Second LeCosPA International Symposium

Everything about Gravity

Second LeCosPA International Symposium

Everything about Gravity

LeCosPA, NTU, Taiwan, 14 – 18 December 2015

editor

Pisin Chen

LeCosPA, National Taiwan University

World Scientific

NEW JERSEY • LONDON • SINGAPORE • BEIJING • SHANGHAI • HONG KONG • TAIPEI • CHENNAI • TOKYO

Published by

World Scientific Publishing Co. Pte. Ltd.
5 Toh Tuck Link, Singapore 596224
USA office: 27 Warren Street, Suite 401-402, Hackensack, NJ 07601
UK office: 57 Shelton Street, Covent Garden, London WC2H 9HE

Library of Congress Cataloging-in-Publication Data
Names: LeCosPA International Symposium (2nd : 2015 : Taipei, Taiwan), author. | Chen, Pisin, 1950– editor. | Guo li Taiwan da xue. Liang Cizhen yu zhou xue yu li zi tian wen wu li xue zhong xin.
Title: Everything about gravity : Second LeCosPA International Symposium, NTU, Taipei, Taiwan, 14–18 December 2015 / editor, Pisin Chen (National Taiwan University, Taiwan).
Other titles: Second LeCosPA International Symposium
Description: Singapore ; Hackensack, NJ : World Scientific, [2017] | Includes bibliographical references.
Identifiers: LCCN 2016053882| ISBN 9789813203945 (hard cover ; alk. paper) | ISBN 9813203943 (hard cover ; alk. paper)
Subjects: LCSH: Quantum gravity--Congresses. | Gravity--Congresses. | General relativity (Physics)--Congresses.
Classification: LCC QC178 .L35 2015 | DDC 539.7/54--dc23
LC record available at https://lccn.loc.gov/2016053882

British Library Cataloguing-in-Publication Data
A catalogue record for this book is available from the British Library.

Desk Editor: Ng Kah Fee

Typeset by Stallion Press
Email: enquiries@stallionpress.com

Printed in Singapore

Preface

To celebrate the centenary of Einstein's General Relativity (GR), the Leung Center for Cosmology and Particle Astrophysics (LeCosPA) of National Taiwan University organized the 2nd LeCosPA International Symposium "Everything about Gravity" on December 14–18, 2015 in Taipei under the auspices of the Division of Astrophysics-Cosmology-Gravity (DACG), Association of Asia-Pacific Physical Societies (AAPPS). 149 participants from 19 countries attended this symposium.

During the 5-day meeting, we examined all aspects of gravity, including but not limited to mathematical structures of GR, Big Bang and black hole singularities, modified gravity, quantum gravity, emergent gravity, cosmology, inflation, dark matter and dark energy, AdS/CFT correspondence, information loss paradox, gravitational waves from Big Bang and other sources, experimental and observational gravity, etc. Altogether, there were 11 Parallel Sessions organized that covered all these topics.

This Symposium Proceedings collects 78 papers contributed by the symposium's Plenary Session and Parallel Session speakers. Organizers of the Parallel Sessions have in addition prepared a summary for their own session. The topics range from quasi-local energy in GR in the presence of gravitational radiations, a gauge theory perspective of gravity, naked black hole firewalls related to the black hole information loss paradox, a new theory of spacetime quantization, relations between the Schwinger effect and the Hawking radiation and Unruh effect, conformal frames in cosmology, surprises in nonrelativistic naturalness, inflation and tensor fluctuations, emergent spacetime for quantum gravity, understanding strongly coupled magnetism through holographic principle, to the detections of dark matter, ultra-high energy cosmic neutrinos and cosmic rays. Last but not least, the closing remark delivered by John Ellis raised the following questions: Does cosmological inflation require a modification of Einstein's gravity?

It may be premature to answer these questions, but one thing is certain. After 100 years of remarkable success of Einstein's general relativity, including the recent LIGO discovery of gravitational waves in 2016, and the tremendous advancement in our understanding of the other three fundamental interactions through quantum field theory, the development of a successful quantum theory of gravity has become a major goal in physics in the 21st century. This symposium was therefore not just for looking back, but also for looking forward.

In addition to the exciting scientific program, the participants and their companions also enjoyed a chain of wonderful social activities. The Opening Night Concert was performed by the NTU Orchestra of Chinese Instruments. Many participants who never heard of traditional Chinese music in the symphonic format were impressed by its breadth and depth. The excursion on the third day not only included several tourist attractions in the city of Taipei, but also the coastal county of Yiland, where the symposium banquet was held in the beautiful Pearl Restaurant. Another unforgettable event was the Christmas party on Thursday, where Alexei Starobinsky, John Ellis, Don Page, Bill Unruh, and I put up Santa's hats (and three of them actually did have white beards) and led the participants to sing Christmas carols. It was fun!

The success of this symposium owed tremendously to the generous supports of Asia-Pacific Center for Theoretical Physics (APCTP), Taiwan's Ministry of Science and Technology, Ministry of Foreign Affairs, Bureau of Foreign Trade, Taipei City Government, National Center for Theoretical Sciences (NCTS), Physical Society of the Republic of China, Gravitational Society of Republic of China, Academia Sinica Institute of Astronomy and Astrophysics (ASIAA), Academia Sinica Institute of Physics (ASIOP), National Taiwan University's Center for Condensed Matter Sciences (CCMS) and Center for Theoretical Science (CTS).

Pisin Chen
Taipei, Taiwan
August 25, 2016

Contents

X. Black Hole Evaporation and Information Loss Paradox 545

XI. Mathematical Structure of GR 597

Second LeCosPA International Symposium on

Everything about Gravity

– Celebrating the Centenary of Estine's General Relativity –
Leung Center for Cosmology and Particle Astrophysics,
National Taiwan University. Taipei, Taiwan
December 14–18, 2015

Under the Auspices of

Associa ion of Asia-Pacific Physical Societies (AAPPS) –
Division of Astrophysics, Cosmology and Gravity (DACG)

Sponsoring Organizations and Institutions.

Asia Pacific Center for Theoretical Physics (APCTP)
Academia Sinica Institute of Astronomy and Astrophysics (ASIAA)
Institute of Physics, Academia Sinica (ASIOP)
Center for Condensed Matter Science, National Taiwan University
Bureau of Foreign Trade (BOFT)
Department of Physics, National Taiwan University
Ministry of Foreign Affairs (MOFA)
Ministry of Science and Technology (MOST)
The Department of Information and Tourism, Taipei City Government (TP EDOIT)
The Gravitational Society of The Republic of China (Taiwan)
The Physical Society of The Republic of China (Taiwan) (PSROC)

MOFA BOFT MOST TPEDOIT APCTP AAPPS-DACG

ASIAA CCMS CTS CIS GSROC NCTS

PLENARY SESSION

Quasi-Local Energy in Presence of Gravitational Radiation*

Po-Ning Chen* and Mu-Tao Wang†
Department of Mathematics, Columbia University,
2990 Broadway, New York, NY 10027, USA
**pnchen@math.columbia.edu*
†*mtwang@math.columbia.edu*

Shing-Tung Yau
Department of Mathematics, Harvard University,
1 Oxford Street, Cambridge, MA 02138, USA
yau@math.columbia.edu

We discuss our recent work [P.-N. Chen, M.-T. Wang and S.-T. Yau, Quasi-local mass in the gravitational perturbations of black holes, to appear.] in which gravitational radiation was studied by evaluating the Wang–Yau quasi-local mass of surfaces of fixed size at the infinity of both axial and polar perturbations of the Schwarzschild spacetime, à la Chandrasekhar. [S. Chandrasekhar, *The Mathematical Theory of Black Holes*, Oxford Classic Texts in the Physical Sciences (Oxford University Press, New York, 1998).]

Keywords: Quasi-local energy; quasi-local mass; gravitational perturbation; black hole.

PACS Number(s): 04.20.Cv, 04.30.Nk

We compute the Wang–Yau quasi-local mass[1,2] of "spheres of unit size" at null infinity to capture the information of gravitational radiation. The set-up, following Chandrasekhar,[3] is a gravitational perturbation of the Schwarzschild solution, which is governed by the Regge–Wheeler equation (see below). We take a sphere of a fixed areal radius and push it all the way to null infinity. The limit of the geometric data is that of a standard configuration and thus the optimal embedding equation[1,2,4] can be solved.

Let us first consider the axial perturbations. The metric perturbation is of the form:

$$-\left(1-\frac{2m}{r}\right)dt^2+\frac{1}{1-\dfrac{2m}{r}}dr^2+r^2d\theta^2+r^2\sin^2\theta(d\phi-q_2dr-q_3d\theta)^2.$$

*This article first appeared in *Int. J. Mod. Phys. D* **25** (2016) 1645001.

The linearized vacuum Einstein equation is solved by a separation of variable ansatz in which q_2 and q_3 are explicitly given by the Teukolsky function and the Legendre function.

In particular,

$$q_3 = \sin(\sigma t) \frac{C_\mu(\theta)}{\sin\theta} \frac{(r^2 - 2mr)}{\sigma^2 r^4} \frac{d}{dr}(rZ^{(-)}),$$

for a solution of frequency σ and a separation of variable constant μ. Here $C_\mu(\theta)$ is related to the μth Legendre function P_μ by

$$C_\mu(\theta) = \sin\theta \frac{d}{d\theta}\left(\frac{1}{\sin\theta}\frac{dP_\mu(\cos\theta)}{d\theta}\right).$$

After the change of variable

$$r_* = r + 2m\ln\left(\frac{r}{2m} - 1\right),$$

$Z^{(-)}$ satisfies the Regge–Wheeler equation:

$$\left(\frac{d^2}{dr_*^2} + \sigma^2\right) Z^{(-)} = V^{(-)} Z^{(-)},$$

where

$$V^{(-)} = \frac{r^2 - 2mr}{r^5}[(\mu^2 + 2)r - 6m],$$

and μ is a separation of variable constant.

On the Schwarzschild spacetime

$$-\left(1 - \frac{2m}{r}\right) dt^2 + \frac{1}{1 - \frac{2m}{r}} dr^2 + r^2 d\theta^2 + r^2 \sin^2\theta d\phi^2,$$

we consider an asymptotically flat Cartesian coordinate system (t, y_1, y_2, y_3) with $y_1 = r\sin\theta\sin\phi$, $y_2 = r\sin\theta\cos\phi$, $y_3 = r\cos\theta$. Given $(d_1, d_2, d_3) \in \mathbb{R}^3$ with $d^2 = \sum_{i=1}^3 d_i^2$, consider the 2-surface

$$\Sigma_{t,d} = \left\{(t, y_1, y_2, y_3) : \sum_{i=1}^3 (y_i - d_i)^2 = 1\right\}.$$

We compute the quasi-local mass of $\Sigma_{t,d}$ as $d \to \infty$.

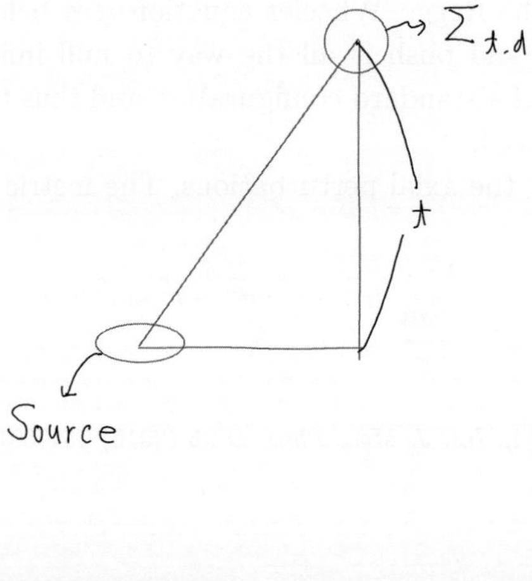

Denote

$$A(r) = \frac{(r^2 - 2mr)}{\sigma^2 r^3} \frac{d}{dr}(rZ^{(-)}).$$

The linearized optimal embedding equation of $\Sigma_{t,d}$ is reduced to two linear elliptic equations on the unit 2-sphere S^2:

$$\Delta(\Delta + 2)\tau = [-A''(1 - Z_1^2) + 6A'Z_1 + 12A]Z_2Z_3,$$

$$(\Delta + 2)N = (A'' - 2A'Z_1 + 4A)Z_2Z_3,$$

where τ and N are the respective time and radial components of the solution, and Z_1, Z_2, Z_3 are the three standard first eigenfunctions of S^2. A' and A'' are derivatives with respect to r, and r^2 is substituted by $r^2 = d^2 + 2Z_1 + 1$ in the above equations.

The quasi-local mass of $\Sigma_{t,d}$ with respect to the optimal isometric embedding is then

$$E(\Sigma_{t,d}) = C^2\{\sin^2(\sigma t)E_1 + \sigma^2 \cos^2(\sigma t)E_2\} + O\left(\frac{1}{d^3}\right),$$

where E_1 and E_2 are two integrals on the standard unit 2-sphere, that depend on the solution τ and N of the optimal isometric embedding equation. Explicitly,

$$E_1 = \int_{S^2} (1/2)[A^2Z_2^2(7Z_3^2 + 1) + 2AA'Z_1Z_3^2(3Z_2^2 - 1) - N(\Delta + 2)N],$$

$$E_2 = \int_{S^2} [A^2Z_2^2Z_3^2 - \tau\Delta(\Delta + 2)\tau].$$

In particular,

$$\partial_t E(\Sigma_{t,d}) = \frac{\sigma \sin(2\sigma t)C^2(\theta)}{d^2}\{E_1 - \sigma^2 E_2\} + O\left(\frac{1}{d^3}\right).$$

Let us compare the quasi-local mass on the small spheres $\Sigma_{t,d}$ along a certain direction to the quasi-local mass of the large coordinate spheres $S_{t,r}$.

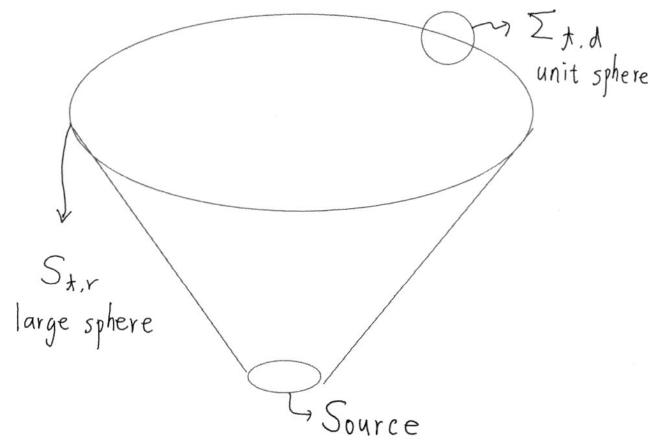

Naively, one may expect to recover $\partial_t E(S_{t,r})$ by integrating the energy radiated away at all directions $\partial_t E(\Sigma_{t,d})$. However, our calculation indicates that there are nonlinear correction terms from the quasi-local energy that should be taken into account.

We can also consider the polar perturbation of the Schwarzschild spacetime in which the metric coefficients g_{tt}, g_{rr}, $g_{\theta\theta}$, and $g_{\phi\phi}$ are perturbed in

$$-\left(1-\frac{2m}{r}\right)dt^2+\frac{1}{1-\frac{2m}{r}}dr^2+r^2d\theta^2+r^2\sin^2\theta d\phi^2.$$

The gravitational perturbation is governed by the Zerilli equation

$$\left(\frac{d^2}{dr_*^2}+\sigma^2\right)Z^{(+)}=V^{(+)}Z^{(+)},$$

where

$$V^{(+)}=\frac{2(r^2-2mr)}{r^5(nr+3m)^2}[n^2(n+1)r^3+3mn^2r^2+9m^2nr+9m^3],$$

and n is the separation of variable constant. Again, we compute the quasi-local mass of spheres of unit-size at null infinity. The calculation is similar to the axial perturbation case but the result is different as the leading term is of the order $\frac{1}{d}$ (as opposed to $\frac{1}{d^2}$ for axial-perturbation) with nonzero coefficients. If such a linear perturbation can be realized as an actual perturbation of the Schwarzschild spacetime, the result would contradict the positivity of the quasi-local mass.[1,2,5] From this, we deduce the following conclusion: There does not exist any gravitational perturbation of the Schwarzschild spacetime that is of purely polar type in the sense of Chandrasekhar.[3]

For an actual gravitational perturbation of the Schwarzschild solution, the vanishing of the $\frac{1}{d}$ gives a limiting integrand that integrates to zero on the limiting 2-sphere at null infinity. In fact, the quasi-local mass density ρ (see [Eq. (2.2) of Ref. 6]) of $\Sigma_{t,d}$ can be computed at the pointwise level. Up to an $O(\frac{1}{d^3})$ term

$$\rho=\left(K-\frac{1}{4}|H|^2\right)-\frac{(|H|-2)^2}{4}+\frac{1}{d^2}\left\{\frac{1}{2}|\nabla^2N|^2+((\Delta+2)N)^2-\frac{1}{4}(\Delta N)^2\right.$$
$$\left.-\frac{1}{4}(\Delta\tau)^2+\frac{1}{2}[\nabla^a\nabla^b(\tau_a\tau_b)-|\nabla\tau|^2-\Delta|\nabla\tau|^2]\right\},$$

where K is the Gauss curvature of $\Sigma_{t,d}$. The first line, which integrates to zero, is of the order of $\frac{1}{d}$ and is exactly the mass aspect function of the Hawking mass.[7] The $\frac{1}{d^2}$ term of the quasi local mass $\int_{\Sigma_d}\rho d\mu_{\Sigma_{t,d}}$ has contributions from the second and third lines (of the order of $\frac{1}{d^2}$), the $\frac{1}{d^2}$ term of the first line, and the $\frac{1}{d}$ term of the area element $d\mu_{\Sigma_{t,d}}$. The above integral formula is obtained after performing integrations by parts and applying the optimal embedding equation several times.

To each closed loop on the limiting 2-sphere at null infinity, we can thus associate a nonvanishing arc integral that is of the order of $\frac{1}{d}$, where d is the distance from the source. We expect the freedom in varying the shape of the loop can increase the detectability of gravitational waves.

Acknowledgments

P.-N. Chen is supported by NSF grant DMS-1308164, M.-T. Wang is supported by NSF grant DMS-1405152 and S.-T. Yau is supported by NSF grants PHY-0714648 and DMS-1308244. This work was partially supported by a grant from the Simons Foundation (#305519 to Mu-Tao Wang). Part of this work was carried out when P.-N. Chen and M.-T. Wang were visiting the Department of Mathematics and the Center of Mathematical Sciences and Applications at Harvard University.

References

1. M.-T. Wang and S.-T. Yau, *Phys. Rev. Lett.* **102** (2009) 021101.
2. M.-T. Wang and S.-T. Yau, *Commun. Math. Phys.* **288** (2009) 919.
3. S. Chandrasekhar, *The Mathematical Theory of Black Holes*, Oxford Classic Texts in the Physical Sciences (Oxford University Press, New York, 1998).
4. P.-N. Chen, M.-T. Wang and S.-T. Yau, *Commun. Math. Phys.* **308** (2011) 845.
5. C.-C. M. Liu and S.-T. Yau, *J. Amer. Math. Soc.* **19** (2006) 181.
6. P.-N. Chen, M.-T. Wang and S.-T. Yau, *Commun. Math. Phys.* **338** (2015) 31.
7. S. W. Hawking, *J. Math. Phys.* **9** (1968) 598.
8. P.-N. Chen, M.-T. Wang and S.-T. Yau, Quasi-local mass in the gravitational perturbations of black holes, to appear.

Gravity: A Gauge Theory Perspective*

James M. Nester*,†,‡,§,¶ and Chiang-Mei Chen*,†,‖

*Department of Physics, National Central University,
Chungli 32001, Taiwan

†Center for Mathematical and Theoretical Physics,
National Central University, Chungli 32001, Taiwan

‡Graduate Institute of Astronomy,
National Central University, Chungli 32001, Taiwan

§Leung Center for Cosmology and Particle Astrophysics,
National Taiwan University, Taipei 10617, Taiwan

¶nester@phy.ncu.edu.tw

‖cmchen@phy.ncu.edu.tw

The evolution of a generally covariant theory is under-determined. One hundred years ago such dynamics had never before been considered; its ramifications were perplexing, its future important role for all the fundamental interactions under the name gauge principle could not be foreseen. We recount some history regarding Einstein, Hilbert, Klein and Noether and the novel features of gravitational energy that led to Noether's two theorems. Under-determined evolution is best revealed in the Hamiltonian formulation. We developed a covariant Hamiltonian formulation. The Hamiltonian boundary term gives covariant expressions for the quasi-local energy, momentum and angular momentum. Gravity can be considered as a gauge theory of the local Poincaré group. The dynamical potentials of the Poincaré gauge theory of gravity are the frame and the connection. The spacetime geometry has in general both curvature and torsion. Torsion naturally couples to spin; it could have a significant magnitude and yet not be noticed, except on a cosmological scale where it could have significant effects.

Keywords: Hamiltonian; quasi-local energy; Poincaré gauge theory.

PACS Number(s): 04.20.Cv, 04.20.Fy

1. Introduction

There have long been disputes about all of the *principles* used by Einstein for his gravity theory. Kretschmann in 1917 argued that *general covariance* has no real physical content and no connection to an extension of the principle of relativity.[1,2] From a different perspective general covariance has deep fundamental ramifications.

General relativity (GR) with general covariance is the premier gauge theory. The consequences of this, especially regarding gravitational energy and

*This article first appeared in *Int. J. Mod. Phys. D* **25** (2016) 1645002.

under-determined evolution, were long perplexing. The Hamiltonian approach clarifies these issues. Gravity can be understood as a gauge theory of the local Poincaré symmetries of spacetime.

2. Some Historical Background

In 1915 Einstein made presentations to the Prussian Academy of Sciences on November 4, 11, 18 and 25 (published one week later). The last had his generally covariant equations along with energy conservation. Hilbert presented his "Foundations of physics" on November 16 and 20; he submitted his first note on Nov. 19 (published in March 1916).[3] His first theorem is central to our concerns:

Theorem I ('Leitmotiv'[a]). *In the system of n Euler–Lagrange differential equations in n variables obtained from a generally covariant variational integral such as in Axiom I, 4 of the n equations are always a consequence of the other $n-4$ in the sense that 4 linearly independent combinations of the n equations and their total derivatives are always identically satisfied.*

Some discussions of Hilbert's work have appeared recently,[3–5] often it has been viewed from the Einstein GR perspective (e.g. Ref. 6). An alternative view[7] of Hilbert's agenda argues that a main aim was to reconcile the tension between *general covariance* and its inevitable consequence: *a lack of unique determinism*[b] — this is *the essence of gauge theory.* Dynamical equations obtained from a variational principle had formerly had deterministic Cauchy initial value problems, but for GR there was a differential identity connecting the evolution equations; they were not independent and could not give uniquely determined evolution. Later, it was found that this is best addressed using the Hamiltonian approach.[8,9]

We note some excerpts from Einstein's correspondence, from Vol. 8 in Ref. 10.

"In your paper everything is understandable to me now except for the energy theorem. Please do not be angry with me that I ask you about this again. ... How is this cleared up? It would suffice, of course, if you would charge Miss Noether with explaining this to me." (Doc. 223 to Hilbert 30 May 1916)

"The only thing I was unable to grasp in your paper is the conclusion at the top of page 8 that ε^σ was a vector." (Doc. 638 to Klein 22 Oct 1918)

"... Meanwhile, with Miss Noether's help, I understand that the proof for the vector character of ε^σ from "higher principles" as I had sought was already given by Hilbert on pp. 6, 7 of his first note, ..." (Doc. 650 from Klein 10 Nov 1918)

Briefly, after a couple of years Klein clarified Hilbert's energy–momentum "vector"; he related it to Einstein's pseudotensor, but he disagreed with Einstein's physical interpretation of divergenceless expressions.[c] Enlisted by Hilbert and Klein,

[a]Guiding theme.
[b]Einstein had struggled with this in connection with his hole argument.
[c]Things were not as easy then; in particular the Bianchi identity and its contracted version were not known to these people[11,12]; they each had to rediscover an equivalent identity on their own.

it was Emmy Noether who resolved the primary puzzle regarding gravitational energy.

3. Automatic Conservation of the Source and Gauge Fields

In 1916 Einstein showed that local coordinate invariance plus his field equations gives material energy momentum conservation, without using the matter field equations (see Doc. 41 in Vol. 6 of Ref. 10). This is referred to as *automatic conservation of the source* (see Sec. 17.1 in Ref. 13); it uses a Noether second theorem local (gauge) symmetry type of argument to obtain current conservation. Hermann Weyl argued in this way for the electromagnetic current in his papers of 1918 (the name *gauge theory* comes from this work) and 1929,[d] whereas modern field theory generally uses Noether's first theorem for current conservation.[15]

The *essence* of gauge theory is *a local symmetry*, consequently: (i) a differential identity, (ii) under-determined evolution, (iii) restricted type of source coupling, (iv) automatic conservation of the source. Yang–Mills is only one special type. Our gauge approach to gravity does not try to force it into the Yang–Mills mold, but rather simply recognizes the natural symmetries of spacetime geometry.

4. Noether's 1918 Contribution

One word well describes 20th century physics: *symmetry.* Most of the theoretical physics ideas involved symmetry — essentially they are applications of Noether's two theorems.[16] The first associates conserved quantities with global symmetries. The second concerns local symmetries: it is the foundation of the modern gauge theories.

Why did Noether make her investigation? Klein was looking into the relationship between Einstein's pseudotensor and Hilbert's energy vector. He published a paper based on his correspondence with Hilbert. We quote some excerpts[16]:

Klein: "You know that Miss Noether advises me continually regarding my work, and that in fact it is only thanks to her that I have understood these questions."

Hilbert: "I fully agree in fact with your statements on the energy theorems: Emmy Noether, on whom I have called for assistance more than a year ago to clarify this type of analytical questions concerning my energy theorem, found at that time that the energy components that I had proposed — as well as those of Einstein — could be formally transformed, ... using the Lagrange differential equations (4) and (5) of my first note, into expressions whose divergence vanishes identically, ... Indeed, I believe that in the case of general relativity, i.e. in the case of the general invariance of the Hamiltonian function, the energy equations which in your opinion correspond to the energy equations of the theory of orthogonal invariance do not exist at all; I can even call this fact a characteristic of the general theory of relativity."

[d]An English translation of Weyl's seminal papers can be found in Ref. 14.

This is why Noether wrote her paper. After presenting her two famous theorems she uses them to draw the conclusion that clarifies the situation[16] :

"Given I invariant under the group of translations, then the energy relations are improper if and only if I is invariant under an infinite group which contains the group of translations as a subgroup.... As Hilbert expresses his assertion, the lack of a proper law of energy constitutes a characteristic of the "general theory of relativity." For that assertion to be literally valid, it is necessary to understand the term "general relativity" in a wider sense than is usual, and to extend it to the aforementioned groups that depend on n arbitrary functions."

Her result regarding the lack of a proper law of energy applies not just to Einstein's GR, but to all geometric theories of gravity. The modern view is that energy–momentum is *quasi-local*, associated with a closed 2-surface.[17]

5. Energy–Momentum Pseudotensors and the Hamiltonian

The Einstein–Lagrangian differs from Hilbert's by a total divergence:

$$2\kappa\mathcal{L}_{\mathrm{E}}(g_{\alpha\beta},\partial_\mu g_{\alpha\beta}) := -\sqrt{-g}g^{\beta\sigma}\Gamma^{\alpha}{}_{\gamma\mu}\Gamma^{\gamma}{}_{\beta\nu}\delta^{\mu\nu}_{\alpha\sigma} \equiv \sqrt{-g}R - \mathrm{div}. \tag{1}$$

The Einstein *pseudotensor* is the associated canonical energy–momentum density:

$$\mathfrak{t}^{\mu}_{\mathrm{E}\nu} := \delta^{\mu}_{\nu}\mathcal{L}_{\mathrm{E}} - \frac{\partial\mathcal{L}_{\mathrm{E}}}{\partial\partial_\mu g_{\alpha\beta}}\partial_\nu g_{\alpha\beta}. \tag{2}$$

Using $\sqrt{-g}G^{\mu}{}_{\nu} = \kappa\mathfrak{T}^{\mu}{}_{\nu}$ one gets a conserved total energy–momentum:

$$\partial_\mu(\mathfrak{T}^{\mu}{}_{\nu} + \mathfrak{t}^{\mu}_{\mathrm{E}\nu}) = 0, \iff \sqrt{-g}G^{\mu}{}_{\nu} + \kappa\mathfrak{t}^{\mu}_{\mathrm{E}\nu} = \partial_\lambda\mathfrak{U}^{[\mu\lambda]}{}_{\nu}. \tag{3}$$

The superpotential was found by Freud in 1939[18]: $\mathfrak{U}_{\mathrm{F}}^{\mu\lambda}{}_{\nu} := \sqrt{-g}g^{\beta\sigma}\Gamma^{\alpha}{}_{\beta\gamma}\delta^{\mu\lambda\gamma}_{\alpha\sigma\nu}$. Other pseudotensors likewise follow from different superpotentials. They are all inherently reference frame dependent. Thus, there are two big problems: (1) which pseudotensor? (2) which reference frame? The Hamiltonian approach has answers. With constant Z^μ, the energy–momentum within a region is

$$\begin{aligned} -Z^\mu P_\mu(V) &:= -\int_V Z^\mu(\mathfrak{T}^{\nu}{}_{\mu} + \mathfrak{t}^{\nu}{}_{\mu})\sqrt{-g}d^3\Sigma_\nu \\ &\equiv \int_V \left[Z^\mu\sqrt{-g}\left(\frac{1}{\kappa}G^{\nu}{}_{\mu} - T^{\nu}{}_{\mu}\right) - \frac{1}{2\kappa}\partial_\lambda(Z^\mu\mathfrak{U}^{\nu\lambda}{}_{\mu})\right] d^3\Sigma_\nu \\ &\equiv \int_V Z^\mu\mathcal{H}^{\mathrm{GR}}_\mu + \oint_{S=\partial V}\mathcal{B}^{\mathrm{GR}}(Z) \equiv H(Z,V). \end{aligned} \tag{4}$$

$\mathcal{H}^{\mathrm{GR}}_\mu$ is the covariant expression for the Hamiltonian density. The *boundary term* 2-surface integral is determined by the superpotential. The value of the pseudotensor/Hamiltonian is *quasi-local*, from just the boundary term, since by the initial value constraints the spatial volume integral vanishes.

6. The Hamiltonian Approach

Noether's work can be combined with the Hamiltonian formulation. In Hamiltonian field theory, the conserved currents are the generators of the associated symmetry. For spacetime translations (infinitesimal diffeomorphisms), the associated current expression (i.e. the energy–momentum density) *is* the Hamiltonian density — the canonical generator of spacetime displacements. Because it can be varied, one gets a handle on the conserved current ambiguity. The Hamiltonian variation gives information that tames the ambiguity in the boundary term — namely boundary conditions. Pseudotensor values are values of the Hamiltonian with certain boundary conditions.[19] Thus, Problem (1) is under control.

The Hamiltonian approach reveals certain aspects of a theory. The constrained Hamiltonian formalism was developed by Dirac[8] and by Bergmann and coworkers. It was applied to GR by Pirani, Schild and Skinner[20] and by Dirac.[21] Later the ADM approach[22] became dominant. For the Poincaré gauge (PG) theory of gravity the Hamiltonian approach was developed by Blagojević and coworkers.[23]

7. The Covariant Hamiltonian and Its Boundary Term

From a first-order Lagrangian formulation, $\mathcal{L} = d\varphi \wedge p - \Lambda$, which gives pairs of first-order equations for an f-form φ and its conjugate p, we developed a 4D-*covariant* Hamiltonian formalism.[19,24–28] The Hamiltonian generates the evolution of a spatial region along a vector field. The Hamiltonian density is the first-order translational Noether current 3-form, it is linear in the displacement vector plus a total differential:

$$\mathcal{H}(Z) := \pounds_Z \varphi \wedge p - i_Z \mathcal{L} =: Z^\mu \mathcal{H}_\mu + d\mathcal{B}(Z), \tag{5}$$

and is a conserved "current" *on shell* (i.e. when the field equations are satisfied):

$$-d\mathcal{H}(Z) \equiv \pounds_Z \varphi \wedge \frac{\delta \mathcal{L}}{\delta \varphi} + \frac{\delta \mathcal{L}}{\delta p} \wedge \pounds_Z p. \tag{6}$$

Furthermore, from *local* diffeomorphism invariance, it follows that $\mathcal{H}_\mu$ is linear in the Euler–Lagrange expressions. Hence, the translational Noether current conservation reduces to a differential identity. This is an instance of Noether's second theorem, exactly the case to which Hilbert's "lack of a proper energy law" remark applies. The value of the Hamiltonian is *quasi-local* (associated with a closed 2-surface):

$$-P(Z,V) = H(Z,V) := \int_V \mathcal{H}(Z) = \oint_{\partial V} \mathcal{B}(Z). \tag{7}$$

The Hamiltonian boundary term has two important roles: (i) it gives the quasi-local values, (ii) it gives the boundary conditions. The boundary term can be adjusted to match suitable boundary conditions. We were led to a set of general boundary terms which are linear in $\Delta\varphi := \varphi - \bar{\varphi}$, $\Delta p := p - \bar{p}$, where $\bar{\varphi}, \bar{p}$ are reference values:

$$\mathcal{B}(Z) := i_Z \begin{Bmatrix} \varphi \\ \bar{\varphi} \end{Bmatrix} \wedge \Delta p - (-1)^f \Delta\varphi \wedge i_Z \begin{Bmatrix} p \\ \bar{p} \end{Bmatrix}. \tag{8}$$

The associated variational Hamiltonian boundary term is

$$\delta\mathcal{H}(Z) \sim d\left[\begin{Bmatrix} i_Z\delta\varphi \wedge \Delta p \\ -i_Z\Delta\varphi \wedge \delta p \end{Bmatrix} + (-1)^f \begin{Bmatrix} -\Delta\varphi \wedge i_Z\delta p \\ \delta\varphi \wedge i_Z\Delta p \end{Bmatrix}\right]. \tag{9}$$

Here *for each bracket independently* one may choose either the upper or lower term, which represent essentially a choice of Dirichlet (fixed field) or Neumann (fixed momentum) boundary conditions for the space and time parts of the fields separately.[e]

For asymptotically flat spaces the Hamiltonian is *well defined*, i.e. the boundary term in its variation vanishes and the quasi-local quantities are well defined at least on the phase space of fields satisfying Regge–Teitelboim like asymptotic conditions:

$$\Delta\varphi \approx \mathcal{O}^+\left(\frac{1}{r}\right) + \mathcal{O}^-\left(\frac{1}{r^2}\right), \quad \Delta p \approx \mathcal{O}^-\left(\frac{1}{r^2}\right) + \mathcal{O}^+\left(\frac{1}{r^3}\right). \tag{10}$$

Also the formalism has natural boundary term related energy flux expressions.[27]

8. Gauge and Geometry

For the history of gauge theory see Ref. 14. Gravity as a gauge theory was pioneered by Utiyama (1956, 1959), Sciama (1961) and Kibble (1961). For accounts of gravity as a spacetime symmetry gauge theory, see Hehl and coworkers,[29–32] Mielke[33] and Blagojević.[34] A comprehensive reader with summaries, discussions and reprints has recently appeared.[35] For the observational constraints on torsion see Ni.[36]

GR can be seen as the original gauge theory: the first physical theory where a local gauge freedom (general covariance) played a key role. Although the electrodynamics potentials with their gauge freedom were known long before GR yet this gauge invariance was not seen as having any important role in connection with the nature of the interaction, the conservation of current, or a differential identity — until the seminal work of Weyl, which post-dated (and was inspired by) GR.

We also note the developments of the concept of a connection in geometry by Levi-Civita, Weyl, Schouten, Cartan, Eddington and others. Riemann–Cartan geometry (with a metric and a metric compatible connection, having both curvature and torsion) is the most appropriate for a dynamic spacetime geometry theory: its local symmetries are just those of the local Poincaré group. The conserved quantities, energy–momentum and angular momentum/center-of-mass momentum are associated with the Minkowski spacetime symmetry, i.e. the Poincaré group.

[e]There are more complicated possibilities, "mixed" choices involving some linear combination of the upper and lower expressions.

9. Riemann–Cartan Geometry and PG Dynamics

It is natural to consider gravity as a gauge theory of the local Poincaré group. The spacetime geometry that suits this perspective is Riemann–Cartan geometry, which has a (Lorentz signature) metric and a *metric compatible connection*: $Dg_{\mu\nu} \equiv 0$.

The translation and Lorentz gauge potentials are, respectively, the *coframe* $\vartheta^\alpha = e^\alpha{}_k dx^k$ and *connection* $\Gamma^\alpha{}_\beta = \Gamma^\alpha{}_{\beta k} dx^k$ one-forms. The associated field strengths are the *torsion* and *curvature* 2-forms:

$$T^\alpha := D\vartheta^\alpha := d\vartheta^\alpha + \Gamma^\alpha{}_\beta \wedge \vartheta^\beta = \frac{1}{2} T^\alpha{}_{ij} dx^i \wedge dx^j, \tag{11}$$

$$R^\mu{}_\nu := d\Gamma^\mu{}_\nu + \Gamma^\mu{}_\lambda \wedge \Gamma^\lambda{}_\nu = \frac{1}{2} R^\mu{}_{\nu ij} dx^i \wedge dx^j. \tag{12}$$

The first and second Bianchi identities are $DT^\alpha \equiv R^\alpha{}_\beta \wedge \vartheta^\beta$ and $DR^\alpha{}_\beta \equiv 0$. The Ricci identity, $[\nabla_\mu, \nabla_\nu]V^\alpha = R^\alpha{}_{\beta\mu\nu} V^\beta - T^\gamma{}_{\mu\nu} \nabla_\gamma V^\alpha$, reflects the holonomy and the Lorentz and translational field strengths. For an orthonormal frame $g_{\mu\nu} =$ const. and $\Gamma^{\alpha\beta}$ is *anti-symmetric.*

The PG dynamics has been discussed in detail in Ref. 28 including (i) the Lagrangian, both second and first-order, (ii) the Noether symmetries, conserved currents and differential identities, (iii) the covariant Hamiltonian including the generators of the local PG symmetries, (iv) our preferred Hamiltonian boundary term, (v) the quasi-local energy–momentum and angular momentum/center-of-mass moment obtained therefrom and (vi) the choice of reference in the boundary term.

10. Preferred Hamiltonian Boundary Terms and Reference

For the PG and GR our preferred Hamiltonian boundary terms are

$$\mathcal{B}_{\rm PG}(Z) = i_Z \vartheta^\alpha \tau_\alpha + \Delta\Gamma^\alpha{}_\beta \wedge i_Z \rho_\alpha{}^\beta + \bar{D}_\beta Z^\alpha \Delta\rho_\alpha{}^\beta, \tag{13}$$

$$\mathcal{B}_{\rm GR}(Z) = \frac{1}{2\kappa}(\Delta\Gamma^\alpha{}_\beta \wedge i_Z \eta_\alpha{}^\beta + \bar{D}_\beta Z^\alpha \Delta\eta_\alpha{}^\beta),$$

$$\eta^{\alpha\beta\ldots} := *(\vartheta^\alpha \wedge \vartheta^\beta \wedge \cdots). \tag{14}$$

Like many other choices, at spatial infinity the latter gives the standard values for energy–momentum and angular momentum/center-of-mass momentum.

Our preferred GR expression has some special virtues:(i) at null infinity it gives the Bondi–Trautman energy and the Bondi energy flux, (ii) it is covariant, (iii) it is positive — at least for spherical solutions and large spheres, (iv) for small spheres it is a positive multiple of the Bel–Robinson tensor, (v) first law of thermodynamics for black holes, (vi) for spherical solutions it has the hoop property, (vii) for a suitable choice of reference it vanishes for Minkowski space.

Regarding the second ambiguity inherent in our quasi-local energy–momentum expressions: the choice of reference. Minkowski space is the natural choice, but one needs to choose a specific Minkowski space. Recently, we proposed (i) 4D *isometric matching on the boundary,*[f] and (ii)*energy optimization* as criteria for selecting the "best matched" reference on the boundary of the quasi-local region. A detailed discussion of our covariant Hamiltonian boundary terms and our reference choice proposal was presented in the MS parallel session[39] and in Ref. 40. They have been tested on spherically symmetric and axisymmetric spacetimes.[41]

11. The Poincaré Gauge Theory of Gravity

The standard PG Lagrangian density has a quadratic field strength form[g]:

$$\mathcal{L}_{\rm PG} \sim \frac{1}{\kappa}(\Lambda + \text{curvature} + \text{torsion}^2) + \frac{1}{\varrho}\,\text{curvature}^2. \tag{15}$$

Varying ϑ, Γ gives quasi-linear second-order dynamical equations for the potentials:

$$\kappa^{-1}(\Lambda + \text{curv} + D\ \text{tor} + \text{tor}^2) + \varrho^{-1}\text{curv}^2 = \text{energy–momentum}, \tag{16}$$

$$\kappa^{-1}\text{tor} + \varrho^{-1}D\ \text{curv} = \text{spin}. \tag{17}$$

The general theory has 11 scalar plus 7 pseudoscalar parameters, but there is one even parity and two odd total differentials; effectively 15 "physical" parameters.[42,43]

Torsion couples to intrinsic spin, not orbital angular momentum.[44] But highly polarized spin density is practically nonexistent in the present day universe. So on ordinary scales matter hardly excites or responds directly to torsion. Torsion could have a significant magnitude and yet be hardly observable: "dark torsion".

At very high densities, it is a different story; at around 10^{52} gm/cm^3 the nucleon spin density is comparable to the material energy density, and beyond that the spin-torsion interaction dominates gravity in the PG. So one can expect major effects in the early universe. But even in the present day, while being hardly noticeable on the lab, solar system, or galactic scale, the gravitational effects of torsion (like Λ) could well have measurable effects on the cosmological scale.

12. General PG Homogeneous and Isotropic Cosmologies

The general PG homogeneous and isotropic cosmology has been considered recently.[45] For such cosmologies the general PG has an effective Lagrangian. From this with $\dot{a} = aH$, 6 first-order equations for a, H, the scalar and pseudoscalar curvatures R, X and the two "scalar" torsion components u, x were obtained:

$$\begin{aligned} -\frac{w_{4+6}}{2}\dot{R} - \frac{\mu_{3-2}}{4}\dot{X} = -[&-3\tilde{a}_2 - w_{4+6}R - \frac{\mu_{3-2}}{2}X]u \\ &+ [6\tilde{\sigma}_2 - \frac{\mu_{3-2}}{2}R + w_{2+3}X]x \\ &+ w_{4-2}[2X - 24(H-u)x]x, \end{aligned} \tag{18}$$

[f]The hardest part of 4D isometric matching is the embedding of the 2D surface S into Minkowski space; Yau and coworkers have extensively investigated this.[37,38]
[g]$\kappa := 8\pi G/c^4$ and ϱ^{-1} has the dimensions of action. Λ is the cosmological constant.

$$-\frac{\mu_{3-2}}{4}\dot{R}+\frac{w_{2+3}}{2}\dot{X}=-[6\tilde{\sigma}_2-\frac{\mu_{3-2}}{2}R+w_{2+3}X]u$$
$$+[12\tilde{a}_3+w_{4+6}R+\frac{\mu_{3-2}}{2}X]x$$
$$-w_{4-2}(2R-12[(H-u)^2-x^2+ka^{-2}])x, \qquad (19)$$
$$\dot{H}-\dot{u}=\frac{R}{6}-2H^2+3Hu-u^2+x^2-ka^{-2}, \qquad (20)$$
$$a_2\dot{u}=\frac{1}{3}(-a_0R-\tilde{\sigma}_2X+\rho-3p+4\Lambda)$$
$$+a_2(u^2-3Hu)-4a_3x^2, \qquad (21)$$
$$\dot{x}=\frac{X}{6}-3Hx+2xu \qquad (22)$$

Here the material energy density satisfies a generalized Friedmann relation:

$$\rho=-\Lambda+3a_0[(H-u)^2-x^2+ka^{-2}]$$
$$-\frac{3}{2}a_2(u^2-2Hu)+6a_3x^2+6\tilde{\sigma}_2x(H-u)$$
$$+\frac{w_{4+6}}{24}[R^2-12R\{(H-u)^2-x^2+ka^{-2}\}]$$
$$+\frac{\mu_{3-2}}{24}[RX-6X\{(H-u)^2-x^2+ka^{-2}\}-12Rx(H-u)]$$
$$-\frac{w_{2+3}}{24}[X^2-24Xx(H-u)]. \qquad (23)$$

The above equations are *the most general — they include all the quadratic PG cosmologies.* Generically, the model has, in addition to the usual metric scale factor, effectively two dynamical "scalar" torsion components carrying spin 0^+ and 0^-. Typically they, and the Hubble expansion rate, show damped oscillations. But the model has other types of behavior in the various degenerate special cases, including GR.[45]

Acknowledgments

This was prepared while J.M.N. was visiting the Morningside Center of Mathematics, Chinese Academy of Sciences, Beijing 100190, China. C.M.C. was supported by the Ministry of Science and Technology of the R.O.C. under the grant MOST 102-2112-M-008-015-MY3.

References

1. J. D. Norton, *Rep. Prog. Phys.* **56** (1993) 791.
2. J. D. Norton, General covariance, gauge theories and the Kretschmann objection, in *Symmetries in Physics: Philosophical Reflections*, eds. K. Brading and E. Castellani (Cambridge University Press, Cambridge, 2003), pp. 110–123.
3. T. Sauer, *Arch. Hist. Exact Sci.* **53** (1999) 529.

4. D. E. Rowe, The Göttingen response to general relativity and Emmy Noether's theorems, in *The Symbolic Universe: Geometry and Physics 1890–1930*, ed. J. Gray (Oxford University Press, Oxford, 1999), pp. 189–233.
5. L. Corry, *David Hilbert and the Axiomatization of Physics (1898–1918): From Grundlagen der Geometrie to Grundlagen der Physik* (Kluwer Academic Publishers, Dordrecht/Boston/London, 2004).
6. J. Renn and J. Stachel, Hilbert's foundation of physics: From a theory of everything to a constituent of general relativity, in *The Genesis of General Relativity, Vol. 4. Gravitation in the Twilight of Classical Physics: The Promise of Mathematics* eds. J. Renn and M. Schemmel (Springer, Berlin, 2007), pp. 858–973.
7. K. Brading and T. A. Ryckman, *Stud. Hist. Philos. Sci. B, Mod. Phys.* **39** (2008) 102.
8. P. A. M. Dirac, *Lectures on Quantum Mechanics* (Belfer, Yeshiva University, New York, 1964).
9. J. Earman, Tracking down gauge: An ode to the constrained Hamiltonian formalism, in *Symmetries in Physics: Philosophy and Reflections*, eds. K. Brading and E. Castellani (Cambridge University Press, Cambridge, 2003), pp. 140–162.
10. The Collected Papers of Albert Einstein: http://einsteinpapers.press.princeton.edu.
11. A. Pais, *Subtle is the Lord: The Science and Life of Albert Einstein* (Clarendon, Oxford, 1982).
12. D. E. Rowe, *Math. Intelligencer* **24** (2002) 57.
13. C. W. Misner, K. S. Thorne and J. A. Wheeler, *Gravitation* (W. H. Freeman, San Francisco, 1973).
14. L. O'Raifeartaigh, *The Dawning of Gauge Theory* (Princeton University Press, Princeton, 1997).
15. K. Brading, *Stud. Hist. Philos. Mod. Phys.* **33** (2002) 3.
16. Y. Kosmann-Schwarzbach, *The Noether Theorems: Invariance and Conservation Laws in the Twentieth Century* (Springer, New York, 2011).
17. L. B. Szabados, *Living Rev. Rel.* **12** (2009) 4.
18. Ph. Freud, *Ann. Math.* **40** (1939) 417.
19. C.-C. Chang, J. M. Nester and C.-M. Chen, *Phys. Rev. Lett.* **83** (1999) 1897.
20. F. Pirani, A. Schild and R. Skinner, *Phys. Rev.* **87** (1952) 452.
21. P. A. M. Dirac, *Proc. Roy. Soc.* (London) *A* **246** (1958) 326.
22. R. Arnowitt, S. Deser and C. W. Misner, The dynamics of general relativity, in *Gravitation: An Introduction to Current Research*, ed. L. Witten (Wiley, New York, 1962), pp. 227–265; reprinted in *Gen. Relativ. Gravit.* **40** (2008) 1997.
23. M. Blagojević and I. A. Nikolić, *Phys. Rev. D* **28** (1983) 2455; I. A. Nikolić, *Phys. Rev. D* **30** (1984) 2508.
24. C.-M. Chen and J. M. Nester, *Class. Quantum Grav.* **16** (1999) 1279.
25. C.-M. Chen and J. M. Nester, *Grav. Cosmol.* **6** (2000) 257.
26. C.-M. Chen, J. M. Nester and R.-S. Tung, *Phys. Lett. A* **203** (1995) 5.
27. C.-M. Chen, J. M. Nester and R.-S. Tung, *Phys. Rev. D* **72** (2005) 104020.
28. C.-M. Chen, J. M. Nester and R.-S. Tung, *Int. J. Mod. Phys.* **24** (2015) 1530026; in *One Hundred Years of General Relativity: From Genesis and Empirical Foundations to Gravitational Waves, Cosmology and Quantum Gravity*, Vol. 1, ed. W.-T. Ni (World Scientific, Singapore, 2015).
29. F. W. Hehl, P. von der Heyde, G. D. Kerlik and J. M. Nester, *Rev. Mod. Phys.* **48** (1976) 393.

30. F. W. Hehl, Four lectures on Poincaré gauge theory, in *Proc. 6th Course of the Int. School of Cosmology and Gravitation on Spin Torsion and Supergravity*, eds. P. G. Bergmann and V. de Sabbatta (Plenum, New York, 1980).
31. F. W. Hehl, J. D. McCrea, E. W. Mielke and Y. Ne'eman, *Phys. Rep.* **258** (1995) 1.
32. F. Gronwald and F. W. Hehl, On the Gauge Aspects of Gravity, in *Proc. 14th Course of the School of Gravitation and Cosmology* (*Erice*), eds. P. G. Bergman, V. de Sabbata and H. J. Treder (World Scientific, Singapore, 1996), pp. 148–198.
33. E. W. Mielke, *Geometrodynamics of Gauge Fields* (Akademie-Verlag, Berlin, 1987).
34. M. Blagojević, *Gravitation and Gauge Symmetries* (Institute of Physics, Bristol, 2002).
35. M. Blagojević and F. W. Hehl, *Gauge Theories of Gravitation* (Imperial College Press, London, 2013).
36. W.-T. Ni, *Rep. Prog. Phys.* **73** (2010) 056901.
37. P.-N. Chen, M.-T. Wang and S.-T. Yau, *Commun. Math. Phys.* **338** (2015) 31.
38. M.-T. Wang and S.-T. Yau, *Phys. Rev. Lett.* **102** (2009) 021101; *Commun. Math. Phys.* **288** (2009) 919.
39. C.-M. Chen, J.-L. Liu, J. M. Nester and G. Sun, Covariant Hamiltonian boundary term: Reference and quasi-local quantities, to appear in *Proc. Second LeCosPA Meeting: Everything about Gravity, MS Parallel Session*, National Taiwan University, Taipei, Taiwan, December 14–18, 2015 World Scientific, edited by Pisin Chen.
40. G. Sun, C.-M. Chen, J.-L. Liu and J. M. Nester, *Chin. J. Phys.* **53** (2015) 110107.
41. G. Sun, C.-M. Chen, J.-L. Liu and J. M. Nester, *Chin. J. Phys.* **52** (2014) 111.
42. P. Baekler, F. W. Hehl and J. M. Nester, *Phys. Rev. D* **83** (2011) 024011.
43. P. Baekler and F. W. Hehl, *Class. Quantum Grav.* **28** (2011) 215017.
44. F. W. Hehl, Yu. N. Obkhov and D. Peutzfeld, *Phys. Lett. A* **377** (2013) 1775.
45. F.-H. Ho, H. Chen, J. M. Nester and H.-J. Yo, *Chin. J. Phys.* **53** (2015) 110109.

From Binaries to Backgrounds: Pulsar Timing and General Relativity

M. Kerr

CSIRO Astronomy and Space Science,
Australia Telescope National Facility,
PO Box 76, Epping NSW 1710, Australia
matthew.kerr@gmail.com

Pulsar timing provides two avenues for testing the predictions and implications of the theory of general relativity. High precision timing experiments of individual systems — particularly of pulsars in binaries, but also of isolated pulsars — allow stringent tests of equivalence principles and the dynamical predictions of general relativity. They provide the strongest constraints on the majority of the parameters in the parameterized post-Newtonian formalism. Pulsar timing arrays — experiments measuring the arrival times of pulses from an ensemble of millisecond pulsars with high precision — are kiloparsec-scale gravitational wave detectors sensitive to the low-frequency (nHz) emission expected from the mergers of supermassive black holes. In this paper, I will give a brief overview of both techniques and of recent results.

Keywords: Gravity; gravitational waves; pulsars; pulsar timing; general relativity.

1. Introduction

In their *tour de force* paper,[1] Hewish *et al.* simultaneously announced the discovery of pulsars and established the method of pulsar timing. In that work, the authors compared pulse arrival times (TOAs) at different frequencies to measure the dispersion due to electrons along the line of sight, establishing that the new source was well outside of the Solar System. They also used the variation in the TOAs due to the earth's motion around the sun to precisely determine the position of the pulsar. Along with the value and stability of the precisely measured signal period, these properties were key in identifying neutron stars as the physical counterpart to pulsars.

Since that time, pulsar timing has continued to play a critical role in astrophysics. While many studies focus on timing as a direct probe of the structure of neutron stars and their magnetospheres, e.g. through the studies of pulsar glitches and timing noise, it is the serendipity of having precise, nearly pointlike clocks in a variety of dynamical systems that has had the greatest impact. Most famously, the discovery of PSR B1913+16 by Hulse and Taylor[2] in a < 8 hr orbit around another neutron star yielded the first indirect evidence of gravitational waves,[3] in precise agreement with the predictions of general relativity (GR). As I will discuss below, this timing campaign is now one of several providing direct, dynamical tests of GR at the $<1\%$ level.

In addition to such "direct" tests, both isolated pulsars and pulsars in binary systems with other neutron stars or white dwarfs can be used to place constraints on non-GR theories of gravity. These include additional couplings of matter to gravity (e.g. Brans–Dicke theory) and theories requiring violations of equivalence principles or Lorentz invariance. Many such theories can be embedded in the parametrized post-Newtonian formalism (PPN[4]), a prescription associating additional terms in the Lagrangian with 9 parameters, 8 of which are independent. Non-GR values of the parameters predict new dynamics. For example, a velocity relative to a preferred frame (Lorentz invariance violation) will cause a spinning neutron star to precess, in turn altering the pulse profile observed at the earth.

Finally, the promise of pulsar timing arrays (PTAs) — ensembles of millisecond pulsars (MSPs) timed with high precision — as detectors of low-frequency gravitational waves is beginning to be realized. Already, stringent upper limits on a stochastic background of gravitational waves from the cosmological population of merging, supermassive binary black holes (e.g. Shannon *et al.*[5]) have begun to challenge predictions based on our understanding of galaxy merger rates, black hole mass functions, and dynamics in the innermost parsecs of merged galaxies. More sensitive radio telescopes, favourable scaling of sensitivity with time, and international collaboration between the three major PTAs (the Parkes Pulsar Timing Array, the European Pulsar Timing Array, and NANOGrav) via the International Pulsar Timing Array, mean PTAs may soon join LIGO in the direct detection of gravitational waves.

In this paper, I will give an overview of the state-of-the art of dynamical tests of GR, constraints on PPN parameters, and the search for nHz gravitational waves. It is meant primarily as a record of the presentation communicated at the 2nd LeCosPA Symposium and can only scratch the surface both in terms of results and literature review. For the former two subjects, I strongly encourage the interested reader to refer to the excellent review articles from Stairs,[6] Will,[7,8] and Manchester.[9] The paper by Manchester also provides an overview of timing arrays.

2. Dynamical Tests of General Relativity

Typical "spectroscopic" binary systems can be completely described by five "Keplerian" parameters, viz. the orbital period (P_b), projected semi-major axis ($a \sin i$, with i the inclination of the orbital plane to the plane of the sky), the epoch of periastron (T_0), the longitude of the ascending node (Ω), and the eccentricity (e). If relativistic corrections are important — e.g. for sub-day orbital periods and relative velocities of 100s of $\mathrm{km\,s^{-1}}$ — then their effects can be computed by adding the appropriate relativistic terms to the Lagrangian. At low order, these terms give rise to five additional constants of the motion, the so-called "post-Keplerian" parameters, which can be expressed in terms of the Keplerian parameters and the object masses m_1, m_2, $M = m_1 + m_2$. They are the rate of periastron advance

$$\dot{\omega} = 3 \left(\frac{P_b}{2\pi} \right)^{-5/3} (T_\odot M)^{2/3} (1 - e^2)^{-1}, \tag{1}$$

the gravitational redshift

$$\gamma = e \left(\frac{P_b}{2\pi}\right)^{1/3} T_\odot^{2/3} M^{-4/3} m_2 (m_1 + 2m_2), \tag{2}$$

the orbital period decay due to gravitational wave emission

$$\dot{P}_b = -\frac{192\pi}{5} \left(\frac{P_b}{2\pi}\right)^{-5/3} \left(1 + \frac{73}{24}e^2 + \frac{37}{96}e^4\right) (1 - e^2)^{-7/2} T_\odot^{5/3} m_1 m_2 M^{-1/3}, \tag{3}$$

and the two parameters of the Shapiro delay $r = T_\odot m_2$ and $s = \sin i$. Here, $T_\odot \equiv GM_\odot/c^3 \approx 4.93\mu$s.

The measurement of both Keplerian and post-Keplerian parameters is performed by fitting a model that accounts for the binary motion, the orientation and motion of the earth around the Solar System barycenter, and any other "nuisance" parameters, such as proper motion of the pulsar, to the observed TOAs. Sensitivity to the parameters increases with the length of the experiment. E.g., a discrepancy in P_b causes a deviation in rotational phase with a linear envelope, which can thus be measured with a precision t^{-1}. Continued timing provides an additional $t^{-1/2}$ from noise reduction, so that the P_b precision scales as $t^{-3/2}$. By similar arguments, the precision of $\dot{P}_b \propto t^{-5/2}$, strongly favouring long experiments.

PSR B1913+16, with a data set spanning more than 30 years, is a prime example of the favourable scaling. The parameters γ and $\dot{\omega}$ are well measured, and together give precise estimates of the two neutron star masses. Together with the (very) well measured Keplerian parameters, they uniquely predict $\dot{P}_b$ in the GR framework. The increasing precision on the measured $\dot{P}_b$ then yields a constraint on any violations of the GR prediction. The current ratio of measured to predicted values is 0.997 ± 0.002,[10] a value that is now limited by the unknown Galactic kinematic contribution to $\dot{P}_b$. Because the system is at modest inclinations, the Shapiro parameters are not measurable, and only one "independent" test of GR is available. Because it involves $\dot{P}_b$, it is necessarily a radiative one.

The double pulsar system PSR J0737-3039A/B, on the other hand, is almost perfect for dynamical tests of GR. It is extremely relativistic — in a 2.4 hr orbit with a periastron advance of 16.9°/yr! — and nearly edge-on (88.7°), allowing all 5 post-Keplerian parameters to be well-measured. The presence of two pulsars likewise provides two independent estimates of the mass function. All four "independent tests" from combinations of the parameters agree with GR at the $\leq$1% level, and some combinations offer non-radiative constraints. Because of the serendipitous alignment with the line-of-sight, the Shapiro parameter s is currently the best-measured, allowing the most stringent test of GR at 0.99987 ± 0.00050[11]! But, because the Shapiro parameters do not produce a cumulative effect on pulse phase, their precision only scales as $t^{-1/2}$. Fortunately, in another bout of serendipity, the double pulsar is much closer to the earth than PSR B1913+16, so that the uncertain kinematic effects are sub-dominant, and $\dot{P}_b$ will ultimately offer the most sensitive test. In 20 years, second-order post-Newtonian corrections to $\dot{\omega}$ may become measurable, allowing a measurement of the neutron star moment of inertia.

There are additional systems with measured post-Keplerian parameters, each yielding important constraints on neutron star masses. In particular, PSR B1534+12 yields a 0.17% test[12] via γ, s, and $\dot{\omega}$. Unfortunately, its $\dot{P}_b$ is dominated by kinematic effects, so it will not become substantially more precise without more sensitive observations. No other system is competitive with B1913+16 or the double pulsar for such GR tests, and new tests will rely on future discoveries and monitoring with new, sensitive, widefield telescopes.

3. Constraints on Post-Newtonian Parameters

Neutron stars are almost ideal for investigating the presence of new dynamical effects. Because they are compact, they can often be treated as test masses, even in tight binaries. Their orbital dynamics are typically in the weak-field régime, allowing for powerful null tests. Their high spin rates and stability make them ideal gyroscopes, and changes in the pulse profile provide a convenient way to measure or constrain changes in the relative angle of the line-of-sight to the star. These properties are ideal for searching for violations of the equivalence principles (EPs), particularly the Einstein EP.

Although such violations can be specific to a particular theory of gravity, many can be encapsulated in the PPN framework. This framework allows computation of the dynamical effects, and thus observational constraints on the parameters. In brief, these parameters are γ, the coupling between matter and curvature (not to be confused with the post-Keplerian parameter), β, nonlinearity in superposition of sources, ξ, governing Lorentz-violating preferred location effects, α_{1-3} giving preferred frame effects, and α_3, ζ_{1-4}, violating conservation of momentum. In GR, γ and β are unity, while all other parameters vanish. In previous decades, measurements of these parameters have been furnished largely by Solar System experiments, and these are summed up in Table 4 of the original version of Will's review paper.[7] There, pulsars provide the best limits on two parameters, α_3 and ζ_2. However, in the intervening decade, precision pulsar measurements have come to dominate the field, and in the revised paper,[8] pulsars provide the best limits on 5 parameters! As we discuss below, the discovery of a pulsar in a triple system will also soon provide the best measurement of the key parameter β in the règime where self-gravity is important. (In general, we caution that the relation between the PPN parameters in systems where self-gravity is important and those where it is not are theory dependent; parameters measured with pulsars inevitably fall in the former group. See the paper of Will[8] for more details.)

Often, a variety of parameters can be tested from a single physical phenomenon, and in this brief overview we will consider two such situations. In the first example, a general consequence of some EP violations is a polarization of the orbit, i.e. an induced eccentricity. By connecting an upper limit on such an eccentricity with theory, the same experiment can probe parameters that predict the polarization from different effects — e.g., through an SEP violation or through precession about a preferred velocity. One of the most well known of these predictions is a gravitational equivalent of the Stark effect, the "Nortvedt effect" in which the orbit of a binary of two objects with different compositions is polarized along the direction of an external gravitational force. This can be tested with good accuracy with lunar laser ranging, but neutron stars in wide, circular binaries with white dwarfs offer competitive limits. Such orbits are naturally produced during many "recycling scenarios" that produce MSPs. Binaries with long periods and small eccentricities thus place stringent limits on η (SEP violation in the Galactic potential), α_1 (polarization along direction of motion due to Lorentz invariance violation), and α_3 (Lorentz invariance and momentum conservation violations).

In order to cast the observed limits into constraints, the binary systems must generally satisfy stringent criteria on their age, P_b, and $\dot{\omega}$ (relative to the Galactic rotation), limiting the number of available systems. Because the limits require a marginalization over some unknown geometric parameters, the small number of available systems substantially degrades the limits, and η is constrained about $100\times$ more poorly than with lunar laser ranging experiments. For the Nortdvedt effect / η, these problems stem largely from invoking the Galactic potential as the polarizer; ideally, a binary should be equipped with a strong, measurable external potential! This hope has recently been realized with the discovery of the hierarchical triple system J0337+1715,[13] a 2.7 ms pulsar in tight (1.6 d) orbit around a white dwarf with an additional white dwarf in a 327 d orbit with the inner pair. The gravitational potential due to the outer star at the inner pair is 10^6 stronger than the Galactic potential, and this system will yield the best measurement of η for strong field objects. Interestingly, the MSP of the triple system is both close to the earth and relatively radio bright, so future pulsar surveys may turn up many similarly interesting systems at larger distances.

Another interesting case is the use of isolated pulsars as gyroscopes. Here, an EP violation induces precession about a particular direction, e.g. about the preferred velocity vector in the case of a Lorentz-invariance violating non-zero α_2, or about the net Galactic gravitation (essentially towards the Galactic centre) in case of non-zero ξ. In both cases, solitary MSPs can set stringent limits, as shown in two papers by Shao *et al.*[14,15] There, the authors used sensitive pulsar timing array observations (see below) of two isolated MSPs, J1744−1134 and B1937+21, to limit any variation in the profile over a span of 15 years. By assuming a particular model for the radiation pattern of the pulsar and marginalizing over unknown geometrical parameters, the authors were able to extract limits more sensitive than terrestrial and Solar System limits by orders of magnitude. PTA work continues and these limits will be improved substantially both with longer observing spans, more suitable pulsars, and better models for pulsar radiation patterns.

Finally, we note that sensitive limits on other quantities — e.g., a $\dot{P}_b$ contribution due to dipolar gravitational radiation — can both test equivalence principles and limit parameters in specific alternative (e.g. scalar-tensor) theories of gravity. The best systems are those with long data sets, little timing noise, and in tight binaries, e.g. PSR J1738+0333, in an 8.5 hr orbit. Limits on variations on Newton's constant are generally about an order of magnitude worse than Solar System tests, but constraints on dipole radiation are competitive. See the paper from Freire *et al.*[16] for more details.

4. The Search for Low-frequency Gravitational Waves

Finally, we briefly discuss recent results using PTAs to search for 10^{-9}–10^{-7} Hz gravitational waves (GW). The fundamental idea is to use the earth and pulsars as test masses in a Galactic-scale interferometer.[17] A source of monochromatic plane waves will retard/advance the arrival times of pulses according to the relative positions of the earth, source, and pulsar. The effect can be understand in terms of an "earth term" evaluated at the present epoch and a "pulsar term" evaluated at the time the electromagnetic pulses were emitted.[18] Because distances to pulsars are not generally known to <1 ly, the signal from each pulsar term enters with a random phase, and searches for GW generally are accomplished by searches for peaks in power spectra rather than through a coherent search (c.f. LIGO). Moreover, sources typically evolve over the $\sim 10^3$ yr light travel time across the array, such that the pulsars in the array will each generally see a different GW frequency, further dephasing the array. On the other hand, the "earth term" is in phase for all pulsars and its amplitude is determined only by the source-earth-pulsar geometry. By searching for correlations in TOAs from multiple pulsars, the GW signature can be detected.[18]

The most promising source of GW in this frequency range is from the merger of supermassive black hole binaries (SMBHs) following major galaxy mergers. Because very massive mergers are rare, it is expected the first detection will not be of a single source but rather of the summed contributions of the cosmological population of mergers. If merging binaries have negligible eccentricity and if GW dominates the evolution of the SMBHs from orbital periods comparable to PTA data spans, the observational signature is particularly simple — a power law of strain versus frequency $h(f) = A\,(f/\mathrm{yr}^{-1})^{-2/3}$. As discussed above, this strain spectrum will manifest in the time series of PTAs through correlations between pulsars with a power spectral density $S(f) = \Gamma_{ab}/12\pi^2 A^2\,(f/\mathrm{yr}^{-1})^{-13/3}\,\mathrm{yr}^3$, where Γ_{ab} depends on the separation of the pulsar pair a, b on the sky. Predictions of A are uncertain, as they depend sensitively on the black hole mass function, the efficiency of <pc-scale processes driving the black holes to merger, and the degree to which eccentricities / merger rates may be modified by the black hole environments. Values of $A \sim 10^{-15}$ are typical of recent calculations,[19] and detection and measurement of A is a primary goal of PTA experiments.

Three PTAs are working to detect both this background and additional sources of GW, as well as to provide sensitive measurements of Solar System dynamics and realizations of a cosmic time-scale. NANOGrav[20] and the European Pulsar Timing Array[21] primarily monitor northern hemisphere sources, while the Parkes Pulsar Timing Array (PPTA[22,23]) makes use of the 64-m Parkes radio telescope to monitor sources at mid- and southern latitudes. The southern hemisphere is generally advantageous for observing Galactic sources. Over time, each of the PTAs have built up longer data sets and improved the precision with which pulse TOAs can be measured. In particular, improved digital signal recorders are intrinsically stable and take advantage of the rapid development of computational capabilities, allowing observing over wider bandwidths and advanced signal processing such as coherent dedispersion to remove the dispersive effects of the interstellar medium.

To date, there is no strong evidence for a GW background. Instead, upper limits have been growing steadily more stringent. In 2005, the best limit[24] was about $A < 11 \times 10^{-15}$. Ten years later, the limit has improved an order of magnitude, with recent work[5] ruling out a (SMBH) power law background of $A > 10^{-15}$. The improvement is impressive, and will grow more so as the limit goes like $t^{-3/2}$, all else remaining equal. This is largely due to the steep spectrum of the assumed model. The lowest accessible GW frequency for a PTA is about $2/T$, because fitting for pulsar spin periods and spin-down rates removes the lowest-frequency signals from the data. The steep spectrum means the lowest frequencies carry ever more power. Moreover, the limit is dominated by the "best" pulsars, those with the least contamination by delays due to the interstellar medium and those that are intrinsically the best clocks (the most regular pulse emission). Indeed, the Shannon *et al.* limit is dominated by a single pulsar, J1909−3744, and uses a single electromagnetic "band" of data, the highest radio frequency recorded by the PPTA. Such high-frequency observations are less susceptible to changes in the line of sight through the interstellar medium, but they are generally less sensitive as pulsars become fainter at higher radio frequencies. Generally, however, the strategy for improving the upper limit is simply to observe the "best" pulsars for as long as possible.

On the other hand, this strategy can fail for two reasons. First, the stringency of the limits depends critically on the assumption about the spectral shape at low frequencies. The most recent limits challenge such predictions for A, making the fixed spectral shape suspect. Environmental effects and finite eccentricity could cause the spectrum to "turn over", i.e. have less power at the low frequencies to which PTAs are most sensitive. NANOGrav have recently explored the consequences of such a turn over and explored the inferences that can be made about merger environments and black hole mass functions.[25] Currently, the data do not strongly constrain any of the proposed mechanisms for weakening the low-frequency GW signal.

The second reason applies in the happy outcome of a GW detection, as here the statistical situation becomes much different. Once the GW background begins to be detectable (comparable to the statistical precision of the observations), it becomes the dominant source of noise in the data. This self-noise means the sensitivity scales only (roughly) as $t^{-1/2}$, and to substantially improve the sensitivity of the array, one must add additional baselines (pulsars). Thus, depending on its amplitude, the detection of the GWB will certainly require a large investment of telescope time. Strides in this direction are being made under the auspices of the International Pulsar Timing Array, a data-sharing collaboration between the three major PTAs. The merged data set offers substantial improvements in sky coverage and telescope capability relative to any individual PTA. However, if the GWB amplitude is low enough, its detection may ultimately require both the large collecting area of future telescopes such as FAST and the Square Kilometer Array and the detection of new, fainter MSPs to add to timing arrays.

5. To the Future

In summary, pulsar timing is a fantastic tool for testing GR and probing the low-frequency GW universe. Already, pulsar timing experiments provide the most sensitive constraints on five of 8 PPN parameters. With discovery and monitoring of the triple pulsar, this will soon become six! In terms of dynamical tests, current systems confirm the accuracy of the predictions of GR to levels approaching $1{:}10^4$. Both favourable scaling with time and improvements in observational capabilities will continue to improve the constraints. New pulsar surveys with current and future telescopes hold forth the promise of finding new systems for both EP and dynamical tests: the holy grail is detection of a pulsar/black hole binary, or of a pulsar orbiting the supermassive black hole Sgr A*. Pulsar timing arrays can contribute to this effort, particularly by ruling out precession of a wide array of neutron stars, thereby constraining PPN parameters. And, of course, the primary goal and promise of detection of nHz GW remains. With such developments, it seems likely the next century of general relativity will be as bright as the first!

References

1. A. Hewish, S. J. Bell, J. D. H. Pilkington, P. F. Scott and R. A. Collins, Observation of a Rapidly Pulsating Radio Source, *Nature* **217**, 709 (February 1968).
2. R. A. Hulse and J. H. Taylor, Discovery of a pulsar in a binary system, *ApJ* **195**, L51 (January 1975).
3. J. H. Taylor, L. A. Fowler and P. M. McCulloch, Measurements of general relativistic effects in the binary pulsar PSR 1913+16, *Nature* **277**, 437 (February 1979).
4. C. M. Will and K. Nordtvedt, Jr., Conservation Laws and Preferred Frames in Relativistic Gravity. I. Preferred-Frame Theories and an Extended PPN Formalism, *ApJ* **177**, p. 757 (November 1972).
5. R. M. Shannon, V. Ravi, L. T. Lentati, P. D. Lasky, G. Hobbs, M. Kerr, R. N. Manchester, W. A. Coles, Y. Levin, M. Bailes, N. D. R. Bhat, S. Burke-Spolaor, S. Dai, M. J. Keith, S. Oslowski, D. J. Reardon, W. van Straten, L. Toomey, J.-B. Wang, L. Wen, J. S. B. Wyithe and X.-J. Zhu, Gravitational waves from binary supermassive black holes missing in pulsar observations., *Science* **349**, 1522 (2015).

6. I. H. Stairs, Testing General Relativity with Pulsar Timing, *Living Reviews in Relativity* **6** (September 2003).
7. C. M. Will, The Confrontation between General Relativity and Experiment, *Living Reviews in Relativity* **9** (March 2006).
8. C. M. Will, The Confrontation between General Relativity and Experiment, *Living Reviews in Relativity* **17** (June 2014).
9. R. N. Manchester, Pulsars and gravity, *International Journal of Modern Physics D* **24**, p. 1530018 (March 2015).
10. J. M. Weisberg, D. J. Nice and J. H. Taylor, Timing Measurements of the Relativistic Binary Pulsar PSR B1913+16, *ApJ* **722**, 1030 (October 2010).
11. M. Kramer, I. H. Stairs, R. N. Manchester, M. A. McLaughlin, A. G. Lyne, R. D. Ferdman, M. Burgay, D. R. Lorimer, A. Possenti, N. D'Amico, J. M. Sarkissian, G. B. Hobbs, J. E. Reynolds, P. C. C. Freire and F. Camilo, Tests of General Relativity from Timing the Double Pulsar, *Science* **314**, 97 (October 2006).
12. E. Fonseca, I. H. Stairs and S. E. Thorsett, A Comprehensive Study of Relativistic Gravity Using PSR B1534+12, *ApJ* **787**, p. 82 (May 2014).
13. S. M. Ransom, I. H. Stairs, A. M. Archibald, J. W. T. Hessels, D. L. Kaplan, M. H. van Kerkwijk, J. Boyles, A. T. Deller, S. Chatterjee, A. Schechtman-Rook, A. Berndsen, R. S. Lynch, D. R. Lorimer, C. Karako-Argaman, V. M. Kaspi, V. I. Kondratiev, M. A. McLaughlin, J. van Leeuwen, R. Rosen, M. S. E. Roberts and K. Stovall, A millisecond pulsar in a stellar triple system, *Nature* **505**, 520 (January 2014).
14. L. Shao, R. N. Caballero, M. Kramer, N. Wex, D. J. Champion and A. Jessner, A new limit on local Lorentz invariance violation of gravity from solitary pulsars, *Classical and Quantum Gravity* **30**, p. 165019 (August 2013).
15. L. Shao and N. Wex, New limits on the violation of local position invariance of gravity, *Classical and Quantum Gravity* **30**, p. 165020 (August 2013).
16. P. C. C. Freire, N. Wex, G. Esposito-Farèse, J. P. W. Verbiest, M. Bailes, B. A. Jacoby, M. Kramer, I. H. Stairs, J. Antoniadis and G. H. Janssen, The relativistic pulsar-white dwarf binary PSR J1738+0333 - II. The most stringent test of scalar-tensor gravity, *MNRAS* **423**, 3328 (July 2012).
17. S. Detweiler, Pulsar timing measurements and the search for gravitational waves, *ApJ* **234**, 1100 (December 1979).
18. R. W. Hellings and G. S. Downs, Upper limits on the isotropic gravitational radiation background from pulsar timing analysis, *ApJ* **265**, L39 (February 1983).
19. A. Sesana, Insights into the astrophysics of supermassive black hole binaries from pulsar timing observations, *Classical and Quantum Gravity* **30**, p. 224014 (November 2013).
20. M. A. McLaughlin, The North American Nanohertz Observatory for Gravitational Waves, *Classical and Quantum Gravity* **30**, p. 224008 (November 2013).
21. M. Kramer and D. J. Champion, The European Pulsar Timing Array and the Large European Array for Pulsars, *Classical and Quantum Gravity* **30**, p. 224009 (November 2013).
22. G. Hobbs, The Parkes Pulsar Timing Array, *Classical and Quantum Gravity* **30**, p. 224007 (November 2013).
23. R. N. Manchester, G. Hobbs, M. Bailes, W. A. Coles, W. van Straten, M. J. Keith, R. M. Shannon, N. D. R. Bhat, A. Brown, S. G. Burke-Spolaor, D. J. Champion, A. Chaudhary, R. T. Edwards, G. Hampson, A. W. Hotan, A. Jameson, F. A. Jenet, M. J. Kesteven, J. Khoo, J. Kocz, K. Maciesiak, S. Oslowski, V. Ravi, J. R. Reynolds, J. M. Sarkissian, J. P. W. Verbiest, Z. L. Wen, W. E. Wilson, D. Yardley, W. M. Yan and X. P. You, The Parkes Pulsar Timing Array Project, *PASA* **30**, p. 17 (January 2013).

24. F. A. Jenet, G. B. Hobbs, W. van Straten, R. N. Manchester, M. Bailes, J. P. W. Verbiest, R. T. Edwards, A. W. Hotan, J. M. Sarkissian and S. M. Ord, Upper Bounds on the Low-Frequency Stochastic Gravitational Wave Background from Pulsar Timing Observations: Current Limits and Future Prospects, *ApJ* **653**, 1571 (December 2006).
25. Z. Arzoumanian, A. Brazier, S. Burke-Spolaor, S. Chamberlin, S. Chatterjee, B. Christy, J. Cordes, N. Cornish, P. Demorest, X. Deng, T. Dolch, J. Ellis, R. Ferdman, E. Fonseca, N. Garver-Daniels, F. Jenet, G. Jones, V. Kaspi, M. Koop, M. Lam, J. Lazio, L. Levin, A. Lommen, D. Lorimer, J. Luo, R. Lynch, D. Madison, M. McLaughlin, S. McWilliams, C. Mingarelli, D. Nice, N. Palliyaguru, T. Pennucci, S. Ransom, L. Sampson, S. Sanidas, A. Sesana, X. Siemens, J. Simon, I. Stairs, D. Stinebring, K. Stovall, J. Swiggum, S. Taylor, M. Vallisneri, R. van Haasteren, Y. Wang and W. Zhu, The NANOGrav Nine-year Data Set: Limits on the Isotropic Stochastic Gravitational Wave Background, *ArXiv e-prints* (August 2015).

Unclothed Firewalls*

Pisin Chen

Leung Center for Cosmology and Particle Astrophysics,
National Taiwan University, Taipei, Taiwan 10617
Kavli Institute for Particle Astrophysics and Cosmology,
SLAC National Accelerator Laboratory,
Stanford University, California 94305, USA
pisinchen@phys.ntu.edu.tw

Yen Chin Ong

Nordita, KTH Royal Institute of Technology and Stockholm University,
Roslagstullsbacken 23, SE-106 91 Stockholm, Sweden
yenchin.ong@nordita.org

Don Nelson Page

Theoretical Physics Institute, University of Alberta,
Edmonton, Alberta Canada T6G 2E1
profdonpage@gmail.com

Misao Sasaki

Yukawa Institute for Theoretical Physics,
Kyoto University, Kyoto 606-8502, Japan
misao@yukawa.kyoto-u.ac.jp

Dong-Han Yeom

Leung Center for Cosmology and Particle Astrophysics,
National Taiwan University, Taipei, Taiwan 10617
innocent.yeom@gmail.com

We have previously argued that fluctuations of the Hawking emission rate can cause a black hole event horizon to fluctuate inside the location of a putative firewall, rendering the firewall naked. This assumes that the firewall is located near where the horizon would be expected based on the past evolution of the spacetime. Here, we expand our previous results by defining two new estimates for where the firewall might be that have more smooth temporal behavior than our original estimate. Our results continue to contradict the usual assumption that the firewall should not be observable except

*This article first appeared in *Int. J. Mod. Phys. D* **25** (2016) 1645003.

by infalling observers. This casts doubt about the idea that firewalls are the 'most conservative' solution to the information loss paradox.

Keywords: Black holes; firewalls; Hawking emission.

PACS Number(s): 04.70.Dy, 04.60.−m

1. The AMPS Argument

Almheiri, Marolf, Polchinski and Sully (AMPS)[1,2] argued that local quantum field theory, unitarity, and no-drama (the assumption that infalling observers should not experience anything unusual at the event horizon if the black hole is sufficiently large) cannot all be consistent with each other for the Hawking evaporation of a black hole with a finite number of quantum states given by the Bekenstein–Hawking entropy.

AMPS suggested that the 'most conservative' resolution to this inherent inconsistency between the various assumptions is to give up no-drama. Instead, an infalling observer would be terminated once he or she hits the so-called firewall. This seems rather surprising, because the curvature is negligibly small at the event horizon of a sufficiently large black hole, and thus one would expect nothing special but low energy physics.

1.1. *The heart of the AMPS argument*

Assuming unitarity, the information contained inside a black hole should eventually be recovered from the Hawking radiation. The late-time radiation purifies the earlier radiation, so the late-time radiation should be maximally entangled with the earlier radiation.[3–5]

By the monogamy of quantum entanglement, the late-time radiation cannot also be maximally entangled with the interior of the black hole. This means that the field configuration across the event horizon is generically not continuous, which leads to a divergent local energy density. More explicitly, the quantum field Hamiltonian contains terms like $(\partial_r \varphi)^2$. The derivative is divergent at some $r = R$ if the field configuration is not continuous across R. This is the firewall.

1.2. *The location of a firewall*

Usually, it is thought that firewalls lie on the black hole event horizons. Of course, in quantum mechanics there are no sharp boundaries, and the positions of event horizons should be uncertain, up to perhaps fluctuations of the order of the Planck length. That is to say, firewalls are presumably like the stretched horizons,[6] with the crucial difference that anything that hits a firewall gets incinerated instead of just passing right through.

By the nature of the event horizon, if the firewall lies either inside or exactly on the horizon, then it is completely invisible to the observers outside. For firewalls that are not too far outside the event horizons, it is still doubtful that they are perceptible to far-away observers, since it would seem that such firewalls are well hidden inside the Planckian region of the *local* thermal atmosphere.

2. Causality Assumption for Firewalls

Here, we extend the implications of our previous argument[7] that a firewall, if it exists, has a location determined by the past history of the Hawking evaporating black hole spacetime and is near where the event horizon would be if the evaporation rate were smooth, without quantum fluctuations. Then, we show that quantum fluctuations of the evaporation rate in the future can move the event horizon to the inside of the firewall location, rendering it naked.

2.1. *Vaidya metric assumption*

For simplicity, we approximate the metric near the horizon of an evaporating black hole by the Vaidya metric with a negative energy influx:

$$ds^2 = -\left(1 - \frac{2M(v)}{r}\right) dv^2 + 2dvdr + r^2 d\Omega^2. \tag{1}$$

Here $M(v)$ is the mass of the black hole, which is decreasing as a function of the advanced time v. For a smooth evaporation rate of a spherical black hole emitting mainly photons and gravitons, we shall take (in Planck units)

$$\dot{M} \equiv \frac{dM}{dv} = -\frac{\alpha}{M^2}, \tag{2}$$

where α is a constant that has been numerically evaluated[8–12] to be about 3.7474×10^{-5}.

2.2. *Location of the event horizon with smooth evaporation*

The apparent horizon is located at $r_{\mathrm{ApH}} = 2M(v)$, whereas the event horizon is generated by radially outgoing null geodesics,

$$\dot{r} \equiv \frac{dr}{dv} = \frac{1}{2}\left(1 - \frac{2M(v)}{r}\right), \tag{3}$$

on the boundary of such null geodesics reaching out to future null infinity, instead of falling into the singularity that is believed to be inside the black hole. For a smooth evaporation rate $\dot{M} \equiv dM/dv = -\alpha/M^2$, the event horizon is given by the solution to Eq. (3) such that it does not diverge exponentially far away from the apparent horizon in the future. If we define $u \equiv 1 - r/(2M)$ and $p \equiv -4\dot{M}$ and assume that $n \equiv -d\ln p/d\ln M$ is constant, then we noted[7] that the event horizon is at

$$u = p + (n-2)p^2 + (n-1)(2n-5)p^3 + (6n^3 - 28n^2 + 37n - 14)p^4 + O(p^5). \tag{4}$$

2.3. *The adiabatic horizon*

For a smooth Hawking evaporation into massless particles with $p \equiv -4\dot{M} = 4\alpha/M^2$, so that $n = 2$, one finds that the event horizon is at

$$r_{\rm EH} = 2M\left[1 - \frac{4\alpha}{M^2} + O\left(\frac{\alpha^3}{M^6}\right)\right]. \tag{5}$$

For a general spherical metric, the covariant generalization of $\dot{f}$ along an outward null direction toward the future is $\dot{f} \equiv df/dv \equiv N^\alpha \partial f/\partial x^\alpha$, with outward null vector N^α normalized so that $\dot{r} \equiv dr/dv \equiv N^\alpha r_{,\alpha} = (1/2)\nabla r \cdot \nabla r = (1/2) - M/r$. Note that $dM/dv = -\alpha/M^2$ implies that $d^2(M^3)/dv^2 = 0$, but since $r_{\rm EH} \approx 2M$, we have $d^2(r_{\rm EH}^3)/dv^2 \approx 0$ as well. We therefore defined[7] an *adiabatic horizon* at $r_{\rm AdH}$ by the outer solution to

$$\frac{d^2}{dv^2}(r_{\rm AdH}^3) \equiv N^\alpha \frac{\partial}{\partial x^\alpha}\left(N^\beta \frac{\partial}{\partial x^\beta} r_{\rm AdH}^3\right) = 0. \tag{6}$$

2.4. *The location of the adiabatic horizon*

The location of the adiabatic horizon is very near where the event horizon would be if the evolution of the event horizon followed the adiabatic mass evolution law $dM/dv = -\alpha/M^2$. We pointed out[7] that $r_{\rm AdH}$ is equivalent to the location where the gradient vector of $(1/4)\nabla(r^2) \cdot \nabla(r^2) = r^2 \nabla r \cdot \nabla r \equiv r^2 - 2Mr$ (which, incidentally, defines M) is in the outward null direction, or $N^\alpha(r^2\nabla r \cdot \nabla r)_{,\alpha} = 0$, which gives $\dot{M} = (1/2) - (3/2)(M/r) + (M/r)^2$ and

$$r_{\rm AdH} \equiv \frac{4M}{3 - \sqrt{1 + 16\dot{M}}}. \tag{7}$$

When $dM/dv = -\alpha/M^2$ holds, this expression agrees with $r_{\rm EH} = 2M[1 - 4\alpha/M^2 + O(\alpha^3/M^6)]$ to the order given.

2.5. *The fourth-root horizon*

One disadvantage of our definition of the adiabatic horizon[7] for a black hole with fluctuating emission is that it directly depends on the mass emission rate $-\dot{M} \equiv -dM/dv$, so that a sudden fluctuation of this emission rate away from α/M^2 can lead to a sudden change in $r_{\rm AdH}$. If one knows what α is, one could avoid sudden jumps in $r_{\rm AdH}$ by replacing $\dot{M}$ in Eq. (7) with $-\alpha/M^2$, so that at any advanced time $r_{\rm AdH}$ just depends on the mass M at that value of v. Then, the time derivative of $r_{\rm AdH}$ can jump suddenly, but not the value of $r_{\rm AdH}$ itself if $-\dot{M}$ is always finite.

However, even this has the disadvantage that near the final evaporation of the black hole, when $M < 4\sqrt{\alpha}$, Eq. (7) with $\dot{M}$ replaced by $-\alpha/M^2$ gives an imaginary square root in the denominator, so real values of this modified adiabatic horizon no longer exist. A modification that would give real and positive values for all positive M but which still matches the first two terms of Eq. (5) is what one might call the *fourth-root horizon*,

$$r_{4\mathrm{rH}} \equiv \sqrt{\sqrt{(2M)^4 + (32\alpha)^2} - 32\alpha}. \tag{8}$$

The inverse of this relation is $M = (1/2)(r_{4\mathrm{rH}}^4 + 64\alpha r_{4\mathrm{rH}}^2)^{1/4}$. If one had a model evaporating black hole for which the event horizon (say denoted by r) were exactly at the fourth-root horizon, so that $r = r_{4\mathrm{rH}}$, then using Eqs. (3) and (8) and the definition $\epsilon \equiv 8\alpha/M^2$ leads to $-M^2\dot{M}/\alpha = 2\sqrt{1+\epsilon^2}\big(1 - \sqrt{\sqrt{1+\epsilon^2} - \epsilon}\big)/\epsilon$, which is very close to 1 for $M \gg 4\sqrt{\alpha}$ ($\epsilon \ll 1$), but then it decreases to a minimum near 0.954142 at $\epsilon \approx 0.398463$ before increasing to 2 as the mass goes to zero and ϵ diverges, with $-M^2\dot{M}/\alpha$ always staying finite and remaining within a factor of 2 of unity and hence giving $\dot{M} \sim -\alpha/M^2$ as M decreases to zero.

2.6. *The smoothed horizon*

One can get an even smoother approximation for the location of the event horizon, and hence for where the firewall might be if its location depends only on the past behavior of the spacetime, by taking the *smoothed horizon* to be at the radius given by the following weighted integral of the mass $M(v')$ over a positive variable y at advanced times $v' = v - 4M(v)y$ before the time v:

$$r_{\mathrm{SmH}}(v) \equiv 2\int_0^\infty dy\, M(v - 4M(v)y)\left(\frac{1+2a}{a^2}\exp\Big(-\frac{y}{a}\Big) - \frac{1+a}{2a^2}\exp\Big(-\frac{y}{2a}\Big)\right), \tag{9}$$

for a positive smoothing parameter a. Then if $M(v)$ had a linear dependence on v over the range where the integrand is significant, this would give the extrapolation $r_{\mathrm{SmH}}(v) = 2M(v + 4M(v))$, the radius of the apparent horizon at a time greater than v by four times the mass at v (i.e. by a time greater than v by the inverse of the approximate expression $1/(4M)$ for the surface gravity of the black hole at the advanced time v), which is what the first two terms of Eq. (5) give for constant $\dot{M}$. Assuming that the Hawking radiation has significant fluctuations only over a timescale of the order of $M(v)$, when $1 \ll a \ll M(v)^2$ the fluctuations in $M(v)$ will be nearly smoothed out and yet the integral will be dominated by a linear behavior in v of the smoothed $M(v)$ to give a good estimate of where the event horizon would be if $M(v)$ actually had the smoothed form. Of course, we do not know whether or not the firewall would be at or near this smoothed horizon radius $r_{\mathrm{SmH}}(v)$, but one might think that if the smoothing parameter were taken to be $a \sim \ln(M(v)) \gg 1$, giving smoothing over roughly the scrambling time, then $r_{\mathrm{SmH}}(v)$ could be a plausible guess for where the firewall might be if it really existed.

2.7. *Location of the firewall near the smoothed horizon*

We shall assume[7] that the firewall, if it exists, is close to where the event horizon would be if the black hole evolves smoothly and adiabatically according to

$dM/dv = -\alpha/M^2$, as perhaps well estimated by the new definition of the smoothed horizon for some suitable value of the smoothing parameter a. However, the actual event horizon depends on the future evolution of the spacetime, and not just on that of its past. Therefore, quantum fluctuations in the future spacetime can lead the event horizon to deviate significantly from the smoothed horizon. If the mass loss rate exceeds the adiabatic formula, then the event horizon will be inside the smoothed horizon. As a result a firewall located at the smoothed horizon would become naked, visible from future null infinity. A schematic Penrose diagram for an unclothed firewall is given in Fig. 1.

3. Quantum Fluctuation for the Mass Evaporation

From $\dot{r} \equiv dr/dv = (1/2) - M/r$, one can write[7] the mass $M = M(v)$ in the Vaidya metric in terms of the event horizon radius $r = r(v) \equiv r_{\rm EH}(v)$ as $M = \frac{1}{2}r - r\dot{r}$. Let M_1, r_1 and M_2, r_2 be the mass and radius of the unperturbed black hole and of its fluctuations, respectively, with the total mass $M = M_1 + M_2$ and the event horizon radius $r = r_1 + r_2$. The unperturbed black hole horizon radius r_1 is near where the smoothed horizon would be in the limit of very large smoothing parameter a.

Now suppose that the unperturbed mass loss would give $M = M_1 = M_1(v) = (1/2)r_1 - r_1\dot{r}_1$, such that $\dot{M}_1 \approx -\alpha/M_1^2$, and that quantum fluctuations $M_2 = M_2(v)$ and $r_2 = r_2(v)$ are small compared with the total mass and the event horizon radius, respectively. Then $M = M_1 + M_2 = (1/2)r - r\dot{r} = (1/2)(r_1 + r_2) - (r_1 + r_2)(\dot{r}_1 + \dot{r}_2) \approx M_1 + (1/2)r_2 - r_1\dot{r}_2$.

For simplicity, we are making the highly idealized assumption that even with quantum fluctuations, the metric remains spherically symmetric and Vaidya near the event horizon.

3.1. *Departure of the event horizon from the unperturbed horizon*

Now for some particular advanced time $v = v_0$, let us ignore quantum fluctuations before this time, so that $M_2(v) = 0$ for $v < v_0$, and let us define the constant $M_0 = M(v_0) = M_1(v_0)$. To leading order in $M_0 \gg 1$ and $|v - v_0| \ll M_0^3$, the fractional decay of the black hole over the advanced time $v - v_0$ is small, and the negative of the coefficient of $\dot{r}_2$ in $M \approx M_1 + \frac{1}{2}r_2 - r_1\dot{r}_2$ may be written as $r_1 \approx 2M_1 \approx 2M_0$. Then, one gets $(1/2)r_2 - 2M_0\dot{r}_2 \approx M_2(v)$. The solution of this differential equation that has no exponentially growing departure of the event horizon $r(v) = r_1 + r_2$ from the unperturbed horizon $r_1(v)$ at late times is[7]

$$r_2 \approx \exp\left(\frac{v - v_0}{4M_0}\right) \int_v^\infty dv' \frac{M_2(v')}{2M_0} \exp\left(\frac{v_0 - v'}{4M_0}\right). \quad (10)$$

3.2. *A particular fluctuation for the evaporation rate*

Since the adiabatic evolution gives $\dot{M}_1 \approx -\alpha/M_0^2$ for $M_0 \gg 1$ and $|v - v_0| \ll M_0^3$, let us consider[7] a quantum mass fluctuation that gives, with $\theta(v - v_0)$ the Heaviside

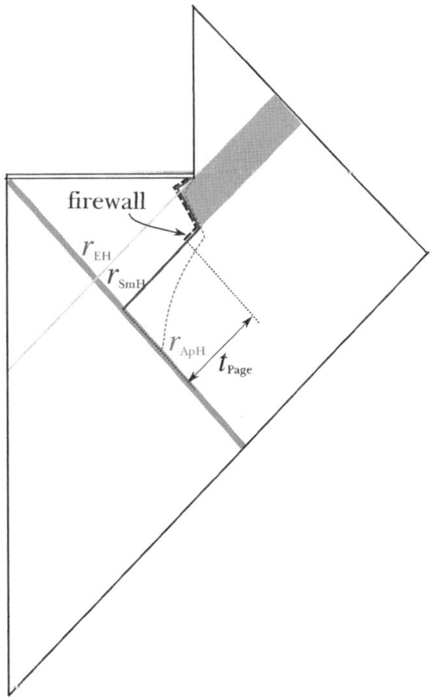

Fig. 1. A Penrose diagram illustrating the formation of a Schwarzschild black hole from a collapsing null shell, and its subsequent Hawking evaporation.[7] The event horizon (r_{EH}) has been shifted inward some distance from the smoothed horizon (r_{SmH}) due to a quantum fluctuation. This renders the firewall (denoted by the dashed curve that appears after the Page time t_{Page}) naked. The apparent horizon (r_{ApH}) is also shown for comparison, but light rays can escape from inside it, since the black hole is shrinking.

step function,

$$\dot{M}_2 = -\theta(v - v_0)\frac{\alpha}{M_0^2} f \exp\left(-\frac{\beta(v - v_0)}{4M_0}\right), \tag{11}$$

which has two constant parameters, namely f for how large the quantum fluctuation in the energy emission rate is relative to the adiabatic emission rate $-\alpha/M^2$ (with f assumed to be positive so that the quantum fluctuation increases the emission rate above the adiabatic value), and β for how fast the quantum fluctuation in the energy emission rate decays over an advanced time of $4M_0$ (the inverse of the surface gravity κ of the black hole).

3.3. *Results for the mass and horizon fluctuation*

Then with $M_2(v) = 0$ for $v < v_0$, one gets

$$M_2 \approx -\theta(v - v_0)\frac{4\alpha}{\beta M_0} f \left[1 - \exp\left(-\frac{\beta(v - v_0)}{4M_0}\right)\right]. \tag{12}$$

Plugging this back into Eq. (10) then gives[7]

$$r_2 \approx -\,\theta(v_0 - v)\frac{8\alpha f}{(1+\beta)M_0}\exp\left(\frac{v - v_0}{4M_0}\right) - \theta(v - v_0)\frac{8\alpha f}{\beta(1+\beta)M_0}\left[1 + \beta - \exp\left(\frac{-\beta(v - v_0)}{4M_0}\right)\right]. \tag{13}$$

3.4. *Consequences of this fluctuation*

This particular form of the emission rate fluctuation implies[7] that the total mass fluctuation from the adiabatic evolution is $M_2(\infty) = -4\alpha f/(\beta M_0)$. Then the radial fluctuation in the event horizon radius at the advanced time $v = v_0$, when $-r_2(v)$ has its maximum value, is $r_2(v_0) \approx [2\beta/(1+\beta)]M_2(\infty)$. This means that if the quantum fluctuation in the energy emission rate is very short compared with $4M_0$ (decaying rapidly in comparison with the surface gravity of the black hole), so that $\beta \gg 1$, then $r_2(v_0) \approx 2M_2(\infty)$, twice the total mass fluctuation. However, we shall just assume that β is of the order of unity and hence get $r_2(v_0) \sim M_2(\infty)$ as an order-of-magnitude relation. Note that the reduction in the radius of the event horizon at $v = v_0$, where the fluctuation in the mass emission rate starts, occurs *before* there is any decrease in the mass below the unperturbed value $M_1(v)$, and hence before any decrease in the radius of the adiabatic horizon,[7] fourth-root horizon, or smoothed horizon, because the location of the 'teleological' event horizon is defined by the future evolution of the spacetime.

3.5. *Quantum fluctuations can render a firewall naked*

Therefore, if the putative firewall occurs at a location determined purely causally by the past behavior of the spacetime, and is sufficiently near where the event horizon would be under adiabatic emission thereafter (say near the smoothed horizon), then quantum fluctuations, at later advanced times that reduce the mass of the hole below that given by the adiabatic evolution, would move the actual event horizon inward (even before quantum fluctuations in the mass emission rate begin), so that the event horizon becomes inside the location of the putative firewall. That is, quantum fluctuations that increase the mass emission rate render such a firewall naked, visible to the external universe.[7]

4. A Possible Objection and Answer

One possible objection to this conclusion is that for α, β, and f all of the order of unity, the inward shift in the event horizon is by a change of radius, r_2, of the order of $1/M$, so that the proper distance from the putative firewall that might be close to the smoothed horizon that is near $r = r_1$ to the event horizon at $r = r_1 + r_2$, in the frame of the timelike firewall surface outside the event horizon, is of the order of the Planck length. The proper acceleration of an observer that stays of the order of the Planck length outside the event horizon would be of the order of the Planck acceleration, giving an Unruh temperature of the order of the Planck temperature. One might object that quantum gravity effects at such extreme accelerations would make a naked firewall in practice indistinguishable from a firewall at or inside the event horizon.

However, we claim that an arbitrarily large black hole will with high probability become highly naked.[7] To see this, consider a black hole with initial entropy $S \gg 1$, which emits roughly S particles throughout its long life time. There would be roughly S approximately independent chances for the proper distance fluctuation of the event horizon inside the firewall to achieve some large value $L \gg 1$, so that the probability at any one time needs only be $P(L) \sim 1/S$. The most probable way to produce such a large fluctuation L at time $v = v_0$, when the Hawking temperature is T_0, is to have thereafter the radiation be locally thermal with a time-dependent temperature $T(v) = T_0(z+1)/[z+1-ze^{-(v-v_0)/(4M_0)}]$, where $z = [T(v_0)-T_0]/T_0 \gg 1$ is a constant parameter chosen to give the desired $L = [8M_0(-r_2(v_0))]^{1/2}$. It can be shown that the probability of this fluctuation is $P(L) \sim \exp[-(\pi/2)L^2]$. Setting $P(L) = 1/S$ then yields $L \approx [(2/\pi)\ln S]^{1/2}$, which is indeed arbitrarily large for arbitrarily large S.

5. Disrobing Firewalls by Black Hole Mining

Adam Brown has suggested[13] that one need not wait for the firewall to become naked by a spontaneous quantum fluctuation, but that one could disrobe it by black hole mining. This could cause the event horizon to shrink much further below the adiabatic horizon than would be very probable with spontaneous quantum fluctuations, so that it seems that with suitable mining one could almost certainly make a firewall naked. However, the details of this are work in progress.

6. Firewalls Are Not Conservative

In conclusion, it seems that if putative firewalls are at locations determined by the past structure of the spacetime to be near where the event horizon would be if the black hole evolved adiabatically in the future, quantum fluctuations in the emission rate,[7] or black hole mining, can render the firewalls naked.

Therefore, the firewall is *not* hidden in the region with Planckian local temperature; its presence would truly be at odds with general relativity and the usual quantum field theory.

More specifically, being in the exterior of the event horizon means that the firewall could potentially influence the exterior spacetime, so that even observers who do not fall into the black hole could have a fiery experience. In addition, the presence of a firewall well outside the event horizon could affect the spectrum of the Hawking radiation, which means that the presence of a firewall could be inferred even by asymptotic observers.[14,15] Such a ‘naked firewall,’ i.e. a firewall far outside the event horizon,[7] is therefore problematic, and giving up the no-drama assumption no longer seems like a palatable 'most conservative' option.

Acknowledgments

The work of PC is supported in part by Taiwan's National Center for Theoretical Sciences (NCTS) and Taiwan's Ministry of Science and Technology (MOST), of DNP by the Natural Sciences and Engineering Council of Canada, of MS by MEXT KAKENHI Grant Number 15H05888, and of DY by the Leung Center for Cosmology and Particle Astrophysics (LeCosPA) of National Taiwan University (103R4000). YCO thanks Nordita for various travel supports.

References

1. A. Almheiri, D. Marolf, J. Polchinski and J. Sully, *J. High Energy Phys.* **1302** (2013) 062, arXiv:1207.3123 [hep-th].
2. A. Almheiri, D. Marolf, J. Polchinski, D. Stanford and J. Sully, *J. High Energy Phys.* **1309** (2013) 018, arXiv:1304.6483 [hep-th].
3. D. N. Page, *Phys. Rev. Lett.* **71** (1993) 1291, arXiv:gr-qc/9305007.
4. D. N. Page, *Phys. Rev. Lett.* **71** (1993) 3743, arXiv:hep-th/9306083.
5. D. N. Page, *J. Cosmol. Astropart.* **1309** (2013) 028, arXiv:1301.4995 [hep-th].
6. L. Susskind, L. Thorlacius and J. Uglum, *Phys. Rev. D* **48** (1993) 3743, arXiv:hep-th/9306069.
7. P. Chen, Y. C. Ong, D. N. Page, M. Sasaki and D.-H. Yeom, *Phys. Rev. Lett.* **116** (2016) 161304, arXiv:1511.05695 [hep-th]
8. D. N. Page, *Phys. Rev. D* **13** (1976) 198.
9. D. N. Page, *Phys. Rev. D* **14** (1976) 3260.
10. D. N. Page, *Phys. Rev. D* **16** (1977) 2402.
11. D. N. Page, *Phys. Rev. Lett.* **50** (1983) 1013.
12. D. N. Page, *New J. Phys.* **7** (2005) 203, arXiv:hep-th/0409024.
13. A. Brown, Private communication (2015).
14. D.-il Hwang, B.-H. Lee and D.-H. Yeom, *J. Cosmol. Astropart. Phys.* **1301** (2013) 005, arXiv:1210.6733 [gr-qc].
15. W. Kim, B.-H. Lee and D.-H. Yeom, *J. High Energy Phys.* **1305** (2013) 060, arXiv:1301.5138 [gr-qc].

A Quantized Spacetime Based on $Spin(3,1)$ Symmetry*

Pisin Chen*,†,‡,§,¶, Hsu-Wen Chiang*,†,‖ and Yao-Chieh Hu*,†,‡,**

*Department of Physics, National Taiwan University, Taipei 10617, Taiwan, ROC

†Leung Center for Cosmology and Particle Astrophysics, National Taiwan University, Taipei 10617, Taiwan, ROC

‡Graduate Institute of Astrophysics, National Taiwan University, Taipei 10617, Taiwan, ROC

§Kavli Institute for Particle Astrophysics and Cosmology, SLAC National Accelerator Laboratory, Stanford University, Stanford, CA 94305, USA

¶pisinchen@phys.ntu.edu.tw

‖b98202036@ntu.edu.tw

**r04244003@ntu.edu.tw

We introduce a new type of the spacetime quantization based on the spinorial description suggested by loop quantum gravity. Specifically, we build our theory on a string theory inspired $Spin(3,1)$ worldsheet action. Because of its connection with quantum gravity theories, our proposal may in principle link back to string theory, connect to loop quantum gravity where SU(2) is suggested as the fundamental symmetry, or serve as a Lorentzian spin network. We derive the generalized uncertainty principle and demonstrate the holographic nature of our theory. Due to the quantization of spacetime, geodesics in our theory are fuzzy, but the fuzziness is shown to be much below conceivable astrophysical bounds.

Keywords: Noncommutative geometry; quantum gravity; generalized uncertainty principle.

PACS Number: 04.60.−m

1. Introduction

In both superstring and loop quantum gravity theories, there exist a minimal distance measure, which suggests that the infinitely-differentiable spacetime manifold may be an illusion that ceases to be valid at very small scales. This suggests that spacetime may be quantized. Snyder coined the attempt as "quantized spacetime".[1] Motivated by possible fundamental descriptions of the quantum gravity, we intro-

*This article first appeared in *Int. J. Mod. Phys. D* **25** (2016) 1645004.

duce a new type of the spacetime quantization based on the spinorial description, specifically, a string theory inspired $Spin(3,1)$ worldsheet action.[2]

In the attempts of quantizing spacetime, the quadratic distance function g is usually promoted to its quantum version without modification. A new way of deforming the spacetime algebra was recently proposed by Adler.[3] Instead of taking the spectral square root[4] of g to obtain the spectrum of the distance operator, this new approach takes the route via the Clifford algebra. One then obtains a linear line element $\Delta s_L = \gamma_\mu \Delta x^\mu$ (see footnote a). The price to pay, however, is that the new transfer function γ_μ is a matrix instead of a pure number. To compensate that, Adler reinterprets Δs_L as a quantum mechanical (QM) object $\hat{\Delta s}$ called "linear line element operator", whose eigenvalue is the proper distance and the state describes an eigendirection ϵ_μ along which the measure is completely certain, i.e.,

$$\Delta s = \langle \hat{\Delta} s \rangle = \langle \epsilon | \gamma_\mu \Delta x^\mu | \epsilon \rangle = (|\epsilon\rangle)^\dagger \gamma_\mu \Delta x^\mu |\epsilon\rangle = \epsilon_\mu \Delta x^\mu. \tag{1}$$

The proper distance is thus treated as the addition of generators of the spin group on the classical x-space, similar to the 't Hooft operator on a spacetime lattice, and the infinitely differentiable manifold is replaced by a piecewise-linear one.

An uncertainty is introduced as $var(\Delta s) = \Delta s^2 - (\dot{x}^\mu \epsilon_\mu)^2 \Delta s^2$, where $\dot{x}^\mu$ is the four-velocity. The preferred direction ϵ_μ along which the measurement is completely certain inevitably breaks the Lorentz invariance, unless the theory is nondissipative or is at high temperature limit. The whole purpose of using the Clifford algebra to preserve the Lorentz invariance is therefore defeated. Another problem is that gamma matrices and the coordinate difference are living in different representations, rendering the linear line element representation dependent. In next section, we will show that following Adler's philosophy but recasting the linear line element as $\Delta s = \gamma_\mu \langle \lambda \gamma^\mu \rangle$, a new kind of spacetime quantization can be developed that is fully covariant, representation independent, and with rich structures and salient properties.[2]

2. Constructing Action Through Bosonization

Equation (1) suggests that Adler's approach of spacetime quantization may be recast as a worldline action in terms of bi-fermionic fields that exhibit bosonic behavior. A natural tool to attain this action would be the bosonization in $2d$ quantum field theory (QFT).[5] By considering the worldline as the low energy limit of a cylindrical worldsheet, we may apply the fermionization on the worldsheet action and compactify it back to $1d$. To fix the gauge structure $g(X)$ such that the fermionization procedure can be applied, we utilize the tetrad formalism and obtain a $SO(3,1)$ charged scalar theory. However, since we are aspired to construct

[a]For the sake of readability, IJK are for $SO(3,1)$ tangent bundle coordinate indices, $\mu\nu\lambda\rho$ for 4-d coordinate, ijk for 3-d coordinate, and $(i)(j)(k)$ for site numbers. Gamma matrices γ are defined according to the type of the indices (Minkowskian if unspecified). We also use natural units $c = \hbar = G = 1$, and the signature of g is chosen to be $(+,-,-,-)$.

a spinor theory, we replace the group structure with $Spin(3,1)$, which is effectively equivalent to $SO(3,1)$ classically.

We fermionize the theory by checking the bosonization dictionary[6] and arrive at a trivial charged fermion action describing a 4d spacetime spinor valued 2d worldsheet Majorana-Weyl spinor. From constraints of the worldline action, we can finally obtain the compactified action in terms of the spinor field ψ. Since the worldsheet spinor is conserved, we obtain the quantization condition[2] $\Delta X^I = \lambda v^I/\sqrt{|v^2|}$, where $v^I \equiv dX^I/d\tau$ is the velocity of the original worldline action in terms of tetrads and $\lambda \sim \mathcal{O}$(Planck length) is the characteristic length of the quantized spacetime. We can also show that $\langle \Delta\hat{X}^\mu \rangle = \lambda\bar{\psi}\gamma^\mu\psi = \lambda v^\mu$, where $v^2 = 1, 0, \text{or} -1$. Notice that since here we focus on eigenstates of the velocity operator, $ds^2 = \bar{\psi}\psi$ must vanish. If the constraint is substituted with $ds^2 = \text{const.}$, we immediately obtain $\Delta s^2 = \bar{\psi}\psi = 1, 0, \text{or} -1$. We now have a concrete quantization scheme that can be reproduced explicitly by a fermionized worldline action.

3. Spacetime Interval Operator

Let us introduce the "spacetime interval operator" $\Delta\hat{X}^I$ as

$$\langle \Delta\hat{X}^I \rangle = \bar{\psi}\Delta\hat{X}^I\psi = \bar{\psi}e^I_\mu \Delta\hat{X}^\mu\psi = \psi^\dagger\gamma^0\Delta\hat{X}^I\psi = \lambda\psi^\dagger\gamma^0\gamma^I\psi, \tag{2}$$

$$\Delta s^2 = \eta_{IJ}\langle \Delta\hat{X}^I\Delta\hat{X}^J \rangle = 4\lambda^2\bar{\psi}\psi - \perp 4\lambda^2, \text{or } 0. \tag{3}$$

Here e^I_μ is the tetrad. $\langle \Delta\hat{X}^I \rangle$ is a normalized non-null vector for a non-null state or a null vector with positive time component for a null state. The choice of the normalization can be appreciated by looking at the solution of the Dirac field equation.

This new measurement is neither a distance measure nor a coordinate difference operator, but a difference operator on the tangent bundle of the manifold. A better way to understand it is to treat it as a discretized version of the velocity four-vector. Only when combined with the tetrad does the coordinate difference measure $\Delta\hat{X}^\mu$ reappear. Just like a vector in GR, the components of $\Delta\hat{X}^I$ are not physical. To measure the velocity of a particle we need a two-particle interaction, and the physical object is the inner product $r_I\Delta\hat{X}^I$, where r_I is the classical trajectory of a probe particle expressed in the same representation as $\Delta\hat{X}^I$. Therefore all the derivations and results we obtained are representation independent.

The new choice of interpretation also implies the existence of an underlying minimal distance. But unlike what Adler obtained[3], in our case even uncertainties are Lorentz invariant. One may try to obtain the variances of the interval measures

$$\langle \text{var}(r_I\Delta X^I) \rangle = |\langle (r_I\Delta X^I)^2 \rangle - \langle r_I\Delta X^I \rangle^2| = \lambda^2|r_I r^I n_J n^J - (r_K n^K)^2|, \tag{4}$$

$$\langle \text{var}(\Delta s) \rangle = |\langle \Delta X^I\Delta X_I \rangle - \langle \Delta X^I \rangle\langle \Delta X_I \rangle| = 3\lambda^2|n_I n^I|. \tag{5}$$

Here $n^I = \langle \Delta X^I \rangle$ and r^I is the ruler. Clearly, along the eigen-direction there is no uncertainty. However, along the transverse direction the measurement is completely uncertain. From this point of view, the behavior of the spacetime interval operator is exactly the same as the spin operators in relativistic QM.

One interesting feature of our theory is that from equation (2), the time difference is proportional to the identity operator, rendering its expectation value positive definite. Thus the arrow of time problem is a nonissue without the need to invoke the second law of thermodynamics.

4. Holography, Non-commutativity and Generalized Uncertainty Principle

We can define the position operator and the position eigenstate as $\langle \hat{X}^\mu \rangle = \sum \lambda \bar{\psi}_{(i)} \gamma^\mu \psi_{(i)}$. The position operator can be envisioned as the 't Hooft operator on a $1d$ spin chain, and the associated Fourier transformation operator as the Wilson line operator. This kind of visualization is inherited from the bosonization. It is now straightforward to derive the commutation and anti-commutation relations:

$$\langle [\hat{X}^\mu, \hat{X}^\nu] \rangle = \sum_i \langle [\Delta \hat{X}^\mu_{(i)}, \Delta \hat{X}^\nu_{(i)}] \rangle, \tag{6}$$

$$\frac{1}{2} \langle \{\hat{X}^\mu, \hat{X}^\nu\} \rangle = \langle \hat{X}^\mu \rangle \langle \hat{X}^\nu \rangle + \sum_i (\lambda^2 g^{\mu\nu} - \langle \Delta X^\mu \rangle_{(i)} \langle \Delta X^\nu \rangle_{(i)}). \tag{7}$$

Clearly the spacetime in this theory, contrary to those of most others, is neither commutative nor anti-commutative.

One important aspect of our theory is its holographic nature. Since a N-line-element state resembles a composite spin-$N/2$ charged particle, by Clebsch-Gordan decomposition it is with no doubt that only $\mathcal{O}(N^3)$ different states are allowed within a four-hyperbola, implying that the spacetime information can be written in a three-dimensional quantum language. With the geodesic equation[2], the degree of freedom, i.e., the entropy, is proportional to the surface area of the system.

Another aspect is that it implies the generalized uncertainty principle (GUP). Let us invoke the ADM formalism to reduce the symmetry down to $Spin(3)$. Under this symmetry, the spacetime behaves as a spin chain in the zero temperature limit, with the coordinate measure being the total spin. It can be reinterpreted as a bosonic system with $SU(2)$ symmetry even at the quantum level, as demonstrated in the non-abelian bosonization. An immediate conclusion is the existence of an exterior derivative that can be treated as a momentum operator.[2,7] One can therefore compute the uncertainty relation, and in the low momentum limit and $\langle x \rangle \to 0$ we obtain the GUP in the usual form:

$$\Delta x_i \Delta p_j \geq \frac{1}{2}(\delta_{ij} + 2\lambda^2 \delta_{ij} \langle p_i \rangle \langle p_j \rangle). \tag{8}$$

5. Astrophysical Tests

It is straightforward to apply the quantized geodesic obtained in the previous section as an astrophysical test for this theory through the variance of the geodesic, i.e., the smearing effect. Consider, for example, the null geodesics of photons with characteristic energy $E \sim 100$ keV emitted from a gamma ray burst (GRB) afterglow at high redshift ($z \approx 10$). Because the Lorentz invariance is manifest in our theory, the smearing effect occurs only along the perpendicular direction. Assuming a Gaussian distribution of photons at the source, we find, from equations (4) and (5), that the fuzziness of the GRB image so induced upon arrival on earth is of the order of $\sqrt{\lambda^2 E d_S\,(z \approx 10)} \approx 10^{-13}$ meter, where $d_S = \int_0^z dz'(1+z')/H(z')$ is the rescaled distance taking the rescaling of the variance into account. Clearly, it is impossible to identity such minute effect in the near future.

6. Conclusion

We show that a non-commutative spacetime theory can be constructed on top of Adler's "linear line element", and derive the associated action from the fermionization of the K-K ground state of the Polyakov action on a cylinder. The theory is shown to be holographic, and passes the astrophysical tests on the smearing effect and Lorentz invariance violation. The theory can also be regarded as the 4-D extension to the $SU(2)$-type models, and from there the generalized uncertainty principle is derived.

This theory is quite potent, with many additional implications that deserve further exploration.[2] For example, the coefficient ξ of the Bekenstein-Hawking entropy, $S = \xi A/4$, may be fixed in our setup. Also so far our theory is still purely kinematic. The dynamics of the theory can be introduced by either invoking loop quantum gravity or by putting back the K-K tower of the worldsheet excitations. Interestingly in loop quantum gravity,[8] it has been argued that the length operator should be defined through Dirac operators. It remains curious whether our work can be served as a representation that can describe length, area and volume easily without introducing the spectral decomposition.[4,9]

Acknowledgement

We appreciate useful discussions with Abhay Ashtekar, Ron Adler and Carlo Rovelli. This work is supported by National Center for Theoretical Sciences (NCTS) of Taiwan, Ministry of Science and Technology (MOST) of Taiwan, and the Leung Center for Cosmology and Particle Astrophysics (LeCosPA) of National Taiwan University.

References

1. H. S. Snyder, *Phys. Rev.* **71** (1947) 38.
2. H.-W. Chiang, Y.-C. Hu and P. Chen, *Phys. Rev. D* **93** (2016) 084043, arXiv:1512.03157.
3. R. J. Adler, arXiv:1402.5921.
4. T. Thiemann, *J. Math. Phys.* **39** (1998) 3372, arXiv:gr-qc/9606092.
5. S. Coleman, *Phys. Rev. D* **11** (1975) 2088.

6. E. Witten, *Commun. Math. Phys.* **92** (1984) 455.
7. E. Batista and S. Majid, *J. Math. Phys.* **44** (2003) 107, arXiv:hep-th/0205128.
8. C. Rovelli, arXiv:1408.3260.
9. E. Bianchi, *Nucl. Phys. B* **807** (2009) 591, arXiv:0806.4710.

Schwinger Effect, Hawking Radiation and Unruh Effect*

Sang Pyo Kim
Department of Physics, Kunsan National University,
Kunsan 54150, Korea
sangkim@kunsan.ac.kr

We revisit the Schwinger effect in de Sitter (ds), anti-de Sitter (Ads) spaces and charged black holes, and explore the interplay between quantum electrodynamics (QED) and the quantum gravity effect at one-loop level. We then advance a thermal interpretation of the Schwinger effect in curved spacetimes. Finally, we show that the Schwinger effect in a near-extremal black hole differs from Hawking radiation of charged particles in a nonextremal black hole and is factorized into those in an Ads space and a Rindler space with the surface gravity for acceleration.

Keywords: Schwinger effect; Hawking radiation; Unruh effect; (anti-)de Sitter space; Rindler space.

PACS Number(s): 04.62+v, 04.70.Dy, 12.20.−m

1. Introduction

Spontaneous creation of particles or charged pairs from an external gauge field or a curved spacetime is one of the most prominent phenomena in quantum field theory. A strong electric field produces pairs of charged particles and antiparticles, known as the Schwinger mechanism.[1] The more remarkable phenomenon is the emission of all species of particles from black holes, known as Hawking radiation.[2] Under an influence of strong backgrounds, the vacuum may spontaneously breakdown due to quantum fluctuations and virtual pairs can be separated either by the energy of the fields or the causality of spacetimes. An accelerating detector measures a thermal spectrum of the Unruh temperature determined by the acceleration, known as the Unruh effect.[3] The spectrum and characteristics for these effects are summarized in Table 1.

Heisenberg and Euler found the one-loop effective action for an electron in a constant electromagnetic field[4] and Schwinger introduced the proper-time integral method to express the effective action in scalar and spinor quantum electrodynamics (QED),[1] which is now known as the Heisenberg–Euler or Schwinger effective action.

*This article first appeared in *Int. J. Mod. Phys. D* **25** (2016) 1645005.

Table 1. Comparison among the Schwinger mechanism, Unruh effect and Hawking radiation.

	Schwinger mechanism	Unruh effect	Hawking radiation
Agent	Constant electric field	Accelerating detector	Charged RN black hole
Mean number	$N_{\rm S} = e^{-\frac{m}{2T_{\rm S}}}$	$N_{\rm U} = \frac{1}{e^{\frac{\omega}{T_{\rm U}}} \mp 1}$	$N_{\rm H} = \frac{\Gamma_{j\omega lm}}{e^{\frac{\omega - q_j \Phi}{T_{\rm H}}} \mp 1}$
Temperature	$T_{\rm H} = \frac{1}{2\pi}\frac{qE}{m}$	$T_{\rm U} = \frac{a}{2\pi}$	$T_{\rm S} = \frac{1}{2\pi}\frac{\sqrt{M^2-Q^2}}{M+\sqrt{M^2-Q^2}}$

The most distinct feature of the Heisenberg–Euler or Schwinger action is poles of the proper-time representation of the action in an electric field. Thus, the one-loop effective action has not only the vacuum polarization (the real part) but also the vacuum persistence amplitude (twice the imaginary part). The vacuum persistence amplitude is a consequence of spontaneous production of charged pairs from the Dirac sea. Note that the Schwinger effect is the particle-hole theory, in which virtual particles from the Dirac sea tunnel through a tilted potential barrier due to the electric field, and does not include the Coulomb attraction of pairs due to the homogeneity of produced pairs.

In this paper, we recapitulate the Schwinger effect in curved spacetimes, such as a de Sitter (dS) space, an anti-de Sitter (AdS) space, and an extremal or near-extremal Reissner–Nordström (RN) black hole. One motivation for studying the Schwinger effect in (A)dS is to unveil the interplay between the Maxwell theory as a U(1) gauge and the quantum gravity effect at one-loop level. Another motivation to study QED in (A)dS is the near-horizon geometry of a near-extremal black hole[5] and the scalar S-wave in the Nariai-geometry of a rotating black hole,[6] which are summarized in Table 2. Further, it would be interesting to investigate whether charged black holes may have the Schwinger effect different from Hawking radiation.

We also provide a thermal interpretation of the Schwinger effect in $(\mathrm{A})\mathrm{dS}_2$, which has recently been introduced by Cai and Kim[7,8] and is a QED analog of the Unruh effect in $(\mathrm{A})\mathrm{dS}_2$.[9,10] The Schwinger effect from an extremal RN black hole has the same spectrum as in AdS_2 since the near-horizon geometry of the extremal black hole is $\mathrm{AdS}_2 \times S^2$ as shown in Table 2. The Schwinger formula from the extremal black hole is similarly given a thermal interpretation.[11,12] Interestingly, the Schwinger effect from a near-extremal black hole is factorized into the Schwinger formula in AdS_2 and that in Rinder_2 with the acceleration of the surface gravity due to small nonextremality of black hole. We find the Schwinger formula in dS

Table 2. QED phenomena in the near-horizon geometry of 4D black holes.

Black hole	Region of parameters	Near-horizon geometry
RN black hole	Near-extremality $M \sim Q$	QED in $\mathrm{AdS}_2 \times S^2$
RN black hole	Nonextremality $M \gg Q$	QED in $\mathrm{Rindler}_2 \times S^2$
Rotating black hole in dS	scalar S-wave	QED in dS_2

in any dimension. A passing remark is that the holographic Schwinger effect is the particle picture of charged pairs including the Coulomb attraction of the pairs.[13]

2. QED Action in dS_2

We consider the planar coordinates for a $(d+1)$-dimensional dS space, dS_{d+1},

$$ds^2 = -dt^2 + e^{2Ht} d\mathbf{x}^2. \tag{1}$$

The electromagnetic field in a curved spacetime measured in a local frame $\theta^\alpha = (dt, e^{Ht}dx^1, \ldots, e^{Ht}dx^d)$ is given by the two-form tensor $\mathbf{F} = F_{\alpha\beta}\theta^\alpha \wedge \theta^\beta$. We assume a constant electric field along the x^1-direction measured by a co-moving observer with the $(d+1)$-velocity $u_\alpha = (1, 0, \ldots, 0)$. Then, $\mathbf{F}_{01} = -E = -\mathbf{F}_{10}$ in the local frame, $\mathbf{A} = -[E(e^{Ht} - 1)/H]dx^1$ and $\mathbf{F} = d\mathbf{A}$ in the metric (1). Thus, the vector potential is given by $A_1 = -E(e^{Ht} - 1)/H$, which has the Minkowski limit $H = 0$.

First, in dS_2 the Schwinger formula (mean number) for charged spinless scalars is given by the dimensionless instanton action[14,15]

$$N_S = e^{-S_{dS}}, \quad S_{dS} = 2\pi \left(\sqrt{\left(\frac{qE}{H^2}\right)^2 + \left(\frac{m}{H}\right)^2 - \left(\frac{1}{2}\right)^2} - \frac{qE}{H^2} \right). \tag{2}$$

The Schwinger formula (2) can be interpreted in terms of the effective temperature introduced by Cai and Kim[7]

$$N_S = e^{-\frac{\bar{m}}{T_{CK}}}, \quad T_{CK} = \sqrt{T_U^2 + T_{GH}^2} + T_U, \tag{3}$$

where T_U is the Unruh temperature for accelerating charge, T_{GH} is the Gibbons–Hawking temperature,[16] and $\bar{m}$ is the effective mass in dS_2, which are respectively

$$T_U = \frac{\frac{qE}{\bar{m}}}{2\pi}, \quad T_{GH} = \frac{H}{2\pi}, \quad \bar{m} = \sqrt{m^2 - \left(\frac{H}{2}\right)^2}. \tag{4}$$

It is interesting to compare the effective temperature (3) with the effective temperature for an accelerating observer in dS_2[9,10]

$$T_{NPT-DL} = \sqrt{T_U^2 + T_{GH}^2}, \quad T_U = \frac{a}{2\pi}. \tag{5}$$

By solving the field equation and using the Bogoliubov transformation, the pair-production rate is found[7,8,12,17]

$$\frac{d^2 N_{dS}}{dtdx} = \frac{(2|\sigma| + 1)H^2 S_\mu}{2(2\pi)^2} \left(\frac{e^{-(S_\mu - S_\lambda)} \pm e^{-2S_\mu}}{1 - e^{-2S_\mu}} \right), \tag{6}$$

where the upper (lower) sign is for scalars (fermions) and the dimensionless instanton actions are

$$S_\mu = 2\pi \sqrt{\left(\frac{qE}{H^2}\right)^2 + \left(\frac{m}{H}\right)^2 - \left[\left(\frac{1}{2}\right)^2\right]}, \quad S_\lambda = 2\pi \frac{qE}{H^2}. \tag{7}$$

The prefactor $H^2 S_\mu / 2(2\pi)^2$ is the density of states and the square bracket in Eq. (7) is present only for scalars but vanishes for fermions. Noting $S_{dS} = S_\mu - S_\lambda$, the leading term of Eq. (6) is the Schwinger formula (2) from the instanton action. Without the density of states, the quantity in Eq. (6) is the mean number of created pairs. In the in-out formalism, the vacuum persistence amplitude (integrated action density) is related to the mean number of pairs as

$$2\mathrm{Im}(\mathrm{W_{dS}}) = \pm[\ln(1 \pm e^{-(S_\mu - S_\lambda)}) - \ln(1 - e^{-2S_\mu})], \tag{8}$$

which is the pressure from quantum gas. The first logarithm in Eq. (8) is the standard QED action with the mean number $N = e^{-\bar{m}/T_{\mathrm{CK}}}$ while the second one is a correction due to a charged vacuum in dS and has the character of spinor QED regardless of spins.

Employing the gamma-function regularization in Ref. 18, the QED action density for scalars is[7,8]

$$\mathcal{L}^{\mathrm{sc}}_{\mathrm{dS}} = \frac{H^2 S_\mu}{2(2\pi)^2}\mathcal{P}\int_0^\infty \frac{ds}{s}\left[e^{-(S_\mu - S_\lambda)s/2\pi}\left(\frac{1}{\sin\left(\frac{s}{2}\right)} - \left(\frac{2}{s} + \frac{s}{12}\right)\right)\right.$$
$$\left. - e^{-S_\mu s/\pi}\left(\frac{\cos\left(\frac{s}{2}\right)}{\sin\left(\frac{s}{2}\right)} - \left(\frac{2}{s} - \frac{s}{6}\right)\right)\right]. \tag{9}$$

Similarly, we find the QED action density for fermions

$$\mathcal{L}^{\mathrm{sp}}_{\mathrm{dS}} = -\frac{H^2 S_\mu}{(2\pi)^2}\mathcal{P}\int_0^\infty \frac{ds}{s}(e^{-(S_\mu - S_\lambda)s/2\pi} - e^{-S_\mu s/\pi})\left(\frac{\cos\left(\frac{s}{2}\right)}{\sin\left(\frac{s}{2}\right)} - \left(\frac{2}{s} - \frac{s}{6}\right)\right). \tag{10}$$

Here, $\mathcal{P}$ denotes the principal value and the subtracted terms renormalize the vacuum energy and the charge. The bosonic and fermionic current density is given by

$$J_{\mathrm{dS}} = 2q\frac{d^2 N_{\mathrm{dS}}}{dtdx}. \tag{11}$$

Note that the current density comes from the second quantized field theory in curved spacetime, which is equivalent to (2 charge) × (density of states) × (mean number). In the limit of $mH \leq qE \ll H^2$, i.e. $S_\mu \approx (qE/H^2)$ and $S_{\mathrm{dS}} \approx \pi(mH/qE)^2$, the current density accumulated over the Hubble time $1/H$ increases as $(H^3/E) \times (2\pi m^2)$ and exhibits the infrared hyperconductivity, which has been observed in Ref. 19.

3. QED Action in dS$_{d+1}$

The Schwinger effect in dS$_4$ has been studied in Refs. 20 and 21. We now study the Schwinger effect in $(d+1)$ dimensions, in which the field equation for charged scalars takes the form

$$\ddot{\phi}_{\mathbf{k}} + dH\dot{\phi}_{\mathbf{k}} + \left(m^2 + e^{-2Ht}\left(\bar{k}_1 + \frac{qE}{H}e^{Ht}\right)^2 + \mathbf{k}_\perp^2 e^{-2Ht}\right)\phi_{\mathbf{k}} = 0, \tag{12}$$

where $\bar{k}_1 = k_1 - qE/H$ and $\mathbf{k}_\perp$ is the $(d-1)$-dimensional momentum transverse to the electric field. Then, the positive and negative frequency solutions for the in-vacuum ($t = -\infty$) are given by the Whittaker function as

$$\begin{aligned} \phi^{(+)}_{\text{in},\mathbf{k}}(t) &= \frac{e^{\pi\kappa/2}}{\sqrt{2\bar{k}}} W_{-i\kappa,i\gamma_{\text{dS}}}\left(2e^{-i\pi/2}\frac{\bar{k}}{H}e^{-Ht}\right), \\ \phi^{(-)}_{\text{in},\mathbf{k}}(t) &= \frac{e^{\pi\kappa/2}}{\sqrt{2\bar{k}}} W_{i\kappa,i\gamma_{\text{dS}}}\left(2e^{i\pi/2}\frac{\bar{k}}{H}e^{-Ht}\right), \end{aligned} \tag{13}$$

where

$$\kappa = -\frac{qE}{H^2}\frac{\bar{k}_1}{\bar{k}}, \quad \bar{k} = \sqrt{\bar{k}_1^2 + \mathbf{k}_\perp^2}, \quad \gamma_{\text{dS}} = \sqrt{\left(\frac{qE}{H^2}\right)^2 + \left(\frac{m}{H}\right)^2 - \left(\frac{d}{2}\right)^2}. \tag{14}$$

The positive and negative frequency solutions for the out-vacuum ($t = \infty$) are given by the confluent hypergeometric function as

$$\begin{aligned} \phi^{(+)}_{\text{out},k}(t) &= \frac{e^{-i\pi/4}e^{\pi\gamma_{\text{dS}}/2}}{\sqrt{4\bar{k}\gamma_{\text{dS}}}} M_{i\kappa,i\gamma_{\text{dS}}}\left(2e^{i\pi/2}\frac{\bar{k}}{H}e^{-Ht}\right), \\ \phi^{(-)}_{\text{out},k}(t) &= \frac{e^{-i\pi/4}e^{-\pi\gamma_{\text{dS}}/2}}{\sqrt{4\bar{k}\gamma_{\text{dS}}}} M_{i\kappa,-i\gamma_{\text{dS}}}\left(2e^{i\pi/2}\frac{\bar{k}}{H}e^{-Ht}\right). \end{aligned} \tag{15}$$

The Schwinger effect in a constant electric field in dS_{d+1} should be independent of t and x_1 due the symmetry of the spacetime and the field. In fact, the integration of the longitudinal momentum k_1 of the out-vacuum solution $\phi^{(+)}_{\text{out},k}(t)$ gives the density of states $H^2\gamma_{\text{dS}}/(2\pi)$ independent of dimensions.[8] Therefore, the pair-production rate per spacetime volume is

$$\frac{d^{d+1}N_{\text{dS}}}{dtdx^d} = \frac{(2|\sigma|+1)H^2 S_\mu}{2(2\pi)^2}\int \frac{d^{d-1}\mathbf{k}_\perp}{(2\pi)^{d-1}}\left(\frac{e^{-(S_\mu - S_\lambda)} \pm e^{-2S_\mu}}{1 - e^{-2S_\mu}}\right), \tag{16}$$

where the upper (lower) sign is for charged scalars (fermions) and the dimensionless instanton actions are

$$S_\mu = 2\pi\sqrt{\left(\frac{qE}{H^2}\right)^2 + \left(\frac{m}{H}\right)^2 - \left[\left(\frac{d}{2}\right)^2\right]}, \quad S_\lambda = 2\pi\frac{qE}{H^2}\left(\frac{\frac{qE}{H}}{\sqrt{\left(\frac{qE}{H}\right)^2 + \mathbf{k}_\perp^2}}\right). \tag{17}$$

Here, the square bracket in Eq. (17) vanishes for fermions. In the case of dS_2, the transverse momentum vanishes and the integration becomes unity and the Schwinger formula (6) is recovered. Further, in the limit of the Minkowski spacetime ($H = 0$), Eq. (16) recovers the Schwinger formula ($d = 3$ in Table 1)

$$\frac{d^{d+1}N_{\text{Min}}}{dtdx^d} = \frac{(2|\sigma|+1)qE}{2(2\pi)}\int \frac{d^{d-1}\mathbf{k}_\perp}{(2\pi)^{d-1}} e^{-\pi\frac{m^2+\mathbf{k}_\perp^2}{qE}}. \tag{18}$$

Finally, in the limit of the zero field ($E = 0$), Eq. (16) reduces to the pure dS radiation for scalars (fermions) in the planar coordinates

$$N_{\text{dS}} = \frac{1}{e^{S_{\text{dS}}} \mp 1}, \quad S_{\text{dS}} = 2\pi\sqrt{\left(\frac{m}{H}\right)^2 - \left[\left(\frac{d}{2}\right)^2\right]}. \tag{19}$$

4. QED Action in AdS$_2$

The two-dimensional AdS space has a constant curvature $\mathcal{R}_2 = -2K^2$, the planar coordinates and the Coulomb potential for a constant electric field

$$ds^2 = -e^{2Kx}dt^2 + dx^2, \quad A_1 = -\frac{E}{K}(e^{Kx} - 1). \tag{20}$$

The Schwinger formula for charged scalars is given by another dimensionless instanton action[15,22]

$$N_{\rm AdS} = e^{-S_{\rm AdS}}, \quad S_{\rm AdS} = 2\pi\left(\frac{qE}{K^2} - \sqrt{\left(\frac{qE}{K^2}\right)^2 - \left(\frac{m}{K}\right)^2 - \left(\frac{1}{2}\right)^2}\right), \tag{21}$$

which is an analytical continuation of the instanton action (2) in dS$_2$. Employing the effective temperature by Cai and Kim,[7] one may give a thermal interpretation for the Schwinger formula (21) as

$$N_{\rm AdS} = e^{-\frac{\bar{m}}{T_{\rm CK}}}, \quad T_{\rm CK} = \sqrt{T_{\rm U}^2 + \frac{\mathcal{R}_2}{8\pi^2}} + T_{\rm U}, \tag{22}$$

where the Unruh temperature for accelerating charge and the effective mass in (A)dS$_2$ are

$$T_{\rm U} = \frac{\frac{qE}{\bar{m}}}{2\pi}, \quad \bar{m} = \sqrt{m^2 - \frac{\mathcal{R}_2}{8}}. \tag{23}$$

The binding nature of a pair in the AdS space increases the effective mass while the dS space intrinsically separates the pair and decreases the effective mass.

Similarly, one obtains the density of states and the pair-production rate in AdS$_2$

$$\frac{d^2N_{\rm AdS}}{dtdx} = \frac{(2|\sigma|+1)K^2S_\nu}{2(2\pi)^2}\left(\frac{e^{-(S_\lambda - S_\nu)} - e^{-(S_\lambda + S_\nu)}}{1 \pm e^{-(S_\lambda + S_\nu)}}\right), \tag{24}$$

where the upper (lower) sign is for charged scalars (fermions) and the dimensionless instanton actions are

$$S_\lambda = 2\pi\frac{qE}{K^2}, \quad S_\nu = 2\pi\sqrt{\left(\frac{qE}{K^2}\right)^2 - \left(\frac{m}{K}\right)^2 - \left[\left(\frac{1}{2}\right)^2\right]}. \tag{25}$$

The square bracket vanishes for fermions. Note that pairs are produced when the Breitenlohner–Freedman stability bound $m^2 \geq (qE/K)^2$ is violated. The QED action density is for scalars

$$\mathcal{L}^{\rm sc}_{\rm AdS} = \frac{K^2S_\nu}{2(2\pi)^2}\mathcal{P}\int_0^\infty \frac{ds}{s}e^{-S_\lambda s/2\pi}\cosh\left(\frac{S_\nu s}{2\pi}\right)\left(\frac{1}{\sin\left(\frac{s}{2}\right)} - \left(\frac{2}{s} + \frac{s}{12}\right)\right), \tag{26}$$

and for fermions

$$\mathcal{L}^{\rm sp}_{\rm AdS} = -\frac{K^2 S_\nu}{(2\pi)^2}\mathcal{P}\int_0^\infty \frac{ds}{s}(e^{-(S_\lambda - S_\nu)s/2\pi} - e^{-(S_\lambda + S_\nu)s/2\pi}) \times \left(\frac{\cos\left(\frac{s}{2}\right)}{\sin\left(\frac{s}{2}\right)}\left(\frac{2}{s} - \frac{s}{6}\right)\right). \tag{27}$$

The bosonic and fermionic current density is given by

$$J_{\rm AdS} = 2q\frac{d^2 N_{\rm AdS}}{dtdx}. \tag{28}$$

5. Hawking Radiation from Nonextremal Black Hole

The RN black hole with the mass M and charge Q has the metric and the Coulomb potential

$$ds^2 = -\left(1 - \frac{2M}{r} + \frac{Q^2}{r^2}\right)dt^2 + \frac{dr^2}{1 - \frac{2M}{r} + \frac{Q^2}{r^2}} + r^2 d\Omega_2^2, \quad \Phi = \frac{Q}{r}, \tag{29}$$

which has the outer and inner horizons

$$r_+ = M + \sqrt{M^2 - Q^2}, \quad r_- = M - \sqrt{M^2 - Q^2}. \tag{30}$$

As summarized in Table 2, the near-horizon geometry of the RN black hole (29) is approximately given by the Rindler space, $\text{Rindler}_2 \times S^2$ when $M \gg Q$ while it is accurately approximated by the AdS space, $\text{AdS}_2 \times S^2$ when $M \simeq Q$. Interestingly, in the tortoise coordinate r_* for an extremal black hole $Q = M$, the radial motion of a charged scalar in the spherical harmonics $Y_{lm}(\theta, \varphi)$,[23]

$$\left[\frac{d^2}{dr_*^2} - V_l(r(r_*))\right]\Phi_l = 0, \tag{31}$$

has in the asymptotic region ($r \gg Q$) the effective potential for the Klein–Gordon equation in the Coulomb potential with an angular momentum $\bar{l}$ and a charge $\bar{Q}$

$$V_l(r) = m^2 + \frac{\bar{l}(\bar{l}+1)}{r^2} - \left(\omega - \frac{q\bar{Q}}{r}\right)^2 + \mathcal{O}\left(\frac{1}{r^3}\right), \tag{32}$$

where

$$\bar{Q} = Q\left(1 + \frac{m^2}{q\omega}\right), \quad \bar{l} + \frac{1}{2} = \sqrt{\left(l + \frac{1}{2}\right)^2 + (qQ)^2\left(\frac{m^2}{q\omega} + 1\right)^2 - (qQ)^2}. \tag{33}$$

Thus, the Schwinger formula from the extremal black hole is approximately given by that from a supercritical point charge in the Minkowski spacetime.

In the tunneling picture for Hawking radiation, virtual pairs are created through the horizon. Thus, the near-horizon geometry of the black hole plays an important role in understanding the Schwinger effect or Hawking radiation. For a nonextremal black hole, the near-horizon geometry has the Rindler coordinates[24]

$$ds^2 = -F(\rho)d\tau^2 + \frac{d\rho^2}{G(\rho)} + r^2(\rho)d\Omega_2^2, \tag{34}$$

where $F \simeq (\kappa\rho)^2$ and $G \simeq 1$ near the horizon $\rho \simeq 0$ with the surface gravity $\kappa = (r_+ - r_-)/(2r_+^2)$. A physically intuitive way to understand the pair production from the background field is to investigate the poles of the Hamilton–Jacobi equation for a charged particle. The Hamilton–Jacobi action for each spherical harmonics

$$S_l = \int \frac{d\rho}{\sqrt{FG}} \sqrt{(\omega - q\Phi)^2 - F\left(\frac{\left(l + \frac{1}{2}\right)^2}{r^2(\rho)} + m^2\right)}, \tag{35}$$

can give the particle production as the decay rate of the vacuum. In the phase-integral method, the mean number is given by the leading term[24–26]

$$\mathcal{N}_l = e^{i\mathcal{S}_{\Gamma_l}}, \tag{36}$$

where $\mathcal{S}_{\Gamma_l} = \oint S_l$ is the action evaluated along a contour Γ_l in the complex plane of ρ. The simple pole at the outer horizon $\rho = 0$ $(r = r_+)$ recovers Hawking radiation of charged particles

$$\mathcal{N}_l = e^{-\frac{\omega - q\Phi(r_+)}{T_{\rm H}}}, \tag{37}$$

where the Hawking temperature and the electrostatic potential on the horizon are

$$T_{\rm H} = \frac{r_+ - r_-}{4\pi r_+^2}, \quad \Phi(r_+) = \frac{Q}{r_+}. \tag{38}$$

6. Schwinger Effect in Near-Extremal Black Hole

As shown in Table 2, the near-horizon geometry of a near-extremal black hole has $\mathrm{AdS}_2 \times S^2$ with the metric[5]

$$ds^2 = -\frac{\rho^2 - B^2}{Q^2}d\tau^2 + \frac{Q^2}{\rho^2 - B^2}d\rho^2 + Q^2 d\Omega_2^2, \tag{39}$$

where coordinates are stretched by smallness parameter ϵ as

$$r - Q = \epsilon Q, \quad t = \frac{\tau}{Q}, \quad M - Q = \frac{(\epsilon B)^2}{2Q}. \tag{40}$$

The Hawking radiation for the parametrization (40) is given by

$$N_{\rm H} = \frac{1}{e^{\frac{\omega - q\Phi(B)}{T_{\rm H}}} - 1}, \quad T_{\rm H} = \frac{\sqrt{\frac{1}{2} + \frac{M}{2Q}}}{4\pi\left(M + \epsilon B\sqrt{\frac{1}{2} + \frac{M}{2Q}}\right)}\epsilon B. \tag{41}$$

Except for particles with energy close to the chemical potential, i.e. $\omega - q\Phi(B) \propto \epsilon B$, Hawking radiation is exponentially suppressed to zero. By taking the limit $\epsilon = 0$

the near-horizon region can be stretched infinitely to $\mathrm{AdS}_2 \times S^2$. This procedure differs from the amplified near-horizon geometry of a nonextremal black hole as $\mathrm{Rindler}_2 \times S^2$. This means that the Schwinger effect from $\mathrm{AdS}_2 \times S^2$ is very accurate for the near-extremal black hole.

The action for the Hamilton–Jacobi equation for the radial motion after separating the spherical harmonics is given by

$$S_l(\rho) = \int d\rho \sqrt{\frac{(q\rho - \omega Q)^2 Q^2}{(\rho^2 - B^2)^2} - \frac{m^2 Q^2 + \left(l + \frac{1}{2}\right)^2}{\rho^2 - B^2}}. \tag{42}$$

Note that there are two finite poles at $\rho = B$, the outer horizon, and $\rho = -B$, the inner horizon as well as an infinite pole at $\rho = \infty$. Then, the contour integral gives the leading term for the Schwinger effect[5]

$$N_{\mathrm{NBH}} = e^{-(S_a - S_b)}, \quad S_a = 2\pi q Q, \quad S_b = 2\pi \sqrt{(q^2 - m^2) Q^2 - \left(l + \frac{1}{2}\right)^2}, \tag{43}$$

where S_a is the sum of residues at $\rho = \pm B$ and S_b is the residue at $\rho = \infty$.

From the field equation, the exact Schwinger formula from the geometry (39) is given by[5]

$$N_{\mathrm{NBH}} = \left(\frac{e^{-(S_a - S_b)} - e^{-(S_a + S_b)}}{1 \pm e^{-(S_a + S_b)}} \right) \times \left(\frac{1 \mp e^{-(S_c - S_a)}}{1 + e^{-(S_c - S_b)}} \right), \tag{44}$$

where $S_c = 2\pi(\omega Q^2/B)$. The mean number for the Schwinger effect has the thermal interpretation[11,12]

$$N_{\mathrm{NBH}} = \left(\frac{e^{-\frac{\bar{m}}{T_{\mathrm{CK}}}} - e^{\frac{\bar{m}}{T_{\mathrm{CK}}}}}{1 \pm e^{-\frac{\bar{m}}{T_{\mathrm{CK}}}}} \right) \times \left(\frac{1 \mp e^{-\frac{\omega - q\Phi}{T_{\mathrm{H}}}}}{1 + e^{-\frac{\omega - q\Phi}{T_{\mathrm{H}}} - \frac{\bar{m}}{T_{\mathrm{CK}}}}} \right), \tag{45}$$

where

$$T_{\mathrm{CK}} = T_{\mathrm{U}} + \sqrt{T_{\mathrm{U}}^2 - \left(\frac{1}{2\pi Q}\right)^2}, \quad \bar{T}_{\mathrm{CK}} = T_{\mathrm{U}} - \sqrt{T_{\mathrm{U}}^2 - \left(\frac{1}{2\pi Q}\right)^2} \tag{46}$$

with $T_{\mathrm{U}} = (qE_{\mathrm{H}}/\bar{m})/2\pi$, the Unruh temperature for accelerating charge on the horizon. Note that T_{CK} and $\bar{T}_{\mathrm{CK}}$ are the temperature for the Schwinger effect in AdS_2 in Sec. 4. The Schwinger effect from the near-extremal black hole is the product of the Schwinger effect in AdS_2 and a correction due to the Hawking radiation from the nonextremality. It has been further observed that the Schwinger effect has the factorization[11,12]

$$N_{\mathrm{NBH}} = e^{\frac{\bar{m}}{T_{\mathrm{CK}}}} \times \underbrace{\left(\frac{e^{-\frac{\bar{m}}{\bar{T}_{\mathrm{CK}}}} - e^{-\frac{\bar{m}}{T_{\mathrm{CK}}}}}{1 \pm e^{-\frac{\bar{m}}{T_{\mathrm{CK}}}}} \right)}_{\text{Schwinger effect in AdS}_2} \times \underbrace{\left(\frac{e^{-\frac{\bar{m}}{T_{\mathrm{CK}}}} (1 \mp e^{-\frac{\omega - q\Phi}{T_{\mathrm{H}}}})}{1 + e^{-\frac{\omega - q\Phi}{T_{\mathrm{H}}} - \frac{\bar{m}}{T_{\mathrm{CK}}}}} \right)}_{\text{Schwinger effect in Rindler}_2}. \tag{47}$$

The reason for the Schwinger effect in $\mathrm{Rindler}_2$ is that the measure of the near-extremality is in fact a measure for the Rindler space. The Schwinger effect of scalar QED in the Rindler space has been studied in Ref. 27.

7. Discussion and Conclusion

We have studied the spontaneous creation of charged pairs from a constant electric field in (A)dS and from a charged RN black hole, which is a nonperturbative quantum effect at one-loop level. The QED action in (A)dS exhibits the interplay of QED and the quantum gravity effect. The Schwinger formula has a thermal interpretation of the Unruh temperature for accelerating charge and the Gibbons–Hawking temperature for dS_2 and the corresponding curvature for AdS_2. We have found the Schwinger formula in dS space of any dimension, which has the correct limit of the Minkowski spacetime and the pure dS space. The near-horizon geometry of charged RN black hole is $AdS_2 \times S^2$ for an extremal black hole while it is $\text{Rindler}_2 \times S^2$ for a nonextremal black hole. Thus, the Schwinger effect for extremal black hole has the same thermal interpretation as in AdS_2. The near-extremal black hole has, however, an additional Schwinger effect in Rindler_2.

Acknowledgments

The author would like to thank Rong-Gen Cai, Chiang-Mei Chen, W-Y. Pauchy Hwang, Hyun Kyu Lee, Don N. Page, Christian Schubert, and Yongsung Yoon for useful discussions and Pisin Chen for warm hospitality during the second LeCosPA Symposium. The participation of LeCosPA Symposium was supported by Asia Pacific Center for Theoretical Physics (APCTP). This research was supported by Basic Science Research Program through the National Research Foundation of Korea(NRF) funded by the Ministry of Education (15B15770630).

References

1. J. Schwinger, *Phys. Rev.* **82** (1951) 664.
2. S. W. Hawking, *Nature* **248** (1974) 30; *Commun. Math. Phys.* **43** (1975) 199.
3. W. G. Unruh, *Phys. Rev. D* **14** (1976) 870.
4. W. Heisenberg and H. Euler, *Z. Phys.* **98** (1936) 714, arXiv:physics/0605038.
5. C.-M. Chen, S. P. Kim, I.-C. Lin, J.-R. Sun and M.-F. Wu, *Phys. Rev. D* **85** (2012) 124041; C.-M. Chen, J.-R. Sun, F.-Y. Tang and P.-Y. Tsai, *Class. Quantum Grav.* **32** (2015) 195003.
6. D. Anninos and T. Hartman, *J. High Energy Phys.* **03** (2010) 096; D. Anninos and T. Anous, *J. High Energy Phys.* **08** (2010) 131.
7. R.-G. Cai and S. P. Kim, *J. High Energy Phys.* **09** (2014) 072.
8. S. P. Kim, Scalar QED action density and Schwinger pair production in $(A)dS_2$, to appear in *Proc. MG14*, arXiv:1512.03142.
9. H. Narnhofer, I. Peter and W. Thirring, *Int. J. Mod. Phys. B* **10** (1996) 1507.
10. S. Deser and O. Levin, *Class. Quantum Grav.* **14** (1997) L163.
11. S. P. Kim, H. K. Lee and Y. Yoon, arXiv:1503.00218.
12. S. P. Kim, *Int. J. Mod. Phys. A* **30** (2015) 1545017.
13. W. Fischler, P. H. Nguyen, J. F. Pedraza and W. Tangarife, *Phys. Rev. D* **91** (2015) 086015.
14. J. Garriga, *Phys. Rev. D* **49** (1994) 6343.
15. S. P. Kim and D. N. Page, *Phys. Rev. D* **78** (2008) 103517.

16. G. W. Gibbons and S. W. Hawking, *Phys. Rev. D* **15** (1977) 2738.
17. S. Haouat and R. Chekireb, arXiv:1504.08201.
18. S. P. Kim, H. K. Lee and Y. Yoon, *Phys. Rev. D* **78** (2008) 105013; *Phys. Rev. D* **82** (2010) 025015.
19. M. B. Fröb, J. Garriga, S. Kanno, M. Sasaki, J. Soda, T. Tanaka and A. Vilenkin, *J. Cosmol. Astropart. Phys.* **04** (2014) 009.
20. T. Kobayashi and N. Afshordi, *J. High Energy Phys.* **10** (2014) 166.
21. C. Stahl and E. Strobel, *AIP Conf. Proc.* **1693** (2015) 050005; C. Stahl, E. Strobel and S.-S. Xue, *Phys. Rev. D* **93** (2016) 025004.
22. B. Pioline and J. Troost, *J. High Energy Phys.* **03** (2005) 043.
23. S. P. Kim, *J. Korean Phys. Soc.* **65** (2014) 907.
24. S. P. Kim, *J. High Energy Phys.* **11** (2007) 048.
25. S. P. Kim and D. N. Page, *Phys. Rev. D* **75** (2007) 045013.
26. S. P. Kim, *Phys. Lett. B* **725** (2013) 500; *Phys. Rev. D* **88** (2013) 044027; *JPS Conf. Proc.* **1** (2014) 013110; *Grav. Cosmol.* **20** (2014) 193.
27. Cl. Gabriel and Ph. Spindel, *Ann. Phys.* **284** (2000) 263.

Conformal Frames in Cosmology*

Guillem Domènech* and Misao Sasaki†
Yukawa Institute for Theoretical Physics,
Kyoto University, Kyoto 606-8502, Japan
**guillem.domenech@yukawa.kyoto-u.ac.jp*
†misao@yukawa.kyoto-u.ac.jp

From higher dimensional theories, e.g. string theory, one expects the presence of nonminimally coupled scalar fields. We review the notion of conformal frames in cosmology and emphasize their physical equivalence, which holds at least at a classical level. Furthermore, if there is a field, or fields, which dominates the universe, as it is often the case in cosmology, we can use such notion of frames to treat our system, matter and gravity, as two different sectors. On one hand, the gravity sector which describes the dynamics of the geometry and on the other hand, the matter sector which has such geometry as a playground. We use this interpretation to build a model where the fact that a curvaton couples to a particular frame metric could leave an imprint in the cosmic microwave background (CMB).

Keywords: Cosmology; inflation; scalar–tensor theory.

PACS Number(s): 04.62.+v, 98.80.−k

1. Introduction: Conformal Frames

In cosmology, it is certainly often assumed that some field(s) rules the evolution of the universe, i.e. dictates metric dynamics, while the rest, say matter, plays out its dynamics in such a given geometry. There is no reason, at least on cosmological scales, to *a priori* assume that the playground geometry for matter, from now on we call it *matter metric*, is exactly equal to that given by the dominating fields, for example if those fields are nonminimally coupled to gravity. These metrics could be nontrivially related through a function of the fields, the simplest case being the so-called conformal transformation, that is a rescaling of the metric. Needless to say, these two geometries must be almost indistinguishable on solar system scales, where very strong constraints apply.

One may wonder if such model, i.e. two different geometries, could be derived from a theoretical model rather than be postulated. In fact, without going into much details, it is fairly easy to see that in higher dimensional spacetimes, this

*This article first appeared in *Int. J. Mod. Phys. D* **25** (2016) 1645006.

is not surprising at all.[1] For example, consider a D-dimensional space where the metric is given by

$$\begin{pmatrix} g^{(D)}_{\mu\nu}(x,y) & g^{(D)}_{\mu B}(x,y) \\ g^{(D)}_{A\nu}(x,y) & g^{(D)}_{AB}(x,y) \end{pmatrix}, \tag{1}$$

where $\mu, \nu = 1 \ldots 4$, $A, B = 5 \ldots D$ and x and y are, respectively, four and $D-4$-dimensional coordinates. As matter is concerned, we know it lives in a four-dimensional spacetime. How is the dimensionality reduced or, in other words, which is the effective metric for matter, covers many possibilities. To gather some examples, one could have

$$g_{\mu\nu}(x) \stackrel{?}{=} \begin{cases} \langle g^{(D)}_{\mu\nu} \rangle_{D-4}, \\ f(x)\langle g^{(D)}_{\mu\nu} \rangle_{D-4}, \\ g^{(D)}_{\mu\nu}(x,0), \\ \ldots, \end{cases} \tag{2}$$

where brackets mean that we integrated out the $D-4$ extra dimensions. Under a dimensional reduction, the initial D tensor modes translate into four tensor, four vector and four scalar modes. Most likely, such dimensional reduction ends up involving dilatonic scalars, that is scalars fields which nonminimally couple to matter and/or gravity. In that sense, it is clear that there is no unique natural conformal or physical frame *a priori*.

For concreteness sake, let us show two typical conformal frames in cosmology and let us work in Planck units ($\hbar = c = M_{\rm pl} = 1$). On one hand, the Jordan frame[2,3] where the action is given by

$$S = \int d^4x \sqrt{-\bar{g}} \{F(\phi)\bar{R} + \mathcal{L}(\phi) + \mathcal{L}_m(\psi, A, \ldots)\}, \tag{3}$$

where matter fields, e.g. fermions ψ and vectors A, minimally couple to the metric $\bar{g}$ but there is a nonminimal coupling between a scalar field ϕ and gravity. On the other hand, the Einstein frame action is

$$S = \int d^4x \sqrt{-g} \{R + \mathcal{L}(\phi) + G(\phi)\mathcal{L}_m(\phi, \psi, A, \ldots)\}, \tag{4}$$

where this time the scalar field ϕ minimally couples to the metric g but there appears a coupling between matter fields and the scalar field ϕ. In the Jordan frame, one usually assumes that matter is *universally* coupled, which for baryons is experimentally consistent. If there were a nonuniversal coupling, we could write the matter Lagrangian as

$$\sum_A G_A(\phi)\mathcal{L}_A(\phi, Q_A), \tag{5}$$

where $Q_A = \{\psi, A, \ldots\}$. However, it should be noted that in the latter case the definition of Jordan frame, i.e. where matter is minimally coupled, is not clear and perhaps one could define a Jordan frame for each Q_A, if any.

A note is in order. Throughout the foregoing discussion, we did not emphasize any frame as more physical than others and, furthermore, from our dimensional reduction point of view there is no reason to believe that such a frame exist. In next section, we review some examples of how physics does not change but interpretations do. Before that, let us remind the reader below how quantities transform under a conformal transformation as we will make use of them.

Transformation rules. Under a conformal transformation given by

$$\tilde{g}_{\mu\nu} = \Omega^2 g_{\mu\nu}, \tag{6}$$

the Ricci scalar transforms as

$$\tilde{R} = \Omega^{-2}\left[R - (D-1)\left(2\frac{\Box\Omega}{\Omega} - (D-4)g^{\mu\nu}\frac{\nabla_\mu\Omega\nabla_\nu\Omega}{\Omega^2}\right)\right], \tag{7}$$

and if one can neglect the dynamics of the dilaton field at short scales, one can find that matter fields transform as

$$\begin{aligned} \tilde{\chi} &= \Omega^{-(D-2)/2}\chi \quad \text{for scalars,} \\ \tilde{A}_\mu &= \Omega^{-(D-4)/2}A_\mu \quad \text{for vectors,} \\ \tilde{\psi} &= \Omega^{-(D-1)/2}\psi \quad \text{for fermions.} \end{aligned} \tag{8}$$

Let us show in more detail the fermion and gauge field cases, as we use this results below. The action for a Dirac fermion is given by

$$S = \int d^4x\sqrt{-g}\left[-i\bar{\psi}\gamma^\mu D_\mu\psi - m\bar{\psi}\psi - \frac{1}{4}F_{\mu\nu}F^{\mu\nu}\right], \tag{9}$$

where $D_\mu = \partial_\mu + ieA_\mu - \frac{1}{4}\omega_{ab\mu}\Sigma^{ab}$, γ^μ are the gamma matrices, $\Sigma^{ab} = \frac{1}{2}[\gamma^a, \gamma^b]$ and $\omega_{ab\mu} = e_{a\nu}\nabla_\mu e^\nu_b$, m is the mass of the fermion and $F_{\mu\nu} = \nabla_\mu A_\nu - \nabla_\mu A_\nu$. After the conformal transformation (6), one obtains

$$S = \int d^4x\sqrt{-\tilde{g}}\left[-i\bar{\tilde{\psi}}\tilde{\gamma}^\mu D_\mu\tilde{\psi} - \tilde{\mathbf{m}}\bar{\tilde{\psi}}\tilde{\psi} - \frac{1}{4}F_{\mu\nu}F^{\mu\nu}\right], \tag{10}$$

where $\tilde{\gamma}^\mu = \Omega^{-1}\gamma^\mu$, $\tilde{\psi} = \Omega^{-3/2}\psi$, $\tilde{m} = \Omega^{-1}m$ and the gauge field A_μ is conformal invariant in four dimensions. We highlighted the mass in bold as it is the main effect of a conformal transformation; the mass of the fermion becomes spacetime-dependent.

2. Frame Independence of Observables

The physical equivalence of conformal frames has been extensively discussed in the literature,[4–14] and here we shall just give a brief review. See Refs. 15–21 for different points of view. We show here with an illustrative example[6] how the observables do not depend on the frame one computes them. In particular, we consider an essential quantity in cosmology, the redshift.

For simplicity's sake, let us assume that matter is minimally coupled to an expanding background. The line element is given by

$$ds^2 = -dt^2 + a^2(t)d^2\sigma_{(K)}, \tag{11}$$

where $a(t)$ is the scale factor, $d^2\sigma_{(K)}$ is an homogeneous and isotropic three-dimensional space and $K = \pm 1, 0$. The expansion of the universe is dictated by the Friedmann equation

$$\left(\frac{\dot{a}}{a}\right)^2 \equiv H^2 = \frac{8\pi G}{3}\rho - \frac{K}{a^2}, \tag{12}$$

and due to such expansion, one observes a cosmological redshift

$$E_{\text{obs}} = \frac{E_{\text{emit}}}{1+z}. \tag{13}$$

This is regarded as a "proof" of expansion, or at least this is how we *interpret* the data.

However, as we are interested in conformal frames, let us go to a special frame by choosing $\Omega = 1/a$ and working with conformal time $d\eta = dt/a$. By doing so, we go to a frame where the line element is given by

$$d\tilde{s}^2 = \Omega^2 ds^2 = -d\eta^2 + d\sigma^2_{(K)}, \tag{14}$$

which is a *static universe* and, as such, photons do not redshift. Then, one may wonder whether this frame is unphysical. Let us show that it is as physical as any other.

Recall that the mass of a fermion, e.g. an electron, is rescaled as (10)

$$\tilde{m} = \Omega^{-1} m = \frac{m}{1+z}, \tag{15}$$

where we used the usual definition of redshift, i.e. $a^{-1} = 1 + z$. Consequently, the Bohr radius scales as $\propto \tilde{m}^{-1}$ and, hence, the atomic energy levels scale inversely proportional, i.e. $\propto \tilde{m}$. Equation (15) yields time-dependent energy levels, namely

$$\tilde{E}_n = \frac{E_n}{1+z}, \tag{16}$$

where $\tilde{E}_n$ is the energy levels in the static frame and E_n is the energy levels in the Jordan (matter) frame. Thus, it is clear that frequency of photons emitted at a time $z(\eta)$ from a level transition $n \to n'$ is

$$\tilde{E}_{nn'} = \frac{E_{nn'}}{1+z}, \tag{17}$$

which is exactly what we observe as Hubble's law. Then, how do we interpret the cosmic microwave background (CMB) photons in this static frame? First, the universe was in thermal equilibrium at $T = 2.725\,\text{K}$ when the electron mass was more than 10^3 times smaller, that is at a time $z > 10^3$. Afterwards, CMB photons have never redshifted. How about the rate of scattering/interaction? The Thomson cross-section $\tilde{\sigma}_T$ scales as $\propto \tilde{m}^{-2}$ and the electron density $\tilde{n}_e$ is indeed constant in the static frame. Relating them to the matter frame quantities, we respectively have:

$$\tilde{\sigma}_T = \sigma_T(1+z)^2 \quad \text{and} \quad \tilde{n}_e = \frac{n_e}{(1+z)^3}. \tag{18}$$

Therefore, the rate of scattering per unit of proper time remains unchanged,

$$\tilde{\sigma}_T \tilde{n}_e d\eta = \frac{\sigma_T\, n_e}{1+z} d\eta = \sigma_T\, n_e\, dt. \tag{19}$$

So far, we showed that observed physics, e.g. redshift, is independent of the frame but interpretations of such observables most likely differ. In this way, it might be more appropriate to call them "representations" rather than frames.[6] One last interesting consequence of the previous example is that the metric itself is nonobservable.

3. Conformal Frame "Dependence" of Inflation

In this section, let us put to good use all of the previously discussed. It is hopefully clear that there is no frame more physical than others and that observables do not depend on the frame one computes them. Bearing this in mind, let us consider a two field inflationary model with the action given by

$$S = \int d^4x \sqrt{-g} \left\{ \frac{1}{2} R + \mathcal{L}_{\rm inf}(\phi) \right\} + \sqrt{-\bar{g}} \mathcal{L}_m(\chi), \tag{20}$$

where ϕ and χ are scalar fields and g and $\bar{g}$ are related by a conformal transformation, i.e.

$$\bar{g}_{\mu\nu} = F^{-1}(\phi) g_{\mu\nu}. \tag{21}$$

This action is not neither in the Jordan frame nor Einstein frame form, but emphasizes the idea of our model. Basically, we assume that ϕ drives inflation and χ is an spectator matter field, so-called the curvaton.[22–24] Thus, the curvaton does not contribute to the inflationary dynamics and need not be in an accelerated expanding universe, i.e. metric g, but feels the metric $\bar{g}$, which might not be inflating. The same description applies for matter fields universally coupled to the metric $\bar{g}$. Since we define inflation as an accelerated expansion, we refer to the fact that matter might be in a very different universe as *conformal "dependence" of inflation*.[25]

In order to avoid any confusion, the action in the Jordan (matter) frame is given by

$$S = \int d^4x \sqrt{-\bar{g}} \left\{ \frac{1}{2} F(\phi) \bar{R} + \bar{\mathcal{L}}_{\rm inf}(\phi) - \frac{1}{2} \bar{g}^{\mu\nu} \nabla_\mu \chi \nabla_\nu \chi - \frac{1}{2} \bar{m}^2 \chi^2 \right\}, \tag{22}$$

while the Einstein frame counterpart is given by

$$S = \int d^4x \sqrt{-g} \left\{ \frac{1}{2} R + \mathcal{L}_{\rm inf}(\phi) - \frac{1}{2F(\phi)} g^{\mu\nu} \nabla_\mu \chi \nabla_\nu \chi - \frac{1}{2} m^2 \chi^2 \right\}, \tag{23}$$

where we redefined $m \equiv F^{-1/2} \bar{m}$ and $\mathcal{L}_{\rm inf}(\phi)$ is the Einstein frame counterpart of $\bar{\mathcal{L}}_{\rm inf}(\phi)$, as in general the functional form changes. Note that in the Einstein frame, there is an explicit coupling between inflaton and curvaton while in the Jordan frame, the curvaton is minimally coupled to the metric $\bar{g}$. Thus, the "natural" frame of the curvaton is the Jordan frame while the "natural" frame of inflation, understood as an accelerated expansion, is the Einstein frame.

At this point, a note is in order. We did not specify the functional form of $F(\phi)$ or, in other words, the matter metric $\bar{g}$. A particular choice of $F(\phi)$ corresponds to fixing the model. Nevertheless, this is not an obstacle as we assume inflation in the Einstein frame and the curvaton is assumed to be subdominant. Thus, the exact form of $F(\phi)$ is not relevant for the dynamics of inflation. The main point is that if the field χ contributes to the total curvature power spectrum, as it is the case for the curvaton model, then $F(\phi)$ may become extremely important. It leaves an observable imprint of the matter point of view.

For the sake of simplicity, we considered a simple inflationary model, so-called power-law inflation,[26] as an illustrative example. We proceed as follows: (i) we compute the curvature power spectrum due to the inflaton in the Einstein frame while completely neglecting the curvaton, (ii) we investigate how does $\bar{g}$ behaves for some choice of $F(\phi)$ and (iii) we compute the power spectrum of the curvaton in the Jordan frame in some interesting examples.

Brief review of power-law (Einstein frame). The inflaton Lagrangian in the power-law model[26] is given by

$$\mathcal{L}_{\rm inf}(\phi) = -\frac{1}{2}g^{\mu\nu}\nabla_\mu\phi\nabla_\nu\phi - V_0 e^{-\lambda\phi}. \tag{24}$$

An exact solution to the equations of motion yields

$$a = a_0\left(\frac{t}{t_0}\right)^p, \quad H \equiv \frac{\dot{a}}{a} = \frac{p}{t}, \tag{25}$$

$$\phi = \frac{2}{\lambda}\ln\left(\frac{t}{t_0}\right), \quad (0 < t < \infty), \tag{26}$$

where $p = 2/\lambda^2$ and $\lambda^2 V_0 t_0^2 = 2(3p-1)$. Slow roll inflation occurs for $p \gg 1$. The curvature and tensor power spectrum at horizon crossing ($k = aH$) are found to be,[26–29] respectively,

$$\mathcal{P}_{\mathcal{R}_c} = \frac{p}{8\pi^2}\frac{H^2}{M_{\rm pl}^2} = \frac{p}{8\pi^2}\frac{H_0^2}{M_{\rm pl}^2}\left(\frac{k}{k_0}\right)^{\frac{-2}{p-1}}, \quad \mathcal{P}_T = \frac{2}{\pi^2}\frac{H^2}{M_{\rm pl}^2}, \tag{27}$$

where we recovered the units and from which one can extract the spectral index, that is

$$n_s^\phi - 1 = \frac{d\ln\mathcal{P}_{\mathcal{R}_c}}{d\ln k} = \frac{-2}{p-1} = n_T, \tag{28}$$

where the superscript ϕ emphasize that it is the contribution from the inflaton. If there is no other fields contributing to the power spectrum, the former result is observable through the CMB. Let us move on to the curvaton.

3.1. *Matter point of view (Jordan frame)*

We choose two different couplings to the curvaton. The first one leads us to another power-law but with a different power-law index and the second one yields a bouncing matter metric.

Jordan frame power-law. Let us choose that the curvaton metric is related to the inflationary metric by

$$F(\phi) = \mathrm{e}^{\gamma\lambda\phi} = (t/t_0)^{2\gamma}, \tag{29}$$

where γ is a free parameter. Plugging the former to Eq. (21) yields

$$\bar{a} = a_0 \left(\frac{\bar{t}}{\bar{t}_0}\right)^{\bar{p}}, \quad \bar{H} \equiv \frac{1}{\bar{a}}\frac{d\bar{a}}{d\bar{t}} = \frac{\bar{p}}{\bar{t}}, \tag{30}$$

where

$$\bar{p} > 1 (0 < \bar{t} < \infty), \quad \bar{p} < 1 (-\infty < \bar{t} < 0), \tag{31}$$

and

$$\bar{p} - 1 = \frac{p-1}{1-\gamma}, \quad \bar{t}_0 = \frac{t_0}{1-\gamma}, \tag{32}$$

which is another power-law type universe. Interestingly, although there is an accelerated expansion in the Einstein frame, the curvaton might feel a different evolution, depending on the choice of γ. Concretely for $\bar{p} > 1$ $(\gamma < 1)$, we have inflation, $0 < \bar{p} < 1$ $(\gamma > p)$ leads to a decelerated contracting universe and for $\bar{p} < 0$ $(1 < \gamma < p)$, it is a superinflationary universe. The latter is the most interesting case as it yields a blue tilted curvaton power spectrum, i.e.

$$\tilde{n}_\chi - 1 = \frac{-2}{\tilde{p} - 1} > 0, \quad (\bar{p} < 0), \tag{33}$$

where we assumed that the curvaton instantly reheats the universe.[22–24] It has exactly the same form as the inflaton power spectrum (28) but with $\bar{p}$ instead of p, as one expects from a power-law universe. The total power spectrum is shown in Fig. 1(a). It should be noted that once we fix γ, we fix the curvaton Jordan frame, which obviously leads to different models, different observational results.

Jordan frame bounce. Finally, we give another example, a bouncing Jordan metric, where this time

$$F(\phi) = (1 + \mathrm{e}^{-\gamma\lambda\phi/2})^{-2} = (1 + (t/t_0)^{-\gamma})^{-2}. \tag{34}$$

Again, plugging this form into Eq. (21) roughly yields (for $\bar{p} < 1$)

$$\bar{a} \approx \begin{cases} a_0(-\bar{t}/\bar{t}_0)^{\bar{p}}, & |\bar{t}| \gg \bar{t}_0, \quad (\bar{t} < 0), \\ a_0(\bar{t}/\bar{t}_0)^{p}, & \bar{t} \gg \bar{t}_0. \end{cases} \tag{35}$$

Initially, the curvaton is in a contracting universe which bounces and catches up with the Einstein frame power-law. The total power spectrum is shown in Fig.1(b). The main feature in this case is a blue tilt at long scales, which if the curvaton dominates at short scales leads to a suppression of the power spectrum for large k.

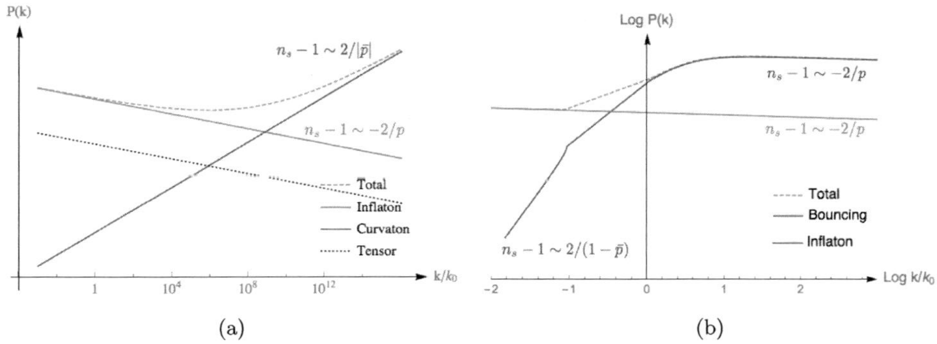

Fig. 1. (Color online) Total power spectrum. (a) Jordan power-law curvaton. In red, the inflaton curvature perturbation; in blue, the one from the curvaton and in magenta, the sum. A black line shows the tensor power spectrum. (b) Jordan bounce curvaton. In red, inflaton curvature perturbation and in blue, the contribution from the curvaton with an initial contracting phase. Lastly, the magenta dashed line is the total power spectrum in the bouncing case.

4. Discussion and Conclusions

We argued that in higher dimensional theories, the presence of dilatonic scalar fields, i.e. that nonminimally couple to gravity and/or matter, is usually expected. When such a nonminimal coupling is present, the notion of conformal frames (or "representations") appears, in particular, the Jordan (matter) frame and the Einstein frame. We reviewed the physical equivalence of observables between frames by showing an illustrative example where the redshift instead of being interpreted as "proof" of expansion is due to a time dependent electron mass.

We then considered a two-field model, where the second field is an spectator field, a curvaton, and it is minimally coupled to a metric which is not inflating. We define inflation in the Einstein frame as an accelerated expansion and we call the fact that the curvaton need not be in such a metric *conformal "dependence" of inflation.* Thus, we easily get a blue tilt contribution from the curvaton to the curvature power spectrum by minimally coupling the curvaton to a superinflationary universe. Furthermore, we considered another example where the curvaton feels a bouncing universe.

In this way, we emphasized the *physical equivalence* between conformal frames in cosmology while paying special attention to how *matter couples* to gravity, which might be observationally relevant if for example, the curvaton mechanism is present. Thus, without specifying the matter frame a model is not completely defined. Moreover, the physical equivalence between frames related by a generalized transformation, called disformal transformation,[30] has also been shown.[31–38]

As a final remark, let us consider the importance of the matter coupling in a more general theory, such as Horndeski[39,40] and its extensions.[41–44] Horndeski model implicitly assumes that matter is minimally coupled, namely

$$S = \int d^4x\sqrt{-g}\{\mathcal{L}_H(\phi, g, R) + \mathcal{L}_{\text{matter}}(\psi, A)\}, \tag{36}$$

where $\mathcal{L}_H$ is a general function of a nonminimally coupled scalar field ϕ which yields second-order differential equations of motion. On the other hand, if the matter sector is minimally coupled to a different metric, i.e.

$$S = \int d^4x\sqrt{-g}\mathcal{L}_H(\phi, g, R) + \sqrt{-\bar{g}}\,\mathcal{L}_{\text{matter}}(\psi, A), \tag{37}$$

we clearly have different model which indeed yields different features. In particular, if $\bar{g}$ is related to g by a derivative-dependent disformal transformation, the latter theory falls in the category of beyond Horndeski theory and may experience a breaking of the Vainshtein mechanism inside astrophysical bodies.[45–47] For these reasons, further study on how matter could couple to gravity might be an interesting direction.

Acknowledgments

GD would like to thank J. Fedrow, J. O. Gong, G. Leung, R. Namba, T. Qiu and J. White for useful discussions on the conformal "dependence" of inflation and N. Deruelle and R. Saito for interesting discussions on matter couplings in Horndeski. This work was supported in part by MEXT KAKENHI Grant No. 15H05888.

References

1. Y. Fujii and K.-I. Maeda, *The Scalar–Tensor Theory of Gravitation* (Cambridge University Press, UK, 2003).
2. P. Jordan, *Z. Phys.* **157** (1959) 112.
3. C. Brans and R. H. Dicke, *Phys. Rev.* **124** (1961) 925.
4. N. Makino and M. Sasaki, *Prog. Theor. Phys.* **86** (1991) 103.
5. V. Faraoni and S. Nadeau, *Phys. Rev. D* **75** (2007) 023501, arXiv:gr-qc/0612075.
6. N. Deruelle and M. Sasaki, *Springer Proc. Phys.* **137** (2011) 247, arXiv:1007.3563 [gr-qc].
7. J.-O. Gong, J.-C. Hwang, W. I. Park, M. Sasaki and Y.-S. Song, *J. Cosmol. Astropart. Phys.* **2011** (2011) 023, arXiv:1107.1840.
8. L. Järv, P. Kuusk, M. Saal and O. Vilson, *Phys. Rev. D* **91** (2015) 024041, arXiv:1411.1947 [gr-qc].
9. P. Kuusk, L. Järv and O. Vilson, *Int. J. Mod. Phys. A* **31** (2016) 1641003, arXiv:1509.02903 [gr-qc].
10. R. Catena, M. Pietroni and L. Scarabello, *Phys. Rev. D* **76** (2007) 084039, arXiv:astro-ph/0604492.
11. T. Chiba and M. Yamaguchi, *J. Cosmol. Astropart. Phys.* **2013** (2013) 040, arXiv:1308.1142.
12. T. Chiba and M. Yamaguchi, *J. Cosmol. Astropart. Phys.* **2008** (2008) 021, arXiv:0807.4965.
13. T. Qiu, *J. Cosmol. Astropart. Phys.* **2012** (2012) 041, arXiv:1204.0189.
14. M. Li, *Phys. Lett. B* **736** (2014) 488, arXiv:1405.0211.
15. S. Capozziello, S. Nojiri, S. Odintsov and A. Troisi, *Phys. Lett. B* **639** (2006) 135, arXiv:astro-ph/0604431.
16. S. Nojiri and S. D. Odintsov, *Phys. Rev. D* **74** (2006) 086005, arXiv:hep-th/0608008.
17. F. Briscese, E. Elizalde, S. Nojiri and S. Odintsov, *Phys. Lett. B* **646** (2007) 105, arXiv:hep-th/0612220.

18. S. Nojiri and S. D. Odintsov, *Phys. Rep.* **505** (2011) 59, arXiv:1011.0544.
19. C. Wetterich, *Phys. Dark Univ.* **2** (2013) 184, arXiv:1303.6878 [astro-ph.CO].
20. C. Wetterich, *Phys. Rev. D* **90** (2014) 043520, arXiv:1404.0535 [gr-qc].
21. C. Wetterich, *Phys. Lett. B* **736** (2014) 506, arXiv:1401.5313 [astro-ph.CO].
22. T. Moroi and T. Takahashi, *Phys. Rev. D* **66** (2002) 063501, arXiv:hep-ph/0206026.
23. K. Enqvist and M. S. Sloth, *Nucl. Phys. B* **626** (2002) 395, arXiv:hep-ph/0109214.
24. D. H. Lyth and D. Wands, *Phys. Lett. B* **524** (2002) 5, arXiv:hep-ph/0110002.
25. G. Domènech and M. Sasaki, *J. Cosmol. Astropart. Phys.* **1504** (2015) 022, arXiv:1501.07699 [gr-qc].
26. F. Lucchin and S. Matarrese, *Phys. Rev. D* **32** (1985) 1316.
27. H. Kodama and M. Sasaki, *Prog. Theor. Phys. Suppl.* **78** (1984) 1.
28. V. F. Mukhanov, H. A. Feldman and R. H. Brandenberger, *Phys. Rep.* **215** (1992) 203.
29. D. H. Lyth and E. D. Stewart, *Phys. Lett. B* **274** (1992) 168.
30. J. D. Bekenstein, *Phys. Rev. D* **48** (1993) 3641, arXiv:gr-qc/9211017 [gr-qc].
31. M. Minamitsuji, *Phys. Lett. B* **737** (2014) 139, arXiv:1409.1566.
32. D. Bettoni and S. Liberati, *Phys. Rev. D* **88** (2013) 084020, arXiv:1306.6724.
33. N. Deruelle and J. Rua, *J. Cosmol. Astropart. Phys.* **2014** (2014) 002, arXiv: 1407.0825.
34. Y. Watanabe, A. Naruko and M. Sasaki, *Europhys. Lett.* **111** (2015) 39002, arXiv: 1504.00672 [gr-qc].
35. H. Motohashi and J. White, arXiv:1504.00846.
36. S. Tsujikawa, *J. Cosmol. Astropart. Phys.* **1504** (2015) 043, arXiv:1412.6210 [hep-th].
37. G. Domènech, A. Naruko and M. Sasaki, *J. Cosmol. Astropart. Phys.* **1510** (2015) 067, arXiv:1505.00174 [gr-qc].
38. G. Domènech, S. Mukohyama, R. Namba, A. Naruko, R. Saitou and Y. Watanabe, *Phys. Rev. D* **92** (2015) 084027, arXiv:1507.05390 [hep-th].
39. G. W. Horndeski, *Int. J. Mod. Phys.* **10** (1974) 363.
40. T. Kobayashi, M. Yamaguchi and J. Yokoyama, *Prog. Theor. Phys.* **126** (2011) 511, arXiv:1105.5723.
41. J. Gleyzes, D. Langlois, F. Piazza and F. Vernizzi, *J. Cosmol. Astropart. Phys.* **2015** (2015) 18, arXiv:1408.1952.
42. J. Gleyzes, D. Langlois, F. Piazza and F. Vernizzi, *Phys. Rev. Lett.* **114** (2015) 211101, arXiv:1404.6495 [hep-th].
43. X. Gao, *Phys. Rev. D* **90** (2014) 081501, arXiv:1406.0822 [gr-qc].
44. C. Deffayet, X. Gao, D. Steer and G. Zahariade, *Phys. Rev. D* **84** (2011) 064039, arXiv:1103.3260 [hep-th].
45. T. Kobayashi, Y. Watanabe and D. Yamauchi, *Phys. Rev. D* **91** (2015) 064013, arXiv:1411.4130 [gr-qc].
46. K. Koyama and J. Sakstein, *Phys. Rev. D* **91** (2015) 124066, arXiv:1502.06872 [astro-ph.CO].
47. R. Saito, D. Yamauchi, S. Mizuno, J. Gleyzes and D. Langlois, *J. Cosmol. Astropart. Phys.* **1506** (2015) 008, arXiv:1503.01448 [gr-qc].

Surprises with Nonrelativistic Naturalness*

Petr Hořava
Berkeley Center for Theoretical Physics and Department of Physics, University of California, Berkeley, CA 94720-7300, USA
Theoretical Physics Group, Lawrence Berkeley National Laboratory, Berkeley, CA 94720-8162, USA
horava@berkeley.edu

We explore the landscape of technical naturalness for nonrelativistic systems, finding surprises which challenge and enrich our relativistic intuition already in the simplest case of a single scalar field. While the immediate applications are expected in condensed matter and perhaps in cosmology, the study is motivated by the leading puzzles of fundamental physics involving gravity: the cosmological constant problem and the Higgs mass hierarchy problem.

Keywords: Technical naturalness; nonrelativistic Nambu–Goldstone bosons; resistivity in strange metals.

PACS Number(s): 11.10.Gh, 11.30.Qc, 72.10.Di, 74.25.Kc

1. Puzzles of Naturalness

Some of the most fascinating open problems in the fundamental physics of the Universe can be phrased as puzzles of technical naturalness

- **The cosmological constant problem.** Why is the cosmological constant so small compared to the Planck scale M_P of gravity?
- **The Higgs mass hierarchy problem.** Why is the Higgs mass so small compared to the anticipated high scale M in particle physics? (Here M could be a GUT scale, or some other natural high scale of beyond-Standard Model physics — but it is logical to take $M \sim M_P$, since M_P is the only such high scale that we know to exist for sure).

In cosmology, if one subscribes to slow-roll inflation, another puzzle of naturalness emerges: Why is the η parameter of the inflaton potential so small? The requirement of technical naturalness[1] follows from some of our most elementary and robust principles of quantum mechanics, quantum field theory and many-body physics, such as

*This article first appeared in *Int. J. Mod. Phys. D* **25** (2016) 1645007.

causality and spacetime locality. If violated in nature, it would suggest disconcerting limitations to the basic principles which have so far been enormously successful across a remarkable range of scales of physical phenomena, from microphysics to condensed matter.

Notably, all the naturalness puzzles mentioned above involve gravity; it is also intriguing that they can all be phrased as questions about the natural size of various terms appearing in the potential of scalar fields. It is possible that it is quantum gravity which behaves in ways different from conventional many-body physics, and is ultimately responsible for the apparent violations of technical naturalness. Before accepting or rejecting this possibility, however, it is worthwhile to adopt technical naturalness as our guiding principle, to look for new symmetries and mechanisms that respect this principle, and to explore and map out the "landscape of naturalness". This is the strategy adopted in this brief review, and in our papers[2–5] whose main results this paper is based on.

Puzzles of technical naturalness are not confined only to particle physics, gravity and cosmology. A beautiful example exists in the field of high-T_c superconductivity:

- ***T*-linear resistivity of strange metals.** In the "strange metal" phase above the critical temperature, the resistivity behaves linearly with temperature, across a wide range of scales (at least by condensed matter standards).

As was explained beautifully by Polchinski in Ref. 6, no known interaction could explain this linear dependence in a technically natural way.[a]

1.1. *Technical naturalness*

Following a period of intuitive developments, the concept of technical naturalness was clearly formulated by G. 't Hooft in 1979 in Ref. 1. The philosophical basis for this principle is clear from the very first sentence of Ref. 1:

> *The concept of causality requires that macroscopic phenomena follow from microscopic equations.*

Then, 't Hooft further formalizes this philosophy as follows:

> *We now conjecture that the following dogma should be followed:*
> — *At any energy scale μ, a physical parameter or set of physical parameters $\alpha_i(\mu)$ is allowed to be very small only if the replacement $\alpha_i(\mu) = 0$ would increase the symmetry of the system.*–

A classic example, which will be quite useful in the following, is the first one discussed in Ref. 1: The relativistic $\lambda\phi^4$ scalar field theory in $3+1$ dimensions, with

[a]At least at the time; since then, the situation has improved somewhat, specifically with the scenarios involving quantum critical points, and particularly those amenable to holographic dual descriptions. However, the puzzle still persists and awaits its conclusive resolution.

action

$$S = \int d^4x \left(\frac{1}{2}\partial_\mu \phi \partial^\mu \phi - \frac{1}{2} m^2 \phi^2 - \frac{\lambda}{4!}\phi^4 \right). \tag{1}$$

Can the mass m be naturally small? Yes, but only if λ is also correspondingly small

$$m^2 \sim \varepsilon \mu^2, \quad \lambda \sim \varepsilon \tag{2}$$

with $\varepsilon \ll 1$ and with μ, a natural scale which makes the dimensions correct. In this example, the symmetry protecting the mutual smallness of m^2 and λ is the celebrated "constant shift" symmetry, generated by

$$\delta\phi(x^\mu) = a \tag{3}$$

with a a real constant, and broken by a small amount at the high scale μ. This "naturalness scale" μ can also be easily extracted from (2):

$$\mu \sim \frac{m}{\sqrt{\lambda}}. \tag{4}$$

Thus, if we wish for μ to be much larger than the physical particle mass m, the theory must be kept at very weak coupling. Or, put differently for a fixed m, the theory is technically natural up to the scale μ, but not much higher.

Our third and final quote from[1] sounds both quite visionary and rather ominous:

> *Pursuing naturalness beyond* 1000 GeV *will require theories that are immensely complex compared with some grand unified theories.*

Some of the GUTs are known to be quite complex; is this statement a premonition of the complexity of the landscape of string vacua and the multi-verse?

2. Naturalness in Nonrelativistic Systems

Somewhat paradoxically, gravity is perhaps the least-understood of all the fundamental interactions. The possibility that gravity at short distances could behave nonrelativistically[7,8] has generated some interest, leading to a new perspective on the possible reconciliation of gravity with quantum mechanics, and on its short-distance completion in the quantum theory. Since gravity plays a central role in the puzzles mentioned above, we are motivated to investigate technical naturalness for nonrelativistic systems. As it turns out, this investigation reveals many surprises, even before we include gravity. Let us start with a single scalar field in a flat nonrelativistic spacetime. (After all, the interesting puzzles of gravity also involve scalars.) Surely, quantum field theories of a single scalar field must have been exhaustively understood in the literature a long time ago? Not so, as we will now see!

2.1. *Towards the classification of nonrelativistic Nambu–Goldstone modes*

To focus our search for novel phenomena of naturalness, we choose the time-honored problem which universally emerges in the context of spontaneous symmetry breaking: the problem of classifying the types of Nambu–Goldstone (NG) modes implied by the breaking of global continuous internal symmetries.

In the relativistic case, the number and behavior of such NG modes is controlled by the simple, elegant and precise statement of Goldstone's theorem: one independent massless NG mode exists for each broken symmetry generator. In nonrelativistic systems, with or without Galilean invariance, a similarly complete classification is not yet available. Goldstone's theorem still exists, but it only says that there are NG modes and that they are gapless.

A sensible strategy for this problem was proposed in Refs. 9–11: the classification will be accomplished if we can classify all effective field theories available to describe the NG mode dynamics. Let us illustrate this method by focusing on nonrelativistic field theories in an "Aristotelian spacetime" M, of $D+1$ spacetime dimensions. M is topologically $\mathbf{R}^{D+1}$, and it is equipped with the flat metric but compared to either Lorentzian or Galilean spacetimes, the symmetries of M are reduced to the "Aristotelian symmetries": The SO(D) of spatial rotations, and the $\mathbf{R}^{D+1}$ of space and time translations.[b] The effective action, in terms of scalars $\pi^I(t,\mathbf{x})$, $I=1,\ldots,K$, will be

$$\int dt\, d^D\mathbf{x}(\Omega_J(\pi)\dot{\pi}^J + g_{IJ}(\pi)\dot{\pi}^I\dot{\pi}^J - h_{IJ}(\pi)\partial_i\pi^I\partial_i\pi^J + \cdots). \tag{5}$$

The Ω term leads to a natural pairing of some of the π^I's into canonical pairs; those will be called the "Type B" modes. The other π^I's do not participate in the Ω term, and they behave akin to the relativistic modes. If this is the generic low-energy action, the classification of NG modes is simple:

- Type A modes, characterized by dispersion relation $\omega \sim k$ with each such mode associated with a single broken symmetry generator.
- Type B modes with dispersion $\omega \sim k^2$. One such mode exists per *pair* of broken symmetry generators paired by Ω_I.

Based on our intuition developed in relativistic systems, one may be tempted to conjecture that (5) is generic, and therefore everything else is fine-tuning. This intuition is incorrect. One can find explicit examples of interacting theories, in which the dispersion relations start at higher orders than quadratic, and the vanishing (or the smallness) of the lower-order terms is not spoiled by quantum corrections.

[b]Such spacetimes have often been called "Lifshitz spacetimes" in the hep-th literature of recent years, but perhaps the term "Aristotelian spacetime" (as first introduced for spacetimes of this symmetry by Penrose[12] in 1968) is more appropriate. We adopt this more accurate terminology from now on.

Consider for example the super-renormalizable linear sigma model of the scalar Φ in the N representation of $O(N)$, in $2+1$ dimensions, defined by turning on a relevant self-interaction at the $z=2$ fixed point,[c] with action

$$S_{\mathrm{LSM}} = \frac{1}{2}\int dt\, d^2\mathbf{x}\left(\dot{\Phi}\cdot\dot{\Phi} - \partial_i\partial_j\Phi\cdot\partial_i\partial_j\Phi - c^2\partial_i\Phi\cdot\partial_i\Phi - m^4\Phi\cdot\Phi - \lambda(\Phi\cdot\Phi)^2\right). \tag{6}$$

Set $c^2=0$ at tree level. The leading quantum contribution to c^2 comes from two loops, and was evaluated in Ref. 2 to be $\delta c^2 = A(\lambda^2/m^4)$, where A is a constant smaller than one ≈ 0.125. This is quite small; how small? Assume that there is a hidden symmetry, which is — similar to the constant shift symmetry in (2) — broken by some small amout ε at some high momentum scale μ, and protects the smallness of c^2. Then, we expect in complete analogy with (2),

$$m^4 \sim \varepsilon\mu^4, \qquad \lambda \sim \varepsilon\mu^3, \qquad c^2 \sim \varepsilon\mu^2. \tag{7}$$

We can again obtain μ in terms of the physical couplings, $\mu \sim m/\lambda$, which predicts the small value of $c^2 \sim \lambda^2/m^4$ if (7) is to hold, but this is precisely the behavior we found by the explicit loop calculation reported above! Hence, if technical naturalness is upheld by this system, there must be a hidden symmetry behind (7).

2.2. *Polynomial shift symmetries*

This hidden symmetry, first identified in Ref. 2, is the linear shift symmetry,

$$\delta\phi(t,\mathbf{x}) = a + a_i x^i. \tag{8}$$

The c^2 term is invariant under this symmetry, up to a total derivative, and therefore protected from (large) quantum corrections.[d]

Equation (8) can be easily generalized to higher polynomial shift symmetries[e]

$$\delta_P\phi(t,\mathbf{x}) = a + a_i x^i + \cdots + a_{i_1\ldots i_P} x^{i_1}\cdots x^{i_P}. \tag{9}$$

This hierarchy of polynomial shift symmetries of degree P leads to a refinement of the classification of NG modes. We obtain, in a technically natural way, two *towers of multi-critical NG modes*[2]

[c]Other examples, leading to the same conclusions, can be found in Ref. 2.
[d]This symmetry (8) is reminiscent of Galileons, but there are clear differences: Galileons[13] are relativistic theories, with scalars invariant under shifts linear in *spacetime* coordinates. Since shifts linear in time are a part of the symmetry of the Galileons, their Hamiltonians and hence their dynamics are quite different from our nonrelativistic theories. Moreover, in our nonrelativistic systems, the linear shift symmetries can be naturally extended to higher polynomial shifts; similar extensions in the relativistic context of the Galileons would lead to apparent violations of perturbative unitarity.
[e]When the degree P is greater than one, we can in principle split $a_{i_1\ldots i_P}$ into irreducible representations of SO(D), and consider strictly smaller subgroups of polynomial shifts of degree P. We will not deal with this refinement of the symmetry structure here.

- Type A_n sequence, with dispersion $\omega \sim k^n$, with one mode per broken symmetry generator.
- Type B_{2n} sequence, with dispersion $\omega \sim k^{2n}$, with one mode per *pair* of broken symmetry generators.

Each of the two towers is labeled by an integer $n = 1, 2, \ldots,$. In addition, in Type A case, we run into the multi-critical generalization of the Coleman–Hohenberg–Mermin–Wagner (CHMW) theorem. There is a "lower critical dimension" of space at which the scalar becomes dimensionless, and its two-point function sensitive to an infrared regulator — the NG mode with $n = D$ does not quite exist as a quantum object. This would appear to limit the sensible universality classes that can represent NG modes of spontaneous symmetry breaking to $n < D$; see, however, the next paragraph for a way out from this limitation.

In Type B case, there is no analog of the CHMW theorem, the dimension of the scalar field is always non-negative, and all the Type B_{2n} modes are free to serve as NG modes associated with spontaneous symmetry breaking.

2.3. *Naturalness of cascading hierarchies*

So far we focused on the classification of NG modes characterized by one Gaussian fixed point, protected by a polynomial shift symmetry of some degree P. Since the polynomial shifts of degree $P' < P$ are contained in the shifts of degree P, the symmetries can be broken step by step, in a hierarchical fashion — first, at some high scale μ from degree P to degree P'; then at some lower scale μ' from P' to P''; and so on, possibly all the way to the breaking of the constant shift symmetry. This hierarchy $\mu \ll \mu' \ll \mu'' \cdots$ produces a hierarchical sequence of crossover scales, which separate consecutive regimes with Types A_n, $A_{n'}$, $A_{n''}$, ... behavior with $n > n' > n'' > \cdots$ (and similarly for Type B). Thus, the system exhibits a pattern of "hierarchical symmetry breaking", cascading naturally from higher to lower values of n. Importantly, the technical naturalness of this mechanism will be protected even in the interacting theories by the pattern of partial breakings of the polynomial shift symmetries.

Note that this mechanism allows us to avoid the naive consequences of the multi-critical version of the CHMW theorem: The $n = D$ NG mode does exhibit a logarithmic propagator, and is sensitive to an IR cutoff, but unlike in the relativistic case, the system can have a natural physical IR cutoff, represented by any of the scales of crossover from $n = D$ to $n < D$, without losing the constant shift symmetry required of the NG modes. In this sense, spontaneous symmetry breaking associated with such naturally regulated $n = D$ NG modes is possible.

3. Interacting Theories with Polynomial Shift Symmetries

Any generic interaction will generally break all polynomial shift symmetries all the way to the constant shift guaranteed by Goldstone's theorem. This is the case in the example (6), and those of Ref. 2. Now that we know that the new kind of symmetry is possible and can be realized by free-field fixed points, we can turn the argument around and ask: Given a polynomial shift symmetry of degree P in $D+1$ spacetime dimensions, what is the most general Lagrangian which respects this symmetry? It is natural to organize this question by the growing dimensions of the scalar composite operators that can appear in the Lagrangian. The first question is: given P, D and the number of fields n, what is the lowest-dimension operator that can appear in the Lagrangian, and is invariant up to a total derivative?

3.1. *Invariants of polynomial shift symmetries*

The questions just posed are essentially cohomological in nature. The classification of possible terms $\mathcal{L}$ in the Lagrangian, made of n copies of ϕ, containing 2Δ spatial derivatives, satisfying the invariance relation

$$\delta_P \mathcal{L} = \partial_i \mathcal{L}_i \tag{10}$$

(with δ_P acting as in (9)), and not equivalent by integration by parts, defines a sequence of cohomology groups, labeled by the integers P, D, n and Δ. The evaluation of these cohomology groups is an intriguing mathematical problem; some partial answers can be found in Ref. 3.

3.2. *Polynomial shift invariants and graph theory*

The operations defining this sequence of cohomology groups has an intriguing representation in the language of graph theory. Using this representation, one thus defines a new infinite sequence of cohomology groups associated with operations on graphs. There is a larger set of graph theory tools needed to set-up (10) and to classify its solutions (see Ref. 3); fortunately, only a pleasingly simple toolset is needed to state the results. Each monomial in $\mathcal{L}$ is made of products of $\partial_i \ldots \partial_j \phi$'s, and can be depicted as follows: Represent each copy of ϕ by a dot (i.e. a vertex of the graph), and each pair $\partial_i \ldots \partial_i \ldots$ of contracted derivatives by a line. The line is attached to the two vertices representing the ϕ's that the ∂_i's are acting on. This language is surprisingly efficient not only for deriving theorems about our invariants, but also for stating the results.

3.3. *Examples*

We begin with invariants with linear shift symmetries: At n points and in dimensions $D > n-1$, the lowest-dimension invariant occurs at $\Delta = n-1$, and it is unique. This is proven in Ref. 3, for any n and without any additional simplifying assumptions common in the analogous statements in the Galileon literature. Moreover, the invariant has an intriguing and surprisingly simple representation in the graph theory language: it is simply given by the sum over all trees with equal weight with the vertices treated as labeled or distinguishable. For example, the lowest-dimension four-point $P = 1$ invariant has six derivatives, and is given by

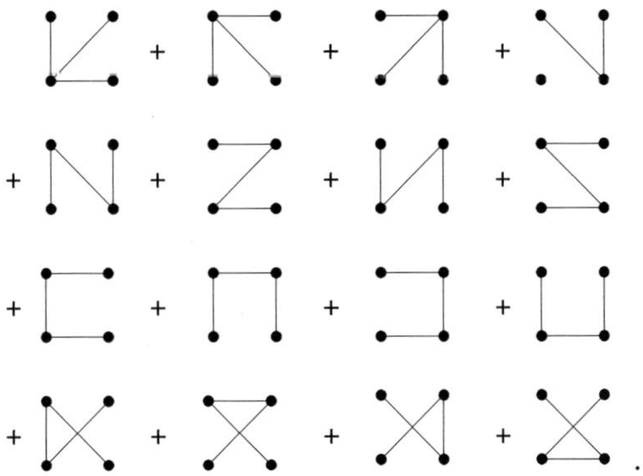

(11)

When we treat the vertices as indistinguishable, this sum becomes

$$4\,\text{[graph]} + 12\,\text{[graph]} = 4\partial_i\partial_j\partial_k\phi\,\partial_i\phi\,\partial_j\phi\,\partial_k\phi + 12\partial_i\phi\,\partial_i\partial_j\phi\,\partial_j\partial_k\phi\,\partial_k\phi. \qquad (12)$$

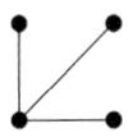

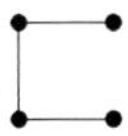

It is at present unknown what underlies this surprising simplicity of the lowest-dimension n-point one-invariants in the language of graph theory.

The graph theory techniques become indispensable for higher values of P and Δ. To illustrate, here is the unique, lowest-derivative four-point, $P = 5$ invariant with vertices treated as indistinguishable:

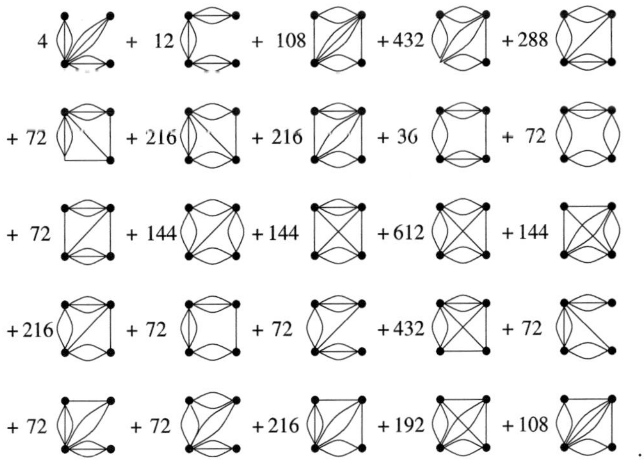

(13)

Each of the $16^3 = 4096$ terms depicted in this invariant has 18 derivatives distributed aroud the four ϕ's. In the graph language, this invariant is formed via a graph theory superposition of three copies of the $P = 1$ $\Delta = 3$ invariant; see Ref. 3 for details and proofs.

4. Back to Physics

Having seen a glimpse of the surprisingly rich mathematical structure of polynomial shift symmetries and their invariants, we now return back to physics, and look for applications involving our technically natural NG modes.

In Sec. 1 we mentioned the puzzle of linear resistivity in the strange metal phase above T_c in high-T_c superconductors. Can we find a mechanism which causes resistivity in metals to scale linearly with temperature in a technically natural way?

In the standard BCS theory, superconductivity is caused by the electron–phonon interactions. In the metallic regime above T_c, the resistivity caused by these interactions is given by the Bloch–Grüneisen formula, leading to the well-known T^5 scaling at sufficiently low T. In contrast, the temperature dependence of resistivity $\rho(T)$ in many high-T_c superconductors looks qualitatively like this:

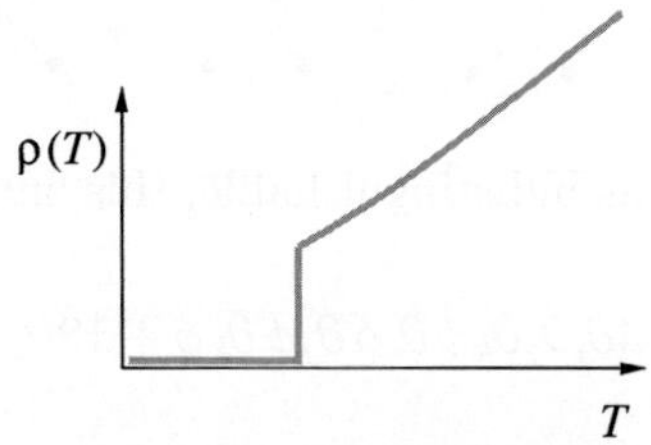

This behavior persists over a surprisingly large range of scales, and it is also quite robust under various deformations of the system, such as doping. How to explain this, in a technically natural way? And what is the "superglue" that serves to create pairing in these systems? The electron–phonon interactions yield $\rho(T) \sim T^5$, and appear to be out. Electron–electron interactions give T^2 and the interactions with impurities lead to $\rho(T) \sim$ Const. Again, none of them give resistivity linear in temperature.[f] Perhaps some very novel, exotic mechanism is taking place?[g]

This is a situation not dissimilar to the puzzles of naturalness involving gravity, which are also often interpreted as a sign that revolutionary new ideas about quantum gravity are needed. However, it is interesting to go against this flow, and revisit the traditional mechanisms with the added understanding of nonrelativistic systems. For gravity, this strategy was initiated in Refs. 7 and 8. Following this path in the case of superconductivity leads us to revisit the time-tested BCS theory, and to return to the electron–phonon interactions as the cause of electron pairing.

Needless to say, acoustic phonons represent a truly classic example of a NG mode, associated with breaking of translational invariance. Having seen that NG modes with higher-order dispersion relations are technically natural, we now ask

[f] $\rho(T) \sim T$ does emerge naturally and simply from the electron–ion interactions in the *high-temperature* regime with T above the Debye temperature, but here we are interested in a much lower range of T, well below the Debye temperature.

[g] Among the most popular such ideas is quantum criticality, which can robustly lead to linear resistivity given the absence of scales near the purported critical point.

what happens in the BCS theory when the role of the phonons is played by Type A_n NG modes of Refs. 2 and 4, with $n > 1$.

4.1. *Multi-critical phonons and the Debye model*

We will address this question in the simplest, "jellium" model of a metal in the Aristotelian spacetime of $3+1$ dimensions, replacing the ionic lattice by a uniform medium, taking the Fermi surface of the electrons to be spherically symmetric. In this model, all umklapp processes are neglected; only the longitudinal component of the phonons remains coupled and behaves as a true scalar. With this scalar being of Type A_1, this would be the simplest model of a conventional BCS superconductor. We now make the crucial assumption that the phonons are NG modes of Type A_n, with $n > 1$.[h]

In the Gaussian approximation, our phonons will be described by a "multi-critical Debye model".[14] Using the hydrodynamic description, we write the fluid density as $\rho(t, \mathbf{x}) = \rho_0 + Q(t, \mathbf{x})$, where Q is the deviation from a uniform isotropic background ρ_0. We assume the velocity $\mathbf{v}(t, \mathbf{x})$ of the fluid to describe a gradient flow,

$$v^i(t, \mathbf{x}) = \partial_i f(t, \mathbf{x}). \tag{14}$$

This ensures that the transverse phonons are absent; the purely longitudinal phonons are then described by the canonical pair f and Q, with $[f(t, \mathbf{x}), Q(t, \mathbf{x}')] = -i\delta^3(\mathbf{x}-\mathbf{x}')$. In the Gaussian approximation, the Hamiltonian is

$$H_{\mathrm{ph}} = \int d^3\mathbf{x} \left(\frac{1}{2}\rho_0 \mathbf{v}^2 + \mathcal{V}(Q) \right). \tag{15}$$

In order for this system to describe the multi-critical Type A_n phonon, we assume that $\mathcal{V}$ is dominated by

$$\mathcal{V} = \frac{1}{2}\zeta_n^2\, \partial_{i_1} \dots \partial_{i_{n-1}} Q\, \partial_{i_1} \dots \partial_{i_{n-1}} Q. \tag{16}$$

This yields phonon excitations with Type A_n dispersion relation $\omega^2 = \zeta_n^2 k^{2n}$.[i]

In the standard Debye model, there is a natural short-distance cutoff: the momenta cannot exceed k_D, and the frequencies are correspondingly cutoff at the Debye frequency ω_D. In the multi-critical Debye model with Type A_n phonons, similarly, the momenta cannot exceed k_D, and ω goes only up the modified Debye frequency $\tilde{\omega}_D$, whose value depends on the coefficients of the phonon dispersion relation. In both cases, the value of ω_D or $\tilde{\omega}_D$ is determined by requiring that the total number of modes in the system is equal to that of the underlying lattice (cf. Fig. 1).

[h] The scalar collective mode that we refer to as the "phonon" does not have to come literally from the lattice vibrations — it could be, for example, magnetic in origin. We refer to this mode as the "phonon" for simplicity, regardless of its microscopic origin.

[i] Terms with lower integer values of n are also allowed, as long as they are sufficiently small so that they only become important at momenta much lower than those of interest.

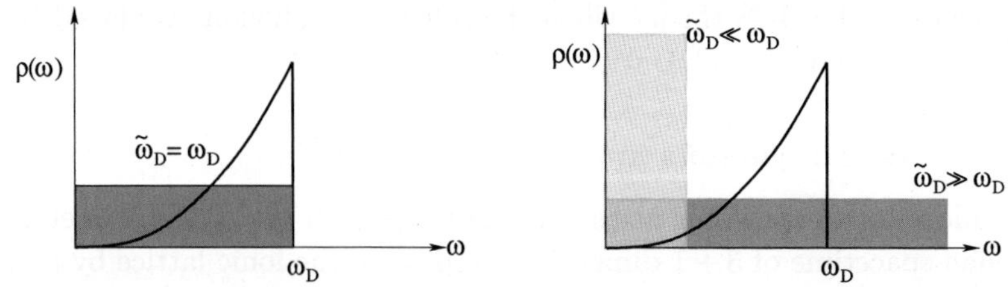

Fig. 1. The density of states $\varrho(\omega)$ as a function of frequency in the multi-critical Debye model with $n = 3$ in $3 + 1$ dimensions (the lower critical case), compared to the density of states in the standard Debye model. (a) the case of $\tilde{\omega}_D = \omega_D$; (b) the regimes with $\tilde{\omega}_D \ll \omega_D$ and $\tilde{\omega}_D \gg \omega_D$, assuming fixed k_D.

An interesting special case is $n = D$, the case of the lower critical dimension.

In the conventional systems with phonons, the terms of lowest order in the k-dominate low-energy dispersion relation, with the size of all higher-order terms generically suppressed hierarchically. How can we turn this into a system with multi-critical phonons? There are two interesting options how a sizable hierarchy can be opened: either by suppressing the lower-order terms, or by enhancing the higher-order term. Phenomenologically, the second option appears interesting for superconductivity, suggesting the following behavior of the density of states

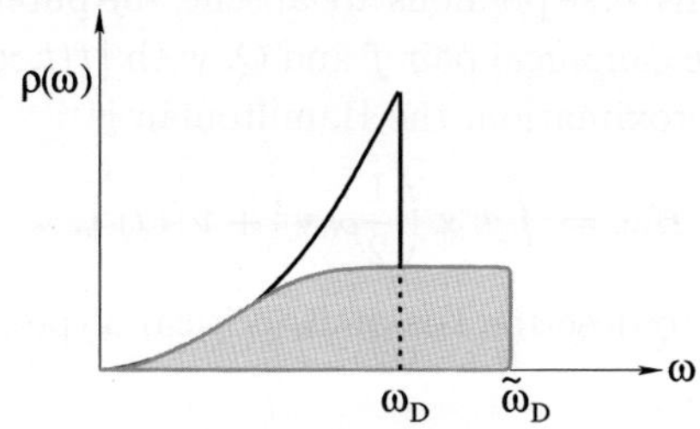

We will adopt this as our model of the acoustic phonons in a metal. Such phonons are an example of the cascading hierarchy of scales discussed in Sec. 2.3 (and in Ref. 4). The $n = 3$ dispersion dominates in the regime below $\tilde{\omega}_D$, and changes to $n = 1$ at a hierarchically much smaller crossover scale $\omega_\otimes \ll \tilde{\omega}_D$ (possibly also with an intermediate crossover to $n = 2$).

4.2. *Coupling to the Fermi surface*

The electrons $\Psi(t, \mathbf{x})$ are assumed for simplicity to have the spherical Fermi surface. We take the coupling of the phonons to the fermions to be

$$g \int dt\, d^3\mathbf{x}\, Q \Psi^\dagger \Psi \equiv g \int dt\, d^3\mathbf{x}\, \partial_i Q_i \Psi^\dagger \Psi. \tag{17}$$

This is the standard coupling between the electron Fermi liquid and the phonon excitations of the lattice, given in terms of the displacement field $Q_i(t, \mathbf{x})$, with our $Q = \partial_i Q_i$. This standard coupling breaks the polynomial shift symmetry of the phonon system all the way to the constant shifts of Q_i, but does so in a way which does not violate the hierarchy of scales $\omega_\otimes \ll \tilde{\omega}_D$.

4.3. *Resistivity in strange metals*

We use Bloch–Boltzmann kinetic theory of transport to calculate the resistivity due to the interaction between the electrons and our multi-critical phonons. The result generalizes the Bloch–Grüneisen formula, which gives resistivity ρ in terms of the *relaxation time* $\tau_{\rm tr}(\varepsilon_F)$, evaluated at the Fermi surface (see, e.g.[15] Chap. 8 in Ref. 15)

$$\rho \sim \frac{1}{\tau_{\rm tr}(\varepsilon_F)} \sim \int_{\mu_{\rm IR}}^{k_D} |g_k|^2 n(k) k^2\, k\, {\rm d}k. \tag{18}$$

Here $n(k) = 1/[\exp(\omega_k/T) - 1]$ is the phonon distribution function, $g_k \sim g\,k/\sqrt{\omega_k}$ the electron–phonon vertex, and $\mu_{\rm IR}$ is an infrared regulator, which can be set to zero in the absence of infrared divergences.[j]

In the range of temperatures of interest, well below the Debye temperature Θ, this formula gives resistivity which depends on T and n as

$$\rho(T) \sim T^{(6-n)/n}. \tag{19}$$

In this standard case of $n = 1$, this reproduces the standard Bloch–Grüneisen scaling $\rho \sim T^5$. More interestingly, in the case of the multi-critical phonons at their lower critical dimension, $n = 3$, we obtain $\rho \sim T$ (accompanied by $T \log T$ corrections) — the electron–phonon interaction is capable of explaining resistivity linear in temperature!

The calculation can be repeated in any number of spatial dimensions D, generalizing (19) to

$$\rho(T) \sim T^{(3+D-n)/n}. \tag{20}$$

The interesting cases are those with $D = 2$ and 3: Type A_n multi-critical phonons with $n \leq D$ suggest resistivity going as T^5, T^2 (which matches the scaling of the electron–electron contribution) or T in three spatial dimensions, and T^4 or $T^{3/2}$ in two dimensions.

5. Conclusions and Future Prospects

Investigating the role of technical naturalness in nonrelativistic systems with Aristotelian spacetime symmetries has led to many surprises, even in the simplest cases of single scalar fields, often challenging our naive intuition based on relativistic systems. These surprises should be finding their first applications in those fields of physics which naturally respect the Aristotelian symmetries: the most obvious

[j]When the phonons are at the lower critical dimension, there are IR divergences that are automatically regulated with $\omega_\otimes > 0$ serving as the natural IR regulator, much as in Sec. 2.3

is condensed matter, where we have found a technically natural explanation for resistivity linear in temperature in metals with multi-critical phonons. Another field which shares similar spacetime symmetries with a preferred rest frame is cosmological inflation; applications in this area are therefore also likely.

Eventually, we are hoping that the lessons learned in the nonrelativistic arena will give new insights to the "fundamental" puzzles of naturalness in high-energy physics and gravity: the cosmological constant problem and the Higgs mass hierarchy problem. If this many surprises and such unexpectedly rich mathematical and physical structure appear already in the case of a single scalar, one can only imagine how many surprises are still hidden in nonrelativistic systems with gauge symmetries, in particular in nonrelativistic gravity.

Acknowledgments

It is a pleasure to thank Tom Griffin, Kevin Grosvenor, Ziqi Yan and Chris Mogni for the many fruitful and enjoyable collaborations on the results summarized in these notes. I wish to thank Pisin Chen and the entire organizing team of the second LeCosPA Symposium on *Everything About Gravity* at National Taiwan University for their flawless hospitality in Taipei. This work has been supported in part by NSF Grant No. PHY-1521446 and by the Berkeley Center for Theoretical Physics.

References

1. G. 't Hooft, *NATO ASI Ser. B* **59** (1980) 135.
2. T. Griffin, K. T. Grosvenor, P. Hořava and Z. Yan, *Phys. Rev. D* **88** (2013) 101701, arXiv:1308.5967.
3. T. Griffin, K. T. Grosvenor, P. Hořava and Z. Yan, *Commun. Math. Phys.* **340** (2015) 985, arXiv:1412.1046.
4. T. Griffin, K. T. Grosvenor, P. Hořava and Z. Yan, *Phys. Rev. Lett.* **115** (2015) 241601, arXiv:1507.06992.
5. K. T. Grosvenor, P. Hořava, C. J. Mogni, and Z. Yan, *Nonrelativistic Renormalization of Scalar Fields with Polynomial Shift Symmetries*, to appear (2016).
6. J. Polchinski, arXiv:hep-th/9210046.
7. P. Hořava, *J. High Energy Phys.* **3** (2009) 20, arXiv:0812.4287.
8. P. Hořava, *Phys. Rev. D* **79** (2009) 084008, arXiv:0901.3775.
9. H. Watanabe and H. Murayama, *Phys. Rev. Lett.* **108** (2012) 251602, arXiv:1203.0609.
10. H. Watanabe and H. Murayama, *Phys. Rev. Lett.* **110** (2013) 181601, arXiv:1302.4800.
11. H. Watanabe and H. Murayama, *Phys. Rev. X* **4** (2014) 031057, arXiv:1402.7066.
12. R. Penrose, Structure of space-time, in *Battelle Rencontres*, C. DeWitt and J. A. Wheeler (eds). (Benjamin, New York, 1968).
13. A. Nicolis, R. Rattazzi and E. Trincherini, *Phys. Rev. D* **79** (2009) 064036, arXiv:0811.2197.
14. T. Griffin, K. T. Grosvenor, P. Hořava and Z. Yan, *Multicritical Phonons and Resistivity of Strange Metals*, to appear (2016).
15. N. Pottier, *Nonequilibrium Statistical Physics: Linear Irreversible Processes* (Oxford University Press, Oxford, NY, 2010).

Quantum Geometrodynamics with Intrinsic Time Development*

Chopin Soo
Department of Physics, National Cheng Kung University,
No. 1, University Rd., Tainan 70101, Taiwan
cpsoo@mail.ncku.edu.tw

Quantum geometrodynamics with intrinsic time development is presented. Paradigm shift from full spacetime covariance to spatial diffeomorphism invariance yields a nonvanishing Hamiltonian, a resolution of the 'problem of time' and gauge-invariant temporal ordering in an ever expanding universe. Einstein's general relativity is a particular realization of a wider class of theories; and the framework prompts natural extensions and improvements with the consequent dominance of Cotton–York potential at early times when the universe was small.

Keywords: Geometrodynamics; problem of time; quantum gravity.

PACS Number(s): 04.70.Bw, 04.20.Jb, 04.70.Dy

1. Introduction

All of us experience the passage of time. But is time an illusion of our perception, an emergent semi classical entity, or is it present at the fundamental level even in quantum gravity? General relativity (GR), Einstein's theory of classical spacetime, ties space and time to (pseudo-)Riemannian geometry. But in quantum gravity, spacetime is a concept of 'limited applicability'.[1] That semi-classical spacetime is emergent begs the question what, if anything at all, plays the role of 'time' in quantum gravity? Wheeler went as far as to claim we have to forgo time-ordering, and to declare 'there is no spacetime, there is no time, there is no before, there is no after'.[1] But without 'time-ordering', how is 'causality', which is requisite in any 'sensible physical theory', enforced in quantum gravity? Furthermore, a resolution of the 'problem of time' in quantum gravity cannot be deemed complete if it fails to account for the intuitive physical reality of time and does not provide satisfactory correlation between time development in quantum dynamics and the passage of time in classical spacetimes.

Wheeler also emphasized it is three-geometry, rather than four-geometry, which is fundamental in quantum geometrodynamics. The call to abandon four-covariance

*This article first appeared in *Int. J. Mod. Phys. D* **25** (2016) 1645008.

is not new. Simplifications in the Hamiltonian analysis of GR, and the fact that the physical degrees of freedom (d.o.f) involve only the spatial metric, lead Dirac to conclude that 'four-dimensional symmetry is not a fundamental property of the physical world'.[2] A key obstacle to the viability of GR as a perturbative quantum field theory lies in the conflict between unitarity and spacetime general covariance: renormalizability can be attained with higher-order curvature terms, but spacetime covariance requires time as well as spatial derivatives of the same (higher) order, thus compromising unitarity. Relinquishing four-covariance to achieve power-counting renormalizability through modifications of GR with higher-order *spatial*, rather than spacetime, curvature terms was Horava's bold proposal.[3]

Geometrodynamics bequeathed with positive-definite spatial metric is the simplest consistent framework to implement fundamental commutation relations (CR) predicated on the existence of spacelike hypersurfaces. In the Arnowitt–Deser–Misner (ADM) description of the spacetime metric, $ds^2 = -N^2dt^2 + q_{ij}(dx^i + N^i dt)(dx^j + N^j dt)$, these are labeled as constant-t hypersurfaces.[4] Quantum states however do not depend on t; in his seminal work, DeWitt observed that the wave function of the universe can depend only on the three-geometry, and time must be determined intrinsically.[5] The canonical action of GR may be written as

$$S = \int \tilde{\pi}^{ij}\dot{q}_{ij}d^3x dt - \int (NH + N^i H_i)d^3x dt + \text{boundary term}, \tag{1}$$

wherein the super-Hamiltonian $H = \frac{2\kappa}{\sqrt{q}}[G_{ijkl}\tilde{\pi}^{ij}\tilde{\pi}^{kl} + \mathcal{V}(q_{ij})]$, and $H_i = -2q_{ij}\nabla_k\tilde{\pi}^{kj} = 0$ is the super-momentum constraint which generates spatial diffeomorphisms of the variables. The DeWitt supermetric, $G_{ijkl} = \frac{1}{2}(q_{ik}q_{jl} + q_{il}q_{jk}) - lq_{ij}q_{kl}$, with deformation parameter l, has signature $(\text{sgn}[\frac{1}{3} - l], +, +, +, +, +)$. For Einstein's theory, $l = \frac{1}{2}$ and the potential $\mathcal{V} = -\frac{q}{(2\kappa)^2}(R - 2\Lambda_{\text{eff.}})$. The symplectic potential decomposes as $\int \tilde{\pi}^{ij}\delta q_{ij} = \int \bar{\pi}^{ij}\delta\bar{q}_{ij} + \tilde{\pi}\delta \ln q^{\frac{1}{3}}$. So clean separation of the conjugate pair, $(\ln q^{\frac{1}{3}}, \tilde{\pi})$, consisting of (one-third of) the logarithm of the determinant of the spatial metric and the trace of the momentum, from $(\bar{q}_{ij} := q^{-\frac{1}{3}}q_{ij}, \bar{\pi}^{ij} := q^{\frac{1}{3}}(\tilde{\pi}^{ij} - \frac{1}{3}q^{ij}\tilde{\pi}))$, the unimodular spatial metric and traceless part of the momentum, allows a deparametrization of the theory wherein $\ln q^{\frac{1}{3}}$ plays the role of the intrinsic time variable when $\beta^2 := l - \frac{1}{3} > 0$.

A framework for quantum geometrodynamics without the paradigm of spacetime covariance, and equipped with spatial diffeomorphism-invariant physical Hamiltonian, has been advocated in a series of works.[6–10] It resolves 'the problem of time' and bridges the deep divide between quantum mechanics and conventional canonical formulations of quantum gravity with a Schrödinger equation which describes first-order evolution and time-ordering in global intrinsic time.

2. Classical GR and Intrinsic Time Development

The super-Hamiltonian constraint factorizes in an interesting way as

$$\frac{\sqrt{q}}{2\kappa}H = G_{ijkl}\tilde{\pi}^{ij}\tilde{\pi}^{kl} + \mathcal{V}(q_{ij}) = -(\beta\tilde{\pi} - \bar{H})(\beta\tilde{\pi} + \bar{H}) = 0, \tag{2}$$

wherein $\bar{H} = \sqrt{\bar{G}_{ijkl}\bar{\pi}^{ij}\bar{\pi}^{kl} + \mathcal{V}(q_{ij})} = \sqrt{\frac{1}{2}[\bar{q}_{ik}\bar{q}_{jl} + \bar{q}_{il}\bar{q}_{jk}]\bar{\pi}^{ij}\bar{\pi}^{kl} + \mathcal{V}(q_{ij})}$. This constrains $\bar{H} = \pm\beta\tilde{\pi}$. In Einstein's theory when $\mathcal{V} = -\frac{q}{(2\kappa)^2}[R - 2\Lambda_{\text{eff.}}]$ and $l = \frac{1}{2}$, the constraints form a first class algebra; and the lapse function, N, which is a Lagrange multiplier, is *a priori* an arbitrary function. But the physical meaning of N is revealed *a posteriori* by the equations of motion (EOM) and constraints. In particular, the lapse function is related to $\partial_t \ln q$ and $\bar{H}$ (which assumes the role of Hamiltonian density in intrinsic time geometrodynamics) through

$$\begin{aligned}
\frac{\partial \ln q^{\frac{1}{3}}(x)}{\partial t} &= \left\{\ln q^{\frac{1}{3}}(x), \int N(y)H(y)d^3y\right\}_{\text{P.B.}} \\
&= \int N(y)\{\ln q^{\frac{1}{3}}(x), H(y)\}_{\text{P.B.}}d^3y \\
&= -\frac{4}{\sqrt{q}}N(x)\kappa\beta^2\tilde{\pi}(x) \\
&= \mp 4\kappa\beta \underset{\sim}{N}(x)\bar{H}(x),
\end{aligned} \tag{3}$$

wherein the Poisson bracket $\{\ln q^{\frac{1}{3}}(x), \tilde{\pi}(y)\}_{\text{P.B.}} = \delta(x-y)$ has been used to arrive at the intermediate step, and the Hamiltonian constraint in the last step (even though N was *a priori* arbitrary). This is the precise form of the *emergent* ADM lapse function N in Einstein's theory.

Dispensing with the Hamiltonian constraint, the EOM of the physical d.o.f. (which lie in the transverse traceless excitations of $(\bar{q}_{ij}, \bar{\pi}^{ij})$) can equivalently be captured by the nontrivial Hamiltonian, $H_{\text{ADM}} := \frac{1}{\beta}\int \frac{\partial \ln q^{\frac{1}{3}}(x)}{\partial t}\bar{H}(x)d^3x$, which generates evolution w.r.t. ADM coordinate time t. To wit,

$$\begin{aligned}
\dot{\bar{q}}_{ij}(x) &= \left\{\bar{q}_{ij}(x), \frac{1}{\beta}\int \frac{\partial \ln q^{\frac{1}{3}}(y)}{\partial t}\bar{H}(y)d^3y\right\}_{\text{P.B.}} \\
&= \int \frac{1}{\beta}\frac{\partial \ln q^{\frac{1}{3}}(y)}{\partial t}\{\bar{q}_{ij}(x), \bar{H}(y)\}_{\text{P.B.}}d^3y \\
&= \int \mp\underset{\sim}{N}(y)4\kappa\bar{H}(y)\{\bar{q}_{ij}(x), \bar{H}(y)\}_{\text{P.B.}}d^3y \\
&= \int \mp\underset{\sim}{N}(y)2\kappa\{\bar{q}_{ij}(x), \bar{H}^2(y)\}_{\text{P.B.}}d^3y \\
&= \int \mp N(y)\{\bar{q}_{ij}(x), H(y)\}_{\text{P.B.}}d^3y.
\end{aligned} \tag{4}$$

This final step can be reached bearing in mind $\frac{\sqrt{q}}{2\kappa}H = \bar{H}^2 - \beta^2\tilde{\pi}^2$ and $\tilde{\pi}$ commutes with $(\bar{q}_{ij}, \bar{\pi}^{ij})$. This is thus equivalent to the evolution generated through conventional $\int \mp N(y)\,\{\bar{q}_{ij}(x), H(y)\}_{\text{P.B.}}\,d^3y$ with the *a posteriori* relation (3) between $N, \partial_t \ln q$ and $\bar{H}$. The $\mp$ sign accompanying N does not affect the resultant classical ADM four-metric which depends on the square of N. The EOM for $\bar{\pi}^{ij}$ can be similarly demonstrated. Adding $\int N^i(y)H_i(y)d^3y$ to the Hamiltonian merely leads to modification of the EOM by Lie derivatives of the variables w.r.t. N^i, with the resultant lapse function (3) generalized to[a]

$$N = \frac{\sqrt{q}(\partial_t \ln q^{1/3} - \frac{2}{3}\nabla_i N^i)}{4\beta\kappa\bar{H}}. \tag{5}$$

In fact this ensures the Hamiltonian constraint is satisfied classically in the form $\frac{K^2}{9} = \frac{4\kappa^2\beta^2}{q}\bar{H}^2$. Thus the Hamiltonian H_{ADM}, *which is no longer constrained to vanish*, equivalently captures the physical content and EOM of Einstein's theory. This framework however allows the potential $\mathcal{V}$ *to depart from that of Einstein's theory without leading to inconsistencies in the constraint algebra.*

3. Global Intrinsic Time Development and Temporal Order

While q is a tensor density, the multi-fingered intrinsic time interval, $\delta \ln q^{\frac{1}{3}} = \frac{1}{3}\frac{\delta q}{q} = \frac{q^{ij}}{3}\delta q_{ij}$, is a scalar entity.[b] Hodge decomposition for any compact Riemannian manifold without boundary yields

$$\delta \ln q^{\frac{1}{3}} = \delta T + \nabla_i \delta Y^i, \tag{6}$$

wherein the gauge-invariant part of $\delta \ln q^{\frac{1}{3}}$ is $\delta T = \frac{2}{3}\delta \ln V$, which is proportional to the three-dimensional diffeomorphism invariant (3dDI) logarithmic change in the spatial volume V. This can be seen from $\int (\delta \ln q^{\frac{1}{3}})\sqrt{q}d^3x = \int (\delta T + \nabla_i \delta Y^i)\sqrt{q}d^3x = (\delta T)\int \sqrt{q}d^3x = (\delta T)V$ (which implies that δT is the average value of $\delta \ln q^{\frac{1}{3}}(x)$ over V), and also $\int (\delta \ln q^{\frac{1}{3}})\sqrt{q}d^3x = \frac{1}{3}\int \frac{(\delta\sqrt{q}^2)}{\sqrt{q}^2}\sqrt{q}d^3x = \frac{2}{3}\delta(\int \sqrt{q}d^3x) = \frac{2}{3}\delta V$, thus yielding $(\delta T)V = \frac{2}{3}\delta V$. Being the logarithmic change of a tensor density, $\delta \ln q^{\frac{1}{3}}$ also has the distinct advantage (over, for instance, the use of simple scalar fields as time) that the $\nabla_i \delta Y^i$ piece in its Hodge decomposition is in fact a Lie derivative of $\ln q^{\frac{1}{3}}$ (in particular, $\mathcal{L}_{\delta\vec{N}} \ln q^{\frac{1}{3}} = \frac{2}{3}\nabla_i \delta N^i$) and can thus be gauged away completely with spatial diffeomorphisms.

Instead of describing the evolution of quantum states w.r.t. gauge-dependent multi-fingered time $\ln q^{\frac{1}{3}}(x)$, it is eminently more meaningful to ask how the wave function of the universe, Ψ, changes w.r.t to 3dDI *global* intrinsic time variable T. With T displacing the physically less concrete ADM coordinate time t in H_{ADM}, this dynamics is determined by the Schrödinger equation,

$$i\hbar\frac{\delta\Psi}{\delta T} = H_{\mathrm{Phys}}\Psi, \quad H_{\mathrm{Phys}} := \int \frac{\bar{H}(x)}{\beta}d^3x, \tag{7}$$

[a]The determinant of the ADM metric is equal to Nq, so the lapse is nonvanishing unless there is degeneracy. In, for instance, Minkowski spacetime, both the numerator and denominator vanish, but we should consider instead de Sitter spacetime with compact slicings, and N is well-defined in the limit of vanishing positive cosmological constant.

[b]In DeWitt's seminal work,[5] the intrinsic time variable was $q^{\frac{1}{4}}$ instead of $\ln q^{\frac{1}{3}}$.

wherein $H_{\rm Phys}$ is the physical Hamiltonian generating evolution in global intrinsic time. This *replaces, and circumvents, the naive imposition of the Hamiltonian constraint as a Schwinger–Tomonaga equation with multi-fingered time* $\ln q^{\frac{1}{3}}(x)$, which would have looked like $i\hbar\frac{\delta\Psi}{\delta\ln q^{\frac{1}{3}}(x)}\Psi = \frac{\bar{H}(x)}{\beta}\Psi$ in the metric representation.

The formalism is applicable to the full theory of quantum gravity,[8,10] and assumptions of mini-superspace models have not been invoked. $\beta > 0$ is required for the Hamiltonian to be bounded from below, and global intrinsic time increases monotonically with our ever expanding universe.

In this framework dictated by first-order Schrödinger evolution in global intrinsic time, the Hamiltonian $H_{\rm Phys}$ is 3dDI provided $\bar{H} = \sqrt{\bar{\pi}^{ij}\bar{G}_{ijkl}\bar{\pi}^{kl} + \mathcal{V}}$ is a scalar density of weight one; and Einstein's GR (with its particular value of β and $\mathcal{V}$) is a particular realization of this wider class of theories. The crucial time development operator can be derived by integrating the Schrödinger equation, yielding $\Psi(T) = U(T, T_0)\Psi(T_0)$, with T-ordered operator $U(T, T_0) := \mathcal{T}\exp[-\frac{i}{\hbar}\int_{T_0}^{T} H_{\rm Phys}(T')dT']$, or in the form of a time-ordered Dyson series,

$$\begin{aligned} U(T,T_o) &= I - \frac{i}{\hbar}\int_{T_o}^{T} dT_1 H_{\rm Phys}(T_1) \\ &+ \left(\frac{-i}{\hbar}\right)^2 \int_{T_o}^{T} dT_2 \int_{T_o}^{T_2} dT_1 H_{\rm Phys}(T_2) H_{\rm Phys}(T_1) + \cdots \\ &+ \left(\frac{-i}{\hbar}\right)^n \int_{T_o}^{T} dT_n \int_{T_o}^{T_n} dT_{n-1}\cdots \\ &\times \int_{T_o}^{T_2} dT_1 H_{\rm Phys}(T_n) H_{\rm Phys}(T_{n-1})\cdots H_{\rm Phys}(T_1) + \cdots . \end{aligned} \tag{8}$$

Unitary and diffeomorphism-invariant $U(T, T_0)$ follows if the 3dDI $H_{\rm Phys}$ is self-adjoint. Since dT is also unchanged under spatial diffeomorphisms, the temporal ordering in $U(T, T_0)$ will be reassuringly gauge-invariant. Time-ordered integrability of the Schrödinger equation is feasible without ambiguity because δT is 'one-dimensional', more precisely, spatially-independent *rather than many-fingered*, and $H_{\rm Phys}$ is a spatial integral, rather than a local Hamiltonian density. Moreover, the necessity of 'time-ordering', which underpins the notion of causality, emerges, because quantum fields and the Hamiltonian do not commute at different 'times'.[c] What is paramount to causality is not the dimension of time, but the sequence, or time-ordered development, of the physical state.

Wheeler's provocative call to forgo all time-ordering may ring true intuitively, but it does not hold up to deeper scrutiny. While spacetime events and their ordering have no primary role and spacetime itself is an entity 'of limited applicability' in quantum gravity, the fundamental gauge-invariant entity that can be, and is in fact, time-ordered is the physical quantum state.

[c]The Hamiltonian in fact depends explicitly upon intrinsic time.

4. Modification of the Potential, and Asymptotic Behavior of the Hamiltonian at Early and Late Intrinsic Times

Paradigm shift from four-covariance to 3dDI (or spatial covariance), not only allows a resolution of the 'problem of time', but it also permits the introduction of higher spatial derivative terms[3] in the potential $\mathcal{V}$ in $\bar{H}$ to improve convergence without sacrificing unitarity. Since $\bar{H}$ is a square root, this suggests a positive semi-definite potential,[6,8]

$$\mathcal{V} = \left[\frac{1}{2}(\bar{q}_{ik}\bar{q}_{jl} + \bar{q}_{il}\bar{q}_{jk}) + \gamma\bar{q}_{ij}\bar{q}_{kl}\right]\tilde{\mathcal{W}}^{ij}\tilde{\mathcal{W}}^{kl}, \quad \gamma \geq -\frac{1}{3};$$
$$\tilde{\mathcal{W}}^i_j = [\sqrt{q}(\Lambda' + a'R)\delta^i_j + b\hbar\sqrt{q}\bar{R}^i_j + g\hbar\tilde{C}^i_j], \tag{9}$$

wherein R and $\bar{R}^i_j$ are, respectively, the scalar and traceless parts of the spatial Ricci curvature, while $\tilde{C}^i_j$ is the Cotton–York tensor (density) which is third-order in spatial derivatives and associated with dimensionless coupling constant g. This is compatible with the introduction of higher spatial curvature terms up to $\tilde{C}^i_j\tilde{C}^j_i$ in H for power-counting renormalizability.[3]

Hodge decomposition for $\delta \ln q^{\frac{1}{3}}$ and its Heisenberg equation of motion lead to $\frac{d}{dT}\ln q^{\frac{1}{3}}(x,T) = \frac{\partial}{\partial T}\ln q^{\frac{1}{3}} + \frac{1}{i\hbar}[\ln q^{\frac{1}{3}}, H_{\text{Phys}}] = 1$ with solution $\ln[\frac{q(x,T)}{q(x,T_{\text{now}})}] = 3(T - T_{\text{now}})$ and $-\infty < T < \infty$. Moreover, $\delta T = \frac{2}{3}\delta \ln V$, i.e. $T - T_{\text{now}} = \frac{2}{3}\ln(V/V_{\text{now}})$. Explicitly separating out T-dependence from entities (labeled with overline) which depend only on $\bar{q}_{ij}$ yields

$$\begin{aligned}\tilde{\mathcal{W}}^i_j &= [\sqrt{q}(\Lambda' + a'R)\delta^i_j + b'\sqrt{q}\bar{R}^i_j + g\hbar\tilde{C}^i_j] \\ &= [\sqrt{q}(\Lambda' + a'q^{-\frac{1}{3}}\bar{q}^{kl}\overline{R}_{kl})\delta^i_j + b'\sqrt{q}q^{-\frac{1}{3}}\bar{q}^{ik}\overline{\bar{R}}_{kj} + g\hbar\tilde{C}^i_j] \\ &\quad + \sqrt{q}(\partial_i \ln q \,\text{terms})\end{aligned} \tag{10}$$

with q-independent Cotton–York tensor density $\tilde{C}^i_j$ which is conformally invariant. The Hamiltonian is explicitly (intrinsic)time-dependent, and not (intrinsic)time-reversal invariant; furthermore, the exponential scaling behavior of q with intrinsic time implies in the limit $T - T_{\text{now}} \to -\infty, V/V_{\text{now}} \to 0$ (i.e. early times when the universe was very small in volume), the potential $\mathcal{V}$ was dominated by the Cotton–York term, whereas the limit $T - T_{\text{now}} \to \infty, V/V_{\text{now}} \to \infty$ (i.e. late times when the universe becomes large) will be dominated by the cosmological constant term, which is compatible with current observations and understanding of our ever expanding universe.[d] In the middle period curvature and cosmological terms will be comparable in importance.

[d]An exemplar of the determination of global intrinsic time interval is its measurement through dimensionless redshift in homogeneous Friedmann–Lemaitre–Robertson–Walker (FLRW) cosmology.

5. Momentric Variable, and the Free Theory

The Poisson brackets for the $(\bar{q}_{ij}, \bar{\pi}^{ij})$ variables are,

$$\{\bar{q}_{ij}(\mathbf{x}), \bar{q}_{kl}(\mathbf{y})\}_{\text{P.B.}} = 0, \quad \{\bar{q}_{kl}(\mathbf{x}), \bar{\pi}^{ij}(\mathbf{y})\}_{\text{P.B.}} = P^{ij}_{kl}\,\delta(\mathbf{x}-\mathbf{y}),$$
$$\{\bar{\pi}^{ij}(\mathbf{x}), \bar{\pi}^{kl}(\mathbf{y})\}_{\text{P.B.}} = \frac{1}{3}(\bar{q}^{kl}\bar{\pi}^{ij} - \bar{q}^{ij}\bar{\pi}^{kl})\delta(\mathbf{x}-\mathbf{y}) \tag{11}$$

with $P^{ij}_{kl} := \frac{1}{2}(\delta^i_k\delta^j_l + \delta^i_l\delta^j_k) - \frac{1}{3}\bar{q}^{ij}\bar{q}_{kl}$ denoting the traceless projection operator. This set is not thus strictly canonical, and difficulties in implementing $\bar{\pi}^{ij}$ as self-adjoint traceless operator in the metric representation lead us to summon the momentric variable (first introduced by Klauder[11]) which is classically $\bar{\pi}^i_j = \bar{q}_{jm}\bar{\pi}^{im}$. The fundamental CR of the traceless momentric and unimodular part of the spatial metric are

$$[\bar{q}_{ij}(x), \bar{q}_{kl}(y)] = 0, \quad [\bar{q}_{ij}(x), \hat{\bar{\pi}}^k_l(y)] = i\hbar\bar{E}^k_{l(ij)}\delta(x-y),$$
$$[\hat{\bar{\pi}}^i_j(x), \bar{\pi}^k_l(y)] = \frac{i\hbar}{2}(\delta^k_j\hat{\bar{\pi}}^i_l - \delta^i_l\hat{\bar{\pi}}^k_j)\delta(x-y), \tag{12}$$

wherein $\bar{E}^i_{j(mn)} = \frac{1}{2}(\delta^i_m\bar{q}_{jn} + \delta^i_n\bar{q}_{jm}) - \frac{1}{3}\delta^i_j\bar{q}_{mn}$ is the vielbein for the supermetric $\bar{G}_{ijkl} = \bar{E}^m_{n(ij)}\bar{E}^n_{m(kl)}$. Quantum mechanically, the momentric operators and CR can be explicitly realized in the metric representation by

$$\hat{\bar{\pi}}^i_j(x) = \frac{\hbar}{i}\bar{E}^i_{j(mn)}(x)\frac{\delta}{\delta\bar{q}_{mn}(x)} = \frac{\hbar}{i}\frac{\delta}{\delta\bar{q}_{mn}(x)}\bar{E}^i_{j(mn)}(x) = \hat{\bar{\pi}}^{\dagger i}_j(x), \tag{13}$$

which are self-adjoint on account of $[\frac{\delta}{\delta\bar{q}_{mn}(x)}, \bar{E}^i_{j(mn)}(x)] = 0$. These eight momentric variables generate SL(3, R) transformations, which preserve positivity and unimodularity of $\bar{q}_{ij}$. Explicitly, $U^\dagger(\alpha)\bar{q}_{kl}(x)U(\alpha) = (e^{\frac{\alpha(x)}{2}})^m_k\bar{q}_{mn}(x)(e^{\frac{\alpha(x)}{2}})^n_l$, wherein $U(\alpha) = e^{-\frac{i}{\hbar}\int\alpha^i_j\bar{\pi}^j_i d^3y}$. Furthermore, they generate, by themselves, at each spatial point, an SU(3) algebra. In fact, defining $T^A(x) := \frac{1}{\hbar\delta(0)}(\lambda^A)^j_i\hat{\bar{\pi}}^i_j(x)$ with Gell–Mann matrices $\lambda^{A=1,\dots,8}$ lead to

$$[T^A(x), T^B(y)] = if^{AB}{}_C T^C\frac{\delta(x-y)}{\delta(0)} \tag{14}$$

with SU(3) structure constants $f^{AB}{}_C$.[12] This leads to the remarkable observation that in the absence of the potential $\mathcal{V}$ the free theory is characterized by SU(3) invariance generated by the momentric (whereas $\bar{\pi}^{ij}$ generate translations which do not preserve the positivity of the metric), because the Casimir invariant T^AT^A is related to the kinetic operator in $\bar{H}$ through

$$\frac{\hbar^2\delta^2(0)}{2}T^AT^A = \hat{\bar{\pi}}^{i\dagger}_j\hat{\bar{\pi}}^j_i = \hat{\bar{\pi}}^i_j\hat{\bar{\pi}}^j_i = \hat{\bar{\pi}}^{ij}\hat{\bar{G}}_{ijkl}\hat{\bar{\pi}}^{kl}. \tag{15}$$

The upshot is its spectrum can be labeled by eigenvalues of the complete commuting set at each spatial point comprising the two Casimirs $L^2 = T^AT^A, C = d_{ABC}T^AT^BT^C \propto \det(\hat{\bar{\pi}}^i_j)$, Cartan subalgebra T^3, T^8, and isospin $I = \sum^3_{B=1}T^BT^B$.

An underlying group structure has the advantage of the action of momentric on wave functions by functional differentiation can be traded for its well-defined action as generators of SU(3) on states expanded in this basis. In addition, the ground state of the free theory, $|0\rangle$, is thus an SU(3) singlet state annihilated by all the momentric operators (i.e. $T^A|0\rangle = 0$).

6. Early Universe and Cotton–York Dominance

As explained, early global intrinsic times and small volumes correspond to the era of Cotton–York dominance of $\mathcal{V}$ at the beginning of the universe, wherein $\bar{H} = \sqrt{\hat{\bar{\pi}}^{\dagger j}_i \hat{\bar{\pi}}^i_j + g^2\hbar^2 \tilde{C}^j_i \tilde{C}^i_j}$. A number of intriguing facts conspire to simplify and regulate the Hamiltonian: the traceless Cotton–York tensor density is expressible as $\tilde{C}^i_j = \bar{E}^i_{j(mn)} \frac{\delta W}{\delta \bar{q}_{mn}}$, wherein $W = \frac{1}{4}\int \tilde{\epsilon}^{ijk}(\bar{\Gamma}^l_{im}\partial_j \bar{\Gamma}^m_{kl} + \frac{2}{3}\bar{\Gamma}^l_{im}\bar{\Gamma}^m_{jn}\bar{\Gamma}^n_{kl})\, d^3x$ is the 3dDI Chern–Simons functional of the affine connection of $\bar{q}_{ij}$. This leads to the similarity transformation of the momentric,

$$\begin{aligned}\hat{Q}^i_j &= e^{gW}\hat{\bar{\pi}}^i_j e^{-gW} \\ &= \frac{\hbar}{i}\bar{E}^i_{j(mn)}\left[\frac{\delta}{\delta \bar{q}_{mn}} - g\frac{\delta W}{\delta \bar{q}_{mn}}\right] \\ &= \frac{\hbar}{i}\bar{E}^i_{j(mn)}\frac{\delta}{\delta \bar{q}_{mn}} + ig\hbar\tilde{C}^i_j. \end{aligned} \tag{16}$$

Moreover, $[\hat{\bar{\pi}}^i_j, \tilde{C}^j_i] = 0$. Consequently, the Hamiltonian density is simply $\bar{H} = \sqrt{\hat{Q}^{\dagger i}_j \hat{Q}^j_i}$.[e]

Ricci curvature terms become increasingly important in the potential after the initial era of Cotton–York dominance. They can be introduced in a manner which preserves the underlying structure which regulate the Hamiltonian by extending the Chern–Simons action with 3dDI invariants of the spatial metric. This not only guarantees 3dDI invariance; but also makes the Hamiltonian density the square-root of a positive semi-definite and self-adjoint object $\hat{Q}^{\dagger i}_j \hat{Q}^j_i$; and ensures the preservation of all these properties even under renormalization of the coupling constants. In increasing order of spatial derivatives, these invariants are $\Lambda \int \sqrt{q} d^3x$, $EH = b\int \sqrt{q} R d^3x$, and the Chern–Simons functional of the affine connection with dimensionless coupling constant. Even higher derivative curvature invariants will come along with super-renormalizable dimensional coupling constants, while the cosmological constant volume term commutes with $\hat{\bar{\pi}}^i_j$ due to the traceless projector $\bar{E}^i_{j(mn)}$. To wit, only the Einstein–Hilbert action in three dimensions and the Chern–Simons functional are relevant. This corresponds to adopting

[e] While $\hat{Q}^{\dagger i}_j$ and $\hat{Q}^i_j$ are related to $\hat{\bar{\pi}}^i_j$ by $e^{\mp gW}$ similarity transformations, they are non-Hermitian, and generate two unitarily inequivalent representations of the noncompact group SL$(3, R)$ at each spatial point; whereas the momentric $\hat{\bar{\pi}}^i_j = \frac{1}{2}(\hat{Q}^{\dagger i}_j + \hat{Q}^i_j)$ generates a unitary representation of $\prod_x \mathrm{SU}(3)_x$.

$W_T = \frac{g}{4}\int \tilde{\epsilon}^{ijk}(\bar{\Gamma}^l_{im}\partial_j\bar{\Gamma}^m_{kl} + \frac{2}{3}\bar{\Gamma}^l_{im}\bar{\Gamma}^m_{jn}\bar{\Gamma}^n_{kl})\, d^3x + b\int \sqrt{q}Rd^3x$, which leads to

$$\hat{Q}^i_j := e^{W_T}\hat{\bar{\pi}}^i_j e^{-W_T} = \frac{\hbar}{i}\bar{E}^i_{j(mn)}\left[\frac{\delta}{\delta\bar{q}_{mn}} - \frac{\delta W_T}{\delta\bar{q}_{mn}}\right]$$
$$- \frac{\hbar}{i}\bar{E}^i_{j(mn)}\frac{\delta}{\delta\bar{q}_{mn}} + ib\hbar\sqrt{q}\bar{R}^i_j + ig\hbar\tilde{C}^i_j, \quad (17)$$

wherein, again due to the $\bar{E}^i_{j(mn)}$ projector, only the traceless part of the Ricci tensor remains. The Hamiltonian density is then

$$\bar{H} = \sqrt{\hat{Q}^{\dagger i}_j \hat{Q}^j_i}$$
$$= \sqrt{\hat{\bar{\pi}}^{\dagger j}_i\hat{\bar{\pi}}^i_j + \hbar^2(g\tilde{C}^i_j + b\sqrt{q}\bar{R}^i_j)(g\tilde{C}^j_i + b\sqrt{q}\bar{R}^j_i) + [\hat{\bar{\pi}}^i_j, ib\hbar\sqrt{q}\bar{R}^j_i]}\,. \quad (18)$$

Remarkably, the coincident commutator $[\hat{\bar{\pi}}^i_j, ib\sqrt{q}\hbar\bar{R}^j_i] = -\frac{5}{12}b\hbar^2\delta(0)\sqrt{q}(5R - \frac{9}{\epsilon})$.[8] Thus the potential for Einstein's theory, which is the spatial Ricci scalar and a (positive) c-number cosmological constant term, emerges. This means that, instead of the naive positive semi-definite form with $\mathcal{V}$ as in (9), the simple and elegant quantum Hamiltonian density $\sqrt{\hat{Q}^{\dagger i}_j\hat{Q}^j_i}$ (with all its aforementioned advantages) already contains Einstein's GR with cosmological constant. Adopting this, the departures from Einstein's theory, which come from $\bar{R}^i_j$ and the Cotton–York tensor, only appear in the higher-curvature higher-derivative combination $(g\tilde{C}^j_i + b\sqrt{q}\bar{R}^j_i)(gC^i_j + b\sqrt{q}R^i_j)$. These 'non-GR' terms are automatically absent in homogeneous FLRW cosmology, and also in constant curvature slicings of Painlevé–Gullstrand solutions of black holes.[13] Consequently, except for Cotton–York preponderance at very early times,[f] Einstein's GR dominates at low curvatures and long wavelengths in a theory in which 'four-dimensional symmetry is not a fundamental property of the physical world'.[2]

In summary, the final Hamiltonian assumes the elegant form,

$$H_{\text{Phys}} = \hbar\int \sqrt{(Q^A)^\dagger Q^A}\,\frac{\delta(0)}{\sqrt{2}\beta}d^3x, \quad Q^A := e^{W_T}T^A(x)e^{-W_T}, \quad (19)$$

wherein $\frac{\delta(0)}{\beta}d^3x$ is a dimensionless volume element, its divergence to be absorbed by renormalization of β (renormalization of the couplings constants in the theory remains to be studied). With dimensionless fundamental variables, the CR are[8,10]

$$[\bar{q}_{ij}(x), \bar{q}_{kl}(y)] = 0,$$
$$[\bar{q}_{ij}(x), T^A(y)] = \frac{i}{2}((\lambda^A)^k_i\bar{q}_{kj} + (\lambda^A)^k_j\bar{q}_{ki})\frac{\delta(x-y)}{\delta(0)}, \quad (20)$$
$$[T^A(x), T^B(y)] = if^{AB}{}_C T^C\frac{\delta(x-y)}{\delta(0)}.$$

[f] Further discussions on the initial vacuum state, primordial fluctuations, relation to Penrose Weyl conjecture and the 'arrow of time' can be found elsewhere.[8]

In conjunction with Schrödinger evolution and time-ordering in global intrinsic time, this framework presents a new vista to resolve and surmount the many conceptual and technical challenges of quantum gravity.

Acknowledgments

This work was supported in part by the Ministry of Science and Technology (R.O.C.) under Grant No. MOST104-2112-M-006-003. I would like to thank Eyo Eyo Ita III and Hoi-Lai Yu for beneficial discussions.

References

1. J. A. Wheeler, Superpsace and the nature of quantum geometrodynamics, in *Battelle Rencontres*, Lectures in Mathematics and Physics, C. M. DeWitt and J. A. Wheeler (eds.) (W. A. Benjamin, New York, 1968).
2. P. A. M. Dirac, *Proc. R. Soc. Lond. A* **246** (1958) 333.
3. P. Horava, *Phys. Rev. D* **79** (2009) 084008.
4. R. L. Arnowitt, S. Deser and C. W. Misner, *Phys. Rev.* **116** (1959) 1322.
5. B. S. DeWitt, *Phys. Rev.* **160** (1967) 1113.
6. C. Soo and H.-L. Yu, *Prog. Theor. Exp. Phys.* 2014 (2014) 013E01.
7. N. Ó. Murchadha, C. Soo and H.-L. Yu, *Class. Quantum Grav.* **30** (2013) 095016.
8. E. E. Ita III, C. Soo and H.-L. Yu, *Prog. Theor. Exp. Phys.* **2015** (2015) 083E01.
9. H.-C. Lin and C. Soo, *Chin. J. Phys.* **53** (2015) 110102.
10. C. Soo and H.-L. Yu, *Chin. J. Phys.* **53** (2015) 110106.
11. J. R. Klauder, *Int. J. Geom. Methods Mod. Phys.* **3** (2006) 81.
12. M. Gell-Mann and Y. Ne'eman, *The Eightfold Way* (W. A. Benjamin, New York, 1964).
13. C. Y. Lin and C. Soo, *Phys. Lett. B* **671** (2009) 493.

Birth of the Inflationary Universe and Tensor Fluctuations*

Jun'ichi Yokoyama

Research Center for the Early Universe (RESCEU), Graduate School of Science, The University of Tokyo, Tokyo 113-0033, Japan

Department of Physics, Graduate School of Science, The University of Tokyo, Tokyo 113-0033, Japan

Kavli Institute for the Physics and Mathematics of the Universe (Kavli IPMU), WPI, UTIAS, The University of Tokyo, Kashiwa, Chiba 277-8568, Japan

yokoyama@resceu.s.u-tokyo.ac.jp

In order to achieve full understanding of our Universe, we wish to clarify how it was created. In this talk, I discussed several possibilities and relevance to the observation of tensor perturbation.

Keywords: Creation of the Universe; inflation; tensor perturbation.

PACS Number(s): 98.80.Cq, 98.80.Jk, 98.80.Es

1. Introduction

One of the ultimate goal of cosmology is to clarify the evolution history of our Universe from its creation to the present as well as its future including the fate of our Universe. For a long time people thought that the Universe is static and do not change in time. The situation changed after the discovery of cosmic expansion by Hubble[1] and Lemaitre.[2] The classical big bang cosmology[3] thus emerged with which we can understand the three most fundamental observations of cosmology, namely, the homogeneous cosmic expansion, the cosmic microwave background radiation (CMB) discovered by Penzias and Wilson,[4] and the abundance of light elements.[5,6]

On the other hand, the classical big bang cosmology also suffers from fundamental difficulties such as the horizon problem, the flatness problem,[7] and the initial singularity problem. Since all these are related to the initial condition of the Universe and due to the lack of reliable theories to describe the very early Universe, these problems had not been seriously studied for a long time.

*This article first appeared in *Int. J. Mod. Phys. D* **25** (2016) 1645009.

Such a situation was somewhat changed by the introduction of grand unified theories (GUTs) by Georgi and Glashow.[8] Since the unification scale of GUTs, $M_{\rm GUT} = 10^{15-16}$ GeV is on one hand too big to be experimentally accessible but on the other hand is just a step off from the Planckian scale, $M_{\rm Pl} = (\hbar c^5/G)^{1/2} = 1.2 \times 10^{19}$ GeV, where quantum gravitational effects become important, there emerged a trend to use the high temperature and density regime in the early Universe as an arena for its verification, and it stimulated serious study of the birth of the Universe.

At the same time, however, GUTs also brought about a serious difficulty to the classical big bang cosmology, namely, overproduction of magnetic monopoles[9,10] at the GUT phase transition[11] which would over-close the Universe[12] with the density parameter $\Omega_{\rm monopole} \sim 10^{15}$! Fortunately, however, the resolution of this serious problem was already prepared in the GUT itself, as observed independently by Sato[13–15] and Guth.[16] Inflationary cosmology based on unified theories of elementary interactions thus came into being.

These pioneers observed that in the symmetric state of the Higgs field, which is realized thermally shortly after the big bang, the field has a large potential energy density of order of $M_{\rm GUT}^4$, and that if this state is metastable with a long enough lifetime, the Universe would be dominated by the false vacuum energy density which would not decrease as long as the Higgs field remains in that state no matter how much the Universe expands. As a result, an exponential cosmic expansion is realized which was referred to as cosmic inflation. Since both the particle horizon and the curvature radius are exponentially stretched during inflation, it can in principle provide solutions to the horizon and the flatness problems.

Both Sato and Guth assumed that the Universe started with or reached to the thermal equilibrium state before inflation sets in, and that the required initial condition of the Higgs field for inflation is achieved by these thermal properties. In this sense, their models are no different from the classical big bang cosmology as for the initial condition of the Universe.

Apart from the above mentioned approach based on unified theories, there was a longer tradition in the Russian school to pursue theories of the very early Universe. While many of them were speculative, a number of important ideas were also proposed. Among them is the curvature square theory which realized exponential cosmic expansion followed by Friedmann regime now referred to as the Starobinsky model.[17] Note, however, the fact that the curvature square term added to the Einstein–Hilbert action of gravity induces de Sitter expansion that had been shown by Nariai and Tomita already in 1971.[18] Interestingly, the main focus of both Nariai–Tomita and Starobinsky was to avoid the initial singularity.

Nowadays it is believed that the actual inflation was induced according to the slow-rollover scenario.[19,20] In particular, Linde's chaotic inflation scenario[21] is attractive because inflation can start naturally even in the chaotic initial condition at the Planckian time full of quantum fluctuations. On the other hand, chaoatic inflation with a simple polynomial potential is observationally disfavored now. Hence, we must consider the initial condition of the Universe even more seriously than before.

2. Null Energy Condition

So far, we have implicitly assumed that the cosmic energy density decreases in time as the Universe expands. Said differently, as we trace the evolution of the Universe back in time, the energy density increases indefinitely as the size of the Universe becomes smaller and smaller. In fact, however, behind such a picture is a prejudice that the energy density is always a decreasing (or better to say, nonincreasing) function in the expanding universe. This is the case if and only if the null energy condition, which asserts that the energy–momentum tensor $T_{\mu\nu}$ satisfies

$$T_{\mu\nu}\xi^\mu\xi^\nu \geq 0 \tag{1}$$

for any null vector ξ^μ. In the Friedmann Universe, where the energy–momentum tensor takes the form of a perfect fluid with energy density ρ and pressure P, (1) is equivalent to

$$\rho + P \geq 0. \tag{2}$$

As the Hubble parameter satisfies

$$\frac{dH}{dt} = -\frac{3}{2}(\rho + P)H, \tag{3}$$

if the null energy condition is violated, the Hubble parameter may increase in time. Then the Universe may emerge in a state with a lower energy density which may be irrelevant to quantum gravity.

Let us see if violation of the null energy condition is possible in various classes of inflationary cosmology.

2.1. *Slow-roll canonical inflation*

This is the most standard class of inflation models where inflation is induced by a scalar field ϕ, the inflaton, with a canonical kinetic term $X \equiv -\frac{1}{2}g^{\mu\nu}\partial_\mu\phi\partial_\nu\phi$ and a potential $V[\phi]$. Taking the variation of the action

$$S = \int \mathcal{L}\sqrt{-g}d^4x = \int (X - V[\phi])\sqrt{-g}d^4x \tag{4}$$

with respect to the metric tensor, one finds the energy–momentum tensor

$$T_{\mu\nu} = -\frac{2}{\sqrt{-g}}\frac{\delta S}{\delta g^{\mu\nu}} = \partial_\mu\phi\partial_\nu\phi + \mathcal{L}g_{\mu\nu}. \tag{5}$$

For homogeneous field configuration, the energy density, ρ, and the pressure, P, are given by

$$\rho = \frac{1}{2}\dot{\phi}^2 + V[\phi], \quad P = \frac{1}{2}\dot{\phi}^2 - V[\phi], \tag{6}$$

respectively. Hence if the potential is so flat that $V[\phi] > \dot{\phi}^2$ is satisfied, the equation-of-state parameter $w \equiv P/\rho$ is smaller than $-1/3$ and an accelerated expansion can be realized. However, since $\rho + P = \dot{\phi}^2 > 0$, the null energy condition is never violated in this class.

2.2. *k-inflation*

This model makes use of a scalar field with a higher-order kinetic function.[22] If we can realize a state with a constant canonical kinetic function, its energy density can have the same equation-of-state as the cosmological constant to drive exponential inflation. In this model, the Lagrangian is a more complicated function of both X and ϕ which we denote as $\mathcal{L} = K(X, \phi)$. Then calculating the energy–momentum tensor in the same way as (5), we find

$$\rho = 2XK_X - K, \quad P = K \tag{7}$$

in this case. Thus an accelerated expansion is possible even without any potential, if $XK_X < -K$ is satisfied. In particular, in the case $K_X = \partial K/\partial X = 0$ holds, we find de Sitter expansion. This is what is called k-inflation. If we expand the Lagrangian as a power series of X like

$$K(X, \phi) = K_1(\phi)X + K_2(\phi)X^2 \cdots , \tag{8}$$

we immediately find that $K_1(\phi)$ and $K_2(\phi)$ should have an opposite sign in order to realize inflation. Inflation can be terminated as the signature changes to the normal one.

From (7) we find $\rho + P = 2XK_X$, and apparently it is possible to violate the null energy condition if we have $K_X < 0$. Then $\dot{H}$ would be positive. However, the life is not so simple as the second-order action for the comoving linear curvature perturbation ζ is given by

$$S_2 = M_G^2 \int dt d^3x a^3 \left[\frac{\Sigma}{H^2}\dot{\zeta}^2 + \frac{\dot{H}}{H^2}\frac{(\nabla\zeta)^2}{a^2} \right] \tag{9}$$

with $M_G^2\Sigma \equiv XK_X + 2X^2K_{XX}$, where M_G is the reduced Planck mass.[23] Thus if $\dot{H}$ is positive, the sound velocity becomes imaginary which causes anomalous growth of small scale perturbations resulting in the gradient instability. Thus such a solution is not allowed.

2.3. *G-inflation*

The G-inflation includes higher-order derivative terms but it is constructed in such a way that both gravitational and scalar field equations of motion are of second-order differential equations.[24] The original model has the Lagrangian

$$\mathcal{L}_\phi = K(\phi, X) - G(\phi, X)\Box\phi, \tag{10}$$

where K and G are general function of ϕ and X. From this Lagrangian, we can calculate the energy–momentum tensor as

$$T_{\mu\nu} = K_X \nabla_\mu \phi \nabla_\nu \phi + K g_{\mu\nu} - 2\nabla_{(\mu} G \nabla_{\nu)} \phi$$
$$+ g_{\mu\nu} \nabla_\lambda G \nabla^\lambda \phi - G_X \Box \phi \nabla_\mu \phi \nabla_\nu \phi, \tag{11}$$

so that, we find

$$\rho = 2K_X X - K + 3G_X H \dot{\phi}^3 - 2G_\phi X, \tag{12}$$
$$P = K - 2(G_\phi + G_X \ddot{\phi})X. \tag{13}$$

Thus violation of null energy condition is again possible for appropriate choices of the functions.

The curvature perturbation in G-inflation model is governed by the action

$$S_2 = \frac{1}{2} \int d\eta d^3x \, z^2 [\mathcal{G}(\zeta_\phi')^2 - \mathcal{F}(\nabla \zeta_\phi)^2], \tag{14}$$

where

$$z := \frac{a\dot{\phi}}{H - G_X \dot{\phi}^3 / 2M_{\rm Pl}^2}, \tag{15}$$

$$\mathcal{F} := K_X + 2G_X(\ddot{\phi} + 2H\dot{\phi}) - 2\frac{G_X^2}{M_{\rm Pl}^2} X^2$$
$$+ 2G_{XX} X \ddot{\phi} - 2(G_\phi - X G_{\phi X}), \tag{16}$$

$$\mathcal{G} := K_X + 2X K_{XX} + 6G_X H \dot{\phi} + 6\frac{G_X^2}{M_{\rm Pl}^2} X^2$$
$$- 2(G_\phi + X G_{\phi X}) + 6G_{XX} H X \dot{\phi}, \tag{17}$$

and the prime represents differentiation with respect to the conformal time η. The squared sound speed is therefore $c_s^2 = \mathcal{F}/\mathcal{G}$. To avoid ghost and gradient instabilities, we require the conditions $\mathcal{F} > 0$ and $\mathcal{G} > 0$. These conditions are compatible even when $\rho + P$ is negative. Thus, this theory can violate null energy condition without causing any instabilities.

2.4. *The case for a positive tensor spectral index*

Thus among the three mechanisms of inflation we have introduced, only G-inflation can realize stable violation of the null energy condition. It has a profound implication for observation of tensor perturbation.

Let us review the spectrum of quantum tensor perturbation generated in quasi-de Sitter universe.[25] Incorporating the tensor perturbation h_{ij} to the cosmological background, the metric reads

$$ds^2 = a^2(\eta) \left[-d\eta^2 + (\delta_{ij} + h_{ij}) dx^i dx^j \right], \tag{18}$$

and the second-order action for h_{ij} is given by

$$S_{2T} = \frac{M_G^2}{8}\int d\eta d^3x a^2 (h_j^{i'} h_i^{j'} - h_{j,l}^i h_i^{j,l}). \tag{19}$$

This action has the same form as a free scalar field again, so we can quantize it in the same way.

Taking new variables as $z_T \equiv a/2, u_{ij} \equiv M_G z_T h_{ij}$, we find

$$S_{2T} = \frac{1}{2}\int d\eta d^3x \left[u_{ij}'^2 - (\nabla u_{ij})^2 + \frac{a''}{a}u_{ij}^2\right]. \tag{20}$$

In quasi-de Sitter universe with $-\dot{H}/H^2 \equiv \epsilon \ll 1$, the scale factor around an arbitrary time η_* can be expressed as

$$a = -\frac{1}{H_*\eta_*}\frac{1}{1-\epsilon}\left(\frac{-\eta}{-\eta_*}\right)^{\frac{1}{2}-\nu_T}, \quad \nu_T = \frac{3}{2}\frac{1-\dfrac{\epsilon}{3}}{1-\epsilon}, \tag{21}$$

using the Hubble parameter at that time H_*. Therefore, we find the normalized mode function

$$u_{ij}^A = \left(-\frac{\pi\eta}{4}\right)^{\frac{1}{2}} H_{\nu_T}^{(1)}(-k\eta)e_{ij}^A(\boldsymbol{k}), \quad A = +, \times, \tag{22}$$

where $e_{ij}^A(\boldsymbol{k})$ is a polarization tensor that satisfies $\langle e_{ij}^A(\boldsymbol{k}) e^{*ijB}(\boldsymbol{k})\rangle = \delta^{AB}$.

Thus, the power spectrum of long-wave tensor perturbation is given by

$$\mathcal{P}_T(k) \equiv \frac{4\pi k^3}{(2\pi)^3}h_{ij}h^{*ij} = \frac{4\pi k^3}{(2\pi)^3}\frac{u_{ij}^A u^{*ijA}}{M_G^2 z_T^2} = \frac{2H^2}{\pi^2 M_G^2}. \tag{23}$$

Here, the value of the Hubble parameter when k mode left the Hubble radius should be used. The tensor spectral index is therefore given by

$$n_t \equiv \frac{d\ln\mathcal{P}_T(k)}{d\ln k} = -2\epsilon. \tag{24}$$

This means that if future observation of the B-mode polarization of CMB revealed that its spectral index is positive, it would mean that the Hubble parameter was an increasing function of time during inflation, so that only G-inflation would remain an option among the three classes. It is said quite often that a positive spectral index of tensor perturbations would rule out inflation paradigm, but this is not the case; it simply restricts viable class of models significantly.

3. Generalized *G*-Inflation and Beyond

The above feature of G-inflation also applies for the generalized G-inflation model which is the most general single-field inflation model with second-order field equations.[24,26] Its Lagrangian is given by the sum of the following functionals of ϕ

and X.

$$\mathcal{L}_2 = K(\phi, X), \tag{25}$$

$$\mathcal{L}_3 = -G_3(\phi, X)\Box\phi, \tag{26}$$

$$\mathcal{L}_4 = G_4(\phi, X)R + G_{4X}(\phi, X)[(\Box\phi)^2 - (\nabla_\mu\nabla_\nu\phi)^2], \tag{27}$$

$$\mathcal{L}_5 = G_5(\phi, X)G_{\mu\nu}\nabla^\mu\nabla^\nu\phi - \frac{1}{6}G_{5X}(\phi, X)[(\Box\phi)^3 - 3(\Box\phi)(\nabla_\mu\nabla_\nu\phi)^2 + 2(\nabla_\mu\nabla_\nu\phi)^3], \tag{28}$$

where $G_{\mu\nu}$ is the Einstein tensor, $(\nabla_\mu\nabla_\nu\phi)^2 = (\nabla_\mu\nabla_\nu\phi)(\nabla^\mu\nabla^\nu\phi)$, $(\nabla_\mu\nabla_\nu\phi)^3 = (\nabla_\mu\nabla_\nu\phi)(\nabla^\nu\nabla^\lambda\phi)(\nabla_\lambda\nabla^\mu\phi)$ and $G_{iX} = \partial G_i/\partial X$.

The Lagrangian of the Einstein gravity may be automatically included by taking $G_4 = M_G^2/2$, and the extension to a nonminimally coupled scalar field is also immediate. Thus, the total action we consider is given by

$$S = \sum_{i=2}^{5}\int d^4x\sqrt{-g}\mathcal{L}_i. \tag{29}$$

We wished to construct a model of the initial condition of inflation where the Universe started its evolution from Minkowski vacuum state, and then the energy density increased due to violation of the null energy condition. But it turned out that the system suffered from gradient instability at transitions in the expansion law, namely, from the initial asymptotically Minkowski regime to inflationary regime, and from inflation epoch to the reheating. This was also the case in all the previous attempts to our knowledge.[27–29]

We have finally succeeded, however, in constructing a worked example of inflation starting from Minkowski space with stable violation of the null energy condition[30] by virtue of the so-called beyond the Horndeski theory where only spatial covariance is maintained with higher-order derivative terms in space.[31–33] In this theory, too, the tensor spectral index would have a positive value, so the observation of tensor perturbation is very important in discussing the real beginning of the Universe.

4. Alternative View

Soon after he proposed the original inflation model, Sato realized that his theory naturally predicts multi-production of Universes[34] in such a way that the mother universe creates child universes, and grand child universes,etc. Now, multiverse is a topic of active research in superstring theory which predicts a large number of possible vacuum states. Then, our Universe may not be of first generation. We have recently discovered an evaporating black hole which may induce a phase transition to transform itself to a wormhole beyond whose throat there exists an inflating spacetime.[35] Then, the origin of our Universe, too, may be totally different from what we have thought of conventionally.

5. Conclusion

Nowadays, cosmology is driven by precision observations. It is healthy as a falsifyable science, but, on the other hand, it is also important to think about fundamental subjects which may never be observationally verified. Research on the birth of the Universe is among such subjects. Nevertheless, it is intriguing that the observation of tensor perturbation may provide us with useful hints on the topic.

Acknowledgments

I would like to thank Professor Pisin Chen for his hospitality during the symposium. This manuscript is partially based on the author's review article on inflationary cosmology.[36]

References

1. E. Hubble, *Proc. Natl. Acad. Sci.* **15** (1929) 168.
2. G. Lemaître, *Ann. Soc. Sci. Bruxelles* **47** (1927) 49.
3. G. Gamow, *Phys. Rev.* **70** (1946) 572.
4. A. A. Penzias and R. W. Wilson, *Astrophys. J.* **142** (1965) 419.
5. R. Alpher, H. Bethe and G. Gamow, *Phys. Rev.* **73** (1948) 803.
6. C. Hayashi, *Prog. Theor. Phys.* **5** (1950) 224.
7. R. Dicke and P. Peebles, The big bang cosmology — enigmas and nostrums, *in General Relativity, An Einstein Centenary Survey*, Chap. 9, S. W. Hawking and W. Israel (eds.) (Cambridge University Press, Cambridge, 1979).
8. H. Georgi and S. Glashow, *Phys. Rev. Lett.* **32** (1974) 438.
9. G. 't Hooft, *Nucl. Phys. B* **79** (1974) 276.
10. A. M. Polyakov, *JETP Lett.* **20** (1974) 194.
11. T. Kibble, *J. Phys. A* **9** (1976) 1387.
12. J. Preskill, *Phys. Rev. Lett.* **43** (1979) 1365.
13. K. Sato, *Mon. Not. R. Astron. Soc.* **195** (1981) 467.
14. K. Sato, *Phys. Lett. B* **99** (1981) 66.
15. M. B. Einhorn and K. Sato, *Nucl. Phys. B* **180** (1981) 385.
16. A. H. Guth, *Phys. Rev. D* **23** (1981) 347.
17. A. A. Starobinsky, *Phys. Lett. B* **91** (1980) 99.
18. H. Nariai and K. Tomita, *Prog. Theor. Phys.* **46** (1971) 776.
19. A. D. Linde, *Phys. Lett. B* **108** (1982) 389.
20. A. Albrecht and P. J. Steinhardt, *Phys. Rev. Lett.* **48** (1982) 1220.
21. A. D. Linde, *Phys. Lett. B* **129** (1983) 177.
22. C. Armendariz-Picon, T. Damour and V. F. Mukhanov, *Phys. Lett. B* **458** (1999) 209, http://arxiv.org/abs/hep-th/9904075; arXiv:hep-th/9904075 [hep-th].
23. J. Garriga and V. F. Mukhanov, *Phys. Lett. B* **458** (1999) 219, http://arxiv.org/abs/hep-th/9904176; arXiv:hep-th/9904176 [hep-th].
24. T. Kobayashi, M. Yamaguchi and J. Yokoyama, *Phys. Rev. Lett.* **105** (2010) 231302, http://arxiv.org/abs/1008.0603; arXiv:1008.0603 [hep-th].
25. A. A. Starobinsky, *JETP Lett.* **30** (1979) 682.
26. T. Kobayashi, M. Yamaguchi and J. Yokoyama, *Prog. Theor. Phys.* **126** (2011) 511, http://arxiv.org/abs/1105.5723; arXiv:1105.5723 [hep-th].
27. P. Creminelli, M. A. Luty, A. Nicolis and L. Senatore, *J. High Energy Phys.* **12** (2006) 080, http://arxiv.org/abs/hep-th/0606090; arXiv:hep-th/0606090 [hep-th].

28. P. Creminelli, A. Nicolis and E. Trincherini, *J. Cosmol. Astropart. Phys.* **1011** (2010) 021, http://arxiv.org/abs/1007.0027; arXiv:1007.0027 [hep-th].
29. D. Pirtskhalava, L. Santoni, E. Trincherini and P. Uttayarat, *J. High Energy Phys.* **12** (2014) 151, http://arxiv.org/abs/1410.0882; arXiv:1410.0882 [hep-th].
30. T. Kobayashi, M. Yamaguchi and J. Yokoyama, *J. Cosmol. Astropart. Phys.* **1507** (2015) 017, http://arxiv.org/abs/1504.05710; arXiv:1504.05710 [hep-th].
31. J. Gleyzes, D. Langlois, F. Piazza and F. Vernizzi, *Phys. Rev. Lett.* **114** (2015) 211101, http://arxiv.org/abs/1404.6495; arXiv:1404.6495 [hep-th].
32. X. Gao, *Phys. Rev. D* **90** (2014) 081501, http://arxiv.org/abs/1406.0822; arXiv: 1406.0822 [gr-qc].
33. J. Gleyzes, D. Langlois, F. Piazza and F. Vernizzi, *J. Cosmol. Astropart. Phys.* **1502** (2015) 018, http://arxiv.org/abs/1408.1952; arXiv:1408.1952 [astro-ph.CO].
34. K. Sato, H. Kodama, M. Sasaki and K.-I. Maeda, *Phys. Lett. B* **108** (1982) 103.
35. N. Oshita and J. Yokoyama, http://arxiv.org/abs/1601.03929; arXiv:1601.03929 [gr-qc].
36. K. Sato and J. Yokoyama, *Int. J. Mod. Phys. D* **24** (2015) 1530025.

Emergent Spacetime for Quantum Gravity*

Hyun Seok Yang
Center for Quantum Spacetime, Sogang University,
Seoul 121-741, Korea
hsyang@sogang.ac.kr

We emphasize that noncommutative (NC) spacetime necessarily implies emergent spacetime if spacetime at microscopic scales should be viewed as NC. In order to understand NC spacetime correctly, we need to deactivate the thought patterns that we have installed in our brains and taken for granted for so many years. Emergent spacetime allows a background-independent formulation of quantum gravity that will open a new perspective to resolve the notorious problems in theoretical physics such as the cosmological constant problem, hierarchy problem, dark energy, dark matter and cosmic inflation.

Keywords: Emergent spacetime; quantum gravity; noncommutative spacetime.

PACS Number(s): 04.60.−m; 11.25.Tq; 02.40.Gh

1. What Every Physicist Should Know about Noncommutative Spacetime

Since Newton the concept of space and time has gone through various changes. In retrospect, the upheaval of new physics has accompanied the novel concept of space and time. And history is a mirror to the future. The concept of emergent gravity and spacetime recently activated by the AdS/CFT correspondence[1–3] advocates that spacetime is not a fundamental entity existed from the beginning but an emergent property from a primal monad such as matrices. However, physicists have still devoted themselves to the 20th century physics and have been complacent in the contemporary paradigm although the concept of emergent spacetime asks us to go beyond the horizon of the contemporary physics. It is our belief that the beginning of the 21st century is the right time to go beyond the spacetime concept developed for the 20th century physics and the emergent spacetime is a significant new paradigm for the 21st century physics which may shed light on the origin of spacetime and our Universe.

In string theory, two exclusive spacetime pictures exist together although they are conceptually in deep discord with each other. One is founded on the general

*This article first appeared in *Int. J. Mod. Phys. D* **25** (2016) 1645010.

relativity while the other is based on the emergent spacetime. However, these two spacetime pictures are exclusive and irreconcilable each other. Thus it is necessary to formulate string theory entirely based on a unique spacetime picture. Since we think that the emergent spacetime picture is a more fundamental approach, we want to understand string theory based on the novel spacetime picture. Indeed string theory basically implies the emergent spacetime picture. (See the quotation in Sec. 4 in Ref. 4.) However, the current string theory assumes from the outset that strings are vibrating in a pre-existing spacetime and so resides in the territory of general relativity. After all, a spacetime is inserted into the theory "by hand" and thus spacetime does not emerge from the theory. Hence, we need a new approach to string theory in order to realize the concept of emergent spacetime.

String theory is defined by replacing particles (point-like objects) with strings (one-dimensional objects). In order to do this, we need to introduce a *new* constant α' whose physical dimension is (length).[2] It is well-known that the new constant α' introduces a new duality depicted by $R \to R' = \frac{\alpha'}{R}$. This is known as the T-duality in string theory, but it is impossible in particle theories ($\alpha' = 0$). It is important to notice that a new physical constant such as $\hbar$ and α' introduces a deformation of some structure in a physical theory.[4,5] For instance, the Planck constant $\hbar$ in quantum mechanics carries the physical dimension $[\hbar] =$ (length) $\times$ (momentum) and so it deforms the algebraic structure of particle phase-space from commutative to noncommutative (NC) space, i.e.

$$xp - px = 0 \quad \Rightarrow \quad xp - px = i\hbar. \tag{1}$$

An educated reasoning motivated by the fact that $[\alpha'] =$ (length) $\times$ (length) leads to a natural speculation that α' brings about the deformation of the algebraic structure of spacetime itself such that

$$xy - yx = 0 \quad \Rightarrow \quad xy - yx = i\alpha'. \tag{2}$$

From the deformation theory point of view, replacing particles with strings is equivalent to the transition from commutative space to NC space. This may be supported by the fact that the NC space (2) defines only a minimal area whereas the concept of point is doomed as we have learned from quantum mechanics defined by the NC phase-space (1). The minimal surface in the NC space (2) acts as a basic building block of string theory. Remarkably the deformation (2) provides us an important clue for a background-independent formulation of string theory.[6]

Since the mathematical structure of the NC space (2) is essentially the same as quantum mechanics in Eq. (1), we will consider the deformation (2) to be quantization. An important question is whether the quantization (2) in terms of α' can be regarded as an alternative formulation of string theory so that it realizes all essential aspects of string theory. The recent results[4–7] suggest that the answer is yes and moreover it naturally realizes the concept of emergent spacetime which was elusive so far in the usual string theory. In this approach, quantum gravity is formulated as a dynamical theory of NC spacetime and the NC spacetime necessarily implies emergent spacetime.[7] Therefore, the NC space (2) denoted by $\mathbb{R}^2_{\alpha'}$ is much more radical and mysterious than we thought. In order to understand NC spacetime correctly, we need to deactivate the thought patterns that we have installed in our brains and taken for granted for so many years.

The reason is the following. Recall that the NC phase-space (1) in quantum mechanics introduces the wave-particle duality. Indeed the NC space (2) also brings about a radical change of physics since the NC nature of spacetime is responsible for a new kind of duality, known as the gauge-gravity duality. A primary cause of the radical change of physics in quantum mechanics is that the NC phase-space (1) introduces a *complex linear* vector space called the Hilbert space. This is also true for the NC space (2) since its mathematical structure is essentially the same as quantum mechanics. Moreover the NC space $\mathbb{R}^2_{\alpha'}$ admits a nontrivial inner automorphism. For example, for an arbitrary NC field $f(x,y)$, we have the relation given by

$$f(x+a,y) = U(a)^\dagger f(x,y) U(a), \quad f(x,y+b) = U(b)^\dagger f(x,y) U(b), \tag{3}$$

where $U(a) = \exp(-\frac{iay}{\alpha'})$ and $U(b) = \exp(\frac{ibx}{\alpha'})$. A striking feature of the NC space is thus that every points are unitarily equivalent because translations in $\mathbb{R}^2_{\alpha'}$ are simply a unitary transformation acting on the Hilbert space $\mathcal{H}$. As a result, the concept of space is doomed and the classical space is replaced by a state in the Hilbert space $\mathcal{H}$. This fact leads to an important picture that classical spacetime is an emergent concept derived from a NC algebra. In other words, NC spacetime necessarily implies emergent spacetime, if spacetime at microscopic scales should be viewed as NC. Moreover, any dynamical variable defined on $\mathbb{R}^2_{\alpha'}$ becomes an operator acting on the Hilbert space $\mathcal{H}$. In particular, any NC field can be regarded as a linear operator acting on the Hilbert space. Note that the NC space (2) is equivalent to the Heisenberg algebra of harmonic oscillator, i.e. $[a, a^\dagger] = 1$, if the

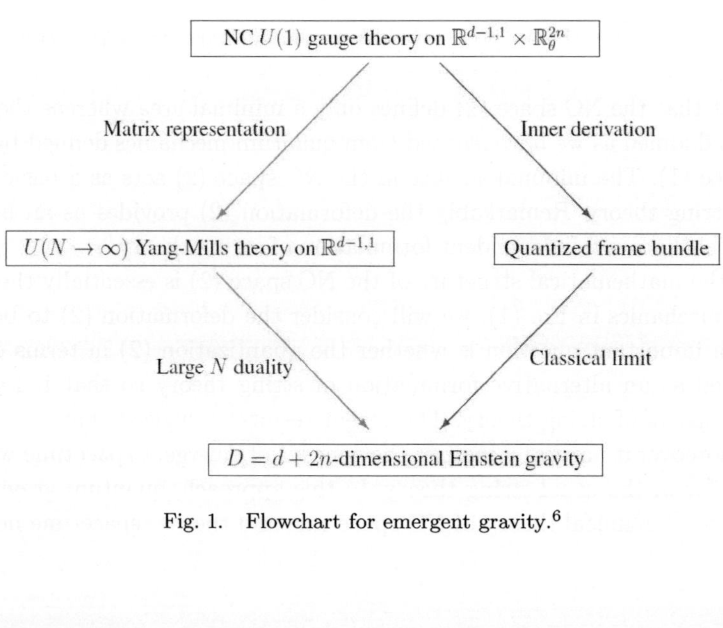

Fig. 1. Flowchart for emergent gravity.[6]

annihilation operator is defined by $a = \frac{1}{\sqrt{2\alpha'}}(x + iy)$. Thus, the Hilbert space for $\mathbb{R}^2_{\alpha'}$ is the Fock space and has a countable basis. Then the representation of NC fields on the Hilbert space $\mathcal{H}$ is given by $N \times N$ matrices where $N = \dim(\mathcal{H}) \to \infty$. Consequently, the NC space (2) leads to an interesting equivalence between a lower dimensional large N gauge theory and a higher dimensional NC $U(1)$ gauge theory.[7]

In next section, we will show that $U(1)$ gauge theory defined on a NC spacetime such as (2) defines quantum gravity according to the large N duality as depicted in Fig. 1. We thus demonstrate how the deformation (2) provides us a background-independent formulation of string theory.[6]

2. Emergent Spacetime from Large N Duality

Let us consider a $2n$-dimensional NC space denoted by $\mathbb{R}^{2n}_\theta$ whose coordinate generators obey the commutation relation

$$[y^a, y^b] = i\theta^{ab}, \quad a, b = 1, \ldots, 2n, \tag{4}$$

where $(\theta)^{ab} = \alpha'(\mathbf{1}_n \otimes i\sigma^2)$ is a $2n \times 2n$ constant symplectic matrix and $l_s \equiv \sqrt{\alpha'}$ is a typical length scale set by the vacuum. The NC space $\mathbb{R}^2_{\alpha'}$ corresponds to the $n = 1$ case. Similarly to the $n = 1$ case, the NC space (4) is equivalent to the Heisenberg algebra of n dimensional harmonic oscillator. Hence the underlying Hilbert space on which the NC $\star$-algebra $\mathcal{A}_\theta$ acts is given by the Fock space defined by

$$\mathcal{H} = \{|\mathbf{n}\rangle \equiv |n_1, \ldots, n_n\rangle |\, n_i \in \mathbb{Z}_{\geq 0},\ i = 1, \ldots, n\}, \tag{5}$$

which is orthonormal, i.e. $\langle \mathbf{n}|\mathbf{m}\rangle = \delta_{\mathbf{n},\mathbf{m}}$ and complete, i.e. $\sum_{\mathbf{n}=0}^{\infty} |\mathbf{n}\rangle\langle\mathbf{n}| = \mathbf{1}_\mathcal{H}$, as is well-known from quantum mechanics. Since the Fock space (5) has a countable basis, it is convenient to introduce a one-dimensional (1D) basis using the "Cantor diagonal method" to put the n-dimensional non-negative integer lattice in $\mathcal{H}$ into one-to-one correspondence with the natural numbers:

$$\mathbb{Z}^n_{\geq 0} \leftrightarrow \mathbb{N} : |\mathbf{n}\rangle \leftrightarrow |n\rangle, \quad n = 1, \ldots, N \to \infty. \tag{6}$$

In this 1D basis, the completeness relation of the Fock space (5) is now given by $\sum_{n=1}^{\infty} |n\rangle\langle n| = \mathbf{1}_\mathcal{H}$.

The Hilbert space for $\mathbb{R}^{2n}_\theta$ is also the Fock space and so the representation of NC fields on the Hilbert space $\mathcal{H}$ is given by $N \times N$ matrices where $N = \dim(\mathcal{H}) \to \infty$. Consider two arbitrary dynamical fields $f(x, y)$ and $g(x, y)$ on a $D = (d + 2n)$-dimensional NC spacetime $\mathbb{R}^{d-1,1} \times \mathbb{R}^{2n}_\theta$ which are elements of the NC $\star$-algebra $\mathcal{A}^d_\theta \equiv \mathcal{A}_\theta\big(C^\infty(\mathbb{R}^{d-1,1})\big) = C^\infty(\mathbb{R}^{d-1,1}) \otimes \mathcal{A}_\theta$. Since the dynamical variables in $\mathcal{A}^d_\theta$ can be regarded as operators acting on the Hilbert space (5), we can represent the operators in the Fock space (5) as $N \times N$ matrices in $\mathrm{End}(\mathcal{H}) = \mathcal{A}^d_N \equiv \mathcal{A}_N\big(C^\infty(\mathbb{R}^{d-1,1})\big) = C^\infty(\mathbb{R}^{d-1,1}) \otimes \mathcal{A}_N$ where $N = \dim(\mathcal{H}) \to \infty$:

$$f(x, y) = \sum_{n,m=1}^{\infty} |n\rangle\langle n|f(x, y)|m\rangle\langle m| := \sum_{n,m=1}^{\infty} F_{nm}(x)|n\rangle\langle m|,$$

$$g(x,y) = \sum_{n,m=1}^{\infty} |n\rangle\langle n|g(x,y)|m\rangle\langle m| := \sum_{n,m=1}^{\infty} G_{nm}(x)|n\rangle\langle m|, \tag{7}$$

where $F(x)$ and $G(x)$ are $N \times N$ matrices in $\mathcal{A}_N^d = \mathrm{End}(\mathcal{H})$. Then, we get a natural composition rule

$$\begin{aligned}(f \star g)(x,y) &= \sum_{n,l,m=1}^{\infty} |n\rangle\langle n|f(x,y)|l\rangle\langle l|g(x,y)|m\rangle\langle m| \\ &= \sum_{n,l,m=1}^{\infty} F_{nl}(x)G_{lm}(x)|n\rangle\langle m|.\end{aligned}$$

The above composition rule implies that the ordering in the NC algebra $\mathcal{A}_\theta^d$ is perfectly compatible with the ordering in the matrix algebra $\mathcal{A}_N^d$. Thus, we can straightforwardly translate multiplications of NC fields in $\mathcal{A}_\theta^d$ into those of matrices in $\mathcal{A}_N^d$ using the matrix representation (7) without any ordering ambiguity. Using the map (7), the trace over $\mathcal{A}_\theta^d$ can also be transformed into the trace over $\mathcal{A}_N^d$, i.e.

$$\int_M \frac{d^{2n}y}{(2\pi)^n|\mathrm{Pf}\theta|} = \mathrm{Tr}_{\mathcal{H}} = \mathrm{Tr}_N. \tag{8}$$

Using the matrix representation (7), we can show[5,8–10] that the $D = (d+2n)$-dimensional NC $U(1)$ gauge theory on $\mathbb{R}^{d-1,1} \times \mathbb{R}_\theta^{2n}$ is exactly mapped to the d-dimensional $U(N \to \infty)$ Yang–Mills theory on $\mathbb{R}^{d-1,1}$:

$$S = -\frac{1}{G_{YM}^2}\int d^DY \frac{1}{4}(\widehat{F}_{AB} - B_{AB})^2, \tag{9}$$

$$= -\frac{1}{g_{YM}^2}\int d^dx \mathrm{Tr}\left(\frac{1}{4}F_{\mu\nu}F^{\mu\nu} + \frac{1}{2}D_\mu\phi_a D^\mu\phi_a - \frac{1}{4}[\phi_a,\phi_b]^2\right), \tag{10}$$

where $G_{YM}^2 = (2\pi)^n|\mathrm{Pf}\theta|g_{YM}^2$ and

$$B_{AB} = \begin{pmatrix} 0 & 0 \\ 0 & B_{ab}\end{pmatrix}.$$

Note that the equivalence between the D-dimensional NC $U(1)$ gauge theory (9) and d-dimensional $U(N \to \infty)$ Yang–Mills theory (10) is not a dimensional reduction but an exact mathematical identity. A remarkable point is that the large N gauge theories described by the action (10) arise as a nonperturbative formulation of string/M theories.[11] For instance, we get the IKKT matrix model for $d = 0$,[12] the BFSS matrix quantum mechanics for $d = 1$[13] and the matrix string theory for $d = 2$.[14] The most interesting case arises for $d = 4$ and $n = 3$ which suggests an engrossing duality[15] that the 10-dimensional NC $U(1)$ gauge theory on $\mathbb{R}^{3,1} \times \mathbb{R}_\theta^6$ is equivalent to the bosonic action of four-dimensional $\mathcal{N} = 4$ supersymmetric $U(N)$ Yang–Mills theory, which is the large N gauge theory of the AdS/CFT duality.[1–3] According to the large N duality or gauge-gravity duality, the resulting large N

gauge theory must be dual to a higher dimensional gravity or string theory as summarized in Fig. 1. Hence, it should not be surprising that the NC $U(1)$ gauge theory should describe a theory of gravity (or a string theory) in the same dimensions. Unfortunately, this important possibility has been largely ignored until recently in spite of the apparent relationship depicted in Fig. 1.

The matrix representation in Fig. 1 is based on the fact that the NC space (4) admits a separable Hilbert space and NC $U(1)$ gauge fields become operators acting on the Hilbert space. Therefore there should be some way to map the NC $U(1)$ gauge theory to the Einstein gravity if we accept the conjectural large N duality. The blue arrows in Fig. 1 shows how to derive D-dimensional Einstein gravity from NC $U(1)$ gauge theory on $\mathbb{R}^{d-1,1} \times \mathbb{R}^{2n}_{\theta}$, which should be expected according to the (conjectural) large N duality. However, we can use the emergent gravity from NC $U(1)$ gauge theory to verify the large N duality by realizing the equivalence between the actions (9) and (10) in a reverse way. It is based on the observation[7,15] that the NC space (4) is a consistent vacuum solution of the d-dimensional $U(N \to \infty)$ Yang–Mills theory (10) in the Coulomb branch, see Fig. 2. The conventional choice of vacuum in the Coulomb branch of $U(N)$ Yang–Mills theory is given by

$$[\phi_a, \phi_b]|_{\text{vac}} = 0 \quad \Rightarrow \quad \langle \phi_a \rangle_{\text{vac}} = \text{diag}((\alpha_a)_1, (\alpha_a)_2, \ldots, (\alpha_a)_N), \tag{11}$$

for $a = 1, \ldots, 2n$. In this case, the $U(N)$ gauge symmetry is broken to $U(1)^N$. If we consider the $N \to \infty$ limit, the large N limit opens a new phase of the Coulomb branch given by

$$[\phi_a, \phi_b]|_{\text{vac}} = -iB_{ab} \quad \Rightarrow \quad \langle \phi_a \rangle_{\text{vac}} = p_a \equiv B_{ab} y^b, \tag{12}$$

where $B_{ab} = (\theta^{-1})_{ab}$ and the vacuum moduli y^a satisfy the Moyal–Heisenberg algebra (4). This vacuum will be called the NC Coulomb branch. Note that the

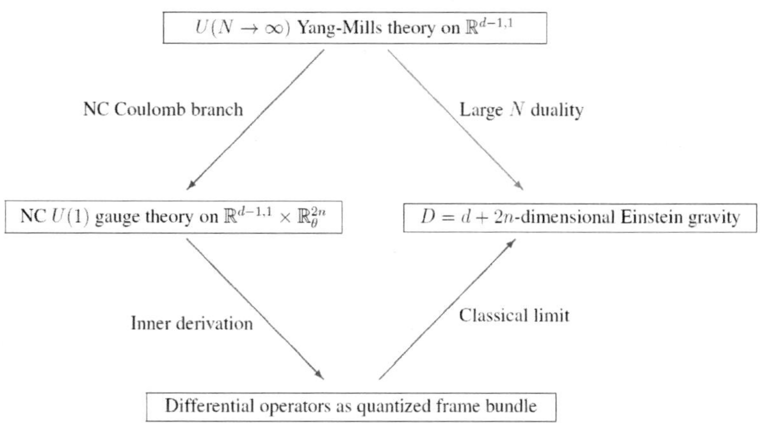

Fig. 2. Flowchart for large N duality.[6]

Moyal–Heisenberg vacuum (12) saves the NC nature of matrices while the conventional vacuum (11) dismisses the property. Suppose that fluctuations around the vacuum (12) take the form

$$D_\mu = \partial_\mu - i\widehat{A}_\mu(x,y), \quad \phi_a = p_a + \widehat{A}_a(x,y). \tag{13}$$

The adjoint scalar fields in Eq. (13) now obey the deformed algebra given by

$$[\phi_a, \phi_b] = -i(B_{ab} - \widehat{F}_{ab}), \tag{14}$$

where

$$\widehat{F}_{ab} = \partial_a \widehat{A}_b - \partial_b \widehat{A}_a - i[\widehat{A}_a, \widehat{A}_b] \tag{15}$$

with the definition $\partial_a \equiv \mathrm{ad}_{p_a} = -i[p_a, \cdot]$. Plugging the fluctuations in Eq. (13) into the d-dimensional $U(N \to \infty)$ Yang–Mills theory (10), we finally get the $D = (d+2n)$-dimensional NC $U(1)$ gauge theory and thus arrive at the reversed version of the equivalence:[7,15]

$$\begin{aligned} S &= -\frac{1}{g_{YM}^2} \int d^d x \mathrm{Tr} \left(\frac{1}{4} F_{\mu\nu} F^{\mu\nu} + \frac{1}{2} D_\mu \phi_a D^\mu \phi_a - \frac{1}{4} [\phi_a, \phi_b]^2 \right) \\ &= -\frac{1}{G_{YM}^2} \int d^D Y \frac{1}{4} (\widehat{F}_{AB} - B_{AB})^2, \end{aligned} \tag{16}$$

where $\widehat{A}_A(x,y) = (\widehat{A}_\mu, \widehat{A}_a)(x,y)$ are $D = (d+2n)$-dimensional NC $U(1)$ gauge fields. It might be emphasized that the NC space (12) is the crux to realize the equivalence (16). If the conventional commutative vacuum (11) were chosen, we would have failed to realize the equivalence (16). Indeed it turns out[15] that the NC Coulomb branch is crucial to realize the emergent gravity from matrix models or large N gauge theories as depicted in Fig. 2.

The relationship between a lower dimensional large N gauge theory and a higher dimensional NC $U(1)$ gauge theory in Figs. 1 and 2 is an exact mathematical identity. Therefore, the next step is to find an isomorphic map from the NC $U(1)$ gauge theory to the Einstein gravity in order to complete the large N duality. To be precise, consider the inverse metric in Einstein gravity given by

$$\left(\frac{\partial}{\partial s} \right)^2 = E_A \otimes E_A = g^{MN}(X) \partial_M \otimes \partial_N, \tag{17}$$

where $E_A = E_A^M(X)\partial_M$ are orthonormal frames on the tangent bundle $T\mathcal{M}$ of a D-dimensional spacetime manifold $\mathcal{M}$. The large N (or gauge-gravity) duality in Figs. 1 and 2 means to realize the vector fields $E_A = E_A^M(X)\partial_M \in \Gamma(T\mathcal{M})$ in terms of NC $U(1)$ gauge fields.

It was shown in Ref. 6 that quantum gravity can be derived from the electromagnetism on NC spacetime by realizing the large N duality in Fig. 2 via the duality chain given by

$$\mathcal{A}_N^d \Rightarrow \mathcal{A}_\theta^d \Rightarrow \mathfrak{D}^d. \tag{18}$$

See the paragraph in Eq. (7) for the definition of $\mathcal{A}_N^d$ and $\mathcal{A}_\theta^d$. The module of derivations $\mathfrak{D}^d$ is a direct sum of the submodules of horizontal and inner derivations[16]:

$$\mathfrak{D}^d = \mathrm{Hor}(\mathcal{A}_N^d) \oplus \mathfrak{D}(\mathcal{A}_N^d) \cong \mathrm{Hor}(\mathcal{A}_\theta^d) \oplus \mathfrak{D}(\mathcal{A}_\theta^d), \tag{19}$$

where horizontal derivation is locally generated by a vector field

$$k^\mu(x,y)\frac{\partial}{\partial x^\mu} \in \mathrm{Hor}(\mathcal{A}_\theta^d). \tag{20}$$

In particular, we are interested in the derivation algebra generated by the dynamical variables in Eq. (13). It is defined by

$$\widehat{V}_A = \{i\,\mathrm{ad}_{D_A} = [D_A, \cdot\,]|D_A(x,y) = (D_\mu, D_a)(x,y)\} \in \mathfrak{D}^d, \tag{21}$$

where $D_A(x,y) = -i\phi_A(x,y)$. It can be shown that the duality chain in Eq. (18) is constructed by the quantization of a (regular) d-contact manifold.[17] The dynamical variables in d-dimensional Yang–Mills gauge theory in Fig. 2 take values in $\mathcal{A}_N^d$ while those in $D = (d+2n)$-dimensional NC $U(1)$ gauge theory take values in $\mathcal{A}_\theta^d$. These two NC algebras $\mathcal{A}_N^d$ and $\mathcal{A}_\theta^d$ are related to each other by considering the NC Coulomb branch for the algebra $\mathcal{A}_N^d$.[6]

In a large-distance limit, i.e. $|\theta| \to 0$, one can expand the NC vector fields $\widehat{V}_A$ in Eq. (21) using the explicit form of the Moyal $\star$-product. The result takes the form

$$\widehat{V}_A = V_A^M(x,y)\frac{\partial}{\partial Y^M} + \sum_{p=2}^{\infty} V_A^{a_1\cdots a_p}(x,y)\frac{\partial}{\partial y^{a_1}}\cdots\frac{\partial}{\partial y^{a_p}} \in \mathfrak{D}^d, \tag{22}$$

where $V_A^\mu = \delta_A^\mu$. Thus, the Taylor expansion of NC vector fields in $\mathfrak{D}^d$ generates an infinite tower of the so-called polyvector fields which describe higher-spin fields with spin $s \geq 2$.[7] Note that the leading term gives rise to the ordinary vector fields that will be identified with a frame basis associated with the tangent bundle $T\mathcal{M}$ of an emergent spacetime manifold $\mathcal{M}$. It is important to perceive that the realization of emergent geometry through the derivation algebra in Eq. (21) is intrinsically local. Therefore, it is necessary to consider patching or gluing together the local constructions to form a set of global quantities. We will assume that local coordinate patches have been consistently glued together to yield global (poly)vector fields. See Refs. 18 and 19 for a global construction of NC $\star$-algebras and Ref. 7 for the globalization of emergent geometry. Let us truncate the above polyvector fields to ordinary vector fields given by

$$\mathfrak{X}(\mathcal{M}) = \left\{ V_A = V_A^M(x,y)\frac{\partial}{\partial X^M} | A, M = 0, 1, \ldots, D-1 \right\}, \tag{23}$$

where $X^M = (x^\mu, y^a)$ are local coordinates on a D-dimensional emergent *Lorentzian* manifold $\mathcal{M}$. The orthonormal vielbeins on $T\mathcal{M}$ are then defined by the relation

$$V_A = \lambda E_A \in \Gamma(T\mathcal{M}) \tag{24}$$

or on $T^*\mathcal{M}$

$$v^A = \lambda^{-1} e^A \in \Gamma(T^*\mathcal{M}). \tag{25}$$

The conformal factor $\lambda \in C^\infty(\mathcal{M})$ is determined by the volume preserving condition

$$\mathcal{L}_{V_A} \nu_t = \big(\nabla \cdot V_A + (2 - d - 2n) V_A \ln \lambda\big) \nu_t = 0, \quad \forall A = 0, 1, \ldots, D-1, \tag{26}$$

where

$$\nu_t \equiv d^d x \wedge \nu = \lambda^2 d^d x \wedge v^1 \wedge \cdots \wedge v^{2n} \tag{27}$$

is a D-dimensional volume form on $\mathcal{M}$. In the end, the Lorentzian metric on a D-dimensional spacetime manifold $\mathcal{M}$ is given by[4,5,7]

$$\begin{aligned} ds^2 &= \mathcal{G}_{MN}(X) dX^M \otimes dX^N = e^A \otimes e^A \\ &= \lambda^2 v^A \otimes v^A = \lambda^2 (\eta_{\mu\nu} dx^\mu dx^\nu + v_b^a v_c^a (dy^b - \mathbf{A}^b)(dy^c - \mathbf{A}^c)), \end{aligned} \tag{28}$$

where $\mathbf{A}^b := A_\mu^b(x, y) dx^\mu$. Therefore, the NC field theory representation of the d-dimensional large N gauge theory in the NC Coulomb branch provides a powerful machinery to identify gravitational variables dual to large N matrices. Therefore, we see that the general large N duality depicted in Fig. 2 can be explained with the map (18).[5–7]

A crucial point is that the NC space (4) is caused by the Planck energy condensate in vacuum and responsible for the generation of flat spacetime. The dynamical origin of flat spacetime in emergent gravity is a tangible difference from general relativity and the crux to clearly resolve the notorious cosmological constant problem.[4,5,20] A remarkable feature is that the huge vacuum energy being a perplexing cosmological constant in general relativity was simply used to generate flat spacetime and thus does not gravitate. Since the Planck energy condensate into vacuum must be a dynamical process, it is natural to expect that the cosmic inflation corresponds to a dynamical mechanism for the instantaneous condensation of vacuum energy to enormously spread out spacetime. Recently, it was shown in Ref. 6 that the cosmic inflation actually arises as a time-dependent solution of the matrix quantum mechanics (i.e. $d = 1$ case in Eq. (16)) describing the dynamical process of the vacuum condensate without introducing any inflaton field as well as an *ad hoc* inflation potential. The emergent spacetime picture admits a background-independent formulation so that the inflation is responsible for the dynamical emergence of spacetime. These results require us to seriously reconsider main sources to introduce the multiverse hypothesis from the standpoint of emergent spacetime. The emergent spacetime picture certainly opens a new perspective that may cripple all the rationales to introduce the multiverse hypothesis such as the cosmological constant problem, chaotic and eternal inflation scenarios, and string landscape.[21]

Although, string theory has a high potential for TOE, it has failed to achieve the goal so far. We think that a main reason for the failure lies on the incapacity for generating space and time through the vibrations of a string without assuming the prior existence of spacetime. We argued that, from the deformation theory point of view, replacing particles (point-like objects) with strings (1D objects) is equivalent to doing commutative space by NC space. Then the gravitational force manifests itself as the deformation of NC algebra via NC $U(1)$ gauge fields, leading to a dynamical NC spacetime and realizing emergent spacetime.[6] The whole picture in Fig. 2, thus implies that NC spacetime can be viewed as a second-quantized string and demonstrates how the deformation (4) provides us a background-independent formulation of string theory.

Acknowledgments

We thank organizers of the 2nd LeCosPA Symposium, especially Pisin Chen for invitation and hospitality. This work was supported by the National Research Foundation of Korea (NRF) grant funded by the Korea government (MOE) (No. 2011-0010597 and No. NRF-2015R1D1A1A01059710).

References

1. J. M. Maldacena, *Adv. Theor. Math. Phys.* **2** (1998) 231.
2. S. S. Gubser, I. R. Klebanov and A. M. Polyakov, *Phys. Lett. B* **428** (1998) 105.
3. E. Witten, *Adv. Theor. Math. Phys.* **2** (1998) 253.
4. J. Lee and H. S. Yang, *J. Korean Phys. Soc.* **65** (2014) 1754.
5. H. S. Yang, *J. High Energy Phys.* **05** (2009) 012.
6. H. S. Yang, arXiv:1503.00712.
7. H. S. Yang, *Int. J. Mod. Phys. A* **30** (2015) 1550016.
8. H. Aoki, N. Ishibashi, S. Iso, H. Kawai, Y. Kitazawa and T. Tada, *Nucl. Phys. B* **565** (2000) 176.
9. N. Seiberg, *J. High Energy Phys.* **09** (2000) 003.
10. H. S. Yang, *Eur. Phys. J. C* **64** (2009) 445.
11. W. Taylor, *Rev. Mod. Phys.* **73** (2001) 419.
12. N. Ishibashi, H. Kawai, Y. Kitazawa and A. Tsuchiya, *Nucl. Phys. B* **498** (1997) 467.
13. T. Banks, W. Fischler, S. H. Shenker and L. Susskind, *Phys. Rev. D* **55** (1997) 5112.
14. R. Dijkgraaf, E. Verlinde and H. Verlinde, *Nucl. Phys. B* **500** (1997) 43.
15. H. S. Yang, *Phys. Rev. D* **90** (2014) 086006.
16. S. Azam, *Commun. Algebra* **36** (2008) 905.
17. I. Vaisman, *J. Math. Math. Sci.* **8** (1985) 521.
18. B. Jurčo, P. Schupp and J. Wess, *Lett. Math. Phys.* **61** (2002) 171.
19. H. Bursztyn and S. Waldmann, *Commun. Math. Phys.* **228** (2002) 103.
20. H. S. Yang, *J. Phys. Conf. Ser.* **343** (2012) 012132.
21. H. S. Yang, arXiv:1504.00464.

Understanding Strongly Coupling Magnetism from Holographic Duality*

Rong-Gen Cai* and Run-Qiu Yang†

Institute of Theoretical Physics,
Chinese Academy of Sciences,
Beijing 100190, China
**cairg@itp.ac.cn*
†aqiu@itp.ac.cn

The unusual magnetic materials are significant in both science and technology. However, because of the strongly correlated effects, it is difficult to understand their novel properties from theoretical aspects. Holographic duality offers a new approach to understanding such systems from gravity side. This paper will give a brief review of our recent works on the applications of holographic duality in understanding unusual magnetic materials. Some quantitative comparision between holographic results and experimental data will be shown and some predictions from holographic duality models will be discussed.

Keywords: Holographic duality; strong coupling; magnetism.

PACS Number(s): 11.25.Tq, 11.27.+d, 04.70.−s

1. Introduction

The anti-de Sitter (AdS)/Conformal field theory (CFT) correspondence or gauge/gravity duality, one of important progresses in quantum gravity, has offered a new viewpoint to understand the connection between the quantum gravity and CFT.[1–4] This correspondence relations a weakly coupling gravity in $(d + 1)$-dimensional AdS space-time to a strongly coupling CFT in d-dimensional space-time. As the correspondence maps a strongly coupling many-body problem into a weakly coupling gravity system in one more dimension, it opens a new approach to deal with the strongly correlated electron systems which are difficult to handle in traditional condensed matter theories. Over the past years, this correspondence has been applied into an abundant of strongly correlated condensed systems such as superconductor/superfuild,[5–11] Fermi/non-Fermi liquids,[12–14] charge density wave and metal/insulator phase transition[15–17] and some systems far from thermal equilibrium.[18–22] In these works, the attentions are focused on the electric transport properties in strongly correlated materials such as conductivity. Though there are a few

*This article first appeared in *Int. J. Mod. Phys. D* **25** (2016) 1645011.

of works involving magnetism in holographic superconductor models, magnetism does not pay the central role. However, some important strongly correlated phenomena including high temperature superconductors and heavy fermion metals are controlled by magnetism of the material. Because of the strong correlation between electrons in the materials, there is still lack of a well theoretical description. The gauge/gravity duality provides an approach and perspective to understand these challenging problems.

As the spontaneous U(1) symmetry breaking, which plays an important role in building holographic superconductor models, the time reversal symmetry breaking is the characteristic of magnetic ordered states. An important property in magnetic materials is the response to external magnetic field. In phenomenology, a holographic model which can describe the phenomena involving spontaneous magnetic ordered state should give time reversal symmetry breaking spontaneously and characteristic response to external magnetic field. In this paper, we will introduce our recent works about using holographic duality to understand strong coupling phenomena controlled by magnetism. We will first give a holographic model to realize the ferromagnetic phase transition and discuss its important properties. Then we will show some singificant results in understanding some novel magnetic materials based on this model, including the coexistence and competition between ferromagnetism and superconductivity, colossal magnetic resistance and antiferromagnetic quantum phase transition (QPT) induced by an external magnetic field.

2. Holographic Ferromagnetic Model

The model, we are considering is Einstein–Maxwell theory in asymptotic AdS spacetime. The first holographic ferromagnetic model was proposed in Ref. 23. However, this mode exists external degree of freedoms which leads ghost or causality violation. Here, we will introduce a modified model proposed in Ref. 24. The action in the bulk is,

$$
\begin{aligned}
S &= \frac{1}{2\kappa^2}\int d^4x\sqrt{-g}(L_1+\lambda^2 L_2),\\
L_1 &= \mathcal{R}+\frac{6}{L^2}-F^{\mu\nu}F_{\mu\nu},\\
L_2 &= -\frac{1}{12}(dM)^2-\frac{m^2}{4}M_{\mu\nu}M^{\mu\nu}-M^{\mu\nu}F_{\mu\nu}-\frac{J}{8}V(M).
\end{aligned}
\tag{1}
$$

where L is the AdS radius and $2\kappa^2 = 16\pi G$ is associated with the gravitational constant in the bulk. We will set them both to be unity in the following. g is the determinant of the bulk metric $g_{\mu\nu}$. A_μ is U(1) gauge potential and $F_{\mu\nu} = (dA)_{\mu\nu}$. dM is the exterior differential of two-form field $M_{\mu\nu}$. m^2, λ and J are three real model parameters with $J < 0$ and $m^2 > 0$ in order the magnetization to happen spontaneously. $V(M_{\mu\nu})$ is a nonlinear potential of the two-form field to describe the self-interaction of the polarization tensor. It should be expanded as the even power of $M_{\mu\nu}$. The choice of nonlinear potential is not unique. Following Ref. 24, we take the following form,

$$V(M_{\mu\nu}) = ({}^*M_{\mu\nu}M^{\mu\nu})^2 = [{}^*(M \wedge M)]^2. \tag{2}$$

Here * is the Hodge-star operator.

In the simple case, the dual boundary is isotropy and homogenous, so all the fields in the bulk are only functions of radius r. Specially, we will use the following ansatz for metric and matter field,

$$ds^2 = -r^2 f(r) e^{a(r)} dt^2 + \frac{dr^2}{r^2 f(r)} + r^2(dx^2 + dy^2), \tag{3}$$

$$M_{\mu\nu} = -p(r) dt \wedge dr + \rho(r) dx \wedge dy, \quad A_\mu = \phi(r) dt + Bx dy \tag{4}$$

with some real functions $f(r)$, $a(r)$, $\phi(r)$, $p(r)$ and $\rho(r)$, which are determined by equations of motions through the variation of action (1).

The bulk field B is a constant magnetic field, which can be regarded as external magnetic field in the dual boundary theory. We will denote the position of the horizon as r_h and the conformal boundary will be at $r \to \infty$. In order to study a dual theory with finite chemical potential or charge density accompanied by a U(1) symmetry, we need to turn on $A_t = \phi(r)$ in the bulk. The magnetic moment density in the dual boundary is expressed as,

$$N = -\frac{\lambda^2}{2} \int_{r_h}^{\infty} \frac{\rho e^{a/2}}{r^2} dr. \tag{5}$$

It has been shown that[25] $\rho \to -\rho$ as $B \to -B$ under the time reversal transformation, which leads to the property of magnetic moment defined in Eq. (5) under the time reversal transformation such as $N \to -N$. This is in agreement with the fact that magnetic field is pseudo-vector in $3+1$ boundary dimensions or pseudo-scalar in $2+1$ boundary dimensions. In the case of $B = 0$, the appearance of nonzero ρ will lead to spontaneous magnetization and time reversal symmetry breaking.

The key physics of this model can be understood by the probe limit when $\lambda \to 0$. In this limit, we neglect the back reaction of polarization field to the gauge field and background geometry. In such a case, the background is just the dyonic Reissner–Nordstrom AdS black hole. In the vicinity of critical temperature and weak external magnetic field, the magnetic part of the grand thermodynamic potential density Ω in grand canonical ensemble is,[25]

$$\Omega \simeq -BN + \frac{2a_0}{\lambda^2 N_1^2}\left(\frac{T}{T_c - 1}\right) N^2 + \frac{-16\tilde{J}_f a_1}{\lambda^6 N_1^4} N^4 + \mathcal{O}(N^6). \tag{6}$$

Here $a_0, a_1 > 0, N_1$ and $\tilde{J}_f < 0$ are all constants, the details of these constants can be found in Ref. 25. We can see that this is just the Ginzburg–Landau (GL) theory of ferromagnetic model. From it, we can obtain the expression of magnetic moment in the ferromagnetic phase when $B = 0$,

$$N/\lambda^2 = \sqrt{\frac{N_1^2 a_0}{-16\widetilde{J}_f a_1}}\left(1 - \frac{T}{T_c}\right)^{1/2}, \quad \text{when } T \to T_c^-. \tag{7}$$

This tells us that the critical exponent 1/2 is an exact result. In the nonzero magnetic field case, the expression (6) gives out correct magnetic susceptibility behavior and hysteresis loop. When the temperature is far way from its critical value, then the expression (6) is not applicable. The spontaneous magnetic moment density and magnetic susceptibility can be determined numerically from the equations of motion. One typical result is shown in Fig. 1. One can find that the model realizes the hysteresis loop of single magnetic domain and that the magnetic susceptibility satisfies the Curie–Weiss law.

The case with full back reaction has also been studied in Ref. 25, which shows that the ferromagnetic phase transition will also happen at low temperature and the phase transition is always a second-order one when we increase the strength of the back reaction.

The holographic ferromagnetic phase transition is a consequence of the repulsion of the two-form field and the attraction from gravity. This can be understood from Fig. 2 schematically. As the high temperature corresponds to strong surface gravity and the gravity dominates the system, the condensation of the two-form field caused by vacuum fluctuation will be attracted into the black hole and disappear. This implies that the thermal fluctuation dominates the materials in high temperature and the ordered magnetic moment will become disordered quickly because of thermal fluctuation. In low temperature, the surface gravity becomes weak and the repulsion dominate the system. The condensation of the two-form field caused by vacuum fluctuation can be held near the horizon. This implies that the thermal fluctuation is weak and cannot destroy the ordered magnetic moment so that the macroscopical magnetic moment appears.

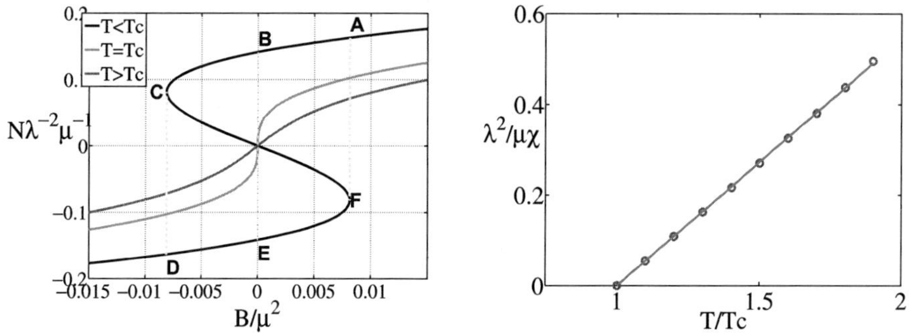

Fig. 1. The magnetic moment N (left panel) as a function of temperature in different magnetic field B and the magnetic susceptibility χ (right panel) as a function of temperature. Here, μ is the chemical potential of dual boundary.

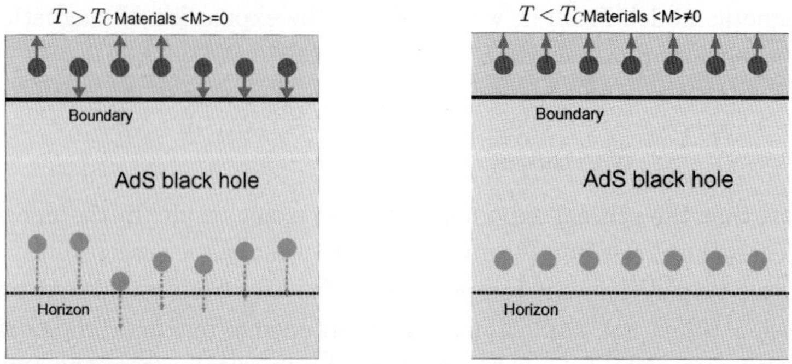

Fig. 2. (Color online) The schematic diagram for the condensation of the two-form field near the black hole horizon. The pink circles stand for the condensation of the spatial part of two-form field caused by vacuum fluctuation. The blue circles stand for microcosmic magnetic moments in materials and the arrows stand for their directions. T is temperature and T_C is the critical temperature.

3. Application in Strongly Correlated Materials

Though ferromagnetic phase transition itself has been understood well in condensed matter theory, the holographic model offers a framework to investigate the strongly correlated phenomena controlled by magnetism. As this framework is independent of the model such as holographic superconductor or Fermi/non-Fermi liquid, we can combine it with them and study spontaneous magnetic order in unconventional superconductors, strange metals or other novel materials.

3.1. *Competition and coexistence of ferromagnetism and superconductivity*

One of the interesting problems is the coexistence of ferromagnetism and superconductivity in heavy Fermi materials such as UGe_2,[26] URhGe[27] and UCoGe.[28] In a very long time, it was thought these two phenomenons are incompatible with each other. In the viewpoint of BCS theory, any magnetic impurity will suppresses the singlet Cooper pair formation, which causes a rapid depression of the superconducting transition temperature. Likewise, the superconductivity can screen off the magnetic field and leads to that the long-range magnetic order is accompanied by the expulsion of superconductivity. However, the new novel materials challenge this belief. In some heavy Fermi materials, the ferromagnetism and superconductivity are found to coexist with each other. The nature of this superconducting state in ferromagnetic materials is currently under debate. The heavy effective mass of valence electrons shows the strong correlation between them, which implies that approximations and conception of free field lose their validness.

In Ref. 30, the authors first study the competition and coexistence of ferromagnetism and superconductivity from holographic duality. By combining a holographic p-wave superconductor model[29] and a holographic ferromagnetism model, Ref. 30 studied the coexistence and competition of ferromagnetism and p-wave superconductivity. It is found that the results depend on the self-interaction of magnetic moment of the complex vector field and which phase appears first. In the case that the ferromagnetic phase appears first, if the interaction is attractive, the system shows the ferromagnetism and superconductivity can coexist in low temperatures. If the interaction is repulsive, the system will only be in a pure ferromagnetic state. In the case that the superconducting phase appears first, the attractive interaction will lead to a magnetic p-wave superconducting phase in low temperatures. If the interaction is repulsive, the system will transform into a pure ferromagnetic phase when the temperature is lowered.

3.2. *Colossal magnetic resistance in holographic duality*

Another interesting application is about the phenomena called "colossal magnetoresistance" (CMR) effect, which appears in the manganites such as $A_{1-x}B_xMnO_3$ (A = La, Pr, Sm, etc. and B = Ca, Sr, Ba, Pb). Such phenomena is among the main areas of research in strongly correlated electron systems.[31–37] In these materials, there is a remarkable magnetoresistivity near the Curie temperature T_C and an insulator (or semiconductor)/metal phase transition associated with a paramagnetic/ferromagnetic phase transition happens. The CMR effect has a completely different physical origin from the "giant" magnetoresistance (CMR) observed in layered and clustered compounds. Because of the progress in computation and mean field studies for realistic models, many important results have been achieved. However, a complete understanding of the CMR effect is still a challenge.

Within massive gravity, Ref. 38 constructs a gravity dual for insulator/metal phase transition and CMR effect. The main results in Ref. 38 can be shown in Fig. 3. From the right panel of Fig. 3, we see that the DC resistance shows an insulator behavior, when the temperature is higher than T_C, while the temperature is less than the Curie temperature T_C, the resistivity decreases, which shows a

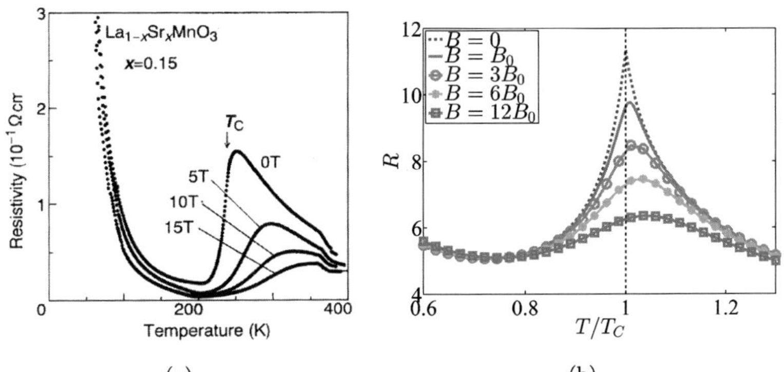

Fig. 3. Left panel: the DC resistivity of $La_{1-x}Sr_xMnO_3$ as a function of temperature for different magnetic field.[31] Right panel: the DC resistivity versus temperature in different external magnetic field.

metal's behavior. There is a distinct peak at the temperature, where spontaneous magnetization begins to appear and an insulator/metal phase transition happens there. This is just one of the characteristic properties of CMR materials in manganese oxides. What's more, when a small magnetic field B is turned on, we find that the resistivity is very sensitive to the external magnetic field. In addition, holographic model shows that MR in the case of weak magnetic field and $T \to T_C^+$ obeys following relation,

$$\mathrm{MR} = \frac{1 - R(B)}{R(B=0)} \propto N^2. \tag{8}$$

It is in agreement with the experimental and theoretical analysis of CMR materials.[31]

3.3. *Antiferromagnetism and quantum phase transition induced by external magnetic field*

The holographic ferromagnetic model has been generalized so that it can describe antiferromagnetism phase transition in Ref. 39. In the antiferromagnetic model, in order to describe the magnetic structures of antiferromagnetism, Ref. 39 introduces two different antisymmetry polarization fields in the bulk. These polarizations couple to U(1) gauge field strength and each other. The system shows a critical temperature T_N if the values of parameters are in some suitable region. In the case of absence of external magnetic field and low temperature, the magnetic moments of these tensor condense in antiparallel manner with the same magnitude, which leads to an antiferromagnetic state. In the weak external magnetic field, the magnetic susceptibility density has a peak at critical temperature and satisfies the Curie–Weiss law in the paramagnetic phase of antiferromagnetic materials.

One unexpected discovery is that the model in Ref. 39 shows critical temperature T_N will be suppressed by external magnetic field. There is a critical magnetic field B_c where the critical temperature will be suppressed into zero. This shows that this holographic antiferromagnetic model can describe a QPT induced by magnetic field. Inspired by this result, Refs. 40 and 41 investigate a holographic model for antiferromagnetic QPT induced by magnetic field. Figure 4 shows the results from holographic model and experimental results of QPT from pyrochlores $(Er_{1-x}Y_x)_2Ti_2O_7$.[42] We can see that, after rescaling the temperature and magnetic field, the antiferromagnetic transition boundary of $(Er_{1-x}Y_x)_2Ti_2O_7$ is insensitive to the doping, which shows a universal behavior. The holographic model can cover well the experimental date. Besides, the holographic model forecasts two scaling relations,

$$\widetilde{\Delta} = C\left(\frac{B}{B_c} - 1\right), \quad \text{with } \widetilde{\Delta} = \frac{\Delta}{k_B T_{N0}}, \tag{9}$$

$$\frac{\widetilde{T}_N}{\ln \widetilde{T}_N} \propto \left(1 - \frac{B}{B_c}\right), \tag{10}$$

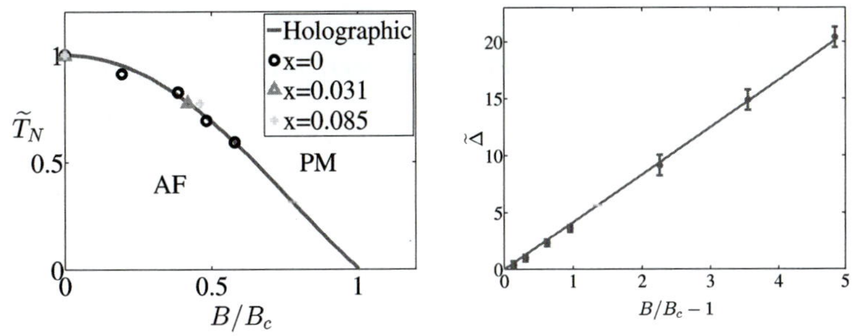

Fig. 4. Left panel: The antiferromagnetic critical temperature T_N versus the external magnetic field B. The solid curve is holographic results and the different dots are experimental data[40,42] for pyrochlore compounds, $Er_{2-2x}Y_{2x}Ti_2O_7$ with three different doping x. T_N is the antiferromagnetic critical temperature and T_{N0} is the value of T_N at $B = 0$. Right panel: The energy gap of the high-field ferrimagnetic state, Δ, of $Er_{2-2x}Y_{2x}Ti_2O_7$ as a function of magnetic dilution (x) and magnetic field B measured in the units of the critical field.[42]

where $\widetilde{T}_N = T_N/T_{N0}$, k_B is Boltzmann constant and C is a constant. These scaling relations can be analytically presented by considering the emergent geometry AdS_2 in the IR limit.[41] The first one has been measured in $(Er_{1-x}Y_x)_2Ti_2O_7$ (see the right panel of Fig. 4) and experimental data give constant $C \approx 5.0$, which is very close to the holographic result $C \approx 4.2$. Because of the difficulty in experiments, it is still an open question for experimenter to check the other scaling relationship in Eq. (10).

4. Summary

Understanding magnetic properties of materials has a very long history. However, it is still mysterious for lots of materials in extreme pressure and low temperature. Magnetism plays a crucial role in Mott transition, heavy fermion, and QPT. In fact, quantum magnetism is one of the manifestations of strong electron–electron interaction, so its treatment should be part of general strongly correlated electron effects. Holographic duality is a very suitable tool for such phenomena. This paper gave a brief review about the applications of holographic duality in understanding unusual magnetic materials. Some quantitative comparison between holographic results and experimental data was made and some predictions from holographic duality models were discussed. Though some works have been done to understand the phenomena in strongly correlated electron systems controlled by magnetism, it is still far from a complete understanding. The most works at present stage are to recover some properties which have been discovered in experiments. Of course, this is important and provides a starting point to build a holographic description for real materials.

Acknowledgments

This work was supported in part by the National Natural Science Foundation of China with Grant Nos.11375247 and 11435006. We thank F. Kusmartsev, Y. B. Wu and C.Y. Zhang for collaborations in some relevant studies.

References

1. J. M. Maldacena, *Int. J. Theor. Phys.* **38** (1999) 1113; *Adv. Theor. Math. Phys.* **2** (1998) 231, arXiv:hep-th/9711200.
2. S. S. Gubser, I. R. Klebanov and A. M. Polyakov, *Phys. Lett. B* **428** (1998) 105, arXiv:hep-th/9802109.
3. E. Witten, *Adv. Theor. Math. Phys.* **2** (1998) 253, arXiv:hep-th/9802150.
4. E. Witten, *Adv. Theor. Math. Phys.* **2** (1998) 505, arXiv:hep-th/9803131.
5. S. A. Hartnoll, C. P. Herzog and G. T. Horowitz, *Phys. Rev. Lett.* **101** (2008) 031601, arXiv:0803.3295 [hep-th].
6. S. A. Hartnoll, C. P. Herzog and G. T. Horowitz, *J. High Energy Phys.* **0812** (2008) 015, arXiv:0810.1563 [hep-th].
7. S. A. Hartnoll, *Class. Quantum Grav.* **26** (2009) 224002, arXiv:0903.3246 [hep-th].
8. C. P. Herzog, *J. Phys. A* **42** (2009) 343001, arXiv:0904.1975 [hep-th].
9. J. McGreevy, *Adv. High Energy Phys.* **2010** (2010) 723105, arXiv:0909.0518 [hep-th].
10. G. T. Horowitz, *Lect. Notes Phys.* **828** (2011) 313, arXiv:1002.1722 [hep-th].
11. R. G. Cai, L. Li, L. F. Li and R. Q. Yang, *Sci. China Phys. Mech. Astron.* **58** (2015) 060401, arXiv:1502.00437 [hep-th].
12. S. S. Lee, *Phys. Rev. D* **79** (2009) 086006, arXiv:0809.3402 [hep-th].
13. H. Liu, J. McGreevy and D. Vegh, *Phys. Rev. D* **83** (2011) 0650209, arXiv:0903.2477 [hep-th].
14. M. Cubrovic, J. Zaanen and K. Schalm, *Science* **325** (2009) 439, arXiv:0904.1993 [hep-th].
15. A. Aperis, P. Kotetes, E. Papantonopoulos, G. Siopsis, P. Skamagoulis and G. Varelogiannis, *Phys. Lett. B* **702** (2011) 181, arXiv:1009.6179 [hep-th].
16. A. Donos and J. P. Gauntlett, *Phys. Rev. D* **87** (2013) 126008, arXiv:1303.4398 [hep-th].
17. Y. Ling, C. Niu, J. Wu, Z. Xian and H. B. Zhang, *Phys. Rev. Lett.* **113** (2014) 091602, arXiv:1404.0777 [hep-th].
18. K. Murata, S. Kinoshita and N. Tanahashi, *J. High Energy Phys.* **1007** (2010) 050, arXiv:1005.0633 [hep-th].
19. M. J. Bhaseen, J. P. Gauntlett, B. D. Simons, J. Sonner and T. Wiseman, *Phys. Rev. Lett.* **110** (2013) 015301, arXiv:1207.4194 [hep-th].
20. A. Adams, P. M. Chesler and H. Liu, *Science* **341** (2013) 368, arXiv:1212.0281 [hep-th].
21. A. M. Garca-Garca, H. B. Zeng and H. Q. Zhang, *J. High Energy Phys.* **1407** (2014) 096, arXiv:1308.5398 [hep-th].
22. P. M. Chesler and L. G. Yaffe, *J. High Energy Phys.* **1407** (2014) 086, arXiv:1309.1439 [hep-th].
23. R. G. Cai and R. Q. Yang, *Phys. Rev. D* **90** (2014) 081901, arXiv:1404.2856 [hep-th].
24. R. G. Cai and R. Q. Yang, arXiv:1504.00855 [hep-th].
25. R. G. Cai, R. Q. Yang, Y. B. Wu and C. Y. Zhang, *J. High Energy Phys.* **1511** (2015) 021, arXiv:1507.00546 [hep-th].
26. G. G. Lonzarich *et al.*, *Nature* **406** (2000) 587.
27. D. Aoki, A. Huxley, E. Ressouche, D. Braithwaite, J. Flouquet, J-P. Brison, E. Lhotel and C. Paulsen, *Nature* **413** (2001) 613.
28. N. T. Huy *et al.*, *Phys. Rev. Lett.* **99** (2007) 067006.
29. R. G. Cai, L. Li and L. F. Li, *J. High Energy Phys.* **1401** (2014) 032, arXiv:1309.4877 [hep-th].
30. R. G. Cai and R. Q. Yang, *Phys. Rev. D* **91** (2015) 026001, arXiv:doi:10.1103/PhysRevD.91.026001, arXiv:1410.5080 [hep-th].

31. A. Urushibara, Y. Moritomo, T. Arima, A. Asamitsu, G. Kido and Y. Tokura, *Phys. Rev. B* **51** (1995) 103.
32. M. Uehara, S. Mori, C. H. Chen and S. W. Cheong, *Nature* **399** (1999) 560.
33. M. B. Salamon and M. Jaime, *Rev. Mod. Phys.* **73** (2001) 583.
34. E. Dagotto, T. Hotta and A. Moreo, *Phy. Rep.* **344** (2001) 1.
35. E. L. Nagaev, *Phys. Rep.* **346** (2001) 387.
36. C. Sen, G. Alvarez and E. Dagotto, *Phys. Rev. Lett.* **105** (2010) 097203.
37. M. Anamitra, S. Cole William, W. Patrick, R. Mohit and T. Nandini, *Phys. Rev. Lett.* **110** (2013) 157201.
38. R. G. Cai and R. Q. Yang, *Phys. Rev. D* **92** (2015) 106002, arXiv:1507.03105 [hep-th].
39. R. G. Cai and R. Q. Yang, *Phys. Rev. D* **91** (2015) 086001, arXiv:1404.7737 [hep-th].
40. R. G. Cai, R. Q. Yang and F. V. Kusmartsev, *Phys. Rev. D* **92** (2015) 086001, arXiv:1501.04481 [hep-th].
41. R. G. Cai, R. Q. Yang and F. V. Kusmartsev, *Phys. Rev. D* **92** (2015) 046005, arXiv:1505.03405 [hep-th].
42. J. F. Niven *et al.*, *Proc. R. Soc. A* **470** (2014) 20140387.

Direct Dark Matter Searches and the CDEX Research Program*

Henry T. Wong
Institute of Physics, Academia Sinica,
Taipei 11529, Taiwan
htwong@phys.sinica.edu.tw

Qian Yue* and Kejun Kang
Department of Engineering Physics,
Tsinghua University, Beijing 100084, China
**yueq@mail.tsinghua.edu.cn*

We sketch the landscape which gives rise to the missing energy density problem and highlight the direct experimental searches of weakly interacting massive particles (WIMPs) dark matter. An overview of the China Jinping underground laboratory as well as the China dark matter experiment (CDEX) dark matter program based on germanium detectors with sub-keV sensitivities is presented. The achieved results, status as well as the R&D and technology acquisition efforts towards a ton-scale experiment are reported.

Keywords: Cosmology; dark matter; laboratory facilities.

PACS Number(s): 98.80.−k, 95.35.+d, 01.52.+r

1. The Grand Landscape

There are fundamental conflicts with our understanding of the Universe when it is comprehended via gravity (with general relativity (GR) describing the macroscopic scales) and via particle physics (with the Standard Modal (SM) describing the microscopic scales).[a] They are formulated under vastly different theoretical framework. Attempts to describe gravity using the language of quantum field theory have not been successful so far.

The incompatibilities between GR and SM is compelling in their interpretation of astronomical observations from galactic to cosmological length scales.[1] The main-stream solution is that the Universe consists of about 25% as "dark matter" and 70% as "dark energy," the exact nature of which are not understood. The

*This article first appeared in *Int. J. Mod. Phys. D* **25** (2016) 1645012.
[a]The theme of this Conference, documented in these Proceedings.

"missing energy density" (i.e. $\Omega_{\rm SM} < \Omega_{\rm Total}$) problem represents a crisis and offers great opportunities in fundamental research. Tremendous interest are generated and huge efforts are being invested in their pursuit. New and significant physics output through a journey with surprising twists and turns can be expected.

The standard paradigm for dark matter[2] is that it is a new form of matter, stable, has no electro-magnetic and strong interactions with SM matter, and is "cold" or nonrelativistic. Among the various particle physics models proposed as cold dark matter, the class receiving the most attention is on weakly interacting massive particles (WIMPs, denoted by χ) with the GeV–TeV mass range. The lightest particles in super-symmetric (SUSY) theories motivated to solve the SM hierarchy problem are good candidates for WIMPs. To be compared with experimental and observational data, WIMPs are taken to be produced via thermal equilibrium in the early Universe. They would be gravitationally bounded in the galactic halo and have Maxwellian velocity distributions.

Several experimental approaches are pursued to probe the possible existence and properties of WIMP dark matter.

(I) **Accelerator searches:** WIMPs may be produced by the collisions of high energy particles and searches are conducted at accelerator experiments such as those at the Large Hadron Collider (LHC), the signatures of which would be distinguished by the missing energy since they would not interact with the detectors.

(II) **Indirect searches:** Astrophysics measurements of positrons, anti-protons, photons and neutrinos by ground-based and satellite experiments are compared with astrophysical models. Anomalous excesses not accounted by known astrophysics sources would require further studies to confirm whether they would be originated from the final-states of WIMP decays or annihilations or interactions.

(III) **Direct searches:** These are low background experiments looking for WIMPs which interact with the detector target while traversing through earth. Consistencies of measurements among different target materials and between the integrated and annual modulated rates may indicate positive signatures.

(IV) **Galactic structures:** Simulation results and observations are compared to place constraints on WIMP interaction models and their parameter space.

Direct WIMP search experiments are typically performed at underground laboratories world-wide which provide effective suppression of cosmic-ray direct and induced background. The favored detection channels are on elastic scattering of WIMPs with the target nucleus $[\chi + N \rightarrow \chi + N]$ with either spin-independent or spin-dependent cross-sections. Different experimental signatures are studied by a wide spectrum of experiments, based on the detection of ionization (charge), bolometry (heat or phonons), or scintillation (photons). Techniques making simultaneous measurements of two of these parameters provide the additional advantages of differentiation between nuclear (signal) versus electron (background) recoil events. Several advances in detector techniques are inspired by direct WIMP searches programs. These include the development of sub-Kelvin cryogenic detectors, two-phase liquid noble gas detectors, as well as the realization of bolometric amplification.

In the subsequent sections, we focus on the China dark matter experiment (CDEX) dark matter program with germanium detectors having sub-keV sensitivities at the China JinPing Underground Laboratory (JPL).

2. China JinPing Underground Laboratory

The CJPL[3] is located in Sichuan, China, and was inaugurated in December 2012. With a rock overburden of about 2400 m, it is the deepest operating underground laboratory in the world. The muon flux is measured to be[4] $(2.0\pm0.4)\times10^{-10}\,\mathrm{cm}^{-2}\,\mathrm{s}^{-1}$, suppressed from the sea-level flux by a factor of 10^{-8}. The ambient gamma and neutron fluxes and radon concentration[5] are, respectively, $6.9\times10^{-2}\,\mathrm{cm}^{-2}\,\mathrm{s}^{-1}$ at 40−2700 keV, $1.75\times10^{-5}\,\mathrm{cm}^{-2}\,\mathrm{s}^{-1}$ below 20 MeV, and $38\pm17\,\mathrm{Bq/m}^3$ under ventilation.

As depicted schematically in Fig. 1, the completed CJPL Phase-I consist of a laboratory hall of dimension 6 m (W) × 6 m (H) × 40 m (L). This space is currently used by the CDEX[6] and PandaX[7] dark matter experiments, as well as for a general

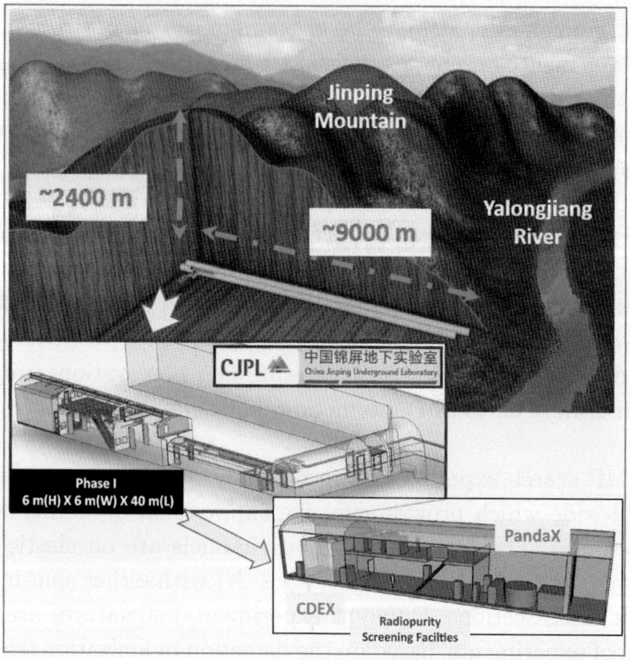

Fig. 1. Schematic diagram of CJPL Phase-I inaugurated in 2012, showing the space allocation to the CDEX and PandaX dark matter experiments, as well as to the radiopurity screening facilities.

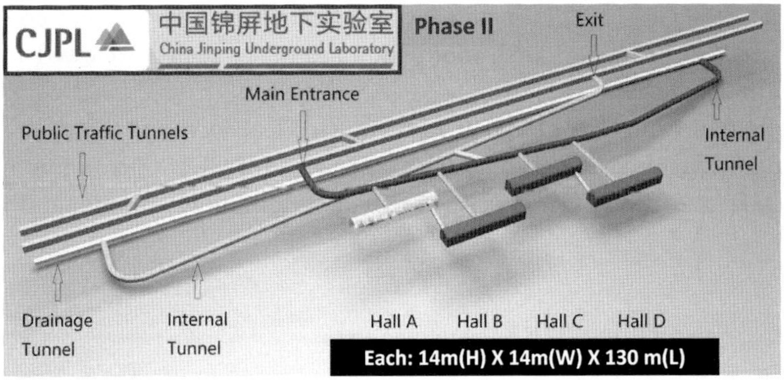

(a)

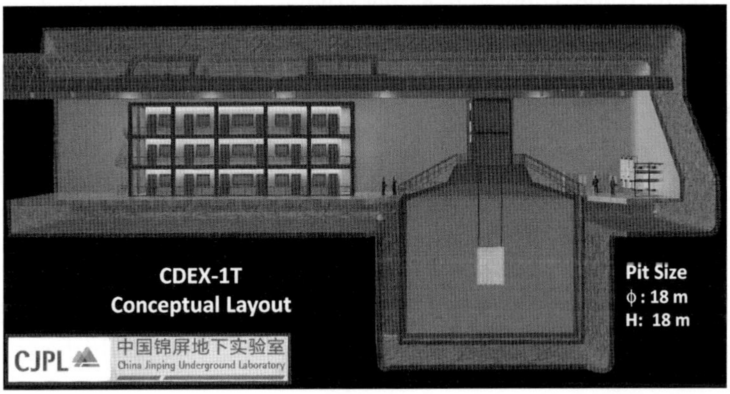

(b)

Fig. 2. (a) Schematic diagram of CJPL Phase-II scheduled to complete by early 2017 and (b) conceptual configuration of a future CDEX-1T experiment at CJPL Phase-II.

purpose low radiopurity screening facility. The drive-in tunnel access can greatly facilitate the deployment of big experiments and large teams. Supporting infrastructures of catering and accommodation, as well as office and workshop spaces, already exist.

Additional laboratory space for CJPL Phase-II, located about 500 m from Phase-I site, is currently under construction. Upon the scheduled completion by early 2017, it will consist of four halls each with dimension 14 m (W) × 14 m (H) × 130 m (L). The tunnel layout is displayed in Fig. 2(a).

3. CDEX Dark Matter Program

Germanium detectors sensitive to sub-keV recoil energy were identified and demonstrated as possible means to probe the "light" WIMPs with mass range

$1\,\text{GeV} < m_\chi < 10\,\text{GeV}$.[8] This inspired development of p-type point-contact germanium detectors (pPCGe) with modular mass of kg-scale,[9] followed by various experimental efforts.[10,12,13]

The unique advantages of CJPL make it an ideal location to perform experiments on dark matter searches. The scientific theme of CDEX[6] is to pursue studies of light WIMPs with pPCGe. It is one of the two founding experimental programs at CJPL.

3.1. *First generation CDEX experiments*

As depicted in Fig. 3, the first-generation experiments adopted a baseline design of single-element "1 kg mass scale" pPCGe enclosed by NaI(Tl) crystal scintillator as anti-Compton detectors adopted in earlier studies of neutrino magnetic moments[11] and dark matter searches.[12] These active detectors are further surrounded by passive shieldings of Oxygen-Free-High-Conductivity (OFHC) copper, boron-loaded polyethylene (PE(B)) and lead, while the detector volume is purged by dry nitrogen to suppress radon contamination.

The pilot CDEX-0 measurement is based on a 20 g prototype Ge detector at 177 (eV_{ee}) threshold with an exposure of 0.784 kg-days.[14] The CDEX-1 experiment adopts a pPCGe detector of mass 1 kg. The first results are based on an analysis threshold of 475 eV_{ee} with an exposure of 53.9 kg-days.[15] After suppression of the anomalous surface background events and measuring their signal efficiencies and background leakage factors with calibration data,[16] all residual events can be

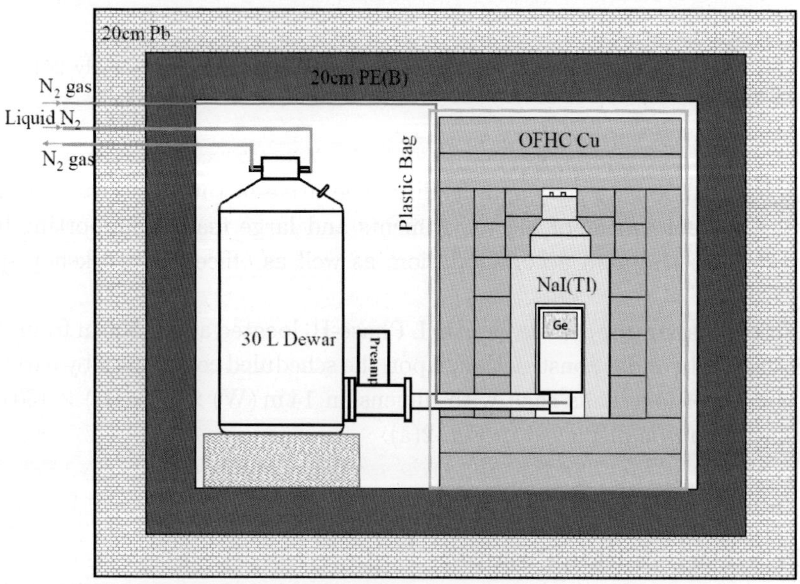

Fig. 3. Schematic diagram of the baseline design of the CDEX-0 and CDEX-1 experiments, using single-element pPCGe detector enclosed by NaI(Tl) crystal scintillator and passive shieldings.

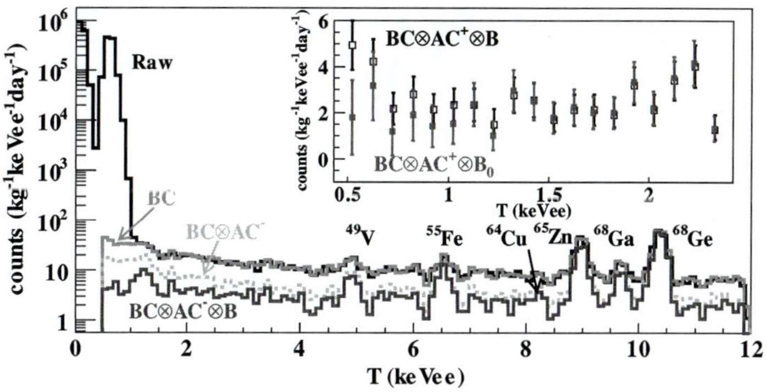

(a)

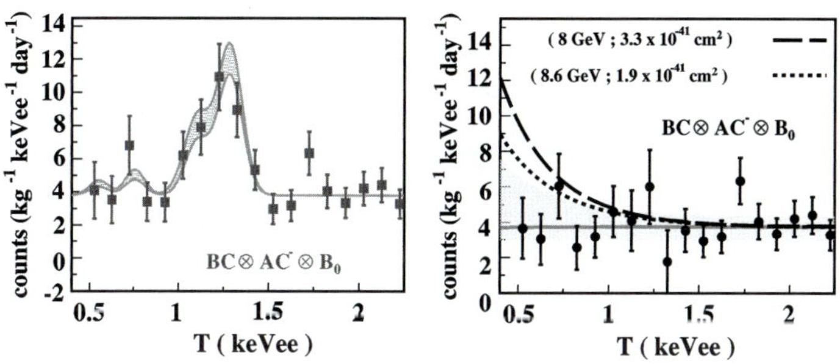

(b)

Fig. 4. Background spectra of the CDEX-1 measurement at their various stages of selection: basic cuts (BC), anti-Compton (AC) and bulk (B) events. All events can be accounted for by known background. Examples of excluded χN recoil spectra are superimposed.

accounted for by known background models, as illustrated in Fig. 4. Dark matter constraints on χN spin-independent cross-sections were derived for both data set, and are displayed in Fig. 5, together with other selected benchmark results.[2] In particular, the allowed region from the CoGeNT[10] experiment is probed and excluded with the CDEX-1 results.

Analysis is currently performed on CDEX-1 data set with year-long exposure. Annual modulation effects as well as other physics channels are being studied. New data is also taken with an upgraded pPCGe with lower threshold.

3.2. *Current efforts and future goals*

The long-term goal of the CDEX program will be a ton-scale germanium experiment at CJPL for the searches of dark matter and of neutrinoless double beta decay ($0\nu\beta\beta$).[17] A pit of diameter 18 m and height 18 m will be built at one of the halls of CJPL-Phase II to house such an experiment, as illustrated in Fig. 2(b).

Towards this end, the "CDEX-10" prototype has been constructed with detectors in array structure having a target mass at the 10 kg range. This would provide

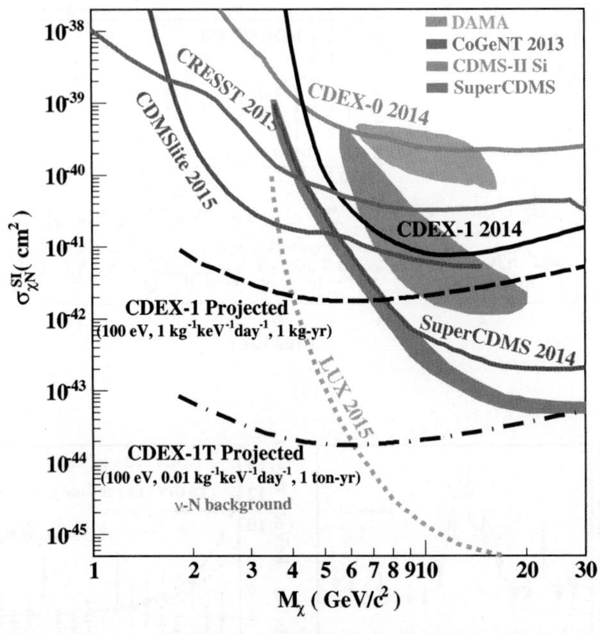

Fig. 5. Exclusion regions derived from the CDEX-0 and CDEX-1 experiments, and comparison with other benchmark results. Projected sensitivities of the current detectors and future projects are superimposed.

a platform to study the many issues of scaling up in detector mass and in improvement of background and threshold. The detector array is shielded and cooled by a cryogenic liquid. Liquid nitrogen is being used, while liquid argon is a future option to investigate, which may offer the additional potential benefits of an active shielding as anti-Compton detector.

In addition, various crucial technology acquisition projects are pursued, which would make a ton-scale germanium experiment realistic and feasible. These include:

(1) detector grade germanium crystal growth;
(2) germanium detector fabrication;
(3) isotopic enrichment of ^{76}Ge for $0\nu\beta\beta$;
(4) production of electro-formed copper, eventually underground at CJPL.

The first detector fabricated by the Collaboration from commercial crystal that matches expected performance will be installed at CJPL in 2016. It allows control of assembly materials placed at its vicinity, known to be the dominant source of radioactive background, as well as efficient testing of novel electronics and readout schemes. The benchmark would be to perform light WIMP searches with germanium detectors with "$0\nu\beta\beta$-grade" background control. This configuration would provide the first observation (or stringent upper bounds) of the potential cosmogenic tritium contaminations in germanium detectors, from which the strategies to suppress such background can be explored.

The projected χN sensitivity for CDEX-1T is shown in Fig. 5, taking a realistic minimal surface exposure of six months. The goal for $0\nu\beta\beta$ will be to achieve sensitivities covering completely the inverted neutrino mass hierarchy.

Acknowledgments

Gravity has played a central role in establishing the missing energy density problem which defines the landscape of basic research today. Gravity *may* continue to be a major player in providing the "Solution". The authors appreciate the efforts of the organizers of this Symposium in bringing together an audience with diversified research background and interest but nevertheless coupled together by this fascinating subject.

References

1. K. A. Olive and J. A. Peacock, *Rev. Part. Phys. Chin. Phys. C* **38** (2014) 327; O. Lahav and A. R. Liddle, *Rev. Part. Phys. Chin. Phys. C* **38** (2014) 345.
2. M. Drees and G. Gerbier, *Rev. Part. Phys. Chin. Phys. C* **38** (2014) 353.
3. K. J. Kang *et al.*, *J. Phys. Conf. Ser.* **203** (2010) 012028; J. M. Li *et al.*, *Phys. Procedia* **61** (2015) 576.
4. Y. C. Wu *et al.*, *Chin. Phys. C* **37** (2013) 086001.
5. Z. Zeng *et al.*, *J. Radioanal. Nucl. Chem.* **301** (2014) 443.
6. K. J. Kang *et al.*, *Front. Phys.* **8** (2013) 412.
7. X. G. Cao *et al.*, *Sci. China Phys. Mech. Astron.* **57** (2014) 1476.
8. Q. Yue *et al.*, *High Energy Phys. Nucl. Phys.* **28** (2004) 877; H. T. Wong *et al.*, *J. Phys. Conf. Ser.* **39** (2006) 266.
9. P. N. Luke *et al.*, *IEEE Trans. Nucl. Sci. B* **36** (1989) 926; P. S. Barbeau, J. I. Collar, and O. Tench, *J. Cosmo. Astropart. Phys. B* **09** (2007) 009.
10. C. E. Aalseth *et al.*, *Phys. Rev. D* **88** (2013) 012002.
11. H. B. Li *et al.*, *Phys. Rev. Lett.* **90** (2003) 131802; H. B. Li *et al.*, *Phys. Rev. D* **75** (2007) 012001.
12. S. T. Lin *et al.*, *Phys. Rev. D* **79** (2009) 061101(R); H. B. Li *et al.*, *Phys. Rev. Lett.* **110** (2013) 261301.
13. G. K. Giovanetti *et al.*, *Phys. Procedia* **00** (2014) 1.
14. S. K. Liu, *Phys. Rev. D* **90** (2014) 032003.
15. K. J. Kang *et al.*, *Chin. Phys. C* **37** (2013) 126002; W. Zhao *et al.*, *Phys. Rev. D* **88** (2013) 052004; Q. Yue *et al.*, *Phys. Rev. D* **90** (2014) 091701(R).
16. H. B. Li *et al.*, *Astropart. Phys.* **56** (2014) 1; A. K. Soma *et al.*, arXiv:1411.4802.
17. S. M. Bilenky and C. Giunti, *Mod. Phys. Lett. A* **27** (2012) 1230015.

Design and Implementation of the TAROGE Experiment*

J. W. Nam*,‡, C.-C. Chen*, C.-H. Chen*, C.-W. Chen*, P. Chen*, Y.-C. Chen*, S.-Y. Hsu*, J.-J. Huang*, M.-H. A. Huang†, T.-C. Liu*, J. Řípa*, Y.-S. Shiao*, M.-Z. Wang* and S.-H. Wang*

*Department of Physics, Institute of Astrophysics, and Leung Center for Cosmology and Particle Astrophysics, National Taiwan University, Sec. 4 No. 1. Roosevelt Road, Taipei 10617, Taiwan

†Department of Energy and Resources, National United University, 2 Lianda Road, Miao-Li 36003, Taiwan

‡jwnam@phys.ntu.edu.tw

Taiwan astroparticle radiowave observatory for geo-synchrotron emissions (TAROGE) is an antenna array on the high mountains of Taiwan's east coast for the detection of ultra-high energy cosmic rays (UHECRs) in an energy above $10^{18.5}$ eV. The antennas point toward the ocean to detect radiowave signals emitted by the UHECR-induced air-shower as a result of its interaction with the geomagnetic field. Looking down from the coastal mountain, the effective area is enhanced by collecting both direct-emission as well as the ocean-reflected signals. This instrument also provides the capability of detecting earth-skimming tau-neutrino through its subsequent tau-decay induced shower. In order to prove the detection concept, initial two stations were successfully built at 1000 m elevation near Heping township, Taiwan, in 2014–2015. Each station consists of 12 log-periodic dipole array antennas for 110–300 MHz. The stations have been operating smoothly for radio survey and optimization of instrumental parameters. In this report, we discuss the design of TAROGE, the performance of the prototype station and the future prospect.

Keywords: Ultra-high energy cosmic rays; extensive air showers; radio detection.

PACS Number(s): 96.50.sd, 07.57.Kp, 84.40.−x

1. Introduction

The radio detection method has recently been established as a promising alternative solution for detecting of ultra-high energy cosmic rays (UHECRs) which is challenging due to the extremely low flux. It would require a detection area several orders of magnitude larger than the current largest cosmic ray detector; Pierre Auger Observatory,[1] to identify their astrophysical sources. Radio emission from

*This article first appeared in *Int. J. Mod. Phys. D* **25** (2016) 1645013.

extensive air showers is a form of geo-synchrotron radiation, which arises from the deflection of electrons and positions in the shower by the Earth magnetic field. Such radiations at radio-band are coherently added during air-shower development because the radio wavelengths are longer than the shower size, which results in intense pulses that can be detected at a distance of several hundred kilometers. Therefore, one can effectively cover a huge detection area with a small number of radio antennas. Furthermore, the radio detection system can be operated all year round and day-and-night, which leads to a duty cycle as high as ~100%. Several frontier experiments such as LOPES, ANITA, and AREA[2–4] have demonstrated powerful capability of the radio detection method for UHECRs. For instance, the ANITA experiment, a NASA balloon-borne antenna array that surveys Antarctica from the altitude of 35 km, reported the observation of 16 events with averaged energy $\langle E \rangle = 1.5 \times 10^{19}$ eV. It turned out that 14 of these 16 events were radio emissions from down-going UHECRs, but reflected off the Antarctic ice surface. ANITA takes an advantage of vast detection area ($\sim 10^6$ km^2) within a horizon distance of ~ 600 km from the 35 km of high altitude. Although ANITA has great advantage of covering a large detection area, there are several important drawbacks: (1) the flight duration of the NASA Antarctic balloons is typically limited to ~30 days. The duty factor further reduces to 10 days/year since the flight opportunity for such balloon projects is roughly once in every three years, (2) ANITA's frequency range is 200–1200 MHz, which is optimized for UHE neutrino detection.[5,6] In contrast, UHE cosmic ray induced geo-synchrotron radiation is significantly enhanced at lower frequencies down to 10 MHz.[7–13] Deployment of low frequency antennas to the ANITA payload is difficult due to its weight and space constraints, (3) because the field strength of signal is proportional to 1/R, when it propagates through a long distance to high altitude, only very high energy events can survive and thus lowers event rate, (4) typical altitude of the shower maximum for UHECRs is 10 km, which is much lower than the ANITA's observation point. Only reflected-off signals are detectable except a small fraction of events coming from near the horizon.

Our approach, Taiwan astroparticle radiowave observatory for geo-synchrotron emissions (TAROGE), is to invoke the proven ANITA detection methodology, but places the antennas at the top of high mountains near the ocean. Figure 1 shows the detection concept of TAROGE. This approach resolves all problems of ANITA described above. Thanks to the ideal geology of Taiwan's east coast, where there are numerous mountains as high as 2–3 km in altitude near ocean. At 2 km altitude the horizon-distance is 160 km, which would provide a 1.3×10^4 km^2 detection area as a reflector with a 60° field-of-view. Although the detection area is smaller than ANITA, TAROGE has a great advantage of year-round operation with a high duty cycle, which compensates the area loss. Furthermore, lowering the detection frequency range to 110 MHz (and possibly below) and reducing the detection altitude to 2 km would enhance the signal strength and would therefore allow TAROGE to lower the energy threshold. The sea water is highly reflective for radio frequency (RF); the reflection coefficient is near 100% for the horizontally polarized wave, and

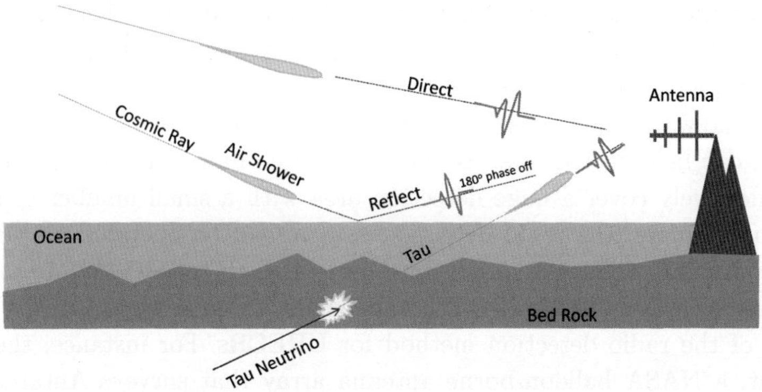

Fig. 1. Detection concept of TAROGE.

higher than 60% for the vertically polarized wave except a small grazing angle. Surface roughness due to ocean wave would be an issue weakening the signal strength. However, this effect is reduced for low frequency range in which the wavelength is longer than the surface roughness for the typical wave condition. TAROGE's receiver configuration that points toward the horizon is sensitive to showers with large zenith angles, which is utilized for UHECRs with energies of $10^{18.5}$–10^{19} eV. The east coastal region of Taiwan is much less populated than other areas of Taiwan, providing a quieter ambient RF environment. Figure 2 shows power spectra obtained from our initial RF survey at multiple locations in the mountain area of eastern Taiwan. Although many frequency bands appear to be polluted by continuous waveform (CW) noises, there are still fairly quite bands where the noise floors are comparable to thermal- and Galactic-noises. Using RF filters and a trigger system which is only sensitive to coherent signal, Taiwan could be a feasible place for this experiment. This detection technique not only can be applied in Taiwan, but may also be extended to other places with better and quieter RF environments in future.

2. The TAROGE-1 Station

As the first step, in order to demonstrate the detection concept of TAROGE and to perform a long-term noise survey, we built a prototype station, TAROGE-1, in Yongshih Mt. at 1000 m elevation near Heping township, Taiwan, in July 2014. This site had an excellent infrastructure including a good accessibility on a paved road, AC power lines, and a 12 m tall truss tower for mounting antennas. Because of a fact that the site has both views of the open ocean and the township in the field of view, it is almost an ideal location for detailed ambient RF studies with various anthropogenic noise sources near the coast as well as cosmic ray signals. Figure 3 shows the overall structure of the TAROGE-1 station. TAROGE-1 consists of 12 of 5 dBi log-periodic dipole array (LPDA) antennas for 110–300 MHz. A dual polarization configuration, the horizontal polarization (H-pol) and the vertical polarization

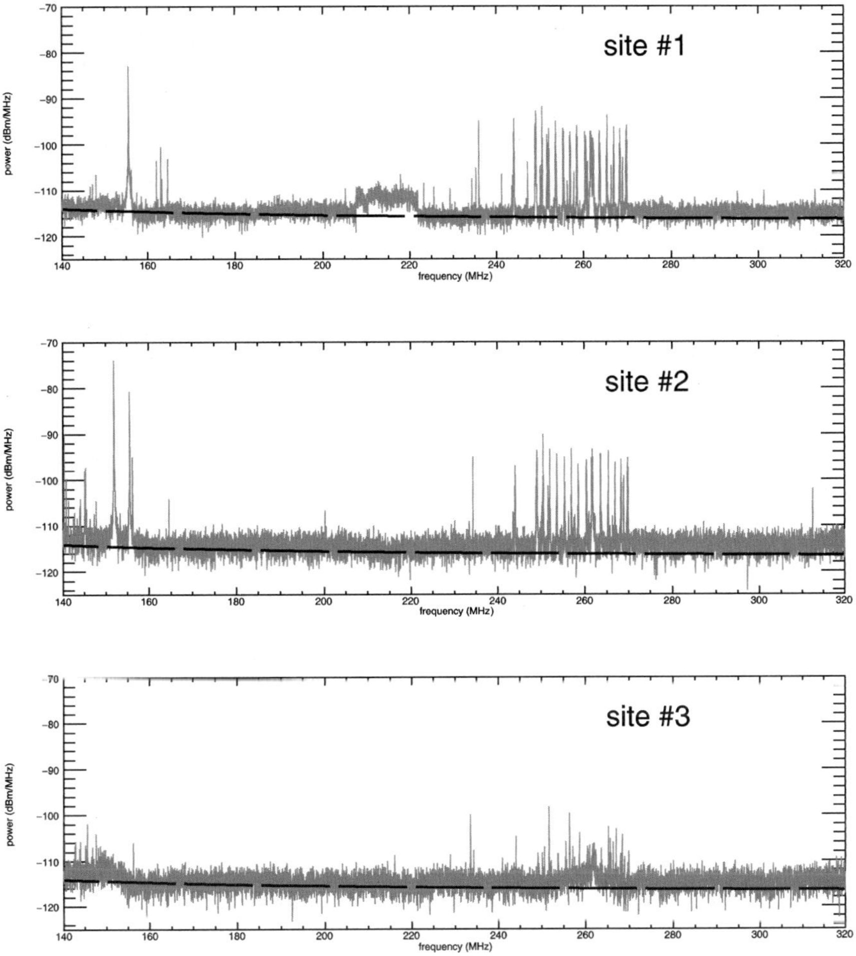

Fig. 2. RF noise spectra in the east mountain area: Data taken in a mountain near ocean (top), in a mountain 10 km away from ocean (middle) and in a deep valley without ocean view (bottom). Dashed lines indicate estimation of the noise floor contributed by thermal- and Galactic-radiations.

(V-pol), is implemented by orientating E-planes of six antennas to the horizontal direction and another to the vertical.

Figure 4 shows the system architecture of TAROGE-1. Received signal in the antenna is fed into the RF front-end before the trigger and digitizer system. The RF front-end consists of a front-end filter, a low noise amplifier (+35 dB), and a second stage amplifier (+30 dB). The RF front-ends are located close to the LPDAs to minimize transfer losses. The signals are transferred via coaxial cables to the data acquisition (DAQ) box where the trigger and digitizer system are placed. The RF signals are split into two paths, the trigger and the digitizer in the DAQ box.

The TAROGE trigger uses the multi-frequency bands coincidence technique, basically similar to ANITA, which provides an effective discriminating power

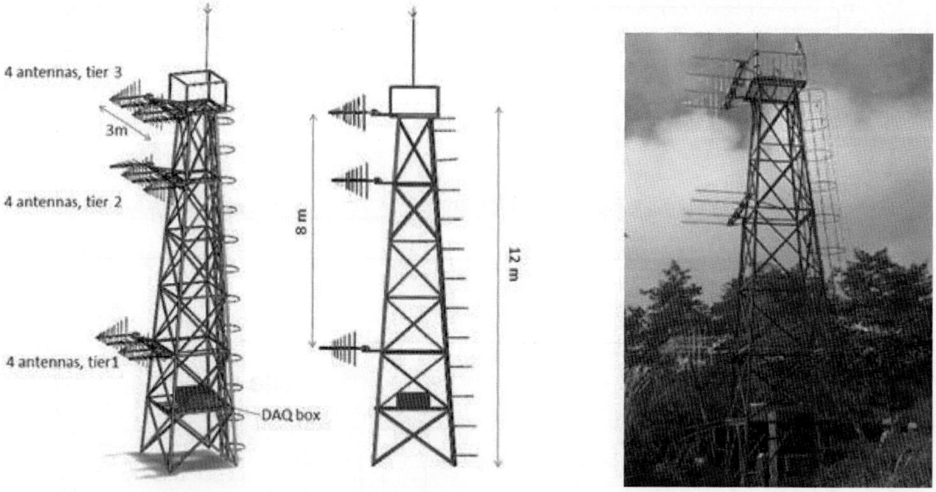

Fig. 3. The TAROGE-1 Station mechanical rendering (left) and figure after the installation completed (right).

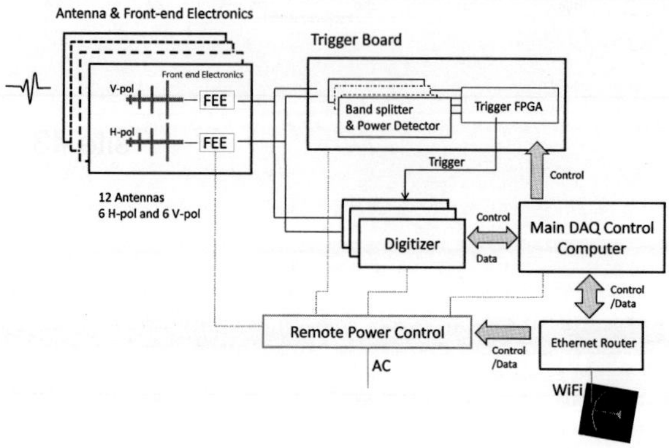

Fig. 4. System architecture of TAROGE-1.

for impulsive geo-synchrotron signals against thermal noises as well as CW anthropogenic-backgrounds. The thermal noise events are further reduced by requiring multiple antennas coincidence. Figure 5 shows the architecture of trigger system. The signal path in the trigger system is fed into four bands; 110–140 MHz, 170–200 MHz, 220–250 MHz and 270–300 MHz. Uncovered frequency regions between the bands are selected to avoid communication CW signals along the east coast of Taiwan. Incident increase of RF power is examined by a power detector and a comparator in each band. Individual thresholds of sub-band can be adjusted by multi-channel DACs through remote commands. The dominated emission power of the geo-synchrotron in Taiwan is V-pol because of a low inclination of Earth magnetic field. However, the Askaryan effect[14,15] in the extensive air-shower still produces a fair amount of power in H-pol.[4] Furthermore, the reflection coefficient

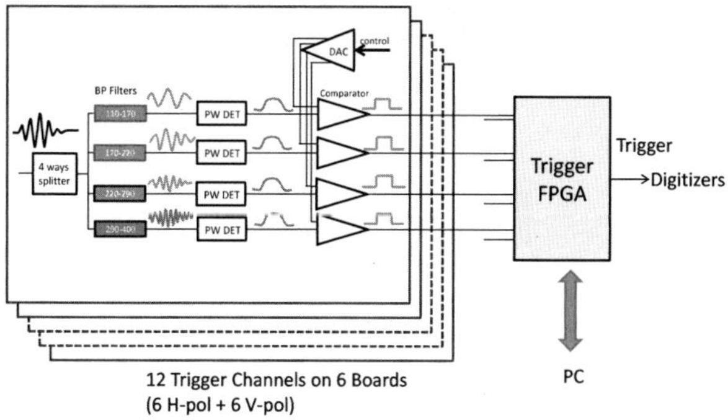

Fig. 5. Schematic diagram of trigger system of TAROGE-1.

of H-pol is much higher than V-pol especially in the low grazing angle which is less than 10°. Therefore, we designed the trigger to perform both in H-pol and V-pol, but independently.

The digitizer system uses affordable commercial digital oscilloscopes with the 300 MHz bandwidth and 8 bit ADCs. We operate the system with 1 GHz sampling rate for a record length of 1 μs. The record length is somewhat longer than the optimal configuration, which causes traffic in data transfer to the main computer, but is limited by the company's firmware. The maximum rate of event readout is ~7 Hz, which is the bottleneck of the TAROGE-1 DAQ. The data size is 50 KB/s (or 4.3 GB/day) after compression. Real-time data transfer and system control is available through the internet. We use a long range Wi-Fi repeater with 24 dBi dish antennas linking between the TAROGE station and the Edu-net at Heping Elementary School in the township. All active elements of the system are placed in Faraday boxes to ensure EMI shielding with shielding power better than −60 dB. Waterproof enclosures are also used to protect electronics and to prevent change of impedance of RF connectors for severe weather conditions.

A precise geometry survey was carried out to measure coordinates of antennas, which is crucial to ensure a good pointing capability. A laser theodolite with an arc-sec angular resolution was used to provide a precision of 1 mm. After completed the installation of station, system validation and calibration were followed using a calibration pulser system which consists of an impulsive pulse generator and a LPDA antenna. The transmitter power was varied using an adjustable attenuator, and polarization angle was varied by changing tip-tilt angle of an antenna mount on a tripod. The pulser signals were sent from multiple locations near the coastal line where its distance to the TAROGE-1 station is 3–5 km. Because the oscilloscope readout reached the saturation rate due to numerous numbers of impulsive noises from the township, triggering on the pulser events was challenging unless a high threshold signal-to-noise ratio (SNR > 10) was required.

Figure 6 (left) displays recorded waveforms of one of the pulser events. The timing calibration was performed using the pulser events. We used the cross-correlation

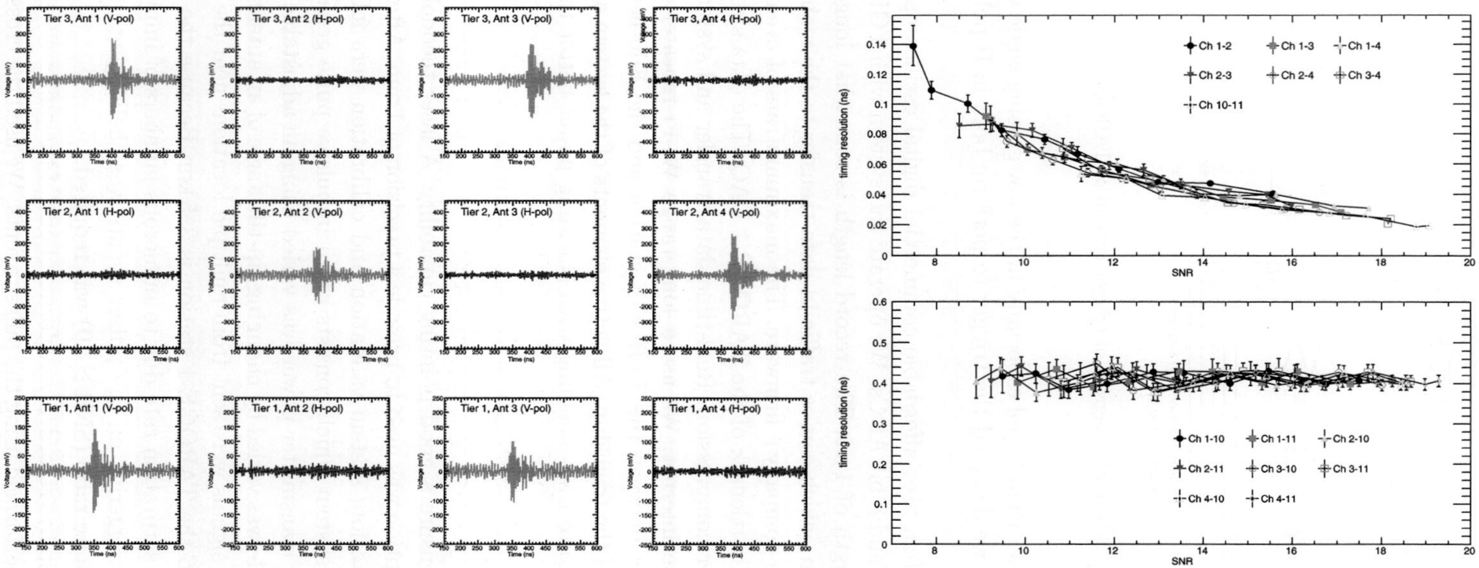

Fig. 6. Left: Pulser event waveforms (V-pol event), Right: Time resolution versus SNR: for channels on the same board (top) and for channels on different boards (bottom).

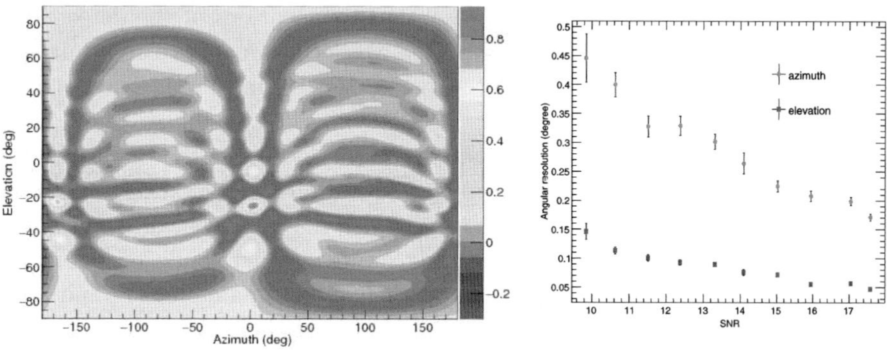

Fig. 7. Left: Interferometry image of the calibration pulser event, Right: Angular resolution versus SNR.

technique[16] to measure arrival time difference between antennas. As shown in Fig. 6 (right), we achieved an excellent timing resolution ~80 ps for SNR = 10 signals for antenna pairs connected to the same oscilloscope. The antenna pairs in different oscilloscopes has ~400 ps time resolution which is dominated by the trigger jitter of the oscilloscope.

We use the interferometry imaging method[17] to determine the arrival direction of the incoming RF signal. Figure 7(left) shows an interferometry image obtained from the calibration pulser event. Figure 7(right) shows an excellent angular resolutions, which is better than 0.15° in elevation angle and 0.45° for horizontal angle, are achieved.

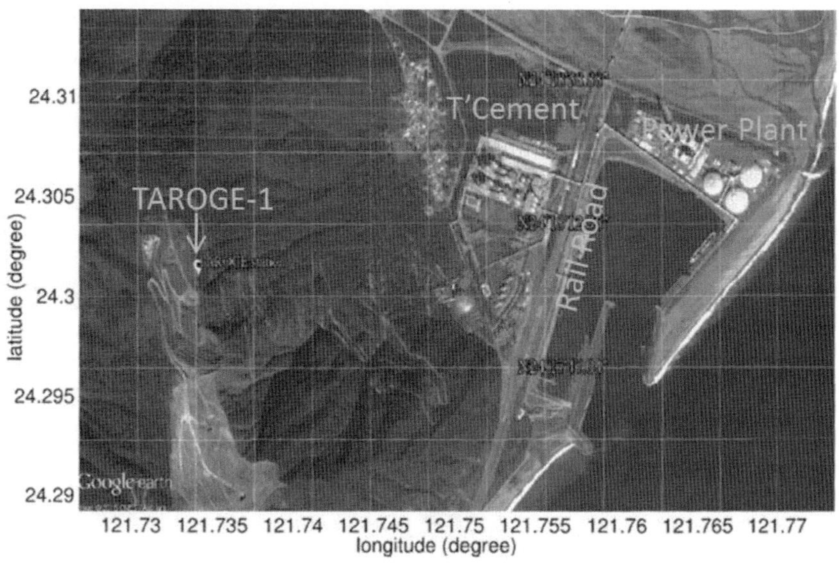

Fig. 8. (Color online) Map of locations of reconstructed impulsive noise events (red mark) superimposed on a map around the TAROGE site.

Figure 8 shows the distribution of reconstructed locations of the detected impulsive noise events in TAROGE-1 data. The event coordinate are determined by extrapolating the arrival directions of signal to the ground level. The result indicates that the most populated event (> 95%) locations are associated to two industrial facilities: one is the cement factory and the other is the power plant. A small portion of events are associated to transportations on the railway and roads.

TAROGE-1 has been operating smoothly since the beginning of August 2014. There are some nonoperating periods during severe weather conditions as like typhoons, power outages due to failure of power line, and maintenances and minor upgrades of the system. The overall duty cycle is more than 65%. The system operated remarkably stable with a duty cycle almost 100% during winter and spring.

3. The TAROGE-2 Station

In December 2015, the TAROGE-2 station was successfully built in the southern part of Yongshih Mt. at 1100 m elevation, located ~1000 m apart from the TAROGE-1 station. The TAROGE-2 site has a great advantage over TAROGE-1 in terms of the ambient RF noise. While TAROGE-1 faces to the Heping Township in where two big industrial facilities are located, TAROGE-2 faces to the ocean without any anthropogenic noise sources in the field of view. Hualian city is located 30 km away from the site in the south direction, or 100° off from the boresight of antennas, where the gain of antenna drops by more than 25 dB. There are three antenna towers in the TAROGE-2 station, which forms triangular radio array. Figure 9 shows the structure of antenna towers and its allocation. One of the towers (tower #1) was placed on the other platform which is located 5 m higher than the base platform. Two dual-polarized LPDA antennas were mounted in each tower with a 2.5 m spacing. Figure 10 shows a photo of TAROGE-2 station.

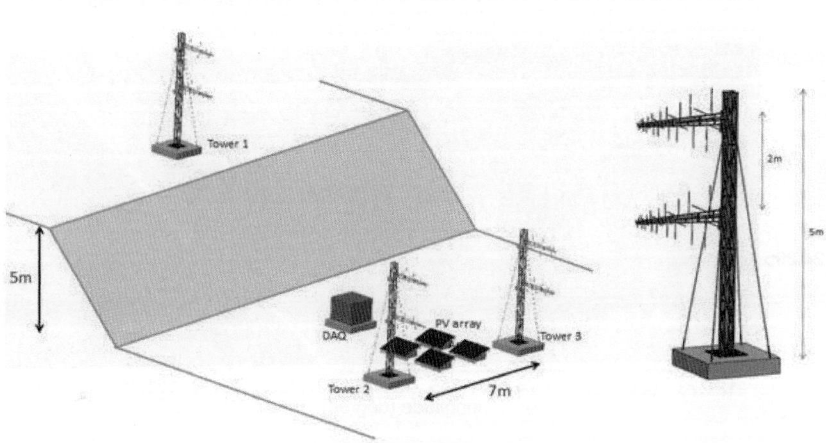

Fig. 9. Rendering of TAROGE-2 station.

Fig. 10. The TAROGE-2 Station after the deployment completed. The Tower-1 is not pictured.

The design of the TAROGE-2 station is conceptually the same as TAROGE-1. However, there are many of remarkable improvements; (1) enhanced antenna gain from 5 dBi to 7 dBi, (2) enlarged bandwidth from 300 MHz to 350 MHz, (3) new design of the trigger board to which is insensitive to temporal appearance of CW noises, (4) improved readout speed by five times allowing upto 40 Hz in DAQ, (5) improved time resolution with a new synchronization technique using the injected sine wave, and (6) improved pointing capability. The TAROGE-2 has a stand-alone power system because the site is located away from the electrical grid. We use a solar photovoltaic (PV) array for the 400 W maximum power.

Like TAROGE-1, the system validation and calibration using the calibration pulser were carried out after the installation completed. TAROGE-2 has been taking data smoothly since late December 2015. A prelimnary analysis of the calibration data shows an improved pointing resolution, 0.1° and 0.05° for the elevation angle and the horizontal angle, respectively. Detailed analyses of the calibration data and physics run data are in progress. The overall duty cycle is about 70% due to the power limit of the PV system. An upgrade using a wind turbine to compensate inefficiency of the PV system during cloudy days is planned for installation in May 2016.

4. Discussions

During 2012–2015, we have successfully completed the construction of two stations, TAROGE-1 for a prototype station which was aimed for systematic studies of RF ambient around eastern high mountain in Taiwan, and TAROGE-2 which was utilized for the cosmic ray detection. The isolated site for TAROGE-2 was found to be a promising location with a wide field of view and the minimal effect of impulsive anthropogenic backgrounds. However, the CW noise from Hualian city is still a remaining issue which degrades the system sensitivity. In order to minimize the effect of the CW noises, several upgrade plans, such as increased number of receivers with high gain antennas, increased number of frequency bands for the trigger, and exploring the low frequency below 80 MHz are under consideration. The initial simulation study expected an event rate as 45 events/(year·station) at the 1000 m observation altitude with assumptions of the 100% duty cycle and no CW effect. A detailed simulation study with the improved model of the geo-synchrotron emission based on the laboratory measurment[18] is in progress. The extension of TAROGE was funded by the Vanguard program of Ministry of Science and Technology (MOST), Taiwan, to build a total of seven stations in five years.

Acknowledgments

This work has been supported by the pioneer program of MOST, Taiwan. We gratefully thank Taiwan Cement Corporation and Heping Elementary school for their invaluable support.

References

1. Pierre Auger Collab., *Nucl. Instr. Methods A* **798** (2010) 172.
2. LOPES Collab. (H. Falcke *et al.*), *Nature* **435** (2005) 313.
3. ANITA Collab. (S. Hoover *et al.*), *Phys. Rev. Lett.* **105** (2010) 151101.
4. Pierre Auger Collab. (T. Huege *et al.*), *Nucl. Instrum. Methods A* **617** (2010) 484.
5. ANITA Collab. (P. Gorham *et al.*), *Phy. Rev. Lett.* **99** (2007) 171101.
6. ANITA Collab. (P. Gorham *et al.*), *Phy. Rev. Lett.* **103** (2009) 051103.
7. T. Huege and H. Falcke, *Astropart. Phys.* **24** (2005) 116.
8. M. Ludwig and T. Huege, *Astropart. Phys.* **34** (2011) 483.
9. O. Scholten, K. Werner and F. Rusydi, *Astropart. Phys.* **29** (2008) 94.
10. J. Alvarez-Muñiz, *et al.*, *Phys. Rev. D* **86** (2012) 123007.
11. T. Huege, M. Ludwig and C.W. James, *AIP Conf. Proc.* **1535** (2013) 128.
12. D. Heck *et al.*, FZKA Report 6019, Forschungszentrum Karlsruhe (1998).
13. T. Huege, R. Ulrich and R. Engel, *Astropart. Phys.* **27** (2007) 392.
14. G. A. Askaryan, *J. Exp. Theor. Phys.* **14** (1962) 441.
15. G. A. Askaryan, *J. Exp. Theor. Phys.* **21** (1965) 658.
16. ANITA Collab. (P. Gorham *et al.*), *Astropart. Phys.* **32** (2010) 10.
17. ANITA Collab. (A. Romero-Wolf *et al.*), *Astropart. Phys.* **60** (2015) 72.
18. T-510 Collab. (K. Belov *et al.*), arXiv:1507.0729.

Time to Move On?*

John Ellis

Theoretical Particle Physics and Cosmology Group,
Physics Department, King's College London,
London WC2R 2LS, United Kingdom

Theoretical Physics Department,
CERN, CH-1211 Geneva 23, Switzerland
John.Ellis@cern.ch

Cosmology and particle physics have long been dominated by theoretical paradigms: Einstein's general theory of relativity in cosmology and the standard model of particle physics. The time may have come for paradigm shifts. Does cosmological inflation require a modification of Einstein's gravity? Have experiments at the LHC discovered a new particle beyond the Standard Model? It is premature to answer these questions, but we theorists can dream about the possibilities.

Keywords: Cosmology; particle physics; inflation; LHC.

PACS Number(s): 04.25.dg, 04.30.Nk, 04.65.+e, 12.0.Dm, 12.60.Jv, 95.30.Sf, 98.80.Es

1. Introduction

It is a privilege to have been invited to give this talk at the end of a meeting celebrating the centenary of Einstein's general theory of relativity. The organizers undoubtedly expected me to talk about models of cosmological inflation, and most of this talk will indeed be about this subject, and whether it suggests a modification of the minimal Einstein–Hilbert action such as an R^2 term[1] or supergravity.[2] However, during this conference, we received dramatic news from CERN of hints of the possible existence of what might, perhaps be a new massive particle beyond the standard model.[3] In the words of the CERN Director-General, Fabiola Gianotti, we theorists "are allowed to be slightly excited",[4] and I will try to share some of this excitement at the end of my talk.[a]

*This article first appeared in *Int. J. Mod. Phys. D* **25** (2016) 1645014.

[a]The discovery of gravitational waves was announced[5] while I was writing up this talk, and I have added a note on constraints this discovery imposes on possible modifications of Einstein's theory.

2. Cosmological Inflation

For almost two decades now, the cosmological paradigm has been the ΛCDM model, based on the general theory of relativity with the addition of a cosmological constant Λ. Einstein supposedly regarded the introduction in 1917 of Λ as "the biggest blunder of his life",[6] but nowadays we would beg to differ. Its presence is a possibility allowed by the general principles of relativity, and hence is probably compulsory. The present-day cosmological constant has depressing implications for astronomers in the far future, who will only be able to study our local gravitationally-bound group of galaxies. However, according to the hypothesis of cosmological inflation,[7] the energy density of the very early universe may have been dominated by an (almost) constant term that would have generated an epoch of (near-) exponential expansion that could explain the great size and age of the universe today and, as an added bonus, have been the origin of the structures seen in the universe today, from clusters of galaxies to Donald Trump.

The most important evidence for cosmological inflation, and constraints on models, come from observations of fluctuations in the cosmic microwave background radiation (CMB), e.g. by the Planck satellite.[8] Together with other cosmological and astrophysical data, these CMB measurements are consistent with the cosmological concordance model with the following energy densities as fractions of that of a universe that is geometrically flat: $\Omega_\Lambda \simeq 0.69, \Omega_{\text{matter}} \simeq 0.31$, with most of the latter being cold dark matter (CDM).[9] I now discuss aspects of the inflationary paradigm and how it is constrained by CMB data.

2.1. *Slow-Roll inflationary models*

Most interpretations of the CMB data adopt the framework of slow-roll inflationary models,[7] in which the near-exponential expansion of the universe would have been driven by the almost constant field energy V of some inflaton scalar field ϕ:

$$\epsilon \equiv \frac{1}{2} M_P^2 \left(\frac{V'}{V}\right)^2, \quad \eta \equiv M_P^2 \left(\frac{V''}{V}\right) \ll 1. \tag{1}$$

Quantum fluctuations in ϕ would have led to scalar and tensor perturbations in the CMB, and the principal observables are the power-law tilt in the scalar spectrum n_s, and the tensor-to-scalar ratio r, which are given in the slow-roll approximation by

$$n_s = 1 - 6\epsilon + 2\eta, \quad r = 16\epsilon, \tag{2}$$

respectively. It is a generic prediction of slow-roll models that the CMB fluctuations should be Gaussian to a very good approximation.

There was great excitement when the BICEP2 experiment reported[10] the measurement of B-mode polarization in the CMB, which might have been generated by primordial tensor (quantum gravitational-wave) perturbations. However, these are now thought to be consistent with pollution by dust,[11] and a recent combined

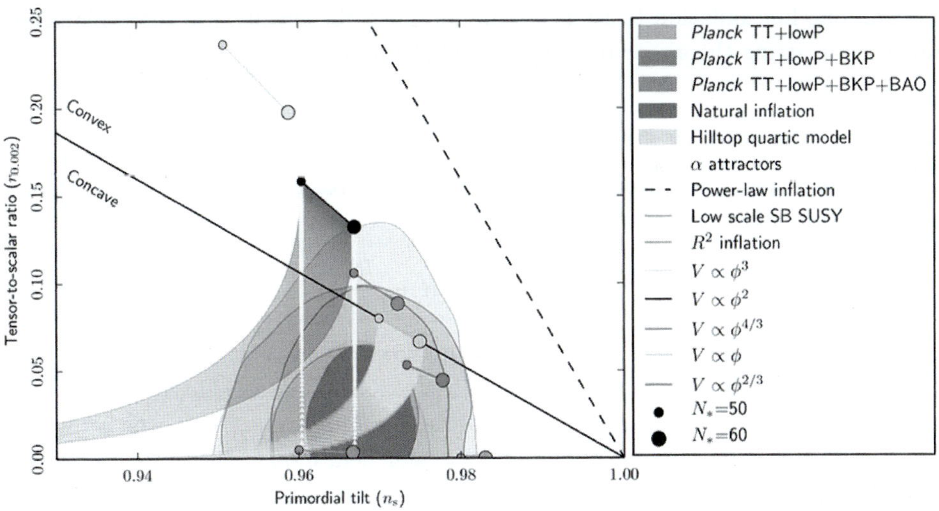

Fig. 1. The marginalized joint 68% and 95% CL regions for the tilt in the CMB scalar perturbation spectrum, n_s, and the relative magnitude of the tensor perturbations, r, obtained from the Planck 2015 data and their combinations with BICEP2/Keck Array and/or BAO data,[12] confronted with the predictions of some inflationary models.

analysis of BICEP2/Keck array and Planck data yields only an upper limit on $r < 0.1$,[12] as seen in Fig. 1, where we also see that $n_s \sim 0.97$. We also see in Fig. 1 that monomial single-field potentials $\propto \phi^n$ are disfavored by the data, whereas the original model of Starobinsky with an $R + R^2$ gravitational action[1] is highly compatible with the data, as would be inflation driven by the Higgs field[13] (with a nonmimimal coupling to gravity) or in no-scale supergravity models,[2] as discussed later. We also note that the experimental restriction on n_s, in particular, starts to provide an interesting constraint on the number of e-folds of inflationary expansion, N_*, which is sensitive to how the inflaton decayed into conventional matter,[14] as also discussed later.

2.2. *Challenges for inflationary models*

It is already a challenge to write down a simple model of inflation that is compatible with the CMB data, with simple power-law potentials being excluded as seen in Fig. 1, but there are other, deeper challenges.

How can one link inflation to 'low-energy' physics at colliders? Within the Standard Model of particle physics, the only candidate for the inflaton is the Higgs boson,[13] but naive extrapolation of standard model measurements to high energies indicates that the Higgs potential probably turns negative,[15] which is unsuitable for inflation.

Can one link inflation to some other well-motivated extension of the standard model? For example, could the inflaton be a supersymmetric partner of a right-handed (singlet) neutrino field,[16] or some sort of axion.[17]?

Can one link inflation to Planck-scale physics? Specifically, are there plausible inflatons in string theory, or in compactified string models?

2.3. *The Starobinsky model*

Let us first review the Starobinsky model,[1] which is based on a nonminimal action for general relativity:

$$S = \frac{1}{2}\int d^4x\sqrt{-g}\left(R + \frac{R^2}{6M^2}\right), \tag{3}$$

where M is some mass parameter that is unknown *a priori*. It would seem superficially that the action (3) does not contain a scalar inflaton field. However, when one makes a conformal transformation to the Einstein frame,[18] (in which the dependence on R is purely linear) one find a scalar field model with a characteristic potential:

$$S = \frac{1}{2}\int d^4x\sqrt{\tilde{g}}\left[\tilde{R} + (\partial_\mu\phi)^2 - \frac{3}{2}M^2(1 - e^{-\sqrt{\frac{2}{3}}\phi})^2\right]. \tag{4}$$

Scalar CMB perturbations were first calculated in this model by, Ref. 19, and their small magnitude requires $M \sim 10^{13}$ GeV $\ll M_P$, i.e. a coefficient of the R^2 term in (3) that is $\gg 1$ in natural units. The value of the scalar tilt n_s and the magnitude of the tensor-to-scalar ratio $r \sim 3 \times 10^{-3}$ are highly compatible with the available CMB data, as already seen in Fig. 1.

2.4. *Higgs inflation*

William of Occam might have suggested using the (only established) Higgs scalar h as the inflaton, which would require a nonminimal coupling to gravity[13]:

$$S = \frac{1}{2}\int d^4x\sqrt{-g}\left[-(1 + \xi h^2)R + (\partial_\mu h)^2 - \frac{\lambda}{2}(h^2 - v^2)^2\right], \tag{5}$$

where ξ is an unknown parameter that must be $\gg 1$, λ is the quartic Higgs self-coupling and v is the Higgs vev. When $1 \ll \xi \ll 10^{17}$, upon transforming to the Einstein frame one finds

$$S = \frac{1}{2}\int d^4x\sqrt{-\tilde{g}}\left[\tilde{R} + (\partial_\mu\chi)^2 - \frac{\lambda}{2\xi^2}(1 + e^{-\sqrt{\frac{2}{3}}\chi})^{-2}\right]. \tag{6}$$

The resulting effective potential gives successful inflation when $\chi \gg 1$, where the potential is similar to that of the Starobinsky model (though clearly different at small χ).

The required value of $\xi \sim 15000$, and one may ask how natural that is. However, a more serious practical problem is that one needs $\lambda > 1$ for field values beyond the Planck scale, whereas the best available calculations indicate that probably $\lambda < 0$

because of renormalization by the top quark.[15] One accurate calculation indicates that λ turns negative at a scale Λ:

$$\log \frac{\Lambda}{\text{GeV}} = 11.3 + 1.0\left(\frac{M_h}{\text{GeV}} - 125.66\right) - 1.2\left(\frac{m_t}{\text{GeV}} - 173.10\right)$$
$$+ 0.4\left(\frac{\alpha_3(M_Z) - 0.1184}{0.0007}\right). \quad (7)$$

Inserting the current experimental values of the Higgs mass $M_h = 125.09 \pm 0.24\,\text{GeV}$,[20] the top mass $m_t = 173.34 \pm 0.27 \pm 0.71\,\text{GeV}$[21] and the strong-interaction coupling $\alpha_3(M_Z) = 0.1177 \pm 0.0013$,[22] one estimates that the Higgs self-coupling λ turns negative at $\ln(\Lambda/\text{GeV}) = 10.0 \pm 1.0$.

Negative λ leads to metastability of the electroweak vacuum and the problem that most of the early universe would have been sucked into the region where the potential is negative.[23] Physics beyond the standard model is needed to avoid these issues, and supersymmetry is one of the possibilities.

2.5. *Inflation cries out for supersymmetry and (no-scale) supergravity*

There are many other motivations for supersymmetry, and inflationis one of them.[24] Models of inflation involve a scalar inflaton field that is elementary, at least at energies $\ll M_P$, and obtaining the right magnitude of perturbations requires a mass $\ll MP$ (typically $\sim 10^{13}\,\text{GeV}$) and/or a small inflaton self-coupling $\lambda \ll 1$. Both of these are technically natural in supersymmetric theories.

The first supersymmetric models of inflation were based on global supersymmetry. However, experience (general relativity, the standard model) teaches us that the only good symmetry is a local symmetry, and the same should apply to supersymmetry. The local version of supersymmetry necessarily involves local spacetime coordinate transformations, and is supergravity. Clearly also, any discussion of early Universe cosmology needs gravity, and supergravity is the way to combine gravity and supersymmetry.

However, there is an issue with generic supergravity models when matter is included, which is that their effective potentials typically have 'holes' with depths $\sim \mathcal{O}(M_P^2)$. These make any treatment of cosmology problematic, let alone inflation. There is, however, an exception, namely the class of no-scale supergravity models.[25] The effective potentials in this class of supergravity models have flat directions, and the scalar potential resembles that in globally supersymmetric models, with controlled corrections.

An added motivation for considering no-scale supergravity models is that they arise naturally in compactifications of string theory.[26] Therefore, no-scale models of inflation may offer some prospects for making contact with some underlying variant of string theory.

2.6. *No-scale supergravity models of inflation*

The first no-scale supergravity models of inflation appeared in the 1980s,[27,28] largely motivated by the 'unholy' form of the effective potential. This approach was revived in 2013[2] in response to the new, high-precision CMB data from Planck. Supergravity models are characterized by a Hermitian Kähler potential K that specifies the internal geometry of the scalar fields ϕ, whose kinetic terms are $(\partial^2 K/\partial\phi^i\partial\phi_j^*)\partial_\mu\phi^i\partial^\mu\phi_j^*$. The simplest no-scale inflationary model contains two complex scalar fields:

$$K = -3\ln\left(T + T^* - \frac{|\phi|^2}{3}\right) \tag{8}$$

and the potential interactions between the scalar fields are characterized by a superpotential W, which we assumed to have the form

$$W = \frac{\mu}{2}\phi^2 - \frac{\lambda}{3}\phi^3, \tag{9}$$

as first studied by Wess and Zumino.[30] We assumed that the field T, which could be interpreted as a modulus of string compactification, is fixed by string dynamics: $T = c/2$, in which case the effective Lagrangian for the inflaton field ϕ is

$$\mathcal{L}_{\text{eff}} = \frac{c}{\left(c - \frac{|\phi|^2}{3}\right)^2}|\partial_\mu\phi|^2 - \frac{\hat{V}}{\left(c - \frac{|\phi|^2}{3}\right)^2} : \quad \hat{V} = \left|\frac{\partial W}{\partial\phi}\right|^2. \tag{10}$$

We note that the prefactors in the Lagrangian terms in (10) modify the form of the effective potential that one would have obtained in the globally-supersymmetric case, when one transforms to a canonically-normalized inflaton field.

The Wess–Zumino superpotential gives good inflation already in the global case,[29] and we found this to be possible also in its no-scale supergravity incarnation for $\lambda/\mu \sim 1/3$,[2] as seen in Fig. 2. Indeed, for the specific case $\lambda = \mu/3$, shown as the black line in Fig. 2, the effective scalar potential is *identical* with that in the original Starobinsky model ((3), (4)).

Looking beyond the Starobinsky limit, one can consider generic potentials of the form[31]

$$V = A(1 - \delta e^{-B\phi} + \mathcal{O}(e^{-2B\phi})). \tag{11}$$

In such models, the most prominent inflationary predictions are

$$n_s = 1 - 2B^2\delta e^{B\phi_*}, \quad r = 8B^2\delta^2 e^{-2B\phi_*}, \quad N_* = \frac{1}{B^2\delta}e^{B\phi_*}, \tag{12}$$

where Φ_* is the value of the inflaton field that yields N_* e-folds of inflation. It is clear from (12 that n_s is in a one-to-one relationship with N_*, and that r may be adjusted, for any fixed value of N_*, by varying the coefficient B in the exponential in (11).[b]

[b]Similar results are found in the so-called α-attractor models.[32]

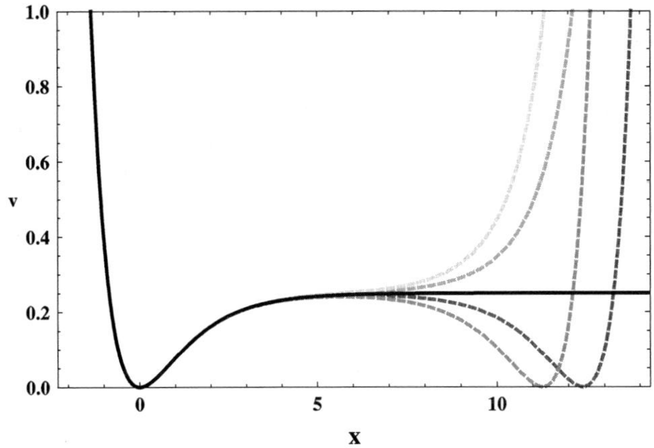

Fig. 2. (Color online) The inflationary potential V in the Starobinsky $R+R^2$ model[1] (solid black line) compared with its form in various no-scale models[2] (dashed colored lines).

As an example of how this freedom could be exploited, one can consider models in which the no-scale property is shared among several moduli fields[31]:

$$K \ni -\sum_i \ln(T_i + T_i^*) : \quad N_i > 0, \ \sum_i N_i = 3, \tag{13}$$

as is characteristic of generic string compactifications. In this case, if the inflaton is identified with T_i one has

$$B = \sqrt{\left(\frac{2}{N_i}\right)}, \quad r = \frac{4N_i}{N_*^2} \tag{14}$$

with the specific choice of a single modulus with $N_i = 3$ corresponding to the Starobinsky model (3, 4). A measurement of r could therefore, within this no-scale supergravity framework, provide a unique window on the phenomenology of string compactification.

2.7. *How many e-folds of inflation?*

The discussion above shows that the number of e-folds, N_*, is a key to the phenomenology of inflationary models, and we also see in Fig. 1 that the CMB data are beginning to provide interesting constraints on N_*.[14] This can be related to the potential V_* at ϕ_* corresponding to the scale k_*:

$$N_* = 67 - \ln\left(\frac{k_*}{a_0 H_0}\right) + \frac{1}{4}\ln\left(\frac{V_*^2}{M_P^2 \rho_{\rm end}}\right) + \frac{1 - 3w_{\rm int}}{12(1 + w_{\rm int})}\ln\left(\frac{\rho_{\rm reh}}{\rho_{\rm end}}\right) - \frac{1}{12}\ln g_{\rm th}, \tag{15}$$

where $\rho_{\rm end,reh}$ are the densities at the end of inflation and after reheating, respectively, $w_{\rm int}$ is the average equation-of-state parameter during inflaton decay and reheating, and $g_{\rm th}$ is the number of degrees of freedom after reheating.

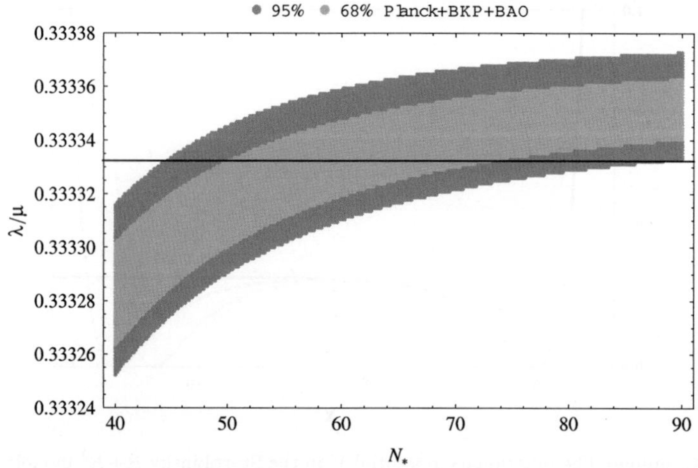

Fig. 3. (Color online) The 68% and 95% CL regions (yellow and blue, respectively) in the $(N_*, \lambda/\mu)$ plane for the no-scale inflationary model[2] with a matter inflaton field and the Wess–Zumino superpotential.[14] The horizontal black line is for $\lambda/\mu = 1/3$, the value that reproduces the inflationary predictions of the Starobinsky model.[2]

2.8. *Constraints on inflaton decay*

The value of N_* and hence r, in particular, is sensitive to the inflaton decay rate Γ_ϕ, since the ratio $\rho_{\rm reh}/\rho_{\rm end} \propto \Gamma_\phi^2$. Thus a measurement of r can constrain models of inflaton decay. This point is illustrated in Fig. 3 for the Starobinsky-like no-scale supergravity models of inflation whose potentials are shown in Fig. 2. In particular, for the special case $\lambda/\mu = 1/3$ that reproduces exactly the Starobinsky model, we see that $N_* \geq 44$ at the 95% CL and ≥ 50 at the 68% CL. If the dominant inflaton decay is via a trilinear superpotential coupling y into two-body final states, the inflaton decay rate $\Gamma_\phi = |y|^2/8\,\pi m_\phi$, and these constraints on N_* correspond to the constraints $y > 10^{-16}$ at the 95% CL and $> 10^{-7}$ at the 68% CL.[14]

These bounds are to be compared with those due to upper limits on y from the production in inflaton decays of gravitinos with mass $m_{3/2}$.[14] We find

$$|y| \lesssim 1.5 \times 10^{-6} \quad \text{for } m_{3/2} = 6\,\text{TeV}, \tag{16}$$

from requiring that long-lived gravitinos do not mess up the success of standard Big-Bang nucleosynthesis (BBN) calculations, weakening for larger $m_{3/2}$ and disappearing altogether if the gravitino is sufficiently heavy to decay before BBN. There is also a constraint

$$|y| < 2.7 \times 10^{-5} \times \left(\frac{100\,\text{GeV}}{m_{\rm LSP}}\right), \tag{17}$$

from the upper limit on the contribution of the lightest supersymmetric particle (LSP) to the cosmological cold dark matter. The latter constraint, in particular, suggests that $|y| \lesssim 10^{-5}$ for LSP masses in the range of a few hundred GeV,

The constraint on the inflaton decay coupling y from the CMB measurement of r is not yet competitive with the other constraints (16, 17), but this can be expected to change in the future.

3. Meanwhile, Back at the LHC

During 2015, the LHC restarted operations, making collisions at a center-of-mass energy of 13 eV, much higher than the 7 Tev and 8 TeV during its first run. The two experiments looking for new high-mass physics, ATLAS and CMS, have each accumulated ~3/fb of high-energy data. They presented their first preliminary data during this conference, and created a sensation.

Both experiments reported statistical excesses in their $\gamma\gamma$ invariant-mass distributions that could perhaps, maybe, be hints of a possible new particle X with mass ~750 GeV decaying into two photons.[3] The significances in the experiences are only $\sim 3\,\sigma$, and are diluted by the 'look-elsewhere' effect that the experiments examined many $\gamma\gamma$ invariant-mass distributions (and many other distributions!), so the overall significance falls far short of the 5-σ 'gold standard' required to claim discovery of a new particle, particularly one so unexpected. If it exists, the X particle would herald a layer of new physics beyond the standard model.

3.1. *Interpretation of the $X(750)$ hint*

The CMS and ATLAS data suggest a product of cross-section and $\gamma\gamma$ branching ratio ~6/fb, and most analyses assume that it is a scalar or pseudoscalar. The most plausible possibility is that it decays to $\gamma\gamma$ via triangular loops of heavy charged particles, most likely fermions weighing $\gtrsim$400 GeV.[33] The production of the $X(750)$ is also often thought to be via loops of massive colored fermions. These could not be a fourth generation of standard model particles, and minimal supersymmetric models are also unable to explain the magnitude of the X 'signal'. There would have to be a whole new world out there!

Four simple possibilities for these extra fermions are: (i) a single vector-like charge 2/3 quark, (ii) an isospin doublet of vector-like charge 2/3 and -1/3 quarks, (iii) an isodoublet and two singlet charge 2/3 and $-1/3$ quarks, and (iv) a complete vector-like generation including leptons as well as quarks.[33] Figure 4 illustrates the possible couplings λ in these four models of X to the vector-like fermions (assumed, for simplicity, to be universal) as functions of their masses (also assumed, for simplicity, to be universal) if they are to reproduce the possible 'signal' (assuming that the dominant X decays are those mediated by fermion loops). It is worth noting that model (iv) contains a weakly-interacting neutral fermion that is a possible dark matter candidate.

In each of these models, triangular diagrams generate decays of the X particle to other pairs of vector bosons: gluon–gluon, $Z\gamma$, ZZ and W^+W^-, at rates relative to $\gamma\gamma$ that are easily calculable and depend on the model.[33] It is premature to exclude any of these models, though model (ii) is under pressure from searches at

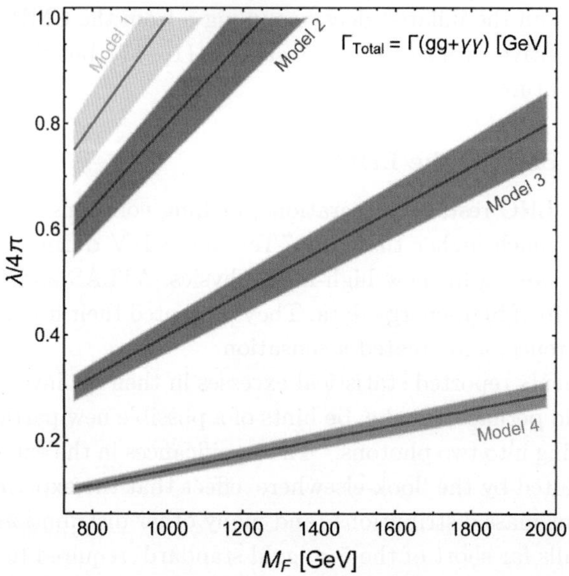

Fig. 4. Values of the vector-like fermion mass m_F and coupling λ (both assumed to be universal) required models (i), (ii), (iii) and (iv) to accommodate the possible $gg \to \Phi \to \gamma\gamma$ signal reported by CMS and ATLAS.[3] The black lines are for the central value of the cross-section, 6 fb, and the colored bands represent the 1-σ uncertainties. Figure from Ref. 34, adapted from Ref. 33.

the LHC. In addition to these decays, the LHC experiments should also look for the heavy fermions that are required. There will be theoretical and experimental work for a generation, *if* the $X(750)$ exists. Since the LHC should make many more collisions in 2016, we can expect to know soon whether the $X(750)$ 'signal' is real. In the meanwhile, we theorists are "allowed to be a little excited".[4,c]

3.2. *What's in it for inflation?*

The appearance of another scalar particle would provide another candidate for the inflaton, and one can consider inflation via a nonminimal X coupling to gravity,[35] just as was discussed for Higgs inflation in Sec. 2.4. The appearance of the X particle could also be used to revive Higgs inflation, or even lead one to consider a two-field model of inflation. If it is real, the $X(750)$ particle could revolutionize inflationary model-building as well as collider physics.

4. Summary

As we saw in Sec. 2, many popular models of inflation postulate some modification of the minimal Einstein–Hilbert action: maybe an R^2 term as in the Starobinsky,

[c]However, we should also remember the words of Laplace, "Plus un fait est extraordinaire, plus il a besoin d'être appuyé de fortes preuves." ("The more extraordinary a claim, the stronger the proof required to support it.")

maybe a nonminimal coupling to a scalar field as in Higgs (or X!) inflation, or maybe promote general relativity to supergravity. It is ironic that, 100 years after the proposal of general relativity, at a time when experiment finally confirms its key prediction of gravitational waves, the time may have come to move on from Einstein. As we saw in Sec. 3, if the possible hint, maybe, of a new particle X with mass $\sim$750 GeV would, if confirmed, require us to move on from the standard model of particle physics, potentially also with interesting implications for inflationary model-building.

5. Future Study

The discovery of gravitational waves,[5] which was announced while this write-up was being completed, is a tremendous vindication of Einstein's theory of general relativity.[36] Perhaps this and similar future observations will give us hints how to extend it? In the meantime, the gravitational-wave data already provide some constraints on possible modifications of Einstein's theory, including an upper limit on the graviton mass[36] $\sim 10^{-22}$ eV, limits on Lorentz violation[37,38] and the violation of the principle of equivalence,[39] and the possible observation of a γ-ray flash (if confirmed) would tells us that electromagnetic and gravitational waves travel at the same velocity[37,40] to within 10^{-17}. *Une affaire à suivre!*

Acknowledgments

The author was supported partly by the London Centre for Terauniverse Studies (LCTS), using funding from the European Research Council via the Advanced Investigator Grant 26732, and partly by the STFC Grant ST/L000326/1. The author thanks Pisin Chen and his team for their kind hospitality during the Second LeCosPA Symposium, and thanks the Universidad de Antioquia, Medellín, for its hospitality while writing up this talk, using grant FP44842-035-2015 from Colciencias (Colombia).

References

1. A. A. Starobinsky, *Phys. Lett. B* **91** (1980) 99.
2. J. Ellis, D. V. Nanopoulos and K. A. Olive, *Phys. Rev. Lett.* **111** (2013) 111301; *Phys. Rev. Lett.* **111** (2013) 12, 129902, arXiv:1305.1247 [hep-th].
3. J. Olsen, CMS physics results from Run 2 presented on Dec. 15th, 2015; M. Kado, ATLAS physics results from Run 2 presented on Dec. 15th, 2015, https://indico.cern.ch/event/442432/; ATLAS Collab., ATLAS-CONF-2015-081, https://atlas.web.cern.ch/Atlas/GROUPS/PHYSICS/CONFNOTES/ATLAS-CONF-2015-081/; CMS Collab., CMS PAS EXO-15-004, https://cds.cern.ch/record/2114808/files/EXO-15-004-pas.pdf.
4. F. Gianotti, talk to CERN staff, 18th January (2016).
5. LIGO Scientific Collab. and Virgo Collab. B. P. Abbott *et al.*, *Phys. Rev. Lett.* **116** (2016) 6, doi:10.1103/PhysRevLett.116.061102, arXiv:1602.03837 [gr-qc]
6. G. Weinstein, arXiv:1310.1033 [physics.hist-ph].

7. J. Martin, C. Ringeval and V. Vennin, *Phys. Dark Univ.* **5–6** (2014) 75–235, arXiv:1303.3787 [astro-ph.CO].
8. Planck Collab. (P. A. R. Ade *et al.*), arXiv:1502.02114 [astro-ph.CO].
9. Particle Data Group Collab. (K. A. Olive *et al.*), *Chin. Phys. C* **38** (2014) 090001, doi:10.1088/1674-1137/38/9/090001.
10. BICEP2 Collab. (P. A. R. Ade *et al.*), *Phys. Rev. Lett.* **112** (2014) 24, doi:10.1103/PhysRevLett.112.241101, arXiv:1403.3985 [astro-ph.CO].
11. Planck Collab. (R. Adam *et al.*), *Astron. Astrophys.* **586** (2016) A133, doi:10.1051/0004-6361/201425034, arXiv:1409.5738 [astro-ph.CO].
12. BICEP2 and Keck Array Collab. (P. A. R. Ade *et al.*), arXiv:1510.09217 [astro-ph.CO].
13. F. L. Bezrukov and M. Shaposhnikov, *Phys. Lett. B* **659** (2008) 703, arXiv:0710.3755 [hep-th].
14. J. Ellis, M. A. G. Garcia, D. V. Nanopoulos and K. A. Olive, *J. Cosmol. Astropart Phys.* **1507** (2015) 07, doi:10.1088/1475-7516/2015/07/050, arXiv:1505.06986 [hep-ph].
15. D. Buttazzo, G. Degrassi, P. P. Giardino, G. F. Giudice, F. Sala, A. Salvio and A. Strumia, *J. High Energy Phys.* **1312** (2013) 089, arXiv:1307.3536 [hep-ph].
16. H. Murayama, H. Suzuki, T. Yanagida and J. Yokoyama, *Phys. Rev. Lett.* **70** (1993) 1912.
17. F. C. Adams, J. R. Bond, K. Freese, J. A. Frieman and A. V. Olinto, *Phys. Rev. D* **47** (1993) 426, arXiv:hep-ph/9207245.
18. K. S. Stelle, *Gen. Relativ. Gravit.* **9** (1978) 353; B. Whitt, *Phys. Lett. B* **145** (1984) 176.
19. V. F. Mukhanov and G.V. Chibisov, *JETP Lett.* **33** (1981) 532.
20. ATLAS and CMS Collab. (G. Aad *et al.*), *Phys. Rev. Lett.* **114** (2015) 191803, doi:10.1103/PhysRevLett.114.191803, arXiv:1503.07589 [hep-ex].
21. ATLAS, CDF, CMS and D0 Collabo., arXiv:1403.4427 [hep-ex].
22. D. d'Enterria and P. Z. Skands (eds.), arXiv:1512.05194 [hep-ph].
23. M. Fairbairn and R. Hogan, *Phys. Rev. Lett.* **112** (2014) 201801, doi:10.1103/PhysRev Lett.112.201801, arXiv:1403.6786 [hep-ph]; A. Hook, J. Kearney, B. Shakya and K. M. Zurek, *J. High Energy Phys.* **1501** (2015) 061, doi:10.1007/JHEP01(2015)061, arXiv:1404.5953 [hep-ph].
24. J. R. Ellis, D. V. Nanopoulos, K. A. Olive and K. Tamvakis, *Phys. Lett. B* **118** (1982) 335; R. Holman, P. Ramond and G. G. Ross, *Phys. Lett. B* **137** (1984) 343.
25. E. Cremmer, S. Ferrara, C. Kounnas and D. V. Nanopoulos, *Phys. Lett. B* **133** (1983) 61.
26. E. Witten, *Phys. Lett. B* **155** (1985) 151.
27. A. S. Goncharov and A. D. Linde, *Class. Quantum. Gravit.* **1** (1984) L75.
28. J. R. Ellis, K. Enqvist, D. V. Nanopoulos, K. A. Olive and M. Srednicki, *Phys. Lett. B* **152** (1985) 175; Erratum-*ibid.* **156B** (1985) 452.
29. D. Croon, J. Ellis and N. E. Mavromatos, *Phys. Lett. B* **724** (2013) 165, arXiv:1303.6253 [astro-ph.CO].
30. J. Wess and B. Zumino, *Phys. Lett. B* **49** (1974) 52, doi:10.1016/0370-2693(74)90578-4.
31. J. Ellis, D. V. Nanopoulos and K. A. Olive, *J. Cosmol. Astropart. Phys.* **1310** (2013) 009, arXiv:1307.3537.
32. R. Kallosh and A. Linde, *J. Cosmol. Astropart. Phys.* **1307** (2013) 002, doi:10.1088/1475-7516/2013/07/002, arXiv:1306.5220 [hep-th].
33. J. Ellis, S. A. R. Ellis, J. Quevillon, V. Sanz and T. You, arXiv:1512.05327 [hep-ph].
34. A. Djouadi, J. Ellis, R. Godbole and J. Quevillon, arXiv:1601.03696 [hep-ph].

35. M. Dhuria and G. Goswami, arXiv:1512.06782 [hep-ph]; Y. Hamada, T. Noumi, S. Sun and G. Shiu, arXiv:1512.08984 [hep-ph]; S. F. Ge, H. J. He, J. Ren and Z. Z. Xianyu, arXiv:1602.01801 [hep-ph].
36. LIGO Scientific and Virgo Collab. (B. P. Abbott *et al.*), arXiv:1602.03841 [gr-qc].
37. J. Ellis, N. E. Mavromatos and D. V. Nanopoulos, arXiv:1602.04764 [gr-qc].
38. A. Kostelecky and M. Mewes, arXiv:1602.04782 [gr-qc].
39. E. O. Kahya and S. Desai, arXiv:1602.04779 [gr-qc].
40. X. Li, F. W. Zhang, Q. Yuan, Z. P. Jin, Y. Z. Fan, S. M. Liu and D. M. Wei, arXiv:1602.04460 [astro-ph.HE].

PARALLEL SESSION

I. Cosmology — Connections with Fundamental Theories

Summary of Parallel Session: "Cosmology — Connections with Fundamental Theories"

Yoji Koyama

Physics Division, National Center for Theoretical Sciences, Hsinchu, 30013, Taiwan

In the past years, there has been a great deal of activities trying to connect cosmology with fundamental aspects of physical laws. Some of the more interesting works have been presented in this parallel session.

1. Introduction

Cosmological inflation is a leading paradigm in the study of very early universe. It not only provides a solution to the flatness and the horizon problems for the observed homogeneity and isotropy of the current universe, it also gives prediction to an almost scale-invariant, Gaussian primordial power spectrum which is in good agreement with the recent precise CMB measurements.

From the viewpoint of theoretical particle physics, inflation provides a possible connection between cosmology and fundamental theory of particle physics which is presumed to describe the universe at its very beginning. However, this is not without challenges. First of all, the standard slow-roll inflation scenario requires almost flat inflaton potential. The slow-roll conditions suffer from the fine-tuning problem due to huge quantum corrections at Planck scale physics, e.g. the eta problem and this shows strong UV sensitivity of particle physics model of inflation. Second, if the primordial gravitational wave is detected, the tensor to scalar ratio fixes the energy scale of inflation which is far above the energy scale that the LHC can reach and possibly around the GUT energy scale. An important consequence of the large tensor to scalar ratio is that, in single field slow-roll inflation, it requires super-Planckian inflaton field excursion via the Lyth bound. This poses a challenge for constructing viable inflation models, since it is difficult to protect the flatness and the smallness of the potential from quantum corrections over super-Planckian field range in effective field theory framework. Thus the large tensor-to-scalar ratio might require the knowledge of physics near the Planck scale.

The non-Gaussianity parameter will soon be measured to the accuracy of $\delta f_{NL} \simeq 1$. This will further restrict the theory space of possible inflation models. With a great advantage of inflationary cosmology that it will be probed in more detail by the future observations, it would help us to improve our understandings of fundamental theory of our universe.

2. Summary of the CF session

In the CF session, the following topics were discussed: string theory and inflation, anisotropic inflation, the infrared quantum effects in de-Sitter space and in a general accelerating universe, and phase transition in inflationary universe. Here we list the speakers of the CF session in order of appearance.

- Yoji Koyama: Dilaton, IR effect in dS space and IR-driven inflation
- Toshifumi Noumi: Effective Field Theory of Anisotropic Inflation
- Fang Ye: Enhancing the Axion Field Range through Mixings
- Ryo Namba: Scale-dependent gravitational waves from a rolling axion
- Hiroyuki Kitamoto: Stochastic Dynamics of Infrared Fluctuations in Accelerating Universe
- Sichun Sun: Echoes of Inflationary Particle Phase Transition in the CMB

Due to the strong UV sensitivity of inflation, there has been extensive studies to construct inflation models based on string theory. The talks of Ye and Namba were about axion. Axion, which is ubiquitous in string theory, has several implications in cosmology. Especially, string axion is an interesting candidate which could source the inflation since the potential of axion is protected from radiative corrections by the shift symmetry.

Motivated by the BICEP results of having a large tensor to scalar ratio r in 2014, Ye proposed a new mechanism to enlarge the axionic field range through mixings to reconcile large field inflation which corresponds to models with super-Planckian inflaton field and string theory where the axionic field range is restricted to be sub-Planckian. This mechanism is effective in the presence of the Stueckelberg mixing.

Namba presented a very interesting model where an axion field is coupled to a chiral gauge field. His question is how robust the relation $P \propto H^2$ for the gravitational power spectrum would be upon detection. He pointed that the axionic coupling to a chiral gauge field would be the another source of gravitational wave in addition to vacuum fluctuations of graviton. He discussed the visibility of the gravitational wave produced by his mechanism in the observation.

Our current universe is highly isotropic over the current horizon scale and hence inflation is usually assumed to be isotropic. However, there is still a room for tiny anisotropy in the inflationary observable such as the scalar power spectrum according to the Planck observation. Noumi presented a general framework of effective field theory to deal with anisotropic inflation. He developed the effective field theory description for spacetime symmetry breaking for anisotropic inflation. It would be interesting to derive the corresponding low energy theorem, similar to the low energy theorem for soft pion scattering amplitudes in field theory.

During inflation, spacetime is de-Sitter like. While inflation would be a phenomenon at very high energy scale, one of the most important features of inflation is from the infrared fluctuations of massless fields on superhorizon scale which generate almost scale invariant power spectra. Koyama and Kitamoto looked at the infrared effect of quantized gravity in de-Sitter space.

Koyama presented an interesting model where the cosmological constant changed, due to the infrared quantum effect of de-Sitter space on the vacuum expectation value of dilaton, slowly with time. Unlike standard slow-roll inflation model which suffers from the eta problem due to huge UV correction, the UV effect in his model is orthogonal to the IR effect and the slow-roll inflation driven by the infrared effect is free from the eta problem and does not rely on a slow-rolling inflaton potetial.

Kitamoto looked at another aspect of the infrared dynamics. It is known that the leading late time behaviour of the secular growing infrared effects in de-Sitter space can be captured by the stochastic method. Kitamoto extended the framework of the stochastic method in de-Sitter space to a general accelerating universe and he was able to derive the corresponding Langevin equation. Moreover he found that semiclassical description for a scalar field leads to the same stochastic equation as in the de Sitter case, provided an appropriate time coordinate is adopted.

In a usual inflationary scenario it is assumed that during inflation the equation of state is that of vacuum with zero temperature and that cosmological phase transition and thermalization of the universe have occurred in the end of reheating era after inflation. Sun presented a scenario of cosmological evolution where there happened cosmological phase transitions around the inflation era. She showed that these inflationary phase transitions leave its imprint on the primordial gravitational power spectrum and provide a new way to probe the electroweak and GUT dynamics through the CMB B-mode polarization.

3. Conclusion

In this parallel session, the connections of cosmology, especially inflationary cosmology, with fundamental theory of particle physics have been discussed from various aspects. String theory, as a consistent theory of quantum gravity, provides much of the motivation and inspiration for new effects that may resolve some of the long standing issues in cosmology. Some of these issues have been addressed in this session by making use of string theory inspired models. It is gratifying to see so many new and promising developments have been reported and shared here.

Acknowledgments

The author would like to thank Chong-Sun Chu for proofreading the manuscript.

Dilaton, IR Effect in dS Space and IR-Driven Inflation

Chong-Sun Chu[1,2] and Yoji Koyama[1]

[1]*Physics Division, National Center for Theoretical Sciences, National Tsing-Hua University, Hsinchu, 30013, Taiwan*

[2]*Department of Physics, National Tsing Hua University, Hsinchu, 30013, Taiwan*

It is known that infrared (IR) quantum fluctuations in de Sitter (dS) space could break the dS symmetry and generate time dependent effects. In this talk, we consider a dilaton-gravity theory. We find that gravitational IR effects lead to a time dependent shift on the vev of the dilaton and results in a screening (temporal) of the cosmological constant/Hubble parameter. In the Einstein frame, taking the dilaton as inflaton, we obtain an inflationary expansion of the slow roll kind. This inflation is driven by the IR quantum effects of dS gravity and does not rely on the use of a slow roll potential. As a result, our model is free from the eta problem which baffle the standard slow roll inflation models.

Keywords: Inflation, time-dependent effect in dS space, non-minimally-coupled gravity, dynamical renormalization group.

1. Introduction

It is widely believed that the universe has underwent a period of accelerated expansion in the early cosmology. Such a period of inflation not only solves the flatness and horizon problems of the standard cosmology, but also, with the introduction of an almost flat inflaton potential, predicts a nearly scale invariant density perturbation. This picture is in excellent agreement with observational results of the cosmic microwave background (CMB) and the large scale structure of the universe. However this simple picture is not without problem as it has proven extremely difficult to bring together the inflationary paradigm with fundamental particle physics. For example, Planck mass suppressed corrections to the inflaton potential generally lead to large corrections to the inflaton mass, resulting in a large slow roll parameter η which renders prolonged slow roll inflation impossible. One may resort to symmetries such as supersymmetry, global symmetries or higher dimensional gauge symmetries to protect the potential.

Dark energy presents another deep mystery of the universe. The current cosmological constant is of the order of $10^{-120}M_P^2$. The deep mystery of the cosmological constant problem[1] is to understand why is there such a huge hierarchy of scales between the current cosmological constant and the Planck scale. It is natural to suspect that a good understanding of the quantum properties of dS space would be necessary in order to tackle this problem properly. It has been conjectured some time ago that IR quantum effects in dS space could lead to a kind of screening to the cosmological constant[2,3] and provide a resolution to the cosmological constant problem. Explicit dS breaking IR effect has been identified[4–7] and demonstrated to lead to a weakening effect (over time) on the cosmological constant.[8]

In this work, we have studied the IR quantum effect from graviton loop in a dilator-gravity theory to see time-dependent effect on the cosmological constant. We realized that instead of trying to use the IR effects of graviton to explain the magnitude of potential, we can also use it to provide an explanation of the slope of the potential, explaining its "slow-rollness of inflation", with the advantage that the eta problem is avoided completely.

2. Key idea

We consider a dilator-gravity theory described by the action[9]

$$S = \int \sqrt{-g} d^4x \left[\frac{M^2}{2} e^{-2\phi/\eta} R - \frac{1}{2} g^{\mu\nu} \partial_\mu \phi \partial_\nu \phi - V(\phi) \right]. \tag{1}$$

Here M is the fundamental scale of the theory and η controls the coupling of dilaton ϕ to gravity. The dilaton has a potential but its detailed form is not important to us. We will only need to assume that the theory admits a minimum where the dilaton is taking a vacuum expectation value (vev) $\phi = v, \quad v \in \mathbf{R}$, and the corresponding potential energy is positive $V(v) > 0$; thereby giving rises to a dS metric.

As is well known for dS space, the quantum fluctuations of a massless minimally coupled (MMC) scalar field grows linearly with time and breaks the dS symmetry.[4] In fact the propagator of a MMC scalar is IR divergent and breaks the dS symmetry explicitly.[5–7] Part of the graviton excitation modes are MMC and so the time dependent IR effects inherent in them could generate time dependent effects on other physical quantities of theory. We identify such an effect on the vev of the dilaton field. We find that the IR effects of the gravitational loops (one loop order) induce a time dependence in the vev $v \longrightarrow v_{\text{eff}} := v + \delta v(t)$, where $\delta v \propto (H_0 t)^2$ and H_0 is the Hubble parameter.

In Einstein frame, this time dependent vev modifies the cosmological constant, $S_E \sim \int d^4x \sqrt{-g_E} V(v) e^{-v_{\text{eff}}/\eta}$. This brings in a time dependence (screening) in the cosmological constant.

3. Perturbative analysis

We consider the action in string frame given by (1). The dilaton potential $V(\phi)$ admits a vacuum at $\phi = v$ such that $V'(v) + \frac{4}{\eta} V(v) = 0$ and defines a dS background:

$$g_{\mu\nu} = a_0^2 \eta_{\mu\nu}, \qquad a_0 = -\frac{1}{H_0 \tau}, \qquad H_0^2 = \frac{1}{3M^2} e^{2v/\eta} V(v) \tag{2}$$

where H_0 is the Hubble constant and τ is the conformal time $\tau = -\frac{1}{H_0} e^{-H_0 t}$.

We evaluate the graviton one-loop corrections to the vev of the dilaton scalar field in the dS background (2). The diagrams to be considered here are shown in Fig. 1. We have expanded the scalar field around v, $\phi = v + \tilde{\phi}$, then the one-loop

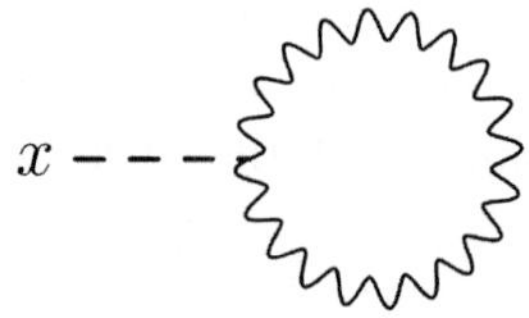

Fig. 1. The one-loop tadpole diagrams due to the graviton loops. The broken line denotes the scalar propagator and the wavy line denotes the graviton propagator.

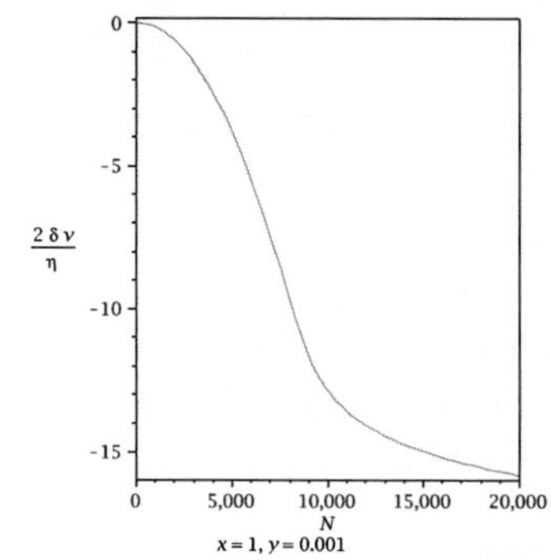

Fig. 2. Plot of $\frac{2\delta v}{\eta}$ against N.

tadpole diagrams generating the vev of $\tilde{\phi}$ correspond to a shift of v,

$$\delta v = \langle \Omega | \tilde{\phi}(x) | \Omega \rangle. \tag{3}$$

Graviton perturbations in dS space can be categorized into MMC modes and conformally coupled modes.[10,11] We are interested in the time dependent part of the correction (3). For this we need only to consider the propagation of MMC mode of the graviton in the loop, which generates time dependent IR logarithm, $\ln a_0$.

We will use the in-in formalism (Schwinger-Keldysh formalism)[12–15] to calculate the vev of $\tilde{\phi}$. One loop result for the effective vev is obtained as

$$v_{\text{eff}} = v + \delta v = v - \frac{\zeta}{\eta}\frac{3H_0^2}{4\pi^2} \ln^2 a_0(\tau) + \text{sub-leading}, \tag{4}$$

where $\zeta = \left(1 + 6\frac{M^2}{\eta^2}e^{-2v/\eta}\right)^{-1/2}$ and sub-leading terms include $\ln a_0, 1/a_0$, etc.

It is clear that the 1-loop result (4) is valid only for $|\epsilon \ln^2 a_0| \ll 1$, $\epsilon = \frac{\zeta^2}{v\eta}\frac{3H_0^2}{4\pi^2}$. The perturbative calculation eventually breaks down at the late time and a resummation of the secular terms is called for. We currently do not have a good enough understanding of the technology to determine precisely the behaviour of the secular IR effects of graviton loops. In this work we will employ the dynamical renormalization group (DRG) method[16,17] to improve the result (4). The DRG effectively resums the leading order time-dependent corrections and requires only small ϵ in order to be valid. The result of DRG is

$$v_{\text{eff}}(\tau) = r + \frac{\eta}{2}W\left(\frac{6M^2}{\eta^2}e^{-2r/\eta}\right), \qquad r := v - \frac{3M^2}{\eta}e^{-2v/\eta} - \frac{3H_0^2N^2}{4\pi^2\eta}. \tag{5}$$

Here N is defined by $N := \ln a_0 = H_0 t$ and $W(z)$ is the Lambert-W function.

The one loop IR effects of the graviton loops induces a time dependent shift δv in the vev and modifies the Hubble parameter in the string frame to [a]

$$H^2 = H_0^2 e^{\frac{2\delta v}{\eta}} \tag{6}$$

[a]Note that we have taken into account the time dependence only in the vev in the exponent,

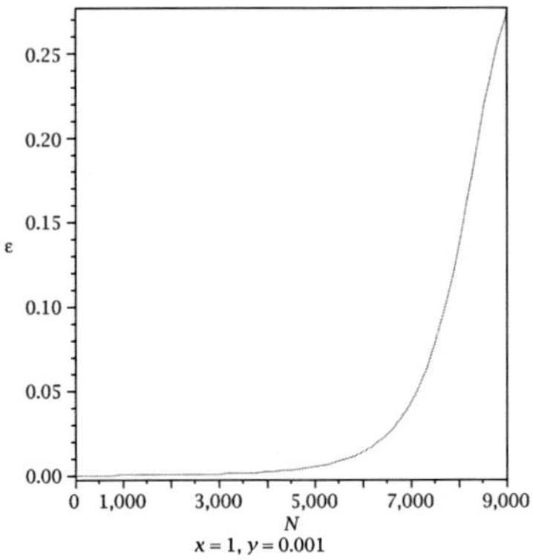

Fig. 3. Plot of ϵ_S against N.

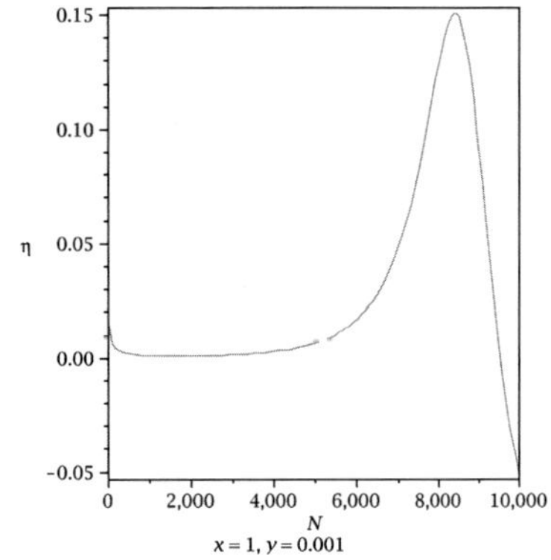

Fig. 4. Plot of η_S against N.

The time dependent exponent $2\delta v/\eta$ is parametrized by

$$x := \frac{Me^{-v/\eta}}{H_0}, \qquad y := \frac{Me^{-v/\eta}}{\eta}. \tag{7}$$

A typical plot of $2\delta v/\eta$ as a function of N is shown in Fig. 2.

Since H changes with time due to the backreaction from the graviton IR effect on the background, we can compute the "slow roll" parameters

$$\epsilon_S := \frac{d}{dt}H^{-1}, \qquad \eta_S := \frac{1}{H}\frac{d\ln\epsilon_S}{dt} \tag{8}$$

to measure the size of the backreaction. We need

$$\epsilon_S, \eta_S \ll 1 \tag{9}$$

in order to trust the quantum field theory computation. Given x and y, ϵ_S and η_S stay very small for a large range of N from the initial time, and then increase rapidly at around N of the order of $N \sim xy^{-1}\ln y^{-1}$. Typical plots of ϵ_S and η_S against N are shown in the Figures 3 and 4.

4. Slow-roll inflation from IR effects

To examine the physical effects of the time dependent vev, we need to go back to the Einstein frame by performing a Weyl scaling of the string frame metric

$$g_{\mu\nu} = g^E_{\mu\nu}e^{2(\frac{\varphi}{\eta}+\beta)}, \tag{10}$$

and we have neglected that in the dilaton potential denoted. Typically, the potential receives a contribution of order (UV cut off)^4 from the zero point energy fluctuation. This term is much bigger than the other UV or IR corrections to the potential. So the dominant time dependence comes from the exponential prefactor.

where φ denote the dynamical part of ϕ above the vev, $\phi = v_{\text{eff}} + \varphi$ and $\beta = \delta v/\eta$. We obtain the Planck mass $M_P = Me^{-\frac{v}{\eta}}$ and the Hubble parameter

$$H_E = H_0 e^{\frac{2\delta v}{\eta}} \tag{11}$$

As δv is always negative for $\eta > 0$, we get a screening of the Hubble constant and the Cosmological constant.

The Einstein frame slow-roll parameters are obtained as

$$\epsilon_E := \frac{d}{dt_E} H_E^{-1} = 2e^{\frac{\delta v}{\eta}} \epsilon_S, \qquad \eta_E := \frac{1}{H_E \epsilon_E} \frac{d\epsilon_E}{dt_E} = e^{\frac{\delta v}{\eta}} (\eta_S - \epsilon_S). \tag{12}$$

Note that N is equal to the number of e-folding N_E in the Einstein frame $N_E := H_E dt_E = N$. The Einstein slow roll parameters will be small as long as we can trust the quantum field theory calculation in dS space. Therefore we are in a regime of parameters (9) that we can trust the quantum field theory computations, the corresponding cosmology in the Einstein frame describes a slow roll inflation with $\epsilon_E, \eta_E \ll 1$.

5. Summary and Discussion

Our result is summarized as follows:

- Slow roll inflation is achieved without a slow roll potential: In our model, the dilaton field is sitting at the minimum of the quantum corrected effective potential. Change of vacuum energy is not due to a rolling of the dilaton field as in model with an inflaton, but is due to the time dependent IR effect of graviton loops on the position of the minimum. In contrast to the simplest slow roll inflation, here inflation is driven by the IR effects of the gravitons themselves.
- No η problem in our model: Inflaton moved slowly by IR effects of graviton and the IR effect is orthogonal to the UV corrections. η problem is resolved in our model.

To discuss the ending of inflation and reheating in our model, it is necessary to include in our model the coupling of the dilaton field to matter fields. We speculated that the IR effect on the dilaton-matter coupling may acts effectively like a damping term and provide a mechanism for the ending and reheating of the inflation.

The dilaton-gravity sector of our model is specified by the Planck mass scale and two dimensionless parameters, ratio of initial Hubble constant with respect to the Planck mass, and the ratio of the dilaton gravity coupling constant with respect to the Planck mass. It is interesting to study the other signatures of the model, such as the tensor scalar ratio r and other features of the power spectrum.

Acknowledgments

We would like to thank Toshiaki Fujimori, Satoshi Iso, Hiroshi Isono, Shoichi Kawamoto, Yoshihisa Kitazawa, Richard Woodard and Jackson Wu for valuable discussions. This work is supported in part by the National Center of Theoretical Science (NCTS) and the grant 101-2112-M-007-021-MY3 of the Ministry of Science and Technology of Taiwan.

References

1. See for example, S. Weinberg, "The Cosmological Constant Problem," Rev. Mod. Phys. **61** (1989) 1.
   J. Polchinski, "The Cosmological Constant and the String Landscape," hep-th/0603249.
2. A. M. Polyakov, "Phase Transitions And The Universe," Sov. Phys. Usp. **25** (1982) 187 [Usp. Fiz. Nauk **136** (1982) 538].
   A. M. Polyakov, "dS space and eternity," Nucl. Phys. B **797** (2008) 199 [arXiv:0709.2899 [hep-th]].
3. L. H. Ford, "Quantum Instability of dS Space-time," Phys. Rev. D **31** (1985) 710.
4. A. Vilenkin and L. H. Ford, "Gravitational Effects upon Cosmological Phase Transitions," Phys. Rev. D **26** (1982) 1231.
5. N. A. Chernikov and E. A. Tagirov, "Quantum theory of scalar fields in dS space-time," Annales Poincare Phys. Theor. A **9**, 109 (1968).
6. B. Allen and A. Folacci, "The Massless Minimally Coupled Scalar Field in dS Space," Phys. Rev. D **35**, 3771 (1987).
7. B. Allen, "The Graviton Propagator in Homogeneous and Isotropic Space-times," Nucl. Phys. B **287**, 743 (1987).
8. N. C. Tsamis and R. P. Woodard, "Quantum gravity slows inflation," Nucl. Phys. B **474**, 235 (1996) [hep-ph/9602315].
9. M. Gasperini, F. Piazza and G. Veneziano, "Quintessence as a runaway dilaton," Phys. Rev. D **65** (2002) 023508 [gr-qc/0108016].
   T. Damour, F. Piazza and G. Veneziano, "Violations of the equivalence principle in a dilaton runaway scenario," Phys. Rev. D **66** (2002) 046007 [hep-th/0205111].
10. N. C. Tsamis and R. P. Woodard, "The Structure of perturbative quantum gravity on a dS background," Commun. Math. Phys. **162** (1994) 217.
11. H. Kitamoto and Y. Kitazawa, "Soft Gravitons Screen Couplings in dS Space," Phys. Rev. D **87**, 124007 (2013) [arXiv:1203.0391 [hep-th]].
12. K. c. Chou, Z. b. Su, B. l. Hao and L. Yu, "Equilibrium and Nonequilibrium Formalisms Made Unified," Phys. Rept. **118**, 1 (1985).
13. E. Calzetta and B. L. Hu, "Closed Time Path Functional Formalism in Curved Space-Time: Application to Cosmological Back Reaction Problems," Phys. Rev. D **35**, 495 (1987).
14. R. D. Jordan, "Effective Field Equations for Expectation Values," Phys. Rev. D **33**, 444 (1986).
15. S. Weinberg, "Quantum contributions to cosmological correlations," Phys. Rev. D **72**, 043514 (2005) [hep-th/0506236].
16. D. Boyanovsky, H. J. de Vega, R. Holman and M. Simionato, "Dynamical renormalization group resummation of finite temperature infrared divergences," Phys. Rev. D **60** (1999) 065003 [hep-ph/9809346].
    D. Boyanovsky and H. J. de Vega, "Dynamical renormalization group approach to relaxation in quantum field theory," Annals Phys. **307** (2003) 335 [hep-ph/0302055].
17. C. P. Burgess, L. Leblond, R. Holman and S. Shandera, "Super-Hubble dS Fluctuations and the Dynamical RG," JCAP **1003** (2010) 033 [arXiv:0912.1608 [hep-th]].

Stochastic Dynamics of Infrared Fluctuations in Accelerating Universe

Gihyuk Cho and Cook Hyun Kim
School of Physics and Astronomy, Seoul National University, Seoul 151-747, Korea
whrlsos@snu.ac.kr and protozerta@snu.ac.kr

Hiroyuki Kitamoto
Division of Physics and Astronomy, Graduate School of Science, Kyoto University, Kyoto 606-8502, Japan
kitamoto@tap.scphys.kyoto-u.ac.jp

In the presence of massless and minimally coupled scalar fields in accelerating universes, quantum fluctuations at super-horizon scales make vacuum expectation values of field operators growing with time. From a semiclassical viewpoint, it was proposed that these infrared effects can be described by a Langevin equation. In de Sitter space, it has been proved that the stochastic approach is equivalent to the leading power resummation of the growing time dependences. We extend the resummation derivation of the Langevin equation in a general accelerating universe. The resulting Langevin equation contains a white noise term as a common feature in accelerating universes, while the coefficient of each term is modified by the slow-roll parameter. Furthermore we show that the same stochastic equation can be derived from the semiclassical description of the scalar fields as far as we adopt an appropriate time coordinate.

Keywords: Accelerating universe, super-horizon, Langevin equation.

1. Introduction

In accelerating universes, the degrees of freedom at super-horizon scales increase with time. This increase gives rise to a growing time dependence to the propagator of a massless and minimally coupled scalar field. It comes from the fact that quantum fluctuations of the scalar field are frozen at super-horizon scales.

In interacting field theories with the scalar field, quantum corrections acquire time dependences through the propagator. We call them the Infrared (IR) effects in accelerating universes. The IR effects at each loop level manifest as polynomials in a factor growing with time. At late times, the leading IR effects come from the leading powers of the factor at each loop level. This fact indicates that perturbation theories are eventually broken if accelerated expansion continues long enough.

From a semiclassical description of the scalar field, A. A. Starobinsky and J. Yokoyama proposed that the IR effects can be evaluated nonperturbatively by a Langevin equation.[1] As for the IR effects in de Sitter (dS) space, N. C. Tsamis and R. P. Woodard proved that the stochastic approach is equivalent to the leading power resummation of them.[2]

We extend the resummation formula of the leading IR effects in a general accelerating universe.[3] Although there were some attempts to derive the stochastic equation on a general background,[4] the consistency with the resummation formula was not verified completely. We find that in a general accelerating universe, the Yang–Feldman equation is reduced to a Langevin equation with a white noise term up to the leading IR effects. Furthermore we show that the choice of time coordinate is crucial to derive the same equation from the semiclassical description of the scalar field.

2. Free scalar field

In this study, we consider a homogeneous, isotropic and spatially flat universe:

$$ds^2 = -dt^2 + a^2(t)d\mathbf{x}^2, \tag{1}$$

where the dimension of the spacetime is four and the Hubble parameter is positive: $H \equiv \dot{a}/a > 0,\ \dot{} = \partial_t$. To describe deformations from dS space, we introduce the slow-roll parameter whose parameter region is given by

$$\epsilon \equiv \frac{-\dot{H}}{H^2}, \quad 0 \leq \epsilon < 1. \tag{2}$$

The lower limit corresponds to dS space and the upper bound means an accelerating universe.

For a massless and minimally coupled scalar field, the quadratic action is

$$S_2 = \int \sqrt{-g}d^4x \left[-\frac{1}{2}g^{\mu\nu}\partial_\mu\varphi\partial_\nu\varphi \right]. \tag{3}$$

The free scalar field φ_0 can be expanded by the annihilation and the creation operators $a_{\mathbf{p}}$, $a^\dagger_{\mathbf{p}}$ as

$$\varphi_0(x) = \int \frac{d^3p}{(2\pi)^3} \left[a_{\mathbf{p}}\phi_{\mathbf{p}}(x) + a^\dagger_{\mathbf{p}}\phi^*_{\mathbf{p}}(x) \right]. \tag{4}$$

We focus on contributions from super-horizon scales where curved spacetimes-specific effects are dominant. At the IR scales, the wave function behaves as[5]

$$\phi_{\mathbf{p}}(x) \simeq -i\frac{2^{\nu_*-1}\Gamma(\nu_*)}{\sqrt{\pi}} \frac{\{(1-\epsilon_*)H_*a^{\epsilon_*}\}^{\frac{1}{1-\epsilon_*}}}{p^{\nu_*}} e^{+i\mathbf{p}\cdot\mathbf{x}}, \tag{5}$$

where $_*$ means the time of horizon crossing: $H_*a_* = p$ and $\nu = \frac{3}{2} + \frac{\epsilon}{1-\epsilon} \geq \frac{3}{2}$. Since each parameter is evaluated at the time of horizon crossing, the leading IR term of the wave function depends on p rather than on t. This fact is consistent with the freezing of quantum fluctuations at super-horizon scales.

The contribution from super-horizon scales to the propagator is written as

$$\langle \varphi_0^2(x) \rangle \simeq \int_{p_0}^{Ha} dp \ \frac{2^{2\nu_*-3}\Gamma^2(\nu_*)}{\pi^3} \frac{\{(1-\epsilon_*)H_*a^{\epsilon_*}\}^{\frac{2}{1-\epsilon_*}}}{p^{2\nu_*-2}}. \tag{6}$$

The upper bound is set on the horizon because the spectrum is dominant at super-horizon scales. We fix the minimum value of the comoving momentum. Physically speaking, a/p_0 means a size of the Universe and it expands with the scale factor. Super-horizon scales can be represented by the inequality: $p_0 < Ha = \dot{a}$. During accelerated expansion: $\ddot{a} > 0$, the degrees of freedom at super-horizon scales increase with time. Since the spectrum is constant, this increase directly leads to a growing time dependence of the propagator.

Changing variables: $p = H'a'$, the propagator (6) can be rewritten as

$$\langle \varphi_0^2(x) \rangle = \int_{t_0}^{t} dt' \; \frac{(2-2\epsilon')^{2\nu'-3}\Gamma^2(\nu')}{\pi^3}(1-\epsilon')^3 H'^3, \tag{7}$$

where each parameter with $'$ is evaluated at t'. The integrand is positive during accelerated expansion and then the time integral induces a growing time dependence.

3. Resummation formula

By employing the Schwinger–Keldysh formalism,[6] we can investigate interacting field theories on a time dependent background. Each vertex is accompanied by one retarded propagator: $G^R(x,x') = \theta(t-t')[\varphi_0(x), \varphi_0(x')]$ starting from it, and the other propagators are the Wightman functions.

The Wightman functions are growing with time as reviewed in the previous section. The retarded propagator has no secular growth in itself but the vertex integral with it induces a secular growth. Even though coupling constants are small, quantum corrections are amplified by these growing factors and then perturbative investigations break down eventually.

As a nonperturbative approach, we extend N. C. Tsamis and R. P. Woodard's study[2] in a general accelerating universe. In an interacting field theory whose nonlinear term is given by a potential, the Yang–Feldman equation is written as

$$\varphi(x) = \varphi_0(x) - i\int \sqrt{-g'}d^4x' \; G^R(x,x')\frac{\partial}{\partial\varphi}V(\varphi(x')). \tag{8}$$

This equation describes quantum effects from all regions of momentum scale. From it, we extract the equation which is valid up to the leading IR effects.

As seen in (5)–(6), the secular growths of the Wightman functions originate in

$$\begin{aligned}\varphi_0(x) \simeq \bar{\varphi}_0(x) \equiv \int \frac{d^3p}{(2\pi)^3}\;\theta(Ha-p)\\ \times\Big[-i\frac{2^{\nu_*-1}\Gamma(\nu_*)}{\sqrt{\pi}}\frac{\{(1-\epsilon_*)H_*a^{\epsilon_*}\}^{\frac{1}{1-\epsilon_*}}}{p^{\nu_*}}e^{+i\mathbf{p}\cdot\mathbf{x}}a_{\mathbf{p}} + (\text{h.c.})\Big].\end{aligned} \tag{9}$$

Here we introduced the step function which is nonzero at super-horizon scales.

For the retarded propagator, its leading IR behavior is given by

$$G^R(x,x') \simeq -i\theta(t-t')\Big[\int_{t'}^{t} dt'' \; a''^{-3}\;\Big]\delta^{(3)}(\mathbf{x}-\mathbf{x}'). \tag{10}$$

The contribution comes from the product of 1 and $\int^t dt' \, a'^{-3}$, which are independent solutions of the equation of motion at $p \to 0$: $\ddot{u} + 3H\dot{u} = 0$.

Substituting (9) and (10), the Yang–Feldman equation (8) is reduced to

$$\varphi(x) = \bar{\varphi}_0(x) - \int^t dt' \left(a'^3 \int_{t'}^{\infty} dt''\ a''^{-3}\right) \frac{\partial}{\partial \varphi} V(\varphi(t', \mathbf{x})). \tag{11}$$

Here we neglected $\int^t dt' \left(a'^3 \int_t^{\infty} dt''\ a''^{-3}\right) \partial V/\partial\varphi(\varphi(t', \mathbf{x}))$, which is subdominant at late times. The remaining integral induces a secular growth since $\left(a'^3 \int_{t'}^{\infty} dt''\ a''^{-3}\right)$ does not decay rapidly at far past. Thus the time dependence of the nonlinear term appears only in the upper bound of the integral. Differentiating both sides with respect to t, we obtain the Langevin equation with the white noise term:

$$\dot{\varphi}(x) = \dot{\bar{\varphi}}_0(x) - \left(a^3 \int_t^{\infty} dt'\ a'^{-3}\right) \frac{\partial}{\partial \varphi} V(\varphi(x)),$$
$$\langle \dot{\bar{\varphi}}_0(t, \mathbf{x}) \dot{\bar{\varphi}}_0(t', \mathbf{x}) \rangle = \frac{(2 - 2\epsilon)^{2\nu - 3} \Gamma^2(\nu)}{\pi^3} \left\{(1 - \epsilon) H\right\}^3 \delta(t - t'). \tag{12}$$

The white noise term comes from the fact that the time dependence of $\bar{\varphi}_0$ appears only through the step function as seen in (9).

The Langevin equation can be translated to the Fokker-Planck equation:

$$\dot{\rho}(t, \phi) = \frac{AH^3}{2} \frac{\partial^2}{\partial \phi^2} \rho(t, \phi) + \left(a^3 \int_t^{\infty} dt'\ a'^{-3}\right) \frac{\partial}{\partial \phi} \Big(\rho(t, \phi) \frac{\partial}{\partial \phi} V(\phi)\Big),$$
$$A = \frac{(2 - 2\epsilon)^{2\nu - 3} \Gamma^2(\nu)}{\pi^3} (1 - \epsilon)^3 > 0. \tag{13}$$

The positivity of the coefficient of the second derivative term means that the free propagator is growing with time. Using the probability density ρ, the vacuum expectation value (vev) of an arbitrary field operator can be evaluated as

$$\langle F(\varphi(x)) \rangle = \int_{-\infty}^{\infty} d\phi\ \rho(t, \phi) F(\phi), \quad F\text{: arbitrary function}, \tag{14}$$

where ϕ is not an operator but a c-number.

The most interesting objects are eventual contributions to vevs of field operators. To evaluate them, we need to solve the Fokker-Planck equation for a final state. If the time variations of H, ϵ are negligible during the IR effects grow, the final state solution is evaluated as

$$\rho_{\infty}(t, \phi) = N^{-1} \exp\Big(- \frac{2}{AH^3} \left(a^3 \int_t^{\infty} dt'\ a'^{-3}\right) V(\phi)\Big), \tag{15}$$

where N is a normalization factor.

4. Semiclassical description

We show that the Langevin equation (12) can be derived also from the semiclassical description of the scalar field. The basic procedure follows the previous studies[1,4] but there is a slight modification.

We start with the equation of motion:

$$\Big(\frac{\partial}{\partial t^2}+3H\frac{\partial}{\partial t}-\frac{1}{a^2}\frac{\partial^2}{\partial \mathbf{x}^2}\Big)\varphi(x)=-\frac{\partial}{\partial\varphi}V(\varphi(x)). \tag{16}$$

Since our interest is about physics at late times, we may neglect the second derivative terms. It should be emphasized that the process depends on the choice of time coordinate.

For a general choice of time coordinate: $dT = \mathcal{H}dt$, the temporal part of d'Alembert operator is written as

$$\frac{\partial}{\partial t^2}+3H\frac{\partial}{\partial t}=\mathcal{H}^2\Big(\frac{\partial^2}{\partial T^2}+\frac{\dot{\mathcal{H}}+3H\mathcal{H}}{\mathcal{H}^2}\frac{\partial}{\partial T}\Big). \tag{17}$$

To compare $\frac{\partial}{\partial T}\varphi$ with $\frac{\partial^2}{\partial T^2}\varphi$ directly, we choose the time coordinate T as its friction coefficient is constant

$$\frac{\dot{\mathcal{H}}+3H\mathcal{H}}{\mathcal{H}^2}=\mu_0\text{: const.} \tag{18}$$

A solution of the equation is given by

$$\mathcal{H}=\frac{1}{\mu_0}\big(a^3\int_t^\infty dt'\ a'^{-3}\big)^{-1}. \tag{19}$$

Here the boundary condition is set as the amplitude of T monotonically increases with time. In contrast to the previous studies,[1,4] we neglect the second derivative terms of (16) with respect to T, $\mathbf{x}$ rather than t, $\mathbf{x}$:

$$\mu_0\mathcal{H}^2\frac{\partial}{\partial T}\varphi(x)=\big(a^3\int_t^\infty dt'\ a'^{-3}\big)^{-1}\frac{\partial}{\partial t}\varphi(x)=-\frac{\partial}{\partial\varphi}V(\varphi(x)). \tag{20}$$

Let us recall that in accelerating universes, more degrees of freedom go outside of the horizon with cosmic evolution. Based on this perspective, we divide the scalar field into super-horizon modes and sub-horizon modes: $\varphi(x)=\bar{\varphi}(x)+\varphi_{\rm UV}(x)$, and identify sub-horizon modes $\varphi_{\rm UV}$ as sources of super-horizon modes $\bar{\varphi}$. To be more precise, we set $\varphi_{\rm UV}$ as

$$\begin{aligned}\varphi_{\rm UV}(x)\equiv\int\frac{d^3p}{(2\pi)^3}\ \theta(p-Ha)\\ \times\left[-i\frac{2^{\nu_*-1}\Gamma(\nu_*)}{\sqrt{\pi}}\frac{\{(1-\epsilon_*)H_*a^{\epsilon_*}\}^{\frac{1}{1-\epsilon_*}}}{p^{\nu_*}}e^{+i\mathbf{p}\cdot\mathbf{x}}a_{\mathbf{p}}+(\text{h.c.})\right],\end{aligned} \tag{21}$$

and keep its time derivative in (20):

$$\big(a^3\int_t^\infty dt'\ a'^{-3}\big)^{-1}\frac{\partial}{\partial t}\{\bar{\varphi}(x)+\varphi_{\rm UV}(x)\}=-\frac{\partial}{\partial\bar{\varphi}}V(\bar{\varphi}(x)). \tag{22}$$

It should be noted that the sign of the step function variable in (21) is opposite to that in (9). Considering the identity: $\dot{\varphi}_{\rm UV}=-\dot{\bar{\varphi}}_0$, we obtain

$$\dot{\bar{\varphi}}(x)=\dot{\bar{\varphi}}_0(x)-\big(a^3\int_t^\infty dt'\ a'^{-3}\big)\frac{\partial}{\partial\bar{\varphi}}V(\bar{\varphi}(x)). \tag{23}$$

This is the same result derived by the resummation formula.

5. Conclusion

We make the leading power resummation of the IR effects in a general accelerating universe. Extracting the leading IR behaviors of the linear term and the retarded propagator, the Yang–Feldman equation is reduced to the Langevin equation with the white noise term (12). The difference from the Langevin equation in dS space appears as the slow-roll parameter dependence of the coefficient of each term.

The white noise term is a consequence of the fact that quantum fluctuations of a massless and minimally coupled scalar field are frozen at super-horizon scales. The nonlinear term is expressed as the local operator. That is because up to the leading IR effects, the retarded propagator is spatially local and the time dependence of the vertex integral appears only in its upper bound.

The Langevin equation can be translated to the Fokker-Planck equation (13). The final state solution of it is necessary to evaluate eventual contributions to vevs of field operators. We find the final state solution (15) in the situation that the time variations of the background parameters are negligible during the IR effects grow.

In fact, the semiclassical description of the scalar field leads to the same Langevin equation. We choose the time coordinate T as its friction coefficient is constant. Eliminating the second derivative terms with respect to T, $\mathbf{x}$ from the equation of motion, the nonlinear term of the Langevin equation is consistent with the resummation formula including its coefficient. As for the noise term, we can obtain the same one by identifying sub-horizon modes as sources of super-horizon modes.

Acknowledgments

H. K. would like to thank Prof. Chong-Sun Chu, Prof. Gary Shiu and the Organizing Committee for the kind invitation to Second LeCosPA International Symposium 'Everything about Gravity'.

References

1. A. A. Starobinsky, Lect. Notes Phys. **246**, 107 (1986). A. A. Starobinsky and J. Yokoyama, Phys. Rev. D **50**, 6357 (1994).
2. N. C. Tsamis and R. P. Woodard, Nucl. Phys. B **724**, 295 (2005).
3. G. Cho, C. H. Kim and H. Kitamoto, arXiv:1508.07877 [hep-th].
4. F. Finelli, G. Marozzi, A. A. Starobinsky, G. P. Vacca and G. Venturi, Phys. Rev. D **79**, 044007 (2009), Phys. Rev. D **82**, 064020 (2010), AIP Conf. Proc. **1446**, 320 (2010), V. Vennin and A. A. Starobinsky, Eur. Phys. J. C **75**, 413 (2015).
5. N. C. Tsamis and R. P. Woodard, Class. Quant. Grav. **20**, 5205 (2003).
6. J. S. Schwinger, J. Math. Phys. **2**, 407 (1961). L. V. Keldysh, Zh. Eksp. Teor. Fiz. **47**, 1515 (1964) [Sov. Phys. JETP **20**, 1018 (1965)].

Inflationary Electroweak/Particle Phase Transitions and New Classical Gravitational Waves on CMB *

Sichun Sun
Institute for Advanced Study, Hong Kong University of Science and Technology, Clear Water Bay, Hong Kong,
sichun@ust.hk
http://www.ust.hk/

Particle phase transitions in the early universe including electroweak and grand unification ones are well-studied subjects. We point out that there are new possible particle phase transitions around inflation. Those new inflationary particle phase transitions, if of the first order, may yield low-frequency gravitational waves (GWs) due to bubble dynamics, leaving imprints on the cosmic microwave background (CMB). In contrast to the nearly scale-invariant primordial GWs caused by vacuum fluctuation, these bubble-generated GWs are classical and have scale dependent B-mode spectra. If decoupled from inflaton, the electroweak phase transition during inflation may serve as a mirror image of the one after reheating where the baryon asymmetry could be generated via electroweak baryogenesis (EWBG). The second new electroweak phase transition may also be the source for EWBG.

Keywords: Gravitational waves, CMB, electroweak physics, baryogenesis.

1. Introduction

Particle phase transitions happened in the early universe are rich subjects closely tied to beyond the standard model physics and particlegenesis in the early universe. With the advent of LHC and the discovery of Higgs, the questions around the electroweak phase transition have intensified. The direct detection experiments for GWs and collider searches are two major ways to probe particle phase transitions.

Inflation is the leading paradigm in the very early cosmology to provide the seeds for CMB fluctuations. In the old inflation scenarios of the early days, the inflaton went through a tunneling process similar to a strong first order phase transition, to drive inflation to happen. This inflaton phase transition scenario is however not successful because the phase transition bubbles cannot collide. If one goes beyond the single field inflation scenario (which is actually a natural step after considering the presence of the standard model Higgs and many popular beyond the standard model scalar extensions[2]), the phase transitions of the other fields during inflation remain valid. The energy density during inflation is very high in almost all inflation models to drive inflation, whereas the temperature during inflation is roughly a constant, so-called the Gibbons-Hawking temperature $T_{GH} = H/(2\pi)$, which can be as low as 10^{-24}GeV. One can find the experimentally allowed inflation models with different Hubble constant H in Fig. 1.

*This talk is mainly based on Ref. 1.

We would like to stress here that our scenario is very different from Higgs inflation[2,3] and Coleman-de Luccia bubble phase transitions.[4] The energy density difference inside and outside the phase transition bubbles is not too large in our scenarios, so as inflation can happen both inside and outside bubbles. This way bubbles can collide and stir up turbulence in the thermal plasma to generate gravitational waves.

The temperature contribution from radiation drops quickly at the beginning of inflation when all the particle contents get diluted, left with only contribution from curvature, T_{GH}. With the lower scale inflation models, $H \sim T_{GH} \leq 10^2$GeV, the Higgs field/phase transition field stay at the origin of the potentials and tunnel to the spontaneous breaking phase due to the decreasing of the temperature, See Fig. 2. Notice that even when $H \leq 10^2$GeV, the energy density in that early universe is still as high as $\rho \sim (10^{10}\text{GeV})^4$. Moreover, the GUT phase transition, if exists, will happen in almost all the inflation scenarios.

The normal phase transitions happened in particle physics after reheating produce gravitational waves that direct detection experiments are currently looking for by Advanced LIGO, Advanced Virgo and LISA. The particle phase transitions here happened *around inflation* in the much earlier universe. They left an imprint on much larger scales in the universe today. If we assume a strong first order phase transition in particle physics (for electroweak phase transition it is preferred by baryogenesis), the drastic bubble nucleation process produced in phase transition will produce gravitational waves. This type of gravitational waves, produced around the same time as the primordial gravitational wave, but having a different yet scale-dependent spectrum, characterized by the bubble sizes at the time of collisions.

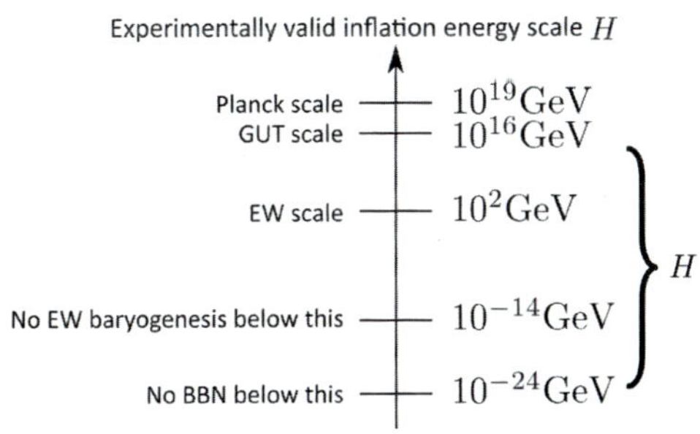

Fig. 1. In various inflation models, Hubble constant during inflation can take different values from 10^{-24} GeV up to 10^{14} GeV. The upper bound is set by the latest experiment. When the Hubble constant is below 10^{-14} GeV, the universe can not reheat above 100 GeV later, where the EW baryogenesis can be hardly achieved. Below 10^{-24} GeV, the reheated universe is too cool to have big bang nucleosynthesis.

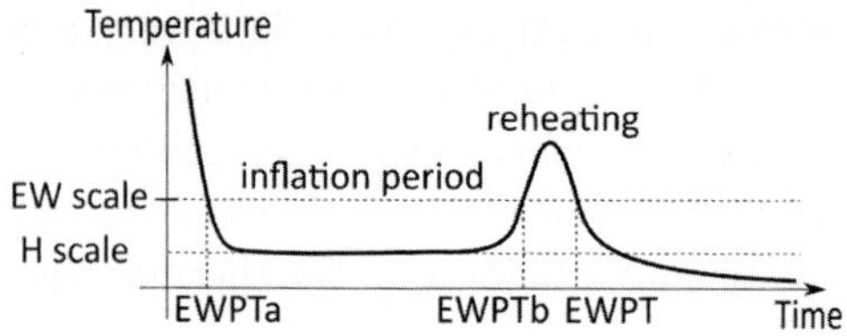

Fig. 2. The thermal temperature vs time in the early universe. Notice that the temperature during inflation comes mainly from the curvature contribution of the de sitter space, which is proportional to the Hubble constant. Two more electroweak phase transitions are proposed here, namely EWPTa and EWPTb.

2. Gravitational wave spectra

We derive the power spectrum P_γ coming from scale dependent GWs produced in de-Sitter space here. During inflation, the action can be written as two independent copies of probe field γ_s,

$$S = \sum_{s=+,\times} \int dt \frac{d^3\boldsymbol{k}}{(2\pi)^3} a^3 \left[\frac{1}{2} \dot{\gamma_s}(\boldsymbol{k}) \dot{\gamma_s}(-\boldsymbol{k}) - \frac{k^2}{2a^2} \gamma_s(\boldsymbol{k}) \gamma_s(-\boldsymbol{k}) \right] . \tag{1}$$

We can then quantize the γ_s fields: $\gamma_s(\boldsymbol{k},\tau) = v_k(\tau) a_{\boldsymbol{k}s} + v_k^*(\tau) a^\dagger_{-\boldsymbol{k}s}$. Solving the equation of motion, we can get the mode function v_k:

$$v_k(\tau) = \frac{H}{\sqrt{2k^3}} \Big(c_1(k)(1 + ik\tau) e^{-ik\tau} + c_2(k)(1 - ik\tau) e^{ik\tau} \Big) , \tag{2}$$

where the coefficients $c_1(k), c_2(k)$ are subject to the consistency condition of quantization $|c_1|^2 - |c_2|^2 = 1$.

The energy density of GWs is essentially the Hamiltonian density, given by

$$\rho_{GW} = \int \frac{dk}{k} \frac{k^3}{2\pi^2} \left(|\dot{v}_k|^2 + \frac{k^2}{a^2} |v_k|^2 \right) . \tag{3}$$

Furthermore, the gravitational energy spectrum can be obtained as

$$\Omega_{GW}(k,\tau) = k \frac{d\rho_{GW}}{dk} / \rho_{\rm tot} \tag{4}$$

$$= \frac{1}{3H^2 M_p^2} \frac{k^3}{2\pi^2} \frac{|v_k'(\tau)|^2 + k^2 |v_k(\tau)|^2}{a(\tau)^2} , \tag{5}$$

where we assume that the universe is spatially flat, meaning that $\rho_{\rm tot} = \rho_{\rm critical} = 3H^2 M_p^2$. Particularly note that during inflation $\rho_{\rm tot} = \rho_{\rm inflaton} + \rho_{\rm rad} + \rho_{\rm higgs}$.

We can also calculate the power spectrum of GWs which measures the two point correlation: $P_\gamma(k,\tau) = \frac{4k^3}{\pi^2 M_p^2} |v_k(\tau)|^2$. This power spectrum of GWs contributes both temperature fluctuations and polarizations on the CMB.

We consider the classical limit for these GWs, where $c_1 \approx c_2 \gg 1$. Inserting the mode functions, we can get the relations for sub-horizon and super-horizon modes respectively,

$$P_\gamma(k,\tau_{\rm obs}\to 0)=\begin{cases}24\left(\frac{a(\tau_*)H}{k}\right)^4\Omega_{GW}(k,\tau_*), & |k\tau_*|\gg 1\\ 24\left(\frac{a(\tau_*)H}{k}\right)^2\Omega_{GW}(k,\tau_*), & |k\tau_*|\ll 1\end{cases}. \tag{6}$$

3. Gravitational waves by the bubbles

The sub-horizon case in our inflationary scenarios is similar to the previous semi-numerical studies on the EW phase transitions such as Refs. 6 and 8. Usually the phase transition is a rapid process compared to the Hubble time and thus the effect of expansion of the universe can be ignored even during inflation.

We can then finally determine the power spectrum of GW generated by bubbles. Using Eq. (6), we arrive at:

$$P_\gamma(k,\tau_{\rm obs}\to 0)=P_\gamma^{\rm crit}\left(\frac{k_{\rm crit}}{k}\right)^4\frac{(a+b)k_{\rm crit}^b k^a}{bk_{\rm crit}^{a+b}+ak^{a+b}}, \tag{7}$$

where $P_\gamma^{\rm crit}$ is the power spectrum at the critical point: $P_\gamma^{\rm crit}=24\left(\frac{a(\tau_*)H}{k_{\rm crit}}\right)^4\Omega_{GW}^{\rm crit}$. An estimation yields: $P_\gamma^{\rm crit}\sim\left(\frac{H}{\beta}\right)^6\left(\frac{\rho_{\rm higgs}}{\rho_{\rm tot}}\right)^2$ for the sub-horizon case.

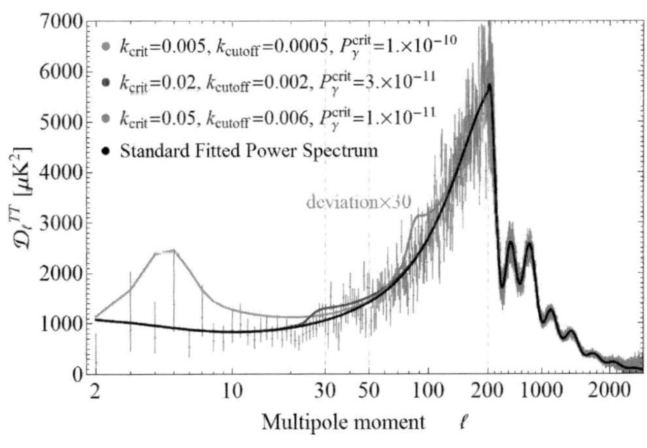

Fig. 3. CMB temperature fluctuation power spectrum. The gray points and error bars are the experimental data of *Planck* 2015 while the black curve is the fitted CMB spectrum using the *Planck* 2015 parameters. (The unit of the momentum k is Mpc^{-1}.)

4. Imprints on CMB

The CMB spectrum can be obtained by inputting the power spectrum into the CLASS[5] where the transfer function is calculated. As we see from Eq. (7), the

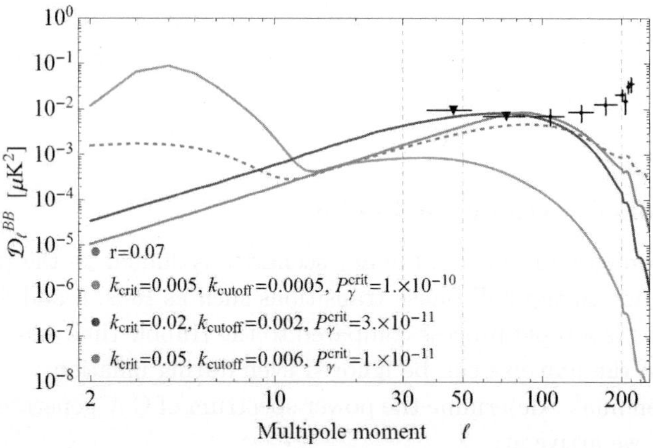

Fig. 4. B-mode power spectrum generated by phase transition bubbles. For comparison, the primordial GWs caused by quantum vacuum fluctuations are showed in dashed line with tensor-to-scalar ratio $r = 0.07$, which is the strongest upper bound to date on primordial GWs. The black ones are the CMB component bandpowers obtained from BICEP2 & *Keck Array* experiment.

power spectrum diverges when $k \to 0$. This divergence is unphysical as the GW generating formulas break down at super horizon scale. We can introduce the horizon scale as a natural cut-off, yielding $k_{\text{cutoff-physical}} = k_{\text{cutoff}}/a_* \sim H$. The critical physical momentum is related to the bubble size via $k_{\text{crit}}/a_* \sim R_b^{-1}$. The corresponding comoving momenta are $k_{\text{crit}} \sim \frac{1}{v_b}\frac{\beta}{H}e^{N_*}k_0$, $\quad k_{\text{cutoff}} \sim e^{N_*}k_0$, where N_* is the e-folding number of phase transition counting from the time that the largest mode k_0 exits the horizon. We choose the scale factor today a_0 as one, thus the largest physical mode today is $k_0 = 0.0002\text{Mpc}^{-1}$ as the inverse of observed universe size. Approximately we can find k_0 corresponding to the position at CMB multipole $\ell_0 \sim 2$. Then we arrive at relations: $\ell_{\text{crit}} \sim 2e^{N_*}\beta/(v_b H), \ell_{\text{cutoff}} \sim 2e^{N_*}$.

- GUT scenarios: The GUT phase transition happens around 10^{16} GeV.[9,10] If choosing $(H/\beta)^6 \sim 10^{-10}, \rho_{\text{GUT}} \sim (10^{16}\text{GeV})^4$, and $H \sim 10^{14}$GeV, then we have $P_\gamma^{\text{crit}} \sim 10^{-10}$.
- EW scenarios: Typically the EWPT temperature is around 10^2GeV $\sim$ 10^3GeV, which requires $T_{GH} \sim H < 10^2$GeV. A first-order EWPT can be achieved in various theories beyond the standard model, e.g.[11] If taking $(H/\beta)^6 \sim 10^{-6}, \rho_{\text{higgs}} \sim (10^3\text{GeV})^4$, and $H \sim 10^{-11}$GeV, then we have $P_\gamma^{\text{crit}} \sim 10^{-10}$. The baryogenesis requires the universe going out of equilibrium, sometimes a strong first-order phase transition. Although the phase transition during inflation is not suited for generating baryons because all the particle generated during inflation will be diluted. The second symmetry restoration (EWPTb or a contrived GUTb) we proposed here might achieve that goal(see a related model[12]).

5. Discussions

In this talk we introduce an indirect approach to probe the first-order CPTs, such as the GUT phase transition and the EWPT, by detecting the GWs through the CMB. These GWs are generated during inflation via bubble collisions or by the turbulence caused by bubble motion in the thermal plasma, characterized with a scale-dependent power spectrum.

The large-scale scalar power spectrum caused by the first-order CPTs during inflation might be suppressed due to the relatively large values of the slow-roll parameter when the thermal radiation is diluted. The further discussion on the density fluctuation depends on the details of the specific inflation models.

Acknowledgments

Sichun Sun was supported by the CRF Grants of the Government of the Hong Kong SAR under HUKST4/CRF/13G.

References

1. H. Jiang, T. Liu, S. Sun and Y. Wang, "Echoes of Inflationary Particle Phase Transitions in the CMB", arXiv:1512.07538 [astro-ph.CO].
2. Y. Hamada, T. Noumi, S. Sun and G. Shiu, "An O(750) GeV Resonance and Inflation," arXiv:1512.08984 [hep-ph].
3. F. L. Bezrukov and M. Shaposhnikov, "The Standard Model Higgs boson as the inflaton," Phys. Lett. B **659** (2008) 703 ,arXiv:0710.3755 [hep-th]; D. Buttazzo, G. Degrassi, P. P. Giardino, G. F. Giudice, F. Sala, A. Salvio and A. Strumia, "Investigating the near-criticality of the Higgs boson," JHEP **1312** (2013) 089 ,arXiv:1307.3536 [hep-ph].
4. S. R. Coleman and F. De Luccia, Phys. Rev. D **21**, 3305 (1980). doi:10.1103/PhysRevD.21.3305
5. Blas, D., Lesgourgues, J., & Tram, T., JCAP, 7, 034 (2011) [arXiv:1104.2933 [astro-ph]]
6. S. J. Huber and T. Konstandin, "Gravitational Wave Production by Collisions: More Bubbles," JCAP **0809**, 022 (2008) [arXiv:0806.1828 [hep-ph]].
7. G. W. Gibbons and S. W. Hawking, Phys. Rev. D **15**, 2738 (1977).
8. M. Kamionkowski, A. Kosowsky and M. S. Turner, "Gravitational radiation from first order phase transitions," Phys. Rev. D **49**, 2837 (1994) [astro-ph/9310044].
9. H. Georgi, H. R. Quinn and S. Weinberg, Phys. Rev. Lett. **33**, 451 (1974).
10. A. H. Guth and S. H. H. Tye, "Phase Transitions and Magnetic Monopole Production in the Very Early Universe," Phys. Rev. Lett. **44**, 631 (1980) [Phys. Rev. Lett. **44**, 963 (1980)].
11. A. G. Cohen, D. B. Kaplan and A. E. Nelson, "Progress in electroweak baryogenesis," Ann. Rev. Nucl. Part. Sci. **43**, 27 (1993) [hep-ph/9302210].
12. A. Dolgov, K. Freese, R. Rangarajan and M. Srednicki, "Baryogenesis during reheating in natural inflation and comments on spontaneous baryogenesis," Phys. Rev. D **56**, 6155 (1997) doi:10.1103/PhysRevD.56.6155 [hep-ph/9610405].

PARALLEL SESSIONS

II. Emergent Gravity and Quantum Gravity

Summary of Parallel Session: "Emergent Gravity and Quantum Gravity"

A. Marcianò

Department of Physics, Fudan University,
Shanghai, 200433, China
marciano@fudan.edu.cn
http://phys.fudan.edu.cn/f7/88/c7605a63368/page.htm

We summarize the presentations of the parallel session "Emergent Gravity and Quantum Gravity", and outline thoughts on the open problems and future prospects of the field.

1. Introduction

Several different approaches to the problem of quantizing gravity were represented at the LeCosPa II Symposium, including Non-commutative Geometry, Regge Calculus, Euclidean Quantum Gravity, Horava-Lifshitz Gravity, Asymptotic Safety, AdS/CFT conjecture of String Theory, Loop Quantum Gravity. In particular, within the topical parallel session on "Emergent Gravity and Quantum Gravity" talks concerned mainly Loop Quantum Gravity, Non-commutative Geometry and Matrix models, Asymptotic Safety and Horava-Lifshitz Gravity. Questions that were faced during the parallel session can be summarized as follows:

(i) *"What are we quantizing? What should we quantize?"*
(ii) *"What are the symmetries of GR? What is time?"*
(iii) *"Is gravity emergent? Should we quantize GR, or some other fundamental mother theory?"*
(iv) *"Can we understand better QG by adding matter to our investigations?"*
(v) *"Is there any relation between gravity and thermodynamics? Can the latter encode signatures of Planck-scale modifications to spacetime symmetries?"*

These questions were addressed relying on the novel confidence that, given the lack of experimental data for QG, we should proceed along two precise directions:

- *Existing many different approaches, it is beneficial not to focus only on one of them in particular, but rather to develop specific attempts while continuing to explore what alternative models can provide as useful.*
- *Cross-fertilization must be encouraged: the lack of experimental guidance, forces us not only to promote a humble dialogue between different approaches and scientific communities, but also to develop theories proposed toward falsification.*

2. On the two different strategies represented

The problem of emergent and quantum gravity was approached in the parallel session along two main directions, the top-down and the bottom-up approaches. These were often co-present in the very same presentation. Among the attempts to be listed in the top-down approach, we can in turn distinguish two main directions of research that were proposed, which depart from posing very fundamental questions, or eventually from emphasizing more technical lessons:

(i) When considering **fundamental questions**, the focus was on:
   (a) *understanding the very nature of space-time, addressing the nature of time and symmetries of GR: Hoi-Lai Yu (Chopin Soo in the plenary session);*
   (b) *regarding GR as an emergent theory and moving toward a quantization of a non-commutative pre-geometry (Hsu-Wen Chiang, Hyun Seok Yang).*

(ii) Looking at **more technical lessons** attained from the models under scrutiny:
   (a) *quantization methods of GR recast in terms of SU(2) variables (Abbay Ashtekar, Jerzy Lewandowski, Antonino Marciano);*
   (b) *Quantization of new formulations of GR (Yifu Cai, Yosuke Misonoh).*

Investigations along the bottom-up line were oriented in order to emphasize the link between QG and the quantization of matter fields on side, and in order to develop a clear phenomenological approach, on the other side, able to derive Planck scale corrections to physical observables quantities that might be in principle tested.

(i) When considering the link between **Matter and Quantum Gravity**, the focus was on:
   (a) recovering viable dynamics of QG from the existence of matter, and ultraviolet modifications of GR from anisotropic scaling (Pietro Donà, Yosuke Misonoh, and Petr Horava in the plenary session);
   (b) understanding back-ground independent quantization of metric d.o.f. using YM fields (Antonino Marcianò).

(ii) While refining attempts to develop a **phenomenology at the Planck scale**, studies were deepening:
   (a) cosmological implications of QG and extended GR (Abbay Ashtekar, Yifu Cai, Jerzy Lewandowski, Antonino Marcianò, Hyun Seok Yang);
   (b) implications of some ri-examination of statistical mechanics (Nikos Kalogeropoulos).

Below we review and sketch the seminars communicated during the parallel session, going thru the details of the presentations.

3. Seminars communicated

Abbay Ashtekar emphasized that the fundamental structure of Loop Quantum Gravity (LQG) deals with polymer like, 1D excitations, that enter the definition of the kinematical Hilbert space. These states are expanded in the spin-network basis, and are supported on 1D *graphs*. While the covariant evolutions of these states within the Hilbert space of LQG makes use of the spin-foam approach, which has support on *two-complexes*, the deployment of Hamiltonian methods have led to the development of applications to cosmology: these are the well celebrated Loop Quantum Cosmology (LQC) approach, and the newly formulated Reduced Loop Quantum Gravity (RLQG) approach. Within the full theory of LQG developments have concerned the derivation of N-point functions, of propagators for the gravitational field, on the recovering in LQC of the Big Bounce picture of early cosmology etc... The author emphasized the existence of a regime motivated by LQG — in its mini-superspace applications to cosmology, namely LQC — in which an energy gap Δ appears that is motivated by he discretization of space-geometry in LQG. The first Friedmann equation get modified into $\dot{a}/a = (8\pi G\rho/3)[1 - \rho/\rho_{\rm sup}]$, $a(t)$ denoting the scale factor of a homogenous isotropic metric, $\dot{a}$ its derivative with respect to time t, ρ the energy density of the matter content of the Universe, and $\rho_{\rm sup}$ a critical density $\rho_{\rm sup} = (18\pi)/(G^2\hbar\Delta^3) \sim 0.41\rho_{\rm Pl}$. Pre-inflationary dynamics matters in this framework,[1] since it fully reproduces cosmological perturbations and accounts for novel falsifiable predictions, such as the on the low-ℓ anomalies of the CMB radiation. Indeed, pre-inflationary era modes whose wavelength exceed the curvature radius are excited and populated at the onset of inflation, so that they can leave imprints on CMB and naturally lead to low ℓ-anomalies.

Pietro Donà presented a study[2] that deploys FRG non-perturbative methods in order to account for possible effects on the running of the Newton constant in GR as induced from the quantum dynamics of matter. In particular, matter affects the gravitational fixed point through the running of the beta-function of the Newton constant, namely $\beta_G = 2G + \frac{G^2}{6\pi}(N_S + 2N_D - 4N_V - 46)$, N_S labeling the number of scalar d.o.f, N_D the number of fermionic d.o.f., and N_D the number of vector fields d.o.f. As a consequence, while the results derived for the Standard Model (SM) and its "small extensions" within a truncated RG flow picture are shown to be consistent with the theoretical analysis, extra dimensions are severe restricted, as well as Minimal Supersymmetric extension of the SM, and the SU(5) and SU(10) Great Unification theories. Furthermore, the Newton coupling G defined from gravity-matter vertices exhibits asymptotic safety in the pure-gravity case and in certain approximation also for a small number of scalars.

Hsu-Wen Chiang presented a work[3] that elaborates on making covariant a recent proposal by Ronald J. Adler[4] on a quantum theory of distance along a curve, in which a Spin(3,1) world-line action from KK_0 states has been considered, reading $S_{\rm spin} = 1/(8\pi\ell_s^2)\int d^2\sigma||\gamma_\mu\partial_\alpha X^\mu||^2$. Introducing the concepts of re-fermionization for the action $S_F = \int d^2\sigma\imath\bar{c}\gamma^\alpha D_\alpha c$ — ψ is a fermion field, while γ are ordinary Dirac matrices — and bosonization $c \propto \sqrt{\gamma^0\not{\psi}}\,\psi_0$, a notion of non-commutative space-time can be recovered. The latter is endowed with spin operators, defined by $\langle \Delta \hat{X}^I \rangle = \bar{\psi}\lambda\gamma^I\psi$. Space-time operators naturally undergo non-commutativity in this picture, while the space-time element reads $ds^2 = \langle \Delta \hat{X}^I \Delta \hat{X}_I \rangle = 4\lambda\bar{\psi}\psi$. Generalized uncertainty relations can be then recovered.

Jerzy Lewandowski started his seminar summarizing the Hamiltonian formulation of GR recast in the Ashtekar variables. He then provided a short review of deparametrization within this latter framework, while accounting for a Klein-Gordon field ϕ that plays the role of time $\phi = t$. Using the scalar constraint for $\hat{\phi}(x)$, the physical Hamiltonian $H_{\rm phys}$ is recovered from the expression of the conjugated momentum $\hat{\pi}(x)$ to $\hat{\phi}(x)$, thus is proportional to the quantity $\hat{\pi}(x) = \pm[-\hat{\phi}_{,a}\hat{\phi}_{,b}\hat{E}^a_i\hat{E}^b_i - 2V(\hat{\phi}(x))|\det\hat{E}| - 2|\det\hat{E}|^{1/2}\hat{C}_{\rm gr}]^{1/2}$, where E^a_i denotes the Ashtekar electric field and $C_{\rm gr}$ the scalar constraints inhering the gravitational d.o.f. Accordingly, observable quantities $\mathcal{O}$ will evolve in the Hamiltonian picture satisfying $d\mathcal{O}/dt = \{\mathcal{O}, H_{\rm phys}\}$. The author then focused on an explicit calculation performed in the radial gauge,[5] recovering the Dirac brackets for the Hamiltonian constrained system under scrutiny. As a result, matter fields are recovered not to commute in the space-time radial gauge. Interestingly, one can speculate on possible applications of this framework to the AdS/CFT scenario.

Antonino Marcianò presented a path toward the understanding of the semiclassical limit of LQG, inserting matter (specifically gauge fields) in the analysis. Even though the semiclassical limit remains a longstanding problem in LQG, the author showed that within the framework of RLQG[6] one can address the semiclassical limit of the coupled system of matter and gravity in a meaningful way. The d.o.f. are captured by spin-network states supported on graphs Γ that are colored with U(1) projected assignments $\{m_l, \imath_v\}$ of the gravitational SU(2) group and representations $\mathcal{R}_{\rm YM}$ of the Yang-Mills (YM) gauge group under scrutiny. States then belong to the total kinematical Hilbert space $\mathcal{H}^{\rm (tot)}_{\rm kin} = \mathcal{H}^{\rm (GR)}_{\rm kin} \otimes \mathcal{H}^{\rm (YM)}_{\rm kin}$, and are labelled, within the language of RLQG, by $|\Gamma; m_l, \imath_v; \mathcal{R}_{\rm YM}\rangle_R = |\Gamma; m_l, \imath_v\rangle_R \otimes |\Gamma; \mathcal{R}_{\rm YM}\rangle$. Within this framework, the action of matter and gravity can be dealt with by resorting to Hamiltonian methods. This enables the computation of matrix-elements of the Hamiltonian operator in the semi-classical limit, and then provides the evolution for the YM Wilson-loops variables supported on the space-like cubulation deployed in RLQG. Expressing the polymer states of the YM theory in terms of the states of the Fock space, which are actually close to experimental set up in the measurement processes, it is possible to recover n-point functions in the semiclassical limit from the non-perturbative approach. Furthermore, the author commented on recovering constraints provided by the Casimir effect, since this would help shedding light on the semi-classical limit of gravity.

Hoi-Lai Yu commented on the necessity of understanding the role of time and gauge symmetries of GR, moving from its ADM Hamiltonian formulation. The crucial observation is that the lapse function N can be fixed a posteriori in any slicing, which implies that the scalar constraint of the ADM Hamiltonian of gravity is not a symmetry generator.[7] Introducing two real parameters δ and β, the lapse function writes $N\,dt := \sqrt{q}(\delta \ln q^{1/3} - \mathcal{L}_{\vec{N}dt} \ln q^{1/3})/(4\beta\kappa\bar{H})$, in which $\kappa = 8\pi G$, while q is the determinant of the three-metric on the spatial hypersurfaces of the ADM foliation, and $\bar{H}$ is related to the trace of the extrinsic curvature by $\sqrt{q}\mathrm{Tr}K = 6\kappa\beta\bar{H}$. As a consequence, 4D covariance is not a symmetry of space-time, and thus one should rather deal with 3D geometries. On the other hand, a metric already delivers a notion of "simultaneity" and a common moment of a rudimentary "time". A notion of intrinsic time can be then introduced, and must be proportional to $\ln q^{1/3}$. Intrinsic time is therefore extracted from the 3D metric, and $\delta \ln q^{1/3}$ turns out to be the Tomonaga-Schwinger time.

Hyun Seok Yang argued that non-commutative spacetime implies emergent spacetime[8] and provided the desired machinery for the background-independent formulation of cosmic inflation.[9] Geometric entities are substituted by algebraic quantities, and the problem of QG is assimilated to the treatment of electromagnetism on non-commutative spacetime. The latter is parametrized using the mathematical tools of the Moyal-plane, undergoing central algebra like commutation relations between the coordinates of a space-time point $[x^\mu, x^\nu] = \imath\theta^{\mu\nu}$. Cosmic Inflation, which is past incomplete and suffers initial singularity in which space and time cease to exist, is then derived in an emergent space-time scenario originates by BFSS Matrix models of non-commutative space-time. Their action, $S_{\mathrm{BFSS}} = (4g^2)^{-1} \int dt\, \eta^{AC}\eta^{BD}\mathrm{Tr}([\phi_A, \phi_B][\phi_C, \phi_D])$, encodes a covariant derivative $D_0 = d/dt - \imath A_0$ and a multiplet ϕ_a that recombine into $\phi_A = (\phi_0, \phi_a)$. Cosmic Inflation then arises as a vacuum solution of the Matrix model fields, provided that $\langle\phi_a\rangle_{\mathrm{vac}} = e^{Ht}p_a$ and $\langle A_0\rangle_{\mathrm{vac}} = a_0(y) = H \int_0^1 d\sigma\, \frac{dy^\mu}{d\sigma}\, p_\mu(\sigma)$.

Yifu Cai presented cosmological consequences of extended theories cast in the language of tele-parallel gravity.[10] The author started showing that GR can be recast in terms of purely torsional Weitzenböck connections $\Gamma^\lambda_{\nu\mu} = e^\lambda_A \partial_\mu e^A_\nu$, in which e^A_μ denotes a vier-bein, if we consider its contortion tensor $T^\lambda_{\nu\mu} = \Gamma^\lambda_{\nu\mu} - \Gamma^\lambda_{\mu\nu}$. Then the Einstein-Hilbert density Lagrangian reads $L := T = 1/4T^{\rho\mu\nu}T_{\rho\mu\nu} + 1/2T^{\rho\mu\nu}T_{\nu\mu\rho} - T^\rho_{\rho\mu}T^{\nu\mu}_\nu$. Extended version of GR can be now recovered in the novel form of $f(T)$, which captures inequivalent theories than usual $f(R)$ ones. The author then showed that $f(T)$ models satisfy cosmological observations at current time, for ad hoc choices of the $f(T)$ function. Dark Energy behavior can be parametrized in terms of $f(T)$ and its first and second functional derivatives $f_T(T)$ and $f_{TT}(T)$ by defining the energy density $\rho_{\mathrm{DE}} = 3/(8\pi G)[-f/6 + T/3f_T]$ and the equation of state parameter $w_{\mathrm{DE}} = -(f - Tf_T + 2T^2 f_{TT})[1 + f_T + 2Tf_{TT}]^{-1}[f - 2Tf_T]^{-1}$. Forthcoming investigations will clarify whether $f(T)$ cosmologies can be a better approximation to the Universe dynamics than ΛCDM.

Nikos Kalogeropoulos focused on a research line that aims at reconstructing foundations of statistical mechanics. To achieve this goal, among several different entropic functionals the author considered $\mathcal{S}_q = (q-1)^{-1}(1-\sum_{i\in I} p_i^q)$, in which p_i denotes a distribution of probability, and which enjoys the dependence on a deformation parameter q. Entropy turns out then to be no more additive, namely for two distributions A and B one finds $\mathcal{S}_q(A\star B)=\mathcal{S}_q(A)+\mathcal{S}_q(B)+(1-q)\mathcal{S}_q(A)\mathcal{S}_q(B)$. As a remarkable consequence, which may provide further phenomenological insights, modified equilibrium states are recovered of the form $Z_q = \sum_{i\in I} e_q^{-\beta_q(E_i-U)}$, in which β_q parametrizes q-deformed temperature's values, E_i denote different energy-levels, U stands for the total internal energy of the statistical system under scrutiny, and $\log_q(x) := (1-q)^{-1}(x^{1-q}-1)$. Through forthcoming investigations, this result might be also connected to modifications of elementary particles' statistics.

Yosuke Misonoh considered the back-reaction to Black Hole (BH) that arises from ultra-violet corrections in Horava-Lifshitz gravity.[11] The author depend a specific framework in which ultraviolet corrections that depend on higher derivatives of degree $\partial^4_{x^i}$ and $\partial^6_{x^i}$ are included. A specific parametrization is addressed in terms of two constants β_1 and β_2 that appear in the action $L_{\rm UV} = -m_{\rm Pl}^{-2}[\beta_1(\dot{u}_\alpha\dot{u}^\alpha)^2+\beta_2\mathcal{R}^2]$, in which the aether time-like (fixed-norm and twistless) four-vector u^α satisfies $\dot{u}^\alpha := u^\mu\nabla_\mu u^\alpha$, $m_{\rm Pl}$ stands for the Planck mass, and $\mathcal{R}_{\mu\nu}$ denotes the three-Ricci tensor. Black hole solution with universal horizon which is a event horizon can be constructed including only the term proportional to $(\dot{u}_\alpha\dot{u}^\alpha)^2$. Nonetheless, inclusion of the $\mathcal{R}^2$ term implies the in-existence of BH, and leads to the statement that universal horizons are physical singularities, without however meaning a necessary violation of the Cosmic Censorship.

4. Partial lessons

Concluding, we can summarize as follows lessons derived during the parallel session:

(i) the lack of guidance from experimental data precludes a final understanding of the problem of quantum gravity (Ashtekar, Yu, Donà, Cai, Marcianò);
(ii) we need to make contact with known physics: particle physics, cosmology, etc... (Donà, Ashtekar, Cai, Marcianò, Lewandowski...);
(iii) we need to develop new concepts and mathematical tools, and mostly we need synergy, and cross-fertilization among different research areas (Ashtekar, Kalogeropoulos, Hsu-Wen Chiang, Hyun Seok Yang, Yosuke Misonoh).

References

1. A. Ashtekar and B. Gupt, Phys. Rev. D **92**, no. 8, 084060 (2015) [arXiv:1509.08899 [gr-qc]].
2. P. Donà, A. Eichhorn and R. Percacci, Phys. Rev. D **89**, no. 8, 084035 (2014) [arXiv:1311.2898 [hep-th]].
3. H. W. Chiang, Y. C. Hu and P. Chen, arXiv:1512.03157 [hep-th].

4. R. J. Adler, arXiv:1402.5921.
5. N. Bodendorfer, P. Duch, J. Lewandowski and J. Swiezewski, JHEP **1601**, 047 (2016) [arXiv:1510.04154 [gr-qc]].
6. E. Alesci and F. Cianfrani, Phys. Rev. D **87**, no. 8, 083521 (2013) [arXiv:1301.2245 [gr-qc]].
7. C. Soo and H. L. Yu, PTEP **2014**, no. 1, 013E01 (2014) doi:10.1093/ptep/ptt109 [arXiv:1201.3164 [gr-qc]].
8. H. S. Yang, arXiv:1504.00464 [hep-th].
9. H. S. Yang, arXiv:1503.00712 [hep-th].
10. Y. F. Cai, S. Capozziello, M. De Laurentis and E. N. Saridakis, arXiv:1511.07586 [gr-qc].
11. Y. Misonoh and K. i. Maeda, Phys. Rev. D **92**, no. 8, 084049 (2015) [arXiv:1509.01378 [gr-qc]].

Gauge Fields in Loop Quantum Gravity and their Semiclassical Limit

J. Bilski
Department of Physics, Fudan University,
Shanghai, 200433, China
jakubbilski14@fudan.edu.cn

F. Cianfrani
Instytut Fizyki Teoretycznej, Uniwersytet Wrocławski,
pl. Maksa Borna 9, 50-204 Wrocław, Poland
francesco.cianfrani@ift.uni.wroc.pl

P. Donà
Department of Physics, Fudan University,
Shanghai, 200433, China
pietro_dona@fudan.edu.cn

A. Marcianò
Department of Physics, Fudan University,
Shanghai, 200433, China
marciano@fudan.edu.cn
http://phys.fudan.edu.cn/f7/88/c7605a63368/page.htm

We summarize work in progress on the semiclassical limit of gauge theories quantized within thin the framework of Loop Quantum Gravity (LQG). We specify the theoretical framework and introduce Quantum Reduced Loop Gravity (QRLG), discussing its role in recovering the semiclassical limit of LQG. We then insert matter, namely gauge fields, in the analysis of the semiclassical limit of LQG, an derive consequences for observational effects. In this brief presentation, we focus on the case of abelian gauge theories, summarizing partial results so far obtained.

Keywords: Quantum gravity.

1. Introduction

The semiclassical limit remains a longstanding problem in LQG. Nonetheless, few attempts have been developed in the last decade, and in particular a novel framework known as QRLG[3] has been advocated in addressing the semiclassical limit of LQG. Within such a framework states of LQG have been selected that corresponds to Bianchi I solutions. This has established a bridge between the full theory of LQG and models of reduced symmetry, namely Loop Quantum Cosmology (LQC).

We aim here at developing an extension of QRLG able to address the semiclassical limit of the coupled system of matter and gravity in a meaningful way. In QRLG the gravitational d.o.f. are captured by spin-network states supported on graphs Γ that are colored with U(1) projected assignments $\{m_l, \imath_v\}$ of the gravitational

SU(2) group. In order to insert gauge fields and their excitations in this picture, we consider representations $\mathcal{R}_{\rm YM}$ of the Yang-Mills (YM) gauge group under scrutiny. The states that we are considering then belong to the total kinematical Hilbert space $\mathcal{H}_{\rm kin}^{\rm (tot)} = \mathcal{H}_{\rm kin}^{\rm (gr)} \otimes \mathcal{H}_{\rm kin}^{\rm (YM)}$, and are labelled, within the language of QRLG, by $|\Gamma; m_l, \imath_v; \mathcal{R}_{\rm YM}\rangle_R = |\Gamma; m_l, \imath_v\rangle_R \otimes |\Gamma; \mathcal{R}_{\rm YM}\rangle$.

Within this framework, the action of matter and gravity can be dealt with by resorting to Hamiltonian methods. This enables the computation of matrix-elements of the Hamiltonian operator in the semi-classical limit, and then provides the evolution for the YM Wilson-loops variables supported on the space-like cubulation deployed in QRLG. Expressing the polymer states of the YM theory in terms of the states of the Fock space, which are actually close to the experimental set up in the measurement processes, it is possible to recover n-point functions in the semiclassical limit from the non-perturbative approach. In the full theory, this has been achieved using the so called constructions of the shadow states,[32] which in the abelian case looks actually trivial to be implemented.

2. Theoretical Framework

At the classical level, in QRLG the spatial part of the metric and its dreibein are gauge-fixed to a diagonal form.[1–3,9] At the quantum level, QRLG is recovered from LQG[10–12] by imposing weakly gauge-fixing conditions on the kinematical Hilbert space of the latter. The cosmological implementation of QRLG then dramatically differs from Loop Quantum Cosmology (LQC),[13–15] in that the latter quantization is performed on minisuperspaces, while the former on the phase space of the full theory. The semiclassical limit of QRLG reproduces the quantum constraint adopted in LQC[6] and leads to enhanced inverse volume corrections.[7,8] Inhomogeneous extension of the Bianchi I model haves been so far considered,[3,4] as well as an attempt to extend QRLG to matter has been developed for scalar fields.[9]

It is remarkable that both LQC and QRLG replace the initial singularity with a Big Bounce scenario.[16] This prediction can be tested by means of the appearance of modifications to the scalar spectrum of the cosmological perturbations.[17–19] These modifications may arise due to quantum corrections to the dynamics of the gravitational and inflaton fields. A realistic model that involves electromagnetic radiation in the theoretical framework is then required to approach the problem from a more meaningful physical perspective.

A procedure consistent with LQG has been developed so far in order to quantize Yang-Mills fields.[20] The basic quantum variables are point holonomies, and together with their smeared momenta define a polymer representation.[21–24] The scalar constraint is quantized once a regularization of the classical expression is implemented: the operator is written entirely in terms of SU(2) holonomies, together with the corresponding smeared fluxes of electric field. A similar formulation has been adopted in QRLG, in which the volume operator is analytic. We focus our attention on states based at graphs having six-valent nodes,[8] which allows for the cubulation of the whole spatial manifold, as introduced by Thiemann.[20] States are of QRLG are a subset of states of LQG, and fare labelled by representations of U(1), in stead representations of SU(2), as in general states of SU(2).

The signature of the metric is fixed to $(-,+,+,+)$, the gravitational coupling constant is $\kappa = 16\pi G$ and $c = 1$. Metric tensor is defined as $g_{\mu\nu} = e^I_\mu e^I_\nu$, where e^I_μ are vierbein fields. Dreibeins are denoted as e^i_a, where lowercase latin indexes $a, b, .. = 1, 2, 3$ label coordinate on each Cauchy hypersurface constructed by ADM decomposition,[25] while $i, j, .. = 1, 2, 3$ are $\mathfrak{su}(2)$ internal indices.

2.1. *Loop Quantum Gravity*

The phase space of LQG is described by holonomies of Ashtekar-Barbero connections,[26] smeared along some curve γ,

$$h_\gamma := \mathcal{P}\exp\left(\int_\gamma A^j_a(\gamma(s))\tau^j\dot{\gamma}^a(s)\right) \tag{1}$$

and by fluxes of densitized triads across some surface S,

$$E(S) := \int_S n_j \epsilon_{abc} E^a_j dx^b \wedge dx^c. \tag{2}$$

The kinematical Hilbert space of the theory is constructed as the direct sum of the space of cylindrical functions of connections along each graph Γ,

$$\mathcal{H}^{(gr)}_{kin} := \bigoplus_\Gamma \mathcal{H}^{(gr)}_\Gamma = L_2(\mathcal{A}, d\mu_{AL}), \tag{3}$$

where $\mathcal{A}$ is the space of connections, $d\mu_{AL}$ denotes the Ashtekar-Lewandowski measure,[27] while the states are cylindrical functions of all links $l_i \in \Gamma$ and they are defined as $\Psi_{\Gamma,f}(A) := \langle A|\Gamma, f\rangle := f\big(h_{l_1}(A), h_{l_2}(A), ..., h_{l_L}(A)\big)$ for some continuous function $f : SU(2)^L \longrightarrow \mathbb{C}$.

The basis of the spin-network states are supported on graphs Γ, and are labelled by irreducible representations of SU(2) group elements. Wigner matrices $D^{j_l}(h_l)$ of spin j are valued on holonomies along the link ls, and contract with intertwiner invariant tensors i_v, which implement SU(2) gauge invariance at each node v. This construction entails spin-network states, whose expression reads

$$\Psi_{\Gamma, j_l, i_v}(h) = \langle h|\{\Gamma, j_l, i_v\}\rangle = \prod_{v\in\Gamma} i_v \cdot \prod_l D^{j_l}(h_l), \tag{4}$$

where the product $\prod_l$ extends over all the links l protruding from v and the $\cdot$ denotes contraction of the SU(2) indexes.

2.2. *Quantum Reduced Loop Gravity*

QRLG implements the restriction to diagonal spatial metric tensor and triads along some fiducial directions, where we pick up coordinates x, y, z. The metric tensor reads

$$dl^2 = a_1^2 dx^2 + a_2^2 dy^2 + a_3^2 dz^2, \tag{5}$$

where the three scale factors are functions of the cosmological time and of all the spatial coordinates.

The graph Γ now contains only three three kinds of links. These are the sets of links l_i along the three fiducial directions. Inverse densitized triads are fixed to be diagonal,

$$E^i_a = p^i \delta^i_a, \qquad |p^i| = \frac{a_1 a_2 a_3}{a_i}, \tag{6}$$

(indices are not summed in this expression), and this implies having imposed SU(2) gauge-fixing conditions in the internal space. Such a gauge-fixing is realized by the projection of SU(2) group elements, which are located on the links l_i, onto U(1) group representations obtained by stabilizing the SU(2) group along the internal directions $\vec{u}_l = \vec{u}_i$, with

$$\vec{u}_1 = (1,0,0), \quad \vec{u}_2 = (0,1,0), \quad \vec{u}_3 = (0,0,1). \tag{7}$$

The kinematical Hilbert space of QRLG now reads:

$$^R\mathcal{H}^{(gr)}_{kin} := \bigoplus_{\Gamma} {}^R\mathcal{H}^{(gr)}_{\Gamma}, \tag{8}$$

in which Γ is a cuboidal graph, while $^R\mathcal{H}^{(gr)}_{\Gamma}$ denotes the reduced Hilbert space on a fixed (reduced) graph. The basis states in $^R\mathcal{H}^{(gr)}_{\Gamma}$ are obtained by projecting SU(2) Wigner matrices on the state of maximum or minimum magnetic number $m_l = \pm j_l$, for the angular momentum component $J_l = \vec{J} \cdot \vec{u}_l$ along the link l:

$$^lD^{j_l}_{m_l m_l}(h_l) = \langle m_l, \vec{u}_l | D^{j_l}(h_l) | m_l, \vec{u}_l \rangle, \qquad h_l \in \mathrm{SU}(2). \tag{9}$$

Then reduced states, called reduced spin network states, are expressed as:

$$^R\Psi_{\Gamma, m_l, i_v}(h) = \langle h | \{\Gamma, m_l, i_v\} \rangle = \prod_{v \in \Gamma} \langle j_l, i_v | m_l, \vec{u}_l \rangle \cdot \prod_l {}^lD^{j_l}_{m_l m_l}(h_l), \;\; m_l = \pm j_l, \tag{10}$$

where $\langle j_l, i_v | m_l, \vec{u}_l \rangle$ are reduced (one-dimensional) intertwiners.

Elements of $^R\mathcal{H}^{(gr)}_{\Gamma}$ can be constructed out of those of the full theory by replacement of the SU(2) basis elements with the following objects

$$\langle j_l, m | m'', \vec{u}_l \rangle \langle m'', \vec{u}_l | D^{j_l}(h_l) | m'', \vec{u}_l \rangle \langle m'', \vec{u}_l | j_l, m' \rangle, \quad m'' = \pm j_l. \tag{11}$$

In a similar way, reduction of canonical variables to $^R h_{l_i}$ and $^R E(S)$ can be easily obtained by smearing along links of reduced, cuboidal graphs Γ and across surfaces S perpendicular to these links, respectively.

If one neglects the scalar curvature term, the scalar constraint operator can be obtained from that of LQG by only considering its euclidean part, and by replacing LQG operators with reduced ones. Its action on three-valent and six-valent nodes has been studied.[3,8]

3. U(1) Gauge Fields in Quantum Reduced Loop Gravity

We address in this section the quantization of abelian gauge fields within the framework of QRLG. For simplicity, we will limit this presentation to this simple illustrative case.

The Hilbert space of the total theory decomposes into

$$\mathcal{H}_{\text{kin}}^{(\text{tot})} = \mathcal{H}_{\text{kin}}^{(\text{gr})} \otimes \mathcal{H}_{\text{kin}}^{(\text{YM})}, \tag{12}$$

namely spin-network states are now the product of cylindrical functions of the gravitational SU(2) group and of the abelian U(1) gauge group. In order to introduce the latter quantities, we first have to individuate the canonical variables of the theory: the position variable is the U(1) gauge connection A_a, which is smeared *à la loop*

$$h_\gamma := \mathcal{P}\exp\Big(\int_\gamma A_a(\gamma(s))\dot{\gamma}^a(s)\Big) = \exp\Big(\int_\gamma A_a(\gamma(s))\dot{\gamma}^a(s)\Big), \tag{13}$$

while the electric fields, which are the conjugated momenta, act as left invariant derivatives in the electromagnetic connection

$$E^a = -\imath \frac{\delta}{\delta A_a}, \tag{14}$$

and in turn are smeared as fluxes.

States of the full theory are represented as $|\Gamma; m_l, \imath_v; \mathcal{R}_{\text{U}(1)}\rangle_R = |\Gamma; m_l, \imath_v\rangle_R \otimes |\Gamma; \mathcal{R}_{\text{U}(1)}\rangle$, where

$$\Psi_{\Gamma,\mathcal{R}_{\text{U}(1)}}(h) = \langle h|\Gamma; \mathcal{R}_{\text{U}(1)}\rangle = \prod_l e^{n_l\theta_l} = e^{\sum_l n_l\theta_l}, \tag{15}$$

where we have introduced the quantity $\theta_l = \int_{\gamma_l} A_a(\gamma(s))\dot{\gamma}^a(s)$, and recognized that $\mathcal{R}_{\text{U}(1)} = n \in \mathbb{Z}$. Finally, we label

$$\Psi_{\Gamma,n_l}(h) = e^{\sum_l n_l\theta_l}. \tag{16}$$

Flux operators then simply act as derivatives in θ_l

$$E_{l'}(S) = -\imath \frac{\delta}{\delta\theta_{l'}}, \tag{17}$$

and carry eigenvalues n_l, the U(1) representations on the links l. The cylindrical function $\Psi_0(h) \equiv \Psi_{\Gamma,0}(h) = 1$, which in this construction corresponds to the Ashtekar-Lewandowski vacuum of LQG, represents a state with infinite uncertainty on the magnetic field, which around a squared loop with infinitesimal measure $||l|| = \epsilon$ can be expressed from

$$h_l = \exp \epsilon^2 B^a \hat{l}_a \tag{18}$$

(having denoted orthogonal directions on the cubulation as $\hat{l}_a$), and zero uncertainty on the electric field, since

$$E_l(S)\Psi_0 = 0. \tag{19}$$

The vacuum $|\Psi_0\rangle$ of the flux-holonomy basis for the U(1) gauge field then differs dramatically from the vacuum $|0\rangle$ of the Fock basis. Nonetheless the relation between the two "vacua" can be easily derived for the case of U(1). We first introduce the decomposition of the electromagnetic field in Fourier modes, namely

$$\varphi^a_{\vec{k}} = \sum_{x^l} A^a_{\vec{x}}\, e^{\imath \vec{k}\cdot\vec{x}}, \tag{20}$$

in which $\vec{x}$ is a vector that individuates the position of the node v on the cubulation, and then calculate

$$\Psi_0(h) = \langle\Psi_0|0\rangle = \prod_{\vec{k}} \exp\left(-\frac{1}{2}\varphi^a_{\vec{k}}\varphi^a_{-\vec{k}}\right) = N\exp\left(-\frac{1}{2}\sum_{l\,l'}\frac{\ln h_l \ln h_{l'}}{||l||\,||l'||}\right). \tag{21}$$

with N normalization constant.

The construction of the shadow states[32] in LQG generalizes this map between states belonging to the loop representation and states belonging to the Fock representation.

Thanks to the map recovered, we can calculate n-point function of the U(1) gauge theory, implementing temporal evolution through the calculation the matrix elements of the Hamiltonian operator in the QRLG semiclassical limit.[33]

4. Conclusions

While seeking the semiclassical limit of gauge theories quantized within thin the framework of LQG, we have focused on the illustrative case of U(1) gauge theories, showing relation between the basis of the loop states, and the standard Fock basis, usually deployed while analyzing QED.

References

1. E. Alesci and F. Cianfrani, Europhys. Lett. **104**, 10001 (2013) [arXiv:1210.4504 [gr-qc]].
2. E. Alesci and F. Cianfrani, arXiv:1303.0762 [gr-qc].
3. E. Alesci and F. Cianfrani, Phys. Rev. D **87**, no. 8, 083521 (2013) [arXiv:1301.2245 [gr-qc]].
4. E. Alesci and F. Cianfrani, arXiv:1506.07484 [gr-qc].
5. E. Alesci, F. Cianfrani and C. Rovelli, Phys. Rev. D **88**, 104001 (2013) [arXiv:1309.6304 [gr-qc]].
6. E. Alesci and F. Cianfrani, Phys. Rev. D **90**, no. 2, 024006 (2014) [arXiv:1402.3155 [gr-qc]].
7. E. Alesci and F. Cianfrani, arXiv:1410.4788 [gr-qc].
8. E. Alesci and F. Cianfrani, arXiv:1506.07835 [gr-qc].
9. J. Bilski, E. Alesci and F. Cianfrani, arXiv:1506.08579 [gr-qc].
10. A. Ashtekar and J. Lewandowski, Class. Quant. Grav. **21**, R53 (2004) [gr-qc/0404018].
11. C. Rovelli, Cambridge, UK: Univ. Pr. (2004) 455 p
12. T. Thiemann, Cambridge, UK: Cambridge Univ. Pr. (2007) 819 p [gr-qc/0110034].
13. M. Bojowald, Lect. Notes Phys. **835**, pp.1 (2011).

14. A. Ashtekar and P. Singh, Class. Quant. Grav. **28**, 213001 (2011) [arXiv:1108.0893 [gr-qc]].
15. K. Banerjee, G. Calcagni and M. Martin-Benito, SIGMA **8**, 016 (2012) [arXiv:1109.6801 [gr-qc]].
16. A. Ashtekar, T. Pawlowski and P. Singh, Phys. Rev. Lett. **96**, 141301 (2006) [gr-qc/0602086].
17. M. Bojowald, G. Calcagni and S. Tsujikawa, Phys. Rev. Lett. **107**, 211302 (2011) [arXiv:1101.5391 [astro-ph.CO]].
18. I. Agullo, A. Ashtekar and W. Nelson, Phys. Rev. Lett. **109**, 251301 (2012) [arXiv:1209.1609 [gr-qc]].
19. G. Calcagni, Annalen Phys. **525**, no. 5, 323 (2013) [Erratum-ibid. **525**, no. 10-11, A165 (2013)] [arXiv:1209.0473 [gr-qc]].
20. T. Thiemann, Class. Quant. Grav. **15**, 1281 (1998) [gr-qc/9705019].
21. A. Ashtekar, S. Fairhurst and J. L. Willis, Class. Quant. Grav. **20**, 1031 (2003) [gr-qc/0207106].
22. A. Ashtekar, J. Lewandowski and H. Sahlmann, Class. Quant. Grav. **20**, L11 (2003) [gr-qc/0211012].
23. W. Kaminski, J. Lewandowski and M. Bobienski, Class. Quant. Grav. **23**, 2761 (2006) [gr-qc/0508091].
24. W. Kaminski, J. Lewandowski and A. Okolow, Class. Quant. Grav. **23**, 5547 (2006) [gr-qc/0604112].
25. R. L. Arnowitt, S. Deser and C. W. Misner, Gen. Rel. Grav. **40**, 1997 (2008) [gr-qc/0405109].
26. A. Ashtekar, Phys. Rev. Lett. **57**, 2244 (1986).
27. A. Ashtekar and J. Lewandowski, J. Math. Phys. **36**, 2170 (1995) [gr-qc/9411046].
28. T. Thiemann, Class. Quant. Grav. **15**, 839 (1998) [gr-qc/9606089].
29. M. Gaul and C. Rovelli, Class. Quant. Grav. **18**, 1593 (2001) [gr-qc/0011106].
30. C. Rovelli, F. Vidotto, *Covariant Loop Quantum Gravity: An Elementary Introduction to Quantum Gravity and Spinfoam Theory*, Cambridge University Press, 2014.
31. M. Domagala, M. Dziendzikowski and J. Lewandowski, arXiv:1210.0849 [gr-qc].
32. A. Ashtekar and J. Lewandowski, Class. Quant. Grav. **18**, L117 (2001) doi:10.1088/0264-9381/18/18/102 [gr-qc/0107043].
33. J. Bilski, F. Cianfrani, P. Donà and A. Marcianò, *in progress.*

Intrinsic Time Quantum Gravity

Hoi Lai Yu

Institute of Physics,
Academia Sinica, Taiwan
hlyu@gate.sinica.edu.tw

Correct identification of the true gauge symmetry of General Relativity being 3d spatial diffeomorphism invariant(3dDI) (not the conventional infinite tensor product group with principle fibre bundle structure), together with intrinsic time extracted from clean decomposition of the canonical structure yields a self-consistent theory of quantum gravity. A new set of fundamental commutation relations is also presented. The basic variables are the 8 components of the unimodular part of the spatial dreibein and 8 $SU(3)$ generators which correspond to Klauder's momentric variables that characterize a free theory of quantum gravity. The commutation relations are not canonical, but have well defined group theoretical meanings. All fundamental entities are dimensionless; and the quantum wave functionals are preferentially in the dreibein representation. The successful quantum theory of gravity involves only broad spectrum of knowledge and deep insights but no exotic ideas.

Keywords: Intrinsic time, momentric; Cotton-York tensor; 3-d diffeomorphism invariant.

1. Intrinsic time and physics of the Hamiltonian constraint

We had shown in[1,2] that the symplectic potential

$$\int \tilde{\pi}^{ij}\delta q_{ij} = \int \bar{\pi}^{ij}\delta \bar{q}_{ij} + \tilde{\pi}\delta \ln q^{\frac{1}{3}},$$

can be cleanly separated into the conjugate pair, $(\ln q^{\frac{1}{3}}, \tilde{\pi})$, consisting of (one-third of) the logarithm of the determinant of the spatial metric and the trace of the momentum, from $(\bar{q}_{ij}, \bar{\pi}^{ij})$, the unimodular part of the spatial metric with traceless conjugate momentum that allows a deparametrization of the theory wherein $\ln q^{\frac{1}{3}}$ plays the role of the intrinsic time variable for $\beta^2 = l - \frac{1}{3} > 0$ with l being the deformation parameter in the DeWitt supermetric; $G_{ijkl} = \frac{1}{2}(q_{ik}q_{jl} + q_{il}q_{jk}) - lq_{ij}q_{kl}$. This decomposition and identification of the intrinsic time variable point to a paradigm shift in the symmetries of Gravitation/space-time.

The fundamental symmetries of Einstein's General Relativity(GR) can be revealed by carrying out its full canonical analysis. Since space-time is not inert in Gravitation, the canonical analysis which has to be done without a fixed background had to wait until the work of Dirac[3] and Arnowitt-Deser-Misner. The 4-d metric and the action can be expressed, essentially without loss of generality as,

$$ds^2 = g_{\mu\nu}dx^\mu dx^\nu = -N^2dt^2 + q_{ij}(dx^i + N^i dt)(dx^j + N^j dt),$$

$$S = \int dtd^3x\,(\tilde{\pi}^{ij}\frac{dq_{ij}}{dt} - N^i H_i - NH) + \text{boundary term};$$

wherein several important features are revealed: only the spatial or 3-metric q_{ij} is dynamical with conjugate momentum $\tilde{\pi}^{ij}$, and the fields N^i and N (which are respectively called the shift and lapse functions, and they parametrize how the same spatial point is deformed from one hypersurface to the next) play the role of Lagrange multipliers associated, respectively, with the super-momentum H_i and super-Hamiltonian H constraints,

$$H_i = -2q_{ik}\nabla_j \tilde{\pi}^{jk}\,(=0),$$

$$H = \frac{2\kappa}{\sqrt{q}}[G_{ijkl}\tilde{\pi}^{ij}\tilde{\pi}^{kl} + V(q_{ij})]\,(=0),$$

and for GR($l = \frac{1}{2}$ and $V(q_{ij}) = -\frac{q}{(2\kappa)^2}[R-2\Lambda_{eff}]$). By the usual method of counting in canonical analysis, there are thus 6 degrees of freedom (d.o.f.) associated with the canonical pairs $(q_{ij}, \tilde{\pi}^{ij})$ for GR in four-dimensions, subject to 4 constraints, resulting in 2 net field d.o.f. in the theory. The constraints and the algebra, or commutation relations they obey (which for GR is also called the Dirac algebra) also shed light on the fundamental symmetries of the theory. Under the action of the super-momentum constraint (also called diffeomorphism constraint for reasons that will soon become clear), the fundamental variables changes by

$$\{f(q_{ij}, \tilde{\pi}^{ij}), H_k[N^k]\}_{\text{P.B.}} = \mathcal{L}_{\vec{N}} f, \qquad H_i[N^i] \equiv \int N^i H_i d^3x;$$

i.e. by a diffeomorphism.

So although it is possible to interpret the constraint H_i causes the same effect on the fundamental variables as the change induced by an infinitesimal general coordinate transformation, it should be emphasized that, 1)the true symmetry of the theory is dictated by the form of the constraints and the precise transformations they generate, and not by changes in the integration dummy spatial coordinate variable which is integrated over in both the action and in the Hamiltonian; 2)H_i generates diffeomorphisms which are changes in the dynamical fields evaluated at the *same* coordinate point, for instance, $\delta_{\vec{N}} q_{ij}(x) = \{q_{ij}(x), H_k[N^k]\}_{P.B.} = \mathcal{L}_{\vec{N}} q_{ij}(x)$, and in this aspect there is no distinction between usual Yang-Mills gauge transformations which are deemed 'internal' while GR is often naively associated with 'external' or space-time *coordinate* transformations. What is important is that the symmetries are not realized through 'transformation of coordinates' but through the transformation of fields in which the coordinate labels are inert and play no active role. Thus as far as spatial diffeomorphisms are concerned, they are genuine *gauge symmetries* of Einstein's theory but with spatial metric being the basic variable and are on equal footing with usual Yang-Mills 'internal' gauge symmetries which describes symmetries under changes at the same coordinate point with connection variables.

For GR, the interpretation of the changes generated by the super-Hamiltonian or Wheeler-DeWitt constraint, H, is not so straightforward. It can be shown that

$$\delta_N q_{ij} = \{q_{ij}, H[N]\} = \frac{4N\kappa}{\sqrt{q}} G_{ijkl}\tilde{\pi}^{kl} = \mathcal{L}_{Nn^\mu} q_{ij}\ [\text{modulo. EOM}],$$

$$\delta_N \tilde{\pi}^{ij} = \{\tilde{\pi}^{ij}, H[N]\} = \frac{N}{2} q^{ij} H - N\sqrt{q}(q^{ki}q^{lj} - q^{ij}q^{kl})R_{kl} + \mathcal{L}_{Nn^\mu}\tilde{\pi}^{ij}.$$

Thus the constraint generates diffeomorphism or Lie derivative, $\mathcal{L}_{Nn^\mu}$, in the direction normal to the Cauchy surface *only on-shell* on the constraint surface (i.e. when the constraints hold) and also when the equations of motion (EOM) are imposed (which for Einstein's theory in vacuum is the vanishing of the spacetime Ricci curvature tensor $R_{\mu\nu}$). This implies that in the quantum context, GR does not possess the full gauge symmetry of four-dimensional diffeomorphisms, but spatial diffeomorphisms will remain intact. This is corroborated by the fact that the Dirac algebra is not the algebra of 4-dimensional diffeomorphisms. The call to abandon 4-covariance is not new. In the simplification of the Hamiltonian analysis of GR, the fact that only the spatial metric is dynamical lead Dirac to conclude that 'four-dimensional symmetry is not a fundamental property of the physical world'.[3] In his seminal article, Wheeler had emphasized that space-time is a concept of 'limited applicability', and it is 3-geometry, not 4-geometry, which is fundamental in Quantum Geometrodynamics.[4]

In quantum field theories (QFT), true gauge symmetries have clear interpretation in the canonical context in Dirac quantization. Physical quantum states are annihilated by the constraints, and this have the direct interpretation that the state is invariant under symmetry transformations of the fundamental configuration variable. For instance in Yang-Mills gauge theories, the Gauss Law constraint is $G^b(\mathbf{x})\Psi[A_{ia}] = \nabla_j\hat{\pi}^{jb}(\mathbf{x})\Psi[A_{ia}] = 0$. Realizing the conjugate momentum of A_{ia} realized in the connection representation by operator, $\hat{\pi}^{ia} = \frac{\hbar}{i}\frac{\delta}{\delta A_{ia}}$ and Taylor expansion leads to

$$\Psi[A_{ia} + \delta_{gauge}A_{ia}] = \Psi[A_{ia}] + \int(\delta_{\text{gauge}}A_{jb}(\mathbf{x}))\frac{\delta\Psi}{\delta A_{jb}(x)}d^3\mathbf{x}$$

$$= \Psi[A_{ia}] + \frac{i}{\hbar}(\int \eta_b G^b d^3\mathbf{x})\Psi[A_{ai}] = \Psi[A_{ia}],$$

wherein $\delta_{\text{gauge}}A_{jb} = -\nabla_j\eta_b$, together with integration by parts over compact Cauchy surface without boundary, or with assumed vanishing of gauge parameter η_b on the boundary. This *linear dependents in canonical momentum*, manifests the *invariance under local gauge transformations* of all physical states in QFT. In precisely the same manner in GR, with regard to spatial diffeomorphisms,[5]

$$\Psi[q_{ij} + \delta_{\text{Diffeo.}}q_{ij}] = \Psi[q_{ij}] + \int(\delta_{\text{Diffeo.}}q_{ij})\frac{\delta\Psi}{\delta q_{ij}}d^3\mathbf{x} = \Psi[q_{ij}] + \frac{i}{\hbar}(\int N^i H_i d^3\mathbf{x})\Psi[q_{ij}],$$

since $\delta_{\text{Diffeo.}}q_{ij} = \mathcal{L}_{\vec{N}}q_{ij} = \nabla_i N_j + \nabla_j N_i$, the momentum operator $\hat{\tilde{\pi}}^{ij} = \frac{\hbar}{i}\frac{\delta}{\delta q_{ij}}$, and the diffeomorphism constraint $H_i\Psi[q_{ij}] = 0$. In Table 1, we enlist the differences in gauge structures between 3dDI and conventional Yang-Mills gauge theories.

Now, one can easily see that the *quadratic canonical momentum dependence* of the Hamiltonian constraint will deprive its role of generating temporal gauge transformations; but only constraints

$$0 \cong \frac{\sqrt{q}}{2\kappa}H = \bar{G}_{ijkl}\bar{\pi}^{ij}\bar{\pi}^{kl} - \beta^2\tilde{\pi}^2 + V(q_{ij}) = (\bar{H} - \beta\tilde{\pi})(\bar{H} + \beta\tilde{\pi}) \Rightarrow \bar{H} = \pm\beta\tilde{\pi}; \quad (1)$$

wherein $\bar{H}(\bar{\pi}^{ij}, \bar{q}_{ij}, q) := \sqrt{\bar{G}_{ijkl}\bar{\pi}^{ij}\bar{\pi}^{kl} + V(\bar{q}_{ij}, q)}$ plays the role of local Hamiltonian density that generates real intrinsic time evolution. Although the constraints form a first class algebra, and the lapse function N is *a priori* arbitrary; classically, the physical meaning of N is fixed *a posteriori* by the EOM and constraints,[1]

$$N = \frac{\sqrt{q}(\partial_t \ln q^{1/3} - \frac{2}{3}\nabla_i N^i)}{4\beta\kappa\bar{H}}.$$

Paradigm shift in the role of the Hamiltonian constraint and the identification of intrinsic time variable from the spatial metric collude to yield a theory of quantum gravity with underlying 3dDI gauge symmetry dictated by the spatial metric. The

Table 1. Comparison of the two different gauge structures

	Diffeomorphism Gauge Structures	Yang-Mills Gauge Structures
Basic Variables	Spatial metric tensor q_{ij}	Gauge connection A_{ia}
Symmetry Generators	$H_i(\mathbf{x}) = -2q_{ik}\nabla_j\pi^{jk}(\mathbf{x})(=0)$	$G^a(\mathbf{x}) = \nabla_i\pi^{ia}(\mathbf{x})(=0)$
Gauge transformation	$[q_{ij}(\mathbf{x}), H_k[N^k]] = \mathcal{L}_{\vec{N}}q_{ij}(\mathbf{x})$; $H_i[N^i] = \int N^i H_i d^3\mathbf{x}$	$[A_{ia}(\mathbf{x}), G^b[\eta_b]] = -\nabla_i\eta_a(\mathbf{x})G^b[\eta_b]$; $G^b[\eta_b] = \int \eta_b G^b d^3\mathbf{x}$
Commutation Relations	$[H_i(\mathbf{x}), H_j(\mathbf{y})]$ $= H_j(\mathbf{x})\partial_i\delta(\mathbf{x}-\mathbf{y}) + H_i(\mathbf{y})\partial_j\delta(\mathbf{x}-\mathbf{y})$	$[G^a(\mathbf{x}), G^b(\mathbf{y})]$ $= if^{ab}{}_c G^c(\mathbf{x})\delta(\mathbf{x}-\mathbf{y})$
	Not product of identical group(i.e. $SL(3R)$)	Infinite tensor product group $\prod_x G$;
Locality & Dimension	at each spatial point of base manifold	G=finite dimensional Lie group
	i.e. not of principle fibre structure	=usually referred to as the 'gauge group'

physical contents of this new theory are equivalent to GR,[6] however, with the great advantages of being capable of modifying the potential, i.e. to include the conformal structures through the Cotton-York tensor to achieve renormalizability of the theory without encountering any inconsistency in the constraint algebra.

Before passing, note that while q is a tensor density, the multi-fingered intrinsic time interval, $\delta \ln q^{\frac{1}{3}} = \frac{q^{ij}}{3}\delta q_{ij}$, is a scalar entity although being non-integrable. Hodge decomposition for any compact Riemannian manifold without boundary yields, $\delta \ln q^{\frac{1}{3}} = \delta T + \nabla_i \delta Y^i$, wherein the integrable gauge-invariant part of $\delta \ln q^{\frac{1}{3}}$ is $\delta T = \frac{2}{3}\delta \ln V_{\text{spatial}}$ proportional to the 3dDI logarithmic change in the spatial volume.[6] Upon quantization in (1), the Schrodinger equation can be derived,

$$i\hbar\frac{\delta\Psi}{\delta T} = \int i\hbar\frac{\delta\Psi}{\delta \ln q^{\frac{1}{3}}(x)}\frac{\delta \ln q^{\frac{1}{3}}(x)}{\delta T}d^3x = \left[\int \frac{\bar{H}(x)}{\beta}d^3x\right]\Psi = H_{\text{Phys}}\Psi,$$

H_{Phys} is the physical Hamiltonian generating evolution in global intrinsic time T.

2. Momentric and commutation relations of geometrodynamics

In usual quantum theories, the fundamental *canonical* CR, $[Q(\mathbf{x}), P(\mathbf{y})] = i\hbar\delta(\mathbf{x}-\mathbf{y})$, implies $\frac{P}{\hbar}$ is the generator of translations of Q which are also symmetries of the 'free theories' in the limit of vanishing interaction potentials. For geometrodynamics, the corresponding canonical CR is $[q_{ij}(\mathbf{x}), \tilde{\pi}^{kl}(\mathbf{y})] = i\hbar\frac{1}{2}(\delta^k_i\delta^l_j + \delta^l_i\delta^k_j)\delta(\mathbf{x}-\mathbf{y})$. However, neither positivity of the spatial metric is preserved under arbitrary translations in superspace generated by the conjugate momentum; nor is the 'free theory' invariant under translations, because the kinetic part, $G_{klmn}\tilde{\pi}^{kl}\tilde{\pi}^{mn}$, of the Hamiltonian dependents on q_{ij}. In quantum gravity, states which are infinitely peaked at the flat metric, or for that matter any particular metric with its corresponding isometries, cannot be postulated ad hoc; consequently, the underlying symmetry of even the 'free theory' is obscure. The Poisson brackets for the barred variables are,

$$\{\bar{q}_{ij}(\mathbf{x}), \bar{q}_{kl}(\mathbf{y})\} = 0, \quad \{\bar{q}_{kl}(\mathbf{x}), \bar{\pi}^{ij}(\mathbf{y})\} = P^{ij}_{kl}\,\delta(\mathbf{x}-\mathbf{y}),$$
$$\{\bar{\pi}^{ij}(\mathbf{x}), \bar{\pi}^{kl}(\mathbf{y})\} = \frac{1}{3}(\bar{q}^{kl}\bar{\pi}^{ij} - \bar{q}^{ij}\bar{\pi}^{kl})\delta(\mathbf{x}-\mathbf{y}); \tag{2}$$

with $P^{ij}_{kl} := \frac{1}{2}(\delta^i_k\delta^j_l + \delta^i_l\delta^j_k) - \frac{1}{3}\bar{q}^{ij}\bar{q}_{kl}$ denoting the traceless projection operator. This set is not strictly canonical. In the metric representation, the implementation of $\bar{\pi}^{kl}$ as traceless, symmetric, and self-adjoint operators is problematic. Remarkably, these difficulties can be cured by passing to the 'momentric variable' (first introduced by Klauder[7]) which is classically $\bar{\pi}^i_j = \bar{q}_{jm}\bar{\pi}^{im}$. In terms of spatial metric and momentric variables, the fundamental CR postulated (from which the classical Poisson brackets corresponding to (2) can be recovered) are then[8]

$$[\bar{q}_{ij}(\mathbf{x}), \bar{q}_{kl}(\mathbf{y})] = 0, \qquad [\bar{q}_{ij}(\mathbf{x}), \bar{\pi}^k_l(\mathbf{y})] = i\hbar \bar{E}^k_{l(ij)}\delta(\mathbf{x}-\mathbf{y}),$$
$$[\bar{\pi}^i_j(\mathbf{x}), \bar{\pi}^k_l(\mathbf{y})] = \frac{i\hbar}{2}(\delta^k_j\bar{\pi}^i_l - \delta^i_l\bar{\pi}^k_j)\delta(\mathbf{x}-\mathbf{y}); \tag{3}$$

wherein $\bar{E}^i_{j(mn)} := \frac{1}{2}(\delta^i_m\bar{q}_{jn} + \delta^i_n\bar{q}_{jm}) - \frac{1}{3}\delta^i_j\bar{q}_{mn}$ (with $\delta^j_i\bar{E}^i_{j(mn)} = \bar{E}^i_{j(mn)}\bar{q}^{mn} = 0$; $\bar{E}^i_{jil} = \bar{E}^i_{jli} = \frac{5}{3}\bar{q}_{jl}$) is the vielbein for the supermetric $\bar{G}_{ijkl} = \bar{E}^m_{n(ij)}\bar{E}^n_{m(kl)}$. Quantum mechanically, the momentric operators and CR *can be explicitly realized in the metric representation* by $\hat{\bar{\pi}}^i_j(\mathbf{x}) := \frac{\hbar}{i}\bar{E}^i_{j(mn)}(\mathbf{x})\frac{\delta}{\delta\bar{q}_{mn}(\mathbf{x})} = \frac{\hbar}{i}\frac{\delta}{\delta\bar{q}_{mn}(\mathbf{x})}\bar{E}^i_{j(mn)}(\mathbf{x}) = \hat{\bar{\pi}}^{\dagger i}_j(\mathbf{x})$ which are self-adjoint on account of $[\frac{\delta}{\delta\bar{q}_{mn}(\mathbf{x})}, \bar{E}^i_{j(mn)}(\mathbf{x})] = 0$.

3. New commutation relations for quantum gravity

Momentric variables, $\hat{\bar{\pi}}^i_j$, generate $SL(3,R)$ transformations of $\bar{q}_{ij} = \delta^{ab}\bar{e}_{ai}\bar{e}_{bj}$ which preserve its positivity and unimodularity. Moreover, they generate at each spatial point, an $SU(3)$ algebra. In fact, with 3×3 Gell-Mann matrices $\lambda^{A=1,\ldots,8}$, it can be checked that $T^A(\mathbf{x}) := \frac{1}{\hbar\delta(\mathbf{0})}(\lambda^A)^j_i\hat{\bar{\pi}}^i_j(\mathbf{x})$ generates the $SU(3)$ algebra with structure constants $f^{AB}{}_C$. The 5 verse 8 asymmetry in the independent components between $\bar{q}_{ij}$ and the symmetric traceless, $\tilde{\pi}^{ij}$ (also $\bar{\pi}^i_j$) in (2) and (3) is rectified by the unimodular dreibein-traceless momentric variables, ($\bar{e}_{ai} := e^{-\frac{1}{3}}e_{ai}, T^A$); each having 8 independent components and obey the *advocated new fundamental CR*,[9]

$$[\bar{e}_{ai}(\mathbf{x}), \bar{e}_{bj}(\mathbf{y})] = 0, \quad [\bar{e}_{ai}(\mathbf{x}), T^A(\mathbf{y})] = i(\frac{\lambda^A}{2})^k_i \bar{e}_{ak}\frac{\delta(\mathbf{x}-\mathbf{y})}{\delta(\mathbf{0})},$$

$$[T^A(\mathbf{x}), T^B(\mathbf{y})] = i f^{AB}_C T^C \frac{\delta(\mathbf{x}-\mathbf{y})}{\delta(\mathbf{0})}. \tag{4}$$

A number of intriguing features are encoded in this set of CR. Because of (4), the quantum wave functionals are preferentially selected to be in the dreibein representation. It is noteworthy that all entities in (4), including $(\bar{e}_{ai}, T^A)$ and $\frac{\delta(\mathbf{x}-\mathbf{y})}{\delta(\mathbf{0})}$, are dimensionless; and *neither the gravitational coupling constant nor Planck's constant make their appearance.* $\delta(\mathbf{0}) := \lim_{x\to y}\delta(\mathbf{x}-\mathbf{y})$ denotes the coincident limit; so there are no divergences in $\frac{\delta(\mathbf{x}-\mathbf{y})}{\delta(\mathbf{0})}$ which is unity in the coincident limit and vanishing otherwise. The second CR in (4) implies $\exp(i\int \alpha_B T^B \delta(\mathbf{0}) d^3\mathbf{y}')\bar{e}_{ai}(\mathbf{x})\exp(-i\int \alpha_A T^A \delta(\mathbf{0}) d^3\mathbf{y}) = (\exp(\frac{\alpha_A(\mathbf{x})\lambda^A}{2}))^j_i \bar{e}_{ja}(\mathbf{x})$, with $\exp(\frac{\alpha_A(\mathbf{x})\lambda^A}{2})$ being a local $SL(3,R)$ transformation(not gauged) which will break down to the 3dDI(gauged) subgroup when interactions are introduced. $SL(3,R)/3dDI$ generates all 3dDI inequivalent quantum states from any 3dDI initial state. While the CR in (4) states that $T^A(\mathbf{x})$ generates, at each spatial point, a separate $SU(3)$ algebra that characterize a free quantum theory. Further details on intrinsic time quantum geometrodynamics and discussions on the causality and time-ordering can be found in Chopin Soo's plenary talk in this meeting.

Acknowledgments

I would like to thank Eyo Eyo Ita III and Chopin Soo for beneficial discussions.

References

1. C. Soo and H.-L. Yu, 'General Relativity without the paradigm of space-time covariance and resolution of the Problem of Time', Prog. Theor. Phys, 013E01 (2014).
2. N. Ó Murchadha, C. Soo and H.-L. Yu, 'Intrinsic time gravity and the Lichnerowicz–York equation', Class. Quantum Grav. **30**, 095016 (2013).
3. P. A. M. Dirac, Proc. Roy. Soc. A**246**, 333 (1958).
4. J. A. Wheeler, *Superspace and the nature of quantum geometrodynamics*, in *Battelle Rencontres*, edited by C. M. DeWitt and J. A. Wheeler (NY: W. A. Benjamin, 1968).
5. Chopin Soo and Hoi-Lai Yu, 'Translation of the invited semi-popular article in Physics Bimonthly, **36**, 336 (2014)'
6. Huei-Chen Lin and Chopin Soo 'Intrinsic time geometrodynamics: explicit examples', Chin. J. Phys. **53**, 110102 (2015)(Special Issue On the occasion of 100 years since the birth of Einstein's General Relativity).
7. J. R. Klauder, 'Overview of Affine Quantum Gravity', Int. J. Geom. Meth. Mod. Phys. **3**, 81 (2006), and references therein.
8. Eyo Eyo Ita III, Chopin Soo, and Hoi-Lai Yu, 'Intrinsic Time Quantum Geometrodynamics' Prog. Theor. Exp. Phys. 083E01 (2015).
9. Chopin Soo and Hoi Lai Yu, 'New commutation relations for quantum gravity', Chin. J. Phys. **53**, 110106 (2015) (Special Issue On the occasion of 100 years since the birth of Einstein's General Relativity).

Cosmic Inflation from Emergent Spacetime Picture

Hyun Seok Yang

Center for Quantum Spacetime, Sogang University, Seoul 121-741, Korea
hsyang@sogang.ac.kr

We argue that the emergent spacetime picture admits a background-independent formulation of cosmic inflation. The inflation in this picture corresponds to the dynamical emergence of spacetime while the conventional inflation is simply an (exponential) expansion of a preexisting spacetime owing to the vacuum energy carried by an inflaton field. We show that the cosmic inflation arises as a time-dependent solution of the matrix quantum mechanics describing the dynamical process of Planck energy condensate in vacuum without introducing any inflaton field as well as an *ad hoc* inflation potential. Thus the emergent spacetime picture realizes a background-independent description of the inflationary universe which has a sufficiently elegant and explanatory power to defend the integrity of physics against the multiverse hypothesis.

Keywords: Emergent spacetime; inflationary universe; quantum cosmology.

1. Why is Emergent Spacetime Necessary for Cosmic Inflation?

Like black holes, the big bang in the very early universe involves extreme conditions that neither relativity nor quantum theory can explain on its own. If we trace back the history of our universe, we will meet an initial singularity in which matter reached almost infinite density and then, according to the theory of general relativity, the space was contracted to a point and the time flow nearly stopped. Thus the big bang suffers the initial singularity in which space and time cease to exist. This implies that the big bang must be a cosmological event generating space and time as well as matters. The initial singularity cannot be avoided in the inflationary cosmology either because there has to be a definite beginning to an inflationary universe.[1] This means that the inflation is incomplete to describe the very beginning of our universe and some new physics is needed to probe the past boundary of the inflating regions. One possibility is that there must have been some sort of quantum creation event as a beginning of the universe.

The inflation scenario so far has been formulated in the context of effective field theory coupled to general relativity. Hence, in this scenario, the existence of space and time is a priori assumed from the beginning and the scenario only describes what happens in a given spacetime. In other words, the inflationary scenario does not describe any generation (or creation) of spacetime but simply characterizes an expansion of a preexisting spacetime. It never addresses the (dynamical) origin of spacetime. Moreover, in most inflationary models, once inflation happens, it never stops and produces not just one universe, but an infinite number of universes. Therefore the conventional inflation inevitably leads to the multiverse which is not falsifiable since the multiverse cannot be tested experimentally.[2]

In consequence, we need a consistent quantum theory to describe how spacetime was generated through the big bang in order to correctly understand the origin of spacetime and our universe. In particular, we need a background-independent theory which does not assume the prior existence of spacetime background but instead provides a mechanism of spacetime generation such that any spacetime structure including flat spacetime arises as a solution of the theory itself. In other words, we need a background-independent formulation of cosmic inflation in which the inflation corresponds to the dynamical emergence of spacetime. Recently such a background independent formulation was proposed[3] so that the cosmic inflation arises as a time-dependent solution of the matrix quantum mechanics describing the dynamical process of Planck energy condensate in vacuum without introducing any inflaton field as well as an *ad hoc* inflation potential. The underlying mathematical principle is the well-known duality between geometry and algebra. In this scheme, the noncommutative (NC) algebra plays a more fundamental role from which the spacetime geometry is derived and the emergent gravity from NC $U(1)$ gauge theory is basically the large N duality, as depicted in Fig. 1.

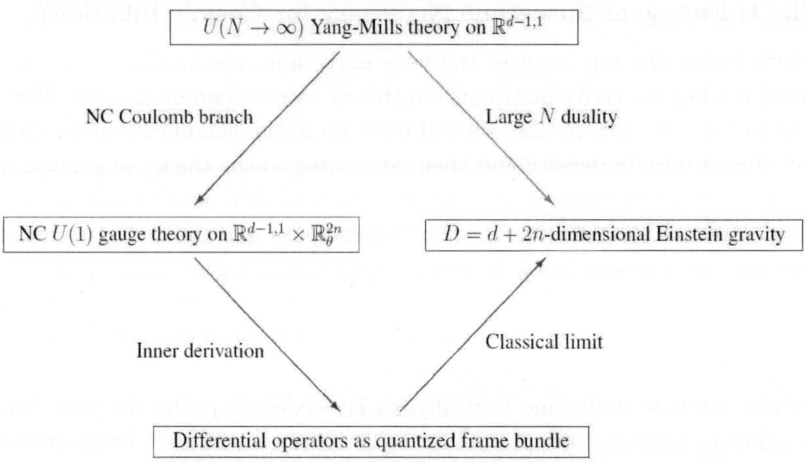

Fig. 1. Flowchart for large N duality.[3]

Since the concept of the multiverse raises deep conceptual issues even to require to change our view of science itself, it should be important to ponder on the real status of the multiverse whether it is simply a mirage developed from an incomplete physics like the ether in the late 19th century or it is of vital importance even in more complete theories. In next section, we will illuminate[3] how the emergent spacetime picture brings about radical changes of physics, especially, regarding to physical cosmology.

2. Inflationary Universe from Emergent Spacetime

Let us consider the matrix quantum mechanics (MQM) to address the background-independent formulation of cosmic inflation. The action in this case is given by[4]

$$S = \frac{1}{g^2}\int dt\, \mathrm{Tr}\Big(\frac{1}{2}(D_0\phi_a)^2 + \frac{1}{4}[\phi_a, \phi_b]^2\Big)$$
$$= \frac{1}{4g^2}\int dt\, \eta^{AC}\eta^{BD}\mathrm{Tr}[\phi_A, \phi_B][\phi_C, \phi_D], \tag{1}$$

where $\phi_0 \equiv iD_0 = i\frac{\partial}{\partial t} + A_0(t)$, $\phi_A(t) = (\phi_0, \phi_a)(t)$ and $\eta^{AB} = \mathrm{diag}(-1, 1, \cdots, 1)$, $A, B = 0, 1, \cdots, 2n$. With the notation of the symbol η^{AB}, it is easy to see that the matrix action (1) has a global automorphism given by

$$\phi_A \to \phi'_A = \Lambda_A{}^B \phi_B + c_A \tag{2}$$

if $\Lambda_A{}^B$ is a rotation in $SO(2n, 1)$ and c_A are constants proportional to the identity matrix. It turns out[3] that the global symmetry (2) is responsible for the Poincaré symmetry of flat spacetime emergent from a vacuum in the Coulomb branch of MQM and so will be called the Poincaré automorphism.

We remark that the time t in the action (1) is not a dynamical variable but a parameter. The concept of time has been introduced in Ref. 3 by considering a one-parameter family of deformations of zero-dimensional matrices which is parameterized by the coordinate t. Then the one-parameter family of deformations can be regarded as the time-evolution of a dynamical system. A close analogy with quantum mechanics implies that the concept of emergent time is connected with the time-evolution of the dynamical system. In this context, the one-dimensional matrix model (1) can be interpreted as a Hamiltonian system of a zero-dimensional (e.g., IKKT) matrix model.[5]

The equations of motion for the matrix action (1) are given by

$$D_0^2\phi_a + [\phi_b, [\phi_a, \phi_b]] = 0, \tag{3}$$

which must be supplemented with the Gauss constraint

$$[\phi_a, D_0\phi_a] = 0. \tag{4}$$

Now we want to apply the large N duality to MQM that is the $d = 1$ case in Fig. 1. It is important to notice that there are two different kinds of vacua in the Coulomb branch if we consider the $N \to \infty$ limit. In addition to the conventional commutative vacuum obeying the property $[\phi_a, \phi_b]|_{\mathrm{vac}} = 0$, there exists a novel coherent vacuum, the so-called NC Coulomb branch,[3] defined by

$$[\phi_a, \phi_b]|_{\mathrm{vac}} = -iB_{ab} \qquad \Rightarrow \qquad \langle\phi_a\rangle_{\mathrm{vac}} = p_a \equiv B_{ab}y^b \tag{5}$$

where the vacuum moduli y^a satisfy the Moyal-Heisenberg algebra given by

$$[y^a, y^b] = i\theta^{ab}, \qquad a, b = 1, \cdots, 2n \tag{6}$$

and $(\theta)^{ab} = (B^{-1})^{ab} = \alpha'(\mathbf{1}_n \otimes i\sigma^2)$ is a $2n \times 2n$ constant symplectic matrix. We emphasize that the NC Coulomb branch (5) together with a constant vacuum energy density given by $\mathcal{E} \equiv \langle A_0(t)\rangle_{\text{vac}}$ is a consistent *vacuum* solution of MQM since it satisfies the equations of motion (3) as well as the Gauss constraint (4). Since $\mathcal{E}$ is proportional to the identity matrix, it plays no role in the temporal covariant derivative D_0 and so it can be dropped without loss of generality. Also note that the coherent vacuum (5) saves the NC nature of matrices while the conventional vacuum dismisses the property.

If we consider fluctuations around the NC vacuum (5) given by

$$D_0 = \frac{\partial}{\partial t} - i\widehat{A}_0(t,y), \qquad \phi_a = p_a + \widehat{A}_a(t,y), \tag{7}$$

the action for the fluctuations is exactly mapped to a $(2n+1)$-dimensional NC $U(1)$ gauge theory as indicated in Fig. 1. If the conventional commutative vacuum were chosen, we would have failed to realize the higher-dimensional NC $U(1)$ gauge theory from MQM. Indeed it turns out[3,5] that the NC Coulomb branch is crucial to realize the emergent gravity from matrix models or large N gauge theories as summarized in Fig. 1. Furthermore the NC Coulomb branch (5) is very different from the conventional commutative vacuum because the former carries a nonzero vacuum energy density in contrast to the latter for which the vacuum energy density identically vanishes. Using the higher-dimensional NC $U(1)$ gauge theory, we can calculate the energy density for the vacuum condensate in the NC Coulomb branch defined by

$$\langle\phi_A\rangle_{\text{vac}} = p_A = \left(i\frac{\partial}{\partial t}, p_a\right). \tag{8}$$

The result is given by

$$\rho_{\text{vac}} = \frac{1}{4G_{YM}^2}|B_{ab}|^2 \tag{9}$$

where $G_{YM}^2(2\pi)^n|\text{Pf}\theta|g^2$ is the $(2n+1)$-dimensional gauge coupling constant.

It was shown in Ref. 3 that quantum gravity can be derived from the electromagnetism on NC spacetime by realizing the large N duality in Fig. 1 via the duality chain given by

$$\mathcal{A}_N^d \quad\Longrightarrow\quad \mathcal{A}_\theta^d \quad\Longrightarrow\quad \mathfrak{D}^d. \tag{10}$$

For the $d=1$ case in Fig. 1, the dynamical variables in MQM take values in $\mathcal{A}_N^1$ while those in $D=(2n+1)$-dimensional NC $U(1)$ gauge theory take values in $\mathcal{A}_\theta^1$. These two NC algebras $\mathcal{A}_N^1$ and $\mathcal{A}_\theta^1$ are related to each other by considering the NC Coulomb branch (8).[3] The module of derivations $\mathfrak{D}^1$ is a direct sum of the submodules of horizontal and inner derivations:[6]

$$\mathfrak{D}^1 = \text{Hor}(\mathcal{A}_N^1) \oplus \mathfrak{D}(\mathcal{A}_N^1) \cong \text{Hor}(\mathcal{A}_\theta^1) \oplus \mathfrak{D}(\mathcal{A}_\theta^1), \tag{11}$$

where horizontal derivation is locally generated by a vector field

$$k(t,y)\frac{\partial}{\partial t} \in \text{Hor}(\mathcal{A}_\theta^1). \tag{12}$$

In particular we are interested in the derivation algebra generated by the dynamical variables in Eq. (7). It is defined by

$$\widehat{V}_A = \{\mathrm{ad}_{\phi_A} = -i[\phi_A, \cdot\,]|\phi_A(t,y) = (iD_0, \phi_a)(t,y)\} \in \mathfrak{D}^1. \tag{13}$$

The large N duality in Fig. 1 says that the gravitational variables such as vielbeins in general relativity arise from the commutative limit of NC $U(1)$ gauge fields via the map (13). Then one may ask where flat Minkowski spacetime comes from. It turns out[7,8] that the $(2n+1)$-dimensional flat Minkowski spacetime is emergent from the vacuum condensate (8) since the corresponding vielbeins and the metric are given by $E_A^{(0)} = \widehat{V}_A^{(0)} = \left(\frac{\partial}{\partial t}, \frac{\partial}{\partial y^a}\right)$ and $ds^2 = -dt^2 + d\mathbf{y}\cdot d\mathbf{y}$. A striking fact is that the flat Minkowski spacetime is emergent from the Moyal-Heisenberg algebra (5) whose energy density is given by Eq. (9). Thus the flat spacetime was originated from the uniform vacuum energy (9) known as the cosmological constant in general relativity. This is a tangible difference from Einstein gravity, in which the energy-momentum tensor identically vanishes, i.e. $T_{AB} = 0$, for the flat spacetime. However, since we have started with the matrix model (1) in which any spacetime structure has not been assumed in advance, the spacetime was not existent at the beginning but simply emergent from the vacuum condensate (5). Therefore the Planck energy condensation into vacuum must be regarded as a dynamical process.

It is not difficult to show[3] that the dynamical process for the vacuum condensate is described by the time-dependent vacuum configuration given by

$$\langle\phi_a(t)\rangle_{\mathrm{vac}} = p_a(t) = e^{\frac{\kappa t}{2}}p_a, \qquad \langle\widehat{A}_0(t,y)\rangle_{\mathrm{vac}} = \widehat{a}_0(t,y), \tag{14}$$

where the temporal gauge field is given by an open Wilson line

$$\widehat{a}_0(t,y) = \frac{\kappa}{2}\int_0^1 d\sigma \frac{dy^a(\sigma)}{d\sigma} p_a(\sigma) \tag{15}$$

along a path parameterized by the curve $y^a(\sigma) = y_0^a + \zeta^a(\sigma)$ where $\zeta^a(\sigma) = \theta^{ab}k_b\sigma$ with $0 \leq \sigma \leq 1$ and $y^a(\sigma = 0) \equiv y_0^a$ and $y^a(\sigma = 1) \equiv y^a$. The constant $H \equiv (n-1)\kappa$ is identified with the inflationary Hubble constant. The $(2n+1)$-dimensional basis for the time-dependent vacuum (14) can easily be calculated using the map (13):

$$V_0 = \frac{\partial}{\partial t} - \frac{\kappa}{2}y^a\frac{\partial}{\partial y^a}, \qquad V_a = e^{\frac{\kappa t}{2}}\frac{\partial}{\partial y^a}. \tag{16}$$

And the dual orthonormal one-forms are given by

$$e^0 = dt, \qquad e^a = e^{Ht}dy_t^a \tag{17}$$

where $y_t^a = e^{\frac{\kappa t}{2}}y^a$. Here we used the relation $V_A = (E_0, \lambda E_a)$, and $\lambda = e^{n\kappa t}$ for the vacuum configuration (14). In the end, the time-dependent metric for the inflating background (14) is given by[3]

$$ds^2 = -dt^2 + e^{2Ht}d\mathbf{y}_t\cdot d\mathbf{y}_t. \tag{18}$$

Note that the temporal gauge field (15) is crucial to satisfy Eqs. (3) and (4) and to get a geodesically complete inflationary spacetime. Moreover the metric (18) is conformally flat, i.e., the corresponding Weyl tensors identically vanish and so describes a homogeneous and isotropic inflationary universe known as the Friedmann-Robertson-Walker metric in physical cosmology.

It is also easy to get a general Lorentzian metric describing $(2n+1)$-dimensional inflating spacetime by considering arbitrary fluctuations around the inflationary background (14).[3] They form a time-dependent NC algebra given by

$$ {}^t\mathcal{A}^1_\theta \equiv \Big\{ \widehat{\phi}_0(t,y) = i\frac{\partial}{\partial t} + \widehat{A}_0(t,y), \quad \widehat{\phi}_a(t,y) = e^{\frac{\kappa t}{2}} \big(p_a + \widehat{A}_a(t,y)\big) \Big\}. \tag{19} $$

We denote the corresponding time-dependent matrix algebra by ${}^t\mathcal{A}^1_N$ which consists of a time-dependent solution of the action (1). Then the general Lorentzian metric describing a $(2n+1)$-dimensional inflationary universe can be obtained by the following duality chain:

$$ {}^t\mathcal{A}^1_N \quad \Longrightarrow \quad {}^t\mathcal{A}^1_\theta \quad \Longrightarrow \quad {}^t\mathfrak{D}^1. \tag{20} $$

The module ${}^t\mathfrak{D}^1$ of derivations of the NC algebra ${}^t\mathcal{A}^1_\theta$ is given by

$$ {}^t\mathfrak{D}^1 = \Big\{ \widehat{V}_A(t) = (\widehat{V}_0, \widehat{V}_a)(t) | \widehat{V}_0(t) = \frac{\partial}{\partial t} + \mathrm{ad}_{\widehat{A}_0}, \quad \widehat{V}_a(t) = e^{\frac{\kappa t}{2}} \mathrm{ad}_{\widehat{\phi}_a} \Big\}, \tag{21} $$

where the adjoint operations are defined by Eq. (13). In the classical limit of the module (21), we get a general inflationary universe described by

$$ ds^2 = -dt^2 + \lambda^2 e^{2Ht} v^a_\alpha v^a_\beta (dy^\alpha_t - \mathbf{A}^\alpha)(dy^\beta_t - \mathbf{A}^\beta) \tag{22} $$

where the conformal factor $\lambda = 1 + \delta\lambda$ is determined by the volume-preserving condition.

In conclusion, the cosmic inflation arises as a time-dependent solution of a background-independent theory describing the dynamical process of Planck energy condensate in vacuum without introducing any inflaton field as well as an *ad hoc* inflation potential. The large N duality in Fig. 1 also implies[3] that cosmic inflation triggered by the Planck energy condensate into vacuum must be a single event. Thus the emergent spacetime is a completely new paradigm so that the multiverse debate in physics circles has to seriously take it into account.[9]

References

1. A. Borde, A. H. Guth and A. Vilenkin, Phys. Rev. Lett. **90** (2003) 151301.
2. Universe or multiverse?, edited by B. Carr (Cambridge Univ. Press, Cambridge, 2007).
3. H. S. Yang, *Emergent spacetime and cosmic inflation*, [arxiv:1503.00712].
4. W. Taylor, Rev. Mod. Phys. **73** (2001) 419.
5. H. S. Yang, Int. J. Mod. Phys. A **30** (2015) 1550016.
6. S. Azam, Commun. Alg. **36** (2008) 905.
7. H. S. Yang, J. High Energy Phys. **05** (2009) 012.
8. J. Lee and H. S. Yang, J. Korean Phys. Soc. **65** (2014) 1754.
9. B. Carr and G. Ellis, Astronomy & Geophysics **49** (2008) 2.29.

Cosmology of $f(T)$ Teleparallel Gravity

Yifu Cai

CAS Key Laboratory for Researches in Galaxies and Cosmology,
Department of Astronomy, University of Science and Technology of China,
Chinese Academy of Science, Hefei, Anhui 230026, China
yifucai@ustc.edu.cn

We briefly review the torsional construction of gravitation dubbed as $f(T)$ gravity, where f is an arbitrary function of the torsion scalar. We review its cosmological applications in different eras of the cosmic history. This construction can provide an interpretation of the late-time acceleration, alternative to a cosmological constant. If one traces back to very early times, inflation can be achieved for a class of extended torsional gravity models. Moreover, the big bang singularity can be avoided in the model of $f(T)$ bounce. A discussion on observational constraints of $f(T)$ gravity was presented focusing on the realization of quintom scenario. This proceeding article is based on an accompanied review[1] and a talk given in the the parallel session "Emergent Gravity and Quantum Gravity" in the 2nd LeCosPA International Symposium – *Everything about Gravity.*

Keywords: Modified gravity theories; teleparallel gravity; f(T) gravity.

1. Introduction

The motivation of modifying Einstein's GR, that possesses GR as a particular limit, can arise from the observational data. In fact, along with the rapid development of observational cosmology since the late 1990s, cosmologists believe that the universe may have undergone two phases of cosmic acceleration. The first phase is called inflation, which is suggested to occur about 10^{-35} second after the Big Bang. The second stage is happening right now, of which the underlying mystery is addressed as an unknown dark energy component. The speeding up of our universe at present was discovered by two independent observational signals on distant Type Ia supernovae (SNIa) in 1998, and later it was confirmed by a number of observations, such as those concerning the large scale structure (LSS), baryon acoustic oscillations (BAO), cosmic microwave background (CMB) radiation, and their cross-correlations.

Although phenomenological studies of these two cosmic accelerations have been successful, the fundamental theory that could explain their microscopic physics remains unknown. One possible approach is to introduce new matter components, such as the inflaton and quintessence. It is phenomenologically viable to construct a particular profile of the potential for the scalar field such that the dynamical evolution can fit to observations. There exists, however, a second approach to explain these accelerating phases, that is to modify the gravity sector. Most of the works in the literature start from the usual, curvature-based formulation, and extend the Einstein-Hilbert action, with the simplest example being the $f(R)$ paradigm.

Nevertheless, it is interesting to start from the Teleparallel Equivalent of General Relativity (TEGR), namely from the torsional formulation of gravity, and try to construct extensions such as the $f(T)$ gravity, in which the Lagrangian is taken to be a non-linear function of the TEGR Lagrangian T. The crucial issue is that although TEGR coincides completely with general relativity at the level of equations, $f(T)$ is different from $f(R)$ gravity, with novel features (amongst others note the significant advantage that the field equations of $f(T)$ gravity are of second order while those of $f(R)$ are of fourth order). The $f(T)$ gravity has interesting cosmological solutions, which provide alternative interpretations for the accelerating phases of the universe.

In this article we review the $f(T)$ gravity and its background dynamics. We refer to[1] for a comprehensive review. In Section 2 we review equations of motion of the $f(T)$ gravity. Then we in Section 3 review its cosmological applications. We show that $f(T)$ gravity can explain the late-time acceleration, alternative to a cosmological constant. Interestingly, if one traces back to very early times, inflation can be obtained for a certain class of extended torsion gravity. Alternatively, the Big Bang singularity can be avoided at even earlier moments due to the appearance of non-singular bounces. We summary our results in Section 4.

2. $f(T)$ gravity

It is well known that, the dynamical variable of the teleparallel gravity, as well as of its $f(T)$ extension, is the vierbein field $\mathbf{e}_A(x^\mu)$. We use Greek indices to run over all space-time coordinates and capital Latin indices to denote the coordinates of the tangent space-time. The vierbein field forms an orthonormal basis for the tangent space at each point x^μ of the manifold, i.e., $\mathbf{e}_A \cdot \mathbf{e}_B = \eta_{AB}$ where $\eta_{AB} = \mathrm{diag}(+1,-1,-1,-1)$ is the Minkowski metric for the tangent space. We analyze the vierbein vector with the use of its component form $\mathbf{e}_A = e^\mu_A \partial_\mu$, and thus the metric tensor can be expressed as

$$g_{\mu\nu}(x) = \eta_{AB} e^A_\mu(x) e^B_\nu(x) \ . \tag{1}$$

Moreover, the vierbein components follow the relations $e^\mu_A e^A_\nu = \delta^\mu_\nu$ and $e^\mu_A e^B_\mu = \delta^B_A$.

In $f(T)$ gravity, similarly to TEGR, one uses the Weitzenböck connection, defined as

$$\hat{\Gamma}^\lambda_{\mu\nu} \equiv e^\lambda_A \partial_\nu e^A_\mu = -e^A_\mu \partial_\nu e^\lambda_A \ . \tag{2}$$

This definition can lead to zero curvature, but nonzero torsion, and accordingly one can write down the torsion tensor as

$$T^\lambda_{\mu\nu} \equiv \hat{\Gamma}^\lambda_{\mu\nu} - \hat{\Gamma}^\lambda_{\nu\mu} = e^\lambda_A \left(\partial_\mu e^A_\nu - \partial_\nu e^A_\mu\right) . \tag{3}$$

Note that the difference between the Levi-Civita and Weitzenböck connections is the contortion tensor $K^{\mu\nu}{}_\rho = -\frac{1}{2}\left(T^{\mu\nu}{}_\rho - T^{\nu\mu}{}_\rho - T_\rho{}^{\mu\nu}\right)$, while it proves useful to define the superpotential as

$$S_\rho{}^{\mu\nu} = \frac{1}{2}\left(K^{\mu\nu}{}_\rho + \delta^\mu_\rho T^{\alpha\nu}{}_\alpha - \delta^\nu_\rho T^{\alpha\mu}{}_\alpha\right) . \tag{4}$$

Using these quantities one can write down the torsion scalar to be,

$$T \equiv S_{\rho}^{\ \ \mu\nu}\, T^{\rho}_{\ \ \mu\nu}. \tag{5}$$

which is also the Lagrangian density of TEGR. Then in the present formalism all the information concerning the gravitational field is included in the torsion tensor $T^{\lambda}_{\mu\nu}$, and the torsion scalar T arises from it in a similar way as the curvature scalar arises from the Riemann curvature tensor. Finally, the torsion scalar gives rise to the dynamical equations for the vierbein, which are comparable with the Einstein equations for the metric. While in teleparallel gravity the action is constructed by the teleparallel Lagrangian density T, the idea of $f(T)$ gravity is to generalize T to an arbitrary function $f(T)$. The action is given by,

$$\mathcal{S} = \int d^4x\, |e| \left[\frac{f(T)}{16\pi G} + L^{(m)}\right], \tag{6}$$

where $|e| = \det(e^A_\mu) = \sqrt{-g}$ and $L^{(m)}$ stands for the matter Lagrangian. For convenience, we would like to rewrite $f(T) = T + F(T)$ and mention that since the Ricci scalar R and the torsion scalar T differ only by a total derivative, the action (6) is equivalent to GR in the case of a vanishing $F(T)$ term, i.e. in the case where $f(T)$ gravity becomes TEGR.

Variation of the action (6) with respect to the tetrad e^A_ν leads to the field equations as

$$\begin{aligned}&e^{-1}\partial_\mu(ee^\rho_A S_\rho^{\ \mu\nu})[1+F_T] - e^\lambda_A T^\rho_{\ \ \mu\lambda} S_\rho^{\ \nu\mu} F_T \\ &+ e^\rho_A S_\rho^{\ \mu\nu}(\partial_\mu T) F_{TT} + \frac{1}{4} e^\nu_A [T + F(T)] = 4\pi G e^\rho_A T^{(m)}{}_\rho^{\ \nu}\,,\end{aligned} \tag{7}$$

where F_T and F_{TT} denote the first and second order derivatives of $F(T)$ with respect to the torsion scalar T, and $T^{(m)}{}_\rho^{\ \nu}$ is the energy-momentum tensor constructed by the matter field Lagrangian. If we assume the background manifold to be a spatially flat Friedmann-Robertson-Walker (FRW) universe then the vierbein takes the form

$$e^A_\mu = \mathrm{diag}(1, a, a, a)\,, \tag{8}$$

with a being the scale factor of the universe (hence the dual vierbein is $e^\mu_A = \mathrm{diag}(1, a^{-1}, a^{-1}, a^{-1})$ and the determinant $e = a^3$). One can immediately see that this choice gives rise to the well-known FRW metric: $ds^2 = dt^2 - a^2(t)\delta_{ij}dx^i dx^j$. Moreover, assuming a perfect fluid for matter, its energy momentum tensor takes

$$T^{(m)}{}_{\mu\nu} = -p_m g_{\mu\nu} - (\rho_m + p_m) u_\mu u_\nu\,, \tag{9}$$

where p_m, ρ_m and u^μ are the pressure, energy density and four velocity of the matter fluid. For the spatially flat FRW background, equations (7) lead to the following effective Friedmann equations:

$$H^2 = \frac{8\pi G}{3}\rho_m - \frac{F(T)}{6} - 2f_T H^2\,, \tag{10}$$

$$\dot{H} = -\frac{4\pi G(\rho_m + p_m)}{1 + F_T - 12H^2 F_{TT}}\,. \tag{11}$$

In the above expressions we have introduced the Hubble parameter $H \equiv \dot{a}/a$, where a dot denotes a derivative with respect to the cosmic time. Additionally, we have a very useful relation for the torsion scalar, namely

$$T = -6H^2 , \tag{12}$$

which can be derived straightforwardly from the definition (5) in the case of the FRW vierbein (8). Usually, one can introduce an equation-of-state parameter $w_m = p_m/\rho_m$ to characterize the dynamics of the matter fluid, where the fluid satisfies the continuity equation

$$\dot{\rho}_m + 3H(1+w_m)\rho_m = 0 . \tag{13}$$

With the above background equations at hand, one can investigate cosmological dynamics governed by any $f(T)$ gravity model.

3. Cosmological applications

One important goal is to construct viable models of $f(T)$ that can accommodate the expanding history of our background universe. In particular, observations including the SNIa, CMB, LSS, and BAO data, have revealed that the universe is currently accelerating. This mystery leads us to consider that, either about 70% of the universe is made up of a substance known as dark energy, about which we have almost no knowledge at all, or that GR is modified at cosmological scales. It is manifest that $f(T)$ gravity can provide a mechanism of realizing the late-time acceleration, alternative to dark energy or curvature-based modified gravity.

3.1. *Late time acceleration*

Let us consider the flat FRW background space-time, of which the metric takes

$$ds^2 = dt^2 - a^2(t)d\vec{x}^2 , \tag{14}$$

and hence, $e^A_\mu = \mathrm{diag}(1, a, a, a)$. In this case the background Friedmann equations were given in (10) and (11). One can reformulate the Friedmann equations to be the same as the ordinary ones in GR, namely

$$H^2 = \frac{8\pi G}{3}(\rho_r + \rho_b + \rho_{dm} + \rho_{DE}) , \tag{15}$$

$$\dot{H} = -4\pi G(\rho_r + \rho_b + \rho_{dm} + \rho_{DE} + P_r + P_{DE}) , \tag{16}$$

by defining the energy density and pressure of an "effective" dark energy component as

$$\rho_{DE} = \frac{2TF_T - F}{16\pi G} , \tag{17}$$

$$P_{DE} = \frac{F - TF_T + 2T^2F_{TT}}{16\pi G(1 + F_T + 2TF_{TT})} . \tag{18}$$

(note that $F = f - T$ as was introduced below Eq. (6)). The above cosmic system involves the radiation ρ_r, the baryon matter ρ_b, the cold dark matter ρ_{dm} and the "effective" dark energy ρ_{DE} arisen from $F(T)$'s contribution. The continuity equations for the various matter components are written as

$$\dot{\rho}_i + 3H(1 + w_i)\rho_i = 0 \ , \tag{19}$$

where the equation-of-state parameters $w_i \equiv \frac{p_i}{\rho_i}$ are defined as the ratio of pressure to energy density. Namely, $w_r = \frac{1}{3}$ for radiation; $w_b = 0$ for baryon matter; $w_{dm} = 0$ for cold dark matter, respectively. For $f(T)$ gravity, one can effectively define:

$$w_{DE} \equiv \frac{P_{DE}}{\rho_{DE}} = -1 + \frac{(F - 2TF_T - T)(F_T + 2TF_{TT})}{(F - 2TF_T)(1 + F_T + 2TF_{TT})} \ . \tag{20}$$

In particular, when $F_T = 0$, one gets $w_{DE} = -1$ and hence this case corresponds to the concordant model, i.e., the ΛCDM. One can further generalize the equation-of-state parameter of the i-th component as a function of the redshift $w_i(z)$, with the redshift given as $1 + z = \frac{a_0}{a}$ (where we can set the present scale factor a_0 to 1). Therefore, the evolutions of various energy densities are described by

$$\rho_i = \rho_{i0} \ \exp\left\{3\int_0^z [1 + w_i(\tilde{z})] d\ln(1 + \tilde{z})\right\} \ . \tag{21}$$

According to today's observations one learns that ρ_{DE} is of the order of 10^{-47} GeV^4 and the equation-of-state parameter is approximately $w = -1$. The energy scale of this dark component is far below any cut-off or symmetry-breaking scales in quantum field theory. Therefore, if dark energy originates from the particle physics nature, one has to search for a delicate mechanism of canceling over large vacuum energy density, but leaving unaffected the rest contribution that slightly deviates from zero, which is known as the famous cosmological constant problem.

In order to perform the stability analysis of the scenario at hand, one can transform the aforementioned dynamical system into its autonomous form $\mathbf{X}' = \mathbf{f(X)}$, where $\mathbf{X}$ is the column vector constituted by suitable auxiliary variables, $\mathbf{f(X)}$ the corresponding column vector of the autonomous equations, and prime denotes derivative with respect to $M = \ln a$. Then one can extract its critical points $\mathbf{X_c}$ satisfying $\mathbf{X}' = \mathbf{0}$, and in order to determine the stability properties of these critical points, one can expand around $\mathbf{X_c}$, setting $\mathbf{X} = \mathbf{X_c} + \mathbf{U}$, with $\mathbf{U}$ the perturbations of the variables considered as a column vector. Thus, for each critical point one can expand the equations for the perturbations up to first order as: $\mathbf{U}' = \mathbf{Q} \cdot \mathbf{U}$, where the matrix $\mathbf{Q}$ contains the coefficients of the perturbation equations. Finally, for each critical point, the eigenvalues of $\mathbf{Q}$ determine its type and stability.

Let us now investigate the dynamics of a cosmic system governed by $f(T)$ gravity. The main goal of this analysis is to demonstrate that there indeed exists a stable accelerating solution at late times, such that the present cosmic speeding up can be explained naturally. For convenience, we consider that the energy components in the universe are only dust matter ρ_m and radiation ρ_r. Then one can rewrite the equations of motion as a dynamical system by introducing the following dimensionless variables:

$$x \equiv -\frac{f}{6H^2}\ , \quad y \equiv \frac{Tf_T}{3H^2}\ , \quad z \equiv \Omega_r \equiv \frac{8\pi G\rho_r}{3H^2}\ , \tag{22}$$

where Ω_r is the dimensionless energy density parameter of radiation. Applying the background Friedmann equations and the continuity equations for all matter components, one can deduce

$$x' = -(2x+y)\frac{z+3-3x-3y}{2my-2+y}\ , \tag{23}$$

$$y' = 2my\frac{z+3-3x-3y}{2my-2+y}\ , \tag{24}$$

$$z' = -4z - 2z\frac{z+3-3x-3y}{2my-2+y}\ , \tag{25}$$

where the prime denotes the derivative with respect to $\ln a$, and where we have defined $m \equiv \frac{TF_{TT}}{F_T}$.

Note that, for a large family of $f(T)$ models, one can express T as a function of y/x, based on their definitions given in (22). Then m can be expressed in terms of y/x, too. To be specific, consider $f(T) = T + \alpha[(-T)^p - \beta]^q$. In this type of models, one can easily calculate that $m = p - 1 + \frac{1-q}{2q}(y/x)$. As a result, for a given form of $f(T)$, the dynamical system governed by equations (23), (24), (25) becomes autonomous. Accordingly, the main Friedmann equation (15) yields

$$\Omega_m \equiv \frac{8\pi G\rho_m}{3H^2} = 1 - x - y - z\ , \tag{26}$$

where Ω_m is the dimensionless density parameter of matter. Also the effective equation-of-state parameter for "dark energy" can be expressed as

$$w_{DE} = -\frac{x + y/2 - my}{(1-y/2-my)(x+y)}\ . \tag{27}$$

In order to extract the dynamical properties of the above autonomous system, we solve the combined equations: $x' = 0$, $y' = 0$, and $z' = 0$. As a consequence, we can obtain two isolated critical points and a continuous critical line, which are provided as follows,

$$\text{Point A}: \ x_c = 0,\ y_c = 0,\ z_c = 1\ ,$$
$$\text{Point B}: \ x_c = 0,\ y_c = 0,\ z_c = 0\ ,$$
$$\text{Line C}: \ x_c = 1 - y_c,\ z_c = 0\ .$$

In the following, we discuss these solutions and examine their stability.

• Point A corresponds to a radiation dominated solution, since at this point one can easily derive that $\Omega_r = 1$. Then one can study the stability of this solution by calculating the eigenvalues of the above linearized system. They are found to be

$$\left(1\ ,\ 2\left(1 - m \pm \sqrt{1 + 2m + m^2 - 2\frac{dm}{d(y/x)}}\right)\right)\ . \tag{28}$$

Hence, this critical point is unstable due to the presence of a positive eigenvalue, which means that the universe evolves away if there exists any small classical perturbation on the background trajectory.

• Point B corresponds to a matter dominated solution, since it exhibits $\Omega_m = 1$. One can further calculate the eigenvalues at this point as

$$\left(-1\ ,\ 2\left(1-m\pm\sqrt{1+2m+m^2-2\frac{dm}{d(y/x)}}\right)\right)\ . \tag{29}$$

Since the cosmological features of this point are not favored by observations, we deduce that for a realistic cosmology one expects that the universe should deviate from it along the background expansion. This can be satisfied if the real part of one of the above eigenvalues is positive definite.

• Line C corresponds to the solution dominated by the $f(T)$ contribution. In this solution, $x_c + y_c = 1$ indicates that both Ω_m and Ω_r are vanishing. Moreover, it is easy to read that $w_{DE} = -1$ and that H becomes constant on this critical line. Therefore, the background evolution on this critical line corresponds to a de Sitter expanding phase. Accordingly, the eigenvalues of the linearized system in this solution are given by $(-4, 0, -3)$, which are either negative or vanishing. This implies that the background evolution described by the last solution is marginally stable. Therefore, the solution of Line C is of cosmological interest in explaining the present acceleration of our universe.

3.2. *Bounce cosmology*

We are interested in examining how cosmological scenarios governed by $f(T)$ gravity can give rise to a non-singular bounce following. We mention that in order to be closer to the convention of the literature on this field, we use G as the gravitational constant, and the connection to other works of the literature is obtained through the known relations $8\pi G = \frac{1}{M_P^2} = \kappa^2$. Whether a universe is expanding or contracting merely depends on the positivity of the Hubble parameter. In the contracting phase that exists prior to the bounce, the Hubble parameter H is negative, while in the expanding one that exists after it we have $H > 0$. By making use of the continuity equations it follows that at the bounce point $H = 0$. Finally, it is easy to see that throughout this transition $\dot{H} > 0$. We refer to[2] for a comprehensive review of phenomenological studies upon bounce cosmology. Having in mind the above general requirements for a cosmic bounce, we can find that such a behavior can be easily obtained in the context of $f(T)$ cosmology. One can start with a specific, desirable form of the bouncing scale factor $a(t)$, and thus can derive explicitly $H(t)$. Concerning the matter fluid content of the universe, with an equation-of-state parameter w_m, its evolution equation (13) straightforwardly gives the solution $\rho_m(t)$ since $a(t)$ is already known. Then, we can insert these relations into (10), and determine the form of $f(T)$, which generates a nonsingular bounce. The above procedure can always be done numerically and exactly. In order to better understand the cosmological implications, we would like to present a specific example of a non-singular bounce that allows the analytical calculation.

To be explicit, we consider a bouncing solution with the scale factor of the universe evolving as

$$a(t) = a_B\left(1 + \frac{3}{2}\sigma t^2\right)^{1/3}, \tag{30}$$

where a_B is the scale factor at the bouncing point, and σ is a positive parameter which describes how fast the bounce takes place. Such an ansatz presents the non-singular bouncing behavior, corresponding to matter-dominated contraction and expansion. In addition, it has the advantage of allowing for semi-analytic solutions. In this ansatz the cosmic time t varies between $-\infty$ and $+\infty$, with $t = 0$ the bounce point. In the present subsection we normalize the bounce scale factor a_B to unity. Using the definition of the Hubble parameter, one can directly derive

$$H(t) = \frac{\sigma t}{(1 + 3\sigma t^2/2)}, \quad T(t) = -\frac{6\sigma^2 t^2}{\left(1 + \frac{3}{2}\sigma t^2\right)^2}. \tag{31}$$

Therefore, provided $-\sqrt{2/3\sigma} \leqslant t \leqslant \sqrt{2/3\sigma}$, the inversion of the above expression yields the following relation

$$t(T) = \pm\left(-\frac{4}{3T} - \frac{2}{3\sigma} + \frac{4\sqrt{T\sigma^3 + \sigma^4}}{3T\sigma^2}\right)^{1/2}, \tag{32}$$

where we have kept the solution pair that gives the correct ($t = 0$ at $T = 0$) behavior. Notice that when $t > \sqrt{2/3\sigma}$ and $t < -\sqrt{2/3\sigma}$, we have assumed that the usual Einstein gravity, i.e. the TEGR, is the prevailing framework, thus negating the need to pursue an $f(T)$ action in that region. Furthermore, we assume the matter content of the universe to be dust, with an equation-of-state parameter $w_m \approx 0$. Inserting this matter fluid into the continuity equation, one can easily arrive at the usual dust evolution, namely $\rho_m = \rho_{mB} a_B^3/a^3$, with ρ_{mB} its value at the bouncing point. Inserting the above expressions into (10) we obtain a differential equation for the reduced form $F(t)$, which can be easily solved analytically as

$$F(t) = \frac{4t}{(2 + 3\sigma t^2)M_P^2} \times \left[\frac{\rho_{mB}}{t} + \frac{6tM_P^2\sigma^2}{2 + 3t^2\sigma} + \sqrt{6\sigma}\rho_{mB}\,\mathrm{ArcTan}\left(\sqrt{\frac{3\sigma}{2}}t\right)\right]. \tag{33}$$

We mention that in the calculation we have set the integration constant to be zero, in order for the solution to be consistent with the Friedmann equation. Thus, the corresponding $f(T)$ expression that generates a bouncing scale factor of the form (30) arises from expression (33) with the insertion of the $t(T)$ relation from (32). Note that the solution (33) is an even function of t, and thus the $\pm$ solutions of (32) correspond to the contraction and expansion phase respectively. Obviously, they give the same form of $f(T)$.

In order to present the above process more clearly, in Fig. 1 we numerically depict the reduced form of $F(T)$ that generates the dust-dominated bouncing solution as desired. We particularly choose the parameters as follows: $a_B = 1$,

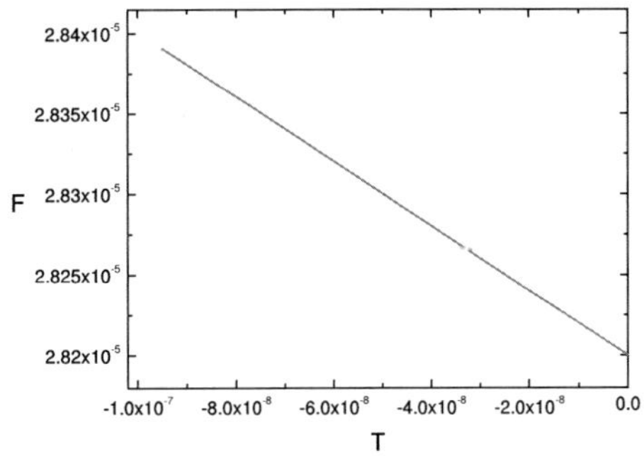

Fig. 1. Numerical result on the reduced form of F as a function of the torsion scalar T in the dust-dominated non-singular bounce cosmology. The model parameters are chosen as: $\sigma = 7 \times 10^{-6} M_P^2$ and $\rho_{mB} = 1.41 \times 10^{-5} M_P^4$, respectively. The graph is plotted in units of M_P.

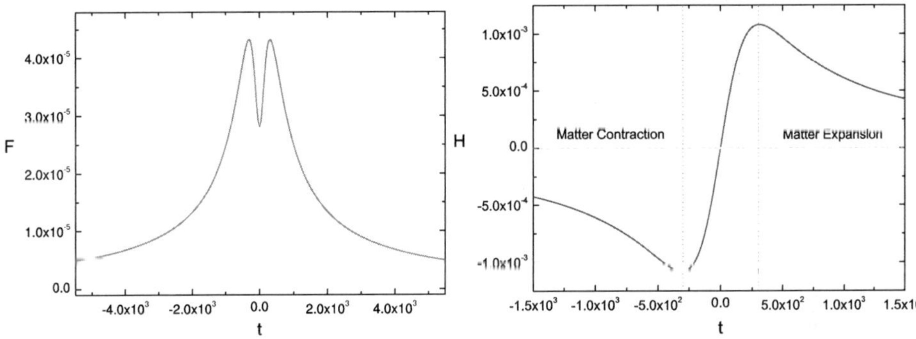

Fig. 2. Numerical result on the evolution of the reduced form F and the Hubble parameter H in terms of the cosmic time t, in the dust-dominated non-singular bounce cosmology. The model parameters are the same as those provided in Fig. 1. The graph is plotted in units of M_P.

$\sigma = 7 \times 10^{-6} M_P^2$, and $\rho_{mB} = 1.41 \times 10^{-5} M_P^4$. We note that the value of σ mainly relies on the amplitude of the CMB spectrum, and that of ρ_{mB} depends on how fast the standard Einstein gravity is recovered in $f(T)$ gravity. Furthermore, we numerically evolve the $F(T)$ and the Hubble parameter H as functions of cosmic time in Fig. 2. In particular, the left panel of Fig. 2 shows that the evolution of $F(T)$ is symmetric with respect to the bouncing moment $t_B = 0$. At t_B $F(T)$ arrives at a minimal value ρ_{mB}/M_P^2, which happens to cancel the contribution of normal matter fields, and thus leads to the non-singular bounce. From the right panel of Fig. 2, one finds that the cosmological evolution follows the usual Einstein gravity away from the bouncing phase, but it is dominated by $F(T)$ in the middle period. These feature are consistent with the analytical designs of the scenario.

4. Discussion

We close this proceeding article by summarizing the analyses and the obtained results. We started with a general discussion of the $f(T)$ gravity. Then we devoted to a study of cosmological applications of $f(T)$ gravity in several situations. We investigated the realization of the late-time acceleration of our universe by virtue of this theory under specific forms and then performed a phase space analysis to examine the classical stability issue. An interesting solution of the early universe was found to be a nonsingular bouncing cosmology under the derivation of $f(T)$ gravity, and hence, it demonstrated that the big bang singularity can be avoided by this theory. Finally, we presented the recent observational constraints on the model parameters.

It is interesting to comment that the $f(T)$ gravity is able to realize the so-called quintom scenario, which is also dubbed as the phantom-divide crossing in the literature. This phenomenological scenario is motivated by the mildly favored observational signal that the equation-of-state parameter of dark energy might have crossed -1 in the recent cosmological past. The difficulty for this scenario is based on the proof of a "No-Go" theorem for dynamical dark energy models, as was reviewed comprehensively in,[3] which states that for dark energy models described by a single perfect fluid or a single scalar field with a Lagrangian of K-essence form, their cosmological perturbations could encounter a severe divergence when the equation-of-state parameter is forced to cross -1.

Acknowledgments

The author is grateful to Pisin Chen and Antonino Marciano for kind invitation and providing the opportunity to present our review work arXiv:1511.07586 in the parallel session "Emergent Gravity and Quantum Gravity" of the 2nd LeCosPA Symposium. He acknowledges to Frederica Wang and Yen-Ling Lee as well as all coordinators for various supports which made this session successful. He also thanks S. Capozziello, M. De Laurentis, E. N. Saridakis and other colleagues for long-termed collaborations in the field of $f(T)$ gravity as well as invaluable friendships. The work is supported in part by the Chinese National Youth Thousand Talents Program, by the USTC start-up funding (Grant No. KY2030000049) and by the NSFC (Grant No. 11421303).

References

1. Y. F. Cai, S. Capozziello, M. De Laurentis and E. N. Saridakis, arXiv:1511.07586 [gr-qc].
2. Y. F. Cai, Sci. China Phys. Mech. Astron. **57**, 1414 (2014) [arXiv:1405.1369 [hep-th]].
3. Y. F. Cai, E. N. Saridakis, M. R. Setare and J. Q. Xia, Phys. Rept. **493**, 1 (2010) [arXiv:0909.2776 [hep-th]].

Non-Additive Entropies in ... Gravity?

Nikos Kalogeropoulos

Weill Cornell Medicine - Qatar,
Education City, PO Box 24144, Doha, Qatar.
nik2011@qatar-med.cornell.edu

We present aspects of entropic functionals relatively recently introduced in Physics which are non-additive, in the conventional sense of the word, some of which have a power-law functional form. We use as an example among them, and to be concrete in this work the "Tsallis entropy", which has, arguably, the simplest functional form. We present some of its properties and speculate about its potential uses in semi-Classical and Quantum Gravity.

Keywords: Non-additive entropy; Tsallis entropy; semiclassical gravity; quantum gravity.

1. Introduction and Motivation

Two of the contemporary trends in Statistical Physics are the re-examination of its foundations; the reasons and range of its applicability, and the construction of "alternative" functionals to that of the Boltzmann/Gibbs/Shannon (BGS) entropy $\mathcal{S}_{BGS}$. As is well-known, for a discrete set of outcomes parametrized by $i \in I$ and corresponding probabilities p_i. and with the Boltzmann constant being indicated by k_B, the BGS entropy is given by

$$\mathcal{S}_{BGS}[\{p_i\}] \ = \ -k_B \sum_{i \in I} p_i \log p_i \qquad (1)$$

The word "alternative" above, is actually a misnomer. The proposed entropic functionals have a different domain of applicability than the domain of $\mathcal{S}_{BGS}$. One such domain is the case of long-range interactions, such as electromagnetism and gravity, the latter being a long-range interaction in the Newtonian framework. It is not at all obvious, even J.W. Gibbs was a skeptic,[1] why $\mathcal{S}_{BGS}$ should describe the collective behavior of systems having such long-range interactions.

One way to deal with issues pertaining to the applicability of $\mathcal{S}_{BGS}$, is to construct other entropic functionals and develop a corresponding "thermodynamic formalism" that can be used to make theoretical predictions based on them, which are testable by experiments. Several such functionals have been proposed during the last 30 years, whose form has either been adapted from Information Theory, or other disciplines, or have been constructed for the purposes of Statistical Mechanics.

One such prominent example, for which the thermodynamic formalism and properties has been largely worked out is the case of the Havrda-Charvát/Daróczy/Vajda/Lindhard-Nielsen/Cressie-Read/Tsallis,[2] henceforth just to be called "Tsallis entropy" $\mathcal{S}_q$ for brevity. Even though the range of applicability of the Tsallis entropy is still subject to discussion, we choose to focus on $\mathcal{S}_q$ in this work due to its simple functional form, which shares common features with other entropic functionals of power-law form that have been recently proposed.

2. Definition and some properties of the "Tsallis entropy"

The Tsallis entropy $\mathcal{S}_q$ is a single-parameter family of functionals that are parametrized by $q \in \mathbb{R}$, even though an extension to $q \in \mathbb{C}$ has been recently proposed.[3] Then, with the notation of (1), $\mathcal{S}_q$ is defined as

$$\mathcal{S}_q[\{p_i\}] \; = \; k_B \cdot \frac{1}{q-1} \cdot \left(1 - \sum_{i \in I} p_i^q\right) \tag{2}$$

The extension to continuous systems is not entirely trivial and still subject to some controversy.[4] Avoiding all the subtleties, we pretend in this work that the naive extension to continuous systems is valid, as most works on this topic tend to assume. Then, let the possible set of outcomes form a sample space Ω endowed with a Borel measure (volume element) $dvol_\Omega$ and a probability distribution function (the Radon-Nikodym derivative of the actual measure μ describing the process, with respect to the volume) $\rho : \Omega \to \mathbb{R}_+$. The Tsallis entropy is then naively defined as

$$\mathcal{S}_q[\rho] \; = \; k_B \cdot \frac{1}{q-1} \cdot \left(1 - \int_\Omega [\rho(x)]^q \; dvol_\Omega\right) \tag{3}$$

We choose units so that $k_B = 1$ in the sequel. Usually Ω is chosen to be a finite dimensional Riemannian manifold $(\mathbf{M}, \mathbf{g})$ with its volume element $dvol_{\mathbf{M}}$ induced by the metric $\mathbf{g}$, and possibly additional data such as a symplectic structure (when $\mathbf{M}$ is the phase space of a system) etc.

Some properties of potential interest of the Tsallis entropy are[1]

(a) The BGS entropy is a member of the family of $\mathcal{S}_q$:

$$\lim_{q \to 1} \mathcal{S}_q \; = \; \mathcal{S}_{BGS} \tag{4}$$

(b) Positivity: $\mathcal{S}_q \geq 0$.

(c) Convexity: $\mathcal{S}_q$ is concave for $q > 0$ and convex for $q < 0$.

(d) Generalized additivity: For two independent occurrences/events A, B, namely if the joint probability $p(A * B)$ of the combined system $A * B$ is expressed in terms of the marginals $p(A)$ and $p(B)$ as $p(A * B) = p(A)p(B)$, then

$$\mathcal{S}_q(A * B) \; = \; \mathcal{S}_q(A) + \mathcal{S}_q(B) + (1-q)\mathcal{S}_q(A)\mathcal{S}_q(B) \tag{5}$$

This is the most significant difference between $\mathcal{S}_{BGS}$ which is just additive for such independent events, and $\mathcal{S}_q$.

(e) Sets of axioms distilling the essential properties of $\mathcal{S}_q$ exist, mirroring the better known Shannon and Khintchine axioms defining $\mathcal{S}_{BGS}$.

3. Accomplishments and critiques

Much like any new proposal, the Tsallis (or any other) of these recently proposed entropic functionals has its strong points but has also its detractors.

We believe that inasmuch as $\mathcal{S}_q$ provides a statistically different way from $\mathcal{S}_{BGS}$ to encode the important aspects of the collective behavior of a system of many degrees of freedom, it is very useful. Naturally, the same can be said about any other entropic functional, especially Renyi's entropy, which has withstood very well the test of time and appears in several disciplines. We should notice that there are many aspects and relations of equilibrium thermodynamics that can be carried over, usually without or with minor modifications, to the case of $\mathcal{S}_q$, which among many things, claims to also describe some non-equilibrium phenomena.[1] Moreover, there are numerous ongoing attempts to apply consequences stemming from $\mathcal{S}_q$ to different disciplines that appear to have only minimal relation to Physics or Information Theory.[1] Hence the scope of $\mathcal{S}_q$ may transcend what is traditionally the domain of Physics, in essence following the case of $\mathcal{S}_{BGS}$.

Some of the above statements, however, may also be seen as possible shortcomings of $\mathcal{S}_q$: the lack of a concrete set of thermodynamic laws obeyed by $\mathcal{S}_q$ is somewhat unnerving. In particular, even definitions of basic concepts such as equilibrium, temperature etc. exist, but they appear, to us at least, to be somewhat artificial.[1] Also: little is actually known about the nature and ab initio calculation of the non-entropic parameter q at both technical and conceptual levels. It is still quite unclear, for instance, what would be the value of q when two or more systems having different values of q interact with each other etc.,[5] nonwithstanding. One may have to resort to using the more general framework of superstatistics,[6] rather than remain within the strict confines of the formalism of $\mathcal{S}_q$ but this is still unclear.

The dynamical foundations of $\mathcal{S}_q$ are really not particularly well-understood. Whereas one can point out that ergodic theory and hyperbolic dynamical systems are underlying the use and widespread validity of $\mathcal{S}_{BGS}$, something similar is not really known for $\mathcal{S}_q$, or any of the other recently proposed entropies. Notice that our comments throughout this work refer only to Hamiltonian systems having many degrees of freedom. The accepted conjecture is that long ("fat-tailed") spatial and temporal correlations in the underlying system should be responsible for a growth rate of the phase space volume which has a power law rather than exponential form, for the system to be described $\mathcal{S}_q$.[7] However very little concrete evidence has been produced to support or refute such claims. And on top of this, someone can always question why it should be $\mathcal{S}_q$ that describes such systems, and not some other entropic functional among the numerous ones that have been constructed which still has a power-law form.

4. Occasional misunderstandings ...

It may appear, from the above arguments, that applying the ideas and the formalism of non-additive entropies, such as $\mathcal{S}_q$, to semi-Classical or Quantum Gravity may not be a particularly fruitful endeavor. Indeed, most problems arising in the development of $\mathcal{S}_q$ have become apparent for relatively "simple" particle systems. Far less is known about how to apply such concepts to field theories, and even more so to Quantum Gravity candidates which have the additional conceptual and technical complications related to diffeomorphism invariance and background-independence.

It is probably due to the numerous subtleties related to the indefinite signature of the space-time metric, to the diffeomorphism invariance of space-time and to the background independence of Quantum Gravity that some misunderstandings have occurred in the part of the literature that attempts to apply $\mathcal{S}_q$ to General Relativity and Quantum Gravity.

A typical case is that of the holographic nature of black hole entropy. This entropy depends on the surface area of a space-like section of the event horizon for let's say the Schwarzschild solution. As a result one might be lead to believe that it the entropy of a Schwarzschild black hole should somehow be related to $\mathcal{S}_q$ for $q \neq 1$ since, after all, $\mathcal{S}_{BGS}$ should be dependent on the "volume" of the "interior" of a black hole rather than its "surface".[8] This apparently reasonable, but rather naive picture, is incorrect for multiple reasons. First, the volume of the interior of a black hole is not defined. The space-time metric is geodesically incomplete, something that really defines the singularity of a space-time.[9] Second, even if one disregards the geodesic incompleteness of the metric , the volume of the "interior" of a Schwarzschild solution depends on the choice of the space-like surface.[10] We could choose such a "volume" to refer to a minimal space-like surface, but this would be an arbitrary and unjustified choice in the classical theory. By contrast, the surface area of any space-like section of the event horizon does not depend on the choice of such section and this is what is used in the entropy formulae. Third, the Schwarzschild solution is static, so it has a Killing vector field which is space-like in the interior part of the solution. Hence, trying to define the volume of the interior of such a solution is like trying to determine the volume of a cylinder: such a concept is physically meaningless due to the existence of the translation invariance along the axis of the cylinder which is an isometry; the volume of sections which are "nice"/"equivariant" under such isometries is what is actually physically relevant.

5. Possible applications to semi-classical and quantum gravity

One cannot help wondering about the possibility of applying $\mathcal{S}_q$ to Gravity. After all, neither black hole entropy nor Quantum Gravity are particularly well-understood, so it may pay off to find "unusual" paths toward resolving such problems. Our current attitude toward this is more favorable than not. Some of the reasons are:

(i) In the Newtonian framework, gravity is a long-range interaction (in 3 and 4 dimensions). Some studies and corresponding arguments involving long-range interactions, such as in space plasmas, seem to point out, according to their authors at least,[11] that the equilibrium distributions that can be derived by applying the Maximum Entropy Principle to $\mathcal{S}_q$, the so-called q-exponentials $e_q(x)$, appear to be a good fit to recently obtained observation data. So, it may be worth considering the possibility of Statistical Mechanics of gravitating systems that relies on $\mathcal{S}_q$.

(ii) It seems that correlations have a significant role to play in understanding the statistical basis of the black hole, and more generally, the holographic nature of "gravitational" entropy. If such correlations turn out to be strong enough to allow the phase space volume to grow in a power-law manner,[7] but not faster, power-law entropies, such as $\mathcal{S}_q$, may be needed for thermodynamics.

(iii) The naive path-integral approach to quantization, which was developed mostly during the 1960s, indicates that gravity is perturbatively (around a Minkowski background, at least) non-renormalizable. The corresponding questions about supergravities[12] or string-inspired "low-energy" actions are still unresolved. So, one is tempted to try to consider path integrals/"partition functions" based on q-exponentials and generalized curvature actions, such as

$$\mathcal{Z}_q = \int_{\mathfrak{M}} e_q^{S_q}\, \mathcal{D}_{\mathfrak{M}} \tag{6}$$

where $\mathfrak{M}$ is the space of pairs of (ultimately Euclidean signature) metrics of the background manifold $\mathbf{M}$ and functions $\phi : \mathbf{M} \rightarrow \mathbb{R}$, with $\mathcal{D}_{\mathfrak{M}}$ being a judiciously chosen measure (such as a q-Gaussian) on $\mathfrak{M}$ and a measure on the space of functions ϕ determining a conformal deformation of the volume $d\nu = e^{-\phi} dvol_{\mathbf{M}}$. The action S_q, familiar as the action of the Brans-Dicke scalar-tensor theory of gravity in the Jordan frame, is

$$S_q = \int_{\mathbf{M}} \mathrm{Scal}_q[\mathbf{g}, \phi]\, e^{-\phi}\, dvol_{\mathbf{M}} \tag{7}$$

where the generalized scalar curvature $\mathrm{Scal}_q[\mathbf{g}, \phi]$ is given by[13]

$$\mathrm{Scal}_q[\mathbf{g}, \phi] = \mathrm{Scal}[\mathbf{g}] - 2 \cdot \frac{\nabla^2 \phi}{\phi} + \left(1 - \frac{1}{q}\right) \cdot \frac{\|\nabla \phi\|^2}{\phi^2} \tag{8}$$

as is suggested by arguments based on optimal transport.[14] This tacitly assumes some form of long-term "memory" in the system which would most likely lead to an integro-differential, rather than Einstein's, equations in the classical limit, something that would have to be appropriately addressed.

(iv) There are very rich analytic and geometric structures[16] associated to $\mathcal{S}_q$ that may be of interest to gravity. In particular, one can provide synthetic definitions of Ricci curvature which are applicable to spaces with low regularity, such as graphs, Alexandrov spaces etc.[15] It is widely believed that the classical "smooth" picture of Riemannian manifolds is emergent from something more fundamental and also far less regular. In this spirit, structures such as generalized convexity conditions, dimension-dependent Poincaré and Sobolev inequalities, Wasserstein spaces etc. which are intimately related to $\mathcal{S}_q$ may be useful in formulating and deriving emergent "smooth" properties of the underlying space-time, especially in the semi-classical limit.

(v) The entropy $\mathcal{S}_q$ appears in Perelman's work on the 3-dimensional Poincaré conjecture.[17] Since space-times are many times assumed to have the form $\Sigma \times \mathbb{R}$ (if one assumes that the Hamiltonian approach has to be applicable and is willing to consider the possibility of excluding space-time topology change) where Σ is a Cauchy hypersurface which is a Riemannian 3-manifold, $\mathcal{S}_q$ may be relevant in further elucidating aspects of geometrization pertinent to gravity.

6. Acknowledgement

We would like to thank Professor Antonino Marciano, for graciously inviting us to orally present our work on aspects of which the present paper relies, in the parallel session that he organized on "Emergent and Quantum Gravity" at this Symposium.

References

1. C. Tsallis, *Introduction to Nonextensive Statistical Mechanics; Approaching a Complex World*, (Springer, New York, USA, 2009).
2. J. Havrda, F. Charvát, *Kybernetika* **3**, 30 (1967); I. Vajda, *Kybernetika* **4**, 105 (1968) (in Czech); Z. Daróczy, *Inform. Control* **16**, 36 (1970); J. Lindhard, V. Nielsen, *Det Kongelige Danske Videnskabernes Selskab Matematisk-fysiske Meddelelser (Denmark)* **38**, 1 (1971); N.A. Cressie, T.R. Read, *J. Roy. Stat. Soc. B* **46**, 440, (1984); C. Tsallis, *J. Stat. Phys.* **52**, 479 (1988).
3. G. Wilk, Z. Włodarczyk, *Physica A* **413**, 53 (2014).
4. S. Abe, *Europhys. Lett.* **90**, 50004 (2010); B. Andresen, *Europhys. Lett.* **92**, 40005 (2010); S. Abe, *Europhys. Lett.* **92**, 40006 (2010); J.P. Boon, J.F. Lutsko, *Phys. Lett. A* **375**, 329 (2011); J.F. Lutsko, J.P. Boon, *Europhys. Lett.* **95**, 20006 (2011); A. Plastino, M.C. Rocca, *Europhys. Lett.* **104**, 60003 (2013).
5. G.-A. Tsekouras, C. Tsallis, *Phys. Rev. E* **71**, 046144 (2005).
6. C. Beck, E.G.D. Cohen, *Physica A* **322**, 267 (2003).
7. R. Hanel, S. Thurner, *Europhys. Lett.* **96**, 50003 (2011).
8. C. Tsallis, L.J.L. Cirto, *Eur. Phys. J. C* **73**, 2487 (2013).
9. J.K. Beem, P.E. Ehrlich, K.L. Easley, *Global Lorentzian Geometry*, 2nd Ed. (Marcel Dekker Inc., New York, NY, USA ,1996).
10. B.S. DiNunno, R.A. Matzner, *Gen. Rel. Grav.* **42**, 63 (2010).
11. G. Livadiotis, D.J. McComas, *Space Sci. Rev.* **175**, 183 (2013).
12. R. Kallosh, *J. High Ener. Phys.* **12**, 009 (2010).
13. J. Lott, *Proc. Amer. Math. Soc.* **135**, 3375 (2007).
14. K.-T. Sturm, *Acta Math.* **196**, 65 (2006); K.-T. Sturm, *Acta Math.* **196**, 133 (2006); J. Lott, C. Villani, *Ann.Math.* **169**, 903 (2009).
15. C. Villani, *Synthetic Theory of Ricci Curvature Bounds*, Takagi Lectures **15**, 81 (2015).
16. C. Villani, *Optimal Transport: Old and New*, Vol. 338, (Springer, Berlin, Germany, 2009).
17. B. Kleiner, J. Lott, *Geom. Topol.* **12**, 2587 (2008).

Black Holes and Thunderbolt Singularities with Lifshitz Scaling Terms

Yosuke Misonoh* and Kei-ichi Maeda

Department of Physics, Waseda University,
Okubo 3-4-1, Shinjuku, Tokyo 169-8555, Japan
misonoh@aoni.waseda.jp

We study a static, spherically symmetric and asymptotic flat spacetime in the hypersurface orthogonal Einstein-aether theory with $z = 2$ Lifshitz scaling terms motivated by Hořava-Lifshitz theory. In our numerical analysis, we find two kinds of interesting solutions: (i) two-parameter family of back hole solutions with a regular universal horizon and (ii) a solution with thunderbolt singularity whose universal horizon turns to be a spacelike singularity.

Keywords: Black holes; Einstein-aether; Lifshitz scaling.

1. Introduction

To establish the fundamental gravitational theory, we have to resolve the difficulty in quantization of gravity. Unlike the other fundamental interactions, gravity is not renormalizable in perturbative approach. In other words, there appears infinite numbers of divergent Feynman diagrams. It means the infinite counter terms are required in order to regularize the gravitational quantum effects.

Recently, Hořava proposed a gravitational theory with Lifshitz scaling.[1] Lifshitz scaling is anisotropic scaling between space and time: $t \to b^{-1}t$, $x^i \to b^{-z}x^i$. Based on this scaling, we can define a scaling dimension such that $[t] = -z$ and $[x^i] = -1$. It is found that if z is set to be the number of the spacial dimension, the scaling dimension of the gravitational coupling constant becomes zero. It means that gravitational interaction acquires the renormalizability at least power-counting level. This gravitational theory with Lifshitz scaling is called Hořava-Lifshitz (HL) theory.

Lifshitz scaling leads unusual dispersion relation $\omega^2 \sim k^{2z}$, and then, the sound speed is given by $c \sim k^{z-1}$. If we consider $z > 1$, the sound speed of the ultimately excited particle diverges. Thus, it is unclear whether the causal boundary can be found or not. This question is partially resolved. In the context of Einstein-aether (æ-) theory, the causal horizon for such an instantaneous propagating mode can be found at least in a static and spherically symmetric spacetime, which is called *universal horizon*.[2,3] Since æ-theory is regarded as an infrared limit of HL theory, one may wonder whether the universal horizon can exist even if the Lifshitz scaling with $z > 1$ are present. In this paper, we will study the black hole solutions in æ-theory with $z = 2$ Lifshitz scaling terms.

2. Einstein-aether with Lifshitz scaling terms

We shall consider the following action:

$$
\begin{aligned}
I &= \frac{1}{16\pi G} \int d^4x\sqrt{-g}\left[\mathcal{L}_{(\mathrm{IR})} + \mathcal{L}_{(\mathrm{UV})}\right] \\
\mathcal{L}_{(\mathrm{IR})} &= R - M^{\mu\nu}{}_{\alpha\beta}\left(\nabla_\mu u^\alpha\right)\left(\nabla_\nu u^\beta\right) , \\
\mathcal{L}_{(\mathrm{UV})} &= -m_{\mathrm{pl}}^{-2}\left(\beta_1 \dot{u}^4 + g_2 \mathcal{R}^2\right) ,
\end{aligned}
\tag{1}
$$

where R is a four dimensional Ricci scalar curvature, G is a gravitational constant, $m_{\mathrm{pl}} := 1/\sqrt{G}$ is a Planck mass which represents a typical Lorentz violating scale and the aether field u^μ is a dynamical unit timelike vector field. The tensor $M^{\mu\nu}{}_{\alpha\beta}$ is defined by

$$
M^{\mu\nu}{}_{\alpha\beta} := c_{13}\delta^\mu{}_\beta\delta^\nu{}_\alpha + c_2\delta^\mu{}_\alpha\delta^\nu{}_\beta - c_{14}u^\mu u^\nu g_{\alpha\beta} , \tag{2}
$$

where where $c_{13} := c_1 + c_3, c_{14} := c_1 + c_4$ with $\{c_i\}$ $(i = 1\text{-}4)$ being the coupling constants in the æ-theory. Since we consider the infrared limit of HL theory, the aether field is described by a gradient of a khronon field φ as

$$
u_\mu := \frac{\nabla_\mu\varphi}{\sqrt{-(\nabla^\alpha\varphi)(\nabla_\alpha\varphi)}} . \tag{3}
$$

The four dimensional spacetime is foliated by three dimensional spacelike hypersurface Σ_φ. Hence we can define a three dimensional Ricci scalar curvature $\mathcal{R}$ in covariant manner:

$$
\mathcal{R} = R - (\nabla_\alpha u_\beta)(\nabla^\beta u^\alpha) + (\nabla_\alpha u^\alpha)^2 + 2\nabla_\alpha[\dot{u}^\alpha - u^\alpha(\nabla_\beta u^\beta)] , \tag{4}
$$

where $\dot{u}^\mu := u^\alpha\nabla_\alpha u^\mu$ is an acceleration of the aether and $\dot{u}^2 := \dot{u}_\mu\dot{u}^\mu$.

The infrared portion $\mathcal{L}_{(\mathrm{IR})}$ corresponds to the action of æ-theory with hypersurface orthogonal aether field, which is equivalent to the infrared limit of non-projectable HL theory. On the other hand, $\mathcal{L}_{(\mathrm{UV})}$ is an ultraviolet modification motivated by HL theory with $z = 2$ Lifshitz scaling, which is composed of the scalar combination of $\mathcal{R}$ and $\dot{u}^\mu$.

Considering quadratic order perturbation around flat spacetime, we obtain the dispersion relations of graviton (helicity-2 polarization) and scalar-graviton (helicity-0 polarization) as follows:

$$
\omega_G^2 = \frac{1}{1-c_{13}}k^2 , \tag{5}
$$

$$
\omega_S^2 = \frac{(c_{13}+c_2)(2-c_{14})}{c_{14}(1-c_{13})(2+c_{13}+3c_2)}k^2 + \frac{8(c_{13}+c_2)g_2}{2+c_{13}+3c_2}\left(\frac{k^2}{m_{\mathrm{pl}}}\right)^2 . \tag{6}
$$

Although the infrared portions which is proportional to k^2 correspond to æ-theory's one, the scalar-graviton acquires additional term which is proportional to k^4 due to $\mathcal{R}^2$ term in action. This additional term causes the divergence of the propagating speed of the scalar-graviton in ultraviolet limit.[4]

To obtain the basic equation, we have to consider the variation with respect to $g^{\mu\nu}$ and φ. However, in static and spherically symmetric spacetime, we can adopt simplified basic equations derived from the variation of $g^{\mu\nu}$ and U^{μ}, where $u^{\mu} = U^{\mu}/\sqrt{-U^{\alpha}U_{\alpha}}$. Since the hypersurface orthogonality of the aether is automatically satisfied, these two set of equations are completely equivalent.[5] Thus, we shall consider the simplified basic equations in the rest of this paper.

3. The static, spherically symmetric and asymptotic flat solutions with Lifshitz scaling terms

In this paper, we focus on a static and spherically symmetric spacetime with asymptotic flatness. The following Eddington-Finkelstein type ansatz is considered in order to avoid a coordinate singularity:

$$ds^2 = -T(r)dv^2 + 2B(r)dvdr + r^2 d\Omega^2 \,, \tag{7}$$

where v is an ingoing null coordinate and $B \geq 0$. The aether in this coordinate system is assumed to be

$$u^{\mu} = (a(r), b(r), 0, 0) \,. \tag{8}$$

Due to the constraint of the aether, the function $b(r)$ is given by

$$b(r) = \frac{a(r)^2 T(r) - 1}{2a(r)B(r)} \,. \tag{9}$$

Since the asymptotic flatness is imposed, the asymptotic values of the above variables are given by

$$T(r) \to 1 \,, \;\; B(r) \to 1 \,, \;\; a(r) \to 1 \,, \;\; b(r) \to 0 \,. \tag{10}$$

To clarify the asymptotic structure of the solution, we shall perform the asymptotic expansion around the frat spacetime:

$$\begin{aligned} T(r) &= 1 + \frac{T_1}{r} + \frac{T_2}{r^2} + \frac{T_3}{r^3} + \frac{T_4}{r^4} + \cdots \,, \\ B(r) &= 1 + \frac{B_1}{r} + \frac{B_2}{r^2} + \frac{B_3}{r^3} + \frac{B_4}{r^4} + \cdots \,, \\ a(r) &= 1 + \frac{a_1}{r} + \frac{a_2}{r^2} + \frac{a_3}{r^3} + \frac{a_4}{r^4} + \cdots \,. \end{aligned} \tag{11}$$

Substituting these series into the basic equations, and solving oder by order, we obtain the asymptotic solution. It is remarkable that every order of these function can be described by two arbitrary coefficient, T_1 and a_2 as in the case of æ-theroy.[6]

The free parameter T_1 is related to a black hole mass M as a Noether charge with respect to time translational symmetry, which is given by

$$M = -\frac{T_1}{2G}\left(1 - \frac{c_{14}}{2}\right) = -\frac{T_1}{2G_{\rm N}} \,, \tag{12}$$

where

$$G_{\rm N} := G\left(1 - \frac{c_{14}}{2}\right)^{-1} \tag{13}$$

is the observed Newton constant. On the other hand, the free parameter a_2 may characterize the distribution of the aether around a black hole.

Note that above asymptotic solution gives us the boundary conditions near spacial infinity. Thus we can obtain a numerical solution to integrate from an appropriate distant spacial point toward the center of a spherical object.

3.1. *Ultraviolet black holes*

To find black hole solution, we have to clarify the definition of a causal horizon in this theory. Since we have an instantaneous scalar-graviton whose propagating speed diverges in ultraviolet limit, the Killing horizon is no longer the event horizon. It is because Killing horizon only means a static limit for a null propagating mode whose sound speed is unity.

In order to define the causal horizon, we focus on the trajectory of such an instantaneous propagating mode. Note that the "light cone" angle of the instantaneous mode will be broaden to 180°. Thus the particle should travel along the three dimensional hypersurface Σ_φ. Then, if there is a spacetime structure such that Σ_φ parallels to a Killing vector ξ^μ associated with time translational symmetry, namely $u \cdot \xi = 0$, any particle cannot escape from this surface as least in a static and spherically symmetric spacetime. This surface is called *universal horizon.*[2,3]

If we assume only $\dot{u}^4$ term ($\beta_1 \neq 0$ and $g_2 = 0$), we can find the black hole solution with universal horizon. Since this solution possesses the causal boundary for ultraviolet particle whose sound speed is infinite, we then call it an *ultraviolet black hole.* The typical example is shown in Fig. 1.

This kind of solution can be found in a finite range of a_2. In other words, this solution depends on a_2 as well as T_1 unlike æ-case.

3.2. *Ultraviolet thunderbolt singularities*

If $\mathcal{R}^2$ term is joined in action, nemely $g_2 \neq 0$, the universal horizon always turns to be a spacelike singularity where the spacetime curvature diverges. The typical example is shown in Fig. 2.

Since the causal horizon itself turns to be physical singularity, this kind of solution cannot be regarded as a black hole solution. However, recall that the universal horizon is a surface defined by $u \cdot \xi = 0$, namely, a static limit for the ultimately excited dispersive particle with $z > 1$ Lifshitz scaling. Thus, the singular information from the universal horizon cannot escape outward. In this sense, this solution does not violate the cosmic censorship conjecture. This property is very similar to *thunderbolt singularity.* The thunderbolt singularity is firstly proposed in the context of quantum black hole evaporation in $(1 + 1)$ dimensional gravitational theory.[7,8]

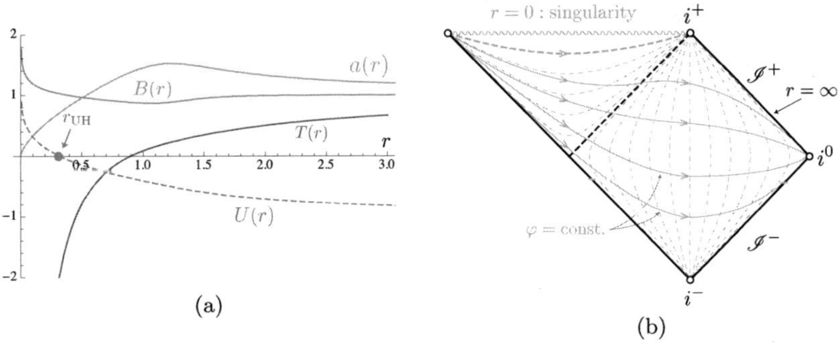

Fig. 1. The typical example of ultraviolet thunderbolt singularity. The left figure (a) represents the evolution of the each components of metric and aether. The coupling constants are set to $c_{13} = 0.100$, $c_2 = -6.135 \times 10^{-4}$, $c_{14} = 0.1000$, and $(a_2, \beta_1/m_{\rm pl}^2, g_2/m_{\rm pl}^2) = (0.200, 1.000, 0)$. The blue, red and green curves indicate the functions $T(r), B(r)$ and $a(r)$, respectively. The dashed brown curves indicate $U(r) := u \cdot \xi$. The universal horizon exists at $r_{\rm UH} = 0.320$. Since there are a universal horizon and a central singularity, this solution is referred to a black hole solution. In the right figure (b), the conformal structure is depicted. The solid green curves indicates Σ_φ, thus, the trajectory of the instantaneous mode. The dashed green curve indicates the universal horizon. The spacetime singularity at the center (the waving green curve).

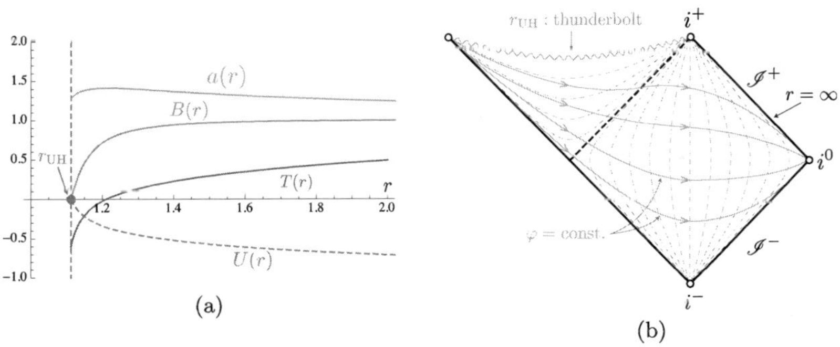

Fig. 2. The typical example of ultraviolet thunderbolt singularity. The left figure (a) represents the evolution of the each components of metric and aether. We choose $(a_2, \beta_1/m_{\rm pl}^2, g_2/m_{\rm pl}^2) = (1.112 \times 10^{-4}, 0, -1.000)$. The rest of the coupling constants are the same as those of Fig. 1. The blue, red and green curves indicate the functions $T(r), B(r)$ and $a(r)$, respectively. The dashed brown curves indicate $U(r) := u \cdot \xi$. The singular universal horizon exists at $r_{\rm UH} = 1.157$. In the right figure (b), the conformal structure is depicted. The solid green curves indicates Σ_φ, thus, the trajectory of the instantaneous mode. The waving green curve indicates the singular universal horizon.

In these case, the physical singularity spreads on null surface. We shall call our solution *ultraviolet thunderbolt singularity.*

4. Conclusion

With $z > 1$ Lifshitz scaling, there will be an instantaneous mode whose sound speed diverges in ultraviolet limit. Thus, the Killing horizon is no longer the event horizon in static and spherically symmetric spacetime. However if there is a surface where $u \cdot \xi = 0$, any particle cannot escape from this boundary even if its sound speed is infinite, which is called universal horizon. In this paper, we have studied the solution with the universal horizon in æ-theory with $z = 2$ Lifshitz scaling.

In our analysis, we have found two types of interesting solutions. One is a black hole solution. This types of solution can be found if we assume only $\dot{u}^4$ term in action, namely $\beta_1 \neq 0$ and $g_2 = 0$. Unlike æ-case, this solution depends on two parameters: T_1 and a_2. T_1 is associated with a black hole mass M, on the other hand, a_2 seems related to the distribution of aether around black hole.

If we consider $\mathcal{R}^2$ term as a ultraviolet modification, we find the ultimate thunderbolt singularity whose universal horizon turns to be physical singularity. Although this solution cannot be regarded as a black hole, the singular information is not detected by outside observer. Hence the cosmic censorship conjecture is not violated. The thunderbolt singularity is known as a final state of black hole evaporation in $(1 + 1)$ dimensional quantum gravity theory. Thus, it is suggestive that such a kind of solution is reproduced in HL theory as a candidate for quantum gravity.

Acknowledgments

This work is supported by a Waseda University Grant for Special Research Projects (project number: 2015S-083) and by Grants-in-Aid from the Scientific Research Fund of the Japan Society for the Promotion of Science (No. 25400276).

References

1. P. Hořava, Phys. Rev. **D 79**, 084008 (2009).
2. E. Barausse, T. Jacobson and T.P. Sotiriou, Phys. Rev. **D 83**, 124043 (2011).
3. D. Blas and S. Sibiryakov, Phys. Rev. **D 84**, 124043 (2011).
4. Y. Misonoh and K. Maeda, Phys. Rev. **D 92**, 084049 (2015).
5. E. Barausse, T.P. Sotiriou, Class. Quantum Grav. **30**, 244010 (2013).
6. C. Elling and T. Jacobson, Class. Quant. Grav. **23**, 5643-5660 (2006).
7. S.W. Hawking and J.M. Stewart, Nucl. Phys. **B400** 393-415 (1993).
8. A. Ishibashi and A. Hosoya, Phys. Rev. **D 66** 104016 (2002).

PARALLEL SESSION

III. Dark Matter and Dark Energy

Charged Q-Balls in Gauge Mediated SUSY Breaking Models

Jeong-Pyong Hong[a], Masahiro Kawasaki[a,b], and Masaki Yamada[a,b]

[a] *Institute for Cosmic Ray Research, The University of Tokyo, 5-1-5 Kashiwanoha, Kashiwa, Chiba 277-8582, Japan*

[b] *Kavli IPMU (WPI), UTIAS, The University of Tokyo, 5-1-5 Kashiwanoha, Kashiwa, 277-8583, Japan*

It is known that after Affleck-Dine baryogenesis, spatial inhomogeneities of Affleck-Dine field grow into non-topological solitons called Q-balls. In gauge mediated SUSY breaking models, sufficiently large Q-balls with baryon charge are stable while Q-balls with lepton charge can always decay into leptons. For a Q-ball that carries nonzero B and L charges, the difference between the baryonic component and the leptonic component in decay rate may induce nonzero electric charge on the Q-ball. This implies that charged Q-ball, also called gauged Q-ball, may emerge in our universe. In this paper, we investigate two complex scalar fields, a baryonic scalar field and a leptonic one, in an Abelian gauge theory. We find stable solutions of gauged Q-balls for different baryon and lepton charges. Those solutions show that a Coulomb potential arises and the Q-ball becomes electrically charged as expected. It is energetically favored that some amount of leptonic component decays, but there is an upper bound on its amount due to the Coulomb force. The baryonic decay also becomes possible by virtue of electrical repulsion, but we find that the evolution itself stops before it occurs mainly due to the Schwinger limit, so that the charged Q-balls eventually survive in the universe.

Keywords: Gauge mediation; charged Q-ball; dark matter; Schwinger effect.

1. Intrduction

After Affleck-Dine baryogenesis,[1] spatial inhomogeneities of Affleck-Dine field grow into non-topological solitons called Q-balls,[2–4] which are typically neutral. We hope these Q-balls survive until present universe and contribute to dark matter. In Gauge Mediated SUSY Breaking Models, Q-balls are stable against decay into nuclei, but may decay into leptons, which are typically lighter than nuclei. Due to the decay into charged leptons, Q-balls can be electrically charged, which we call charged Q-balls,[5] and by virtue of the stability of the baryonic component, charged Q-balls may survive in the universe, so we can expect the charged Q-balls may contribute to the dark matter.

2. Global Q-ball

Q-ball is defined as the lowest energy configuration for a fixed charge in a global U(1) scalar field theory. Using the Lagrange multiplier method, we only need to minimize the following function:

$$E_\omega \equiv E + \omega \left[Q - \frac{1}{2i} \int d^3x (\phi^* \dot{\phi} - \phi \dot{\phi}^*) \right] \tag{1}$$

where E is given by

$$E = \int d^3x \left[\frac{1}{2} \left(|\dot{\phi}|^2 + |\nabla\phi|^2 \right) + V(\phi) \right]. \tag{2}$$

and $\phi \equiv \sqrt{2}\Phi$ is the renormalized scalar field, where Φ is the original complex scalar field. We set the boundary conditions as

$$\frac{d\phi}{dr}(0) = 0, \quad \phi(\infty) = 0. \tag{3}$$

Then it is derived that the solution has the form[6]

$$\phi(x,t) = e^{i\omega t}\phi(x). \tag{4}$$

In gauge mediated SUSY breaking models, it is known that we can approximate $V \sim V_0 =$ const. and there exists an analytic solution[7] given by

$$\phi = \begin{cases} \phi_0 \sin(\omega r)/\omega r, & r < R \equiv \pi/\omega \\ 0, & r > R \end{cases} \tag{5}$$

$$\tag{6}$$

whose energy is then written as

$$E = \frac{4\pi\sqrt{2}}{3} V_0^{1/4} Q^{3/4}. \tag{7}$$

Then,

$$dE/dQ = \sqrt{2}\pi V_0^{1/4} Q^{-1/4}. \tag{8}$$

This indicates that for a large charge, dE/dQ may become small enough. Therefore, for a baryonic Q-ball with a large baryon number, it may hold that

$$\frac{dE}{dB} < m_p, \tag{9}$$

where m_p ($\simeq$ 1GeV) is the proton mass. This means that the Q-ball is stable against the decay into protons. On the other hand, if Q-ball carries a lepton number, it can decay into leptons. This is the motivation of our assumption that baryonic component of Q-ball is stable while leptonic component can decay into leptons.

3. Charged Q-ball (1-scalar)

Since we are interested in electrically charged Q-ball, we must include a U(1) gauge field A_μ as well as the complex scalar field. But we consider only Coulomb potential as the gauge field, which means we neglect the eletrodynamical effects and consider only electrostatic solutions. The spatially localized configuration in this theory is called charged (gauged) Q-ball. We show some typical solutions for gauge mediation model in Fig. 1. We see that the scalar field (blue line) is pushed outward by the coulomb repulsion.

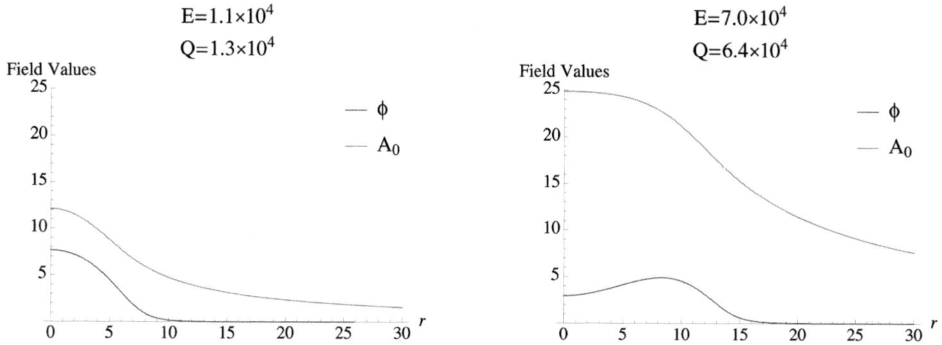

Fig. 1. Gauged Q-balls in gauge mediation-like model with $e^2 = 0.002$ and $m_\Phi = 1$.

4. Charged Q-ball (2-scalar)

Since we need bayronic and leptonic components, we consider a two scalar model. We assume that the decay of the Q-ball always takes place if energetically allowed and that the evolution of the Q-ball can be approximated as a sequence of gauged Q-ball solutions with the same B, due to the stability of the baryonic component. The solutions for $B = 8.4 \times 10^4$ are shown in Fig. 2. We see that as the leptonic component decays, the gauge field, or the Coulomb potential arises. Therefore, it is expected that through decay process, the initially formed neutral Q-ball may evolve into the charged or gauged Q-ball, which means that the charged Q-balls may

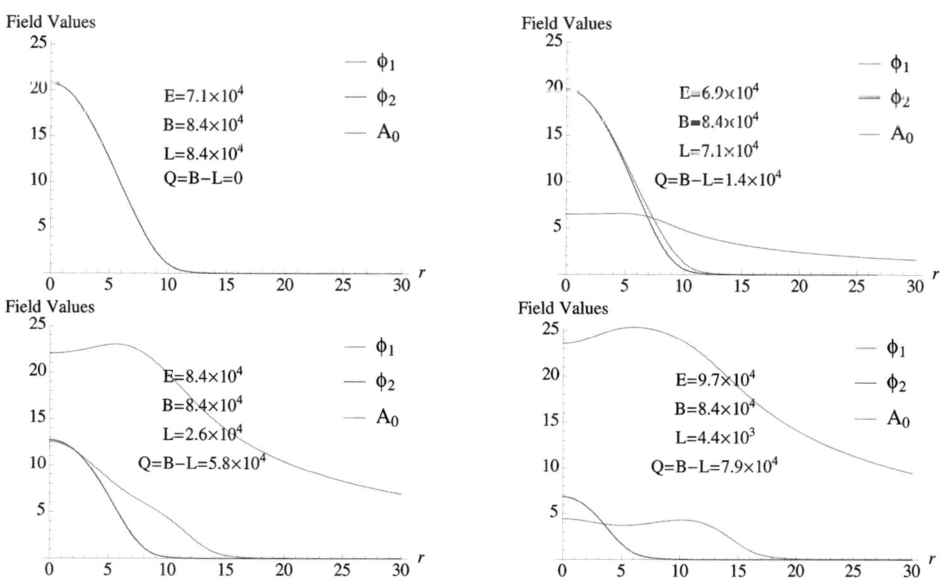

Fig. 2. 2-scalar gauged Q-balls in gauge mediation model with $e^2 = 0.002$, $m_\Phi = 1$ and $B = 8.4 \times 10^4$.

emerge in our universe. Next, if we look at the energy, we see that in Fig. 2, the energy starts to increase in the middle of the decay, which means that the energy of the emitted particle becomes negative. This in turn means that the particle starts to be bound to the Q-ball. This may be understood in the context of quantum mechanics of many-body system as a cloudy bound state which is analogous to that of an atomic system.

When the electric field is large enough, electron-positron pair production occurs from the vacuum, which is called Schwinger process. This prevents further growth of the electric charge of the Q-ball. We use

$$E_S = \frac{m_e^2}{e} \tag{10}$$

as the critical value of electric field for the process. Then, the condition that the electric field at the surface of the Q-ball reaches this value is given by

$$\frac{m_e^2}{e} = \frac{eQ_S}{4\pi R^2} \tag{11}$$

$$= \frac{eQ_S}{4\pi^3}\omega_{0B}^2 \tag{12}$$

$$= \frac{eQ_S}{2\pi}V_0^{1/2}B^{-1/2}. \tag{13}$$

Therefore, the electric charge of the Q-ball can grow until it reaches

$$Q_S \sim \frac{4\pi^2}{e^2}\left(\frac{V_0}{(10^6\text{GeV})^4}\right)^{-1/2}\left(\frac{B}{10^{39}}\right)^{1/2}. \tag{14}$$

Thus, only when

$$B \gtrsim 10^{39}\left(\frac{V_0}{(10^6\text{GeV})^4}\right) \tag{15}$$

applies the discussion in the previous paragragh. Otherwise, Q_S becomes smaller than $4\pi^2/e^2$ and the evoution stops when $Q = Q_S$.

When the size of Q-ball becomes smaller than Compton length $1/m_e$:

$$R \simeq \frac{\pi}{\omega_{0B}} = \frac{1}{\sqrt{2}}V_0^{-1/4}B^{1/4} \tag{16}$$

$$< \frac{1}{m_e}, \tag{17}$$

that is,

$$B \lesssim 10^{38}\left(\frac{V_0}{(10^6\text{GeV})^4}\right), \tag{18}$$

the pair creation does not occur even if the electric field at the surface of the Q-ball becomes E_S, since the pair creation occurs at spatial interval about Compton length. Therefore, the electric charge of the Q-ball can grow until the electric field at $r = 1/m_e > R$ becomes E_S, that is, $Q = 4\pi/e^2$.

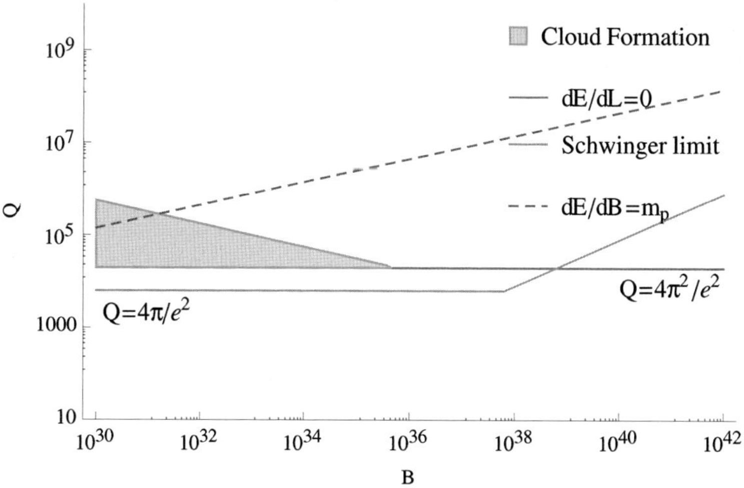

Fig. 3. Various conditions on electric charge of the Q-ball Q. The red line describes the Schwinger limit, and the blue line describes when $(\partial E/\partial L)_B = 0$. In the shaded region, the leptonic cloud is formed, and the upper line of the shaded region describes an upper limit on the electric charge of the Q-ball. The electric charge of the Q-ball can grow until the lowermost line, thus avoiding the baryonic decay, which is illustrated by the dashed line.

5. Conclusion

We illustrate the full results in Fig. 3. The red line describes the Schwinger limit, and the blue line describes when the particles start to be bounded to the Q-ball. In the shaded region, the leptonic cloud is formed, and the upper line of the shaded region describes an upper limit on the electric charge of the Q-ball, which is the condition that the Q-ball size becomes equal to cloud size. The electric charge of the Q-ball can grow until the lowermost line, thus avoiding the baryonic decay, which is illustrated by the dashed line, so we can expect that the charged Q-ball may be stable until the present universe and contribute to the dark matter.

The neutral Q-balls can be detected by Super-Kamiokande[8–10] or IceCube[11] through the absorption of protons, but these are not suited for detection of the (positively) charged Q-balls since charged Q-balls cannot absorb the protons due to the electrical repulsion. Thus, the electric charge is expected to make differences in the experimental signatures of the relic Q-balls. The charged Q-balls are expected to behave as some kind of nuclei, and are known to be detectable by such detectors as MACRO[8,12] and the observational bounds on mass and flux of the relic charged Q-balls are obtained.[9]

Acknowledgments

This work is supported by Grant-in-Aid for Scientific research from the Ministry of Education, Science, Sports, and Culture (MEXT), Japan, No. 25400248 (M.K.), World Premier International Research Center Initiative (WPI Initiative), MEXT, Japan, and the Program for the Leading Graduate Schools, MEXT, Japan (M.Y.). M.Y. acknowledges the support by JSPS Research Fellowships for Young Scientists, No.25.8715.

References

1. I. Affleck and M. Dine, Nucl. Phys. B249, 361 (1985).
2. A. Kusenko and M. Shaposhnikov, Phys. Lett. B418, 46 (1998).
3. S. Kasuya and M. Kawasaki, Phys. Rev. D61, 04130 (2000).
4. S. Kasuya and M. Kawasaki, Phys. Rev. D62, 023512 (2000).
5. K. Lee, J. A. Stein-Schabes, R. Watkins and L. M. Widrow, Phys. Rev. D39, 1665 (1989).
6. S. Coleman, Nucl. Phys. B262, 263 (1985).
7. G. Dvali, A. Kusenko, and M. Shaposhnikov, Phys. Lett. B417, 99 (1998).
8. A. Kusenko, V. Kuzmin, M. E. Shaposhnikov and P. G. Tinyakov, Phys. Rev. Lett. 80, 3185 (1998), [hep-ph/9712212].
9. J. Arafune, T. Yoshida, S. Nakamura, and K. Ogure, arXiv:hep-ph/0005103.
10. Super-Kamiokande, Y. Takenaga *et al.*, Phys. Lett. B647, 18 (2007), [hep-ex/0608057].
11. S. Kasuya, M. Kawasaki, T. Yanagida, arXiv:1502.00715 [hep-ph].
12. MACRO, M. Ambrosio *et al.*, Eur. Phys. J. C13, 453 (2000), [hep-ex/9904031].

Directional Dependence of the Local Estimation of H_0 and the Non-Perturbative Effects of Primordial Curvature Perturbations

Antonio Enea Romano

University of Antioquia, Instituto de Fisica
Medellin, Colombia
antonio.romano@udea.edu.co
www.udea.edu.co

Recent measurements of the cosmic microwave background (CMB) radiation have shown an apparent tension with the present value of the Hubble parameter inferred from local observations of supernovae, which look closer, i.e. brighter, than what is expected in a homogeneous model with a value of H_0 equal to the one estimated from CMB observations. We examine the possibility that such a discrepancy is the consequence of the presence of a local inhomogeneity seeded by primordial curvature perturbations, finding that a negative peak of the order of less than two standard deviations could allow to fit low red-shift supernovae observations without the need of using a value of the Hubble parameter different from H_0^{CMB}. The type of inhomogeneity we consider does not modify the distance to the last scattering, making it compatible with the constraints of the PLANCK mission data. The effect on the luminosity distance is in fact localized around the region in space where the transition between different values of the curvature perturbations occurs, producing a local decrease, while the distance outside the inhomogeneity is not affected.

Keywords: Hubble constant; local structure; inhomogeneities.

1. Introduction

Recent observations[1] of the cosmic microwave background radiation (CMB) have pointed to an apparent discrepancy between the value of the Hubble parameter H_0 deduced from cosmological data and the value inferred form local astrophysical observations.[1,2]

Following Ref. 3 we denote with ${H_0}^{app}$ the value of the Hubble constant estimated from observational data ignoring the presence of an inhomogeneity and with ${H_0}^{true}$ the value obtained taking into account the inhomogeneity. According to this notation, since both the above mentioned estimations were based on the assumption of homogeneity, the apparent tension can be expressed as $H_{0,SN}^{app} \approx 1.09 H_{0,CMB}^{app}$, where $H_{0,SN}^{app}$ and $H_{0,CMB}^{app}$ are the values estimated from fitting respectively low-redshift supernovae and CMB observations. One possible explanation of such a difference could be the effects of the local structure on the analysis of cosmological and astrophysical observations, and we will show what kind of inhomogeneity could resolve the tension, i.e. could give $H_{0,SN}^{true} = H_{0,CMB}^{true}$.

2. Low red-shift supernovae and the H_0 estimation

We will adopt the following notation for the luminosity distance, $D_L^{mod}(z, H_0)$, where we are denoting with a superscript the cosmological model or the observed data, and H_0 is explicitly treated as an argument to avoid confusion. According to the above notation low red-shift supernovae observations are fitted by a homogeneous cosmological model according to

$$D_L^{hom}(z_{SN}, H_{0,SN}^{app}) = D_L^{obs}(z_{SN}), \tag{1}$$

with $H_{0,SN}^{app} \approx 1.09 H_{0,CMB}^{app}$. We will show that if the effects of primordial curvature perturbations are taken into account a local inhomogeneity could resolve the apparent discrepancy, i.e.

$$D_L^{inh}(z_{SN}, H_{0,CMB}^{true}) = D_L^{obs}(z_{SN}). \tag{2}$$

Such an inhomogeneity can arise naturally from primordial curvature perturbations as predicted by inflation and constrained by CMB observations to have a standard deviation of about 5×10^{-5}, providing a simple mechanism for its origin.

3. The effects of primordial curvature perturbations

The metric after inflation near a peak of the primordial curvature perturbation[3,4] can be written as

$$ds^2 = -dt^2 + a_F^2(t) e^{2\zeta(r)}(dr^2 + r^2 d\Omega^2), \tag{3}$$

where $\zeta(r)$ is the primordial curvature perturbation. The spherical symmetry approximation is justified by the property of a Gaussian random field[5] according to which large peaks of a stochastic function tend to have a spherical shape. At a sufficiently early time when both the LTB and the metric in Eq.(3) are valid, they can be matched[3] leading to the following relation

$$k(r) = -\frac{1}{r^2}[(1 + r\zeta')^2 - 1]. \tag{4}$$

The use of the LTB solution allows to study exactly the formation of the structure seeded by $\zeta(r)$, and in particular to go beyond the limitations of perturbation theory. It turns out in fact that such early time tiny curvature perturbations of the order of 10^{-5} can lead to a highly non linear structure today whose effects cannot be neglected.

Eq.(4) is very important because it gives the connection between the early and the present Universe, and it can be used in two ways: given an early Universe curvature perturbations it allows to find the present day structure seeded by it, or vice versa, given a present day structure it can be used to find what primordial curvature perturbation could have originated it. Normally the first approach is adopted,[3,4] but here we will adopt the second one, since it is more convenient to define the model of the inhomogeneity in terms of $k(r)$. In this way we can find a natural explanation for the origin of the present day highly non linear structure whose effect cannot be captured by perturbations theory, while the use of the LTB solution gives the non perturbative evolution of the structure.

4. Effects on the luminosity distance

To compute $D_L^{\Lambda LTB}(z)$ we need to solve the radial null geodesics

$$\frac{dr}{dz} = \frac{\sqrt{1+2E(r(z))}}{(1+z)\dot{R}'[T(r(z)),r(z)]}, \tag{5}$$

$$\frac{dt}{dz} = -\frac{R'[T(r(z),r(z))]}{(1+z)\dot{R}'[T(r(z)),r(z)]}, \tag{6}$$

and then we substitute in the formula for the luminosity distance in a LTB space

$$D_L^{inh}(z, H_0^{true}) = D^{\Lambda LTB}(z) = (1+z)^2 R(t(z), r(z)) \tag{7}$$

where we set the parameters of the LTB solution so that

$$H_0^{LTB} = \frac{2}{3}\frac{\dot{R}(t_0,0)}{R(t_0,0)} + \frac{1}{3}\frac{\dot{R}'(t_0,0)}{R'(t_0,0)} = H_0^{true}. \tag{8}$$

For homogeneous cosmological models we assume a flat ΛCDM solution according to

$$H(z) = H_0^{app}\sqrt{\Omega_M(1+z)^3 + \Omega_\Lambda}\,,$$

$$D_L^{hom}(z, H_0^{app}) = (1+z)\int_0^z \frac{dx}{H(x)}\,. \tag{9}$$

The above definitions are very important since they give an explicit mathematical definition of H_0^{app} and H_0^{true}, which is coherent with the notion of apparent observable we adopted and which can be expressed as

$$D_L^{hom}(z, H_0^{app}) = D_L^{inh}(z, H_0^{true}) = D_L^{obs}(z). \tag{10}$$

We consider solutions with the function $k(r)$ given by a gaussian

$$k(r) = Ae^{-\left(\frac{r-r_0}{\sigma}\right)^2}, \tag{11}$$

centered approximately at the radial comoving coordinate of low-redshift supernovae used for the local estimation of H_0. Different values of r_0 can be chosen to control the shape of the inhomogeneity. This solution corresponds to a compensated inhomogeneity which is asymptotically homogeneous and, if the width of the gaussian is sufficiently small, there is no important difference between the central and asymptotic value of $k(r)$. This property is quite important because it ensures that the Universe is effectively a flat ΛCDM at the center and sufficiently far from it, making sure that the effect on the luminosity distance at the last scattering is very small.

We will adopt a system of units in which $H^{true}_{0,CMB} = 1$, and use the fiducial value $\Omega_\Lambda = 0.692$. The time used as initial condition at the center for the geodesics equations is obtained by integrating the Einstein equation in the scale factor from the big-bang till today.

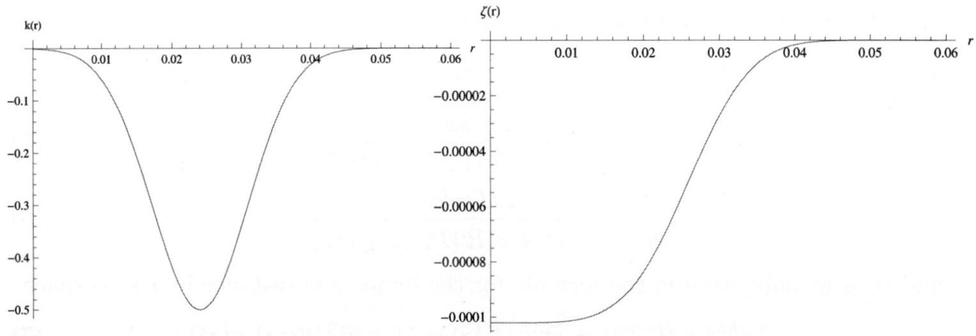

Fig. 1. The primordial curvature perturbation $\zeta(r)$ and the function $k(r)$ are plotted for $A = -0.5, r_0 = r_{SN} * 0.8, \sigma = r_0/2.5$, and $r_{SN} = r^{\Lambda CDM}$. The quantities A and $k(r)$ are in units of $(H^{true}_{0,CMB})^2$, while r_0 is in units of $(H^{true}_{0,CMB})^{-1}$.The effect on the luminosity distance corresponds to the region of transition between different values of the curvature perturbation.

As shown in Fig. 3, since the type of inhomogeneity we have considered is compensated, it has no effect on the luminosity distance outside it, i.e.

$$D^{hom}(H^{true}_{0,CMB}, z_{LS}) \approx D^{inh}(H^{true}_{0,CMB}, z_{LS}), \tag{12}$$

where z_{LS} is the redshift of the last scattering surface, or any redshift sufficiently larger than z_{SN}. This is due to the fact that the central and the asymptotic value of the curvature function is approximately the same. According to our definitions of apparent observable, fitting CMB data with a homogeneous or inhomogeneous model is consequently giving the same estimate for $H^{app}_{0,CMB} \approx H^{true}_{0,CMB}$.

For the case of low redshift supernovae instead it can be seen in Fig. 2 that

$$D^{inh}(H^{true}_{0,CMB}, z_{SN}) \approx D^{hom}(H^{app}_{0,SN}, z_{SN}) = D^{obs}_{z_{SN}}, \tag{13}$$

which implies that

$$H^{app}_{0,SN} > H^{true}_{0,SN} = H^{true}_{0,CMB}. \tag{14}$$

This shows how the effects of the inhomogeneity are able to resolve the apparent disagreement between cosmological and local estimations of H_0, since the luminosity distance is modified only around z_{SN}.

As seen in Figs. 2 and 3 the effect of the structure seeded by this primordial curvature perturbation profile corresponds to and underdense central region followed by an overdense shell, which is what we expect for a compensated inhomogeneity, due to the fact that the function $k(r)$ is asymptotically zero. The luminosity distance is decreased respect to the homogeneous case both in the under and over dense

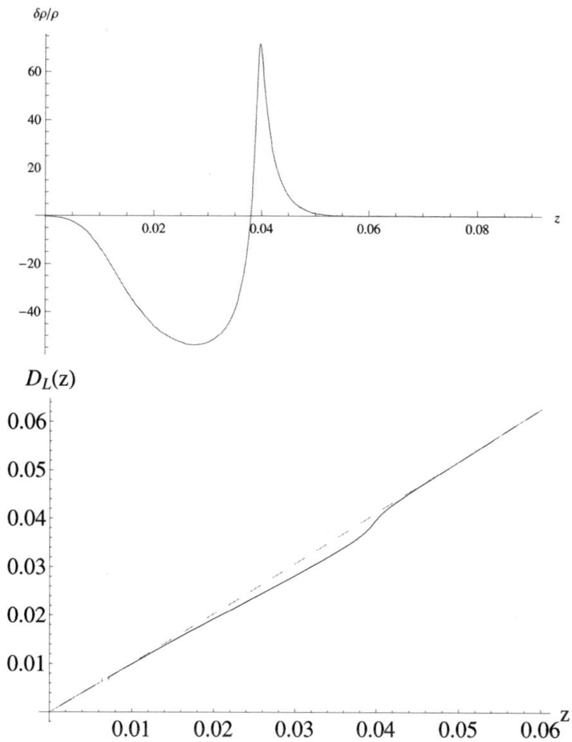

Fig. 2. On the top the percentage density contrast $\frac{\delta\rho}{\rho} = 100(1 - \frac{\rho^{LTB}(z)}{\rho^{\Lambda CDM}(z)})$ is plotted as a function of the redshift, showing how contrary to the linear theory approximation, when non perturbative effects are taken into account, not only underdense regions but also overdense regions can be associated to the decrease of the luminosity distance necessary to explain observations. On the bottom we plot in units of $(H^{true}_{0,CMB})^{-1}$ the luminosity distance for a ΛCDM model and an inhomogeneous model. In both the top and bottom plot the value of the Hubble parameter is the same for the homogeneous and inhomogeneous models, $H_0 = H^{true}_{0,CMB}$. The dashed line is the plot of $D^{hom}(H^{true}_{0,CMB}, z)$ while the solid line is for $D^{inh}(H^{true}_{0,CMB}, z)$. The luminosity distance is approximately the same outside the inhomogeneity, for $z > z_{SN}$, implying that the distance to the last scattering is not affected by the inhomogeneity, and consequently $H^{true}_{0,CMB} \approx H^{app}_{0,CMB}$. This shows that the analysis of the CMB data performed under the assumption of homogeneity is not introducing any misestimation for H_0, contrary to its local estimation, which is based on luminosity distance observations affected by it as shown in Fig. 2.

regions, a relativistic effect associated to large gradient terms which are normally neglected in Newtonian or perturbative calculations as shown for example in Ref. 6. This inhomogeneity is seeded by a curvature perturbation of less than two standard deviations as shown in Fig. 1, and the effect on the luminosity distance is associated to the region of transition between different values of $\zeta(r)$. Such a perturbation can occur quite naturally according the inflationary scenario, and smaller amplitude perturbations could still induce some important effects.

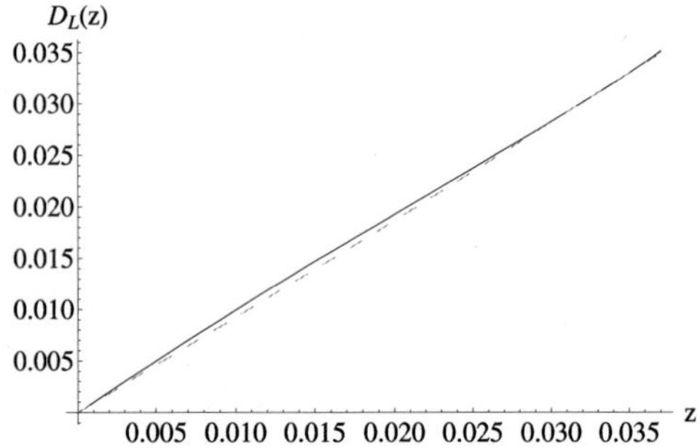

Fig. 3. According to the definition in eq.(12-14), the luminosity distance $D^{hom}(H^{app}_{0,SN}, z)$ for a homogeneous model and the luminosity distance $D^{inh}(H^{true}_{0,CMB}, z)$ for an inhomogeneous model are plotted in units of $(H^{app}_{0,CMB})^{-1}$ respectively with a dashed and solid line. Despite the difference in the value of H_0 for the two models, $D^{inh}(H^{true}_{0,CMB}, z_{SN}) \approx D^{hom}(H^{app}_{0,SN}, z_{SN})$, implying that the observed luminosity distance $D^{obs}_L(z_{SN})$ can be fitted using $H^{true}_{0,CMB} \approx H^{app}_{0,CMB}$ instead of $H^{app}_{0,SN}$ if the effect of the inhomogeneity is taken into account by using D^{inh} instead of D^{hom}, resolving the apparent discrepancy between local and cosmological observations.

The density profile shown in Fig. 3 is compatible with the results of the analysis of galaxy surveys.[7] Quite remarkably the Cepheids used for the calibration of the supernovae luminosity distance for the local estimation of H_0 in Ref. 2 are inside the region with the largest overdensity(Subregion 2 in Ref. 7) with a density profile compatible with our model in the region of available data, suggesting a possible directional dependence[a] of the local structure effects on the H_0 estimation which could be partially attributed to the presence in the region of the Sloan Great Wall.[8]

It would be interesting though to check carefully if using supernovae in different directions away from Sloan Great Wall the H_0 estimation could change significantly from the one obtained by Ref. 2.

References

1. Planck Collaboration, P. Ade *et al.*, (2013), arXiv:1303.5076.
2. A. G. Riess *et al.*, Astrophys.J. **730**, 119 (2011), arXiv:1103.2976.
3. A. E. Romano, M. Sasaki, and A. A. Starobinsky, Eur.Phys.J. **C72**, 2242 (2012), arXiv:1006.4735.
4. A. E. Romano, S. Sanes, and M. Sasaki, (2013), arXiv:1311.1476.
5. J. M. Bardeen, J. R. Bond, N. Kaiser and A. S. Szalay, Astrophys. J. **304** (1986) 15.
6. K. Bolejko *et al.*, Phys.Rev.Lett. **110**, 021302 (2013), arXiv:1209.3142.
7. R. C. Keenan, A. J. Barger and L. L. Cowie, Astrophys. J. **775**, 62 (2013) [arXiv:1304.2884 [astro-ph.CO]].
8. J. R. Gott, III, M. Juric, D. Schlegel, F. Hoyle, M. Vogeley, M. Tegmark, N. A. Bahcall and J. Brinkmann, Astrophys. J. **624**, 463 (2005) [astro-ph/0310571].

[a]Supernovae are in the same right ascension range of Subregion 2, but span a wider declination range, so here we are assuming a similar density profile would be obtained if the galaxy survey analysis was extended to the same declination range of the supernovae, or alternatively if H_0 was estimated only using the supernovae within the same declination range of Subregion 2.

The Fate of Dark Energy with Non-Minimal Coupling to Gravity

Chao-Qiang Geng

Chongqing University of Posts & Telecommunications,
Chongqing 400065, China
Department of Physics, National Tsing Hua University,
Hsinchu 30087, Taiwan
National Center for Theoretical Sciences,
Hsinchu 30087, Taiwan
geng@phys.nthu.edu.tw

Chung-Chi Lee

National Center for Theoretical Sciences,
Hsinchu 30087, Taiwan
g9522545@oz.nthu.edu.tw

Yi-Peng Wu

Institute of Physics, Academia Sinica,
Taipei 11529, Taiwan
ypwu@phys.sinica.edu.tw

We study the cosmological evolution of a quintessence field which has a non-minimal coupling to gravity (extended quintessence). Current observational constraints suggest that the mass of the extended quintessence field have to be extremely light so that the fraction of quintessence density invoked by such a non-minimal coupling is highly suppressed. We show that this tiny gravity-induced dark energy density may control the future cosmological evolutions, leading to attractor solutions depending on the value of the coupling constant ξ. After confronting the massless quintessence scenario with Solar System constraints, we find that the background equation of state of our universe may drastically deviate from -1 some 30 (180) billion years later for $\xi \simeq 1$ (0.1).

Keywords: Dark energy; quintessence; non-minimal coupling.

1. Introduction

Quintessence[1] is proposed as a candidate of dark energy, the source energy that drives the late-time cosmic acceleration. The dynamics of quintessence may be described by a scalar field evolving in a potential, leading to time-dependent equation of state and also spatially inhomogeneous perturbations of the density. One of the main advantage to consider a quintessence field ϕ is the existence of tracker solutions,[2] which can bypass the cosmic coincidence problem arised in the class of dark energy models that predict a nearly constant equation of state along the expansion history. However, the tracking basin of attraction is significantly narrowed down by current observational constraints[3–5] so that a viable tracker model inevitably reproduces the cosmic coincidence problem.

In the extended scenario of quintessence where ϕ exhibits non-minimal couplings with gravity,[6] the cosmic coincidence problem is reduced via the gravity-induced tracker solutions.[7] To describe a suitable cosmology, the extended quintessence field must be effectively massless, where the field potential $V(\phi)$ have to be sufficiently flat and the gravitationally coupled dark energy density ρ_ϕ^{nc} is strictly constrained. We show that this class of massless fields can result in interesting future cosmic evolution with equation of state depending on the strength of the non-minimal coupling. This is due to the attractor solutions invoked by the existence of ρ_ϕ^{nc}, even if this density fraction is highly suppressed today. In this paper, we present summary results of the (nearly) massless quintessence models from detail discussion given by Ref. 8.

2. Extended Quintessence: Dark Energy with Non-Minimal Coupling to Gravity

In the extended quintessence scenario,[6] the quintessence field ϕ is non-minimally coupled to the scalar curvature R through the action

$$\mathcal{S} = \int d^4x\sqrt{-g}\left[F(\phi)R - \frac{1}{2}(\nabla\phi)^2 - V(\phi)\right] + \mathcal{S}_M\,, \tag{1}$$

where $\mathcal{S}_M$ is the action for the relativistic and nonrelativistic matter component. This action can be recast into the Brans-Dicke theory through a field redefinition $F(\phi) = \Phi$, so that

$$\mathcal{S}_\phi \equiv \mathcal{S}_\Phi = \int d^4x\sqrt{-g}\left[\Phi R - \frac{\omega}{\Phi}(\nabla\Phi)^2 - V(\Phi)\right]\,, \tag{2}$$

where $\omega = F/(2F_\phi^2)$ and $F_\phi \equiv dF/d\phi$.

As a typical model, we shall specify the form of $F(\phi)$ to

$$F(\phi) = \frac{1}{2\kappa^2} + \frac{1}{2}\xi\phi^2, \tag{3}$$

where $\kappa^2 = 8\pi G$ and ξ denotes the non-minimal coupling parameter. By definition, the field equations of our theory then read

$$\left(1+\xi\kappa^2\phi^2\right)\left(R^\mu_\nu - \frac{\delta^\mu_\nu}{2}R\right) - \kappa^2\xi\left(\nabla^\mu\nabla_\nu - \delta^\mu_\nu\nabla^2\right)\phi^2 - \kappa^2\nabla^\mu\phi\nabla_\nu\phi$$
$$+\delta^\mu_\nu\kappa^2\left(\frac{1}{2}(\nabla\phi)^2 + V(\phi)\right) = \kappa^2 T^\mu_\nu\,, \tag{4}$$

where T^μ_ν is the energy-momentum tensor.

Taking the metric in the flat Friedmann-Robertson-Walker (FRW) background $ds^2 = -dt^2 + a^2(t)d\vec{x}^2$ with $a(t)$ the scale factor, we obtain the Friedmann equations

$$H^2 = \frac{\kappa^2}{3}\left(\rho_M + \rho_\phi\right)\,, \quad \dot{H} = -\frac{\kappa^2}{2}\left(\rho_M + \rho_\phi + P_M + P_\phi\right)\,, \tag{5}$$

where ρ_M (P_M) is the energy density (pressure) of matter, and ρ_ϕ (P_ϕ) is the effective energy density (pressure) of the scalar field, given by

$$\rho_\phi = \frac{\dot{\phi}^2}{2} + V - 3\xi H^2\phi^2 - 6\xi H\phi\dot{\phi}\,, \tag{6}$$

$$P_\phi = \frac{\dot{\phi}^2}{2} - V + \xi\left(2\dot{H} + 3H^2\right)\phi^2 + 2\xi\left(\ddot{\phi}\phi + \dot{\phi}^2 + 2H\phi\dot{\phi}\right)\,, \tag{7}$$

respectively. On the other hand, the Klein-Gordon equation reads

$$\Box\phi + \xi R\phi - V_\phi = 0, \tag{8}$$

where $\Box \equiv \nabla^\mu\nabla_\mu$ and $V_\phi \equiv dV/d\phi$. Since $R = 6\dot{H} + 12H^2$ in the homogeneous FRW background, we find

$$\ddot{\phi} + 3H\dot{\phi} + V_\phi - 6\xi\left(\dot{H} + 2H^2\right)\phi = 0\,. \tag{9}$$

In terms of the e-folding $N = \ln a$ and rescale ϕ by $\kappa\phi$, Eq. (9) becomes

$$\phi'' + \frac{3}{2}(1-w)\phi' - 3\xi\left(1-3w\right)\phi + \frac{\kappa V_\phi}{H^2} = 0. \tag{10}$$

where the primes are e-folding derivatives, and we have applied the relation $2H' = -3(1+w)H$. For the background equation of state we will take $w = 1/3$ during radiation domination (RD) and $w = 0$ during matter domination (MD).

3. Cosmological Evolution with a Nearly Constant Potential

Let us consider a nearly constant potential $V(\phi) = V_0 + V_\phi(\phi - \phi_*) + ...$, where $V_\phi \ll \kappa V_0$, and the Klein-Gordon equation (10) is reduced to

$$\phi'' + \frac{3}{2}(1-w)\phi' - 3\xi\left(1-3w\right)\phi = 0, \tag{11}$$

during the epoches $\kappa^2 V_0 \sim H_0^2 \ll H^2$. Assuming that w is a constant, Eq. (11) has the exact solution

$$\phi(N) = C_+ e^{L_+ N} + C_- e^{L_- N}, \tag{12}$$

where $C_\pm$ are constants to be determinated by initial conditions and

$$L_\pm(w,\xi) = -\frac{3}{4}(1-w) \pm \sqrt{\frac{9}{16}(1-w)^2 + 3\xi(1-3w)}. \tag{13}$$

In what follows we set $C_- = 0$ so that $\phi = C_+ e^{L_+ N}$, assuming that ϕ is released from rest.

For $\xi < 0$, we have $L_+ \leq 0$, and if $\rho_\phi \gg V_0$ the energy density Eq. (6)

$$\rho_\phi \approx \rho_\phi^{nc} = \frac{1}{\kappa^2}\left(\frac{L_+^2}{2} - 3\xi - 6\xi L_+\right) H^2\phi^2. \tag{14}$$

Given that $\rho_\phi \propto H^2\phi^2 \propto a^{2L_+ - 3(1+w)}$ decays faster than the matter density ($\propto a^{-3}$) as ϕ approaches zero. The value of ϕ is frozen around zero and the quintessence field eventually reproduces the behavior of a cosmological constant as $\rho_\phi \to V_0$.

For $\xi > 0$, the initial density of ϕ may be negative and $L_+ > 0$ holds during RD and MD so that $e^{L_+ N}$ is always increasing with time. If $|\rho_\phi| \gg V_0$, the energy density takes the form

$$\rho_\phi = \frac{1}{\kappa^2}\left(\frac{L_+^2}{2} - 3\xi - 6\xi L_+\right) H^2\phi^2, \tag{15}$$

where it can be checked that $\rho_\phi < 0$ in both RD and MD. One may deduce the equation of state $w_\phi = P_\phi/\rho_\phi$ from Eq. (15) to find that

$$w_\phi = \frac{1}{3}\ \text{(RD)}\quad \text{and} \quad w_\phi = \frac{1}{2} - \sqrt{\frac{1}{4} + \frac{4}{3}\xi}\ \text{(MD)}, \tag{16}$$

where $-1 \le w_\phi < 0$ in MD for $0 < \xi \le 3/2$.

Once the density of V_0 is dominant, the universe will undergo a de Sitter expansion. Taking $w = -1$ with a constant $H = H_{\rm ds}$ to Eq. (11), we obtain $\phi'' + 3\phi' - 12\xi\phi = 0$, and the solution (keeping only the increasing mode) reads

$$\phi_{\rm ds} = C_{\rm ds} e^{L_{\rm ds} N}, \quad L_{\rm ds} = -\frac{3}{2} + \sqrt{\frac{9}{4} + 12\xi}, \tag{17}$$

where the energy density Eq. (6) is of the form

$$\rho_\phi = V_0 + \frac{1}{\kappa^2}\left(\frac{L_{\rm ds}^2}{2} - 3\xi - 6\xi L_{\rm ds}\right) H_{\rm ds}^2\phi_{\rm ds}^2. \tag{18}$$

Note that $L_{\rm ds} > 0$ since $\xi > 0$ and therefore, ϕ remains increasing with time during the V_0 domination epoch.

Since $\phi_{\rm ds}$ is increasing with time, the subdominant component of ρ_ϕ (the second term in the right-hand side of Eq. (18)) eventually becomes comparable with V_0, where the de Sitter expansion is interrupted. We assume that $V_\phi = 0$ and that the asymptotic universe has a constant equation of state $w = w_f$, which satisfies

$$\phi'' + \frac{3}{2}(1 - w_f)\phi' - 3\xi\,(1 - 3w_f)\,\phi = 0. \tag{19}$$

Again, by neglecting the decreasing mode, the solution is given by

$$\phi_f = C_f e^{L_f N}, \quad L_f = -\frac{3}{4}(1 - w_f) + \sqrt{\frac{9}{16}(1 - w_f)^2 + 3\xi(1 - 3w_f)}. \tag{20}$$

Provided that the scalar field is the dominant species of the universe, the Friedmann equation yields $3H^2 = \kappa^2\rho_\phi$, where the energy density

$$\rho_\phi = V_0 + \frac{1}{\kappa^2}\left(\frac{L_f^2}{2} - 3\xi - 6\xi L_f\right) H^2\phi_f^2. \tag{21}$$

For a sufficiently large N such that $\phi_f \gg 6/(L_f^2 - 6\xi - 12\xi L_f)$, the asymptotic solution shows a fine cancellation between V_0 and ρ_ϕ^{nc} (the second term in the right-hand side of Eq. (21)). This implies $\phi_f^2 \sim H^{-2} \propto a^{3(1+w_f)}$, which leads to

$$w_f = \frac{-3 + 2\xi}{3(1 + 2\xi)}, \quad \text{and} \quad L_f = \frac{4\xi}{1 + 2\xi}, \quad (0 < \xi \le 3/2). \tag{22}$$

Given that $-1 < w_f \leq 0$ with $0 < \xi \leq 3/2$, the scalar field will keep overtaking the matter density after the epoch of the de Sitter expansion.

Note that if the initial value of ρ_ϕ is too small such that it remains negative during MD, Eq. (16) shows that $w_\phi < -1$ for $\xi > 3/2$, and therefore ρ_ϕ will always be dominated by the nonminimal coupling term (ρ_ϕ^{nc}) as given by Eq. (15). In this case, ρ_ϕ is always negative, and V_0 can never be important so that there exhibits no cosmic acceleration.

If ρ_ϕ becomes positive before MD, V_0 will dominate the universe and drive a de Sitter expansion. Similarly, this de Sitter expansion shall be interrupted with the growths of ϕ, and the resulting equation of state $w = w_\phi = (-3+2\xi)/[3(1+2\xi)]$. However, since $w_\phi > 0$ for $\xi > 3/2$, the matter density ρ_M may eventually catch up with ρ_ϕ and become one of the dominant density component.

Let us assume that the asymptotic solution for $\xi > 3/2$ has a constant equation of state w_f with ϕ_f taken the form of Eq. (20). For arbitrary initial conditions, the final attractor is a fine cancellation between ρ_M and ρ_ϕ^{nc}, leading to the lowest order Friedmann equation $\rho_M + \rho_\phi^{nc} = 0$. Since $\rho_\phi^{nc} \propto H^2\phi_f^2$, the Friedmann equation indicates $\phi_f^2 \sim H^{-2}a^{-3} \propto a^{3w_f}$, which results in

$$w_f = \frac{4\xi}{3(1+4\xi)}, \quad \text{and} \quad L_f = \frac{2\xi}{1+4\xi}, \quad (\xi > 3/2). \tag{23}$$

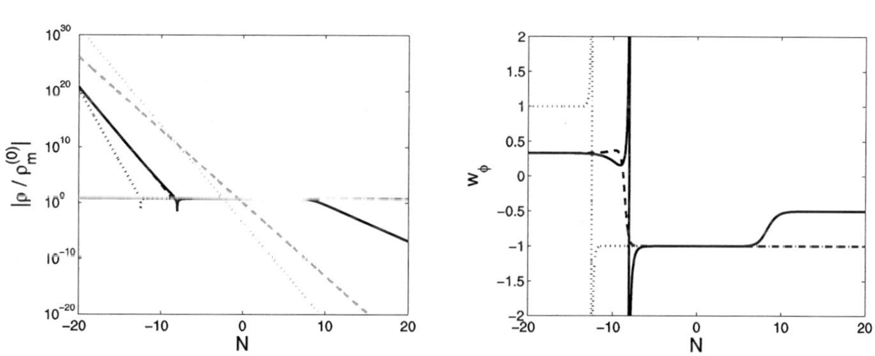

Fig. 1. Evolutions of the ratio of the energy density to the dust-like matter density today (left panel) and the equation of state (right panel) in the constant potential model. In both panels, the initial condition $|\rho_\phi| \gg V_0$ is shown by $\xi = 0.3$ (black solid line), $\xi = -0.3$ (black dashed line) and $\xi = 0.3$ with $C_+ = 0$ (black dotted line). In the left panel, the gray dotted (dashed) line is the energy density of radiation (dust-like matter), respectively. The gray solid line in the left panel corresponds to the constant value $V_0/\rho_m^{(0)}$.

4. Observational Constraints and Results

It is useful to define $\epsilon \equiv 1 - V(\phi)/\rho_\phi$ to parametrize the fraction of the quintessence density additional to its potential. The density fraction induced by the nonminimal coupling is then given by $\Omega_\phi^{nc} \equiv \epsilon\,\Omega_\phi$, where $\Omega_{\phi 0} \approx 0.69$.

For $\xi > 0$, ϵ is negative at present even if the initial condition is $\epsilon = 0$. Given that the background curvature is now decreasing with the increase of $|\epsilon|$, we define the time $N = N_f$ such that

$$3H^2 = \left|\frac{L_{\rm ds}^2}{2} - 3\xi - 6\xi L_{\rm ds}\right| H^2 \phi_{\rm ds}^2(N_*)\,, \tag{24}$$

where $\phi_{\rm ds}$ and $L_{\rm ds}$ are given by Eq. (17), and we take $C_{\rm ds} = \phi_{\rm ds}(N=0) \approx \phi_0$. The onset of the power-law expansion is thus

$$N_f = \frac{1}{2L_{\rm ds}}\left[\ln\frac{6}{|L_{\rm ds}^2 - 6\xi - 12\xi L_{\rm ds}|} - 2\ln C_{\rm ds}\right]. \tag{25}$$

Observational data from CMB, BAO, Type-Ia supernovae can constrain the value of ξ (see Ref. 8). The most important constraint to our scenario comes from the Solar System tests,[9,10] which gives $\omega \geq 40000$ to the action (2) and therefore $F(\phi_0)/F_\phi^2(\phi_0) \geq 8\times 10^4$ (95% C.L.). This indicates that today's field value $\kappa^2\phi_0^2 \leq 1/(160000\xi^2 - \xi)$, where the lower bounds for ϵ are $\epsilon_0 = (-1.4\times 10^{-4}, -4.2\times 10^{-5})$ for $\xi = (0.1, 1)$, respectively. As a consequence, we find a minimum $N_* \approx 13.0$ ($N_* \approx 2.3$) and a maximum $|\Omega_{\phi 0}^{nc}| \approx 9.4\times 10^{-5}$ (2.9×10^{-5}) for $\xi = 0.1$ ($\xi = 1$), which implies that the onset of the ξ-induced power-law expansion can occur at some 179.0 (31.7) billion years later. For the time unit, we take $H_0 dt = d\ln a = dN$ with $H_0^{-1} \approx 13.8\times 10^9$ years.

References

1. R. R. Caldwell, R. Dave, and P. J. Steinhardt, Phys. Rev. Lett. **80**, 1582 (1998).
2. I. Zlatev, L. Wang, and P. J. Steinhardt, Phys. Rev. Lett. **82**, 896 (1999); P. J. Steinhardt, L. Wang, and I. Zlatev, Phys. Rev. D **59**, 123504 (1999).
3. S. Bludman, Phys. Rev. D **69**, 122002 (2004).
4. T. Chiba, Phys. Rev. D **81**, 023515 (2010); T. Chiba, A. D. Felice, and S. Tsujikawa, Phys. Rev. D **87**, 083505 (2013).
5. P. Y. Wang, C. W. Chen and P. Chen, JCAP **1202**, 016 (2012).
6. F. Perrotta, C. Baccigalupi, and S. Matarrese, Phys. Rev. D **61**, 023507 (2000); C. Baccigalupi, S. Matarrese, and F. Perrotta, Phys. Rev. D **62**, 123510 (2000).
7. S. Matarrese, C. Baccigalupi, and F. Perrotta, Phys. Rev. D **70**, 061301 (2004); T. Chiba, M. Siino, and M. Yamaguchi, Phys. Rev. D **81**, 083530 (2010).
8. C. Q. Geng, C. C. Lee, and Y. P. Wu, [astro-ph/1215.04019].
9. B. Bertotti, L. Iess, and P. Tortora, Nature (London), **425**, 374 (2003).
10. L. Perivolaropoulos, Phys. Rev. D **81**, 047501 (2010).

Cosmology with an Adapted System of Coordinates

F. Nugier

Leung Center for Cosmology and Particle Astrophysics,
National Taiwan University, Taipei, Taiwan 10617
fnugier@ntu.etu.tw
https://lecospa.ntu.edu.tw/

Most of the observables in cosmology are light-propagated quantities. I will present a system of coordinates, the "geodesic light-cone (GLC) coordinates", which is adapted to the propagation of null-like signals observed by a geodesic observer. This system is fully general and adapted to calculations in inhomogeneous geometries. I will show that its properties make it useful for a large spectrum of applications, from evaluating the distance-redshift relation, calculating averages on our past light cone, estimating the effect of the large-scale structure on the Hubble diagram, to calculating weak lensing quantities.

Keywords: Inhomogeneous cosmology; general relativity; gravitational lensing.

1. Motivations

Except for planet Earth from which we have geological information, the vast majority of observations available to us about our Universe comes from our past light cone. It is thus natural to use coordinates adapted to it. We present coordinates which were first developed to simplify averages of scalars on our past light-cone,[1] next used to estimate the effect of inhomogeneities on luminosity distance[2–5] and Hubble diagram,[6,7] and recently applied to compute lensing quantities.[8]

2. The geodesic light-cone coordinates

Our light-cone adapted metric consists of 6 arbitrary functions (Υ, U^a, γ_{ab}) and is close to "*observational coordinates*"[9,10] (but different[11]) and totally gauged fixed:

$$ds^2_{\rm GLC} = \Upsilon^2 {\rm d}w^2 - 2\Upsilon {\rm d}w {\rm d}\tau + \gamma_{ab}({\rm d}\widetilde{\theta}^a - U^a {\rm d}w)({\rm d}\widetilde{\theta}^b - U^b {\rm d}w) \ . \tag{1}$$

This metric uses a time coordinate τ corresponding to the proper time of an observer in geodesic motion, a null coordinate w describing its past light cones, and finally angles $\tilde{\theta}^a$ $(a = 1, 2)$ that photons keep along their path orthogonal to a 2-spheres $\Sigma(w, \tau)$ of constant time in our past light-cone (see Fig. 1). A possible simple choice in the FLRW limit is: $w = \eta + r$ (conformal time + radius), $\tau = t$ (cosmic time), $(\tilde{\theta}^1, \tilde{\theta}^2) = (\theta, \phi)$, $\Upsilon = a(t)$, $U^a = 0$, $\gamma_{ab} = a^2 r^2 {\rm diag}(1, \sin^2\theta)$. Υ can be seen as an inhomogeneous scale factor (lapse function), U^a is similar to a shift-vector and γ_{ab} is the metric inside $\Sigma(w, \tau)$.[11] The computation of the distance-redshift relation in GLC coordinates is facilitated by two direct simplifications:

$$\text{Redshift:} \quad (1 + z_s) = \Upsilon(w_o, \tau_o, \widetilde{\theta}^a)/\Upsilon(w_o, \tau_s, \widetilde{\theta}^a) \ , \tag{2}$$

$$\text{Angular distance:} \quad d_A = \gamma^{1/4}(\sin\tilde{\theta}^1)^{-1/2} \ \text{ with } \gamma \equiv \det(\gamma_{ab}) = \frac{|\det(g_{\rm GLC})|}{\Upsilon^2} \ . \tag{3}$$

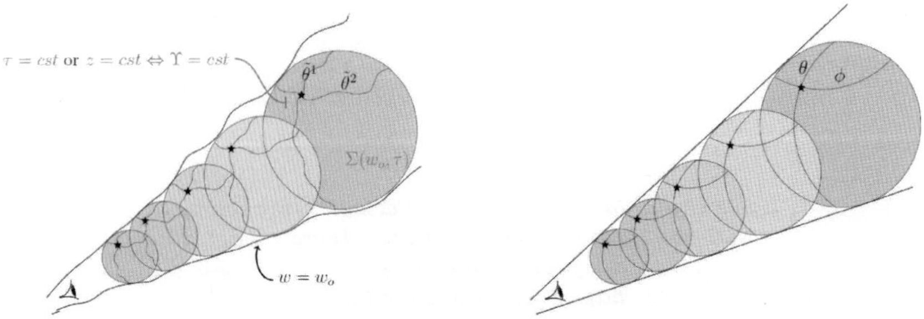

Fig. 1. *Left:* Inhomogeneous light-cone parametrized by GLC coordinates. *Right:* Homogeneous light-cone or an adapted system (like GLC coordinates) in an inhomogeneous geometry.

3. Simplification of light-cone averages

The general expression of the light-cone average[1] of a scalar S (like d_L, d_L^{-2}) is given by $\langle S\rangle (V_0, A_0) = I(S; V_0, A_0)/I(1; V_0, A_0)$. We define the average integral such that it is gauge invariant, invariant under reparametrisations of the form $(A, V) \to (\tilde{A}(A), \tilde{V}(V))$ and under general coordinate transformations $x \to y$:

$$I(S; V_0, A_0) = \int_{\mathcal{M}_4} d^4x\sqrt{-g}\, D(V_0 - V)D(A - A_0)\, \mathcal{N}(V, A, \partial_\mu)\, S(x) \ , \quad (4)$$

where $\mathcal{N}(V, A, \partial_\mu)$ is a normalization, $D(X) = \delta_D(X)$ or $\Theta(X)$ (Heaviside function). From that definition, three configurations are possible (null, spatial, spherical):

Average $\overset{\Theta}{\delta_D}$	$\langle S\rangle^{A_0}_{V_0}$	$\langle S\rangle^{V_0}_{A_0}$	$\langle S\rangle_{V_0,A_0}$
Illustration	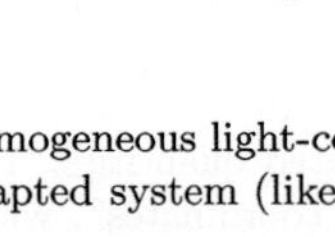	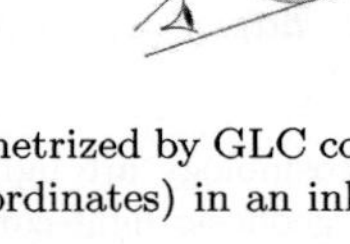	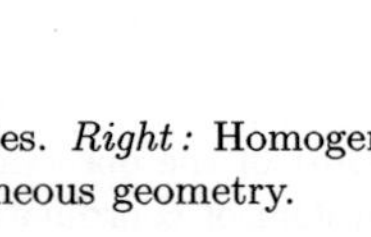
$\mathcal{N}(V, A, \partial_\mu)$	$\frac{\lvert\partial_\mu V\partial^\mu A\rvert}{\sqrt{-\partial_\nu A\partial^\nu A}}$	$\sqrt{-\partial_\mu A\partial^\mu A}$	$\lvert\partial_\mu V\partial^\mu A\rvert$

Among these three types of averages, $\langle S\rangle_{V_0,A_0}$ is closer to physical observables as it averages over the deformed 2-sphere embedded in the light-cone $V = V_0$ and a spatial hypersurface $A = A_0$. In GLC coordinates (where $V \to w$, $A \to \tau$) we can simplify the average and use Eq. (2) to get (with $\tau_z \equiv \tau(z_s, w_o, \widetilde{\theta}^a)$):

$$\langle S\rangle_{w_o,z_s} = \left(\int d^2\widetilde{\theta}\sqrt{\gamma(w_o, \tau_z, \widetilde{\theta}^b)}\, S(w_o, \tau_z, \widetilde{\theta}^b)\right) \Big/ \left(\int d^2\widetilde{\theta}\sqrt{\gamma(w_o, \tau_z, \widetilde{\theta}^b)}\right) \ , \quad (5)$$

allowing us to average scalars on the observed sky, at a certain redshift z_s.

4. Distance-redshift relation at $\mathcal{O}(2)$

The GLC metric enables the computation of the luminosity distance $d_L(z)$ at $\mathcal{O}(2)$ in perturbations within the well-known Newtonian gauge (NG):

$$ds^2_{\rm NG} = a^2(\eta)\left[-(1 + 2\Phi)d\eta^2 + (1 - 2\Psi)(dr^2 + r^2[d\theta^2 + \sin^2\theta\, d\phi^2])\right] \quad (6)$$

with $\Phi = \psi + \frac{1}{2}\phi^{(2)}$, $\Psi = \psi + \frac{1}{2}\psi^{(2)}$ the gauge invariant Bardeen potentials (assuming no matter shear at $\mathcal{O}(1)$). Doing the full $\mathcal{O}(2)$ transformation between GLC and NG coordinates brings the relations $(\tau, w, \tilde{\theta}^1, \tilde{\theta}^2) = \text{func}(\eta, r, \theta, \phi)$ and $(\Upsilon, U^a, \gamma^{ab}) = \text{func}(\psi, \psi^{(2)}, \phi^{(2)})$. These last allow to compute $d_L = (1+z)^2 d_A$, using Eq. (3). The first order is given by different terms containing ψ_s (at the source), (Integrated) Sachs-Wolfe ([I]SW), Doppler, and lensing (convergence) effects. Similarly obtained, the $\mathcal{O}(2)$-corrections contain[2,6,11–13]: two dominant terms (Doppler)^2, (Lensing)^2; combinations of $\mathcal{O}(1)$-terms like ψ_s^2, ([I]SW)^2, [I]SW $\times$ Doppler; genuine $\mathcal{O}(2)$-terms such as $\psi_s^{(2)}$, Lensing$^{(2)}$; new integrated effects, angle deformations; redshift perturbations from Eq. (2) (including transverse peculiar velocity); and terms like Lens-Lens coupling and corrections to the Born approximation.

5. Effect of large-scale structure on the Hubble diagram

In Sec. 4 the distance d_L was expressed in terms of $(\psi, \psi^{(2)}, \phi^{(2)})$, hence it needs a description of the Bardeen potentials at $\mathcal{O}(1,2)$. We expand the first order gravitational potential ψ in Fourier modes and denote by $\overline{(...)}$ the ensemble (or stochastic) average over perturbations. The $\mathcal{O}(2)$ potentials can be related to ψ by[14]: $\psi^{(2)}, \phi^{(2)} \propto \nabla^{-2}(\partial_i\psi\partial^i\psi), \partial_i\psi\partial^i\psi$. Hence the combination of light-cone and stochastic averages enables us to study the effect of matter perturbations on the whole sky (and sky-averaged observables like d_L). For example, the (trivial) average of ψ^2 gives: $\overline{\langle\psi_s^2\rangle} = \int_0^\infty \frac{dk}{k}\mathcal{P}_\psi(k)$ where $\mathcal{P}_\psi(k) \equiv (k^3/2\pi^2)|\psi_k(\eta)|^2 = (3/5)^2 A(k/k_0)^{n_s-1}T^2(k)\, g^2(z)$ is the power spectrum describing perturbations. At linear order, with A, n_s, k_0 taken from WMAP, $T(k)$ is a transfer function[15] including a baryonic component (Silk damping), and $g(z)$ is the growth factor describing the recent time evolution of perturbations.[11] In CDM we get exactly the spectral coefficients coming from each correction of $d_L(z_s, \theta^a)$ described in Sec. 4: $\overline{\langle d_L\rangle} = \int_0^\infty \frac{dk}{k}\mathcal{P}_\psi(k)\, C(k\Delta\eta)$. We do the same in ΛCDM, with reasonable assumptions to simplify integrations,[7] using also a non-linear power spectrum.[16,17]

Like d_L one can also average the flux $\Phi = L/(4\pi d_L^2) \simeq \Phi_0 + \Phi_1 + \Phi_2$. We get:

- $\overline{\langle d_L^{-2}\rangle} \equiv (d_L^{FLRW})^{-2}\,[1 + f_\Phi(z)]$ where $f_\Phi(z) \simeq f(z)\int_0^\infty \frac{dk}{k}\left(\frac{k}{\mathcal{H}_0}\right)^2 \mathcal{P}_\psi(k)$,
- $\overline{\langle d_L\rangle}(z) = d_L^{FLRW}\,[1 + f_d(z)]$ with $f_d = -(1/2)f_\Phi + (3/8)\overline{\langle(\Phi_1/\Phi_0)^2\rangle}$.

The d_L-correction involves the flux variance lead by peculiar velocity and lensing. Similarly, we get the average/dispersion of the distance modulus $\mu = 5\log_{10} d_L$:

$$\overline{\langle\mu\rangle} = \mu^{\text{FLRW}} - 1.25(\log_{10} e)\left[2f_\Phi - \overline{\langle(\Phi_1/\Phi_0)^2\rangle}\right] \quad , \tag{7}$$

$$\sigma_\mu = 2.5(\log_{10} e)\sqrt{\overline{\langle(\Phi_1/\Phi_0)^2\rangle}} \quad . \tag{8}$$

Compared to the Union 2 data and using a non-linear power spectrum in ΛCDM (Fig. 2, Left), we find that peculiar velocities explain well the scatter at small z and

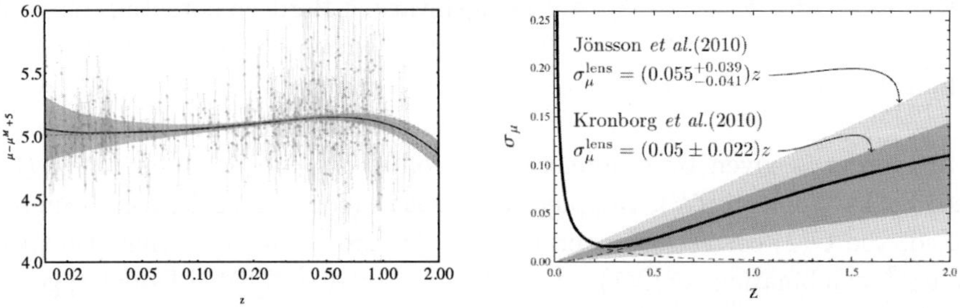

Fig. 2. *Left:* Theoretical dispersion from power spectrum vs data. *Right:* Dispersion σ_μ from peculiar velocities and lensing (solid line) compared to experimental estimates (grey areas).

that lensing only partially explains the scatter at large z. Finally, we can compare our dispersion on the Hubble diagram with the experimental estimations coming from lensing (Fig. 2, Right). We find that the total effect is well fitted by Doppler ($z \leq 0.2$) + lensing ($z > 0.3$) effects and that the lensing prediction, in great agreement with experiments so far,[18,19] should be confirmed by future surveys.

6. Jacobi map and weak lensing

We now consider lensing,[8] motivated by Sec. 5 and recent work[20] on galaxy number counts in GLC. The separation of two light rays emitted at the same time from a source S and converging to an observer O follows the geodesic deviation eq. (GDE):

$$\nabla_\lambda^2 \xi^\mu = R^\mu_{\alpha\beta\nu} k^\alpha k^\nu \xi^\beta \tag{9}$$

with $\nabla_\lambda \equiv \mathrm{D}/\mathrm{d}\lambda \equiv k^\mu \nabla_\mu$, k^μ the photon momentum, λ an affine parameter along the photon path, ξ^μ the orthogonal displacement of the rays. The GDE is projected on the Sachs basis $\{s^\mu_A\}_{A=1,2}$ (two zweibeins with flat index $A = 1,2$): $g_{\mu\nu} s^\mu_A s^\nu_B = \delta_{AB}, s^\mu_A u_\mu = 0, s^\mu_A k_\mu = 0, \Pi^\mu_\nu \nabla_\lambda s^\nu_A = 0$; with u_μ the peculiar velocity of the comoving fluid (S, O comoving too), Π^μ_ν a "screen" projector orthogonal to u_μ and $u_\mu + (u^\alpha k_\alpha)^{-1} k_\mu$. We define the Jacobi map J^A_B, from the observed sky angle $\bar{\theta}^A_o$ to $\xi^A = \xi^\mu s^A_\mu$, by $\xi^A(\lambda) = J^A_B(\lambda, \lambda_o)\, \bar{\theta}^A_o$. The projected quantities ξ^A and $R^A_B = R_{\alpha\beta\nu\mu} k^\alpha k^\nu s^\beta_B s^\mu_A$ (optical tidal matrix) bring us the Jacobi equation[21,22]:

$$\frac{\mathrm{d}^2}{\mathrm{d}\lambda^2} J^A_B(\lambda, \lambda_o) = R^A_C(\lambda)\, J^C_B(\lambda, \lambda_o) \ , \tag{10}$$

$$\text{with I.C.:} \quad J^A_B(\lambda_o, \lambda_o) = 0 \quad , \quad \frac{\mathrm{d}}{\mathrm{d}\lambda} J^A_B(\lambda_o, \lambda_o) = (k^\mu u_\mu)_o\, \delta^A_B \ . \tag{11}$$

A direct resolution of Eq. (10) gives the angular distance of the source $d_A(\lambda_s) \equiv \sqrt{dS_s/d^2\Omega_o} = \sqrt{\det J^A_B(\lambda_s, \lambda_o)}$. Also, the (unlensed) angular position of the source $\bar{\theta}^A_s$ and the observed lensed position $\bar{\theta}^A_o$ (of the image) are given by: $\bar{\theta}^A_s = \left(\xi^A / \bar{d}_A\right)_s$, $\bar{\theta}^A_o = \left(k^\mu \partial_\mu \xi^A / k^\mu u_\mu\right)_o$, where $\bar{d}_A$ is the homogeneous and isotropic background our model refers to. This allows us to define the so-called amplification matrix as:

$$\mathcal{A}^A_B \equiv \frac{\mathrm{d}\bar{\theta}^A_s}{\mathrm{d}\bar{\theta}^B_o} = \frac{J^A_B(\lambda_s, \lambda_o)}{\bar{d}_A(\lambda_s)} = \begin{pmatrix} 1 - \kappa - \hat{\gamma}_1 & -\hat{\gamma}_2 + \hat{\omega} \\ -\hat{\gamma}_2 - \hat{\omega} & 1 - \kappa + \hat{\gamma}_1 \end{pmatrix} , \tag{12}$$

which defines the lensing quantities: κ (convergence), $\hat{\omega}$ (vorticity), $|\hat{\gamma}| \equiv \sqrt{(\hat{\gamma}_1)^2 + (\hat{\gamma}_2)^2} = \sqrt{(1-\kappa)^2 + \hat{\omega}^2 - \mu^{-1}}$ (shear), $\mu \equiv 1/(\det \mathcal{A})$ (magnification).

Let us now turn to the GLC coordinates and express these lensing quantities in it. First, the zweibeins are written as $s^\mu_A = (s^\tau_A, 0, s^a_A)$ and $k^\mu \equiv \omega \Upsilon^{-1} \delta^\mu_\tau$ (where ω is a pure constant). Second, the solution to Eqs. (10) and (11) in GLC is:

$$J^A_B(\lambda, \lambda_o) = s^A_a(\lambda) \left[2u_\tau (\dot{\gamma}_{ab})^{-1} \right]_o s^B_b(\lambda_o) \tag{13}$$

where $(\ldots)^\cdot \equiv \partial_\tau(\ldots)$. The angular distance and the magnification become:

$$d_A = 2u_{\tau_o} (\gamma\gamma_o)^{1/4} / \sqrt{(\det \dot{\gamma}_{ab})_o} \quad , \quad \mu = \left(\bar{d}_A / d_A \right)^2 = \Phi / \bar{\Phi} \quad , \tag{14}$$

involving $\bar{d}_A = a^2(\tau) r^2$ with $r = w - \int \mathrm{d}\tau / a(\tau)$ measured from the observer and Φ ($\bar{\Phi}$) the flux in the inhomogeneous (homogeneous) geometry. The expression of the zweibeins can also be obtained in the GLC coordinates,[8] but it is more convenient to compute the squared lensing quantities, combined with $s^A_a s^A_b = \gamma_{ab}$ and $\epsilon_{AB}\, s^A_a s^B_b = \sqrt{\gamma}\, \epsilon_{ab}$ (ϵ the anti-symmetric symbol), to directly get:

$$\begin{pmatrix} (1-\kappa)^2 + \hat{\omega}^2 \\ \hat{\gamma}_1^2 + \hat{\gamma}_2^2 \end{pmatrix} = \left(\frac{u_{\tau_o}}{\bar{d}_A} \right)^2 \left(\left[\frac{\gamma\, \dot{\gamma}_{ab} \gamma^{bc} \dot{\gamma}_{cd}}{\left(\det^{ab} \dot{\gamma}_{ab} \right)^2} \right]_o \gamma \gamma^{ad} \pm \frac{2\sqrt{\gamma\, \gamma_o}}{\left(\det^{ab} \dot{\gamma}_{ab} \right)_o} \right) . \tag{15}$$

We thus have general lensing quantities expressed with only 3 metric functions (of γ_{ab}), showing the great advantage of using GLC coordinates. Similar expressions for the deformation matrix $\mathcal{S}^A_B \equiv \frac{\mathrm{d}J^A_C}{\mathrm{d}\lambda} (J^{-1})^C_B$ and its elements (optical scalars) simplify in GLC (depending only on γ_{ab}, Υ and their time derivatives). The usefulness of GLC coordinates to compute lensing quantities was illustrated[8] for an off-center observer in a Lemaître-Tolman-Bondi model (considering only the decaying mode).

7. Conclusions

We have briefly described the many advantages in using the GLC coordinates. They are indeed adapted to calculations involving light-propagation. They can also be used for weak lensing (where γ_{ab} acts as a screen), and help to get new predictions on cosmology[23] or study other aspects of weak (and possibly strong) lensing.

References

1. M. Gasperini, G. Marozzi, F. Nugier, and G. Veneziano, "Light-cone averaging in cosmology: Formalism and applications," *JCAP* **1107** (2011) 008, `arXiv:1104.1167 [astro-ph.CO]`.
2. I. Ben-Dayan, M. Gasperini, G. Marozzi, F. Nugier, and G. Veneziano, "Backreaction on the luminosity-redshift relation from gauge invariant light-cone averaging," *JCAP* **1204** (2012) 036, `arXiv:1202.1247 [astro-ph.CO]`.
3. I. Ben-Dayan, M. Gasperini, G. Marozzi, F. Nugier, and G. Veneziano, "Do stochastic inhomogeneities affect dark-energy precision measurements?," *Phys. Rev. Lett.* **110** (2013) 021301, `arXiv:1207.1286 [astro-ph.CO]`.
4. G. Marozzi, "The luminosity distance-redshift relation up to second order in the Poisson gauge with anisotropic stress," *Class. Quant. Grav.* **32** no. 4, (2015) 045004, `arXiv:1406.1135 [astro-ph.CO]`. [Corrigendum: Class. Quant. Grav.32,179501(2015)].

5. G. Fanizza, M. Gasperini, G. Marozzi, and G. Veneziano, "A new approach to the propagation of light-like signals in perturbed cosmological backgrounds," *JCAP* **1508** no. 08, (2015) 020, arXiv:1506.02003 [astro-ph.CO].
6. I. Ben-Dayan, G. Marozzi, F. Nugier, and G. Veneziano, "The second-order luminosity-redshift relation in a generic inhomogeneous cosmology," *JCAP* **1211** (2012) 045, arXiv:1209.4326 [astro-ph.CO].
7. I. Ben-Dayan, M. Gasperini, G. Marozzi, F. Nugier, and G. Veneziano, "Average and dispersion of the luminosity-redshift relation in the concordance model," *JCAP* **1306** (2013) 002, arXiv:1302.0740 [astro-ph.CO].
8. G. Fanizza and F. Nugier, "Lensing in the geodesic light-cone coordinates and its (exact) illustration to an off-center observer in Lemaître-Tolman-Bondi models," *JCAP* **1502** no. 02, (2015) 002, arXiv:1408.1604 [astro-ph.CO].
9. R. Maartens, *PhD Thesis*. PhD thesis, University of Cape Town, 1980.
10. G. F. R. Ellis, S. D. Nel, R. Maartens, W. R. Stoeger, and A. P. Whitman, "Ideal observational cosmology.," *Phys.Rep.* **124** (1985) 315–417.
11. F. Nugier, *Lightcone Averaging and Precision Cosmology*. PhD thesis, 2013. arXiv:1309.6542 [astro-ph.CO].
12. O. Umeh, C. Clarkson, and R. Maartens, "Nonlinear relativistic corrections to cosmological distances, redshift and gravitational lensing magnification: II. Derivation," *Classical and Quantum Gravity* **31** no. 20, (Oct., 2014) 205001, arXiv:1402.1933.
13. O. Umeh, C. Clarkson, and R. Maartens, "Nonlinear relativistic corrections to cosmological distances, redshift and gravitational lensing magnification: I. Key results," *Classical and Quantum Gravity* **31** no. 20, (Oct., 2014) 202001, arXiv:1207.2109.
14. N. Bartolo, S. Matarrese, and A. Riotto, "The full second-order radiation transfer function for large-scale cmb anisotropies," *JCAP* **0605** (2006) 010, arXiv:astro-ph/0512481 [astro-ph].
15. D. J. Eisenstein and W. Hu, "Baryonic features in the matter transfer function," *Astrophys. J.* **496** (1998) 605, arXiv:astro-ph/9709112 [astro-ph].
16. **Virgo Consortium** Collaboration, R. Smith *et al.*, "Stable clustering, the halo model and nonlinear cosmological power spectra," *Mon. Not. Roy. Astron. Soc.* **341** (2003) 1311, arXiv:astro-ph/0207664 [astro-ph].
17. R. Takahashi, M. Sato, T. Nishimichi, A. Taruya, and M. Oguri, "Revising the Halofit Model for the Nonlinear Matter Power Spectrum," *Astrophys. J.* **761** (2012) 152, arXiv:1208.2701 [astro-ph.CO].
18. J. Jonsson, M. Sullivan, I. Hook, S. Basa, R. Carlberg, *et al.*, "Constraining dark matter halo properties using lensed SNLS supernovae," arXiv:1002.1374 [astro-ph.CO].
19. T. Kronborg, D. Hardin, J. Guy, P. Astier, C. Balland, *et al.*, "Gravitational lensing in the Supernova Legacy Survey (SNLS)," arXiv:1002.1249 [astro-ph.CO].
20. E. Di Dio, R. Durrer, G. Marozzi, and F. Montanari, "Galaxy number counts to second order and their bispectrum," arXiv:1407.0376 [astro-ph.CO].
21. P. Fleury, H. Dupuy, and J.-P. Uzan, "Interpretation of the Hubble diagram in a nonhomogeneous universe," *Physical Review D 87,* **123526** (2013) , arXiv:1302.5308 [astro-ph.CO].
22. G. Fanizza, M. Gasperini, G. Marozzi, and G. Veneziano, "An exact Jacobi map in the geodesic light-cone gauge," *JCAP* **1311** (2013) 019, arXiv:1308.4935 [astro-ph.CO].
23. G. Fanizza, M. Gasperini, G. Marozzi, and G. Veneziano, "Time of flight of ultra-relativistic particles in a realistic Universe: a viable tool for fundamental physics?," arXiv:1512.08489 [astro-ph.CO].

PARALLEL SESSION

IV. Gravitational Waves from the Big Bang and Other Sources

Summary of the Parallel Session: "Gravitational Waves from the Big Bang and Other Sources"

Frederico Arroja

Leung Center for Cosmology and Particle Astrophysics, National Taiwan University, No.1, Sec.4, Roosevelt Road, Taipei, 10617 Taipei, Taiwan (R.O.C)

This manuscript summarizes the eight talks of the parallel session "Gravitational Waves from the Big Bang and other sources" of the Second LeCosPA Symposium.

Keywords: Summary; gravitational waves.

1. Introduction

This work is based on a talk given on the 18th of December of 2015 by the author and organizer to summarize the contributions to the parallel session "Gravitational Waves from the Big Bang and other sources" of the 2nd LeCosPA International Symposium "Everything About Gravity – Celebrating the Centenary of Einstein's General Relativity", https://lecospa.ntu.edu.tw/symposium/2015/. The video of the talk can be found online at https://www.youtube.com/watch?v=AZuCbRVV6_U.

One hundred years ago, A. Einstein presented his radical theory of gravity and spacetime. That theory is called General Relativity (GR) and it reconciled Maxwell's theory of electromagnetism with Newtonian mechanics and gravity. It turns out that the latter theory is a good description of reality only for small speeds and weak fields otherwise it has to be modified. During the last century, GR has passed all observational/experimental tests. It made predictions which were confirmed, e.g. the bending and gravitational redshift of light and the perihelion precession of the orbit of Mercury. The theory of GR gave us black holes (BH), still to be discovered but believed by many to exist at the center of most massive galaxies. It is the basis of the so-called ΛCDM model. The latter is considered the standard model of cosmology at present. The theory has practical important applications, e.g. the Global Positioning System. Furthermore, GR predicts the existence of Gravitational Waves (GW). These are ripples of spacetime itself that propagate at the speed of light. They have been indirectly observed by using binary pulsars. In particular, the equations of GR for GW are needed to explain the orbital decay of these binary systems. The missing energy is carried away from these systems by the GW. The predictions of GR agree with observational data to sub-percent level. Despite these successes, we would like a direct detection of these spacetime waves and, at the moment, there are several experiments looking for them. This is the "holy grail" of gravitational physics. On a general note, the theory of GR is a great example of the power of the human brain guided by beauty/simplicity in the laws of nature. Many believe it is not the final theory that describes gravity.

This parallel session was devoted to talks related to GW from both theoretical and observational/experimental perspectives. There were a total of eight speakers and each talk had variable duration. We will in the following sections, briefly summarize the main points of each talk and provide references for the interested reader. Most of the slides of the talks can be found online in the website of the Symposium.

2. Inflationary GW in the effective field theory of modified gravity

The first talk, by S. Tsujikawa, was on inflationary GW using a very general effective field theory approach (EFT). The main reference for this work is Ref. 1. The motivations for the work are to identify the origins of the inflaton, the dark energy and dark matter and quantum gravity as well. In most approaches to the previous problems, one or more scalar fields are involved so it is natural to consider the general framework of EFT to describe them all in an unified and systematic way. For instance, the results of Ref. 1 apply to theories with one scalar field including Horndeski and beyond Horndeski theories as well as Hořava-Lifshitz gravity, a candidate theory for quantum gravity that breaks general diffeomorphism invariance. One of the main results was a second-order action for tensor perturbations where one finds a modified dispersion relation for the gravitons due to the presence of higher spatial derivatives appearing in Hořava-Lifshitz gravity. With this action, assuming slow-roll inflation and a spatial derivative expansion, de Felice and Tsujikawa[1] computed the power spectrum including the first sub-leading corrections. They also showed that in a beyond Horndeski theory one can resort to a disformal transformation to go to the so-called Einstein frame. Both the curvature and tensor perturbations are disformal invariant and the tensor power spectrum is the same as in GR.[2] In the near future, they plan to analyse most single field models present in the literature in a unified way and constrain the parameters of the models with the latest Planck satellite data of the Cosmic Microwave Background (CMB).

3. Gravitational waves from second-order scalar perturbations

Another robust prediction of GR and the standard model of cosmology is that at second-order in perturbation theory, tensor perturbations are generated from the observed scalar fluctuations. Previous estimates in the literature found that these second-order GW will not be detectable with the currently planned experiments, e.g. LISA, LIGO II, BBO. However, the previous estimate did not take into account that in the early universe there may have been an early Matter-Dominated (MD) era. In fact, it is believed that reheating after inflation may be well described by such MD era and if so one may expect large GW on small scales as shown by Assadullahi and Wands '09. An optimistic estimate by the speaker, S. Kuroyanagi, showed that depending on the parameters of the model, i.e. the reheating temperature and the energy scale of inflation, one may be able to observe these GW with for example

DECIGO. However the previous estimate was made in a region of parameter space when perturbation theory breaks down so the speaker solved the necessary equations numerically, also taking into account a smooth transition from the MD era to the Radiation-Dominated (RD) era. She found that the amplitude of GW is suppressed with respect to the optimistic estimate but still may be detectable in the not so-far future. The results were still preliminary and should be made public soon.

4. Primordial black holes as a novel probe of primordial GW

The motivations for this talk are: to investigate inflationary GW on small scales, in fact in a range of scales as large as possible and to provide a different way to probe GW, an alternative to other well-established methods, e.g. Pulsar Timing Arrays and constraints for Big Bang Nucleosynthesis, etc. The main reference for the work is Ref. 3 and a further publication to appear.

It is well-known that primordial black holes (BH) can form during the RD era from the gravitational collapse of over-dense regions. They form shortly after the Fourier mode, k, characteristic of the size of the over-dense regions enters the horizon. Because we have not yet observed any primordial BH one can obtain an upper bound limit on the variance of the primordial curvature perturbation. On CMB and Large Scale Structure scales, the constraint is much weaker than for example the constraint from direct observational of the CMB by the Planck satellite. However on much smaller scales, i.e. for $k > 10^8 \mathrm{Mpc}^{-1}$, the constraint from primordial black holes is actually the most important one. The variance of the primordial curvature perturbations is $\sigma^2 \lesssim 2 \times 10^{-3}$. Following a line of though similar to the previous section but in a somewhat opposite direction, the speaker T. Suyama, argued that at second-order also tensor perturbations produce scalar perturbations. Because these scalar perturbations may form primordial BH and the latter ones have not been observed, from this on can find an upper limit on the variance of the tensor perturbations[3] as $\sigma_T^2 \sim \sigma \lesssim 0.04$.

5. Gravitational waves from bubble collisions in the early universe

The fourth talk by D.-H. Kim, explored the theoretical side of GWs originated from the collision of bubbles in the early universe. The talk was based on Ref. 4.

Bubbles form from a first-order quantum phase transition of a scalar field when it tunnels from the false to the true vacuum state. In this work, the author numerically solved the Einstein field equations together with the Klein-Gordon equation with a quartic potential using double-null coordinates and assuming a system of two identical bubbles. He computed the GWs produced from the bubble collisions. The waveforms were obtained from integration of the energy-momentum tensor. The main results are: in the weak collision regime, it was found that the waveforms show damped monochromatic patterns while in the strong collision regime, the waveforms predominantly exhibit "bumps and spikes", the details of which can be found in Ref. 4. For future works, the authors plans to compute the power spectrum of the GWs, go beyond the thin-wall approximation used in the current computation and include higher-order multipole moments.

6. Resonant primordial gravitational waves amplification

The final "theoretical" talk was given by C. Lin based on the publication 5. The main motivation is: suppose that in the near future we measure the tensor-to-scalar ratio, r, this would imply that $r \sim 10^{-2} > 2 \times 10^{-3}$, which in turn implies we need a way to evade the so-called Boubekeur-Lyth (BL) bound (if $\Delta\phi < M_p$ then $r \lesssim 2 \times 10^{-2}$, where $\Delta\phi$ is the field excursion in field space and M_p is the Planck mass) for GW produced from inflation. This bound is important because it basically is a natural consequence of the validity of the effective field theory.

One way to evade the BL bound is to modify GR. In this talk, the author minimally extended GR to a theory with a non-vanishing graviton mass. He showed that the scalar modes are the same as in standard single field slow-roll inflation, the vector modes remain non-dynamical and the tensor modes receive a mass correction. The coherent oscillations of the inflaton during reheating lead to coherent oscillations of the graviton's mass and this gives rise to the resonant amplification of GW. After reheating one assumes that the graviton's mass approaches zero, the GR value. The main results are: the tensor power spectrum can have a blue tilt and the tensor-to-scalar ratio receives an amplification factor which may be as large as 10^{10} depending on the model parameters: the graviton's mass and the inflaton's decay rate. This amplification factor provides a way to evade the BL bound.

7. Search for gravitational waves from binary black hole mergers: parameter estimation and Fisher matrix

In this section, we turn to more observational related questions. Based on Refs. 6 and 7, H.-S. Cho gave a talk on the pros and cons of the Fisher matrix (FM) formalism versus parameter estimation (PE) in the search for GW from binary BH mergers. He argued that one needs a hybrid model for the waveforms based on both theoretical modeling and numerical relativity simulations around the merger time. Searches for these events in the data use a matched filtering technique, searching through a large template database and then use PE to find from 9 to 15 model parameters. This is a computationally expensive procedure and alternative faster techniques are desirable. One such technique where statistical errors can be done easily for a know signal is the FM approach. The FM procedure requires large signal-to-noise ratios, it assumes Gaussian noise and known signal parameters however it is semi-analytical and fast. PE makes no assumptions, uses real noise, searches the full parameter space but it is computationally expensive or even prohibitive. The author showed that FM error predictions agree well with PE errors, and the errors in mass parameters can be much larger for spinning systems than those for nonspinning systems due to mass-spin degeneracy.

8. Electromagnetic counterparts to GW from binary BH mergers

The talk by F. Zhang, based on Refs. 8 and 9 can be shortly described to be about multi-messenger astronomy. He argued that there is an electromagnetic counterpart to binary BH mergers if there is a surrounding plasma. In numerical simulations one finds that during the pre-merger phase there are two jets, near the merger, the emission is approximately isotropic and in the post-merger phase one observes a single jet. One of the main goals of the talk was to explain the detailed mechanisms driving the radiation. Using an idealized Force Free Electrodynamics (see the above references for all the details) which basically assumes that the plasma cannot experience force, it was shown that the dynamical spacetime squashes and stretches the plasma and produces electromagnetic radiation which is carried out by Alfven and fast-magnetosonic waves.

9. Directed searches for broadband extended gravitational wave emission in energetic core-collapse supernovae

M. van Putten was reporting on a series of works.[10–14] He argued that core-collapse supernovae (CC-Sne) form high-density matter and some form BH instead of neutron stars. There is a major output of energy in GW bursts with ascending and descending chirps. About 20% of all CC-Sne are Type Ib/c and these have very massive progenitors. 1% of these give origin to long gamma-ray bursts. The result of a CC-Sne is a rapidly rotating BH, nearly extremal that can only spin down by interacting (in an intermittent fashion) with its disk. These produce long gamma-ray and GW bursts. Regarding detection, GW burst from SN Ic are competitive with merger systems if just 1% of them is successful. The author searched for a signal from the nearby (only 12 Mpc away) supernovae SN 2010br in LIGO's science run 6 but did not find evidence in the data. This could be due to SN 2010br being a particularly weak event and in addition to that its light curve is poorly known which may have lead to an incorrect estimation of the time of the GW arrival. In the future, the author plans to broaden the search time window to look for this event.

10. Conclusion/Outlook

The parallel session was quite successful in terms of the audience and of the depth and range of topics related to both theoretical and observational considerations that it attracted. The outlook for the (very) near future looks bright. In particular, the Advanced Laser Interferometer GW Observatory (LIGO) is taking and analysing data right now at the time of the writing of this summary. It is expected to detect an event a day but there is a large uncertainty in this estimate (about 3 orders of magnitude). The sensitivity is about 10 times better than the initial LIGO sensitivity. It will detect inspiralling neutron star binaries (of 1.4 solar masses) out to 300 Mpc and binary BH mergers to redshift 0.4. The LISA pathfinder mission to test technologies for the Evolved Laser Interferometer Space Antenna (eLISA), scheduled to be launched in 2034, was successfully launched in December 2015. Other ground-based and balloon based observatories aiming at detecting primordial GW from inflation, e.g. BICEP3/Keck array, ACTPol, SPTPol, EBEX, SPIDER are taking data right now or are planned and approved to be deployed in the next 5 years. Also in the near future we may have a satellite dedicated to this search, it is called LiteBIRD, and aims at a sensitivity on the tensor-to-scalar of the order of 10^{-3}. Let us note that these primordial GW are quantum fluctuations of spacetime and their detection would without doubt provide us significant insights into the theory of the early universe and even possibly on the theory of quantum gravity. We think it is fair to say that we are at the dawn of the gravitational-wave astronomy era. Once a first detection is made, there will be a whole new spectrum to explore and, judging from history, the near future will be a period of incredible discoveries. We dare to say that it will be the most exciting field of physics (perhaps of even science) to be working on, with only a strong competitor from the LHC if the latest results of a diphoton excess at 750 GeV turn out to be real and are the discovery of a new fundamental particle.

11. Acknowledgments

We would like to thank all the speakers of the session for sharing their presentation slides and Pisin Chen and the other members of the Local Organizing Committee for organizing such an enjoyable Symposium. FA is supported by the National Taiwan University (NTU) under Project No. 103R4000 and by the NTU Leung Center for Cosmology and Particle Astrophysics (LeCosPA) under Project No. FI121.

References

1. A. De Felice and S. Tsujikawa, Phys. Rev. **D91**, 103506 (2015).
2. S. Tsujikawa, JCAP **1504**, 043 (2015).
3. T. Nakama and T. Suyama, Phys. Rev. **D92**, 121304 (2015).
4. D.-H. Kim, B.-H. Lee, W. Lee, J. Yang, and D.-h. Yeom, Eur. Phys. J. **C75**, 133 (2015).
5. C. Lin and M. Sasaki, Phys. Lett. **B752**, 84 (2016).
6. R. O'Shaughnessy *et al.*, Phys. Rev. **D89**, 064048 (2014).
7. H.-S. Cho, Class. Quant. Grav. **32**, 235007 (2015).
8. H. Yang and F. Zhang, (2015), 1508.02119.
9. H. Yang, F. Zhang, and L. Lehner, Phys. Rev. **D91**, 124055 (2015).
10. M. H. P. M. van Putten, C. Guidorzi, and F. Frontera, Astrophys. J. **786**, 146 (2014).
11. M. H. P. M. van Putten, G. M. Lee, M. Della Valle, L. Amati, and A. Levinson, (2014), [Mon. Not. Roy. Astron. Soc.444,58(2014)].
12. M. H. P. M. van Putten, Mon. Not. Roy. Astron. Soc. **447**, L11 (2015).
13. A. Levinson, M. H. P. M. van Putten, and G. Pick, Astrophys. J. **812**, 124 (2015).
14. M. H. P. M. van Putten, Astrophys. J. **810**, 7 (2015).

Gravitational Waves from Bubble Collisions in the Early Universe *

Dong-Hoon Kim

Basic Science Research Institute,
Ewha Womans University, Seoul 120-750, Republic of Korea
ki13130@gmail.com

Cosmic bubbles nucleated through the quantum tunneling process would expand and undergo collisions with each other. We focus on collisions of two equal-sized bubbles and compute gravitational waves emitted from the collisions. The mechanism of the collisions can be modeled by means of a real scalar field and its quartic potential. Out of this model we can compute gravitational waves from the collisions by integrating the energy-momentum tensors over the volume of the wave sources; in the quadrupole approximation. Our computational results show that the waveforms are characterized by (i) cusp-like bumps with frequency modulation during the initial-to-intermediate stage of strong collisions and (ii) smooth monochromatic oscillations during the final stage of weak collisions.

Keywords: Gravitational waves, cosmic bubbles, collisions, Coleman-de Luccia instanton, scalar field.

1. Introduction: Bubbles in the early universe and gravitational waves

According to the theory of cosmic "inflation", the early universe experienced an extreme burst of expansion, which lasted a tiny fraction of a second, but smoothed out irregularities–inhomogeneities, anisotropies and the curvature of space, and made the universe appear homogeneous and isotropic.[1] It has been suggested that inflationary models of the early universe most likely lead to a "multiverse".[2] One such model is "eternal inflation":[3] it proposes that many bubbles of spacetime individually nucleate and grow inside an ever-expanding background multiverse. The nucleation and growth of such bubbles can be modeled by a Coleman-de Luccia (CDL) instanton, a type of quantum transition between two classically disconnected vacua at different energies; the higher energy (false vacuum), the lower energy (true vacuum).[4] A scalar field initially in the false vacuum state may tunnel quantum mechanically to the true vacuum state. This nucleates bubbles of the true vacuum (new phase) inside of the false vacuum (old phase) background; through a first-order phase transition. These bubbles then expand and collide with each other. The mechanism of bubble collisions can be effectively modeled by the CDL instanton: as bubbles continue to collide repeatedly, the scalar field transitions back and forth repeatedly between the false vacuum and the true vacuum, eventually settling down in the true vacuum as the collision process is gradually terminated.

*Proceedings for *The 2nd LeCosPA Symposium: Everything about Gravity, Celebrating the Centenary of Einstein's General Relativity.* Talk on December 14, 2015, Taipei, Taiwan.

Gravitational waves (GWs) can be produced from bubble collisions, and we may consider the following scenario. Suppose that we live in a single large true vacuum bubble and that the boundary of our bubble would collide with another bubble that is located outside our observable universe.[5] Then there would exist some signatures of bubble collisions and these could be carried via GWs. In this scenario, the mechanism of bubble collision–GW emission can be well approximated by a two-bubble collision model.

2. Gravity-scalar field dynamics for colliding bubbles

The mechanism of two equal-sized colliding bubbles can be effectively modeled by means of a CDL instanton.[4] One can build a model for this, which consists of gravitation, a real scalar field and a potential of the field. A system of Einstein gravity coupled with a scalar field that is governed by a potential can be described by the following action:

$$\mathcal{S} = \int d^4x\sqrt{-g}\left[\frac{1}{16\pi}R - \frac{1}{2}\nabla_\mu\phi\nabla^\mu\phi - V\left(\phi\right)\right], \tag{1}$$

where R denotes the Ricci scalar, ϕ the scalar field and $V\left(\phi\right)$ the potential of the scalar field.[6] From this system the Einstein equations are derived:

$$R_{\mu\nu} - \frac{1}{2}Rg_{\mu\nu} = 8\pi T_{\mu\nu}, \tag{2}$$

where the energy-momentum tensors on the right-hand side are written as

$$T_{\mu\nu} = \phi_{;\mu}\phi_{;\nu} - \frac{1}{2}\phi_{;\rho}\phi_{;\sigma}g^{\rho\sigma}g_{\mu\nu} - V\left(\phi\right)g_{\mu\nu}. \tag{3}$$

Also, the scalar field equation for the system reads

$$\nabla^2\phi = \frac{dV}{d\phi}. \tag{4}$$

The Einstein equations (2) and the scalar field equation (4) constitute a scalar field model that effectively describes the mechanism of two colliding bubbles in curved spacetime.[6] Given a potential $V\left(\phi\right)$, the scalar field solution ϕ and the geometry solution $g_{\mu\nu}$ should be obtained by solving Eqs. (2) and (4) simultaneously. To this end, we prescribe an ansatz for the geometry $g_{\mu\nu}$ with the hyperbolic symmetry, using the *double-null* coordinates:

$$ds^2 = -\alpha_{\mathrm{h}}^2\left(u,v\right)dudv + r_{\mathrm{h}}^2\left(u,v\right)dH^2, \tag{5}$$

where $dH^2 = d\chi^2 + \sinh^2\chi d\theta^2$ with $0 \leq \chi < \infty$, $0 \leq \theta < 2\pi$,[6] and $\alpha_{\mathrm{h}}\left(u,v\right)$ and $r_{\mathrm{h}}\left(u,v\right)$ are to be determined by solving Eqs. (2) and (4) simultaneously in the coordinates (u, v, χ, θ). In the flat spacetime limit, the double-null coordinates are defined as $u \equiv \tau - x$ and $v \equiv \tau + x$ with $\tau^2 \equiv t^2 - y^2 - z^2$, $t = \tau\cosh\chi$, $y = \tau\sinh\chi\sin\theta$, $z = \tau\sinh\chi\cos\theta$: in our analysis, the x-axis of Cartesian coordinates is chosen to coincide with a line adjoining the centers of the two bubbles, and the y-axis and the z-axis lie in a plane perpendicular to the x-axis.

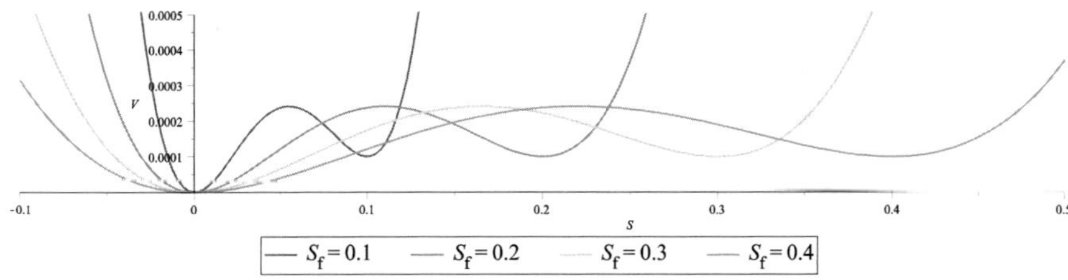

Fig. 1. The potential $V(S) = \frac{3V_f}{\beta S_f^4}S^4 - \frac{2(\beta+3)V_f}{\beta S_f^3}S^3 + \frac{3(\beta+1)V_f}{\beta S_f^2}S^2$; expressed in terms of the false vacuum field $S_f = \sqrt{4\pi}\phi_f$, the vacuum energy of the false vacuum V_f and a free parameter β. The patterns of the potential are shown for $\beta = 0.1$, $V_f = 10^{-4}$ and $S_f = \sqrt{4\pi}\phi_f =$ (1) 0.1, (2) 0.2, (3) 0.3, (4) 0.4.

To build the simplest model of the CDL instanton for two identical colliding bubbles, we consider the potential above to be

$$V(S) = \frac{3V_f}{\beta S_f^4}S^4 - \frac{2(\beta+3)V_f}{\beta S_f^3}S^3 + \frac{3(\beta+1)V_f}{\beta S_f^2}S^2, \qquad (6)$$

where the scalar field, $S \equiv \sqrt{4\pi}\phi$ has been rescaled for computational convenience, and the coefficient for each term on the right-hand side has been appropriately fixed to tune the shape of the potential; in terms of the false vacuum field $S_f = \sqrt{4\pi}\phi_f$, the vacuum energy of the false vacuum V_f and a free parameter β.[6] Given this potential, we are able to solve the scalar field equation (4) simultaneously with the Einstein equations (2), using the ansatz given by Eq. (5), in the coordinates (u, v, χ, θ). Our scalar field solution is independent of the coordinates χ and θ, and hence is expressed in the coordinates (u, v) only; namely, $S(u, v) = \sqrt{4\pi}\phi(u, v)$.[6–8] With the choice of the constants, $\beta = 0.1$, $V_f = 10^{-4}$ and $S_f =$ (1) 0.1, (2) 0.2, (3) 0.3, (4) 0.4, the potential takes the forms as given by Figure 1.[6] With this potential, our numerical solution $S(u, v)$ is obtained as presented by Figure 2.[6]

3. Computation of gravitational waves from bubble collisions

Suppose that a present observer lives in the true vacuum region of one of the two bubbles. Then signatures of bubble collisions which took place in the distant past will be carried to the present observer via GWs. Here we assume that the distance between the center of collision region and the observer can be arbitrarily large (within the size of our universe), and thus that the time for a collision event, which is the retarded time to the present observer, can be quite far in the distant past; namely, $t_R = t_P - r/c \ll t_P$, where t_R denotes the retarded time, t_P the present time and r the distance. Therefore, our bubble collision may be regarded as a localized event, as long as the observer is reasonably far away from the collision region.

The *quadrupole approximation* of GWs is given by

$${}_Q h_{ij}^{TT}(t, \mathbf{x}) = \frac{4}{|\mathbf{x}|}\Lambda_{ij,kl}(\mathbf{n}) \int d^3x' T^{kl}(t_R, \mathbf{x}'), \qquad (7)$$

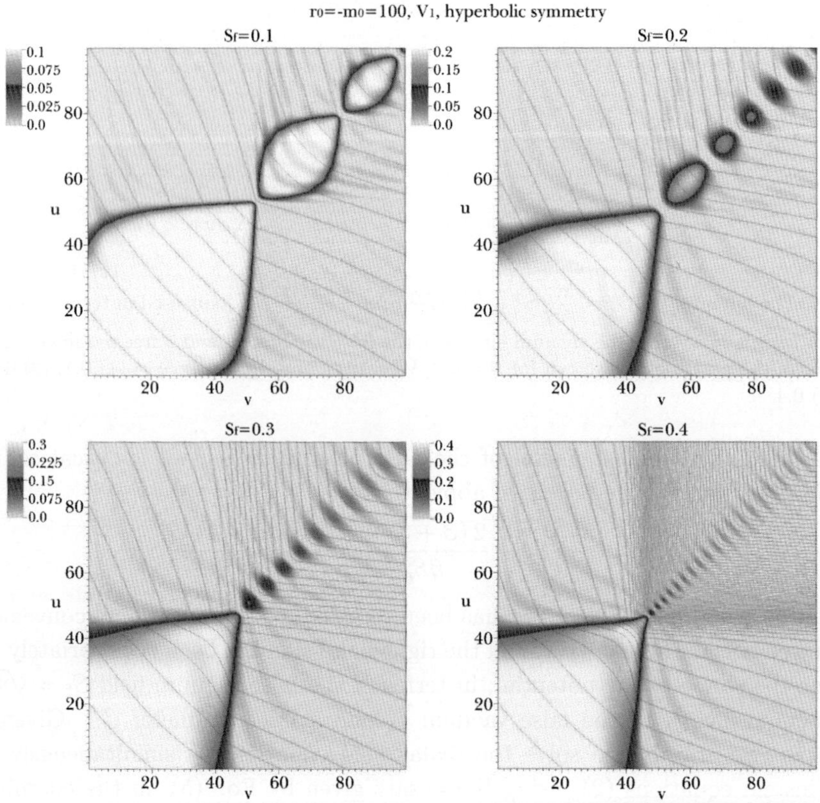

Fig. 2. The numerical solutions $S\,(u,v) = \sqrt{4\pi}\phi\,(u,v)$ obtained with the potential from Figure 1; with various false vacuum field values, $S_{\rm f} = \sqrt{4\pi}\phi_{\rm f} =$ (1) 0.1, (2) 0.2, (3) 0.3, (4) 0.4.

where TT denotes the transverse trace-free gauge, and the projection tensor for gravitational radiation, $\Lambda_{ij,kl}\,(\mathbf{n}) \equiv P_{ik}P_{jl} - \frac{1}{2}P_{ij}P_{kl}$ with $P_{ij} = \delta_{ij} - n_i n_j$ and $\mathbf{n}$ denoting the unit vector along the propagation direction of the waves. Here we have adopted the unit convention $c = G = 1$. Throughout this work our computation is carried out only based on this piece. This is well-suited for our main purpose in this work: to provide *qualitative* insights into patterns of GWs from the colliding two-bubble system in time domain.

3.1. *GWs produced during the initial-to-intermediate stage of strong collisions*

In this regime of strong bubble collisions, our solution for the scalar field, $S\,(u,v) = \sqrt{4\pi}\phi\,(u,v)$ has been obtained and presented by Figure 2.[6] Using this through Eqs. (3) and (7), we are able to compute GWs produced from the collisions; with an appropriate coordinate transformation from (t,x,y,z) to (u,v,χ,θ). After some lengthy analysis, we find the expression:[9]

$$
\begin{aligned}
{}_{\mathrm{Q}}h_{ij}^{\mathrm{TT}}(t,\mathbf{x}) \approx \frac{2\pi}{|\mathbf{x}|}\eta^{2}\left[\delta_{ix}\delta_{jx} - 2\delta_{ix}n_{j}\cos\vartheta + \frac{1}{2}n_{i}n_{j}\left(1+\cos^{2}\vartheta\right) - \frac{1}{2}\delta_{ij}\sin^{2}\vartheta\right] \\
\times\left[\int_{0}^{x_{\mathrm{o}}} du\, T_{uu}\left(t_{\mathrm{R}}\pm u, t_{\mathrm{R}}\mp u\right) - 2\int_{0}^{x_{\mathrm{o}}} du\, T_{uv}\left(t_{\mathrm{R}}\pm u, t_{\mathrm{R}}\mp u\right)\right. \\
\left. + \int_{0}^{x_{\mathrm{o}}} du\, T_{vv}\left(t_{\mathrm{R}}\pm u, t_{\mathrm{R}}\mp u\right) - \frac{1}{2t_{\mathrm{R}}^{2}}\int_{0}^{x_{\mathrm{o}}} du\, T_{\chi\chi}\left(t_{\mathrm{R}}\pm u, t_{\mathrm{R}}\mp u\right)\right],
\end{aligned}
\tag{8}
$$

where $t_{\mathrm{R}} = t - |\mathbf{x}|$, and the wall thickness $\eta \sim \frac{3}{2}\phi_{\mathrm{f}}^{2}/\left(b\epsilon^{4}\right)$, with $2b$ representing the separation of the bubbles and ϵ^{4} the potential difference between the two minima (equivalent to V_{f}), respectively,[7] and the quantity in the first bracket has originated from the projection tensor $\Lambda_{ij,kl}(\mathbf{n})$ with $\mathbf{n} = (\cos\vartheta, \sin\vartheta, 0)$ and ϑ denoting the angle of propagation taken from the x-axis. In order for this computation to be valid, the wall thickness η must be taken sufficiently small; the *thin-wall approximation.*

The numerical results for our GWs (8) are presented by Figure 3;[9] with various false vacuum field values, $S_{\mathrm{f}} = \sqrt{4\pi}\phi_{\mathrm{f}} =$ (1) 0.1, (2) 0.2, (3) 0.3, (4) 0.4, in accordance with the scalar field solutions as presented by Figure 2.

3.2. *GWs produced during the final stage of weak collisions*

Following Ref. 8, the expression for our GWs (8) can be simplified further by neglecting the gravitational effects on the bubbles. Replacing $g_{\mu\nu}$ by $\eta_{\mu\nu}$ through Eq. (3) and after some algebra, the expression (8) can be reduced to[9]

$$
\begin{aligned}
{}_{\mathrm{Q}}h_{ij}^{\mathrm{TT}}(t,\mathbf{x}) \approx \frac{2\pi}{|\mathbf{x}|}\eta^{2}\left[\delta_{ix}\delta_{jx} - 2\delta_{ix}n_{j}\cos\vartheta + \frac{1}{2}n_{i}n_{j}\left(1+\cos^{2}\vartheta\right) - \frac{1}{2}\delta_{ij}\sin^{2}\vartheta\right] \\
\times \int_{0}^{x_{\mathrm{o}}} dx' \left(\frac{\partial\phi}{\partial x'}\right)^{2}_{\tau=t_{\mathrm{R}}}.
\end{aligned}
\tag{9}
$$

Toward the end of the bubble collision process, the scalar field oscillates around the true vacuum state, i.e. $|\phi| \ll 1$, being nearly 'monochromatic'. In this limit, a solution to the scalar field equation (4) turns out to be[9]

$$
\phi(\tau, x) = \phi_{\mathrm{o}}\frac{J_{1}\left(\omega_{\mathrm{t}}\sqrt{\tau^{2}-x^{2}}\right)}{\omega_{\mathrm{t}}\sqrt{\tau^{2}-x^{2}}}, \tag{10}
$$

where J_{n} denotes the Bessel function of the first kind, $\omega_{\mathrm{t}} \equiv \sqrt{6(\beta+1)V_{\mathrm{f}}/\left(\beta S_{\mathrm{f}}^{2}\right)}$ represents the 'terminal' frequency of the bubble collisions, and ϕ_{o} is the amplitude which is determined by the initial conditions of the field.

Substituting Eq. (10) into Eq. (9), our GWs in the regime of weak bubble collisions can be computed. The results are presented by Figure 4;[9] with various false vacuum field values, $S_{\mathrm{f}} = \sqrt{4\pi}\phi_{\mathrm{f}} =$ (1) 0.1, (2) 0.2, (3) 0.3, (4) 0.4.

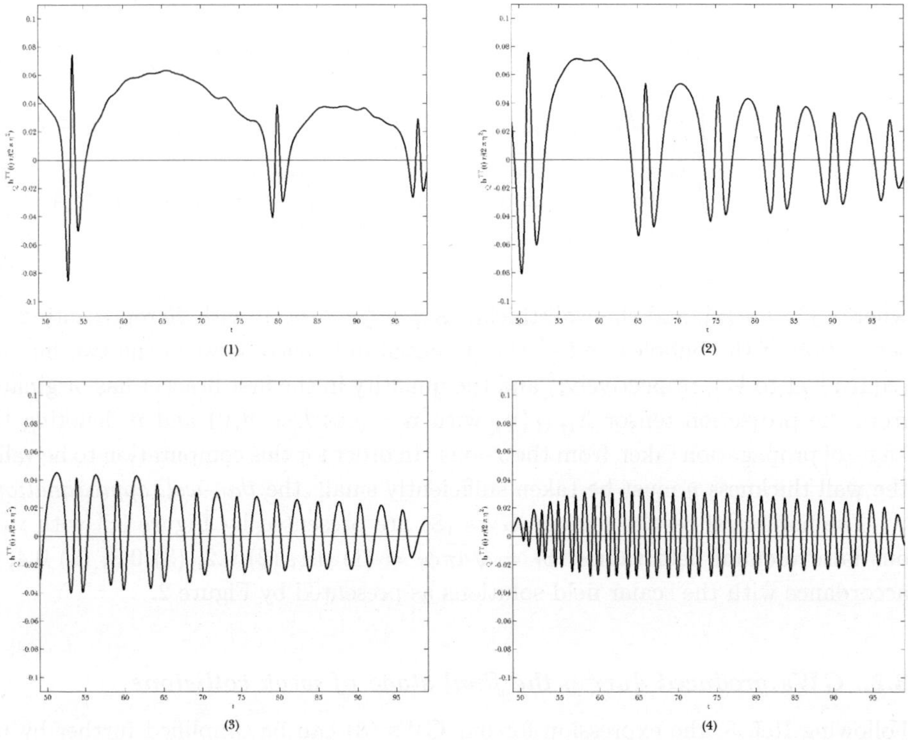

Fig. 3. The numerical plots of ${}_{\mathrm{Q}}h^{\mathrm{TT}}(t)\,|\mathbf{x}|\,/\,(2\pi\eta^2)$ with various false vacuum field values, $S_{\mathrm{f}} = \sqrt{4\pi}\phi_{\mathrm{f}} =$ (1) 0.1, (2) 0.2, (3) 0.3, (4) 0.4.

4. Conclusions

We have computed GWs emitted from collisions of two equal-sized bubbles in time domain. The waveforms have been obtained for (i) the initial-to-intermediate stage of strong collisions and (ii) the final stage of weak collisions, in *full General Relativity* and in the flat spacetime approximation, using numerical and analytical methods, respectively. During (i), the waveforms show the non-linearity of the collisions, characterized by a modulating frequency and cusp-like bumps, whereas during (ii), the waveforms exhibit the linearity of the collisions, featured by a constant frequency and smooth oscillations, as can be checked from Figures 3 and 4, respectively. Also, depending on the false vacuum field value ϕ_{f}, the waveforms have different scales of frequency. During (i), the modulating frequency increases overall as the false vacuum field value ϕ_{f} increases, whereas during (ii), the frequency ω_{t} decreases as the false vacuum field value ϕ_{f} increases in an inversely proportional relationship, i.e. $\omega_{\mathrm{t}} \sim \phi_{\mathrm{f}}^{-1}$. It is interesting to note that the relationship between the false vacuum field value and the frequency during (i) changes almost inversely during (ii).

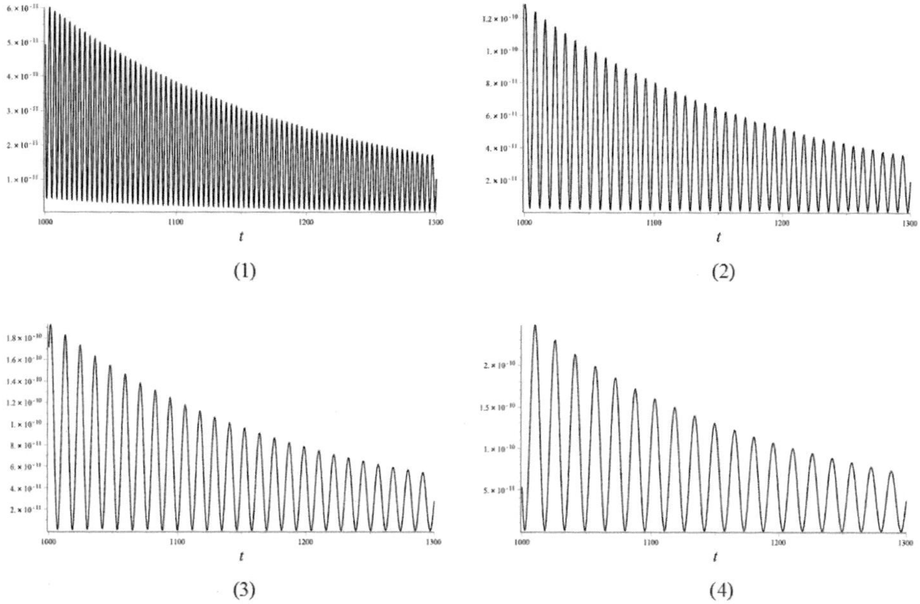

Fig. 4. The plots of ${}_{Q}h^{\mathrm{TT}}(t)\,|\mathbf{x}|\,/\,(2\pi\eta^2)$ with various false vacuum field values, $S_{\mathrm{f}} = \sqrt{4\pi}\phi_{\mathrm{f}} =$ (1) 0.1, (2) 0.2, (3) 0.3, (4) 0.4, with the corresponding frequencies, $\omega_{\mathrm{t}} \simeq$ (1) 0.8124, (2) 0.4062, (3) 0.2708, (4) 0.2031 $\sim$ (1) 1, (2) 1/2, (3) 1/3, (4) 1/4.

The thin-wall and quadrupole approximations served our purpose well in that we were able to gain some *qualitative* insights into the time-domain gravitational waveforms from bubble collisions. However, to obtain more physically reasonable waveforms, taking into account a generic thickness and relativistic motion of bubble wall, the next-to-leading order corrections beyond each approximation will be necessary. Finding the corrections will involve a huge amount of computation, and thus is left for follow-up studies.

Acknowledgment

The author would like to thank Dong-han Yeom and Frederico Arroja for valuable comments and discussions. This work is supported by Basic Science Research Program through the National Research Foundation of Korea (NRF) funded by the Ministry of Education (2013R1A1A2008901).

References

1. A. H. Guth, Inflationary universe: A possible solution to the horizon and flatness problems, Phys. Rev. D **23**, 347 (1981);
A. D. Linde, A new inflationary universe scenario: A possible solution of the horizon, flatness, homogeneity, isotropy and primordial monopole problems, Phys. Lett. B **108**, 389 (1982);

A. Albrecht and P. J. Steinhardt, Cosmology for grand unified theories with radiatively induced symmetry breaking, Phys. Rev. Lett. **48**, 1220 (1982).
2. A. D. Linde, Nonsingular regenerating inflationary universe, Cambridge University preprint, Print-82-0554 (1982);
   A. H. Guth, Eternal inflation and its implications, J. Phys. A **40**, 6811 (2007).
3. A. Vilenkin, The birth of inflationary universes, Phys. Rev. D **27**, 2848 (1983).
4. S. R. Coleman and F. De Luccia, Gravitational effects on and of vacuum decay, Phys. Rev. D **21**, 3305 (1980).
5. S. Chang, M. Kleban and T. S. Levi, When worlds collide, JCAP **0804**, 034 (2008).
6. D. Hwang, B.-H. Lee, W. Lee, and D. Yeom, Bubble collision with gravitation, JCAP **1207**, 003 (2012).
7. S. W. Hawking, I. G. Moss, and J. M. Stewart, Bubble collisions in the very early universe, Phys. Rev. D **26**, 2681 (1982).
8. A. Kosowsky, M. S. Turner, and R. Watkins, Gravitational radiation from colliding vacuum bubbles, Phys. Rev. D **45**, 4514 (1992).
9. D.-H. Kim, B.-H. Lee, W. Lee, J. Yang, and D. Yeom, Gravitational waves from cosmic bubble collisions, Eur. Phys. J. C **75**, 133 (2015).

Search for Gravitational Waves from Binary Back Hole Mergers: Parameter Estimation and Fisher Matrix

H. S. Cho

Korea Institute of Science and Technology Information,
Daejeon 34141, Korea
chohs1439@pusan.ac.kr

The orbits of binary black hole systems shrink due to gravitational radiations. At the last stage of merger, the gravitational wave (GW) signals can be detected by the ground-based detectors, such as LIGO and Virgo. In the data analysis of GWs emitted from merging black hole binaries, the matched filtering method is employed in the detection pipeline to identify GW events. Once a detection is made, the parameter estimation seeks the physical parameters of the GW source. This pipeline repeatedly performs overlap computations by generating theoretical waveforms and matching those to the detector data based on Monte Carlo simulations. In this talk, we briefly review the search and the parameter estimation in GW data analysis, and introduce the Fisher matrix (FM) method that has been mainly used to predict the uncertainties in the parameter estimations analytically. The FM is very easy to compute and has very low computational cost compared to Monte Carlo simulations. Using the FM, we calculate the parameter estimation uncertainties for binary black hole systems.

Keywords: Gravitational waves; binary black holes; parameter estimation; Fisher matrix.

1. Introduction

100 years ago the gravitational waves (GWs) were predicted by the Einstein's general relativity theory. Until recently, no direct detection has been made because the GW strain is extremely weak. At last, however, GWs emitted from a merging binary black hole (BBH) have been directly observed by the two detectors of LIGO (Laser Interferometer Gravitational-Wave Observatory) on September 14, 2015.[1] BBH mergers are among the most promising sources of GW transients for ground-base detectors, such as LIGO[2] and Virgo.[3] In GW data analysis for compact binary coalescences (CBCs), such as BBHs, the search pipeline identifies whether or not the detector data stream includes a real GW signal by matching the theoretically generated waveforms to the detector data.[4] Once a detection is made in search, the next step is to figure out the the physical parameters of the GW source, and this process is performed in the parameter estimation pipeline.[5] The parameter estimation analysis is generally based on the Monte Carlo (MC) simulations. The MC simulation method enables us to explore the whole parameter space, however, this method needs very large computational cost and take long time up to several weeks depending on the waveform model and the detector sensitivity.

On the other hand, for a given signal model with parameters known, the Fisher matrix (FM) method can easily approximate the parameter estimation uncertainties if the wave signal is strong enough.[6] Although several limitations have been addressed in the past works,[7,8] the FM has been still mainly used because it is quite easy to compute and needs very low computational cost compared to the MC simulations. In this work, we briefly review the GW search and parameter estimation in the CBC data analysis. Next, we introduce the FM formalism and apply this method to the parameter estimation for BBHs. We calculate the parameter estimation uncertainties and compare the results between a nonspinning and an aligned-spin BBHs assuming the initial LIGO and the advanced LIGO sensitivities.

2. CBC gravitational wave search and phenomenological waveforms

The matched filtering is the optimal filter for extracting well-modeled signals, such as CBC waveforms, from stationary, Gaussian noise. The matched filter output is expressed by the signal-to-noise ratio ρ as[4]

$$\rho(t) = 4\int_0^\infty df \frac{\tilde{x}(f)\tilde{h}^*(f)}{S_n(f)} e^{2\pi i f t}, \tag{1}$$

where $\tilde{h}^*(f)$ is the complex conjugate of the Fourier-transformed waveform template, $\tilde{x}(f)$ is the Fourier-transformed GW strain data, and $S_n(f)$ is the single-sided noise power spectral density. When the template waveform h has the same shape as the signal waveform s buried in the detector data x, the matched filtering gives the optimal search statistic. It is highly unlikely to obtain $\rho \gg 1$ for purely stationary Gaussian noise, so the detection strategy usually involves setting a lower threshold on ρ to identify event candidates.[9]

Since the match can be most efficiently computed in the Fourier domain, the frequency-domain waveform templates are computationally cheaper. For the low mass systems, where the binary mass is smaller than $\sim 15M_\odot$, only the inspiral phase is likely to have most of the signal power,[10,11] and the two-body dynamics in the inspiral phase can be solved using the post-Newtonian (PN) approximation. However, the gravitational waveforms emitted from BBH mergers generally consist of three phases, inspiral-merger-ringdown (IMR). If the merger-bringdown phases are non negligible in the total signal power, the complete IMR waveforms should be used in the data analysis. Effort to establish the frequency-domain IMR waveforms models have been made over the past years. Those works have been carries out by constructing phenomenological families using hybrid waveforms obtained by combining PN and numerical relativity simulation waveforms. Several phenomenological models have been developed for various BBH configutrations from a nonspinning to a precessing systems. PhenomA has been developed for nonspinning BBH waveforms,[12] while PhenomB[13] and PhenomC[14] were designed to model the waveforms of aligned-spin BBH systems. To include precessing effects, PhenomP has also been developed.[15] Recently, PhenomD has been introduced to improve the accuracies of the previous models for aligned-spin BBHs.[16]

The phenomenological waveform models are defined in the Fourier domain as the form,

$$\tilde{h}(f) = A_{\rm eff}(f) e^{i\Psi_{\rm eff}(f)}, \tag{2}$$

where $A_{\rm eff}$ and $\Psi_{\rm eff}$ are the effective amplitude and effective phase, and those are modeled separately. The model waveforms are parameterized by their physical parameters such as masses and spins. The wave phase contains the arbitrary parameters, the coalescence time t_c and the coalescence phase ϕ_c. However, the inverse Fourier transform can compute the match for all possible coalescence times at once, and by taking the absolute value of the complex number the match can be maximized over the coalescence phase.[4] Consequently, the phenomenological models need only two mass parameters for nonspinning BBHs, and two mass parameters and one or two spin parameters for spinning BBHs. For efficiency, the chirp mass $M_c \equiv (m_1 m_2)^{3/5}/(m_1+m_2)^{1/5}$ and the symmetric mass ratio $\eta \equiv m_1 m_2/(m_1+m_2)^2$ are generally used as mass parameters.

3. Parameter estimation and Fisher matrix

When a GW signal is detected in the CBC search pipeline, the parameter estimation pipeline seeks the physical parameters of the GW source by computing the overlap between the signal s and template waveforms h:

$$\langle s|h\rangle = 4\ {\rm Re} \int_{f_{\rm low}}^{\infty} df \frac{\tilde{s}(f)\tilde{h}^*(f)}{S_n(f)}, \tag{3}$$

where $f_{\rm low}$ is the low frequency cutoff. In CBC parameter estimation, one should deal with from 9 to 15 parameters depending on spins. The template waveforms used in the overlaps are generated continuously varying the parameters, and their result is given by the marginalized posterior probability density functions (PDFs) for the parameters.

In the Bayesian statistics, a posterior probability that the gravitational wave signal in the data stream x is characterized by the parameters λ, can be expressed by $p(\lambda|x) \propto p(\lambda)L(x|\lambda)$ where $p(\lambda)$ is a prior probability, $L(s|\lambda)$ is a likelihood written by[17,18]

$$L(x|\lambda) \propto \exp\left[-\frac{1}{2}\langle x - h(\lambda)|x - h(\lambda)\rangle \right], \tag{4}$$

where $h(\lambda)$ is a template waveform. The data x consists of a GW signal s and a noise n. If the SNR is high enough, the noise can be neglected, then the likelihood can be written as

$$L \propto \exp\left[-\frac{1}{2}\{\langle s|s\rangle + \langle h|h\rangle - 2\langle s|h\rangle\} \right], \tag{5}$$

and the overall shape of L is mainly determined by the third term, i.e., the overlap between s and h. Consequently, assuming a flat prior the Bayesian posterior PDF is obtained by the overlaps between the signal and template waveforms.

On the other hand, the Fisher matrix (FM) has been generally used to predict the uncertainties in parameter estimation.[19,20] The FM does not explore the whole parameter space, that only predicts the magnitudes of the measurement uncertainties for a given signal with known parameter values. For large SNRs, errors in the estimation of parameters $\Delta\lambda^i$ obeys a Gaussian probability distribution of the form[18]

$$p(\Delta\lambda^i) \propto \exp[-\frac{1}{2}\Gamma_{ij}\Delta\lambda^i\Delta\lambda^j], \tag{6}$$

where Γ_{ij} is the Fisher Matrix (FM) defined as[6,21,22]

$$\Gamma_{ij} = -\rho^2 \left\langle \frac{\partial \hat{h}}{\partial \lambda_i} \middle| \frac{\partial \hat{h}}{\partial \lambda_j} \right\rangle \bigg|_{\lambda=\lambda_0}, \tag{7}$$

where $\hat{h} \equiv h/\sqrt{\langle h|h\rangle}$ is the normalized waveform. If the waveform h is a Fourier-domain analytic function of the parameters λ_i, the partial derivatives can be computed analytically, thus the FM is easily obtained by performing one match computation at the true values λ_0 of the parameters. The inverse of Γ_{ij} represents the covariance matrix of the parameter errors, and the error (σ_i) of each parameter is determined as $\sigma_i = \sqrt{(\Gamma^{-1})_{ii}}$. The FM is a semi-analytic approach, thus much more computationally cheap than the MC simulations.

Table 1. Comparison of parameter estimation uncertainties of BBHs.

Detector	initial LIGO		advanced LIGO	
Binary system	Nonspinning	Aligned-spin	Nonspinning	Aligned-spin
$\sigma_{Mc}/M_\odot$	0.030	0.038	0.0040	0.0073
σ_η	0.0034	0.0095	0.0013	0.0078
$\sigma_{\chi_{\rm eff}}$	-	0.037	-	0.021

Using the FM method we calculated statistical uncertainties (σ_i) for a nonspinning and an aligned-spin BBH systems. We assume the BH masses of $(15,5)M_\odot$ for both systems, and the spin of $\chi_{\rm eff} = 0.5$ for the aligned-spin system. The parameter $\chi_{\rm eff}$ indicates the total spin magnitude defined in the range of [-1, 1]. We use the analytic sensitivity curves for the initial LIGO and the advanced LIGO detectors,[23,24] where the low frequency cutoffs are 40 Hz and 10 Hz for the initial LIGO and the advanced LIGO, respectively. The signal strength is assumed to be $\rho = 20$. We use the waveform model PhenomC,[14] and the effective spin is assumed to be 0 for the nonspinning case. In Table 1, the results show comparisons between nonspinning and aligned-spin BBH systems as well as between initial LIGO and advanced LIGO sensitivities. First, we can find that through the improvement of the detector sensitivity from the initial LIGO to the advanced LIGO, the parameter estimation uncertainties can be significantly reduced. This is mainly caused by the difference between the two sensitivity curves at the low frequency region. The GW chirp signals produced by CBCs are likely to contain most of the wave cycles in the low frequency region, thus as the detector is more sensitive in the low frequency the parameter estimation becomes more accurate. Second, we also find that the uncertainties for the mass parameters can be increased if we consider the BH spin. Since the effective spin is more strongly correlated with the symmetric mass ratio than with the chirp mass, this increasing behavior is more pronounced for the parameter η.

4. Summary and discussion

In this work, we briefly reviewed the procedure of the CBC data analysis. The search analysis is basically based on the matched filtering technique which is the optimal filter for well-modeled signals in the stationary, Gaussian noise. The gravitational waveforms emitted from BBH mergers consist of IMR phases, and various IMR waveform models have been developed. In parameter estimation, the template waveforms are continuously generated and the posterior PDF is obtained from the overlaps between the templates and the detector data which includes the signal. The parameter estimation analysis is based on the MC simulations, thus very large computational cost is necessary to perform the full analysis. On the other hand, the FM method has been mainly used to predict the statistical uncertainties in parameter estimation for know strong signals. The FM can be easily computed from the analytic wave function, thus that is much more computationally cheap than the MC simulations. Using PhenomC, which is a frequency-domain gravitational waveform model designed for aligned-spin BBHs, we calculated the parameter estimation uncertainties for both a nonspinning (where the effective spin was set to be 0 in the PhenomC waveforms) and an aligned-spin BBH systems assuming the initial LIGO and the advanced LIGO sensitivities. We found that the mass and spin parameters can be recovered much more accurately for the advanced LIGO. We also found that by taking into account the spin parameter in parameter estimation, uncertainties in the mass parameters can be increased.

References

1. B. P. Abbott *et al.*, (LIGO-Virgo Collaboration), Observation of Gravitational Waves from a Binary Black Hole Merger, *Phys. Rev. Letter* **116**, 061102 (2016).
2. J. Abadie *et al.*, (LIGO Scientific Collaboration), Advanced LIGO, arXiv:1411.4547 (2014).
3. F. Acernese *et al.*, Advanced Virgo: a 2nd generation interferometric gravitational wave detector, *Class. Quantum Grav.* **32** 024001 (2014).
4. B. Allen, W. G. Anderson, P. R. Brady, D. A. Brown and J. D. E. Creighton, FINDCHIRP: An algorithm for detection of gravitational waves from inspiraling compact binaries, *Phys. Rev. D* **85**, 122006 (2012).
5. J. Aasi *et al.*, (LIGO Scientific Collaboration, Virgo Collaboration), Parameter estimation for compact binary coalescence signals with the first generation gravitational-wave detector network, *Phys. Rev. D* **88**, 062001 (2013).
6. M. Vallisneri, Use and abuse of the Fisher information matrix in the assessment of gravitational-wave parameter-estimation prospects, *Phys. Rev. D* **77**, 042001 (2008).

7. I. Mandel, C. Berry, F. Ohme, S. Fairhurst, and W. M. Farr, Parameter estimation on compact binary coalescences with abruptly terminating gravitational waveforms, *Class. Quantum Grav.* **31**, 155005 (2014).
8. H. -S. Cho and C. -H. Lee, Application of the effective Fisher matrix to the frequency domain inspiral waveforms, *Class. Quantum Grav.* **31**, 235009 (2014).
9. Aasi J *et al.*(LIGO Scientific Collaboration and Virgo Collaboration), Search for gravitational waves from binary black hole inspiral, merger, and ringdown in LIGO-Virgo data from 20092010, *Phys. Rev. D* **87**, 022002 (2013).
10. Ajith P, Gravitational-wave data analysis using binary black-hole waveforms, *Class. Quantum Grav.* **25**, 114033 (2008).
11. H. -S. Cho, Parameter estimation using a complete signal and inspiral templates for low mass binary black holes with Advanced LIGO sensitivity, *Class. Quantum Grav.* **32**, 235007 (2015).
12. Ajith P *et al.*, Phenomenological template family for black-hole coalescence waveforms, *Class. Quantum Grav.* **24**, S689 (2007).
13. Ajith P *et al.*, Inspiral-Merger-Ringdown Waveforms for Black-Hole Binaries with Nonprecessing Spins, *Phys. Rev. Letter* **106**, 241101 (2011).
14. Santamaria L *et al.*, Matching post-Newtonian and numerical relativity waveforms: Systematic errors and a new phenomenological model for nonprecessing black hole binaries, *Phys. Rev. D* **82**, 064016 (2010).
15. Hannam M *et al.*, Simple Model of Complete Precessing Black-Hole-Binary Gravitational Waveforms, *Phys. Rev. Letter* **113**, 151101 (2014).
16. Khan S *et al.*, Frequency-domain gravitational waves from non-precessing black-hole binaries. II. A phenomenological model for the advanced detector era, *Phys. Rev. D* **93**, 044007 (2016).
17. C. Cutler and E. É. Flanagan, Gravitational waves from merging compact binaries: How accurately can one extract the binarys parameters from the inspiral waveform?, *Phys. Rev. D* **49** 2658 (1994).
18. L. S. Finn, Detection, measurement, and gravitational radiation, *Phys. Rev. D* **46** 5236 (1992).
19. H. -S. Cho, Search and parameter estimate in gravitational wave data analysis and the fisher matrix, *J. of the Korean Phys. Soc.* **66**, 1637 (2015).
20. H. -S. Cho, Accuracy in measuring the neutron star mass in gravitational wave parameter estimates for nonspinning compact binaries, *J. of the Korean Phys. Soc.* **67**, 960 (2015).
21. P. Jaranowski and A. Królak, Optimal solution to the inverse problem for the gravitational wave signal of a coalescing compact binary, *Phys. Rev. D* **49** 1723 (1994).
22. H. -S. Cho, E. Ochsner, R. O'Shaughnessy, C. Kim, and C. -H. Lee, Gravitational waves from BH-NS binaries: Effective Fisher matrices and parameter estimation using higher harmonics, *Phys. Rev. D* **87**, 024004 (2013).
23. T. Damour, B. R. Iyer, and B. S. Sathyaprakash, Comparison of search templates for gravitational waves from binary inspiral, *Phys. Rev. D* **63** 044023 (2001).
24. P. Ajith and S. Bose, Estimating the parameters of nonspinning binary black holes using ground-based gravitational-wave detectors: Statistical errors, *Phys. Rev. D* **79**, 084032 (2009).

Primordial Black Holes as a Novel Probe of Primordial Gravitational Waves

Teruaki Suyama
Research Center for the Early Universe (RESCEU), Graduate School of Science, The University of Tokyo, Tokyo 113-0033, Japan
suyama@resceu.s.u-tokyo.ac.jp

We propose a novel method to probe primordial gravitational waves by means of primorial black holes (PBHs). When the amplitude of initial tensor perturbations generated during inflation on comoving scales much smaller than those relavant to Cosmic Microwave Background is very large, it induces scalar perturbations due to second-order effects. If the amplitude of resultant scalar perturbations becomes too large, then too many PBHs are produced to a level that is inconsistent with a variety of existing observational constraints on the abundance of PBHs. This consideration leads to upper bounds on the amplitude of primordial tensor perturbations. The resultant PBH upper bounds turn out be tighter than other bounds obtained from Big Bang Nucleosynthesis, Cosmic Microwave Background, LIGO and Virgo.

Keywords: Primordial gravitational wave; primordial black holes.

1. Introduction

Stochastic gravitational wave background (SGWB) on a wide range of scales is thought to have been generated in the early universe. SGWB on largest observable scales have been investigated by Planck[1] and BICEP2.[2] SGWB on smaller scales can be constrained by measuring the value of $N_{\rm eff}$, the effective number of extra degrees of freedom of relativistic species, at Big Bang Nucleosynthesis (BBN) through the current abundance of light elements,[3] or at photon decoupling through the anisotropy of Cosmic Microwave Background (CMB).[4,5] More recently SGWB on small scales has been constrained by CMB spectral distortions as well.[6,7] SGWB can also be directly constrained by gravitational wave detectors (see e.g. Ref. 8). BBN and CMB have played a major role in constraining SGWB since they are applicable on a wide range of scales, noting ground-based laser interferometers tend to target gravitational waves (GWs) on a relatively limited frequency range with high sensitivity. However, it would be worthwhile to note that upper bounds obtained through $N_{\rm eff}$ needs an assumption about the number of relativistic species in the early universe, as is discussed later. In addition, in obtaining BBN or CMB bounds we implicitly assume that any physical mechanisms, both known and unknown, increase $N_{\rm eff}$, making $N_{\rm eff}$ larger than the standard value $N_{\rm eff} = 3.046$.[9] However, in principle it is possible to *decrease* $N_{\rm eff}$ (see e.g. Refs. 10–12). Therefore, it would be desirable to have another independent cosmological method to probe SGWB on a wide range of scales, which does not depend on the assumptions mentioned above.

The very physical mechanism we employ to probe SGWB is the formation of primordial black holes (PBHs), black holes formed in the early universe well before the cosmic structure formation. There are a number of possible mechanisms to create PBHs (see e.g. Ref. 13 and references therein), but one of the simple and natural mechanisms is the direct collapse of radiation overdensity during the radiation-dominated era, which happens when the density perturbation becomes order unity at the moment of the horizon crossing of the perturbation.[14–16] (See also Refs. 17 and 18 for updated discussions of the formation condition and see also Refs. 19–25 for numerical simulations of the formation of PBHs.) There is no conclusive evidence for the existence of PBHs and so upper bounds on the abundance of PBHs on various mass scales have been obtained by various kinds of observations (see e.g. Ref. 13). These upper bounds can be translated into upper bounds on the power spectrum of the curvature perturbation on small scales,[26,27] namely, they can be used to exclude models of the early universe which predict too many PBHs.

We use PBHs to constrain primordial tensor fluctuations. Large amplitude tensor perturbations induce scalar perturbations due to their second-order effects, though tensor and scalar perturbations are decoupled at the linear level. If the typical amplitude of primordial GWs is too large, then the typical amplitude of resultant induced scalar perturbations becomes also too large and there appear too many regions where the amplitude of density perturbation becomes order unity at the horizon crossing to form PBHs. Thus we can obtain upper bounds on the amplitude of primordial tensor modes requiring PBHs are not overproduced as they enter the Hubble radius and so one may place upper bounds on tensor modes also using this direct collapse mechanism. However, the initial condition and dynamics of non-linear gravitational waves originated from SGWB during the radiation-dominated era have not been fully understood and so in this contribution we consider the second-order GWs, noting the dynamics of non-linear radiation density perturbations is better understood.

To calculate induced scalar perturbations one may solve the Einstein equations for scalar perturbations including the source terms which are second-order in tensor perturbations, assuming some form of the initial power spectrum of tensor perturbations. In general, the statistics of induced scalar perturbations are highly non-Gaussian, since they are generated by second-order tensor perturbations. If we assume some statistical properties of initial tensor perturbations, e.g. Gaussianity, in principle one can determine the probability density function (PDF) of induced scalar perturbation, which is necessary to calculate the abundance of PBHs given some initial tensor power spectrum to obtain PBH constraints on primordial GWs. All the details of those analyses will be presented elsewhere,[28] and here we present simple analytic estimations, which correctly reproduce the main results of Ref. 28, as is discussed shortly.

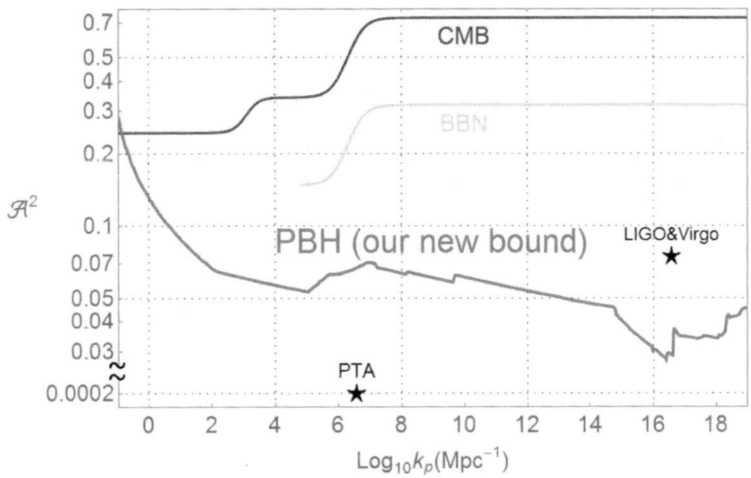

Fig. 1. Comparison of the constraints on the amplitude of the delta-function power spectrum Eq.(2), obtained from PBH, BBN, CMB, PTAs, LIGO and Virgo.

2. PBH upper bound on the tensor perturbations

For simplicity we assume the typical amplitude of tensor fluctuations of only a limited range of wavelength around λ_p, namely we consider a peak in the power spectrum around λ_p. Let us denote by σ_h the root-mean-square amplitude (RMS) or the typical amplitude of tensor fluctuations when the relevant modes under consideration are on super-horizon scales. Note that σ_h^2 roughly corresponds to the dimensionless power spectrum $\mathcal{P}_h$ of primordial GWs. Suppose these large amplitude tensor perturbations whose wavelength is around λ_p reenter the horizon during the radiation-dominated era. It generates radiation density perturbations δ_r whose wavelength is around λ_p and whose RMS σ_{δ_r} at around the horizon crossing is roughly given by $\sigma_{\delta_r} \sim \sigma_h^2$, simply because δ_r is generated by second-order GWs. As confirmed in Ref. 28, this relation indeed holds well at least for the delta-function power spectrum. If $\delta_r > \delta_{r,\mathrm{th}} \sim 0.4$,[17] the induced radiation density perturbations collapse to form PBHs. The probability of formation of PBHs has to be extremely small to be consistent with various observations.[13] Therefore, in order not to overproduce PBHs, one may require $\mathcal{P}_h \sim \sigma_h^2 \sim \sigma_{\delta_r} \lesssim 0.1\delta_{r,\mathrm{th}} \sim 0.04$. This upper bound is applicable to all the scales which reenter the horizon during the radiation-dominated era. Furthermore, this bound is probably conservative on scales which reenter the horizon during the (early as well as late) matter-dominated era, since formation of black holes by direct collapse of primordial perturbations is easier when the universe is (literally or effectively) dominated by pressureless dust. To summarize, we have

$$\mathcal{P}_h \lesssim 0.04 \quad \text{(PBH)}, \tag{1}$$

and this is applicable from the comoving wavelength of $\sim$ 1Gpc all the way down to $\sim$ 1Gpc $\times e^{-60} \sim$ 0.3m, assuming the total number of e-foldings during inflation is sixty.

Our analytic estimations can reproduce the results of more rigorous calculations in Ref. 28, where the delta-function initial tensor power spectrum is assumed:

$$\mathcal{P}_h(k) = \mathcal{A}^2 k_p \delta(k - k_p). \tag{2}$$

The PBH abundance for this power spectrum is calculated and constraints on $\mathcal{A}^2$ as a function of k_p are shown in Fig.1, along with other constraints.

3. Discussion

As already mentioned, PBH constraints are applicable from $\sim$ Gpc all the way down to $\sim$ 0.3m if we assume the number of e-foldings during inflation is sixty. Note that the exclusion of an overproduction of smallest PBHs ($M_{\rm PBH} \lesssim 10^5 g$) depends on the assumption that stable Planck mass relics are left over at the end of Hawking evaporation, which contribute to cold dark matter (see Refs. 13, 29 and references therein). The range of comoving scales corresponding to $M_{\rm PBH} \lesssim 10^5 g$ is roughly $\lesssim$ 50m, and so PBH upper bounds in this range depend on this assumption. If Planck mass relics are not formed, to what extent an overproduction of PBHs with $M_{\rm PBH} \lesssim 10^5 g$ is cosmologically problematic is uncertain. Such an overproduction of smallest PBHs may lead to an early matter-dominated era, during which PBH binaries are formed and emit GWs, or larger PBHs may form due to merger taking place after the collapse of perturbations of PBHs' density, thereby leaving observable traces. Therefore, in principle one may still exclude such an over production of smallest PBHs even without the leftover of Planck mass relics.

We have imposed $\sigma_{\delta_r} \lesssim 0.1\delta_{r,\rm th}$ to obtain PBH constraints, Eq.(1). The factor of 0.1 here is to ensure the sufficient rareness of formation of PBHs. Note that to discuss the probability of PBH formation one needs to know the PDF of induced radiation density perturbation, which is in general highly non-Gaussian since it is generated by second-order tensor perturbations. In Ref. 28 we assume Gaussianity of tensor perturbations on top of the delta-function power spectrum, and numerically confirmed the factor of roughly 0.1 is indeed necessary to avoid an overproduction of PBHs (compare Eq.(1) and the PBH constraints in Fig.1). Strictly speaking, observational constraints on PBHs depend on their mass[13] and so how rare PBH formation has to be also depends on their mass, hence the scale dependence of PBH constraints in Fig.1.[28]

This argument implies PBH constraints on tensor perturbations depend on the statistics of tensor perturbations, determining the statistics of induced density perturbation, just as PBH constraints on scalar perturbations depend on the statistics of scalar perturbations.[30] If high-σ realizations of tensor perturbations are suppressed (enhanced) in comparison to a Gaussian case, PBH constraints on tensor perturbations are tighter (weaker). One may conjecture the dependence of PBH constraints of tensor perturbations on the statistics of tensor perturbations is weaker than the dependence of PBH constraints of scalar perturbations on the statistics of scalar perturbations. This is because in the former case the probability of PBHs is determined by the statistics of induced radiation perturbations, generated from second-order tensor perturbations, and so the PDF of induced radiation perturbations is something similar to a χ^2 distribution.[28] Namely, the probability of high-σ peaks of radiation leading to PBH formation is determined by lower-σ realizations of tensor perturbations, for which non-Gaussianity is less important than it is for higher-σ realizations.

Acknowledgments

The author would like to thank Dr. Frederico Arroja for inviting me to the session "Gravitational Waves from Big Bang and Other Sources". This work was supported by JSPS Grant-in-Aid for Young Scientists (B) No.15K17632 (T.S.), MEXT Grant-in-Aid for Scientific Research on Innovative Areas "New Developments in Astrophysics Through Multi-Messenger Observations of Gravitational Wave Sources" No.15H00777 (T.S.).

References

1. P. Ade *et al.*, Planck 2013 results. XVI. Cosmological parameters, *Astron.Astrophys.* **571**, p. A16 (2014).
2. P. Ade *et al.*, Detection of *B*-Mode Polarization at Degree Angular Scales by BICEP2, *Phys.Rev.Lett.* **112**, p. 241101 (2014).
3. B. Allen, The Stochastic gravity wave background: Sources and detection (1996).
4. T. L. Smith, E. Pierpaoli and M. Kamionkowski, A new cosmic microwave background constraint to primordial gravitational waves, *Phys.Rev.Lett.* **97**, p. 021301 (2006).
5. Y. Kikuta, K. Kohri and E. So, Detecting the relic gravitational wave from the electroweak phase transition at SKA (2014).
6. A. Ota, T. Takahashi, H. Tashiro and M. Yamaguchi, CMB μ distortion from primordial gravitational waves (2014).
7. J. Chluba, L. Dai, D. Grin, M. Amin and M. Kamionkowski, Spectral distortions from the dissipation of tensor perturbations, *Mon.Not.Roy.Astron.Soc.* **446**, p. 2871 (2015).
8. J. Aasi *et al.*, Improved Upper Limits on the Stochastic Gravitational-Wave Background from 2009?2010 LIGO and Virgo Data, *Phys.Rev.Lett.* **113**, p. 231101 (2014).
9. G. Mangano, G. Miele, S. Pastor, T. Pinto, O. Pisanti *et al.*, Relic neutrino decoupling including flavor oscillations, *Nucl.Phys.* **B729**, 221 (2005).
10. K. Ichiki, M. Yahiro, T. Kajino, M. Orito and G. Mathews, Observational constraints on dark radiation in brane cosmology, *Phys.Rev.* **D66**, p. 043521 (2002).
11. P. S. Apostolopoulos and N. Tetradis, The Generalized dark radiation and accelerated expansion in brane cosmology, *Phys.Lett.* **B633**, 409 (2006).
12. R. Maartens and K. Koyama, Brane-World Gravity, *Living Rev.Rel.* **13**, p. 5 (2010).
13. B. Carr, K. Kohri, Y. Sendouda and J. Yokoyama, New cosmological constraints on primordial black holes, *Phys.Rev.* **D81**, p. 104019 (2010).
14. Y. B. Zel'dovich and I. D. Novikov, The Hypothesis of Cores Retarded during Expansion and the Hot Cosmological Model, *Sov. Astron.* **10**, p. 602 (1967).

15. B. J. Carr and S. Hawking, Black holes in the early Universe, *Mon.Not.Roy.Astron. Soc.* **168**, 399 (1974).
16. B. J. Carr, The Primordial black hole mass spectrum, *Astrophys.J.* **201**, 1 (1975).
17. T. Harada, C.-M. Yoo and K. Kohri, Threshold of primordial black hole formation, *Phys.Rev.* **D88**, p. 084051 (2013).
18. T. Harada, C.-M. Yoo, T. Nakama and Y. Koga, Cosmological long-wavelength solutions and primordial black hole formation, *Phys. Rev.* **D91**, p. 084057 (2015).
19. D. K. Nadezhin, I. D. Novikov and A. G. Polnarev, The hydrodynamics of primordial black hole formation, *Sov. Astron.* **22**, 129 (April 1978).
20. M. Shibata and M. Sasaki, Black hole formation in the Friedmann universe: Formulation and computation in numerical relativity, *Phys.Rev.* **D60**, p. 084002 (1999).
21. J. C. Niemeyer and K. Jedamzik, Dynamics of primordial black hole formation, *Phys.Rev.* **D59**, p. 124013 (1999).
22. A. G. Polnarev and I. Musco, Curvature profiles as initial conditions for primordial black hole formation, *Class.Quant.Grav.* **24**, 1405 (2007).
23. A. Polnarev, T. Nakama and J. Yokoyama, Self-consistent initial conditions for primordial black hole formation, *JCAP* **1209**, p. 027 (2012).
24. T. Nakama, T. Harada, A. Polnarev and J. Yokoyama, Identifying the most crucial parameters of the initial curvature profile for primordial black hole formation, *JCAP* **1401**, p. 037 (2014).
25. T. Nakama, The double formation of primordial black holes, *JCAP* **1410**, p. 040 (2014).
26. E. Bugaev and P. Klimai, Constraints on amplitudes of curvature perturbations from primordial black holes, *Phys.Rev.* **D79**, p. 103511 (2009).
27. A. S. Josan, A. M. Green and K. A. Malik, Generalised constraints on the curvature perturbation from primordial black holes, *Phys.Rev.* **D79**, p. 103520 (2009).
28. T. Nakama and T. Suyama, *in preparation* (2015).
29. J. H. MacGibbon, Can Planck-mass relics of evaporating black holes close the universe?, *Nature* **329**, 308 (1987).
30. C. T. Byrnes, E. J. Copeland and A. M. Green, Primordial black holes as a tool for constraining non-Gaussianity, *Phys. Rev.* **D86**, p. 043512 (2012).

Resonant Amplification of Primordial Gravitational Waves

Chunshan Lin and Misao Sasaki
Yukawa Institute for Theoretical Physics, Kyoto University

We propose a mechanism to evade the Lyth bound in models of inflation. We minimally extend the conventional single-field inflation model in general relativity (GR) to a theory with non-vanishing graviton mass in the very early universe. The modification primarily affects the tensor perturbation, while the scalar and vector perturbations are the same as the ones in GR with a single scalar field at least at the level of linear perturbation theory. During the reheating stage, the graviton mass oscillates coherently and leads to resonant amplification of the primordial tensor perturbation. After reheating the graviton mass vanishes and we recover GR.

Keywords: Inflation; primordial gravitational waves; massive gravity.

1. Introduction

Inflation[1] is the leading paradigm of very early universe cosmology, but its physical origin is still mysterious. The generation of primordial gravitational waves is a generic prediction of the inflationary universe. Several next-generation experiments are aimed at measuring primordial gravitational waves down to $r \sim 10^{-3}$.

According to the Lyth bound,[2] the tensor-to-scalar ratio is propotional to the variation of the inflaton field during inflation, i.e. $\Delta\phi/M_p \simeq \int dN\sqrt{r/8}$. The threshold $\Delta\phi = M_p$ then corresponds to $r = 2\times10^{-3}$, assumed that tensor power spectrum is nearly scale-invariant. The sizeable amplitude of the primordial gravitational waves requires a super-Planckian excursion of the inflaton, i.e. $\Delta\phi > M_p$.

In quantum field theory, the naturalness principle tells us that super-Planckian is out of the validity of effective field theory. Therefore, the detection of the primordial tensor perturbation with its amplitude larger than the threshold value $r = 2\times10^{-3}$ has a profound impact on our understanding of fundamental physics. It implies that either quantum field theory or gravity may be modified in the very early universe.

In our work,[3] we proposed a new mechanism to evade the Lyth bound. We consider a minimal extension of GR with a non-vanishing graviton mass term in the very early universe. Then the coherent oscillation of the inflaton induces that of graviton mass and gives rise to resonant amplification of the primordial tensor perturbation. After reheating, the graviton mass vanishes as the inflaton decays and we recover GR.

2. Resonant primordial gravitational waves

In the framework of massive gravity, generally gravitational waves receive a mass correction. The equation of motion reads[3]

$$\ddot{\gamma}_k + 3H\dot{\gamma}_k + \left(\frac{k^2}{a^2} + m_g^2\right)\gamma_k = 0. \tag{1}$$

We assume that the graviton mass is proportional to the inflation during reheating,

$$m_g^2 \simeq \lambda\phi^2, \tag{2}$$

where ϕ is the inflaton. The potential is expanded around the minimum as

$$V(\phi) \simeq \frac{1}{2}M^2\phi^2 + \cdots . \tag{3}$$

where M is the mass of the inflaton during reheating and the dots stand for higher order corrections which are irrelevant at low energy scale. Asymptotically for large $Mt \gg 1$, we have

$$\phi \simeq \frac{\phi_*}{\sqrt{3\pi}Mt}\sin(Mt)\exp\left(-\frac{1}{2}\Gamma Mt\right), \tag{4}$$

To include the effect of decaying inflaton, we have simply added a decaying factor $e^{-\Gamma Mt}$ into the above solution, without specifying the detailed model of reheating.

The equation of motion of gravitational waves (1) becomes a Mathieu-type equation,

$$\frac{d^2\gamma_k}{dx^2} + \frac{2}{x}\frac{d\gamma_k}{dx} + \frac{\xi \cdot e^{-\Gamma x}}{x^2}\sin^2(x)\gamma_k = 0, \tag{5}$$

where $x \equiv Mt$ and $\xi \equiv \frac{\lambda\phi_*^2}{3\pi M^2}$. Noted that we have neglected the spatial gradient term since we are interested in the long wavelength modes. We have numerically solved Eq. (5). The results are plotted in Figs. 1 and 2. The resonant amplification factor depends on the value of ξ and the decay rate Γ. After reheating, $\phi \to 0$, the graviton becomes massless and we recover GR.

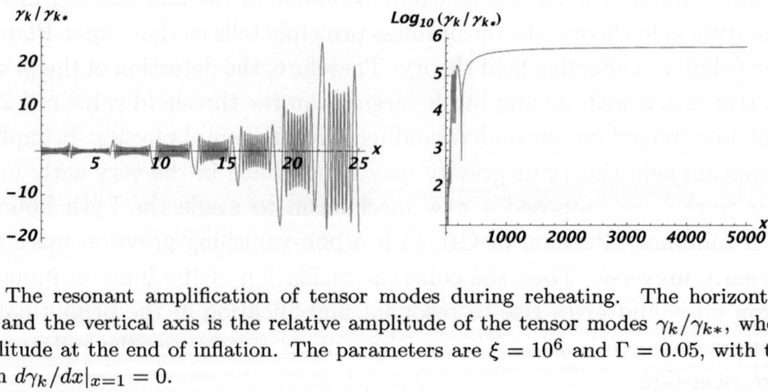

Fig. 1. The resonant amplification of tensor modes during reheating. The horizontal axis is $x \equiv Mt$ and the vertical axis is the relative amplitude of the tensor modes γ_k/γ_{k*}, where γ_{k*} is the amplitude at the end of inflation. The parameters are $\xi = 10^6$ and $\Gamma = 0.05$, with the initial condition $d\gamma_k/dx|_{x=1} = 0$.

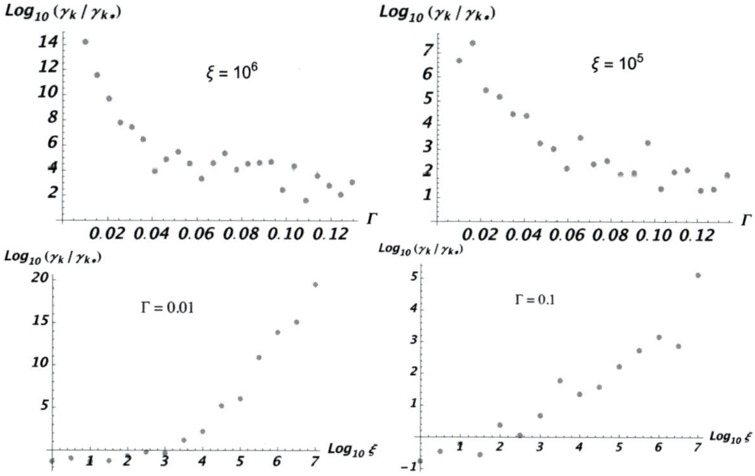

Fig. 2. The parameter dependence of the resonant amplification. The horizontal axis is Γ for the upper panels and ξ for the lower panels. The vertical axes is $Log_{10}(\gamma_k/\gamma_{k*})$ evaluated at $x = 1000$.

In the minimal model of massive gravity, the scalar mode remains the same as the one in GR. Thus the tensor-to-scalar ratio we observe today is

$$r = \frac{A \times P_\gamma}{P_\mathcal{R}} = 16\epsilon \times A, \tag{6}$$

where A is the resonant amplification factor of the tensor modes during reheating. For instance, with the parameters choice in Fig. 1, the factor A could be of the order of 10^{11}. During 60 e-folds, ϕ traverses a distance $\Delta\phi \simeq 15M_p\sqrt{2r/A}$. Hence a sizeable tensor-to-scalar ratio is possible even for a sub-Planckian excursion. We conclude that the Lyth bound can be explictly evaded.

3. Conclusion

we have minimally extended GR to a theory with a non-vanishing graviton mass term and proposed a mechanism to enhance the primordial tensor perturbation from inflation in the the early universe. We find that the Lyth bound can be explicitly evaded due to the parametric resonance of gravitational waves during reheating epoch.

References

1. A. H. Guth, Phys. Rev. D **23**, 347 (1981).
2. D. H. Lyth, Phys. Rev. Lett. **78**, 1861 (1997).
3. C. Lin and M. Sasaki, arXiv:1504.01373 [astro-ph.CO].

Directed Searches for Broadband Gravitational-Wave Emission in Energetic Core-Collapse Supernovae*

Maurice, H. P. M. Van Putten

Room 614, Astronomy and Space Science, Sejong University,
98 Gunja-Dong Gwangin-gu, Seoul 143-747, Korea
mvp@sejong.ac.kr
www.sejong.ac.kr

Core-collapse supernovae are a diverse population with energetic broad-line events, that include their off-spring in ultra-relativistic gamma-ray bursts and a recently discovered highly superluminous event ASASSN-15L (SN2015L). They may be associated with the formation of black holes, rather than magnetars, by their ample baryon-poor energy reservoir in angular momentum desribed by the Kerr metric. Black holes formed herein evolve through various phases that may include direct accretion, Bardeen accretion and spin down to slow rotation, down to a dimensionless spin parameter 0.36 defined by a fixed point in equations of suspended accretion. Evidence may come from gravitational wave emission from associated non-axisymmetric accretion flows, especially from high density matter at the Inner Most Stable Circular Orbit (ISCO) during the latter phase of spin down. We here present a chirp-based spectrogram based on a recently developed "butterfly" filter to probe nearby events by LIGO-Virgo and KAGRA for ascending and descending chirps, making use of embarrassingly parallel computing. Included is a demonstration on the nearby SN2010br Type Ibc ($D \simeq 12$ Mpc) event covered by LIGO S6. Comments on the recently discovered GWB150914 are included.

Keywords: Supernovae; black holes; gravitational waves; matched filtering.

1. Introduction

Core-collapse supernovae (CC-SNe) show a roughly bi-modal distribution in the velocities of their ejecta.[24,32] Events of Type Ib/c further feature a branching ratio of about 1% into normal long gamma-ray bursts (LGRBs) with a true energy in gamma-rays $E_\gamma \simeq 9 \times 10^{51}$ erg.[13,15,16] Typically, the kinetic energies E_k in CC-SNe are representative for their total explosion energy. This is turned on its head, however, by the recently discovered highly luminous event ASSASN-15L (SN2015L) with an output in electromagnetic radiation $E_{rad} \simeq 1.1 \times 10^{52}$ erg on par with E_k that does not derive from nuclear decay of ^{56}Ni.[10]

CC-SNe are believed to be factories of possibly rapidly rotating neutron stars and stellar mass black holes. Since the latter represents the tail of a broad distribution in progenitor mass, the potentially large baryon-free energy reservoir E_{rot} in angular momentum of the latter is an attractive candidate for powering the most extreme events. Hidden from view by the opaque progenitor stellar envelope, direct observation will require a detailed probe by gravitational waves. In fact,

*Invited talk at the 2nd LeCosPa meeting, NTU, Taipei, Taiwan, 2015.

measurement of E_{rot} in gravitational radiation from interactions with high density accretion flows induced by relativistic frame dragging is promising for calorimetric evidence of Kerr black holes.[33]

Black holes of mass M newly formed in core-collapse of relatively massive stars have been widely considered in models of LGRBs by virtue of E_{rot} being baryon-free and E_{rot}/M being large compared to magnetars, see e.g. Ref. 8. A relatively minor fraction of

$$E_{rot} \simeq 6 \times 10^{54}\,\mathrm{erg}\left(\frac{M}{M_\odot}\right) \tag{1}$$

in a rapidly rotating black hole hereby accounts for E_γ, leaving a window for an accompanying major emission unseen in MeV neutrinos and gravitational waves,[34] the latter from non-axisymmetric mass motion at the Inner Most Stable Circular Orbit (ISCO) defined by the Kerr metric.[6,7,20] If so, the latter defines the evolution of black hole mass and spin[34] and may be probed by the gravitational wave observatories LIGO-Virgo and KAGRA.[2–4,11]

For ASASSN-15L, $E_{rad} \sim E_k$ appears to originate from a possibly much more efficient process than nuclear decay, such as dissipation of baryon-poor outflows from a central engine, that may be closely related to the explosion mechanism itself when powered by its angular momentum.[12] Furthermore, the light crossing and cooling time scales of the expanding envelope are on the order of one day, far below the duration of months of its light curve in electromagnetic radiation. Consequently, its light curve provides a tracking in real-time of the long-lived activity of the putative inner engine. Since E_{rad} in ASASSN-15L requires an energy budget that is close to the maximal rotational energy $E_c \simeq 3 \times 10^{52}$[18] (but see Ref. 26) of a neutron star at extremely optimistic efficiencies, it may, like normal LGRBs, represent the transient emission from a rotating black hole.[39]

By the above, core-collapse of relatively massive stars may present telltale light curves of evolving black holes. Any accompanying gravitational radiation will be unsteady in the process of exhausing the angular momentum of the central energy reservoir, that should be apparent in evolving frequencies, i.e., ascending and/or descending chirps on the same time scale as the currently observed light curves in electromagnetic radiation shown in Fig. 1. These considerations suggest the need for a general method to search for broadband extended gravitational wave emission (BEGE).[38]

To implement BEGE, we developed a chirp-based spectram in the $(t, f(t))$ plane by a "butterfly" filter in $(f, df/dt)$ over a choice of intermediate duration τ. In the application to LGRBs, for instance, τ=1 s will capture phase coherence on a time scale of one second. In the application to ASASSN-15L type events, a similar choice may be applicable to an initial failed LGRB hidden in the first tens of seconds, *or* τ =1 d in searches for BEGE associated with aforementioned E_{rad} over the course of months.

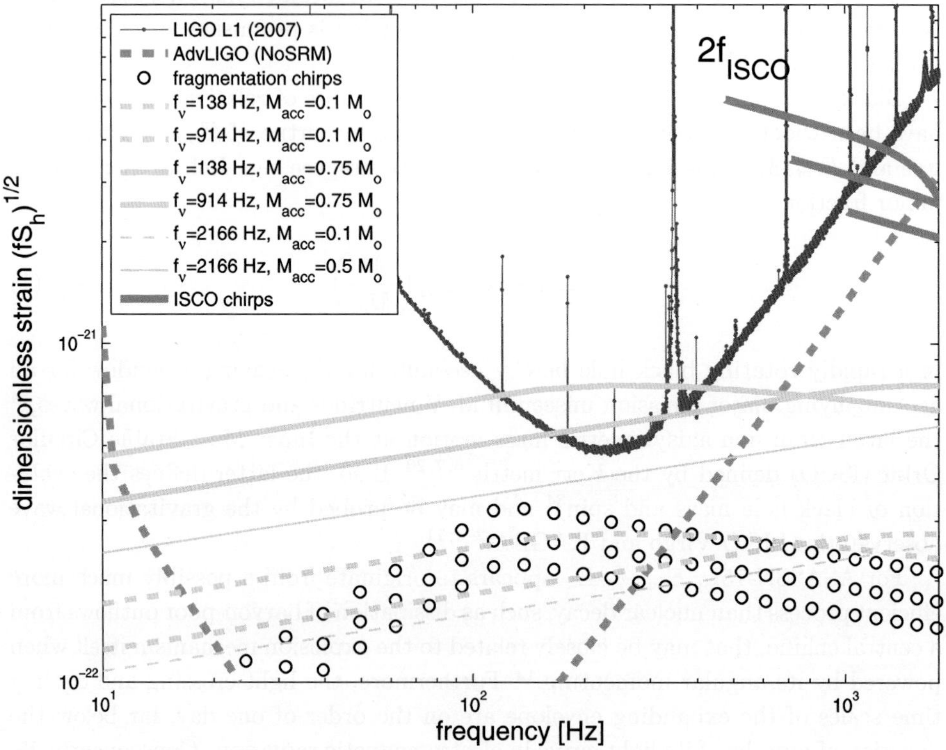

Fig. 1. Overview of some ascending and descending chirps from accretion flows onto new formed black holes in core-collapse supernovae, expressed by characteristic gravitational wave strain $h_{char}(f)$ of quadrupole gravitational radiation at a frequency f at a fiducial source distance $D = 100$ Mpc. For reference, included is the dimensionless strain noise $h_n = \sqrt{fS_h}$ in LIGO S5. The vertical distance of $h_{char}(f)$ to h_n represents the theoretical limit of signal-to-noise ratio in matched filtering. Orientation-averaged results shown are (a) broadband emission from non-axisymmetric accretion flows[25] green solid and dashed curves, (b) fragmentation chirps[29] circles and (c) chirps[38] red solid curves at twice the orbital frequency f_{ISCO} from non-axisymmetric ISCO waves stimulated by feedback from a rapidly rotating black hole. The red curves shown refer to a black hole mass $M = 10M_\odot$ and (a) and $M = 7, 10$ and $15M_\odot$ in (b) and (c). (Adapted from Refs. 25 and 35.)

In Sec. 2, we revisit growth of black holes newly formed in core-collapse of massive stars associated with Fig. 1. In Sec. 3, a general purpose pineline for BEGE is presented. A case study on SN2010br is presented by applying BEGE to LIGO S6 data in Sec. 4. We conclude with outlook on future searches in Sec. 5.

2. Black hole growth in core-collapse of massive rotating stars

Black holes identified in X-ray binaries are believed to originate in core-collapse of massive stars e.g. Ref. 8. What is observed today are remnants of black hole birth and subsequent evolution over the course of an associated supernova, comprising accretion of fall back matter from the progenitor envelope that may pass through several phases, depending on the angular momentum of the progenitor star. While isolated progenitors[40–42] might be slowly rotating, those in intra-day binaries may be rapidly rotating by tidal locking.[28]

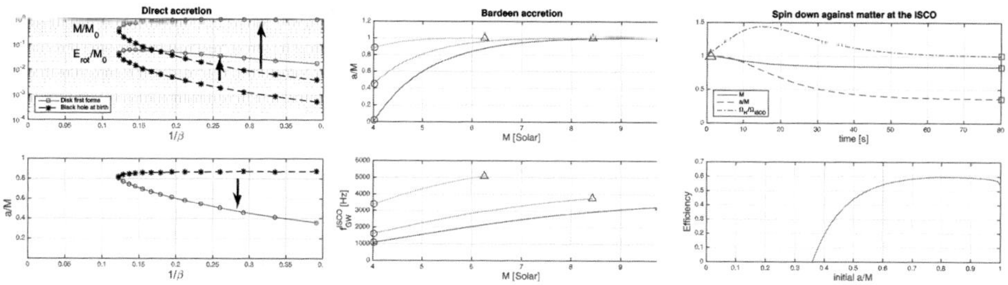

Fig. 2. Three phases of black hole growth following birth in core-collapse (star): direct (left) and Bardeen accretion (middle), followed by spin down against matter at the ISCO (right) producing, respectively, growth in mass to a non-extremal black hole, growth in mass and spin to a near-extremal black hole ($a/M \simeq 1$) and spin down to a slowly rotating black hole ($a/M \simeq 0.36$). While the first phase is on the free fall time scale of the progenitor envelope, the second and third scale can be long lasting by inverse scaling with the density of the accretion flow.

Though details of processes giving rise to black hole birth remain uncertain, newly formed it must satisfy a constraint on its dimensionless spin parameter

$$\frac{a}{M} \leq 1, \tag{2}$$

where $a = J/M$ denotes the specific angular momentum of a black hole, here expressed in geometric units with Newton's constant $G = 1$ and the velocity of light $c = 1$. In a uniformly rotating core, a black hole at birth is near-extremal and of small mass which subsequently evolves through different phases. Revisited in Ref. 39, we have:

(i) Phase I. *Direct accretion* of fall back matter whose angular momentum barrier falls inside the ISCO. The black hole hereby *surges* in mass while decreasing its dimensionless spin parameter (2) on a free fall time scale of tens of seconds of the progenitor envelope;
(ii) Phase II. *Bardeen accretion*[6] from an accretion disk newly formed by fall back matter stalled against an angular momentum barrier outside the ISCO. The black hole now grows in mass and spin on a time scale set by the rate of accretion, generally dependend on density and only moderately perturbed by outflows;[37]
(iii) Phase III. *Spin down* against matter at the ISCO by frame dragging induced Alfvén waves mediated by an inner torus magnetosphere, after the accretion rate has dropped below a critical value.[17,37] (Persist high accretion may result in a time delay between Phase II and Phase III.) The duration of this suspended accretion state scales inversely with the variance in the poloidal magnetic field energy, set by the mass of a torus at the ISCO.[34]

Fig. 2 illustrates Phase I-III. In particular, it shows a black hole remnant of relatively *high mass* relative to the progenitor mass M_0, defined by the high mass $M_\triangle$ near-extremal black hole produced in Bardeen accretion, and *slow spin*, defined by a stable fixed point in the equations suspended accretion when, at late time, the angular velocity of the black hole Ω_H settles down to Ω_T of a torus at the ISCO,

$$\Omega_H = \Omega_T << \frac{1}{2M}, \tag{3}$$

that is approached with essentially exponential decay. According to Figs. 2 and (3) and provided that there is enough mass in the progenitor for a complete Phase II (reaching near-extremal black holes), the remnant satisfies

$$M = 0.84 M_\triangle, \quad a/M = 0.36. \tag{4}$$

Since $M_\triangle$ may reach the progenitor mass M_0 arbitrarily close, the implied *massive slow-spin remnants may explain*[39] *the progenitor black holes of LGWB150914.*[1,22]

3. A butterfly filter for chirp-based spectrograms

Phase II and III open a window to long duration chirps from non-axisymmetric accretion flows down to the ISCO. The latter may produce descending chirps associated with catalytic conversion of black hole spin energy with a gravitational wave frequency $f(t)$ as a function of time and total lifetime of black hole spin T satisfy

$$f_{ISCO}(t) \simeq f_1 + (f_0 - f_1)\, e^{-at/T} \tag{5}$$

where a, f_i $(i = 1, 2)$ are constants, with late-time asymptote[32]

$$f_{GW} \simeq 600\text{–}700\,\text{Hz} \left(\frac{10 M_\odot}{M}\right) \tag{6}$$

for $f_{GW} = 2 f_{ISCO}$. (The lowest frequency corresponds to an initially extremal black hole with maximal energy output.)

To search for ascending and descending gravitational wave chirps over the course of Phase I-III as indicated in Fig. 1, we must take into account that this radiation is phase coherent over at most intermediate time scales $\delta t << \tau << T$. Here, δt ~ms denotes the ms time scale of variability in GRB light curves and $T \simeq T_{90}$ denotes the lifetime of the inner engine, due to the inherently turbulent motions of the underlying accretion flows.

To capture chirps on a time scale τ, we consider extracting spectrograms based on a linear filter by matched filtering, using a bank of chirp templates that covers a range in frequency and time rate-of-change in frequency. The latter represent "butterflies" in Fig. 3. Fig. 4 shows its application to LIGO S6 data.

4. Application to SN2010br

At a distance of about 12 Mpc, SN2010br[9,27] was an extremely nearby Type Ib/c supernava that, as a member of the parent population of normal LGRBs, poses an

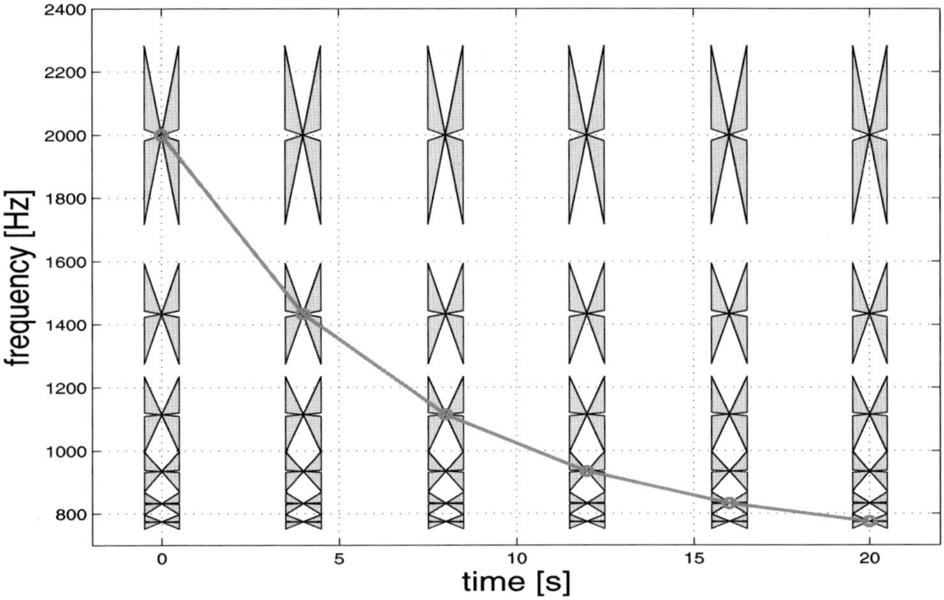

Fig. 3. Chrip-based spectrogram obtained by a butterfly filter, to identify trajectories of long duration chirps. The butterflies obtain from a bank of chirp templates of intermediate duration τ that density cover a range in frequency and time rate-of-change frequency. The vertex and opening angle of each butterfly represent a central frequency f and, respectively, a bandpass of slopes df/dt. (Reprinted from Ref. 38.)

interesting candidate for a probe of any gravitational wave emission. However, its light curve is essentially not resolved, limited to a few days of observation intended to determine its spectral type.

We nevertheless consider it an interesting but challenging example suitable for a case study, in the application of our butterfly filter comprising a total of 64,000 chirp templates, implemented by embarrassengly parrallel computing on a mini-supercomputer with 64 cores. We somewhat arbitrarily chose a two-week data segment starting one month prior to the its discovery. This analysis required about one month of CPU time.

Fig. 5 shows that plenty of SN(f) $>$ 6 are detected in L1 and H1. However, notably absent are (a) SN(f) $>$ 6 events at multiple frequencies within a single 64-second time frame and (b) correlations of SN(f) $>$ 6 values between L1 and H1. On this basis, we disregard the SN(f) $>$ 6 found as spurious.

5. Conclusion and outlook

Modern robotic optical transient surveys promise to provide numerous nearby core-collapse supernovae. The association of energetic events with LGRBs and the highly luminous ASASSN-15L suggest these are of interest for their potentially long-lived compact inner engines possibly in the form of rotating black holes. Given a general

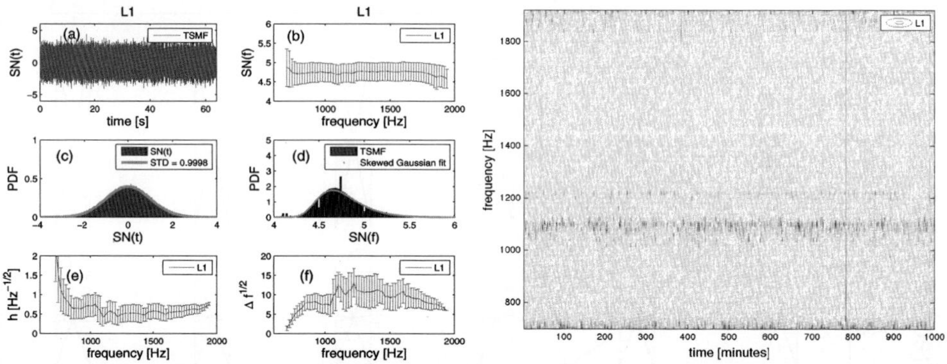

Fig. 4. Application of the butterfly filter of Fig. 3 to LIGO S6 data of the L1 detector. (Left six panels.) (a) Matched filtering against chirp templates of LIGO S6 data produce SN(t) in terms of normalized Pearson coefficients as a function of time t. (b) Maxima SN(f) of SN(t) as a function of frequency define a pseudo-spectrum based on the butterfly filter of Fig. 3 with $\tau = 1$ s. Shown is the mean <SN(f)> over a two week period including its 1 σ variations. (c) SN(t) satisfies a Gaussian distribution, shown normalized to unit variance in the application to a 64 s frames of LIGO data ($n = 2^{18}$ samples). (d) SN(f) satisfies a skewed Gaussian PDF. (e-f) Imperfections in the true spectrum result from non-uniform coverage by our template bank covers of $(f, \Delta f)$, revealed in a small jump around 1100 Hz. (Right panel.) The spectrogram of SN(f) is featureless with no lines, effectively suppressed by the butterflies of Fig. 3. A non-uniformity in our template bank is apparent in a weak noisy feature about 1100 Hz. (Reprinted from Ref. 38.)

outlook on ascending and descending chirps, resulting from conversion of orbital, respectively, spin angular momentum associated with non-axisymmetric accretion flows around rotating black holes (Figs. 1-2), we present a dedicated chirp-based spectrogram based on a butterfly filter in $(f, df/dt)$ space (Figs. 3-4).

For concreteness, we consider SN 2010br, covered by LIGO S6, in a two-week analysis of both L1 and H1 (Fig. 5). We obtained a null-result, possibly because its true time-of-onset, highly uncertain by an ill-resolved light curve, took place before our choice of two weeks data or a larger size template bank is required. (For comparison, 8.64 million templates were used in a recent *BeppoSAX* analysis.[36])

This motivated the development of a two step search, starting with a coarse grained butterfly filter to identify events of interest, here defined by SN(f) > 6. Downselecting by correlating the results from L1 and H1, these events appear to be (numerous) spurious glitches.

Our butterfly filter is proposed as a new pipeline for systematic probes for BEGE from nearby energetic core-collapse events from, e.g., Ref. 5 and modern robotic optical surveys, e.g. Ref. 19 and references therein, in the emerging era of advanced LIGO-Virgo and KAGRA. To further optimize this pipeline, we plan to increase the bank of our chirp templates and to downselect based on SN(f) > 6 at a minimum of two distinct frequencies within each one minute time frame of L1 and H1, before seeking correlations between the two detectors, and, further, to explore implementations on GPU systems and/or Intel's many core CPUs.

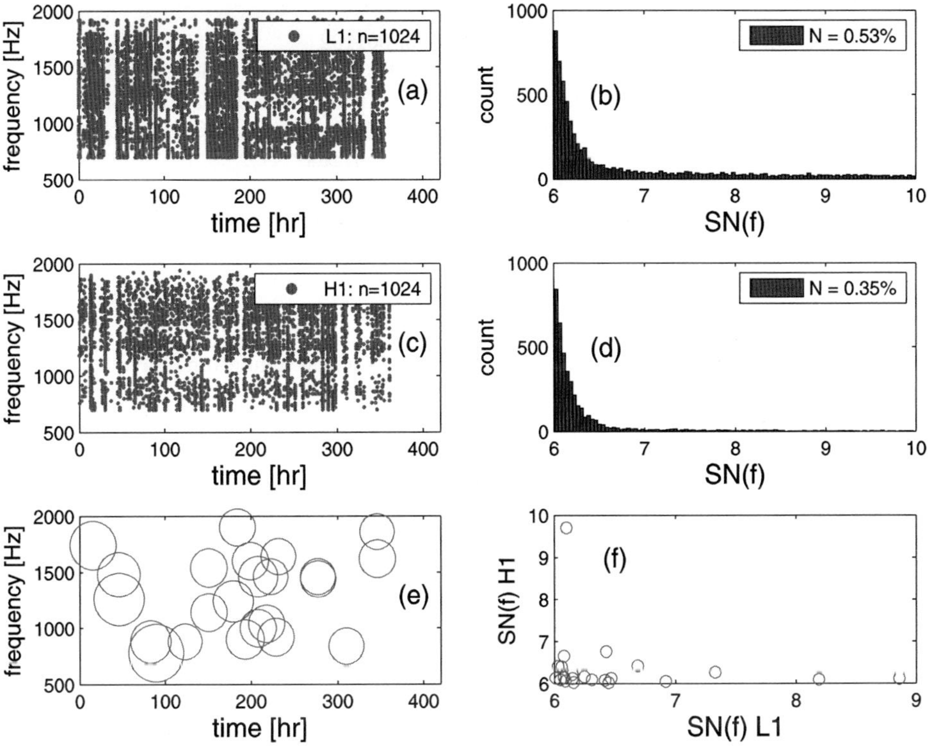

Fig. 5. Scatter plots of the tails SN(f)> 6 in 64 second frames of L1 (a,b) and H1 (c,d) covering a total of two weeks of LIGO S6 data. These high SNR events represent a few tenths of % of the data. Shown are results for 1024 scalings in lifetimes T of a model inner engine, giving template banks of 32 and, respectively, 64 thousand chirps of $\tau = 1$ s duration by time slicing. Coincidences in these tails reduced to 9 (*left*) and 23 (*right*) events (e), where circles indicate the product of the SN(f) values in L1 and H1. However, there are no multiple coincidences in any one minute epoch, nor is there a correlation between the SN(f) of L1 and H1 in (f). These coincidences, coarse grained at one minute resolution, appear to be insignificant. (Reprinted from Ref. 38.)

Acknowledgments

The author thanks the organizers of the 2nd LeCosPa meeting for stimulating discussions. This research is based in part on a Basic Research Grant (2015) from the Korean National Research Foundation and made use of LIGO S6 data from the LIGO Open Science Center (losc.ligo.org), provided by the LIGO Laboratory and LIGO Scientific Collaboration. LIGO is funded by the U.S. National Science Foundation. Some of this work was supported by MEXT, JSPS Leading-edge Research Infrastructure Program, JSPS Grant-in-Aid for Specially Promoted Research 26000005, MEXT Grant-in-Aid for Scientific Research on Innovative Areas 24103005, JSPS Core-to-Core Program, A. Advanced Research Networks, and the joint research program of the Institute for Cosmic Ray Research, University of Tokyo. Parallel computations were performed using a Cloud OS VPGEONET.

References

1. Abbott, B., et al., 2016, Phys. Rev. Lett., 116, 061102
2. Abramovici, A., Althouse, W.E., Drever, R.W.P., et al., 1992, Science, 256, 325
3. Acernese, F. et al. (Virgo Collaboration), 2006, Class. Quantum Grav., 23, S635
4. Acernese, F. et al. (Virgo Collaboration), 2007, Class. Quantum Grav., 24, S381
5. Asiago Supernova Catalogue, `http://graspa.oapd.inaf.it/`
6. Bardeen, J. M. 1970, Nature, 226, 64
7. Bardeen, J. M., Press, W. M., & Teukolsky, S. A. 1972, ApJ, 178, 347
8. Bethe, H.A., Brown, G.E., & Lee, C.-H., eds., 2003, *Formation and Evolution of Black Holes in the Galaxy* (Springer-Verlag)
9. Chomiuk, L., & Soderberg, A., 2010, ATel #2587
10. Dong, S., et al., 2016, Science, 351, 257
11. Somiya, K., (for the KAGRA Collaboration), 2012, Class. Quantum Grav., 29, 124007
12. Bisnovatyi-Kogan, G. S., 1970, Astron. Zh., 47, 813
13. Frail, D.A., et al., 2001, ApJ, 567, L41
14. Frontera, F., Guidorzi, C., Montanari, E., et al. 2009, ApJS, 180, 192
15. Ghirlanda, G., Ghisillini, G., & Firmani, C., 2006, New J. Phys., 8, 123
16. Ghirlanda, G., Ghisillini, G., Salvaterra, R., et al., 2013, MNRAS, 428, 123
17. Globus, N., & Levinson, A. 2014, ApJ, 796, 26
18. Haensel, P., Zdunik, J. L., Bejger, M., et al. 2009, A&A, 502, 605
19. Heo, J.-E., Yoon, S., Lee, D.-S., et al., 2016, NewA, 42, 24
20. Kerr, R.P., 1963, Phys. Rev. Lett., 11, 237
21. Kouveliotou, C., et al., 1993, Astrophys. J., 413, L101
22. LIGO Virgo collaboration 2016, arXiv:1602.03840
23. MacFadyen A. I., & Woosley S. E., 1999, ApJ, 524, 262
24. Maurer, J. I., Mazzali, P. A., Deng, J., et al. 2010, MNRAS, 402, 161
25. Levinson, A., van Putten, M.H.P.M., Pick, G., 2015, ApJ, 812, 124
26. Metzger, B.D., Margalit, B., Kasen, D., & Quataert, E., 2015, MNRAS 454, 3311
27. V. Nevski et al., 2010, CBET # 2245
28. Paczyński, B.P., 1998, ApJ, 494, L45
29. Piro, A.L., & Pfahl, E., 2007, ApJ, 658, 1173
30. Gehrels, N., Chincarini, G., Giommi, P., et al., 2004, ApJ, 611, 1005
31. Swift/BAT, 2014, http://swift.gsfc.nasa.gov/analysis/ threads/batthreads.html
32. van Putten, M.H.P.M., Levinson, A., & Della Valle, M., 2011, A&A, 535, L6
33. van Putten, M.H.P.M., & Levinson, A., 2002, Science, 295, 1874
34. van Putten, M.H.P.M., & Levinson, A., ApJ, 584, 937
35. van Putten, M.H.P.M., Levinson, A., Lee, H.K., Regimbau, T., Punturo, M., & Harry, G.M., 2004, Phys. Rev. D, 2004, 69, 044007
36. van Putten, M.H.P.M., Guidorzi, C., & Frontera, F. 2014, ApJ, 786, 146
37. van Putten, M.H.P.M., 2015, ApJ, 810, 7
38. van Putten, M.H.P.M., 2016, ApJ, 819, 169
39. van Putten, M.H.P.M., & Della Valle, M., 2016, submitted
40. Woosley, S.L., 1993, ApJ, 405, 273
41. Woosley, S.E., & Bloom, J.S., 2006, ARA&A., 44, 507
42. Woosley, S.E., 2010, ApJ, 719, L204

PARALLEL SESSION

V. Relativistic and Particle Astrophysics

Summary of Parallel Session: "Relativistic and Particle Astrophysics"

J. Řípa
Leung Center for Cosmology and Particle Astrophysics, National Taiwan University
No.1, Sec.4, Roosevelt Road, Taipei 10617, Taiwan (R.O.C)
jripa@ntu.edu.tw

1. Summary of the Presentations

The parallel session on "Relativistic and Particle Astrophysics" was held on 14th December 2015. Here a short summary of each presentation of this parallel session is given:

Sadakazu Haino: *Particle Physics and Cosmology with Alpha Magnetic Spectrometer*

- The author summarized the history of direct measurements of cosmic-rays in GeV−TeV region and the latest results from the four years of data taking with the Alpha Magnetic Spectrometer (AMS) placed on the International Space Station (ISS). The physics implications of the results by AMS were also discussed.
- The author presented the resent measurements of proton flux[1] and helium flux[2] in primary cosmic rays.
- Detection of anti-nuclei in cosmic rays would be a strong evidence of primordial anti-matter. Cosmic-ray antiparticles can provide unique opportunity to study fundamental physics such as indirect searches for dark matter and understanding of its nature.
- The author presented the resent measurements of e^+ fraction and e^+/e^- fluxes[3,4] in primary cosmic rays. The AMS measurements of $\bar{p}/p$ ratio were also discussed.
- AMS provides the data in an unprecedented precision with a single instrument.

Keiichi Mase: *The Dawn of Neutrino Astrophysics*

- The author described IceCube,[5] which has recently observed high energy extraterrestrial neutrinos above a few tens of TeV including several PeV neutrinos.[6] Although neutrinos of those energies are predicted to be produced through collisions of hadronic particles accelerated at astronomical sources, they had never been detected before.
- Neutrino observations in combination with observations by other messengers such as cosmic rays and gamma-rays can provide knowledge about the astrophysical sources that produce them, e.g. AGNs, GRBs, SNe, and cosmogenic neutrinos, which are produced through the interactions of ultra-high-energy cosmic

rays (UHECRs) with photons from the cosmic microwave background radiation (CMB). Since neutrinos are weakly interacting particles without electric charge, it is possible to track them back to the source.

- IceCube also contributes to elementary particle physics by searching for neutrinos from dark matter annihilations and by investigating atmospheric neutrino oscillations.[7]
- The author presented the latest IceCube astrophysical neutrino observations.[8,9]
- There are planed extensions of the IceCube experiment: IceCube-Gen2[10] and PINGU.[11]

Tsung-Che Liu: *Profile of Third Flight of the Antarctic Impulsive Transient Antenna (ANITA) in 2015*

- The author gave an overview of the Antarctic Impulsive Transient Antenna (ANITA) experiment,[13] which has been designed to study ultra-high-energy cosmic neutrinos(UHECN) and cosmic rays by detecting the radio pulses emitted by their interactions with the Antarctic ice sheet.[12]
- It detects the coherent radio emission (Askaryan radiation[14,15]) with vertical polarization from neutrino induced particle cascades in ice.
- ANITA also detects UHECRs by geosynchrotron radio emission with predominant horizontal polarization produced in the atmosphere and reflected by the ice.
- The third flight of the ANITA was launched on 2015/12/18 and landed on Antarctica after 23-days-long flight. The author introduced the instrument performance during the flight, and showed examples of the collected data.
- There is a plan for the continuation of this experiment. The ANITA-IV flight is planned in 2016/2017.

Chin-Hao Chen: *Searching Ultra-High Energy Cosmic Neutrinos and Flavor Identification with ARA Detector*

- The author presented the current status of searching UHECNs with Askaryan Radio Array (ARA)[18] in South Pole, and the potential of flavor identification of UHECNs with ARA experiment.
- If UHEN produces an electromagnetic shower which propagates through a dielectric medium (e.g. ice) it leads to the $\sim 20\,\%$ excess negative charge. The excess electrons releases the coherent Cerenkov radiation at about 100 MHz to a few GHz. This is called (Askaryan effect[14,15]) and ARA uses this effect to detect the highest energetic UHENs in ice. With the attenuation length of $\sim 100 - 1500\,\mathrm{m}$, ice is a natural target material for radio detections.
- Detecting UHECNs[16] with energies above $10^{17}\,\mathrm{eV}$, or the GZK neutrinos,[17] is a fundamental problem in neutrino astronomy. By finding GZK neutrinos, not only the GZK process can be verified, but also provides valuable insights of the ultra-high energy cosmic rays. The author presented the latest results.[19]

– Equally important is to identify the flavor of the GZK neutrinos[20,21] because the flavor composition of these neutrinos provides information about their source.

Shih-Hao Wang: *Feasibility of Determining Diffuse Ultra-High Energy Cosmic Neutrino Flavor Ratio through ARA Neutrino Observatory*

– The flavor composition of UHECNs carries precious information about the physical properties of their sources, the nature of neutrino oscillations and possible exotic physics involved during the propagation.
– The author in his work showed the feasibility to constrain the UHECN flavor ratio by measurement of the angular distribution of the incoming events at a neutrino observatory.[22] Simulations were performed assuming the ARA[18] detector configuration.
– The author's results assuming the standard oscillation and the neutrino decay scenarios are summarized in terms of the probability of the flavor ratio extraction and resolution as a functions of the number of observed events and the angular resolution of neutrino directions. By fitting the direction distribution of neutrino events, a preliminary constraint on the UHECNs flavor ratio can be set for the planned ARA37 configuration.

Guey-Lin Lin: *Thermal Transport of Dark Matter in the Sun*

– The author studied the thermal transport occurring in the system of solar captured dark matter (DM) and explore its impact on the DM indirect search signal.[23] He focused on the scenario of self-interacting DM (SIDM).
– The motivation is that Sun is a good target for detecting DM signal through observing neutrinos. DM density in the Sun is expected to be larger than the average DM density in the solar density.
– The author examined the DM temperature evolution and demonstrated that the DM temperature can be higher than the core temperature of the Sun if the DM-nucleon cross section is sufficiently small such that the energy flow due to DM self-interaction becomes relatively important.
– He argued that the correct DM temperature should be used for accurately predicting the DM annihilation rate, which is relevant to the DM indirect detection.

Jakub Řípa: *Gamma-Ray Bursts and Their Relation to Astroparticle Physics and Cosmology*

– The article gives an overview of gamma-ray bursts (GRBs)[24] and their relation to astroparticle physics and cosmology. GRBs are the most powerful explosions in the universe that are characterized by flashes of gamma-rays typically lasting from a fraction of a second to thousands of seconds.
– There were discovered several correlations in the properties of the GRB emission which makes them a potential tool to constrain cosmological parameters[25,26] next to the "traditional" such as SNe Ia or CMB.

- For example, it was suggested that some of the properties of the prompt and the early optical lightcurves of GRBs could be used as standard candles.[27] To confirm that more early optical observations are needed and they can be provided by the Ultra-Fast Flash Observatory pathfinder (UFFO-p),[28] which will be able to start optical observations already few seconds after the trigger.
- GRBs were proposed to be sites for accelerating ultra-high energy cosmic rays and sources of very high energy neutrinos up to $10^{17} \sim 10^{19}$ eV.[29] However, the recent results from the IceCube detector suggests that the efficiency of neutrino production may be much lower than predicted.[30]

Dong-Hoon Kim: *General Relativistic Effects on Pulsar Radiation*

- In this work, the author considers a magnetic dipole model of a pulsar and investigates general relativistic (GR) effects on electromagnetic radiation from the pulsar: the effects on (i) radiation along the magnetic poles and (ii) curvature radiation in the magnetosphere.
- To determine such effects, the author studies electrodynamics of the pulsar in curved spacetime; by solving Maxwell's equations in curved spacetime.[31] In particular, to study electromagnetic radiation near compact objects, e.g. black holes or neutron stars, he applies a "test-particle" method and a "semi-relativistic" approximation.
- The author finds that the major GR effects result from modification of the magnetic field in the strongly curved spacetime of the pulsar. In case (i), modification of the field intensity at the magnetic poles leads to modulation of the frequency of radiation along the poles: e.g. for PSR 1937+214, the field intensity increases by $10.5 \pm 3.7\,\%$ due to the curved spacetime and this will increase the frequency by the same amount. On the other hand, in case (ii), modification of the field lines in the magnetosphere leads to modulation of the frequency of curvature radiation:[32] the radius of curvature of the field lines decreases due to the curved spacetime and this will increase the frequency in an inverse manner.
- The author also investigates GR effects on polarization of the curvature radiation; by computing the Stokes parameters[33] in curved spacetime. The properties of polarization are expected to change due to the curved spacetime: modulation of the radius of curvature leads to modulation of the Stokes parameters, which should cause the ratio between linear and circular polarization to change in curved spacetime. (c.f. Note that Ref. 33 is about curvature radiation in "flat" spacetime, but it is referred to it only for how to compute the Stokes parameters)

Hayashinaka Takahiro: *QED Correction of the Radiation from the Magnetors*

- The author investigated the effect of quantum electrodynamics (QED) on the radiation from the magnetars (strongly magnetized neutron stars). The strength of the magnetic field of magnetars is estimated to be higher than $\sim 10^{13}$ Gauss which is the QED scale determined by electron mass. One has to consider the

QED effect in such a high energy scales because the photon action or the e.o.m. is corrected by loop effect of electron.

- He used the Euler-Heisenberg Lagrangian to find the correction for the magnetic dipole radiation which is the leading radiation term due to the lack of electric dipole radiation. The results were shown for the weak field region ($B \ll 10^{13}$ Gauss) and the strong field region ($B \gg 10^{13}$ Gauss).
- The author found the 1-loop QED correction term to the magnetic dipole radiation explicitly and claimed that when the field strength goes extremely large, there would be no correction.

In addition to the talks in this parallel session, there were two relevant plenary presentations:

- **Jiwoo Nam:** *Ultra-High Energy Neutrinos and Cosmic Rays via Radio; ANITA, ARA, and TAROGE*
- **Henry Wong:** *Direct Experimental Searches of Dark Matter*

References

1. M. Aguilar, D. Aisa, B. Alpat, et al., Precision measurement of the proton flux in primary cosmic rays from rigidity 1 GV to 1.8 TV with the Alpha Magnetic Spectrometer on the International Space Station, *Physical Review Letters* **114**, 171103 (2015).
2. M. Aguilar, D. Aisa, B. Alpat, et al., Precision measurement of the helium flux in primary cosmic rays of rigidities 1.9 GV to 3 TV with the Alpha Magnetic Spectrometer on the International Space Station, *Physical Review Letters* **115**, 211101 (2015).
3. L. Accardo, M. Aguilar, D. Aisa, et al., High statistics measurement of the positron fraction in primary cosmic rays of $0.5 - 500\, GeV$ with the Alpha Magnetic Spectrometer on the International Space Station, *Physical Review Letters* **113**, 121101 (2014).
4. M. Aguilar, D. Aisa, A. Alvino, et al., Electron and positron fluxes in primary cosmic rays measured with the Alpha Magnetic Spectrometer on the International Space Station, *Physical Review Letters* **113**, 121102 (2014).
5. IceCube collaboration, A. Achterberg, M. Ackermann, et al., First year performance of the IceCube neutrino telescope, *Astroparticle Physics* **26**, 155 (2006).
6. M. G. Aartsen, R. Abbasi, Y. Abdou, et al., First observation of PeV-energy neutrinos with IceCube, *Physical Review Letters* **111**, 021103 (2013).
7. M. G. Aartsen, M. Ackermann, J. Adams, et al., Determining neutrino oscillation parameters from atmospheric muon neutrino disappearance with three years of IceCube DeepCore data, *Physical Review D* **91**, 072004 (2015).
8. M. G. Aartsen, M. Ackermann, J. Adams, et al., Observation of high-energy astrophysical neutrinos in three years of IceCube data, *Phys. Rev. Let.* **113**, 101101 (2014).
9. C. Kopper, W. Giang and N. Kurahashi, Observation of astrophysical neutrinos in four years of IceCube data, in *Proceedings of Science (ICRC2015), The 34th International Cosmic Ray Conference*, (SISSA, 2016), p. 1081
10. IceCube-Gen2 collaboration, M. G. Aartsen, M. Ackermann, J. Adams, et al., IceCube-Gen2: A Vision for the Future of Neutrino Astronomy in Antarctica, *arXiv:1412.5106* (2014).
11. The IceCube-PINGU collaboration, M. G. Aartsen, R. Abbasi, M. Ackermann, et al., Letter of Intent: The Precision IceCube Next Generation Upgrade (PINGU), *arXiv:1401.2046* (2014).

12. P. W. Gorham, S. W. Barwick, J. J. Beatty, et al., Observations of the Askaryan effect in ice, *Physical Review Letters* **99**, 171101 (2007).
13. ANITA collaboration, P. W. Gorham, P. Allison, et al., The Antarctic Impulsive Transient Antenna ultra-high energy neutrino detector: design, performance, and sensitivity for the 2006-2007 balloon flight, *Astroparticle Physics* **32**, 10 (2009).
14. G. A. Askar'yan, Excess negative charge of an electron-photon shower and its coherent radio emission, *Soviet Journal of Experimental and Theoretical Physics* **14**, 441 (1962).
15. G. A. Askar'yan, Coherent radio emission from cosmic showers in air and in dense media, *Soviet Journal of Experimental and Theoretical Physics* **21**, 658 (1965).
16. A. M. Hillas, Cosmic rays: recent progress and some current questions, *arXiv:astro-ph/0607109* (2006).
17. R. U. Abbasi, T. Abu-Zayyad, M. Allen, et al., First Observation of the Greisen-Zatsepin-Kuzmin Suppression, *Physical Review Letters* **100**, 101101 (2008).
18. Ara collaboration, P. Allison, J. Auffenberg, et al., Design and initial performance of the Askaryan Radio Array prototype EeV neutrino detector at the South Pole, *Astroparticle Physics* **35**, 457 (2012).
19. ARA Collaboration, P. Allison, R. Bard, et al., Performance of two Askaryan Radio Array stations and first results in the search for ultra-high energy neutrinos, *arXiv:1507.08991* (2015).
20. K.-C. Lai, C.-C. Chen and P. Chen, The strategy of discrimination between flavors for detection of cosmogenic neutrinos, *Nuclear Instruments and Methods in Physics Research A* **742**, 119 (2014).
21. M. G. Aartsen, M. Ackermann, J. Adams, et al., Flavor Ratio of Astrophysical Neutrinos above 35 TeV in IceCube, *Physical Review Letters* **114**, 171102 (2015).
22. S.-H. Wang, P. Chen, J. Nam and M. Huang, Feasibility of determining diffuse ultra-high energy cosmic neutrino flavor ratio through ARA neutrino observatory, *Journal of Cosmology and Astro-Particle Physics* **11**, 062 (2013).
23. C.-S. Chen, G.-L. Lin and Y.-H. Lin, Thermal transport of the solar captured dark matter and its impact on the indirect dark matter search, *arXiv:1508.05263* (2015).
24. C. Kouveliotou, R. A. M. J. Wijers and S. Woosley, Gamma-ray Bursts, *Gamma-ray Bursts* (Cambridge University Press, Cambridge, UK, 2012).
25. F. Y. Wang, Z. G. Dai and E. W. Liang, Gamma-ray burst cosmology, *New Astronomy Reviews* **67**, 1 (2015).
26. L. Amati and M. D. Valle, Measuring cosmol. param. with gamma-ray bursts, in *Proc. of the MG13 Meet. on General Relativity*, (World Scientific, Singapore, 2015), p. 769.
27. A. Panaitescu and W. T. Vestrand, Taxonomy of gamma-ray burst optical light curves: identification of a salient class of early afterglows, *MNRAS* **387**, 497 (2008).
28. I. H. Park, S. Brandt, C. Budtz-Jørgensen, et al., Ultra-Fast Flash Observatory for the obser. of early phot. from gamma-ray bursts, *New Jour. of Phys.* **15**, 023031 (2013).
29. E. Waxman and J. N. Bahcall, High energy neutrinos from cosmological gamma-ray burst fireballs, *Phys. Rev. Let.* **78**, 2292 (1997).
30. R. Abbasi, Y. Abdou, T. Abu-Zayyad, et al., An absence of neutrinos associated with cosmic-ray acceleration in γ-ray bursts, *Nature* **484**, 351 (2012).
31. J. A. Petterson, Magnetic field of a current loop around a Schwarzschild black hole, *Physical Review D* **10**, 3166 (1974).
32. K. Konno and Y. Kojima, General Relativistic Modification of a Pulsar Electromagnetic Field, *Progress of Theoretical Physics* **104**, 1117 (2000).
33. J. A. Gil and J. K. Snakowski, Curvature radiation and the core emission of pulsars, *Astronomy and Astrophysics* **234**, 237 (1990).

Particle Physics and Cosmology with Alpha Magnetic Spectrometer

S. Haino

Institute of Physics, Academia Sinica,
Taipei, 11529, Taiwan
Sadakazu.Haino@cern.ch

The Alpha Magnetic Spectrometer (AMS) is a general purpose high-energy particle physics detector. It was installed on the International Space Station (ISS) on 19 May 2011 to conduct a unique long duration mission (~20 years) of fundamental physics research in space. So far 6 results have been published [1-6] on cosmic-ray positions, electrons, protons and Helium nuclei. Over the last two decades, there has been a strong interest in the cosmic ray positrons in both particle physics and astrophysics. Protons and Helium nuclei in cosmic rays are considered to be mainly produced by Galactic sources such as supernova remnants. Precise knowledge of their spectrum gives important information on the origin, acceleration, and subsequent propagation processes of cosmic rays in the Galaxy.

Keywords: AMS; cosmic rays.

1. Alpha Magnetic Spectrometer

1.1. *Detector*

The Alpha Magnetic Spectrometer (AMS) is a general purpose high-energy particle physics detector. It was installed on the International Space Station (ISS) on 19 May 2011 to conduct a unique long duration mission (~20 years) of fundamental physics research in space. The layout and description of the detector are presented in Ref. [1]. The key elements used in this measurement are the permanent magne, the silicon tracker, four planes of time of flight (TOF) scintillation counters, and an array of 16 anticoincidence counters (ACC). The AMS also contains a transition radiation detector (TRD), a ring imaging Čerenkov detector (RICH), and an electromagnetic calorimeter (ECAL).

The AMS coordinate system is concentric with the magnet and above, below, and downward-going refer to the AMS coordinate system. Timing, location, and orientation are provided by GPS units affixed to the AMS and to the ISS. The detector performance has been steady over time.

The tracker has nine layers, the first (L1) at the top of the detector, the second (L2) above the magnet, six (L3 to L8) within the bore of the magnet, and the last (L9) above the ECAL. L2 to L8 constitute the inner tracker. The tracker accurately determines the trajectory of cosmic rays by multiple measurements of the coordinates. Together, the tracker and the magnet measure the rigidity R of charged cosmic rays. The spatial resolution in each tracker layer is 10 and 7.5 μm, for proton and Helium, respectively, in the bending direction. The maximum detectable rigidity (MDR) is 2 and 3.2 TV, for proton and Helium, respectively, over the 3 m lever arm from L1 to L9.

Each layer of the tracker also provides an independent measurement of the charge Z of the cosmic ray. The charge resolution of the combined inner tracker is $\Delta Z \simeq 0.05$ and 0.07 for Z=1 and 2 particles, respectively.

Two planes of TOF counters are located above L2 and two planes are located below the magnet. The overall velocity ($\beta = v/c$) resolution has been measured to be $\Delta\beta/\beta^2 = 0.04$ and 0.02 for Z=1 and 2 particles. This discriminates between upward- and downward-going particles. The pulse heights of the two upper layers are combined to provide an independent measurement of the charge with an accuracy $\Delta Z \simeq 0.08$ for Z=2. The pulse heights from the two lower planes are combined to provide another independent charge measurement with the same accuracy.

The 16 anticoincidence counters (ACC) form a cylindrical shell between the inner tracker and the magnet. Adjacent counters are combined to provide 8 readout sectors. The ACC have an efficiency of 0.999 99 to reject cosmic rays which enter the inner tracker from the side.

The coincidence of 3 out of the 4 TOF layers with no ACC requirement was used to provide an unbiased trigger. The unbiased trigger, prescaled by 1%, was used to measure the efficiency of the charged particle trigger. The efficiency of the unbiased trigger was estimated directly from the data to be above 99.8% for all rigidities using events in which one of the four TOF layers gave no signal. This allowed the estimation of the efficiency of each TOF layer and, consequently, the efficiency of the unbiased trigger.

Since launch, the detector has been monitored and controlled around the clock. The time, location, and orientation are provided by GPS units affixed to AMS and to the ISS. The detector performance has been steady over time.

1.2. *Beam test*

Before launch, at the CERN SPS, AMS was extensively calibrated with 180 and 400 GeV/c proton beams and beams of positrons, electrons, and pions from 10 to 290 GeV/c. In total, calibrations with 18 different energies and particles at 2000 positions were performed. These data allow the determination of the tracker rigidity resolution function with high precision and the verification of the absolute rigidity scale.

1.3. *Simulation*

Simulated events were produced using a dedicated program developed by the collaboration from the GEANT- 4.9.6 package based on Monte Carlo methods. This program simulates electromagnetic and hadronic interactions of particles in the material of AMS and generates detector responses. The digitization of the signals is simulated precisely according to the measured characteristics of the electronics. The simulated events then undergo the same reconstruction as used for the data. The inverse rigidity for 400 GeV/c protons from the test beam and the Monte Carlo simulation is compared. The resolution has a pronounced Gaussian core with non-Gaussian tails of ~5% and the simulation precisely reproduces the measured resolution including the tails.

2. AMS Results

2.1. *Positron fraction*

Over the last two decades, there has been a strong interest in the cosmic ray positron fraction in both particle physics and astrophysics. The first results from AMS on the positron fraction were reported in [1]. They generated widespread interest. In the second publication [2], we reported new results based on all the data collected during 30 months of AMS operations on the ISS, from 19 May 2011 to 26 November 2013. Because of the excellent and steady performance of the detector, and an increase of the data sample by a factor of 1.7, the measurement of the positron fraction is extended up to 500 GeV with improved precision.

Figure 1 shows the behavior of the positron fraction at low energies, from 1 to 35 GeV. As seen, below ~8 GeV, the positron fraction decreases rapidly as expected from the diffuse production of positrons [16]. Then the fraction begins to increase steadily with energy. The AMS data provide accurate information on the minimum of the positron fraction.

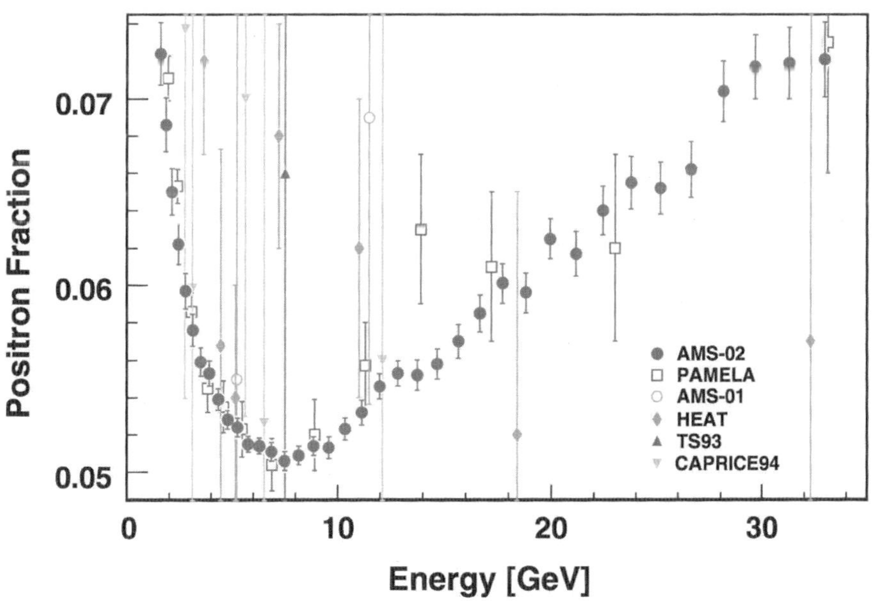

Fig. 1. The positron fraction from 1 to 35 GeV. It shows a rapid decrease from 1 to ~8 GeV followed by a steady increase. The AMS data provide accurate information on the minimum of the positron fraction [2].

Our earlier result [2], in which we observed the increase of the positron fraction with decreasing slope above 20 GeV, is consistent with this new measurement. The increase of the positron fraction has been reported by earlier experiments. The new result extends the energy range to 500 GeV and is based on a significant increase in the statistics by a factor of 1.7. Figure 2 explores the behavior of the positron fraction at high energies (> 10 GeV) and compares it with earlier measurements. We observe that above ~200 GeV the positron fraction is no longer increasing with energy.

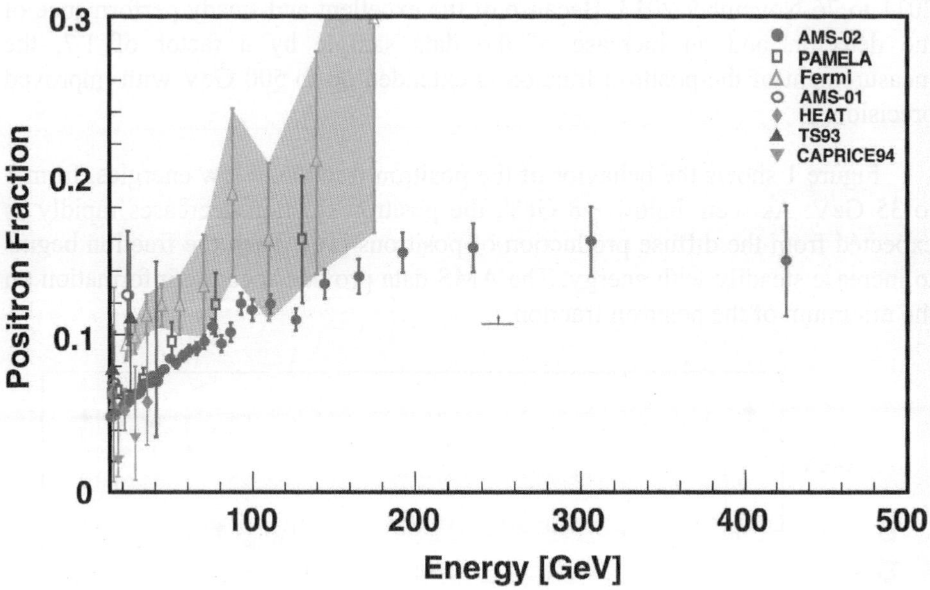

Fig. 2. The positron fraction above 10 GeV, where it begins to increase. The present measurement demonstrates that, above ~200 GeV, the positron fraction is no longer increasing [2].

The new measurement shows a previously unobserved behavior of the positron fraction. The origin of this behavior can only be ascertained by continuing to collect data up to the TeV region and by measuring the antiproton to proton ratio to high energies. These are among the main goals of AMS.

2.2. *Electron and positron flux*

Measurements of the electron flux and the positron flux in primary cosmic rays with AMS is presented in [3]. Precise measurements of the individual electron and positron fluxes as a function of energy provide insight into the origin of cosmic rays. The electron and positron fluxes multiplied by E^3 are presented in Figure 3. Below ~10 GeV, the behavior for both electrons and positrons is affected by solar modulation as seen in our data by variations of the fluxes over this data-taking interval. However, above ~20 GeV the effects of solar modulation are insignificant within the current experimental accuracy. The data show that above ~20 GeV and up to 200 GeV the electron flux decreases more rapidly with energy than the positron flux, that is, the electron flux is softer than the positron flux. This is not consistent with only the secondary production of positrons.

The differing behavior of the spectral indices versus energy indicates that high-energy positrons have a different origin from that of electrons. The underlying mechanism of this behavior can only be ascertained by continuing to collect data up to the TeV region (currently, the largest uncertainties above ~200 GeV are the statistical errors) and by measuring the antiproton to proton ratio to high energies. These are among the main goals of AMS.

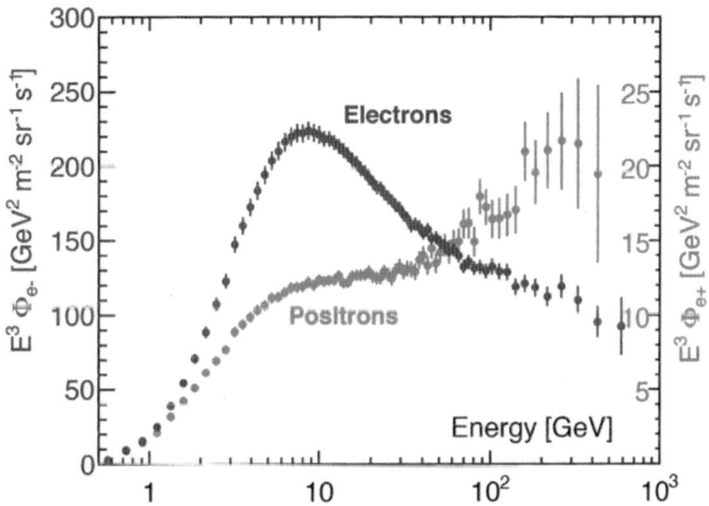

Fig. 3. The electron (blue) and positron (red) flux measured by AMS. Fluxes are scaled by E^3 [3].

2.3. *Proton and helium flux*

Protons and Helium nuclei are the most abundant charged particles in cosmic rays. Knowledge of the precise behavior of the proton and Helium spectra is important in understanding the origin, acceleration, and propagation of cosmic rays. Recent important measurements of the proton flux in cosmic rays have reported different variations of the flux with energy. These measurements generated widespread interest. In particular, the ATIC–2, CREAM, and PAMELA experiments showed deviations of the proton and Helium flux from a single power law. Many models were proposed to account for the hardening of the flux based on different sources, acceleration mechanisms, diffusive propagation effects, and their superposition.

Figure 4 shows the proton and Helium flux as a function of rigidity [5,6]. A power law with a constant spectral index does not fit the flux above 45 GV. The flux deviates from a single power law and the spectral index progressively hardens at rigidities larger than 100 GV. The magnitude of the helium spectral index is different from that of the proton spectral index, but the rigidity dependence is similar for helium and protons. The proton to helium flux ratio spectral index increases with rigidity up to about 45 GV and becomes constant for higher rigidities; i.e., the proton to helium flux ratio is well described by a single power law above 45 GV.

Acknowledgments

AMS is a U.S. DOE sponsored international collaboration. In Taiwan, the research is supported by Academia Sinica and the Ministry of Science and Technology (MOST) under Grants No. 103-2682-M-008- 002 and No. 103-2112-M-001-036.

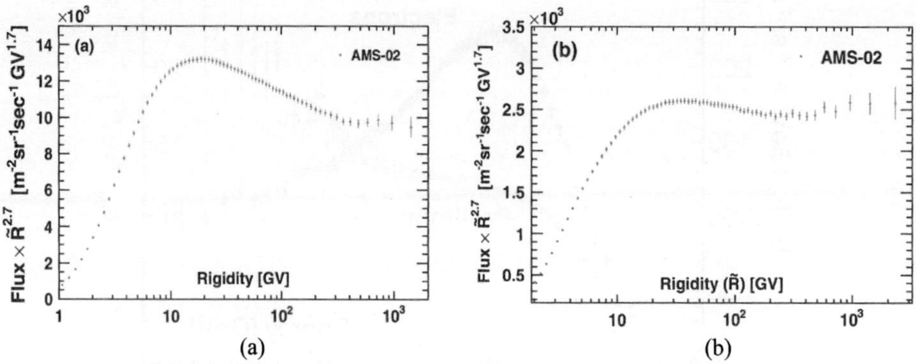

(a) (b)

Fig. 4. (a) Proton and (b) flux measured by AMS. The fluxes are scaled by $R^{2.7}$ [5,6].

References

1. M. Aguilar, *et al.*, Phys. Rev. Lett. **110**, 141102 (2013).
2. L. Accardo, *et al.*, Phys. Rev. Lett. **113**, 121101 (2013).
3. M. Aguilar, *et al.*, Phys. Rev. Lett. **113**, 121102 (2013).
4. M. Aguilar, *et al.*, Phys. Rev. Lett. **113**, 221102 (2013)
5. M. Aguilar, *et al.*, Phys. Rev. Lett. **114**, 171103 (2013).
6. M. Aguilar, *et al.*, Phys. Rev. Lett. **115**, 211101 (2013).

Profile of the Third Flight of the Antarctic Impulsive Transient Antenna (ANITA) during 2014−2015

T. C. Liu for ANITA collaboration

Leung Center for Cosmology and Particle Astrophysics, National Taiwan University, Taipei, 10617, Taiwan(ROC)
tcliu@ntu.edu.tw

The Antarctic Impulsive Transient Antenna (ANITA) experiment is designed to study ultra-high-energy cosmic neutrinos and cosmic rays by detecting the radio pulses that are emitted by their interactions with the Antarctic ice sheet. The third flight of the ANITA was launched on December 18, 2015, and landed in Antarctica after 23 days of flight. Here, we discuss instrument performance during the flight, and show examples of the data.

Keywords: Cosmic ray; extra-high energy neutrino; radio wave.

1. Introduction

Scientists understand the rules of the cosmos by observing the cosmogenic fundamental particles, such as photons, cosmic rays, and neutrinos. Scientific consensus holds that, above 30 TeV, a photon carrying information can interact with the cosmic microwave background (CMB) and lose its original information. Furthermore, cosmic rays, which are charged particles, can be deflected by interstellar B-fields and can interact with CMB photons through the GZK process. By contrast, neutrinos can serve as excellent cosmic messengers for the exploration of the universe.[1] The propagation of neutrino in the universe is quite different from the propagation of other charged particles (cosmic rays); neutrinos cannot be deflected by the interstellar magnetic field. Hence a neutrino keeps its original direction and carries information of very distant cosmic sources.

The Antarctic Impulsive Transient Antenna (ANITA)[2] is a NASA Balloon-Borne observatory designed to investigate various phenomena related to energetic astroparticles, such as ultra-high energy cosmic rays (UHECRs) and ultra-high energy cosmic neutrinos (UHECNs). Recent data of ground-based observatories, such as Hi-Res[3] and Pierre Auger,[4,7] provide evidence of the UHECRs and GZK processes.[8,9] In the standard model, the propagation range of UHECRs ($E_{CR} > 6 \times 10^{19}$ eV) is limited by interaction with the CMB radiation. Greisen, Zatseptin and Kuzmin set a cutoff in the 1960s, that provides the mean path along which UHECRs interact with the CMB. The values of this mean path is approximately 50 Mpc. UHECRs are converted to lower energy hadrons and neutrinos after they interact with the CMB. Berezinsky and Zatsepin predicted that the energy range of the resulting neutrino flux ($E_\nu = 10^{17-20}$ eV) should be 2 orders of magnitude below the GZK energy.[10] In other words, this resulting neutrino flux is strongly constrained by the source models of the UHECRs.[4–6,11–16]

Fig. 1. "Hang test" of ANITA-III at Long Duration Balloon (LDB) site. ANITA-III includes 48 broadband antennas, functioning from 200 to 1200 MHz, and one low frequency antenna, functioning from 30 to 80 Hz. "Hang test" is the final step for qualification. In this picture, people are checking the mechanical structures of the ANITA-III system and testing its communications and operation.

2. Radio Detection Method

In 1962, Gurgen Askaryan postulated that the Askaryan effect, an asymmetrical distribution of positive and negative charged particles (with a 20% excess of electrons over positrons) should occur during the development of high energy showers. While the net charges move faster than the phase velocity of light in a medium, the charged particles emit electromagnetic radiation, called Cherenkov radiation. The coherent Cherenkov emission that are induced by Askaryan effect has been observed in several media, such as sand,[17] rock,[18] salt,[18] and ice[19] in SLAC.

The Askaryan effect has emerged as a feasible method for detecting UHE astroparticles through radio-frequency (RF) radio technology. The RF emission can propagate over long distances in media with low attenuation.[20] For instance, ice is a transparent medium for both visible light and RF, but the attenuation length of visible light (~100m) is an order of magnitude shorter than the attenuation length (~1km) of RF signal. Of these two methods, the RF radio detection of method provides the a more effective means of obtaining a large detection volume for UHE astroparticles.

During the development of the cosmic-ray air showers, the electron-positron pairs of particle cascade are deflected by the Geo-magnetic field and emit radio-frequency radiation. ANITA collaboration performed a beam test experiment, SLAC T-510,[21] to measure the radio-frequency (RF) radiation from charged particle cascades with magnetic field in laboratory. The polarization characteristics of this Geo-magnetic emission help us to verify the observed signal of ANITA.

3. ANITA-I and ANITA-II

ANITA is the first NASA observatory for cosmogenic neutrinos of any flavours. The observatory detects neutrinos with energy levels in the range from EeV to ZeV (10^{18-21} eV). Three ANITA missions launched from NASAs Long Duration Balloon (LDB) site at McMurdo, Antarctica, in the summers of 2006, 2008, and 2014. The first flight module, ANITA-I, consisted of 32 antennas; the payload floated 32 km above Antarctica and collected UHECN events for 31 days. ANITA-I did not detect any neutrino candidates;[2] therefore, researchers moved the most constraining limit of the cosmogenic UHE neutrino flux spectrum from EeV to ZeV (10^{18-21} eV). The ANITA-I flight made the first observation of extensive UHECR showers in the air above Antarctica; it recorded a total of 16 UHECR events with a mean energy 10^{19} eV.[24]

The second flight module, ANITA-II, launched from the LDB site on December 21, 2008, and survived for 28.5 days. ANITA-II had 8 more quad-ridged horn antennas than its predecessor. The total number of antennas was 40. Antennas only triggered when they received fast signals with vertical polarization. The sensitivity to cosmogenic neutrinos of ANITA-II was approximately a fourfold improvement over that of ANITA-I. ANITA-II recorded approximately $\sim$ 27.8 M neutrino events. One neutrino candidate event was discovered by ANITA-II. Because the ANITA-II trigger scheme incorporated modifications, the total number of UHECR events for ANITA-II was expected to be lower than the 16 UHECR events detected by ANITA-I. Hence, ANITA-II only provided additional four isolated UHECR events during its flight.[23,24]

4. ANITA-III

The ANITA-III payload launched (Fig. 2) on December 17, 2014, and landed on January 7th, 2015, near Davis Station (AU) after 22.4 days of flight (which was a relatively short flight duration compared with those of ANITA-I and ANITA-II). Fig. 3 shows the trajectory of the ANITA-III over Antarctica. ANITA-III had a radius of field of view of approximately 600 km, and we stopped its journey before it flew out of the Antarctica airspace. ANITA-III conveyed 48 broadband antennas, operated from 200 to 1200 MHz, and one low frequency antenna, operated from 30

Fig. 2. The BOSS (launch vehicle) hangs the ANITA-III and waits for the completion of inflation. The train with the helium gas cylinders is inflating the balloon, as shown on the right. The ANITA-III payload was launched on December 17, 2014; it floated above Antarctica for only 22 days, a relatively short flight duration compared with those of ANITA-I and ANITA-II.

to 80 Hz. Each antenna had two feeds, one horizontal and one vertical; each detected both polarizations of radiation emissions from air showers. Compared with ANITA-I and ANITA-II, the ANITA-III payload had improved sensitivity to neutrinos and UHECRs, as listed in Table 1. Simulations suggested that the neutrino sensitivity may have been a twofold or threefold improvement. ANITA-III data were recovered in the summer of 2105, and the data analysis is currently underway.

ANITA-III had 48 dual-polarization broadband antennas (with 96 channels in total); those antennas had time windows of approximately 100 ns. The detection frequency of its radio covered a range from 200 to 1200 MHz and was RF transparent in ice. The events were triggered by an impulse radio whose signal was three times stronger than that of thermal noise; events were triggered only if more than three antennas were fired. The expected angular resolution was $\sim 0.2^\circ$ in elevation and $\sim 0.7^\circ$ in azimuth. Fig. 4 shows the triggered event of ANITA-III.

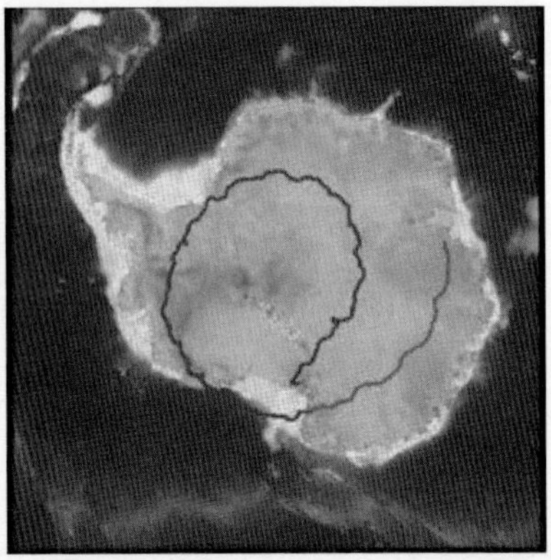

Fig. 3. Trajectory of the ANITA-III above Antarctica. It launched on December 17, 2014, from the SDB site and landed on January 7, 2015 near Davis Station (AU) after 22.4 days of flight. *Credit to CSBF, NASA.*

Table 1. Upgraded Items of ANITA-III.

1. 48 dual polarized horn antennas (8 more horn antennas than ANITA-II)
2. New improved low-noise amplifiers are applied in channels
3. Addition of dedicated H-pol trigger for improved UHECR detection
4. Increased trigger sensitivity and higher trigger throughput
5. Real-time GPU-based event prioritizer and realtime mapper

5. Future plans regarding ANITA

The ANITA-IV flight has been scheduled for a launch window during the winter of 2016-2017 winter. After upgrades to several devices (low-noise amplifiers, real-time 3-bit signal correlator, programmable notch filters, and trigger processor) ANITA-IV is expected to reach a 50%−60% improvement in energy threshold and a 30% improvement in exposure. Overall, ANITA-IV has some chance of achieving a fourfold improvement over ANITA-III.

References

1. D. Seckel and T. Stanev, *Phys. Rev. Lett.* **95** (2005) 141101
2. P. Gorham *et al.*, *Phys. Rev. Lett.* **103** (2009) 051103
3. D. R. Bergman, *Nucl.Phys.Proc.Suppl.* **165** (2007) 19-26
4. J. Abraham *et al.*, *Phys. Rev. Lett.* **101** (2008) 061101
5. C. T. Hill and D. N. Schramm, *Phys. Rev.* D **31** (1985) 564
6. F. W. Stecker, *Phys. Rev. Lett.* **21** (1968) 1016-1018
7. J. Abraham *et al.*, *Phys. Lett.* B **685** (2010) 239-246
8. K. Greisen, *Phys. Rev. Lett.* **16** (1966) 748
9. G. T. Zatsepin and V. A. Kuzmin, JETP **4** (1966) 78
10. V. S. Beresinsky and G. T. Zatsepin, *Phys. Lett.* B **28** (1969) 423

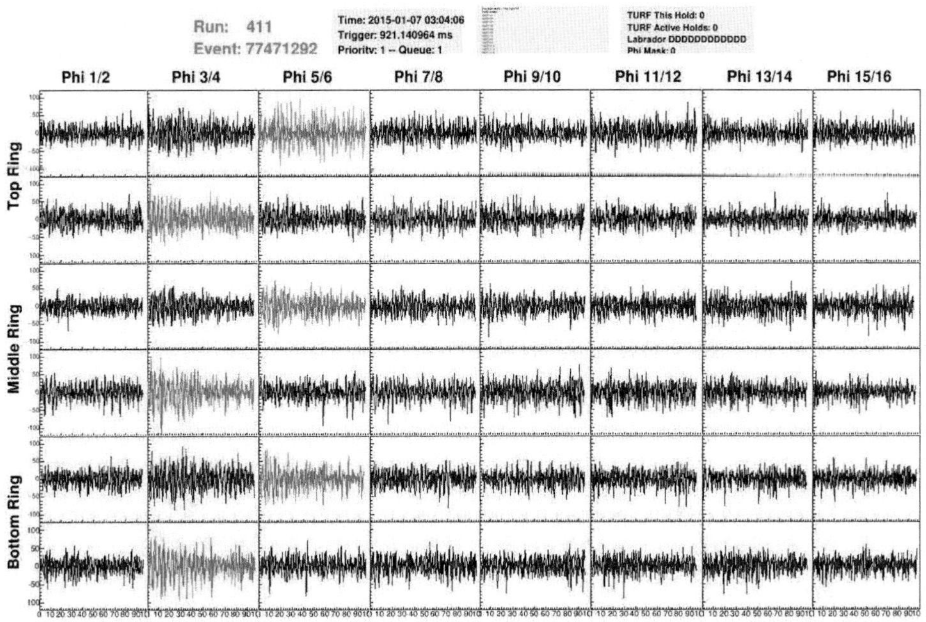

Fig. 4. Example of ANITA-III events. This figure shows 96 waveforms of all channels and the red and green lines denote the triggered waveform of V-pol and H-pol, respectively.

11. J. Linsley, *Phys. Rev. Lett.* **10** (1963) 146
12. D. J. Bird *et al.*, *Phys. Rev. Lett.* **71** (1993) 3401
13. H. Hayashida *et al.*, *Ap. J.* **522** (1999) 225
14. C. C. H. Jui *et al.* , AIP Conf. Proc **516** (2000) 370
15. T. Abu-Zayyad, at. al., *Astrophys. J.* **557** (2001) 686
16. M. Takeda *et al.*, *Astroparticle Physics* **19** (2003) 447-462
17. P. W. Gorham *et al. Phys. Rev.* E **62** (2000) 8590-8605
18. P. W. Gorham *et al. Phys. Rev.* D **72** (2005) 023002
19. P. W. Gorham *et al.* (ANITA Collaboration), *Phys. Rev. Lett.* **99** (2007) 171101
20. T. Barrella, S. Barwick, D. Saltzberg, Journal of Glaciology **57** (2011) 201
21. K. Belov *et al.*, SLAC-PUB-16338
22. I. Kravchenko *et al.*, *Phys. Rev.* D **85** (2012) 062004
23. P. W. Gorham *et al.*, *Phys. Rev.* D **82** (2010) 022004
24. S. Hoover *et al.*, *Phys. Rev. Lett.* **105** (2010) 151101

Searching Ultra-High Energy Cosmic Neutrinos and Flavor Identification with ARA Detector

Chin-Hao Chen
Leung Center for Cosmology and Particle Astrophysics, National Taiwan University, Taipei, 10617, Republic of China
chchen@phys.ntu.edu.tw

Detecting ultra-high energy neutrinos (UHECNs) with energies above 10^{17}eV, or the GZK neutrinos, is a fundamental problem in neutrino astronomy. By finding GZK neutrinos, not only the GZK process can be verified, but also provides valuable insights of the ultra-high energy cosmic rays. Equally important is to identify the flavor of the GZK neutrinos. The flavor composition of these neutrinos provides information about the source of these UHECNs. With these UHECNs travelling astronomical distances, it is also possible to measure neutrino oscillation parameters with extremely long baseline. Here we present the current status of searching UHECNs with Askaryan Radio Array (ARA) in South pole, and the potential of flavor identification of UHECNs with ARA experiment.

Keywords: Ultra high energy neutrino; Askaryan effect; radio detection.

1. UHECR and GZK Cutoff

Cosmic rays are high energy particles originated from the universe. By studying these energetic particles, we can have a deeper understanding on the mechanism of the production of these high energy particles. Fig. 1 [1] is the spectrum of cosmic ray, with the kinetic energy ranging between 10^0 to 10^{12} GeV. The spectrum has a monotonic trend where the number of particles decrease with increasing energy.

In this spectrum, there are two structures. The first one is at energy 10^6 GeV, which is also known as the "knee". The second structure is at energy 10^9 GeV, known as the "ankle". People generally agree that particles with energy below 10^8 is originated within the galaxy. For the particles above ankle, they must have an extragalactic origin.

When the measured cosmic ray spectrum is pushed to even higher energy to above 10^{20} eV or 10^{11} GeV, a new structure has been observed. In Fig. 2,[2] the ankle can be observed at $10^{18.5}$ eV. Then at $E = 10^{19.5}$, there is a sudden cutoff of the spectrum.

This cutoff is first proposed by Greisen,[3] Zatsepin and Kuzmin.[4] Most of the cosmic rays are protons. When a proton with energy above 10^{19} eV traversing the universe, it may collide with a photon coming from cosmic microwave background. This process will create a delta resonance, which subsequently decays into a charged pion and a neutron. The high energy proton will lose energy through this process, or the so called GZK process, which produced this cutoff in the spectrum. In this process, the generated charged pion will decay and produce 2 muon neutrinos and 1

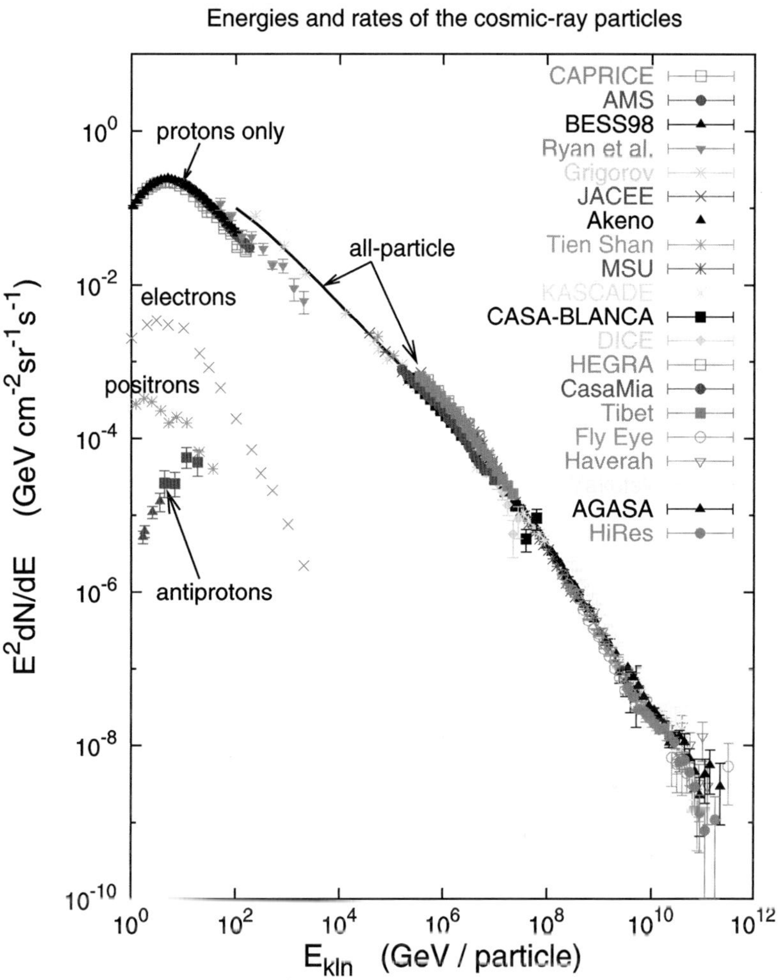

Fig. 1. Spectra of the cosmic ray betweeb E = 10^0 to 10^{12} GeV. [1]

electron neutrino. With the observation of cosmic microwave background and GZK cutoff, finding GZK neutrinos will be a crucial step to validate the GZK process.

By studying these ultra high energy cosmic ray (UHECR) particles, with energy larger than 10^{17} eV, we can understand the mechanism of the cosmic accelerator which generates the particles to such high energies. Most of the UHECR are protons, which carry charges. When a charged particle is travelling in the universe, the trajectory will be bend by the interstellar magnetic fields. Therefore it is difficult to point back to its source.

On the other hand, neutrinos, unlike charged particles, are chargeless, which means its moving direction will not be bent by magnetic fields. Neutrinos are also

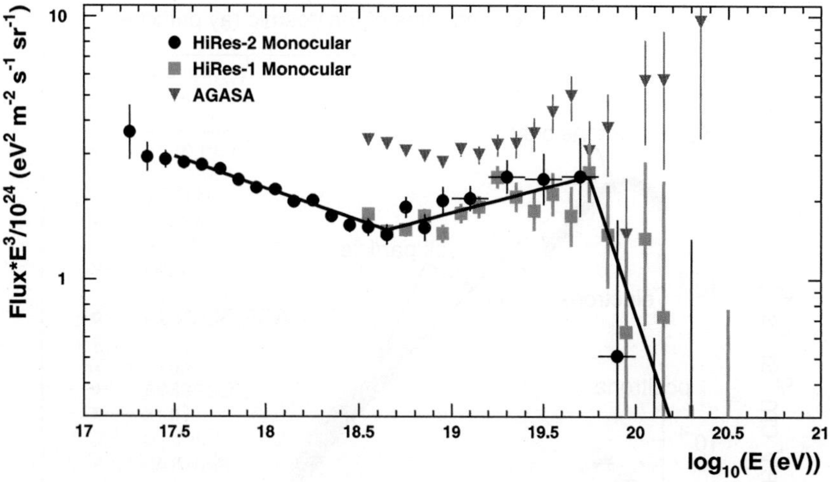

Fig. 2. The cosmic ray energy spectra shows the GZK cutoff at E > $10^{19.5}$ eV.[2]

weak interacting particles, which makes them rarely interact with other matter. Therefore, neutrinos become a great probe for ultra high energy cosmic accelerators. By reconstruct the direction of incoming neutrinos, we can pin point back to its origin.

2. Searching GZK Neutrinos with ARA

Detecting neutrinos are difficult. Because of its small cross-section and a steep spectrum, the volume of the detector must be large. Also, since neutrino is weakly interacting, we can only detect the neutrino via its secondary particles from the interaction. The detection method depends on the type of secondary particles we are looking for.

For neutrinos with energy above 10^{17} eV, the radio detection method is proven to be a highly effective method. The idea is proposed by Askaryan.[5,6] When a ultra high energy neutrino collides in the dielectric medium, such as ice, it will generate a electromagnetic shower. It will have a 20% excess of negative charge inside the dielectric medium. When these charged particles moving at the speed of light in the ice, they will generate a coherent Chrenkov light at the frequency ranging between several hundred MHz to few GHz.

By looking for coherent radio signal emitted by the EM component of a shower induced by the interaction of a high energy cosmic neutrino and some material on Earth, which is the Askaryan effect, we are able to detect these neutrinos. With the attenuation length is 100-1500 m, ice is a natural target material for radio detection.

Askaryan Radio Array (ARA) is an experiment dedicated in detecting ultra high energy neutrinos via Askaryan effect. ARA is located at South Pole. The full array will consist of 37 detector stations, as shown in Fig. 3. Each station is separated by 2 km. Currently three stations are deployed and operating continuously since 2012 (ARA01) and 2013 (ARA02 and ARA03). Each station is designed as an independent detector, as shown in Fig. 4. It consists of four strings with four antennas attached. The antennas are 200 m below the surface of the ice, and each antenna are separated by 10 m. Each string has two vertical polarized (V-pol) two horizontal polarized (H-pol) antennas, with the order from top to bottom is H-pol, V-pol, H-pol, V-pol. In total there are 16 antennas per station. The DAQ rate is about 5Hz. By measuring the timing difference between antennas, we are able to detect the incoming neutrino and reconstruct its direction.

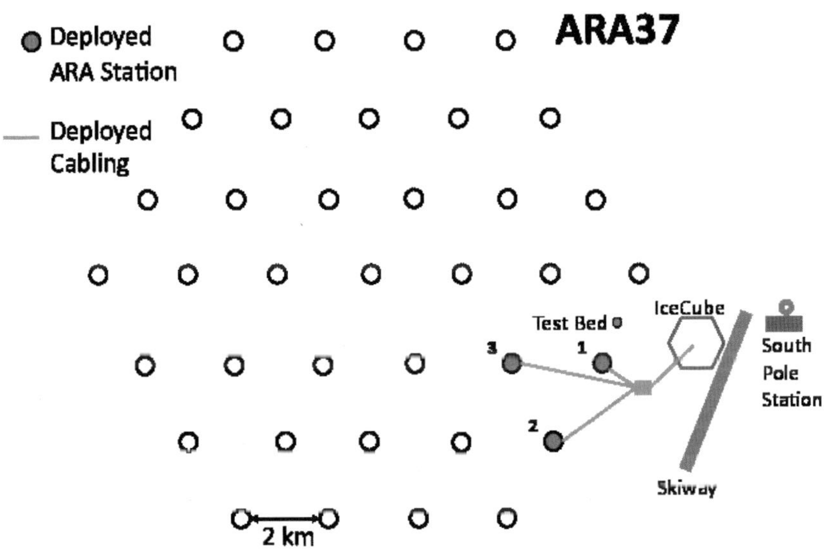

Fig. 3. The area map of the planned ARA detectors in South Pole.[7]

Fig. 5[7] is the result from ARA02 and ARA03, with data taken for 10 months in 2013. During the 10 month data taking period, both ARA02 and ARA03 does not find any UHECN events, therefore the upper limit of the neutrino flux has been set. From the plot, it is easy to see that the ARA is sensitive to neutrinos of energies between 10^{16} and $10^{20.5}$ eV, with the best sensitivity at around 10^{18} eV. The result significantly improved the limit obtained by ARA TestBed [8] by a factor of 10 to 20. It is not as sensitive as other experiments, but with with the current dataset, we are able to project the expected sensitivity when we have the full ARA array with 37 stations taking 3 years worth of data.

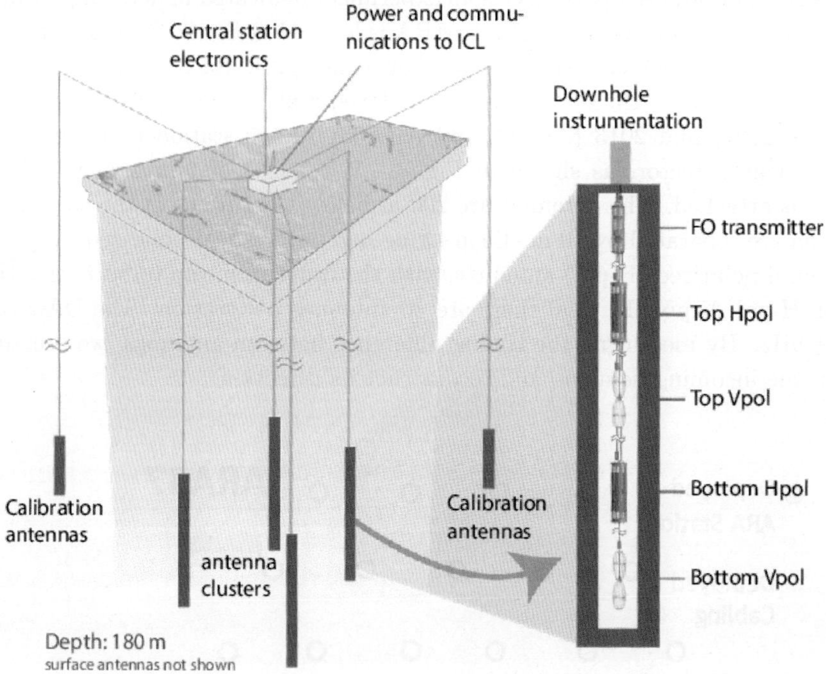

Fig. 4. The baseline design for an ARA station.[7]

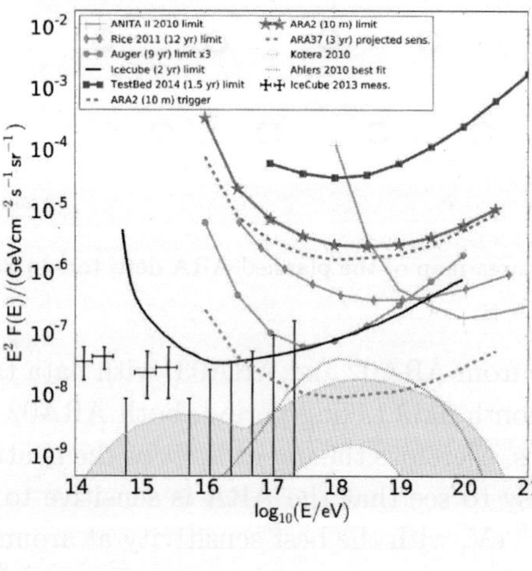

Fig. 5. Neutrino limits and sensitivities of the two ARA stations.[7]

The flavor of the UHECNs is also an important topic. By identifying the flavor of the UNECNs, it will help us to discriminate different theories of the UHECN productions. Recently the flavor ratio of neutrinos above 35 TeV has been measured by IceCube,[9] where the flavor ratio measured on Earth (f_e, f_μ, f_τ) is consistent with (1:1:1), if the neutrinos are generated by charged pion decay and averaged by neutrino oscillations. For neutrinos with energy above 10^{17} eV, the event rate is very low. Therefore event-by-event flavor identification would be preferred. With detailed studies, it might be possible to identify the flavor of the incoming neutrino from the radio signal it released and detected by ARA.

3. Summary

In summary, by detecting the radio signal generated by ultra high energy neutrinos via Askaryan effect in ice, ARA has great potential to detect the high energy neutrinos effectively. With two more stations plan to be deployed in South Pole in 2017-18 and more stations being planned, the capability on neutrino detections will be further enhanced.

References

1. A. M. Hillas, (2006), astro-ph/0607109.
2. R. U. Abbasi *et al.*, *Phys. Rev. Lett.* **100**, p. 101101 (2008).
3. K. Greisen, *Phys. Rev. Lett.* **16**, p. 748 (1966).
4. K. V. A. Zatsepin G. T., *JETP Lett.* **4**, p. 78 (1966).
5. G. A. Askaryan, *JETP* **41**, p. 616 (1962).
6. G. A. Askaryan, *JETP* **48**, p. 988 (1966).
7. P. Allison *et al.*, (2015), 1507.08991.
8. P. Allison *et al.*, *Astroparticle Physics* **70**, p. 62 (2015).
9. M. G. Aartsen *et al.*, *Phys. Rev. Lett.* **114**, p. 171102 (2015).

Feasibility of Determining Diffuse Ultra-High Energy Cosmic Neutrino Flavor Ratio through ARA Neutrino Observatory

Shih-Hao Wang,[1,*] Pisin Chen,[1,2] Jiwoo Nam,[1] and Melin Huang[1]

[1] *Department of Physics, Graduate Institute of Astrophysics, and Leung Center for Cosmology and Particle Astrophysics, National Taiwan University, Taipei 10617, Taiwan, R.O.C.*

[2] *Kavli Institute for Particle Astrophysics and Cosmology, SLAC National Accelerator Laboratory, Menlo Park, CA 94025, U.S.A.*

**wsh4180@gmail.com*

The flavor composition of ultra-high energy cosmic neutrinos (UHECN) carries precious information about the physical properties of their sources, the nature of neutrino oscillations and possible exotic physics involved during the propagation. Since UHECN with different incoming directions would propagate through different amounts of matter in Earth and since different flavors of charged leptons produced in the neutrino-nucleon charged-current (CC) interaction would have different energy-loss behaviors in the medium, measurement of the angular distribution of incoming events by a neutrino observatory can in principle be employed to help determine the UHECN flavor ratio. In this paper we report on our investigation of the feasibility of such an attempt. Simulations were performed, where the detector configuration was based on the proposed Askaryan Radio Array (ARA) Observatory at the South Pole, to investigate the expected event-direction distribution for each flavor. Assuming ν_μ-ν_τ symmetry and invoking the standard oscillation and the neutrino decay scenarios, the probability distribution functions (PDF) of the event directions are utilized to extract the flavor ratio of cosmogenic neutrinos on Earth. The simulation results are summarized in terms of the probability of flavor ratio extraction and resolution as functions of the number of observed events and the angular resolution of neutrino directions. We show that it is feasible to constrain the UHECN flavor ratio using the proposed ARA Observatory.

Keywords: Neutrino astronomy, neutrino experiments, cosmogenic neutrinos, neutrino detectors.

1. Introduction

Ultra-high energy cosmic neutrinos (UHECNs, $E > 10^{17}$eV) can be generated by ultra-high energy cosmic rays (UHECRs, $E > 10^{18}$eV), which properties are still little known, via photo-pion production or proton-proton interaction and the subsequent decays of charged pion and muon, e.g., $\pi^+ \to \nu_\mu + \mu^+ \to \nu_\mu + \bar{\nu}_\mu + \nu_e + e^+$, where the targets can be the intergalactic medium near the astrophysical sources (see Refs. 1, 2 for a review), or the cosmic microwave background (CMB) photons (the Greisen-Zatsepin-Kuzmin process).[3] Neutrinos originating from the GZK process are known as the cosmogenic neutrinos (or GZK neutrinos), which are guaranteed to exist based on the fact that CMB have been well observed and the notion is consistent with the observed GZK cutoff in the cosmic ray spectrum.[4,5]

The relative flux ratio between different UHECN flavors, or briefly, the flavor ratio, can provide information about the physical properties of UHECR sources. For example, the transition of flavor ratio with increasing energies can be used to constrain the strength of magnetic field.[6] Furthermore, during the propagation from the source to the Earth, the flavor composition of a neutrino would oscillate,[7,8] and may even be altered by some new physics beyond the Standard Model, such as the neutrino decay,[9,10] sterile neutrinos with tiny mass differences,[11] the violation of CPT or Lorentz invariance, and the quantum decoherence.[10,12]

As neutrinos interact with nucleons in the medium via the neutral current (NC) interaction, $\nu_l + N \to \nu_l + X$, or via the charged current (CC) interaction, $\nu_l + N \to l^- + X$, where l stands for lepton and X for hadronic debris, showers and charged leptons are generated. The secondary charged leptons have different energy-loss characteristics in the medium, which can manifest themselves in the event topology and in the direction distribution of events as they propagate through different amounts of matter in the Earth. For example, ν_τ can undergo the regeneration process ($\nu_\tau \to \tau \to \nu_\tau$) without being absorbed due to τ decay,[13,14] and as a result can exhibit a higher flux in the up-going directions. Therefore the neutrino distribution can be a useful tool for measuring the flavor ratio and should be applicable to any detector with sufficient angular resolution of event direction.

One of UHECN detection method is to observe the optical Cherenkov lights emitted by secondary charged particles and showers, e.g. IceCube[15] and ANTARES.[16] By combining analysis of direction distribution and event topologies, IceCube recently reported its identification of flavor ratio from 10TeV to PeV range.[17] However, its instrumented volume, currently of cubic kilometer scale for the largest, limits the event rate for UHE neutrinos.

The other very promising method is the radio technique based on the Askaryan effect:[18] neutrino-induced showers propagating in dense media would develop about 20 % of excess negative charges and emit Cherenkov radiation, coherent in the radio frequencies up to a few GHz due to the compact shower size. The radiated power in the coherent regime is roughly proportional to the shower energy, making this technique especially sensitive to UHECNs. This effect has been verified in a series of beam experiments.[19] Another advantage of this approach is the long radio attenuation length in some natural media, e.g. the Polar ice, with lengths typically of order of 0.1 to 1 km, and therefore detectors are able to obtain large target volume and achieve greater sensitivity. Observatories of this type are: ANITA,[20] ARA,[21] and ARIANNA.[22] To demonstrate the feasibility of extracting the flavor ratio from the event direction distribution of radio detection, we constructed a simulation with the detector configuration based on ARA currently under construction at the South Pole. The inference on the flavor ratio at the source is beyond our scope.

2. Simulation Setup

Our simulation is built by integrating existing packages that consists of two parts:

2.1. *Neutrino Generation and Propagation Using MMC*

The Muon Monte Carlo (MMC) package[24] is employed to generate and propagate all types of neutrinos and secondary charged leptons (except electrons, which develop into showers within short distance). For charged leptons, energy losses via ionization, pair production, bremsstrahlung, photonuclear interaction, and decay are taken into account. Taus decay into electron, muon, or hadrons are considered, and hence the $\nu_\tau \to \tau \to \nu_\tau$ regeneration in the propagation.

Neutrinos are generated isotropically at the Earth's surface and start their propagation to the detection volume. The density profile of the Earth is calculated based on Ref. 25 while the composition and the topography on the Earth's crust are not considered. The detection volume, which is the South Polar ice sheet in the vicinity of ARA, is approximated by a cylindrical ice volume, centered at 1 km below the ice surface with a radius of 8 km and a height of 2 km. In the propagation, all neutrinos and charged leptons are tracked until they either are absorbed by the Earth or exit the detection volume; energy losses greater than 1 PeV are treated stochastically, where showers below this cutoff can not trigger the detector efficiently.

2.2. *Neutrino Event Detection Using SADE*

The Simulation of Askaryan Detection and Events (SADE) package is used for the subsequent event detection.[26] Every energy loss exceeding 1 PeV within the detection volume in previous part is converted into electromagnetic or hadronic shower according to its products. In general, a neutrino event can generate multiple showers in the detection volume. Analytic formulae are applied to the calculation of shower characteristics and the frequency spectrum of Cherenkov radiation.[27]

A ray-tracing routine then finds the paths of both the direct and the reflected rays (due to the ice-air interface at the surface and varying refractive index in the ice) connecting each shower to each antenna in the ice, taking into account *in situ* measurements of ice properties such as the index of refraction[28,29] and the radio attenuation length[21,30] at the South Pole. The radiation emitted from each shower, subjected to refraction and attenuation after propagation, is summed up in the time domain.

The detector based on the planned configuration of ARA is a hexagonal array of 37 antenna stations arranged in a triangular grid with 2 km spacing, covering a total area of about 100 km^2. Each station is an autonomously operating cluster of eight vertically polarized and eight horizontally polarized antennas deployed at the depth around 200 m in the ice. The trigger conditions for a detected event require that (1) the received voltage of a triggered antenna should exceed three times of its RMS thermal noise voltage; (2) at least five out of sixteen antennas in a station are triggered; and (3) at least one station is triggered. More details about the setup of simulation parameters can be found in Ref. 31.

3. Angular Distribution of Neutrino Events

In the simulation, monoenergetic neutrino events with initial energies $\log_{10}(E_\nu/\text{eV}) = 17$–$19.5$ are generated. To acquire the expected zenith angle distribution of detected neutrino events for each flavor α, $D_\alpha(E_\nu, \cos\theta)$, isotropic cosmogenic neutrino spectrum for all flavors predicted in Refs. 32, as well as the incident flux ratio among three flavors at the Earth's surface are imposed.

The flavor ratio of neutrinos arriving at the Earth's surface adopted in our analysis is $f_e^E : f_\mu^E : f_\tau^E = 1 : 1 : 1$, as expected in the standard oscillation scenario;[7,8] $6 : 1 : 1$ and $0 : 1 : 1$ predicted in the neutrino decay scenarios with normal and inverted neutrino mass hierarchy, respectively[9] (neutrinos and antineutrinos are counted together because they are hard to be discriminated in thex detection). Throughout this paper these flavor ratios are assumed to be energy-independent over the considered neutrino energy range.

The resulting distribution for each flavor $D_\alpha(\cos\theta)$ in standard scenario is shown in Fig. 1. The zenith angle θ is defined as the angle between the local vertical axis of the detector and the direction of neutrino momentum, and hence events with negative $\cos\theta$ are down-going. The "all-sky" flavor ratio of detected events, $f_e : f_\mu : f_\tau$, is defined as the event ratio among flavors after integrating over all zenith directions. If incident flavor ratio at the Earth's surface, $f_e^E : f_\mu^E : f_\tau^E = 1 : 1 : 1$, are assumed, we find that $f_e : f_\mu : f_\tau = 0.584 : 0.154 : 0.262$. The ν_e events account for the most portion because their secondary electrons have higher detection efficiency, where all their energies are released into showers and strong signals are emitted.

The probability for up-going neutrinos ($\cos\theta > 0$) is suppressed by Earth shadowing, as the neutrino traveling distance is much longer than the interaction length. The common cutoff in down going direction at $\cos\theta \simeq -0.7$ is because there is less material for interaction, and for the Cherenkov cone generated by neutrinos events, having an opening angle $56°$ in ice, travels downward and is less possible to cover the area where the antennas are deployed. The probabilities reach the maximum near the horizontal direction ($\cos\theta \simeq 0$), where the traveling distance of neutrinos becomes comparable to the interaction length and the detection volume has its maximum span. The divergence of distribution shape between flavors here result from secondary charge leptons; unlike electrons, EeV μ and τ leptons can on average propagate distance of order of 10 km before come to a stop[33] and reach the detector volume, even the CC interaction is happened outside. The ν_τ events, in particular, have a different shape compared to the others especially in the horizontal and up-going directions, primarily due to ν_τ regeneration. This feature will be used to extract the flavor composition at the Earth's surface (f^Es) in the next section. However, the resemblance between ν_e and ν_μ distributions will lead to the degeneracy in the extraction. Therefore, the ν_μ-ν_τ symmetry is assumed as an extra constraint.

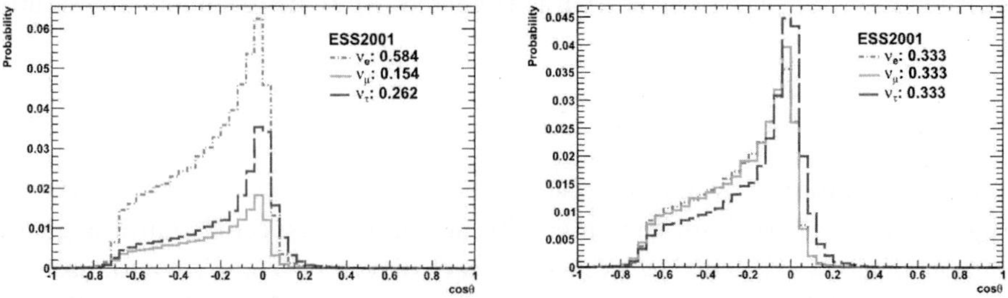

Fig. 1. Left panel: the expected direction distribution of detected neutrino events for each flavor α, $D_\alpha(\cos\theta)$, assuming the ESS flux[32] and the incident flux ratios $f_e^E : f_\mu^E : f_\tau^E = 1/3 : 1/3 : 1/3$. The all-sky flavor ratio of detected events $f_e : f_\mu : f_\tau$ is 0.584 : 0.153 : 0.263. Right panel: the expected direction distributions are scaled to equal fraction for comparing the curve shapes.

4. Pseudo-Observation and Flavor Ratio Extraction

To investigate the discriminating power of the proposed method, an ensemble of pseudo-observation data is generated and fit by the expected direction distribution to extract the flavor ratios.

4.1. *Preparing and Fitting Pseudo-Data Samples*

The pseudo-data sample is prepared in three steps. First, neutrino events with different flavors and directions are randomly generated according to the expected angular distributions, $D_\alpha(\cos\theta)$. Secondly, the zenith angle, θ, of each pick-up event is smeared by adding a Gaussian random number with the standard deviation $\Delta\theta$ equal to an assigned experimental angular error in reconstructed neutrino direction. Finally, this event collection is evenly divided into subsets, where each data set represents a hypothetical experimental data sample with a total number of detected events equal to $N_{\rm obs}$ and these data sets constitute a statistical ensemble.

To fit the hypothetical data samples with angular resolution $\Delta\theta$, the expected direction distribution for each flavor α is convolved with a Gaussian resolution function of standard deviation $\Delta\theta$. The convolved distribution for each flavor, called M_α, is used to construct the probability density function of event direction:

$$P(\cos\theta; f_e, f_2, \Delta\theta) = f_e M_e(\cos\theta) + (1 - f_e) \times [f_2 M_\mu(\cos\theta) + (1 - f_2) M_\tau(\cos\theta)], \quad (1)$$

where f_e and f_2 are unknown fraction coefficients with values between zero and one. This expression ensures f_e and f_2 are independent of each other. However, since the ν_μ-ν_τ symmetry imposed in Sec. 3, the incident ν_μ and ν_τ fluxes on the Earth are equal ($f_\mu^E = f_\tau^E$), and the value of f_2 is known and fixed, with only the fraction of ν_e events, f_e, undetermined. The maximum likelihood estimation is applied for flavor ratio extraction, and the associated best fit value is denoted as $\hat{f}_e$.

4.2. *The Probability of Success and Flavor Ratio Resolution*

The probability of success of flavor ratio extraction is calculated as the number of successful data sets divided by the total number of data sets. In general, the probability increases with the number of observed events and a better angular resolution, as expected. The probability of success for the standard scenario is greater than those for the decay scenarios since the latter have true ν_e ratios so closed to the boundaries that even small statistical fluctuation may result in failure. It was reported in Ref. 21 that fully deployed ARA is able to detect about 50 cosmogenic neutrinos in three years, with an angular resolution of neutrino direction about 6°. The successful probability is about 50 % to 60 % for $N_{\text{obs}} = 50$. For the standard scenario, the probability becomes greater than 70 % for $N_{\text{obs}} \geq 100$ and over 90 % for $N_{\text{obs}} \geq 250$. On the other hand, for the decay scenario with normal hierarchy, the probability is over 50 % once $N_{\text{obs}} \geq 100$. Therefore our method is feasible for ARA to extract the flavor ratio of cosmogenic neutrinos.

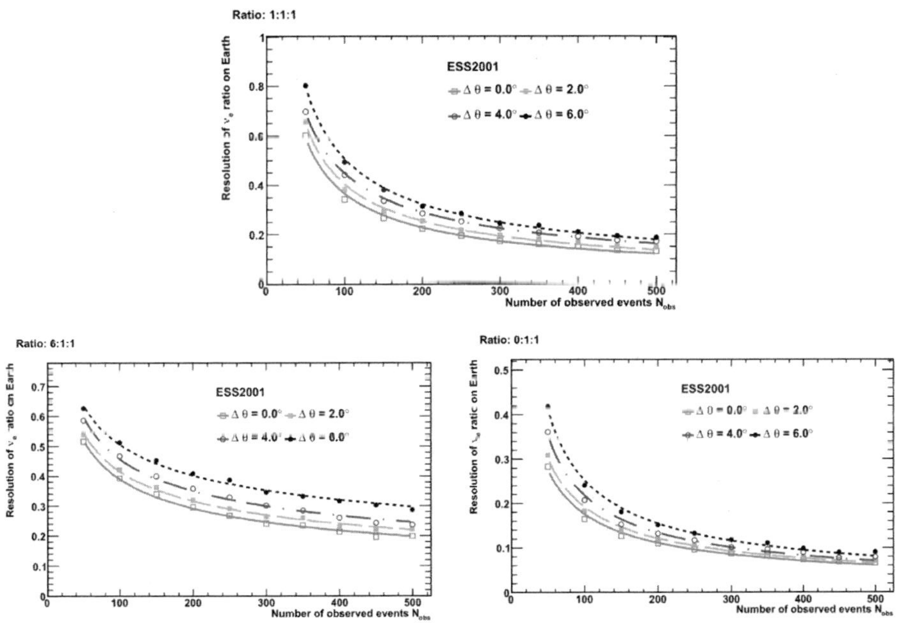

Fig. 2. The resolution of ν_e ratio at the Earth's surface R^E as a function of the number of observed events N_{obs}, assuming initial ratio of $1:1:1$ (top panel), $6:1:1$ (bottom left), and $0:1:1$ (bottom right). Results for different angular resolutions are plotted: $\Delta\theta = 0°$ (the perfect case, red solid curve), 2° (green long dashed), 4° (blue dash-dotted), 6° (black dashed).

We define the resolution of ν_e event fraction as the spread of all fitted values $\hat{f}_e$s in the ensemble with respect to the expected value $f_{e,\text{exp}}$: $R_{\pm}(N_{\text{obs}}, \Delta\theta) = \sqrt{\sum_{i=1}^{N_s}(\hat{f}_{e,i} - f_{e,\text{exp}})^2/(N_s - 1)}$, where N_s is the number of hypothetical experimental data sets. The distribution of $\hat{f}_e$s asymptotically approaches a Gaussian distribution as the number of observed events $N_{\rm obs}$ increases, and therefore the resolution defined here is approximately 1σ uncertainty at 68 % confidence level.

The upper and lower bounds of ν_e event fraction, $f_{e,\rm exp} \pm R$, are then transformed into the corresponding flavor ratios at the Earth's surface, $f^E_{e,\rm true} \pm R^E_\pm$. The resolution of ν_e ratio on the Earth R^E for given $N_{\rm obs}$ and $\Delta\theta$ is plotted in Fig. 2. The lower (upper) resolution R^E_- (R^E_+) is taken in the decay scenario with normal (inverted) hierarchy, while the average value is taken in the standard one.

To measure the separability of a flavor ratio predicted in a "fake" scenario from that extracted from a given "true" scenario (with $f_{e,\rm exp}$ and R), we define the discriminating power of flavor ratios as $|f_{e,\rm model} - f_{e,\rm exp}|/R$, where $f_{e,\rm model}$ is the ν_e event ratio expected by the scenario to be examined. Given $N_{\rm obs} = 100$ and $\Delta\theta = 6°$ for ARA, if the standard scenario is the real one, the discriminating power with respect to the decay scenario with normal (inverted) hierarchy will be at about 0.65σ (1.3σ) level. If the decay scenario with normal (inverted) hierarchy is real, then the discriminating power with respect to the standard scenario will be at about 0.7σ (1.2σ) level, while that to its inverted (normal) counterpart will be at about 2.0σ (1.8σ) level. Hence a preliminary constraint on the neutrino decay can be set by the method proposed in this article. The limit of our method with perfect angular resolution is also shown in Fig. 2 (red solid curves).

5. Conclusion and Future Works

In summary, UHECN observatories using radio Cherenkov technique such as ARA and ARIANNA, are sensitive in the UHE regime and expected to accumulate of order of 10 to 100 cosmogenic neutrinos per year in the near future, and hence provide sufficient statistics for the flavor ratio identification. In this work, the direction distribution of neutrino events is proposed to determine the flavor ratio for such observatories. A simulation is constructed and the expected distribution for ARA is derived, and this method is proved to be able to set a preliminary constraint on the neutrino decay models for ARA.

However, we have not fully taken advantage of all the information possessed by neutrino events. For example, the spatial distribution and the energies of showers, which can be retrieved by incorporating the vertex reconstruction, can help to distinguish electron flavor from the other two; and the characteristics of radio signals may allow the classification of shower types[34] and further the identification of the neutrino flavor event by event. The neutrino spectrum and the cross sections should also be incorporated into the fitting in order to reduce the systematic uncertainties. These will be included in the future simulation, and we expect to get a higher resolution of the neutrino flavor ratio.

Acknowledgments

We thank David Seckel and his coworkers of University of Delaware for sharing with us their simulation code SADE; our investigation would have been incomplete without this code. We also thank Guey-Lin Lin of National Chiao-Tung University for valuable discussions and suggestions. This work is supported by Taiwan National Science Council under Project No. NSC100-2119-M-002-025- and by US Department of Energy under Contract No. DE-AC03-76SF00515.

References

1. F. Halzen and D. Hooper, Reports on Progress in Physics **65**, 1025 (2002).
2. J. K. Becker, Physics Reports **458**, 173 (2008).
3. K. Greisen, Phys. Rev. Lett. **16**, 748 (1966); G. Zatsepin and V. Kuz'Min, JETP Lett. **4**, 114 (1966); V. Beresinsky and G. Zatsepin, Phys. Lett. B **28**, 423 (1969).
4. R. U. Abbasi *et al.* (HiRes Collaboration), Phys. Rev. Lett. **100**, 101101 (2008).
5. J. Abraham *et al.* (Pierre Auger Collaboration), Phys. Rev. Lett. **101**, 061101 (2008).
6. T. Kashti and E. Waxman, Phys. Rev. Lett. **95**, 181101 (2005).
7. J. G. Learned and S. Pakvasa, Astropart. Phys. **3**,267 (1995).
8. H. Athar, M. Jeżabek, and O. Yasuda, Phys. Rev. D **62**, 103007 (2000).
9. J. F. Beacom *et al.*, Phys. Rev. Lett. **90**, 181301 (2003); Phys. Rev. D **69**, 017303 (2004).
10. G. Barenboim and C. Quigg, Phys. Rev. D **67**, 073024 (2003).
11. J. F. Beacom *et al.*, Phys. Rev. Lett. **92**, 011101 (2004).
12. D. Hooper, D. Morgan, and E. Winstanley, Phys. Rev. D **72**, 065009 (2005).
13. E. Bugaev *et al.*, Astropart. Phys. **21**,491 (2004).
14. F. Halzen and D. Saltzberg, Phys. Rev. Lett. **81**, 4305 (1998).
15. F. Halzen, The European Physical Journal C - Particles and Fields **46**, 669 (2006).
16. The ANTARES Collaboration, in 34th ICRC (2015); arXiv:1510.04508 [astro-ph].
17. M. G. Aartsen *et al.* (IceCube Collaboration), Astrophys. J. **809**, 98 (2015).
18. G. A. Askaryan, JETP **14**, 441 (1962); **21**, 658 (1965).
19. D. Saltzberg *et al.*, Phys. Rev. Lett. **86**, 2802 (2001); P. W. Gorham *et al.*, Phys. Rev. D **72**, 023002 (2005); Phys. Rev. Lett. **99**, 171101 (2007).
20. P. Gorham *et al.*, Astropart. Phys. **32**, 10 (2009); Phys. Rev. Lett. **103**, 051103 (2009); Phys. Rev. D **82**, 022004 (2010); erratum: arXiv:1011.5004 [astro-ph].
21. P. Allison *et al.* (ARA Collaboration), Astropart. Phys. **35**,457 (2012).
22. S. W. Barwick, J. Phys. Conf. Ser. **60**, 276 (2007); L. Gerhardt *et al.*, Nucl. Instrum. Meth. **A 624**, 85 (2010); S. W. Barwick *et al.*, Astropart. Phys. **70**, 12 (2015).
23. J. Alvarez-Muñiz and E. Zas, Phys. Lett. **B 411**, 218 (1997).
24. D. Chirkin and W. Rhode, arXiv:0407.075v2 [hep-ph].
25. A. M. Dziewonski and D. L. Anderson, Phys. Earth and Plan. Int. **25**, 297 (1981).
26. D. Seckel, S. Hussain and P. Stengel, private communication (2011).
27. J. Alvarez-Muñiz, R. A. Vázquez, and E. Zas, Phys. Rev. D **62**, 063001 (2000).
28. I. Kravchenko, D. Besson, and J. Meyers, Journal of Glaciology **50**, 522 (2004).
29. H. Landsman *et al.*, Nucl. Instrum. Meth. **A 662**, Supplement 1, S54 (2012).
30. S. Barwick *et al.*, Journal of Glaciology **51**, 231 (2005).
31. S. Wang *et al.*, JCAP **11**, 062 (2013).
32. R. Engel, D. Seckel, and T. Stanev, Phys. Rev. D **64**, 093010 (2001).
33. S. Iyer Dutta *et al.*, Phys. Rev. D **63**, 094020 (2001).
34. J. Alvarez-Muñiz, R. A. Vázquez, and E. Zas, Phys. Rev. D **61**, 023001 (1999).

Gamma-Ray Bursts and their Relation to Astroparticle Physics and Cosmology

J. Řípa

Leung Center for Cosmology and Particle Astrophysics, National Taiwan University
No.1, Sec.4, Roosevelt Road, Taipei 10617, Taiwan (R.O.C)
jripa@ntu.edu.tw

This article gives an overview of gamma-ray bursts (GRBs) and their relation to astroparticle physics and cosmology. GRBs are the most powerful explosions in the universe that occur roughly once per day and are characterized by flashes of gamma-rays typically lasting from a fraction of a second to thousands of seconds. Even after more than four decades since their discovery they still remain not fully understood. Two types of GRBs are observed: spectrally harder short duration bursts and softer long duration bursts. The long GRBs originate from the collapse of massive stars whereas the preferred model for the short GRBs is coalescence of compact objects such as two neutron stars or a neutron star and a black hole. There were suggestions that GRBs can produce ultra-high energy cosmic rays and neutrinos. Also a certain sub-type of GRBs may serve as a new standard candle that can help constrain and measure the cosmological parameters to much higher redshift than what was possible so far. I will review the recent experimental observations.

Keywords: Gamma-ray bursts; astroparticle physics; cosmology.

1. Gamma-Ray Bursts

This section briefly overviews the gamma-ray burst (GRB) phenomena.[1–5] GRBs are one of the most extreme explosive events ever observed. They were discovered in late sixties and reported to the scientific community in early seventies.[6,7] In spite of tremendous effort and numerous observations there are still open questions concerning their detailed physics. They occur roughly once per day and are characterized by flashes of γ-rays typically lasting from a fraction of a second to thousands of seconds, see Fig. 1(a). It has been found that the duration distribution of their γ-ray prompt emission is bimodial[8] which suggested that there were two groups of GRBs, see Fig. 1(b). Later, based on more observations, it has been confirmed that these two GRB groups were two distinct astrophysical populations: I. so called long GRBs with prompt γ-ray emission $\gtrsim 2\,\mathrm{s}$ that has been identified to be gravitational collapses of massive stars due to their association with type Ic core-collapse supernovae, and II. so called short GRBs with prompt γ-ray emission $\lesssim 2\,\mathrm{s}$ that has been suggested to originate in a merger of two compact objects such as NS-NS or NS-BH.[9] It was proposed that they lie at cosmological distances[10] and they are created in collisions of highly relativistic outflow of the accelerated jetted matter.[11,12] In many cases it was found that the prompt γ-ray emission is followed by longer-lasting afterglow in soft X-ray, optical or radio waves[13,14] explained as a result of the interaction of the relativistic outflow with the circum-burst medium. Observations found that their redshifts are up to $z = 9.4$ with $\langle z \rangle \approx 0.5$ for short GRBs and $\langle z \rangle \approx 2.0$ for long GRBs[9] which requires energy release up to an isotropic-equivalent value of $\sim 10^{54}$ erg.

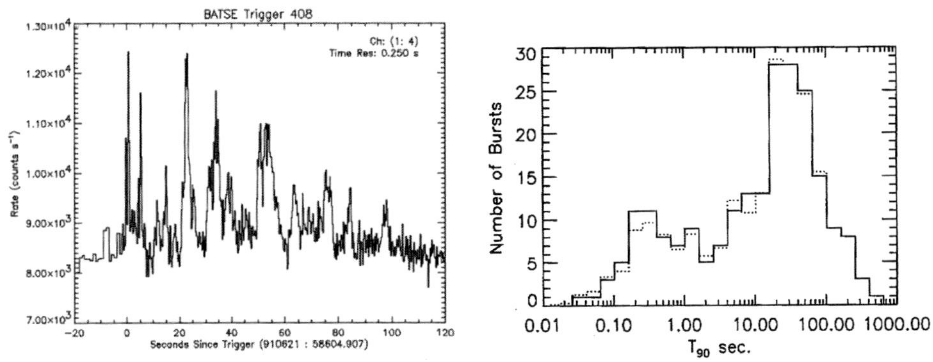

(a) (b)

Fig. 1. (a) An example of GRB prompt gamma flux curve detected by BATSE instrument. From Ref. 15. (b) The distribution of T_{90} durations of GRBs indicating two distinct groups. From Ref. 8.

2. Relation of GRBs to Cosmology

There were discovered several correlations in the properties of the GRB emission: the anti-correlation between isotropic luminosity $L_{\rm iso}$ and spectral lag $\tau_{\rm lag}$ of the prompt γ-ray emission;[16] the correlation between $L_{\rm iso}$ and light-curve variability V;[17] the spectral peak energy of the prompt γ-ray emission $E_{\rm peak}$ and isotropic-equivalent released energy $E_{\rm iso}$ correlation,[18] see Fig. 2(a); $E_{\rm peak}$ and collimated energy E_γ correlation;[19] $E_{\rm peak}$ and peak luminosity $L_{\rm p}$ correlation;[20] correlation between $E_{\rm peak}$, $E_{\rm iso}$ and rest-frame temporal break in optical afterglow light curve $t_{\rm break}$;[21] and a correlation between $L_{\rm iso}$, $E_{\rm peak}$ and prompt 'high-signal' time-scale $T_{0.45}$ (variability).[22] This makes GRBs a potential tool to constrain cosmological parameters,[23–26] see Fig. 3(a).

It was also found that the peak flux $F_{\rm p}$ (at 2 eV and $z = 2$) might anti-correlate with peak time $t_{\rm p}$ in the optical lightcurves,[27–29] see Fig. 2(b). A similar anti-correlation for the peak luminosity in R band of the onset bumps and less steep for the re-brightening bumps in the optical lightcurves was found.[30] Unfortunately for many observations the peak time $t_{\rm p}$ is unknown and further optical observations at earlier times are needed.

More early optical observations with fine sapling of the light curves around the peak times below one minute could be provided by the Ultra-Fast Flash Observatory pathfinder (UFFO-p).[31–33] It is a novel spaceborn instrument dedicated to detect GRBs and rapidly follow their optical/ultraviolet counterparts to provide prompt

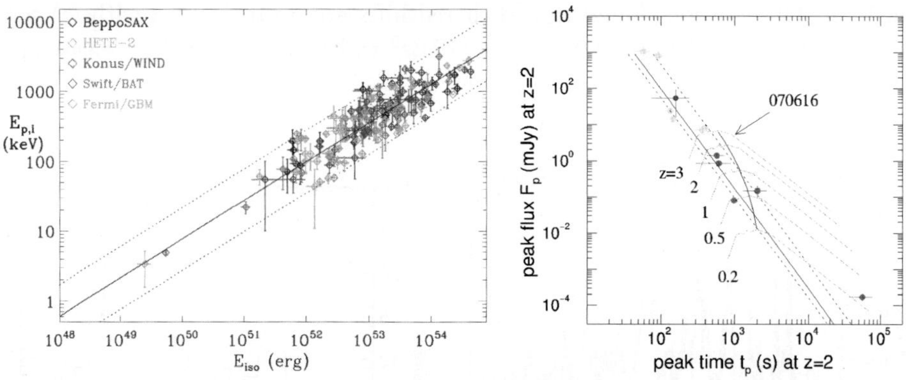

(a) (b)

Fig. 2. (a) The $E_{\mathrm{p,i}} - E_{\mathrm{iso}}$ correlation for long GRBs with marked best-fit power-law (black solid line) for different instruments. From Ref. 26. (b) The anti-correlation of optical light-curve peak flux (at 2 eV) for redshift $z = 2$ and peak time for six fast rising afterglows (light blue points) and five afterglows slow rising (purple points). The solid straight line shows the best fit. From Ref. 27.

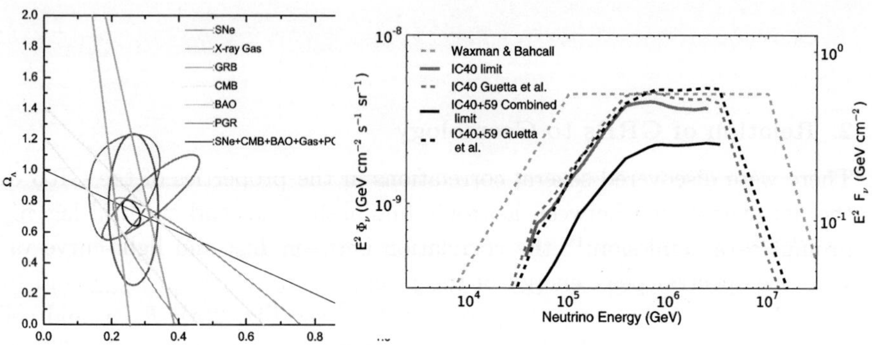

(a) (b)

Fig. 3. (a) Joint 1σ confidence intervals given by constraints from the datasets of galaxy clusters, GRBs, CMB shift parameter, SNe Ia, BAO, and 2dF Galaxy Redshift Survey. From Ref. 25, 24. (b) Comparison of predictions of ν flux from GRBs based on observed γ-ray spectra (dashed lines) with 90 % confidence upper limits obtained from the results of IceCube for 40 and 59 detector strings. From Ref. 50.

optical and early afterglow measurements. It consists of two scientific instruments, see Fig. 4. A GRB location is determined in a few seconds by the first instrument called UFFO Burst Alert & Trigger telescope (UBAT)[34–37] which employs the coded mask imaging technique[38] and the detector comprising of Yttrium Oxyorthosilicate scintillating crystals and multi-anode photomultiplier tubes. It has the energy range of $\approx 10 - 150\,\mathrm{keV}$ with half-coded field of view (HCFOV) $70.4^\circ \times 70.4^\circ$ and angular resolution $\leq 10'$. The second instrument is the Slewing Mirror Telescope (SMT)[39–42] with the field of view of $17' \times 17'$ and UV range of $200 - 650\,\mathrm{nm}$. It consists of a Ritchey-Chrétien telescope with an Intensified Charge-Coupled Device in its focal plane. In front of the telescope there is placed a fast plane slewing mirror which allows to redirect the optical path and start observation within $\approx 1\,\mathrm{s}$ since it receives the target direction. SMT slewing mirror provides approximately the same sky coverage as UBAT's HCFOV. UBAT and SMT have been assembled and integrated with the control electronics on UFFO-p which is planned to be launched on the Lomonosov Moscow State University satellite.[43]

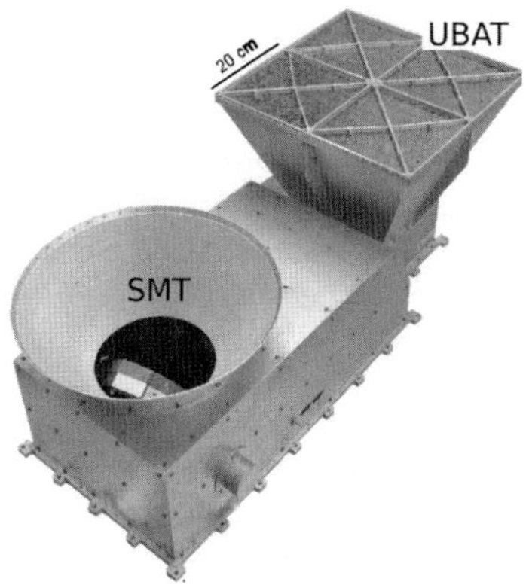

Fig. 4. The photo shows assembled UFFO-p with its two scientific instruments: UBAT and SMT. From Ref. 37.

3. Relation of GRBs to Astroparticle Physics

GRBs were proposed to be sites for accelerating ultra-high energy cosmic rays and sources of very high energy neutrinos up to $10^{17} \sim 10^{19}\,\mathrm{eV}$.[44–48] However, the recent results from the IceCube detector[49] suggests that the efficiency of neutrino production may be much lower than predicted.[50] 4-years data of IceCube put constraints on the prompt neutrino flux from GRBs, see Fig. 3(b). A single low-significance neutrino, compatible with the atmospheric neutrino background, was found in coincidence with one of the 506 observed GRBs.[51]

References

1. T. Piran, The physics of gamma-ray bursts, *Rev. of Mod. Phys.* **76**, 1143 (2004).
2. P. Mészáros, Gamma-ray bursts, *Rep. on Prog. in Phys.* **69**, 2259 (2006).
3. N. Gehrels and S. Razzaque, Gamma-ray bursts in the swift-Fermi era, *Frontiers of Physics* **8**, 661 (2013).
4. G. Vedrenne and J.-L. Atteia, Gamma-Ray Bursts, *Gamma-Ray Bursts: The Brightest Explosions in the Universe* (Springer Praxis Books, Springer-Verlag Berlin, 2009).

5. C. Kouveliotou, R. A. M. J. Wijers and S. Woosley, Gamma-ray Bursts, *Gamma-ray Bursts* (Cambridge University Press, Cambridge, UK, 2012).
6. R. W. Klebesadel, I. B. Strong and R. A. Olson, Observations of gamma-ray bursts of cosmic origin, *ApJ* **182**, L85 (1973).
7. E. P. Mazets, et al., Burst of cosmic gamma-emission from observations on Cosmos 461, *Pisma v Zhurnal Eksperimentalnoi i Teoreticheskoi Fiziki* **19**, 126 (1974).
8. C. Kouveliotou, C. A. Meegan, G. J. Fishman, et al., Identification of two classes of gamma-ray bursts, *ApJ* **413**, L101 (1993).
9. E. Berger, Short-duration gamma-ray bursts, *Annual Review of Astronomy and Astrophysics* **52**, 43 (2014).
10. B. Paczynski, Gamma-ray bursters at cosmological distances, *ApJ* **308**, L43 (1986).
11. M. J. Rees and P. Meszaros, Relativistic fireballs - Energy conversion and time-scales, *MNRAS* **258**, 41 (1992).
12. M. J. Rees and P. Meszaros, Unsteady outflow models for cosmological gamma-ray bursts, *ApJ* **430**, L93 (1994).
13. E. Costa, F. Frontera, J. Heise, et al., Discovery of an X-ray afterglow associated with the γ-ray burst of 28 February 1997, *Nature* **387**, 783 (1997).
14. J. van Paradijs, P. J. Groot, T. Galama, et al., Transient optical emission from the error box of the γ-ray burst of 28 February 1997, *Nature* **386**, 686 (1997).
15. G. J. Fishman, C. A. Meegan, R. B. Wilson, et al., The first BATSE gamma-ray burst catalog, *ApJS* **92**, 229 (1994).
16. Norris, J. P., G. F. Marani and J. T. Bonnell, Connection between energy-dependent lags and peak luminosity in gamma-ray bursts, *ApJ* **534**, 248 (2000).
17. E. E. Fenimore and E. Ramirez-Ruiz, Redshifts for 220 BATSE gamma-ray bursts determined by variability and the cosmol. consequences, *arXiv:astro-ph/0004176* (2000).
18. L. Amati, F. Frontera, M. Tavani, et al., Intrinsic spectra and energetics of BeppoSAX Gamma-Ray Bursts with known redshifts, *A&A* **390**, 81 (2002).
19. G. Ghirlanda, G. Ghisellini and D. Lazzati, The collimation-corrected gamma-ray burst energies corr. with the peak ener. of their νF_ν spect., *ApJ* **616**, 331 (2004).
20. D. Yonetoku, T. Murakami, T. Nakamura, et al., Gamma-ray burst formation rate inferred from the spectral peak energy-peak luminosity relation, *ApJ* **609**, 935 (2004).
21. E.-W. Liang and B. Zhang, Model-independent multivariable gamma-ray burst luminosity indicator and its possible cosmological implications, *ApJ* **633**, 611 (2005).
22. C. Firmani, et. al., Discovery of a tight correlation among the prompt emission properties of long gamma-ray bursts, *MNRAS* **370**, 185 (2006).
23. Ghirlanda, G., G. Ghisellini and C. Firmani, Gamma-ray bursts as standard candles to constrain the cosmological parameters, *New Journal of Physics* **8**, 123 (2006).
24. F. Y. Wang, Z. G. Dai and Z.-H. Zhu, Measuring dark energy with gamma-ray bursts and other cosmological probes, *ApJ* **667**, 1 (2007).
25. F. Y. Wang, Z. G. Dai and E. W. Liang, Gamma-ray burst cosmology, *New Astronomy Reviews* **67**, 1 (2015).
26. L. Amati and M. D. Valle, Measuring cosmol. param. with gamma-ray bursts, in *Proc. of the MG13 Meet. on General Relativity*, (World Scientific, Singapore, 2015), p. 769.
27. A. Panaitescu and W. T. Vestrand, Taxonomy of gamma-ray burst optical light curves: identification of a salient class of early afterglows, *MNRAS* **387**, 497 (2008).
28. A. Panaitescu and W. T. Vestrand, Optical afterglows of gamma-ray bursts: peaks, plateaus and possibilities, *MNRAS* **414**, 3537 (2011).
29. A. Panaitescu, W. T. Vestrand and P. Woźniak, Peaks of optical and X-ray afterglow light curves, *MNRAS* **433**, 759 (2013).
30. E.-W. Liang, L. Li, H. Gao, et al., A Comprehensive study of gamma-ray burst optical

emission. II. Afterglow onset and late re-brightening components, *ApJ* **774**, 13 (2013).
31. P. Chen, S. Ahmad, K. B. Ahn, et al., The UFFO (Ultra Fast Flash Observatory) pathfinder: science and mission, in *Proc. of the 32nd Inter. Cos. Ray Conf.*, Vol. 8 (2011), p. 243.
32. J. Nam, S. Ahmad, K. B. Ahn, et al., The UFFO Slewing mirror telescope for early optical observation from gamma ray bursts, *Mod. Phys. Let. A* **28**, 40003 (2013).
33. I. H. Park, S. Brandt, C. Budtz-Jørgensen, et al., Ultra-Fast Flash Observatory for the obser. of early phot. from gamma-ray bursts, *New Jour. of Phys.* **15**, 023031 (2013).
34. A. Jung, S. Ahmad, K. B. Ahn, et al., Design and fabrication of detector module for UFFO Burst Alert & Trigger Telescope, in *Proc. of the 32nd Inter. Cos. Ray Conf.*, Vol. 8 (2011), p. 235.
35. Y.-Y. Chang, C. R. Chen, P. Chen, et al., Inverted-conical light guide for crosstalk reduc. in tightly-packed scintil. matrix and MAPMT assembly, *NIMPA* **771**, 55 (2015).
36. J. Lee, S. Jeong, J. E. Kim, et al., Design, construction and performance of the detector for UFFO Burst Alert & Trigger Telescope, in *EAS Publication Series, Gamma-ray Bursts: 15 Years of GRB Afterglows*, Vol. 61 (EDP Sciences, UK, 2013), p. 525.
37. J. Řípa, M. B. Kim, S. Jeong, et al., Testing and performance of UFFO Burst Alert & Trigger Telescope, in *Proc. of Science (SWIFT 10), Swift: 10 Years of Discovery*, (SISSA, 2015), p. 102
38. P. H. Connell and V. Reglero, The Ultra Fast Flash Observatory pathfinder - UFFO-p GRB imaging and location with its coded mask X-ray imager UBAT, in *EAS Publication Series, Gamma-ray Bursts: 15 Years of GRB Afterglows*, Vol. 61 (EDP Sciences, UK, 2013), p. 517.
39. S. Jeong, J. W. Nam, K. B. Ahn, et al., Slewing Mirror Telescope optics for the early observation of UV/optical phot. from gamma-ray bursts, *Opt. Expr.* **21**, 2263 (2013).
40. S. Jeong, J. W. Nam, K. B. Ahn, et al., Development of Slewing Mirror Telescope optical system for the UFFO-pathfinder, in *EAS Publication Series, Gamma-ray Bursts: 15 Years of GRB Afterglows*, Vol. 61 (EDP Sciences, UK, 2013), p. 561.
41. S. Jeong, S. Brandt, C. Budtz-Jørgensen, et al., Observation of early photons from gamma-ray bursts with the Lomonosov / UFFO-pathfinder, in *III Workshop on Robotic Autonomous Observatories, RevMexAA*, Vol. 45 (UNAM, 2014), p. 139.
42. J. E. Kim, H. Lim, J. W. Nam, et al., Readout of the UFFO Slewing Mirror Telescope to detect UV/optical photons from gamma-ray bursts *Jour. of Inst.* **8**, P07012 (2013).
43. M. Panasyuk, Moscow State University satellite "Mikhail Lomonosov" - the multi-purpose obs. in space, in *Proc. of the 32nd Int. Cos. Ray Con.*, Vol. 8 (2011), p.313.
44. E. Waxman and J. N. Bahcall, High energy neutrinos from cosmological gamma-ray burst fireballs, *Phys. Rev. Let.* **78**, 2292 (1997).
45. E. Waxman and J. N. Bahcall, Neutrino afterglow from gamma-ray bursts: $\sim 10^{18}$ eV, *ApJ* **541**, 707 (2000).
46. P. Mészáros and E. Waxman, TeV neutrinos from successful and choked gamma-ray bursts, *Phys. Rev. Let.* **87**, 171102 (2001).
47. E., Waxman, Astrophysical sources of high energy neutrinos, *Nuclear Physics B Proceedings Supplements* **118**, 353 (2003).
48. P. Mészáros, Gamma ray bursts as neutrino sources, *arXiv:1511.01396* (2015).
49. D. Guetta, Neutrinos from gamma ray bursts in the IceCube and ARA era, *Journal of High Energy Astrophys.* **7**, 90 (2015).
50. R. Abbasi, Y. Abdou, T. Abu-Zayyad, et al., An absence of neutrinos associated with cosmic-ray acceleration in γ-ray bursts, *Nature* **484**, 351 (2012).
51. M. G. Aartsen, M. Ackermann, J. Adams, et al., Search for prompt neutrino emission from gamma-ray bursts with IceCube, *ApJ* **805**, L5 (2015).

QED Correction to Magnetic Dipole Radiation

Takahiro Hayashinaka and Teruaki Suyama
Research Center for the Early Universe, The University of Tokyo,
7-3-1 Hongo, Bunkyo-ku, Tokyo, 113-0033, Japan
hayashinaka@resceu.s.u-tokyo.ac.jp
suyama@resceu.s.u-tokyo.ac.jp

The first order QED correction term to the classical magnetic dipole radiation formula is explicitly given in terms of local configuration of the electromagnetic current which produces global dipole magnetic field around pulsars. Slow rotation of the star and weakness of the magnetic dipole are assumed to obtain an explicit formula. We also reviewed the analysis in strong magnetic field regime and found that no correction would happen in this regime except for the running of coupling constant effect. We advocate the conventional use of the classical magnetic dipole radiation formula to estimate the magnetic field strength of the pulsars.

Keywords: Quantum electrodynamics; strong magnetic field; dipole radiation.

1. Introduction

Recent observations of pulsars revealed the existence of magnetors[1,2] (neutron stars with strong magnetic field). The measurement of the magnetic field strength of the NS $B_{\rm NS}$ is done indirectly using a P-$\dot{P}$ (period of pulsation and its time derivative) diagram. Magnetic dipole radiation can emit angular momentum and is thought to be responsible for the spin down (breaking) of the pulsars. This is the reason why we can estimate the magnetic field strength of the pulsars from current observation without polarization.

The observed value of the magnetic field strength almost reaches to 10^{16} G which is far beyond the critical value defined by the scale of quantum electrodynamics (QED), $B_{\rm crit} = m^2/e = 4 \times 10^{13}$ G. Here, m is electron mass and e is its charge. Thus it is possible to regard the magnetors as natural proving ground for the strong field QED. Many researches on QED effect on magnetors (or NS) have been made[3–5] so far. However, there seems no explicit formula for the magnetic dipole radiation with QED correction on which the magnetic field measurement depends as far as authors know. It has been argued that this effect is tiny so that one can always neglect it at the (effective) Lagrangian level,[6] but they did not show the contribution to the magnetic dipole radiation itself.

We will, in this paper, consider the QED correction to the magnetic dipole radiation to examine the consistency of the determination of $B_{\rm NS}$ in a plain and direct manner. We limit ourselves to the subcritical regime $|B| < B_{\rm crit}$ at first since full order calculation seems extremely hard. We can make use of Euler-Heisenberg Lagrangian[7–9] to find the correction term. We, then, move on to the strong magnetic field regime $|B| > B_{\rm crit}$ with using an asymptotic expansion of the EH Lagrangian.

This paper is organized as follows. In Sec. 2, we give an introduction to the EH Lagrangian and derive the effective equation of motion for the background EM field. We solve it with a source term (classical current term generated by NS) in Sec. 3. The dipole radiation formula is obtained here. Sec. 4 deals with the strong field limit and its consequence. Finally, conclusion is given in Sec. 5. All the calculation in this paper is performed in a unit with $\hbar = c = 1$ and $\mu_0 = \epsilon_0 = 1$.

2. Euler Heisenberg Lagrangian

The interaction we will consider here is the one between electromagnetic fields $A_\mu(x)$ and Dirac field $\psi(x)$ which is totally depicted in terms of quantum electrodynamics (QED). QED theory has the effective photon self-interaction in consequence of its elementary process (photon-positron–electron interaction) and fermion loop effect. The effective Lagrangian for a constant EM background field up to 1-loop level was found by Euler and Heisenberg[7] and it is called Euler-Heisenberg Lagrangian today. Schwinger also found the same Lagrangian from the QED Lagrangian later.[8] The EH Lagrangian is written in terms of two invariants $\mathcal{F} = \frac{1}{4}F_{\mu\nu}F^{\mu\nu} = -\frac{1}{2}(\boldsymbol{E}^2 - \boldsymbol{B}^2)$ and $\mathcal{G} = \frac{1}{4}F_{\mu\nu}\tilde{F}^{\mu\nu} = -\boldsymbol{E}\cdot\boldsymbol{B}$ as

$$\begin{aligned}\mathcal{L}_{\mathrm{EH}} &= \mathcal{L}_{\mathrm{EM}} + \mathcal{L}_{\mathrm{EH}}^{(1)} \\ &= -\mathcal{F} - \frac{1}{8\pi^2}\int_0^\infty \frac{\mathrm{d}s}{s^3}\mathrm{e}^{-m^2 s}\left[(es)^2\mathcal{G}\frac{\Re\cosh(es\sqrt{2(\mathcal{F}+i\mathcal{G})})}{\Im\cosh(es\sqrt{2(\mathcal{F}+i\mathcal{G})})} - \frac{2}{3}(es)^2\mathcal{F} - 1\right].\end{aligned} \tag{1}$$

We can see separate contributions from the classical Maxwell Lagrangian $\mathcal{L}_{\mathrm{EM}} = -\mathcal{F}$ and the correction of QED $\mathcal{L}_{\mathrm{EH}}^{(1)}$ in Eq. (1). $\mathcal{L}_{\mathrm{EH}}^{(1)}$ includes arbitrary higher order terms of the coupling ($\mathcal{O}(e^n)$ terms) in its expansion. Each $\mathcal{O}(e^n)$ terms has n vector fields, which describes the n-point self-interaction or the nonlinearity of the background photon.

The EH Lagrangian can be applied to the case in which both temporal and spatial scales of changing of the EM fields are enough slower (or larger) than the typical scale of QED which is determined by the electron mass m. The time scale of QED is given by $1/m \sim 8\times 10^{-20}$s, and the spatial scale is given by electron Compton wave length $1/m \sim 3\times 10^{-10}$m. The scale for Neutron stars is designated by rotational angular velocity ω_{NS} and radius R_{NS} of the neutron star. Typical value for ω_{NS} and R_{NS} is roughly given by $\omega_{\mathrm{NS}} \gtrsim 1$ms and $R_{\mathrm{NS}} \sim 10$km, both of which are much larger than $1/m$. Therefore, we can justify the use of the EH Lagrangian to find the correction term for the magnetic dipole radiation.

The modified equation of motion is obtained by variation of Eq. (1) Of course, it is cumbersome to deal with full nonlinearity. Instead, we will take only the first order correction term using perturbation technique. This truncation is justified when all the component of the field strength $F_{\mu\nu}$ is smaller than the critical field strength $F_{\mathrm{crit}} \equiv m^2/e$. The correction to the Lagrangian or the equation of motion seems to be quite small ($\sim \mathcal{O}(10^{-4})$) even if the field strength reaches to the critical value. It is, however, not straightforward to say that this is also the case when the correction term to the radiation is written.

We also consider the source term J^μ which describes the time-varying (rotating) magnetic dipole. If we adopt the Lorenz gauge condition $A^\mu{}_{,\mu} = 0$, we obtain an inhomogeneous nonlinear wave equation

$$\Box A^\mu(x) = J^\mu(x) - \frac{4\alpha^2}{45m^4}(4\mathcal{F}F^{\mu\nu} + 7\mathcal{G}\tilde{F}^{\mu\nu})_{,\nu}, \tag{2}$$

which can be solved perturbatively and $\alpha = e^2/4\pi \sim 1/137$ is the fine structure constant.

3. Magnetic Dipole Radiation and Correction Term

We show the perturbative solution of Eq. (2) in this section.

The first step is to decompose the nonlinear equation (2) into the sequential linear equations. We have to regard the latter (nonlinear) term on the right hand side of Eq. (2) as a small perturbation. The zeroth and the second order reads

$$\Box A^\mu_{(0)} = J^\mu, \quad \Box A^\mu_{(1)} = -\frac{4\alpha^2}{45m^4}(4\mathcal{F}F^{\mu\nu} + 7\mathcal{G}\tilde{F}^{\mu\nu})_{,\nu}\Big|_{(0)} \equiv J^\mu_{(1)}, \tag{3}$$

where $A^\mu = A^\mu_{(0)} + A^\mu_{(1)}$ and the symbol $|_{(0)}$ denotes the substitution $A^\mu \to A^\mu_{(0)}$.

We need to know $\mathcal{O}(r^{-1})$ behaviors of the solutions of (3) to calculate the total energy flux. The electric and magnetic fields are related to each other by a relation $\boldsymbol{E} = \boldsymbol{B} \times \boldsymbol{n}$ for a plane wave propagating along a unit vector $\boldsymbol{n}$. Only the vector potential $\boldsymbol{A}$ in a wave zone (a spatial region in which the plane wave approximation is valid) is needed for our aim since $\boldsymbol{B} = \nabla \times \boldsymbol{A}$.

Let us consider a neutron star located at the origin $\boldsymbol{x} = \boldsymbol{0}$ whose radius and angular velocity are given by $R_{\rm NS}$ and $(0, 0, \omega_{\rm NS})$. The source current $\boldsymbol{J}$ is assumed to be localized around this neutron star. Note that $R_{\rm NS}\,\omega_{\rm NS}$ is much smaller than light speed $c = 1$.

The retarded Green's function $G_{\rm ret}(x, x') = \delta(t - t' - |\boldsymbol{x} - \boldsymbol{x}'|)/4\pi|\boldsymbol{x} - \boldsymbol{x}'|$ is used to express the solution of the first of (3) as

$$\boldsymbol{A}_{(0)}(t, \boldsymbol{x}) = \int_V \mathrm{d}^3x' \frac{\boldsymbol{J}(t - |\boldsymbol{x} - \boldsymbol{x}'|, x')}{4\pi|\boldsymbol{x} - \boldsymbol{x}'|}. \tag{4}$$

Assuming $r = |\boldsymbol{x}| \gg r' = |\boldsymbol{x}'|$, we obtain multipole expansion of the vector potential $\boldsymbol{A}_{(0)}(t, \boldsymbol{x}) \simeq \frac{1}{r}\left(\dot{\boldsymbol{d}}(t - r) + \ddot{\boldsymbol{D}}(t - r) + \dot{\boldsymbol{m}}(t - r) \times \hat{\boldsymbol{x}}\right)$ where $\dot{\boldsymbol{d}}(t) \equiv \int \mathrm{d}^3x' \boldsymbol{J}(t, \boldsymbol{x}')$, $\ddot{\boldsymbol{D}}(t) \equiv \int \mathrm{d}^3x' \nabla \cdot \dot{\boldsymbol{J}}(t, \boldsymbol{x}')\{(\boldsymbol{x}')^2\hat{\boldsymbol{x}} - 3(\hat{\boldsymbol{x}} \cdot \boldsymbol{x}')\boldsymbol{x}')\}$ are the electric dipole/quadrupole moment, and $\dot{\boldsymbol{m}}(t) \equiv \frac{1}{2}\int \mathrm{d}^3x' \boldsymbol{x}' \times \dot{\boldsymbol{J}}(t, \boldsymbol{x}')$ is the magnetic dipole moment. When the electric multipoles vanish, the magnetic dipole moment becomes the leading term. Then, total energy flux per unit time is given by $I = \frac{2}{3}|\ddot{\boldsymbol{m}}|^2$.

Let us move on to the next order. With a new source term $J^\mu_{(1)}$, (3) can be solved in the same way as $\boldsymbol{A}_{(0)}$. We therefore have to calculate the multipoles of

$\boldsymbol{J}_{(1)}$. Before going to the calculation, let us just note the conservation low $J^{\mu}_{(1),\mu} = 0$, and its consequence $A^{\mu}_{(1),\mu} = 0$. We, thus, do not have to care about the breaking of the gauge condition by the perturbation.

The new source term $J^{\mu}_{(1)}$ is written in terms of the zeroth order (induced by J^{μ}) physical fields $F^{\mu\nu}_{(0)}$ and $\tilde{F}^{\mu\nu}_{(0)}$. The field strength is approximately written as $F^{\mu\nu}_{(0)} \simeq F^{\mu\nu}_{\text{MD}} + F^{\mu\nu}_{\text{rad}}$, where $F^{\mu\nu}_{\text{MD}}$ represents the contribution of the rotating magnetic dipole itself which is rapidly decay as $|\boldsymbol{x}| \to \infty$ and $F^{\mu\nu}_{\text{rad}}$ denotes the contribution of the radiation which is tiny compared to $F^{\mu\nu}_{\text{MD}}$ but dominant for the distant region. The magnetic field $\boldsymbol{B}_{(0)}$ involved with $F^{\mu\nu}_{\text{MD}}$ is approximated by rotating a static dipole magnetic field $\tilde{\boldsymbol{B}}$ by a certain angle. The electric field induced by the rotation is given by $\boldsymbol{E}_{(0)}(x') \simeq \boldsymbol{v}' \times \boldsymbol{B}_{(0)}(x')$ (here $\boldsymbol{v}(x')$ is rotating velocity at point x').

We obtain $\mathcal{G}_{(0)} = 0$ since $\boldsymbol{E}_{(0)}$ and $\boldsymbol{B}_{(0)}$ are orthogonal. Note also that the behavior of the $\mathcal{F}_{(0)}$ in distant region ($|\boldsymbol{x}| \to \infty$) is dominated by the radiation part $F^{\mu\nu}_{\text{rad}}$. However, $\mathcal{F}_{(0)}$ vanishes in this limit since $\boldsymbol{E} = \boldsymbol{B} \times \boldsymbol{n}$ holds for the radiation. Thus, we can focus only on the contribution from $F^{\mu\nu}_{\text{MD}}$.

The static dipole magnetic field toward z-axis can be expressed as $\tilde{\boldsymbol{B}}(\boldsymbol{x}) = (A\cos\varphi, A\sin\varphi, B)$ here A, B are functions of ρ and z in cylindrical coordinates. The axial symmetry of the system yields (anti)symmetry of these functions such as

$$A(\rho, z) = -A(\rho, -z), \quad B(\rho, z) = +B(\rho, -z). \tag{5}$$

Rotating $\tilde{\boldsymbol{B}}(\boldsymbol{x})$ by α around y-axis (denoted by $R_y(\alpha)$) then by ωt around z-axis (denoted by $R_z(\omega t)$), we obtain $\boldsymbol{B}_{(0)}(t, \boldsymbol{x}) = \Lambda \tilde{\boldsymbol{B}}(\Lambda^{-1}\boldsymbol{x})$ ($\Lambda = R_z(\omega t) R_y(\alpha)$ is a rotation matrix for this transformation). The rotation velocity of the frame is given by $\boldsymbol{v} = \boldsymbol{\omega} \times \boldsymbol{r} = \omega\rho(-\sin\varphi, \cos\varphi, 0)$.

The electric dipole $\dot{\boldsymbol{d}}_{(1)} = \int \mathrm{d}^3x' \boldsymbol{J}_{(1)}$ is a candidate for the dominant contribution to $\boldsymbol{A}_{(1)}$, and evaluated as $\dot{d}^i_{(1)} = -\frac{16\alpha^2}{45m^4} \int \mathrm{d}^3x' (\mathcal{F}_{(0)} F^{i0}_{(0)})_{,0}$. Form the approximation $2\mathcal{F}_{(0)} \simeq |\boldsymbol{B}_{(0)}|^2 = (A^2 + B^2)$, we found $\dot{d}^i_{(1)} = 0$ after the integration. We can, thus, conclude that there is no quantum contribution to the electric dipole radiation and the leading contribution to $\boldsymbol{A}_{(1)}$ is the electric quadrupole $\ddot{\boldsymbol{D}}_{(1)}$ or the magnetic dipole $\dot{\boldsymbol{m}}_{(1)}$. It can be easily shown that the electric quadrupole is always suppressed by the appearance of the electric field $\boldsymbol{E}_{(0)}$ and the dominant contribution comes from $\mathcal{O}(|\boldsymbol{B}_{(0)}|^3)$ term in the magnetic dipole $\dot{\boldsymbol{m}}_{(1)}$.

After some vector analysis, we found that main contribution to the magnetic dipole moment is given by

$$\boldsymbol{m}_{(1)} = \frac{8\pi\alpha^2}{45m^4} \int_0^\infty \mathrm{d}\rho \int_{-\infty}^{\infty} \mathrm{d}z\, \rho \left(A^2(\rho, z) + B^2(\rho, z)\right) B(\rho, z) \begin{pmatrix} \sin\alpha\cos\omega t \\ \sin\alpha\sin\omega t \\ \cos\alpha \end{pmatrix}. \tag{6}$$

This has to be related to stellar quantities such as the neutron star radius R_{NS} and the surface magnetic field strength B_{NS}. The magnetic field of the neutron star B_{NS} is conventionally linked to the magnetic dipole moment of the star $|\boldsymbol{m}_{\text{NS}}| = B_{\text{NS}} R^3_{\text{NS}}/2$. $\boldsymbol{m}_{\text{NS}}$ is also connected to the normalization of the source current $\boldsymbol{J}$

which is assumed to be localized around the surface of the neutron star. For a ring current I of radius $R_{\rm NS}$, the magnetic moment is given by $|\boldsymbol{m}_{\rm NS}| = I\pi R_{\rm NS}^2$. We adopt this equation to determine the intensity I of the current $\boldsymbol{J}$. Note that, from (4), the functions $A(\rho, z)$ and $B(\rho, z)$ are proportional to the intensity I.

Because $A(\rho, z)$ and $B(\rho, z)$ have mass dimension two, they must be scaled as $A(\rho, z) = I/R_{\rm NS}\,\bar{A}(\rho/R_{\rm NS}, z/R_{\rm NS})$, $B(\rho, z) = I/R_{\rm NS}\,\bar{B}(\rho/R_{\rm NS}, z/R_{\rm NS})$ where $\bar{A}$ and $\bar{B}$ are dimensionless quantities. Therefore (6) can be written as

$$\boldsymbol{m}_{(1)} = \frac{8\pi\alpha^2}{45m^4} I^3 K \begin{pmatrix} \sin\alpha\cos\omega t \\ \sin\alpha\sin\omega t \\ \cos\alpha \end{pmatrix}, \quad K \equiv \int_0^\infty \mathrm{d}\bar{\rho} \int_{-\infty}^\infty \mathrm{d}\bar{z}\, \bar{\rho}\,(\bar{A}^2 + \bar{B}^2)\,\bar{B}, \quad (7)$$

where $\bar{\rho} = \rho/R_{\rm NS}$, $\bar{z} = z/R_{\rm NS}$. The dimensionless integral K is determined by the geometrical configuration of the source current hence can be called form factor.

The total magnetic dipole radiation is given by

$$\frac{2}{3}|\ddot{\boldsymbol{m}}_{\rm tot}|^2 \simeq \frac{2}{3}|\ddot{\boldsymbol{m}}_0|^2 \left\{1 - \frac{\alpha}{45\pi^3} K \left(\frac{B_{\rm NS}}{B_{\rm crit}}\right)^2\right\}, \quad (8)$$

where we used $|\ddot{\boldsymbol{m}}_0| = |\ddot{\boldsymbol{m}}_{\rm NS}|\sin\alpha$ since z component of the magnetic dipole does not contribute to the radiation. This is our main result. We found the correction term suppressed by the fine structure constant $\alpha = e^2/4\pi$ and critical magnetic field strength $B_{\rm crit} = e^2/m$ as expected. We also found the form factor K which carries the information of the stellar scale structure of the source current $\boldsymbol{J}$.

The suppression factor $\alpha/45\pi^3 = 5.2 \times 10^{-6}$ seems too small to cause significant change from the classical theory even for a near critical magnetic field $B_{\rm NS} \sim B_{\rm crit}$, but the value of the form factor might be as large as $K = -2 \times 10^4$ for a ring current with small cross section ($\sim (10^{-2}R_{\rm NS})^2$) which flows right on the equator of the star.

4. Strong Field Limit

In the previous section, the strength of the magnetic field is assumed to be subcritical. We cannot apply the obtained result to the critical or supercritical magnetic field. It seems difficult to solve the full order equation of motion, however, we can analytically investigate the theory in strong field limit $B \gg B_{\rm crit}$. The dominant term of the effective Lagrangian in the strong magnetic field limit[9] is $\mathcal{O}(B^2 \ln B)$ which describes the correction to the kinetic term. It can be regarded as the effect of running coupling constant. We have no correction other than the running of coupling constant. Therefore the classical theory is recovered in this limit.

One might think that the electric field induced by the rotating magnetic dipole $\boldsymbol{E} \sim \boldsymbol{v} \times \boldsymbol{B}$ can be supercritical even though the slow roll condition ($\omega R_{\rm NS} \ll 1$) is assumed, and be concerned about Schwinger effect or particle pair production from the strong electric field. This fear is groundless because $\mathcal{G} = -\boldsymbol{E} \cdot \boldsymbol{B} = 0$. We can use classical magnetic dipole formula for supercritical field without worrying about QED vacuum decay.

5. Conclusion

We have investigated quantum correction to the magnetic dipole radiation which can cause the spin down of pulsars. Euler-Heisenberg Lagrangian (1) and perturbation was used to obtain the correction term (8) directly. We found the suppression factor $\alpha/45\pi^3$, the form factor K depends on the configuration of the magnetic field or the source current generated by the star, and quadratic dependence $(B/B_{\rm crit})^2$ in the correction term. The numerical value of K was obtained for a simple configuration of the source current. It can compensate the suppression factor in case of the highly concentrated (localized) current configuration, but we expect that it cannot be much more significant than astronomical and observational uncertainties for reasonable situation. Should these astronomical and observational uncertainties (such as the angle between the rotational pole and magnetic pole) be removed well, the value of K would carry partial information of the pulsar structure. We also discussed the supercritical regime of QED theory. We confirmed that the classical theory would be recovered with the running coupling constant in strong field limit.

6. Acknowledgement

This work is supported by JSPS KAKENHI, Grant-in-Aid for JSPS Fellows, 15J09390 (T.H.), JSPS Grant-in-Aid for Young Scientists (B) No.15K17632 (T.S.) and No.15K17659, MEXT Grant-in-Aid for Scientific Research on Innovative Areas New Developments in Astrophysics Through Multi-Messenger Observations of Gravitational Wave Sources No.15H00777 (T.S.), JSPS KAKENHI No. 15H05888 (T.S. We would like to thank T. Shigeyama, T. Nakano and M. Masuyama for their great helps from astronomical points of view.

References

1. R. C. Duncan and C. Thompson, Formation of very strongly magnetized neutron stars - Implications for gamma-ray bursts, *apjl* **392**, L9 (June 1992).
2. S. Mereghetti, J. Pons and A. Melatos, Magnetars: Properties, Origin and Evolution, *Space Sci. Rev.* **191**, 315 (2015).
3. J. S. Heyl and L. Hernquist, Qed one-loop corrections to a macroscopic magnetic dipole, *Journal of Physics A: Mathematical and General* **30**, p. 6475 (1997).
4. M. G. Baring and A. K. Harding, Magnetic photon splitting: Computations of proper time rates and spectra, *Astrophys. J.* **482**, p. 372 (1997).
5. K. Kohri and S. Yamada, Polarization tensors in a strong magnetic field, *Phys. Rev.* **D65**, p. 043006 (2002).
6. M. Freytsis and S. E. Gralla, QED Plasma and Magnetars (2015).
7. W. Heisenberg and H. Euler, Folgerungen aus der Diracschen Theorie des Positrons, *Zeitschrift für Physik* **98**, 714 (1936).
8. J. Schwinger, On gauge invariance and vacuum polarization, *Physical Review* **82**, p. 664 (1951).
9. G. V. Dunne, *Heisenberg-Euler effective Lagrangians: Basics and extensions* 2004.

PARALLEL SESSION

VI. Observational/Experimental Gravity

Summary of Parallel Session: "Observational/Experimental Gravity"

Angela Di Virgilio
INFN-Sezione di Pisa, Largo B. Pontecorvo 3, 56124 Pisa, Italy, angela.divirgilio@pi.infn.it

Wei-Tou Ni
National Tsing Hua University, Hsinchu, Taiwan, 30013 ROC, weitou@gmail.com

Sperello di Serego Alighieri
INAF - Osservatorio Astrofisico di Arcetri, 50125 Firenze, Italy

Hung-Yi Pu
Institute of Astronomy & Astrophysics, Academia Sinica, Taipei, Taiwan, 10617 ROC

Sheau-Shi Pan
Center for Measurement Standards, ITRI, Hsinchu, Taiwan, 30011 ROC

We indicate the progress of experimental gravity, present an outlook in this field, and summarise the Observational/Experimental Parallel Session together with a related plenary talk on gravitational waves of the 2nd LeCosPA Symposium.

Keywords: Experimental gravity; observational gravity; equivalence principles; cosmic polarization rotation; CMB; gravitational waves; Lense-Thirring measurement; black hole observation.
PACS numbers: 04.80.Cc, 04.80.Nn, 04.80.-y, 95.55.Ym, 98.80.Es.

1. Progress and Outlook

For one hundred years since the advent of General Relativity (GR), the first thing in observational and experimental gravity is the observation of light deflection during the solar eclipse in 1919. Since then, the 3 classical tests of GR (perihelion advance, gravitational redshift and light deflection) have been verified[1] to $10^{-3} - 10^{-4}$. With the advent of space age in 1957 and the development of space radio communication, Shapiro proposed a fourth test (Shapiro time delay) of GR that electromagnetic wave packets passing near the Sun would be retarded due to gravity curving of spacetime.[2] Shapiro time delay is measured to agree with GR in terms of Eddington parameter γ (equal to 1 for GR) as 1.000021 ± 0.000023 from Cassini spacecraft Doppler tracking.[3] Lense-Thirring frame dragging has been measured to about 10 percent to agree with GR by satellite laser ranging (SLR) of LAGEOS 1 and LAGEOS 2, and by GP-B gyro relativity experiment.[4,5] GP-B experiment has also verified the equivalence principle for rotating body to an ultimate precision.[6]

At present, Lunar Laser Ranging (LLR) test and solar-system radio tests of GR have about the same accuracy. In the near future, interplanetary laser ranging and spacecraft laser ranging in the solar system will improve the accuracy of tests of relativistic gravity by another 3-4 orders of magnitude reaching the second post-Newtonian order.[1] The observation of precise timing of pulses from pulsars is now catching up the solar system observations and will also reach the second post-Newtonian order soon if not earlier.[7] The development of ring laser gyroscopes makes it possible to measure the absolute angular velocity and to estimate the rotation rates relative to the local inertial frame with ultra-precision. This is ideal for measuring the Lense-Thirring frame dragging on Earth. GINGER (Gyroscopes IN GEneral Relativity) is proposed to measure this frame dragging to 1 percent; GINGERino is under implementation (Sec. 2).[8] Technology based on the success of GINGER experiment could be applied to improve the tie between the astronomic reference frame and the solar-system dynamical frame significantly.

Einstein Equivalence Principle is an important cornerstone of GR and metric theories of gravity. From the nonbirefringence of cosmic propagation of electromagnetic wave packet, the constitutive tensor of spacetime for local linear gravity-coupling to electromagnetism must be of core metric form with an axion (pseudoscalar) degree of freedom and a dilaton (scalar) degree of freedom; from observations it is empirically verified to 10^{-38}, i.e. to $10^{-4} \times (M_{\mathrm{Higgs}}/M_{\mathrm{Planck}})^2$.[9] This is significant in constraining the infrared behaviour of quantum gravity. Empirically, axion is constrained by the non-observation of cosmic polarization rotation (Sec. 2); dilation is constrained by the agreement of CMB spectrum to Planck spectrum.[9,10]

Galileo equivalence principle is experimentally verified by Eötvös-type experiments and free-fall experiments[11,12] to about 10^{-13}. In the next mission/mission proposals for testing Galileo equivalence principle, the improvements are aimed at $2-5$ orders of magnitude.[13,14] In a series of papers in the 1970s, Rubin, Ford, and Thonnard measured the rotation curves of a number of disk galaxies and found that rotation speeds were larger than would be expected from the gravitational attraction arising from the visible mass distribution.[15–17] Authors interpreted their findings as evidence for a new dark matter component. Logically this conflict, known as the missing mass problem, could arise from a mass discrepancy, an acceleration discrepancy, or possibly even both. In 1983, Milgrom proposed the phenomenological modified Newtonian dynamics (MOND) law for small accelerations.[18] Under this hypothesis, the gravitational dynamics become modified when the acceleration is smaller than $a_0 \sim 10^{-10}$ m s^{-2}. However, great efforts in finding the missing mass have not been fruitful. Neither the construction of a viable relativistic gravitational theory incorporating the MOND law is fully successful. It remains an open question. It is also interesting to note that $a_0 \sim \Lambda^{1/2}$.[19]

On the cosmological scale, the discovery of cosmic acceleration have indicated the existence of cosmological constant or the cosmological-constant-like dark energy. The search for a microscopic theory behind this phenomenon may give clue to the microscopic origin of gravity.[19] The observation on cosmic structure has confirmed the inflationary scenario that it could have originated from quantum fluctuations in the inflationary period.[20,21] GWs from inflationary period could give imprints of tensor anisotropies and production of B-modes on CMB. CMB polarization observations have constrained tensor to scalar ratios r of inflationary/primordial GWs to be less than $0.07 - 0.1$.[22]

Most of the experimental gravitation community are working on GW experiments/observations on various frequency bands from aHz to THz.[23] Real-time detection is possible above 300 pHz, while below 300 pHz the detection is possible on GW imprints or indirectly. Advanced LIGO has achieved 3.5 fold better sensitivities with a reach to neutron star binary merging event at 70 Mpc and began its first observing run (O1) in the middle of September 2015 searching for GWs. On September 14 the first GW observation has taken place, and the merger of two Black Holes have been recorded.[24] We will see a global network of second generation km-size interferometers for GW detection soon. Another avenue for real-time direct detection is from the PTAs. The PTA bound on stochastic GW background already excludes most theoretical models; this may mean we could detect very low frequency GWs anytime too with a longer time scale. Although the prospect of a launch of space GW is only expected in about 20 years, the detection in the low frequency band may have the largest signal to noise ratios. This will enable the detailed study of black hole co-evolution with galaxies and with the dark energy issue. We will see improvement of a few orders to several orders of magnitude in the GW detection sensitivities over all frequency bands in the next hundred years.[25]

The gravitational deflection has already been applied to astrophysics and cosmology as gravitational lensing to weigh the lensing sources.[26] It becomes an important tool of astrophysics and cosmology. Besides GW observations, electromagnetic observations on black holes have been proposed (Sec. 2). It is well-known that satellite positioning systems need to incorporate GR corrections. With the clock reaching 10^{-18} and beyond, application to measure the Earth gravity and the altitude could be realized.

2. Cosmic Polarization Rotation, Lense-Thirring Frame Dragging and Black Hole Shadow Observation

2.1. *Summary on CPR*

A review about cosmic polarization rotation (CPR), i.e. a rotation of the polarization angle (PA) for radiation traveling over large distances across the universe, was presented by Sperello di Serego Alighieri.[27] CPR is very relevant for this Symposium, since it would be observed if there were the pseudoscalar field coupling to electromagnetism, which Ni[28] found as a unique counter-example to the conjecture that any consistent Lorentz-invariant theory of gravity obeying the weak equivalence principle (WEP) would also obey the Einstein equivalence principle (EEP). In fact, since general relativity (GR) is based on the EEP, our confidence on EEP and GR would be greatly increased if we could show that there is no CPR, because in this case the EEP would be tested to the same high accuracy of the WEP. The search for CPR is important also because it would tell us if and how one of the three elementary pieces of information (direction, energy, and PA), which photons carry to us about the universe, is changed while they travel. Since 1990 CPR has been searched using the polarization of radio galaxies, both in the radio and in the ultraviolet, and, more recently, using the polarization of the cosmic microwave background (CMB). The results of a recent review on CPR[29] and of a few updates were presented. In summary, the results so far are consistent with a null CPR with upper limits of the order of 1°. Two current problems in CPR searches were discussed. The first involves the PA calibration at CMB frequencies, which is becoming the limiting factor, imposing a systematic error of about 1°, which is larger than the statistical errors of the best CMB polarization experiments. Improvements are expected from more precise measurements of the polarization angle of celestial sources at CMB frequencies and a calibration source on a satellite.[30] The second problem results from the unfortunate choice of the CMB community for a PA convention which is opposite to the standard one, adopted by all astronomers for many decades and enforced by the International Astronomical Union.[31] This is causing obvious confusion and misuderstanding, particularly for CPR, for which results use both conventions. A recommendation has been issued that all astronomers, including CMB polarimetrists, use the standard PA convention.[32] Concerning CPR tests, improvements are expected by better targeted high resolution radio polarization measurements of radio galaxies and quasars, by more accurate ultraviolet polarization measurements of radio galaxies with the coming generation of giant optical telescopes and by future CMB polarization measurements. An update[33] of CPR constraint was presented from the analysis of the recent measurements of sub-degree B-mode polarization in the cosmic microwave background from 100 square degrees of SPTpol data.[34] The CPR fluctuation constraint from the joint ACTpol-BICEP2-POLARBEAR polarization data is 23.7 mrad (1.36°).[35] With the new SPTpol data included, the CPR fluctuation constraint is updated to 17 mrad (1°) with the scalar to tensor ratio $r = 0.05 \pm 0.1$.[33]

2.2. *The GINGER Project and the Lense-Thirring Measurement*

GINGER (Gyroscopes IN General Relativity)[36] is based on an array of ring-lasers and is aiming at measuring the Lense-Thirring effect at the level of 1%. Large frame ring-lasers are at present the most sensitive device to measure absolute angular rotation, and has been already demonstrated that they have a sensitivity very close to what is necessary to measure the Lense-Thirring effect.[37] At present it is under discussion the real construction of GINGER. At the same time, experimental activity is in progress; the large frame prototype GINGERino[8] is investigating if the GranSasso underground laboratory is adeguate for an experiment as GINGER, and the prototype called GP2 has been installed in Pisa in order to develop a suitable control strategy to constrain the geometry of the ring-laser and guarantee the long term stability of the scale factor of the ring-laser. The long term stability of the scale factor of the ring-laser is the most challenging experimental problem of the research activity around GINGER.

2.3. *The Greenland Telescope Project and BH Shadow Observation*

The size and shape of the shadow cast by a black hole event horizon is determined by the null geodesics and directly related to the background spacetime metric. Direct imaging of such shadow image and the test of physics in strong gravity is one of the important goals in modern astronomy. The ongoing Greenland Telescope (GLT)[38] project in Academia Sinica Institute of Astronomy and Astrophysics (ASIAA) is devoting itself to this exciting area.[39] The target source of GLT project is the supermassive black hole located at the center of M87. The baseline of the future telescope in Greenland (hence the name Greenland Telescope), the Submillimeter Array in Hawaii, and the Atacama Large Millimeter/submillimeter Array in Chile, will play a key role for future Very Long Baseline Interferometry (VLBI) observation. With longest baseline > 9000 km, the angular resolution can reach $\sim 20\ \mu$as at 350 GHz, high enough to resolve the black hole shadow of M87, which has a estimated angular size $\sim 40\ \mu$as. The first light and related VLBI test of GLT in Thule, a northwest cost of Greenland, will be obtained in 2016. The GLT will then be established at the Summit station in 2018/2019.[40] The GLT project is a collaborative project between ASIAA, Smithsonian Astrophysical Observatory, MIT Haystack Observatory, and National Radio Astronomy Observatory .

References

1. See, e.g., W.-T. Ni, *Int. J. Mod. Phys. D* **25**, 1630003 (2016).

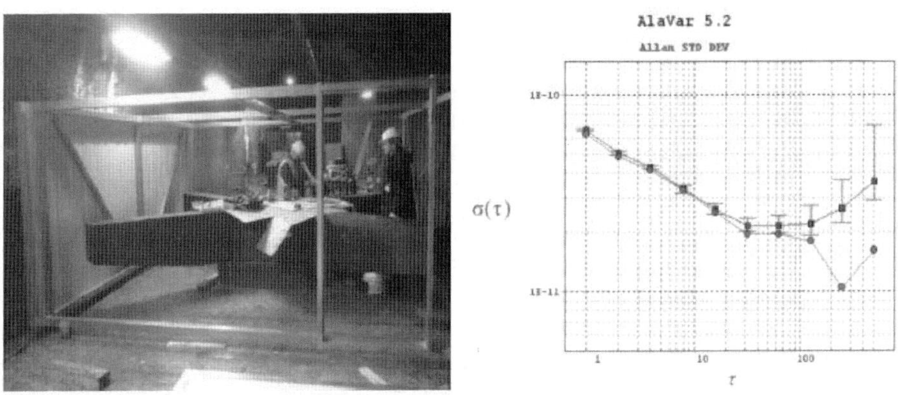

Fig. 1. Left: The construction of GINGERino inside le underground GranSasso laboratory. Right: The preliminary Allan deviation, expressed in rad/s.

2. I. Shapiro, *Phys. Rev. Lett.* **13** 789 (1964).
3. B. Bertotti, L. Iess, and P. Tortora, *Nature* **425**, 374-376 (2003).
4. I. Ciufonlini, and E. C. Pavlis, *Nature* **431**, 958-960 (2004).
5. C. W. F. Everitt, et al., *Phys. Rev. Lett.* **106**, 221101 (2011).
6. W.-T. Ni, *Phys. Rev. Lett.* **106**, 221101 (2011).
7. See, e.g., R. N. Manchester, *Int. J. Mod. Phys. D* **24**, 1530018 (2015).
8. J. Belfi, et al., 'First Results of GINGERino', arXiv:1601.02874.
9. W.-T. Ni, *Phys. Lett. A* **379**, 1297 (2015); and references therein.
10. W.-T. Ni, *Phys. Lett. A* **378**, 3413 (2014).
11. J.G. Williams, S. G. Turyshev, and D. H. Boggs, *Int. J. Mod. Phys. D* **18**, 1129 (2009).
12. S. Schlamminger, et al., *Phys. Rev. Lett.* **100**, 041101 (2008); and references therein.
13. http://smsc.cnes.fr/MICROSCOPE/index.htm.
14. http://www.sstd.rl.ac.uk/fundphys/step/; http://einstein.stanford.edu/.
15. V. C. Rubin, W. K. Ford, Jr , *Astrophys. J.* **159**, 379, (1970).
16. V. C. Rubin, N. Thonnard, and W. K. Ford, Jr., *Astrophys. J.* **225**, L107 (1978).
17. V. C. Rubin, N. Thonnard, and W. K. Ford, Jr, *Astrophys. J.* **238**, 471 (1980).
18. M. Milgrom, *Astrophys. J.* **270**, 365 (1983).
19. See, e.g., M. Bucher and W.-T. Ni, *Int. J. Mod. Phys. D* **24**, 1530030 (2015).
20. See, e.g., M. Davis, *Int. J. Mod. Phys. D* **23**, 1430021 (2014).
21. See, e.g., K. Sato and J. Yokoyama, *Int. J. Mod. Phys. D* **24**, 1530025 (2015).
22. BICEP2/Keck and Planck Collab., *Phys. Rev. Lett.* **114**, 101301 (2015); C.-L. Kuo, this meeting.
23. W.-T. Ni, 'Sensitivities of gravitational-wave detection', *these proceedings.*
24. LIGO and Virgo Collaboration, *Phys. Rev. Lett.* **116**, 6, 061102, 2016
25. See, e.g., W.-T. Ni, *Int. J. Mod. Phys. D* **24**, 1530031 (2015).
26. See, e.g., T. Futamase, *Int. J. Mod. Phys. D* **24**, 1530011 (2015).
27. S. di Serego Alighieri, 'Gaining confidence on General Relativity with Cosmic Polarization Rotation', *these proceedings.*
28. Ni, W.-T., *Phys. Rev. Lett.* **38**, 301 (1977).
29. di Serego Alighieri, S., *Int. J. Mod. Phys. D* **24**, 1530016 (2015).
30. Kaufman, J.P., Keating, B.G., and Johnson, B.R., *MNRAS* **455**, 1981 (2016).
31. IAU Commission 40, *Polarization Definitions*, *Transactions of the IAU*, **Vol. XVB**, p. 166 (1974).
32. IAU Recom., http://www.iau.org/news/announcements/detail/ann16004 (2015).
33. W.-P. Pan et al., 'New constraints on cosmic polarization rotation including SPTpol B-mode polarization observations', *these proceedings.*
34. R. Keisler et al., *Astrophys. J.* **807**, 151 (2015).
35. H.-H. Mei et al., *Astrophys. J.* **805**, 107 (2015).
36. A. Di Virgilio, 'GINGER, An array of Ring Lasers to test General Relativity', *these proceedings*
37. F. Bosi et al., *Phys. Rev. D* **84**, 122002 (2011).
38. H.-Y. Pu, 'Observing the Black Hole Shadow of M87 and the Greenland Telescope Project', *these proceedings.*
39. ASIAA website for the GLT project: url:http://vlbi.asiaa.sinica.edu.tw/project.php
40. Inoue M., et al., *Radio Science* **49**, 564 (2014).

Gaining Confidence on General Relativity with Cosmic Polarization Rotation

Sperello di Serego Alighieri

INAF - Osservatorio Astrofisico di Arcetri
Largo E. Fermi 5, 50125 Firenze, Italy
sperello@arcetri.astro.it
www.arcetri.astro.it/~sperello

The theory of general relativity, for which we celebrate the centennial at this Symposium, is based on the Einstein equivalence principle. This principle could be violated through a pseudoscalar-photon interaction, which would also produce a rotation of the polarization angle for radiation traveling over very long distances. Therefore, if we could show that this cosmic polarization rotation does not exist, our confindence in general relativity would be greatly increased. We review here the astrophysical searches for cosmic polarization rotation, which have been made in the past 26 years using the polarization of radio galaxies and of the cosmic microwave background. So far no rotation has been detected within about 1 degree. We discuss current problems and future prospects for cosmic polarization rotation measurements.

Keywords: General relativity; equivalence principles; polarization; radio galaxies; cosmic microwave background.

1. The importance of cosmic polarization rotation

The possibility that space-time is curved by a gravitational field, as postulated by the theory of general relativity (GR), rests on the extension of the equivalence between an accelerated frame and a gravitational field from experiments involving just moving bodies (the weak or Galilean equivalence principle, WEP[17]) to those concerning all non-gravitational forces, including electromagnetic fields and photons (the Einstein equivalence principle, EEP[16]). In fact, the trajectory of a photon in an accelerated frame is curved because of the finite speed of light, and the same should equivalently occur in a gravitational field. However, this extension from the WEP to the EEP would not be justified if there were a pseudoscalar field ϕ coupling to electromagnetism and leading to a violation of the EEP, while obeying the WEP.[33,34] Such pseudoscalar field, if it existed, would also produce a rotation of the polarization angle (PA) for radiation traveling over large distances across the universe,[10] the so called cosmic polarization rotation (CPR). If we could show that there is no CPR, the EEP would be tested with the same high accuracy of the WEP, greatly increasing our confidence in the EEP and consequently in GR. In fact CPR could also be induced by violations of other fundamental physical principles (see Ref. 36 for a recent review), connected with the breaking of symmetry and parity: clearly, if there were any CPR, i.e. if the CPR angle α were not zero, it should be either positive for a counter-clockwise rotation, or negative for a clockwise rotation,[a] leading to asymmetry.

Searches for CPR are important also to make the best use of the information about the universe which photons carry to us. Almost all the information we have about the universe outside the Solar System is carried to us by photons; a few cosmic rays, several elusive neutrinos, and the just discovered gravitational waves are the only exceptions. Photons carry information about their direction, their energy (or wavelength/frequency), and their polarization. The latter consists essentially in the position angle (PA) of the polarization ellipse, i.e. photons carry throughout the universe an important geometrical information, although our eyes cannot see it. In order to make the best use of this information, it is important to know if and how it is changed while photons travel to us. We know that the direction of photons can be modified by a strong gravitational field, and that their energy is modified by the expansion of the universe. Is the polarization PA also modified while photons travel large distances in vacuum?[b] The searches for CPR deal with this important question.

In the following I will briefly review the tests on CPR carried out with different methods in the past 26 years, and discuss current problems and future prospects for these tests. A more extensive recent review of CPR tests can be found in Ref. 15.

2. Searches for cosmic polarization rotation

CPR tests are simple in principle: they require a distant source of polarized radiation for which the polarization orientation PA_{em} at the emission or at the last interaction can be established. By measuring the observed orientation PA_{obs}, the CPR angle can be calculated:

$$\alpha = PA_{obs} - PA_{em}.$$

The problem is the estimate of PA_{em}, since it is not generally possible to know the geometry of the emission process. Fortunately this problem can be solved when the last interaction was due to scattering and when the scattering geometry can be figured out. In this case we can use the fact that scattered radiation is polarized perpendicularly to the plane containing the incident and scattered rays. This simple physical law has been applied to CPR tests, using both the ultraviolet (UV) radiation of radio galaxies (RG) and the tiny disuniformities of the cosmic microwave background (CMB). The first CPR tests 26 years ago used instead a stastical analysis of the radio polarization in RG.[9] The most accurate CPR tests

[a]We adopt the convention for PA enforced by the International Astronomical Union (IAU): it increases counter-clockwise facing the source, from North through East.[22]

[b]We know that the polarization PA is modified while photons travel in a plasma with a magnetic field, the so called Faraday rotation, which is proportional to the wavelength squared. However we deal here only with possible modifications in vacuum.

obtained with the various methods are summarized in Fig. 1, which is derived from the data of Ref. 15.

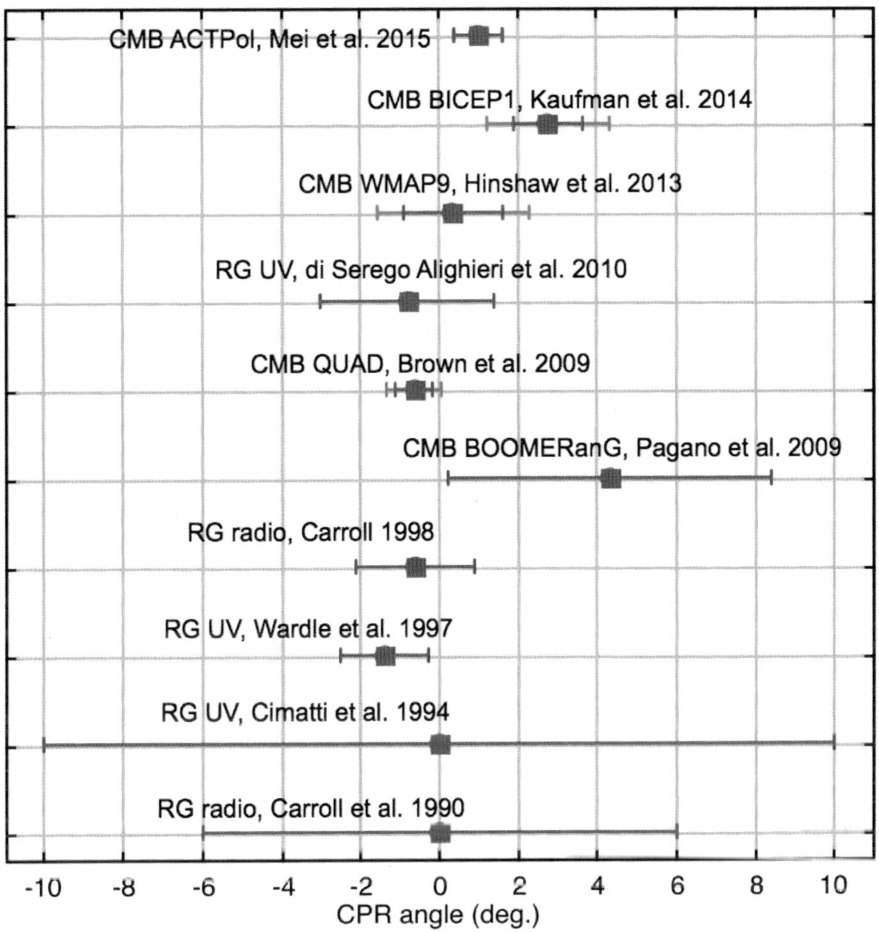

Fig. 1. CPR angle measurements by the various experiments, displayed in chronological order. Blue error bars are for the statistical error, while red ones include also the systematic one, if present. A systematic error should be added to the ATCPol measurement, equal to the unknown difference of the Crab Nebula polarization PA between 146 GHz and 89 GHz.[32]

Since this review[15] some new measurements connected with CPR have become available. The POLARBEAR collaboration[2] reports about a difference of 1.08° in the instrument polarization angle obtained minimizing the EB spectrum and that obtained from the Crab Nebula using the PA measurement of Ref. 6. This corresponds to a measurement of CPR, performed with the effect of a rotation on the EB spectrum and using the Crab Nebula for the PA calibration, and giving a CPR angle $\alpha = 1.08° \pm 0.2°(stat.) \pm 0.5°(syst.)$. The BICEP2 collaboration[3] finds that TB and EB spectra at high monopoles are consistent with a coherent rotation of the polarization angle by 1.1° with a random error of 0.2°. However this is not necessarily a measurement of the CPR angle α, since this TB and EB minimization procedure[27] cannot disentangle α from a PA systematic error due to calibration uncertainties (see the next section). Recently the first results about CPR from the Planck satellite have become available: a CPR angle $\alpha = 0.0° \pm 1.3° (stat.) \pm 1° (syst.)$ has been obtained from an analysis of just TE and EE spectra.[20] However this is not the final word on CPR from Planck, since, when also the BB, TB, and EB spectra will be included in the analysis, it is foreseen that the statistical uncertainty on the CPR angle should decrease to about 0.03°,[20] and hopefully it will also be possible to decrease the systematic errors with an improved PA calibration.

The CPR tests considered above assume that the rotation angle α is uniform across the sky. However it is also possible that α depends on the direction one is looking at[35] and that one can therefore measure its fluctuations $\langle\delta\alpha^2\rangle^{1/2}$. The CPR data obtained on 8 RG[13] have been used to constrain $\langle\delta\alpha^2\rangle^{1/2} \leq 3.7°$.[24] A better constraint $\langle\delta\alpha^2\rangle^{1/2} \lesssim 1°$ has been derived from the WMAP7 data.[18] B-mode CMB polarization data have been used to set a constraint $\langle\delta\alpha^2\rangle^{1/2} \leq 1.56°$.[14]

Attempts have been made to derive constraints on CPR by combining data from various CMB experiments. For example, a CPR angle of 2.12 ±1.14° has been obtained[30] combining the data of BOOMERanG,[37] WMAP9,[21] and BICEP1.[25] However we find it problematic to combine data taken at different frequencies, in different areas of the sky, and particularly in the presence of large and different systematic errors due to uncertainties in the PA calibration.

In summary, the results so far are consistent with a null CPR with upper limits of the order of one degree.

3. Current problems in searches for cosmic polarization rotation

Searches for CPR using the UV polarization of RG have reached the limits allowed by current instrumentation, for the lack of suitable RG, for which the test can be performed and which are bright enough so that their polarization can be measured with the available instruments.

The most accurate results are now obtained with the CMB polarization, averaging over large sky areas, i.e. assuming uniform CPR over these areas. A current problem with CPR searches using the CMB is the calibration of the polarization PA for the lack of sources with precisely known PA at CMB frequencies. This introduces a systematic error, which is similar to the statistical measurement error, of the order of 1 degree (see the red error bars in Fig. 1). Recently the polarization PA of the Crab Nebula (Tau A) for a beam of $5'$ and $10'$ has been measured with an accuracy of 0.2° at 89.2 GHz.[6] However most CMB polarization measurements are made at higher frequencies (100–150 GHz) and the Crab Nebula is not visible from the South Pole, the site of several CMB experiments. In order to overcome the PA calibration problem, some CMB polarization experiments have used a TB and EB nulling procedure.[27] However this procedure would eliminate together the PA systematic error and any CPR angle α, so it cannot be used for CPR tests.

A second problem with CMB tests of CPR is that unfortunately the CMB polarimetrists have adopted the convention that the polarization PA increases clockwise (looking at the source), which is opposite to the standard convention adopted by all other polarimetrists for centuries and enforced by the IAU (PA increases counter-clockwise).[22] This corresponds to a change of the sign of the U Stokes parameter and is obviously producing problems when comparing measurements with different methods, like for CPR tests, also because the CMB convention has not been well documented in the CMB polarization papers. The "CMB convention" has been unfortunately adopted following a software for pixelization on a sphere.[19] However a pixelization software using the IAU convention exists[41] and changing the sign of the U Stokes parameter does not require any sophisticated software and can be done easily.[4] Recently the IAU has issued a recommendation that all astronomers, including those working on the CMB, follow the IAU convention for the PA.[23] It would clearly be desirable that we all use the same convention and avoid parochialism in Science.

4. Summary and future prospects

All results are so far consistent with a null CPR. The CPR test methods have reached so an accuracy of the order of 1° and upper limits to any rotation of the same order.

The different methods are complementary in many ways. They cover different wavelength ranges and the methods at shorter wavelength have an advantage, if CPR effects grow with photon energy, as foreseen in some cases.[28,29] They also reach different distances, and the CMB method obviously reaches furthest. However the relative difference in light travel time between z = 3 and z = 1100 is only 16%.

Improvements can be expected by better targeted high resolution radio polarization measurements of RGs and quasars, by more accurate UV polarization measurements of RGs with the coming generation of giant optical telescopes,[7,12,38] and by future CMB polarization measurements such as those from BICEP3.[5] Also the final results by the Planck satellite[1] are foreseen to bring improvements for CPR measurements because of the small expected statistical error (see Section 2). In any case, in order to exploit its great accuracy, Planck will have to reduce accordingly also the systematic error in the calibration of the polarization angle, which at the moment is of the order of 1° for the CMB polarization experiments. The best prospects to achieve this improvement are likely to be more precise measurements of the polarization angle of celestial sources at CMB frequencies with ATCA[31] and with ALMA,[39] and a calibration source on a satellite.[26]

References

1. Ade, P.A.R. et al. (Planck Collaboration), *A&A* **571**, A16 (2014).

2. Ade, P.A.R. et al. (The POLARBEAR Collaboration), *ApJ* **794**, 171 (2014).
3. Ade, P.A.R. et al. (2015, BICEP2 Collaboration), *ApJ* **814**, 110 (2015).
4. Ade, P.A.R. et al. (Planck Collaboration), *A&A* **586**, A141 (2016).
5. Ahmed, Z., Amiri, M., Benton, S.J., et al., *Proc. SPIE*, Vol. 9153, 1 (2014).
6. Aumont, J., Conversi, L., Thum, C., et al., *A&A* **514**, A70 (2010).
7. Bernstein, R.A., McCarthy, P.J., Raybould, K., et al., *Proc. SPIE*, Vol. 9145, 91451C (2014).
8. Brown, M.L., Ade, P., Bock, J., et al., *ApJ* **705**, 978 (2009).
9. Carroll, S. M., Field, G. B., and Jackiw, R., *Phys. Rev. D* **41**, 1231 (1990).
10. Carroll, S. M., *Phys. Rev. Lett.* **81**, 3067 (1998).
11. Cimatti, A., di Serego Alighieri, S., Field, G. B., and Fosbury, R. A. E., *ApJ* **422**, 562 (1994).
12. de Zeeuw, T., Tamai, R., and Liske, J., *The Messenger* **158**, 3 (2014).
13. di Serego Alighieri, S., Finelli, F., and Galaverni, M., *ApJ* **715**, 33 (2010).
14. di Serego Alighieri, S., Ni, W.-T., and Pan, W.-P., *ApJ* **792**, 35 (2014).
15. di Serego Alighieri, S., *IJMPD* **24**, 1530016 (2015).
16. Einstein, A., *Jahrb. Radioact. Elect.* **4**, 411 (1907).
17. Galilei, G., *Discorsi e dimostrazioni matematiche, intorno a due nuove scienze* (Elsevirii, Leiden, 1638).
18. Gluscevic, V., Hanson, V., Kamionkowski, M., Hirata, C.M., *Phys. Rev. D* **86**, 103529 (2012).
19. Gorski, K.M., Hivon, E., Banday, A.J., et al., *ApJ* **622**, 759 (2005).
20. Gruppuso, A., Gerbino, M., Natoli, P., et al., [`arXiv:1509.04157`] (2015).
21. Hinshaw, G., Larson, D., Komatsu, E., et al., *ApJS* **208**, 19 (2013).
22. IAU Commission 40, *Polarization Definitions, Transactions of the IAU*, Vol. XVB, p. 166 (1974).
23. IAU Recommendation, Paris, December 8th, 2015, `http://www.iau.org/news/announcements/detail/ann16004/` (2015).
24. Kamionkowski, M., *Phys. Rev. D* **82**, 047302 (2010).
25. Kaufman, J.P., Miller, N. J., Shimon, M., et al. (BICEP1 Collaboration), *Phys. Rev. D* **89**, 062006 (2014).
26. Kaufman, J.P., Keating, B.G., and Johnson, B.R., *MNRAS* **455**, 1981 (2016).
27. Keating, B.G., Shimon, M., and Yadav, A.P.S., *ApJL* **762**, L23 (2013).
28. Kostelecký, V.A. and Mewes, M., *Phys. Rev. Lett.* **87**, 251304 (2001).
29. Kostelecký, V.A. and Mewes, M., *Phys. Rev. D* **66**, 056005 (2002).
30. Li, S.-Y., Xia, J.-Q., Li, M., Li, H., Zhang, X., *ApJ* **799**, 211 (2015).
31. Massardi, M., Burke-Spolaor, S.G., Murphy, T., et al., *MNRAS* **436**, 2915 (2013).
32. Mei, H.-H., Ni, W.-T., Pan, W.-P., Xu, L., and di Serego Alighieri, S., *ApJ* **805**, 107 (2015).
33. Ni, W.-T., Montana State Univ., `http://astrod.wikispaces.com/` (1973).
34. Ni, W.-T., *Phys. Rev. Lett.* **38**, 301 (1977).
35. Ni, W.-T., *Chin. Phys. Lett.* **22**, 33 (2005).
36. Ni, W.-T., *Rep. Prog. Phys* **73**, 056901 (2010).
37. Pagano, L., de Bernardis, P., de Troia, G., et al., *Phys. Rev. D* **80**, 043522 (2009).
38. Sanders, G.H., *Jour. Astrophys and Astron.* **34**, 81 (2013).
39. Testi, L. and Walsh, J., *The Messenger* **152**, 2 (2013).
40. Wardle, J. F. C., Perley, R. A., and Cohen, M. H., *Phys. Rev. Lett.* **79**, 1801 (1997).
41. Wallis, C.G.R., Bonaldi, A., Brown, M.L., and Battye, R.A., *MNRAS* **453**, 2058 (2015).

New Constraints on Cosmic Polarization Rotation Including SPTpol B-Mode Polarization Observations

W.-P. Pan

Department of of Physics, National Tsing Hua University,
No. 101, Section 2, Kuang-Fu Road, Hsinchu, Taiwan 30013, ROC
two.joker@gmail.com

S. di Serego Alighieri

Osservatorio Astrofisico di Arcetri, Istituto Nazionale di Astrofisica,
Largo Enrico Fermi 5, I-50125 Firenze. Italia
sperello@arcetri.astro.it

W.-T. Ni

Department of of Physics, National Tsing Hua University,
No. 101, Section 2, Kuang-Fu Road, Hsinchu, Taiwan 30013, ROC
weitou@gmail.com

L. Xu

School of Physics and Optoelectronic Engineering, Dalian University of Technology,
No. 2 Linggong Road, Ganjingzi District, Dalian City, Liaoning Province, China, 116024
lxxu@dlut.edu.cn

We present an update of the cosmic polarization rotation (CPR) constraint from the recent SPTpol measurements of sub-degree B-mode polarization in the cosmic microwave background (CMB) of 100 square degrees of sky. Our previous CPR fluctuation constraint from the joint ACTpol-BICEP2-POLARBEAR polarization data is 23.7 mrad (1.36°). With new SPTpol data included, the CPR fluctuation constraint is updated to 17 mrad (1°) with the scalar to tensor ratio $r = -0.05 \pm 0.1$.

Keywords: Experimental gravity; Observational gravity; Equivalence principles; Cosmic polarization rotation (CPR); CMB; Scalar-to-tensor ratio; Gravitational waves.

PACS Numbers: 04.80.Cc, 04.80.−y, 98.80.Es

1. Introduction

A cornerstone of General Relativity (GR) and metric theories of gravity is the Einstein Equivalence Principle[1] (EEP). EEP is an important tool in the studies of new realms of physics such as early Universe, Black Hole Physics and Quantum Gravity. The precision of its empirical validity is crucial in these studies and rests on the possibility of extending also to the EEP the tests of the weak or Galilean equivalence principle[2] (WEP). However, this extension from the WEP to the EEP would not be fully justified if there were a pseudoscalar field coupling to electromagnetism and leading to a violation of the EEP, while obeying the WEP.[3,4] Such pseudoscalar field, if it existed, would also produce a rotation of the polarization angle (PA) independent of frequency for radiation traveling over large distances across the universe,[3,5] the so called cosmic polarization rotation (CPR).

In the premetric formulation of continuous electrodynamics or spacetime electrodynamics,[6] the field strength F_{kl} ($\boldsymbol{E}$, $\boldsymbol{B}$) and excitation (density with weight +1) H^{ij} ($\boldsymbol{D}$, $\boldsymbol{H}$) are related by the constitutive tensor density

$$H^{ij} = \chi^{ijkl}(F_{kl}). \tag{1}$$

In general $\chi^{ijkl}(F_{kl})$ is a functional of F_{kl}. For local linear medium or for energy much less than Planck energy, (1) is reduced to the following multi-linear (tensor) relation:

$$H^{ij} = \chi^{ijkl}\, F_{kl}. \tag{2}$$

The Lagrangian for electromagnetic field is

$$L_I^{(EM)} = -\,(1/(16\pi))H^{ij}\, F_{ij} = -\,(1/(16\pi))\chi^{ijkl}\, F_{ij}\, F_{kl}. \tag{3}$$

χ^{ijkl} has 21 degrees of freedom with the following permutation symmetry

$$\chi^{ijkl} = -\,\chi^{jikl} = -\,\chi^{ijlk}. \tag{4}$$

In the χ-framework (3), from the nonbirefringence of cosmic propagation of electromagnetic wave packet, the constitutive tensor of spacetime for local linear gravity-coupling to electromagnetism must be of core metric (h^{ik}) form with an axion (pseudoscalar) φ degree of freedom and a dilaton (scalar) ψ degree of freedom:[7]

$$\chi^{ijkl} = ½\,(-h)^{1/2}[h^{ik}\, h^{jl} - h^{il}\, h^{kj}]\psi + \varphi e^{ijkl}. \tag{5}$$

From observations, the nonbirefringence is empirically verified to 10^{-38}, i.e. to $10^{-4} \times (M_{Higgs}/M_{Planck})^2$.[8] This is significant in constraining the infrared behavior of quantum gravity. Empirically, axion is constrained by the non-observation of cosmic polarization rotation (Sec. 2);[8,9,10] dilation is constrained by the agreement of CMB (Cosmic Microwave Background) spectrum to Planck spectrum.[8,11]

If empirically there was no CPR to high accuracy, the EEP would be verified to high accuracy in this aspect. This would greatly increase our confidence in the EEP and consequently in GR. In fact, CPR could be induced also by violations of other fundamental physical principles (see Ref. [12] for a review), connected with the breaking of symmetry and parity: clearly, if there were any CPR, i.e. if the CPR angle were not zero, it should be either positive for a counter-clockwise rotation, or negative for a clockwise rotation, leading to asymmetry. A number of other theoretical models lead to the existence of this frequency-independent CPR also. They all have the same or equivalent effective Lagrangian and are not easy to be distinguished solely from astrophysical and cosmological polarization observation although different models predict different CPR angles. More precise and comprehensive observations are needed to distinguish models if CPR is detected. Feng *et al.*[13] proposed CPT violation and dynamical dark energy models. Liu *et al.*[14] gave constraints on the coupling between the quintessence and the pseudoscalar. Geng *et al.*[15] proposed a new type of effective interaction in terms of the CPT-even dimension-six term equivalent to what we are discussing to generate CPR, and used the neutrino number asymmetry to induce a nonzero polarization rotation angle in the data of the CMB polarization. They found that in their model, the rotation effect can be of the order of magnitude of 10–100 mrad or smaller. In our original pseudoscalar model, the natural coupling strength φ is of order 1. However, the isotropy of our observable universe to 10^{-5} may lead to a change of $\Delta\varphi$ over cosmological distance scale 10^{-5} smaller. Hence, observations to test and measure $\Delta\varphi$ to 10^{-6} will be significant. A positive result may indicate that our patch of inflationary universe has a 'spontaneous polarization' in the fundamental law of electromagnetic propagation influenced by neighboring patches and by a determination of this fundamental physical law we could 'observe' our neighboring patches.

2. New CPR constraints including SPTpol B-Mode polarization observations

Within the last scattering region, three processes can produce B-mode polarization or convert E-mode polarization to B-mode polarization in CMB: (i) local quadrupole anisotropies in the CMB due to large scale gravitational waves (GWs);[16] (ii) primordial magnetic field[17] and (iii) CPR due to pseudoscalar-photon interaction. During propagation, three processes can convert E-mode polarization into B-mode polarization: (i)' gravitational lensing,[18] (ii)' Faraday rotation due to magnetic field (including galactic magnetic field); and (iii)' CPR due to pseudoscalar-photon interaction. The cause of both (i) and (i)' is gravitational deflection; the cause of both (ii) and (ii)' is magnetic field. CPR is independent of frequency while Faraday rotation is dependent on frequency. Therefore Faraday rotation can be corrected for, using observations at different frequencies. However, it is negligible at CMB frequencies and corrections do not need to be applied. The deviation (if there is any) of the speed of gravity propagation from the light speed would change the acoustic peak position of CMB B-mode. From the modeling of observed acoustic peak position of CMB B-mode, this deviation could be measured.[19,20] However, the fitting for velocity of gravity propagation is basically orthogonal to the fitting of other parameters (depends more on the amplitude instead of the acoustic peak position) we are going to consider here.[20] In this paper, we shall assume the speed of gravity propagation is the same as that of light propagation. As to the foreground, the Galactic dust B-mode emission needs to be subtracted in the CMB B-mode polarization measurements (e.g., Ref. 21). At present, there are two quantities that can be measured for CPR α – the mean value $|<\alpha>|$ and the fluctuation amplitude $<(\alpha - <\alpha>)^2>^{1/2}$ in a patch of sky or over all sky.

CPR is currently constrained to be less than about a couple of degrees from the UV polarization observations of radio galaxies and the CMB polarization observations -- 0.02 rad for CPR mean value $|<\alpha>|$ and 0.03 rad for the CPR fluctuations $<(\alpha - <\alpha>)^2>^{1/2}$.[8-10] In 2014, we started to look for the CPR fluctuations from the CMB B-mode observations.[22] In our second study,[23] we use the new Planck dust measurement[21] to update our fits of the BICEP2 CPR fluctuation constraint to be 32.8 mrad (1.88°), and the joint ACTpol-BICEP2-POLARBEAR CPR fluctuation constraint to be 23.7 mrad (1.36°).

Recently, Keisler *et al.*[24] have announced a new measurement of the B-mode polarization power spectrum (the BB spectrum) from 100 deg^2 of sky observed with SPTpol (a polarization-sensitive receiver installed on the South Pole Telescope) during 2012 and early 2013 including data in spectral bands centered at 95 and 150 GHz. They reported the BB spectrum in five bins in multipole space, spanning the range $300 \leqslant \ell \leqslant 2300$, and for three spectral combinations: 95 GHz × 95 GHz, 95 GHz × 150 GHz, and 150 GHz × 150 GHz. The power spectra are consistent with predictions for the BB spectrum arising from the gravitational lensing of E-mode polarization. The preference for lensed B modes is of 4.9 σ after marginalizing over tensor power and foregrounds.

BB power spectrum from SPTpol,[24] ACTpol,[25] BICEP2/Keck,[26] and POLARBEAR[27] are drawn on Fig. 1 (reproduced from Figure 4 of Ref. 24). The solid gray line shows the expected lensed BB spectrum from the Planck+lensing+WP+highL best-fit model.[28] The dotted line shows the nominal 150 GHz BB power spectrum of Galactic dust emission derived from an analysis of polarized dust emission in the BICEP2/Keck field using Planck data.[26] The dash-dotted line shows the sum of the lensed BB power and dust BB power. From Ref. 21, the contribution of dust emission to the BB power spectrum of SPTpol, ACTpol and POLARBEAR could be neglected at this stage.

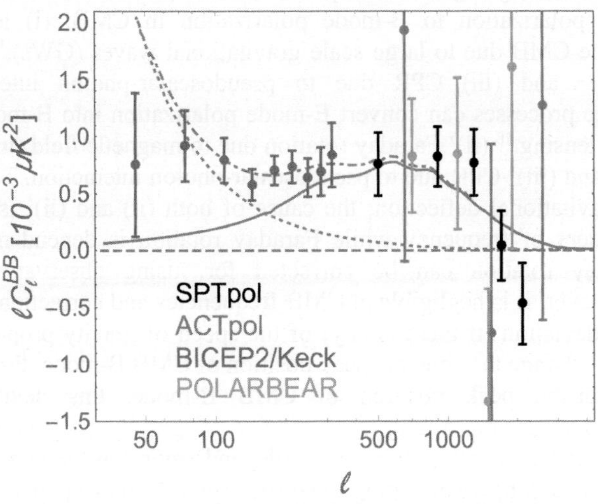

Fig. 1. BB power spectrum from SPTpol, ACTpol, BICEP2/Keck, and POLARBEAR (Ref. 24.)

We fit the B-mode spectrum of various combinations to the two parameters — the tensor to scalar ratio r and the CPR fluctuation power $<(\alpha - <\alpha>)^2>$. The E-mode power spectrum and the lensing B-mode power spectrum are updated from Fig. 3 of Planck 2015 results XIII[29] and Fig. 4 of Planck 2015 results XV[28] respectively. The left column of Figure 2 shows the fitting results of the BICEP2 8 point data; the right column of Figure 2 shows the fitting results of the joint POLARBEAR and BICEP2 8 point data. The left column of Figure 3 shows the fitting results of the joint POLARBEAR, BICEP2 8 point and ACTPOL data; the right column of Figure 2 shows the fitting results of the joint POLARBEAR, BICEP2 8 point, ACTPOL and SPTpol data.

The CPR fluctuation constraint from the joint ACTpol-BICEP2-POLARBEAR polarization data is 23.7 mrad (1.36°).[23] With the new SPTpol data included, the CPR fluctuation constraint is updated to 17 mrad (1°) with scalar to tensor ratio $r = -0.05 \pm 0.1$.

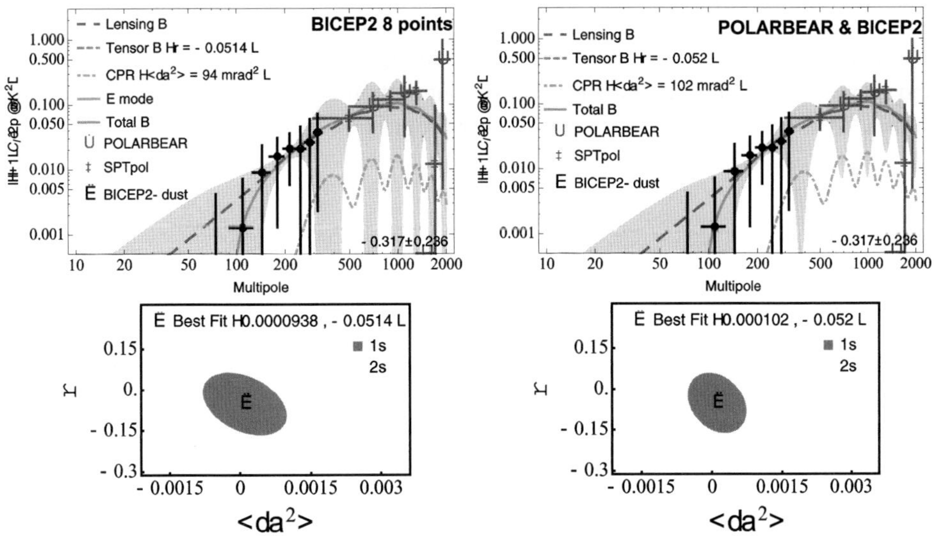

Fig. 2. *Left*: CMB B-Mode power spectrum with data points and models including the tensor-to-scalar ratio *r*, lensing, CPR, and dust contributions. The dust effect is subtracted from the 8 BICEP2 points data by using the PLANCK dust result[x13]. The lowest *l* BICEP2 data point is not included in the fitting because the PLANCK dust result doesn't cover the the full interval of it. The bottom plot shows the 1σ and 2σ contours on the tensor-to-scalar ratio *r* and the rms-sum of the CPR angle due to pseudo-scalar photon interaction for the top one. *Right*: The result of the joint POLARBEAR + BICEP2 (8 points) fitting. The value of the third multipole band of POLARBEAR (*l* from 1300 to 1700) is negative, i.e., $-0.317 \pm 0.236\ \mu K^2$ with the value written on the binning interval just above the horizontal axis.

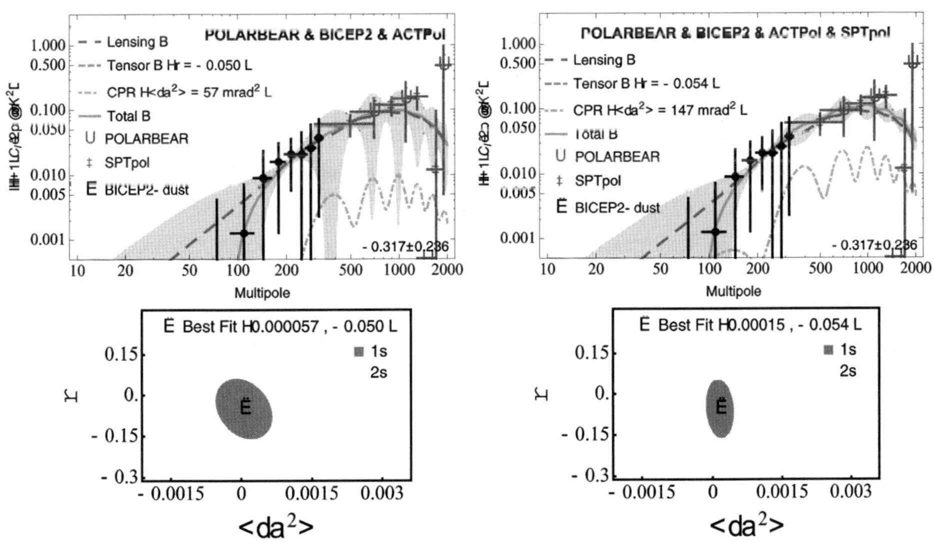

Fig. 3. *Left*: The result of the joint POLARBEAR + BICEP2 (8 points) + ACTPol fitting. *Right*: The result of the joint POLARBEAR + BICEP2 (8 points) + ACTPol + SPTpol fitting. From Fig. 2 and Fig. 3, we can see the higher multipole ($l > 400$) experiments (POLARBEAR/ACTPol/SPTpol) are more sensitive to the CPR angle; the BICEP2 ($l<400$) is more sensitive to the tensor-to-scalar ratio *r*.

Acknowledgements

We would like to thank Carlo Baccigalupi and Jun-Qing Xia for their helpful discussions and comments on how to extract the dust contribution in specific regions of sky from *Planck* CMB maps and on CPT violating cosmological models respectively.

References

1. A. Einstein, *Jahrb. Radioact. Elect.* **4**, 411 (1907).
2. G. Galilei, *Discorsi e dimostrazioni matematiche, intorno a due nuove scienze* (Elsevirii, Leiden, 1638).
3. W.-T. Ni, *Montana State Univ. preprint* (1973), http://astrod.wikispaces.com/.
4. W.-T. Ni, *Phys. Rev. Lett.* **38**, 301 (1977).
5. S. M. Carroll, G. B. Field, and R. Jackiw, *Phys. Rev. D* **41**, 1231 (1990).
6. See, e.g., F. W. Hehl and Y. N. Obukhov, *Foundations of Classical Electrodynamics: Charge, Flux, and Metric* (Birkhäuser: Boston, MA, 2003).
7. W.-T. Ni, Equivalence Principles and Precision Experiments, in *Precision Measurement and Fundamental Constants II*, ed. by B. N. Taylor and W. D. Phillips, Natl. Bur. Stand. (U S) Spec. Publ. **617**, 647 (1984).
8. W.-T. Ni, *Phys. Lett. A* **379**, 1297 (2015) and references therein.
9. S. di Serego Alighieri, *Int. J. Mod. Phys. D* **24**, 1530016 (2015).
10. S. di Serego Alighieri, Gaining confidence on general relativity with cosmic polarization rotation, *these proceedings*.
11. W.-T. Ni, *Phys. Lett. A* **378**, 3413 (2014).
12. W.-T. Ni, *Prog. Theor. Phys. Suppl.* **172**, 49 (2008), arXiv:0712.4082.
13. B. Feng, M. Li, J.-Q. Xia, X. Chen, and X. Zhang, *Phys. Rev. Lett.* **96,** 221302 (2006).
14. G.-C. Liu, S. Lee, and K.-W. Ng, *Phys. Rev. Lett.* **97**, 161303 (2006).
15. C. Q. Geng, S. H. Ho, and J. N. Ng, *J. Cosmol. Astropart. Phys.* **09**, 010 (2007).
16. A. G. Polnarev, *SvA* **29**, 607 (1985).
17. A. Kosowsky, T. Kahniashvili, G. Lavrelashvili and B. Ratra, *Phys. Rev. D* **71**, 043006 (2005).
18. M. Zaldarriaga and U. Seljak, *Phys. Rev. D* **55**, 18309 (1997).
19. L. Amendola, G. Ballesteros, and V. Pettorino, *Phys. Rev. D* **90**, 043009 (2014).
20. M. Raveri, C. Baccigalupi, A. Silvestri, and S.-Y. Zhou, *Phys. Rev. D* **91**, 061501(R) (2015).
21. R. Adam et al. (Planck Collaboration) 2014, Planck intermediate results. XXX. The angular power spectrum of polarized dust emission at intermediate and high Galactic latitudes, *A&A* **586**, A133 (2016), arXiv:1409.5738.
22. S. di Serego Alighieri, W.-T. Ni, and W.-P. Pan, *Ap. J.* **792**, 35 (2014), arXiv:1404.1701.
23. H.-H. Mei, et al., *Astrophys. J.* **805,** 107 (2015), arXiv:1412.8569.
24. R. Keisler, et al., *Astrophys. J.* **807,** 151 (2015), arXiv:1503.02315.
25. S. Naess, M. Hasselfield, J. McMahon, et al., *J. Cosmol. Astropart. Phys.* **1410**, 007 (2014), arXiv:1405.5524.

26. BICEP2/Keck & Planck Collaborations, P. A. R. Ade, et al. 2015, *Phys. Rev. Lett.* **114,** 101301 (2015), arXiv:1502.00612.
27. POLARBEAR Collaboration, *ApJ*, **794**, 171 (2014), arXiv:1403.2369.
28. Planck Collaborations, P. A. R. Ade, et al. 2015, arXiv:1502.01591.
29. Planck Collaborations, P. A. R. Ade, et al. 2015, arXiv:1502.01589.

GINGER

A. D. V. Di Virgilio[1*], N. Beverini [2], J. Belfi[1], A. Ortolan[3]
[1] *INFN Sezione di Pisa, Pisa, Italy*
[2] *Department of Physics University of Pisa, Italy*
[3] *INFN National Laboratory of Legnaro, Legnaro (Padua)*
* *angela.divirgilio@pi.infn.it*

The Gross Ring G is a ring-laser, it is located in the geodetic observatory of Wettzell. It has demonstrated that a ring laser has a sensitivity and stability such to measure the Earth rotational rate with the relative precision of 5 part in 10^9 and a long term stability of 3 hours. The GINGER project is intending to take this level of sensitivity further and to improve the long-term stability. A factor of about 10 improvement in sensitivity could allow the observation of the frame dragging produced by the Earth rotating mass (Lense-Thirring effect), and predicted by General Relativity. At the same time GINGER will provide a local measurement of the Earth rotational rate with a sensitivity comparable to that of the International Earth Rotation and Reference Systems Service (IERS). In GINGER the stability of each ring and the relative alignment of the different rings play a crucial role. Therefore the preliminary goal is the demonstration of the feasibility of a larger gyroscope structure, where the mechanical stability is obtained through an active control of the geometry. A ring-laser prototype (GP-2) has been set up in Pisa in order to develop and test the active control of the ring geometry, while a second apparatus (GINGERino) has been installed inside the Gran Sasso underground laboratory in order to investigate the seismic noise of the site in view of an installation of GINGER. The status of both prototypes is presented.

Keywords: Inertial sensor, general relativity, gravitomagnetism, Lense-Thirring, gravitational waves.

1. Introduction

Ring laser gyroscopes (RLG) have demonstrated to currently be the most sensitive device for testing rotational motion on Earth with respect to an inertial frame.[1] After several years of experimentation carried out by a collaboration between the University of Canterbury (Christchurch NZ) and theTechnische Universität München (TUM) of Munich Germany,[2] the Gross-Ring (G), a 16 m^2 area ring laser, has been constructed at the Geodetic Observatory in Wettzell (Bavaria). It has continuously improved over the years, by more and more accurate control of the environmental conditions, finally achieving a rotational sensitivity of the order of 0.25 $prad/s$ over 10^4 s, which corresponds to resolving the Earth rotation rate to 3 part in 10^9. In G the long term stability is guaranteed by its monolithic structure, which is made in zerodur, a zero expansion coefficient material. In 2008, in the central area of the VIRGO 3km arms interferometric gravitational waves antenna, located in Cascina near Pisa (Italy), a moderate size ring laser of 1.82 m^2 area called G-Pisa was realised by a collaboration between INFN and Department of Physics of the University of Pisa, and with the support of the expertise of Prof. Schreiber (TUM).

The effort was inspired by the desire to establish very small ground tilt movements for the VIRGO interferometer. G-Pisa was first operated with the optical cavity orientated in the local horizontal plane and later in the vertical plane.[3] In contrast to G, G-Pisa is based on an hetero-lithic mechanical structure, it has been the first hetero-lithic ring-laser which has been operated unattempted and without interruption for more than a month, using the active control of the perimeter. It was taking data on March 11th, 2011, when it registered the rotation induced by the seismic Rayleigh waves generated by Fukushima earthquake.[4] Putting together the expertise acquired by the INFN and University of Pisa groups and the other principal groups operating large laser gyros, (TUM and University of Canterbury), a proposal was published in 2011[5] that proposes a ring laser array to observe the Lense-Thirring effect. Later on this project was named GINGER (Gyroscopes IN GEneral Relativity). According to the theory of General Relativity, this effect arises from the metric deformation induced by a massive rotating body. Einstein's equations predicts a 1 *ppb* correction to the Earth rotational rate for an inertial sensor, when measured in an Earth based laboratory with respect to the Earth rotation rate calculated by the quasar positions, measured by the Very Long Baseline Interferometry (VLBI). So far this frame dragging effect has been confirmed in the contest of satellite experiments only.[6,7] The independent measurement in a terrestrial laboratory would give local information, complementary to the satellites ones, which is averaged over the satellite's orbit. GINGER has not been financed yet, at present it is supported by a Memorandum of Understanding between INFN, the TUM University of Munich, the Department of Earth Science of the LMU University of Munich, and the University of Canterbury at Christchurch in NewZealand. In Italy the INFN institute has financed an experiment called G-GranSasso which is taking care of the preparatory phase of the GINGER project, which is under the responsibility of Angela Di Virgilio; the INFN sections of Naples, Padua, and Turin participates to this preparatory phase, they are led by Alberto Porzio of CNR-SPIN of Naples, Giampiero Naletto of the Electronic Engineering Department of the University of Padua, and Angelo Tartaglia of the Polytechnic of Turin. Our experimental work is as well supported by the Italian INGV (Istituto Nazionale di Geofisica e Vulcanologia), with Gilberto Saccorotti of the INGV of Pisa. The future evolution of the GINGER project is at the moment under discussion. In what follows the relativistic effects on the ring-laser is briefly sketched and the two prototypes[8] GINGERino and GP2 are described.

2. Earth rotation rate and relativistic effects

The term gravitomagnetism designates the effects and phenomena predicted by GR which occur when the source of the gravitational field is rotating. Actually, rotational effects are due also to rotating reference frames (i.e. non inertial observers), and the kinematic part of the Sagnac effect is a typical example. If the source of gravity, as for the rotating Earth, can be approximated as a steadily and freely rotating object, the line element for an observer on Earth is given by

$$ds^2 = (c^2 + 2\mathbf{A} \cdot \mathbf{x})dt^2 - 2(\mathbf{\Omega} \wedge \mathbf{x}) \cdot d\mathbf{x}\, dt - d\mathbf{x} \cdot d\mathbf{x} + O(|d\mathbf{x}|^2) \tag{1}$$

where t is the observer proper time, $\mathbf{x}$ the local Fermi coordinates, $\mathbf{A}$ the observer acceleration, and Ω the observer rotation with respect the local inertial frame whose velocity is momentarily the same as that of the accelerating observer.[5,9] Light path must satisfy the null geodetic equation $ds = 0$. In the non relativistic limit, the proper-time difference $\Delta t = t_+ - t_-$ between the clock wise (t_+) and counter clock wise (t_-) propagation times of light along a close path $\mathcal{S}$, depends only on the time-space mixed components $(2\mathbf{\Omega} \wedge \mathbf{x})$ of the metric, as it is the only term that changes sign with the direction of motion, i.e.

$$\Delta t = \frac{2}{c^2} \oint_{\mathcal{S}} 2(\mathbf{\Omega} \wedge \mathbf{x}) \cdot d\mathbf{x} = \frac{4}{c^2} \iint_a \nabla \cdot (\mathbf{\Omega} \wedge \mathbf{x}) d^2x = \frac{4a}{c^2}\, \mathbf{\Omega} \cdot \mathbf{n}\ , \tag{2}$$

where the line integral is calculated on $\mathcal{S}$ with tangent vector $d\mathbf{x}$, and a is the enclosed area. As RL converts time difference into frequency difference, the Sagnac frequency follows from Eq. (2) and the condition of resonance in the optical cavity $t_+ = NT_+$ and $t_- = NT_-$, where T_+ and T_- are the periods of the two counter-propagating waves and $N \in \mathbb{N}$. The resonance condition implies $f_s \equiv 1/T_+ - 1/T_- = c^2 \Delta\tau/(\lambda p) = (4a)/(\lambda p)\, \mathbf{\Omega} \cdot \mathbf{n}$. Under the same assumptions as before, we may work out the metric elements by solving Einstein equations, and the final result reads[5]

$$f_s = \frac{4a}{\lambda p} (\mathbf{\Omega}_\oplus + \mathbf{\Omega}_G + \mathbf{\Omega}_B) \cdot \mathbf{n}\ , \tag{3}$$

where the effective angular velocity vector $\mathbf{\Omega}$ sensed by a RL is the sum of 3 contributions: i) the kinematic Earth rotation vector as measured by IERS or Sagnac term $\mathbf{\Omega}_\oplus \equiv \Omega \mathbf{u}_\oplus$, ii) the pure mass contribution or Geodetic term $\mathbf{\Omega}_G \equiv -2GM/(c^2R)\Omega_\oplus \sin\theta\ \mathbf{u}_\theta$, and iii) the Earth angular momentum contribution or Lense-Thirring term $\mathbf{\Omega}_B = G/(c^2R^3)\,[\mathbf{J}_\oplus - 3(\mathbf{J}_\oplus \cdot \mathbf{u}_r)\mathbf{u}_r]$; here $\mathbf{R} \equiv R\mathbf{u}_r$ is the position of the laboratory with respect to the center of the Earth, $\mathbf{J}_\oplus \simeq (2/5)MR^2\mathbf{\Omega}_\oplus$ is the angular momentum of the Earth, the unit vector $\mathbf{u}_r$ is along the radial direction, and $\mathbf{u}_\theta$ is along the local meridian in the sense of increasing colatitudes. Using Earth mass, rotation rate and radius, we have that Geodetic and Lense-Thirring terms are both of order $\sim 10^{-9}$ with respect to the Sagnac term.

3. The GINGER project

To reach the GINGER goals, it is necessary to use hetero-lithic mechanical structure for the ring-laser, the monolithic structure of G cannot be extended to a three dimensional array. The sensitivity and long term stability already obtained by "G" must be improved along the following lines:[5]

(1) Decreasing the signal to noise ratio, assuming a RL shot noise limited. This can be obtained enlarging the RL side L, since the signal to noise ration increases

with L^2. Note that, with equivalent mirrors quality, the backscatter noise, which is the most severe limitation of RLs, decreases increasing the side L as L^5.

(2) Increasing the long term stability by monitoring and controlling the laser operative parameters and the environment. A deep underground site seems necessary to reduce surface noise due to meteorology and seismic motion.

(3) Improving the scale factor stability and accuracy of the RLs. This implies a strict active control of the ring geometry and of its orientation with respect the Earth rotation axis. A viable solution is to implement an effective control system for the geometry stabilization of the RL optical cavity, and to use a tri-axial array of RLs. The proposed 3-dimensional geometry foresees a set of six mirrors in an octahedral configuration. Probably it will be wise to add at least a fourth ring to this array.

(4) Estimating the nonlinear laser dynamics. To achieve the required accuracy, an estimate of non reciprocal effects resulting in a null-shift term in f_{nr} must be implemented.

Here we present some experimental activities we are carrying out in order to address the issues of line (3), i.e. the geometry stabilization of a single RL cavity (GP2), and the validation of the GranSasso underground location (GINGERino). A fourth prototype called GEMS (GINGER External Metrology System) is in development at the Department of Electronic Engineer of the University of Padua. GEMS is based on the interferometric technique developed for the NASA Stellar Interferometric Mission (SIM). This technique has demonstrated the capability of measuring the distance between two optical retroreflectors with a < 10 pico–m accuracy.[10]

3.1. *The GP2 RL gyroscope*

The scale factor $k_S = \frac{4a}{(\lambda p)}$ must be known and kept constant at the level of one part in 10^{10}, this implies that the geometry must be controlled. The GP2 prototype is conceived to test the control strategies for the RL cavity geometry, with the aim of gaining a sub-nanometer stabilization of the scale factor geometry a/p. It is based on an innovative design, in which the mirrors that constitute the square optical cavity are equipped with a piezo nano-positioning system (at the moment a total of 6 piezo-transucers, which can be extended up to 12), in order to actively control their relative positions. The relevant observables to constraint the geometry are the perimeter of the ring cavity and the lengths of the two diagonal cavities, defined by the mirrors on opposite corners. The use of the diagonal lengths is pioneering in the RL technology. We have demonstrated that, when the lengths of the two diagonals are locked to the same value, the perturbations to the mirror positions affect only quadratically the ring laser perimeter and scale factor.[11,12] In this case, the regular square geometry corresponds to a saddle-point of the perimeter, and so the scale factor fluctuations are minimized. To allow the implementation of this new control strategy, the vacuum chamber of GP2 encloses also the optical paths of an external laser beam between the opposite mirrors of the cavity. To stabilise the absolute lengths of the two diagonal cavities respect to a reference laser, an original interferometric technique has been worked out. It consists in locking the resonance frequencies (f_n) of the two cavities to the same laser carrier frequency and, at the same time, in determining an integer multiple of the two free spectral ranges (FSR). The error signals needed to implement, for each diagonal, a two frequency measurement (f_n and FSR) in closed loop, are recovered by phase-modulating the interrogating laser source with a combination of three independent frequencies, through a single electro-optic modulator. The stabilisation scheme has been tested in a table top experiment on two Fabry-Perot resonators composed by two couples of spherical mirrors. These resonators was mounted on an optical bench to simulate the cavities along the diagonals of a RL gyroscope. With this setup, we were able to demonstrate the capability of stabilizing the two lengths at the level of 1 part in 10^{11}, and setting their difference equal to zero within the precision of 500nm, with residual fluctuations only limited by the laser frequency noise.[13] Recently, this metrology technique has been partially implemented in GP2. Preliminary results demonstrate the compatibility of the scheme with the RL operation, resulting in a sub-nanometer diagonal lengths stabilization with a signal integration time of few hundred seconds.[11]

4. GINGERino, the prototype to validate the GranSasso underground location

The GINGERino structure is the evolution of our first RL prototype G-Pisa[3] to a larger size instrument. It uses the same type of corner boxes of G-Pisa that house the mirrors and the same type of mechanical levers for aligning the ring optical cavity. The resonator was mounted on a cross-shaped granite basis of 4 tons of weight and linked up by vacuum pipes in order to obtain a square optical cavity of $3.60 \times 3.60 m^2$. It is located inside the Gran Sasso underground Laboratories, in a tunnel in the core of the mountain under 1400 m of rock. The granite basis is supported centrally by a concrete monument, which is firmly attached to the tunnel floor. A seismic station is located alongside the interferometer, equipped with two very broad band seismometer (Guralp-360 and Trillium-240, the latter linked up to the Italian national seismic array). A two-dimensional tiltmeter, with nano-rad sensitivity, completes the setup. It was operated from May 2015 onwards. Despite the fact that the mirrors presently mounted in the ring are not as goos as they should be, a level better than 2×10^{-6} (i.e. about 20 pico-rad/s) is achieved in 100 s integration time. Some Earthquakes were registered and analysed during these preliminary runs.[14] The details of GINGERino and the preliminary analysis can be found in Ref. 15. Next improved runs are expected in spring 2016, when the new set of mirrors will be inserted in the cavity.

5. Conclusions

The recent performance of the G ring laser demonstrates the possibility for the ring laser technology to access the resolution required for sensing relativistic effects on rotation. Two main critical points have to be tackled for achieving the required progress. The first is connected to the residual uncontrolled deformation and tilts of the optical cavity of the ring resonator. The second regards the excess of ground rotational noise typically induced by Earth surface environment. Two experiments in Italy are running for the investigation of these two aspects, GP-2 and GINGERino. GP-2 ring laser is designed for the study of active methods for controlling the deformation of the cavity shape, exploiting the stabilization of the diagonal lengths achieving currently a resolution of the order of 1 part in 10^9. The ultimate performance of such an approach is ultimately limited by the frequency stability of the optical frequency reference and thus compatible with the geometrical stability required by GINGER. Preliminary results indicate a good stability of the constrained length of the two resonators. The sensitivity of the geometrical scale factor from not planar deformations remains to be addressed. The GINGERino ring laser is investigating the local noise inside the GranSasso underground laboratory of the INFN. Its stability is currently reaching a level of a few ppm at about 100 s of integration time and is limited by the fluctuation laser parameters connected to a large amount of intracavity backscattered light. A new set of improved mirrors will be available in spring 2016, with this new set of mirrors the backscattering noise should be reduced.

References

1. K. U. Schreiber and J.-P. R. Wells, Invited review article: Large ring lasers for rotation sensing, *Review of Scientific Instruments* **84** (2013).
2. G. E. Stedman, *Light Revolution* (Xlibris Corporation, May 2015).
3. J. Belfi, N. Beverini, F. Bosi, G. Carelli, A. Di Virgilio, D. Kolker, E. Maccioni, A. Ortolan, R. Passaquieti and F. Stefani, Performance of "g-pisa" ring laser gyro at the virgo site, *Journal of Seismology* **16**, 757 (2012).
4. J. Belfi, N. Beverini, G. Carelli, A. Di Virgilio, E. Maccioni, G. Saccorotti, F. Stefani and A. Velikoseltsev, Horizontal rotation signals detected by "g-pisa" ring laser for the m-w=9.0, march 2011, japan earthquake, *Journal of Seismology* **16**, 767 (2012).
5. F. Bosi, G. Cella, A. Di Virgilio, A. Ortolan, A. Porzio, S. Solimeno, M. Cerdonio, J. P. Zendri, M. Allegrini, J. Belfi, N. Beverini, B. Bouhadef, G. Carelli, I. Ferrante, E. Maccioni, R. Passaquieti, F. Stefani, M. L. Ruggiero, A. Tartaglia, K. U. Schreiber, A. Gebauer and J. P. R. Wells, Measuring gravitomagnetic effects by a multi-ring-laser gyroscope, *Physical Review D* **84** (2011).
6. P. C. Ciufolini I, Analysis, *Nature* **431**, 958 (2004).
7. C. W. F. Everitt, D. B. DeBra, B. W. Parkinson, J. P. Turneaure, J. W. Conklin, M. I. Heifetz, G. M. Keiser, A. S. Silbergleit, T. Holmes, J. Kolodziejczak, M. Al-Meshari, J. C. Mester, B. Muhlfelder, V. G. Solomonik, K. Stahl, P. W. Worden, W. Bencze, S. Buchman, B. Clarke, A. Al-Jadaan, H. Al-Jibreen, J. Li, J. A. Lipa, J. M. Lockhart, B. Al-Suwaidan, M. Taber and S. Wang, Gravity probe b: Final results of a space experiment to test general relativity, *Phys. Rev. Lett.* **106**, p. 221101 (May 2011).

8. A. Di Virgilio, M. Allegrini, A. Beghi, J. Belfi, N. Beverini, F. Bosi, B. Bouhadef, M. Calamai, G. Carelli, D. Cuccato, E. Maccioni, A. Ortolan, G. Passeggio, A. Porzio, M. L. Ruggiero, R. Santagata and A. Tartaglia, A ring lasers array for fundamental physics, *Comptes Rendus Physique* **15**, 866 (2014).
9. M. L. Ruggiero and A. Tartaglia, Gravitomagnetic effects, *Nuovo Cimento B* **117**, p. 743 (2002).
10. P. G. Halverson, O. Alvarez-Salazar, A. Azizi, F. Dekens, B. Nemati and F. Zhao, Progress towards picometer accuracy laser metrology for the Space Interferometry Mission - update for ICSO 2004, in *5th International Conference on Space Optics*, ed. B. Warmbein, ESA Special Publication, Vol. 554June 2004.
11. P. T. U. o. S. Santagata, R., Sub-nanometer length metrology for ultra-stable ring laser gyroscopes., (2015).
12. R. Santagata, A. Beghi, J. Belfi, N. Beverini, D. Cuccato, A. Di Virgilio, A. Ortolan, A. Porzio and S. Solimeno, Optimization of the geometrical stability in square ring laser gyroscopes, *Classical and Quantum Gravity* **32** (2015).
13. J. Belfi, N. Beverini, D. Cuccato, A. Di Virgilio, E. Maccioni, A. Ortolan and R. Santagata, Interferometric length metrology for the dimensional control of ultra-stable ring laser gyroscopes, *Classical and Quantum Gravity* **31** (2014).
14. A. Simonelli, J. Belfi, N. Beverini, G. Carelli, A. Di Virgilio, E. Maccioni, G. De Luca and G. Saccorotti, First deep underground observation of rotational signals from an earthquake at teleseismic distance using a large ring laser gyroscope (January 2016).
15. J. Belfi, N. Beverini, F. Bosi, G. Carelli, D. Cuccato, G. De Luca, A. Di Virgilio, A. Gebauer, E. Maccioni, A. Ortolan, A. Porzio, R. Santagata, A. Simonelli and G. Terreni, First results of gingerino, a deep underground ring-laser (January 2016).

Observing the Black Hole Shadow of M87 and the Greenland Telescope Project: GR Test in the Strong Gravity Regime

Hung-Yi Pu

Institute of Astronomy & Astrophysics, Academia Sinica, 11F of Astronomy-Mathematics Building, AS/NTU No. 1, Taipei 10617, Taiwan
hypu@asiaa.sinica.edu.tw

Direct imaging of the "shadow" cast by the black hole event horizon and the test of physics in strong gravity is one of the important goals in modern astronomy. The ongoing Greenland Telescope project in Academia Sinica Institute of Astronomy and Astrophysics (ASIAA) is devoting itself to this exciting area. ASIAA has officially acquired the 12m ALMA (Atacama Large Millimeter/submillimeter Array) Vertex Prototype Telescope in 2011. This telescope, now renamed as the "Greenland Telescope (GLT)", is going to be moved to the Summit Station in Greenland after upgrades. The first light of GLT at the Summit Station is expected to be available in 2018/2019. Together with SMA (Submillimeter Array) in Hawaii and ALMA (Atacama Large Millimeter/submillimeter Array) in Chile, the angular resolution at sub-mm will reach ~20 μarcsec, providing an unique opportunity to observe the shadow of the supermassive black hole at the center of M87, and the accretion/jet structure near the event horizon. The GLT project is a collaborative project between ASIAA, Smithsonian Astrophysical Observatory, MIT Haystack Observatory, and National Radio Astronomy Observatory.

Keywords: Greenland Telescope; black hole; VLBI; radiative transfer; general relativity.

1. Greenland Telescope Project

Base on dynamical or energy arguments, the existence of black holes in the Universe is commonly accepted in modern astronomy. By the help of Very Long Baseline Interferometry (VLBI) technic, it is possible to directly image the the shadow cast by the black hole event horizon for a more direct evidence for the existence of black holes. The Greenland Telescope, a future 12-meter antenna in Greenland, will help to increase the current Very Long Baseline Interferometry (VLBI) baseline length and sensitivity. In future, GLT, the Submillimeter Array in Hawaii, and the Atacama Large Millimeter/submillimeter Array in Chile, will play a key role for VLBI observation (Fig. 1) for the supermassive black hole (SMBH) at the centre of M87.

Located at ~16.7 Mpc away, the shadow size of the SMBH of M87 has the size $\sim 40\mu as$, providing the 2nd largest shadow size observed from Earth. Complementary with the case of the largest black hole shadow ($\sim 50\mu as$), which is cast by the SMBH located at our Milky Way galaxy (Sgr A*), observation of M87 is unique for its location (at northern sky), jet activity, absent of the interstellar scatting effect, and a longer dynamical time-scale.

With longest baseline > 9000 km, the angular resolution can reach $20\mu as$ at 350 GHz, high enough to resolve the black hole shadow. In addition to the VLBI science,

GLT will also contribute to single-dish science at THz band.[3] The (expected) time line of the GLT project is listed below:[1,2]

- 2009 ASIAA/VLBI group established
- 2011 ASIAA and SAO/CfA officially acquired the 12-meter ALMA-NA prototype anttena
- 2015 Antenna retrofitting for cold
- 2016 ship Antenna to Thule, Greenland. Antenna re-assemble and test at Thule, including VLBI test
- 2017–2018(expected) Single-dish and VLBI first light at Thule
- 2019(expected) Transport antenna across ice sheet to Summit Station
- 2019(expected) First light at the Summit Station

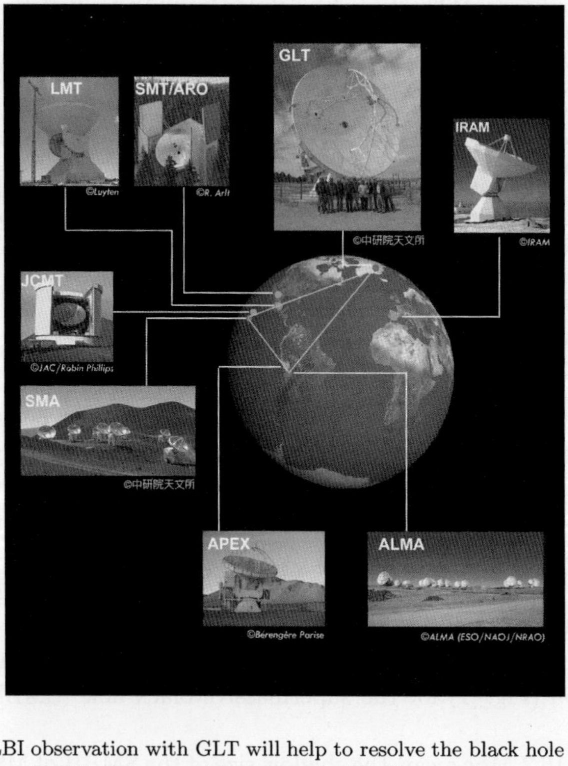

Fig. 1. Future VLBI observation with GLT will help to resolve the black hole shadow cast by the SMBH at the center of M87. The essential stations for VLBI are indicated in red dots, including GLT, SMA (Submillimeter Array), and ALMA (Atacama Large Millimeter/submillimeter Array).

2. Black Hole Shadow

The size and shape of the black hole shadow is determined by the null geodesics. As the geodesics are related to the background spacetime metric, observation of black hole shadow provides a direct test in strong gravity regime. Figure 2 demonstrate the null geodesics and the resulting shadow shape for a non-rotating and a rapidly rotating black hole, assuming the background spacetime is described by the Kerr metric.

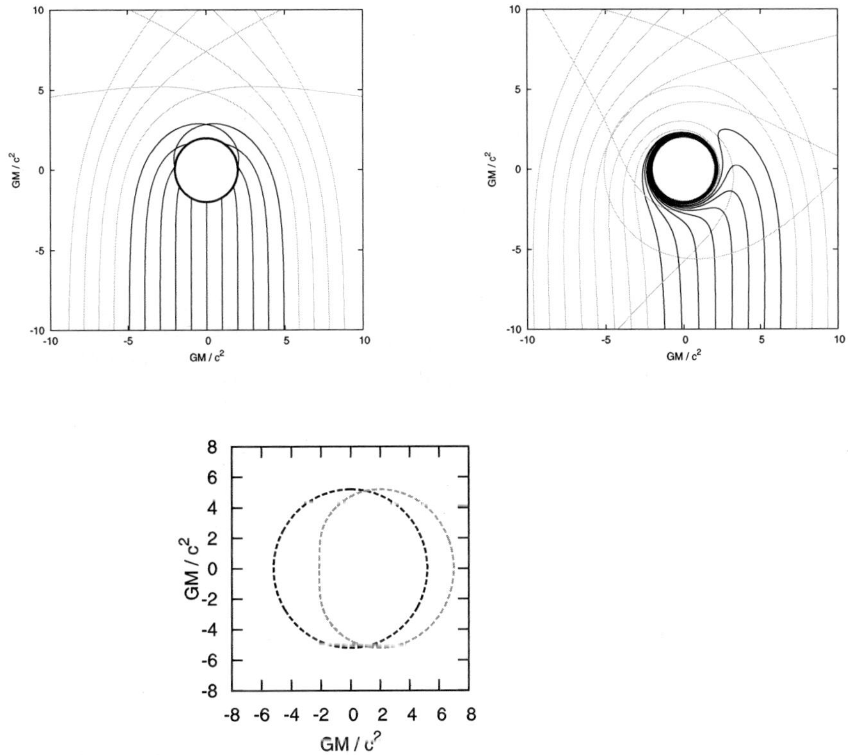

Fig. 2. Observation of black hole shadow as a GR test. Top panels show the null geodesics on the equatorial plane for a non-rotating (left; $a/M = 0$) and rapid rotating (right; $a/M = 0.998$) black hole, respectively. The black geodesics indicate the shadow region. The bottom panel shows the resulting shape of the black hole shadows, when distant observer is located at the equatorial plane. The shadow of a rapid rotating black hole has a "D" shape.

3. Astronomical Considerations

In realistic case, the observed BH shadow image is closely related to the accretion/jet environment. There are several factors related to final image, such as: 1. dynamics of the emitting materials, 2. distribution of emitting materials, 3. emission mechanism. Some examples of how different accretion environment can modify the resulting black hole image demonstrated in Fig. 3. Note that the black hole shadow can only been seen when the surrounding materials become optically

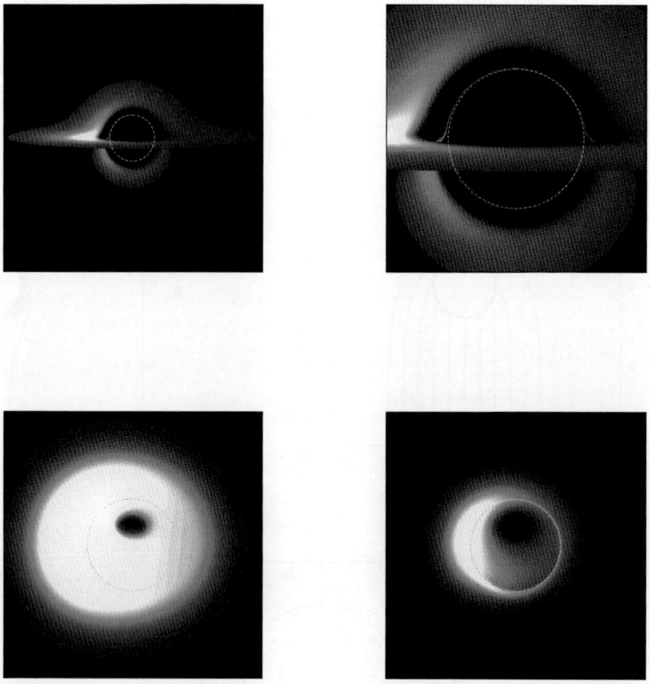

Fig. 3. Examples of the black hole image when a non-rotating black hole is surrounding by different accretion environment. The theoretical black hole shadow shape (as shown in Fig. 2) is indicated by the green dashed circle. Upper panel: The black hole is surrounding by a optically thick, rotating disk, with the inner edge of the disk is located at the innermost stable circular orbit (6 GM/c^2). The inclination angle is 85 degree. The right figure shows the zoom-in plot of the central region. The fine structure surrounding the dashed circle is resulting from the light passing through the gap inside the innermost stable circular orbit, then reach to the observer. Bottom panel: The black hole is surrounding by a rotating shell, with inclination angle of 45 degrees. When the material is optically thick for the observed frequency, we can only see the photosphere of the shell and the shadow is unseen (left). A black region is formed due to that there is less emitting material near the funnel region close to the axis. When the material become optically thin, the shadow can be observed (right).

thin at the observed frequency, so observers can "see through" the materials (accretion/jet) surrounding the central black hole. General relativistic radiative transfer is an essential tool for taking into account the relativistic effects, including the redshift between the fluid rest frame around the black hole and the distant observer, and the integration of emission processes in each of the fluid rest frames along a null geodesic.[4,5] The jet component can further affect the final image. Knowledge for the accretion flow and jet around the black hole is therefore crucial for predicting future Millimeter/submillimeter VLBI observations of M87, and several studies have been devoting to this important topic.[6–8] More details of accretion and jet launching physics will also be revealed by the upcoming VLBI observations.

References

1. ASIAA website for the GLT project: `http://vlbi.asiaa.sinica.edu.tw/project.php`
2. Inoue M. et al. 2014, Greenland Telescope Project: Direct Confirmation of Black Hole with Sub-millimeter VLBI, *RADIO SCIENCE*: **49(7)**, 564-571, 2014-07
3. Hirashita, H. et al. , 2016, First-generation Science Cases for Ground-based Terahertz Telescopes, *Publ. Astron. Soc. Japan*, **68**, R1
4. Younsi, Z., Wu, K., & Fuerst, S. V. 2012, General relativistic radiative transfer: formulation and emission from structured tori around black holes, *A&A*, **545**, A13
5. Pu, H.-Y., Yun K. et al. 2016, Odyssey: a Public GPU-based Code for General-relativistic Radiative Transfer in Kerr Spacetime, *Astrophy. J.*, accepted (arXiv:1601.02063)
6. Broderick, A. E. and Leob, A. 2009 Imaging the Black hole Silhouette of M87: Implications for Jet Formation and Black Hole Spin, *Astrophy. J.* **697**, 1164
7. Dexter J. et al. 2012, The size of the jet launching region in M87, *Mon. Not. R. Astron. Soc.* **421**, 1517
8. Mościbrodzka, M., Falcke, H., & Shiokawa, H. 2016, General relativistic magnetohydrodynamical simulations of the jet in M?87, *A&A*, **586**, A38

PARALLEL SESSION

VII. Alternative and Modified Theories of Gravity

Summary of Parallel Session: "Alternative and Modified Theories of Gravity"

Keisuke Izumi

Departament de Física Fonamental, Institut de Ciències del Cosmos (ICCUB), Universitat de Barcelona, Martí i Franquès 1, E-08028 Barcelona, Spain

We summarize *Alternative and Modified Theories of Gravity* session in *the 2nd LeCosPA International Symposium "Everything about Gravity"*.

Keywords: Modified theory of gravity.

1. Introduction and summary

The great theory of gravity, general relativity (GR), has been verified by many experimental and observational results, and describes them with high precision at distances longer than 0.01mm and shorter than 1Gpc. Beyond these scales, however, the theory of gravity is still shrouded in mystery. At cosmological distances, we are facing with the so-called dark energy problem.[1] Looking at small scale physics, on the other hand, we confront the singularity problem.[2] In the quantum theory it appears as the divergence in loop diagrams, and thus is related to the problem of renormalization. These big mysteries give an indication of the demand to upgrade the theory of gravity, and many alternative and modified theories of gravity have been proposed.

The talks in *Alternative and Modified Theories of Gravity* session cover quite broad topics. Some of speakers proposed new theories of gravity. The observational constraints from cosmology and astrophysical objects were shown. Theoretical discussions of classical theories were done. Quantum aspects of gravity were also discussed. These topics are deeply related to all topics in other parallel sessions in *the 2nd LeCosPA International Symposium "Everything about Gravity"*. We briefly show what each speaker talked about. For the detail, see their proceedings or the original papers.

2. Proposals and reviews

P. M. Ho and his collaborator propose an idea of geometry to describe theories of gravity, *Geometry of Area Without Length.*[3] To define a free string theory, i.e. the Nambu-Goto action, we do not need to define the length and only the notion of area is required. They derive a vacuum equation analogous to the vacuum Einstein equation, and show static spherical solutions, including the Schwarzschild solution.

Poincar gauge theory[4] was review by C. C. Lee. The curvature and torsion tensors can be recognized as the field strength of the gauge fields corresponding to the energy-momentum and angular momentum conservations. C. C. Lee and his collaborators solve the equation of motion with Friedmann ansatz and find an acceleration phase, which may be a solution of the present accelerating expansion of our Universe.[5]

L. W. Luo and his collaborators derive four-dimensional effective theories from five-dimensional teleparallel gravity.[6] Teleparallel gravity is defined on Weitzenböck geometry, which gives zero curvature but non-vanishing torsion. It turns out that GR can be reformulated, called teleparallel equivalent of general relativity[7] (TEGR). In braneworld scenario of TEGR the four dimensional effective theory becomes the same as that from the five dimensional GR, while a different theory from GR can be obtained by Kaluza-Klein reduction in TEGR.

3. Cosmology and astrophysics

C. Y. Chen and his collaborators study the Friedmann universes in (original and modified) Eddington-inspired-Born-Infeld theories and check the asymptotic behavior in the future and past.[8] With some parameters, singularities in both future and past are avoidable. Moreover, accelerating solutions without dark energy exist.

F. Arroja and his collaborators construct general scalar-tensor theories of gravity including, e.g., Horndeski models, which are invariant under the disformal transformations.[9] Non-invertible transformation results in additional equations of motion, which can mimic the cosmological background of a flat Fridmann ansatz with a barotropic perfect fluid with any constant equation of state.

H. Ogawa and his collaborators investigate (in)stability of black hole in shift-symmetric Horndeski theories.[10] Scalar hair of black hole in this theory was pointed out.[11] However, they show that this solution is unstable.

R. Saito and his collaborator point out that, in a recently introduced class of gravitational theories that goes beyond Horndeski, the Vainshtein screening mechanism does not work inside any gravitational source, but does outside.[12] This is a new type of deviation from GR. They solve the modified Lane-Emden equations, and give constraints on the parameters of the theory.

4. Theoretical discussions

General relativity has nontrivial good structures in nonlinear regime. For instance, gravitons propagate at speed of light. However, when we consider the modification from GR, even if the modification is tiny, superluminal modes easily appear.[13–15] With superluminal propagations, the usual causal analysis based on null curves doesn't work and causal structures becomes nontrivial. N. Tanahashi and his collaborators analyze the causal structure in Lovelock gravity,[15] which is the higher curvature theory with the second order differential equation and which is expected to be the low-energy effective theory of the string theory. Although this theory has generically superluminal modes of graviton, they show that any Killing horizons are exactly horizons in the sense of causality. (In Gauss-Bonnet case, it is proved by K. Izumi.[14]) They show that at least one of angular propagations becomes superluminal on black hole solutions and in higher curvature region acausal structures appear. They also point out the shock wave formation, which means the violation of the cosmic censorship.

Y. Yamashita and her collaborator derive the bigravity theory from the braneworld context.[16] Bigravity must be a low-energy effective theory because of the strong coupling and acausality in high energy regime. They consider a two-brane setup in DGP braneworld model and obtain the bigravity model with a nontrivial scalar field.

5. Quantum aspects

C. M. Shen and his collaborators investigate the semiclassical approach of gravity, i.e. quantum field theory on curved spacetime.[17] Calculation of the vacuum polarization of the energy-momentum tensor for non-interacting fields has been well established, but the formula is quite messy. Thus, it hard to solve the dynamics including the vacuum polarization. There is an attempt to describe the effect of the vacuum polarization by auxiliary scalar fields,[18] but there were no natural ways to fix the parameters of solutions of auxiliary scalar fields. They introduce the boundary contributions and show the natural way to fix the parameters in two-dimensional spacetimes. They also demonstrate that their method give consistent results with well known vacuum polarization on concrete spacetimes.

M. Horikoshi and his collaborators calculate the gravitational loop effect in the high-energy Higgs potential.[19] Since Higgs particle has been observed and its mass is fixed, we can calculate the loop corrections of Higgs potential. They calculate the loop corrections including gravitational effect. The results depend on the cutoff scale because of nonrenormalizable property of GR, but with any cutoff scale the potential has a global minimum. They also consider R^2 correction to GR, which can make the renormalizable property better, and show that the potential becomes irrelevant to the cut off.

K. Izumi and his collaborators investigate the relation between unitarity and renormalizability in Lifshitz scaling theory.[20] Unitarity and renormalization are the essential properties of quantum theories, and may require equivalent conditions.[21] Without the Lorentz symmetry, both the unitarity and the renormalizability conditions are modified from those in relativistic theories. For renormalizability, the power counting condition has to be modified from the conventional one. The reference frame dependence of scattering amplitudes result in stronger conditions for unitarity. They show that both unitarity and renormalizability require identical conditions as in the case of conventional relativistic theories.

Acknowledgments

K. I. is supported by FPA2013-46570-C2-2-P, AGAUR 2009-SGR-168, and MDM-2014-0369 of ICCUB (Unidad de Excelencia 'Maria de Maeztu').

References

1. A. G. Riess *et al.* [Supernova Search Team Collaboration], Astron. J. **116**, 1009 (1998). S. Perlmutter *et al.* [Supernova Cosmology Project Collaboration], Astrophys. J. **517**, 565 (1999).
2. S. W. Hawking, G. F. R. Ellis, *"Large scale structure of space-time"*, Cambridge University Press(Cambridge, 1975).
3. P. M. Ho and T. Inami, arXiv:1508.05569 [hep-th].
4. F. W. Hehl, P. Von Der Heyde, G. D. Kerlick and J. M. Nester, Rev. Mod. Phys. **48**, 393 (1976)
5. H. H. Tseng, C. C. Lee and C. Q. Geng, JCAP **1211**, 013 (2012).
   C. Q. Geng, C. C. Lee and H. H. Tseng, Phys. Rev. D **87**, no. 2, 027301 (2013).
6. C. Q. Geng, L. W. Luo and H. H. Tseng, Class. Quant. Grav. **31**, 185004 (2014).
   C. Q. Geng, C. Lai, L. W. Luo and H. H. Tseng, Phys. Lett. B **737**, 248 (2014).
7. K. Hayashi and T. Shirafuji, *New General Relativity*, Phys. Rev. **D 19** (1979) 3524.
8. M. Bouhmadi-Lopez, C. Y. Chen and P. Chen, Eur. Phys. J. C **74**, 2802 (2014).
   M. Bouhmadi-Lpez, C. Y. Chen and P. Chen, Eur. Phys. J. C **75**, 90 (2015).
   M. Bouhmadi-Lopez, C. Y. Chen and P. Chen, Phys. Rev. D **90**, 123518 (2014).
   C. Y. Chen, M. Bouhmadi-Lopez and P. Chen, arXiv:1507.00028 [gr-qc].
9. F. Arroja, N. Bartolo, P. Karmakar and S. Matarrese, JCAP **1509**, 051 (2015).
10. H. Ogawa, T. Kobayashi and T. Suyama, arXiv:1510.07400 [gr-qc].
11. T. Kobayashi and N. Tanahashi, PTEP **2014**, 073E02 (2014)
12. R. Saito, D. Yamauchi, S. Mizuno, J. Gleyzes and D. Langlois, JCAP **1506**, 008 (2015).
13. Y. C. Ong, K. Izumi, J. M. Nester and P. Chen, Phys. Rev. D **88**, 024019 (2013)
    K. Izumi and Y. C. Ong, Class. Quant. Grav. **30**, 184008 (2013)
    S. Deser, K. Izumi, Y. C. Ong and A. Waldron, Phys. Lett. B **726**, 544 (2013)
    K. Izumi, J. A. Gu and Y. C. Ong, Phys. Rev. D **89**, no. 8, 084025 (2014)
    S. Deser, K. Izumi, Y. C. Ong and A. Waldron, arXiv:1312.1115 [hep-th].
    S. Deser, K. Izumi, Y. C. Ong and A. Waldron, Mod. Phys. Lett. A **30**, 1540006 (2015)
14. K. Izumi, Phys. Rev. D **90**, no. 4, 044037 (2014)
15. H. Reall, N. Tanahashi and B. Way, Class. Quant. Grav. **31**, 205005 (2014).
    H. S. Reall, N. Tanahashi and B. Way, Phys. Rev. D **91**, no. 4, 044013 (2015).
16. Y. Yamashita and T. Tanaka, JCAP **1406**, 004 (2014)
    Y. Yamashita and T. Tanaka, arXiv:1510.07551 [hep-th].
17. C. M. Shen, K. Izumi and P. Chen, Phys. Rev. D **92**, no. 2, 024035 (2015) [Phys. Rev. D **92**, no. 4, 049902 (2015)].
18. P. O. Mazur and E. Mottola, Phys. Rev. D **64**, 104022 (2001)
19. Y. Abe, M. Horikoshi and T. Inami, to appear.
20. T. Fujimori, T. Inami, K. Izumi and T. Kitamura, Phys. Rev. D **91**, no. 12, 125007 (2015).
    T. Fujimori, T. Inami, K. Izumi and T. Kitamura, arXiv:1510.07237 [hep-th].

21. C. H. Llewellyn Smith, Phys. Lett. B **46**, 233 (1973).
J. M. Cornwall, D. N. Levin and G. Tiktopoulos, Phys. Rev. Lett. **30**, 1268 (1973) [Phys. Rev. Lett. **31**, 572 (1973)].
J. M. Cornwall, D. N. Levin and G. Tiktopoulos, Phys. Rev. D **10**, 1145 (1974) [Phys. Rev. D **11**, 972 (1975)].

Causality, Hyperbolicity and Shock Formation in Lovelock Theories

Norihiro Tanahashi*, Harvey S. Reall and and Benson Way
DAMTP, Centre for Mathematical Sciences,
University of Cambridge, Wilberforce Road, Cambridge CB3 0WA, UK
* *N.Tanahashi@damtp.cam.ac.uk*

We study gravitational wave propagation in Lovelock theories, which are extensions of Einstein's theory by higher-curvature corrections, to examine if these theories have good properties such as causality and hyperbolicity. We study the propagation on various background spacetime, and find that initial value problem may cease to be well-posed in some case. We also show that the sound speed of gravitational wave depends on the background and it may cause shock formation in these theories. We discuss implications of these phenomena.

Keywords: Classical theories of gravity; black holes.

1. Introduction

Lovelock theories of gravity are the most general metric theory with a covariant Lagrangian that gives equations of motion with derivatives only up to second order. Under these conditions Lovelock showed that the most-general Lagrangian $\mathcal{L}$ in d dimensions is given by[1]

$$\frac{\mathcal{L}}{\sqrt{-g}} = R - 2\Lambda - \sum_{p\geq 2}^{\lfloor\frac{d-1}{2}\rfloor} 2k_p \delta^{c_1\ldots c_{2p}}_{d_1\ldots d_{2p}} R_{c_1c_2}{}^{d_1d_2}\ldots R_{c_{2p-1}c_{2p}}{}^{d_{2p-1}d_{2p}}, \tag{1}$$

where k_p are coupling constants and $\delta^{c_1\ldots c_{2p}}_{d_1\ldots d_{2p}} = (2p)!\delta^{c_1}_{[d_1}\delta^{c_2}_{d_2}\cdots\delta^{c_{2p}}_{d_{2p}]}$ is the generalized Kronecker delta. As stated above, the Euler-Lagrange equation for this Lagrangian is manifestly up to second order in derivatives of the metric:

$$E_{ab} \equiv R_{ab} - \frac{2\Lambda}{d-2}g_{ab} + B_{ab} - \frac{1}{d-2}B^c{}_c g_{ab}, \tag{2}$$

where

$$B^a{}_b \equiv \sum_{p\geq 2}^{\lfloor\frac{d-1}{2}\rfloor} k_p \delta^{ac_1\ldots c_{2p}}_{bd_1\ldots d_{2p}} R_{c_1c_2}{}^{d_1d_2}\ldots R_{c_{2p-1}c_{2p}}{}^{d_{2p-1}d_{2p}}. \tag{3}$$

The theory up to the $p=2$ term is called Einstein-Gauss-Bonnet theory. Thanks to the absence of higher derivatives, the number of degrees of freedom in this theory is the same as that of GR, and also it is protected from the Ostrogradsky instability caused by such higher derivative terms. This property is one of the reasons to focus on these theories among other gravity theories with higher-curvature corrections. Although these theories have such a nice feature, there may be significant modifications to wave propagation and causality due to the Lovelock terms. In our work, we examine causality, hyperbolicity and shock formation in Lovelock theories of gravity to examine what kind of new feature may arise due to the curvature corrections.

2. Characteristics in Lovelock Theories

The field equation (2) has a property that it is quasilinear in second time derivative of the metric in any coordinate chart, hence it may be expressed as

$$E_{ab} = \frac{\partial E_{ab}}{\partial(\partial_0^2 g_{cd})}\partial_0^2 g_{cd} + \cdots = 0, \tag{4}$$

where x^0 is the time coordinate and the ellipses denote terms with x^0 derivatives up to first order. If the coefficient matrix of the $\partial_0^2 g_{cd}$ term is not degenerate on a $x^0 =$ constant surface Σ, we can solve this equation to fix the values of $\partial_0^2 g_{cd}$ and the time evolution of the metric will be uniquely determined. On the other hand, if the matrix is degenerate then $\partial_0^2 g_{cd}$ will not be specified by this equation, hence it is allowed to be discontinuous across Σ. In such a case we call Σ characteristic. Since a disturbance at a spacetime point can propagate on a characteristic surface but not beyond it, a characteristic surface gives a boundary of causal wave propagation. The coefficient matrix in Eq. (4) is written covariantly using the normal $\xi = dx^0$ of Σ as

$$P(x,\xi) = \frac{\partial E_{ab}}{\partial(\partial_e\partial_f g_{cd})}\xi_e\xi_f, \tag{5}$$

which is called the principal symbol of the field equation (4). A surface Σ with normal ξ is characteristic if $\det P(x,\xi) = 0$ is satisfied on Σ, and an eigenvector for a vanishing eigenvalue of P corresponds to a mode propagating on Σ.

We applied this formalism to Lovelock theories of gravity in Ref. 2 to clarify causal structure and various features of wave propagation in this theory. In our approach, the principal symbol is separated into the GR and Lovelock part as

$$(P\cdot t)^a{}_b = (P_{GR}\cdot t)^a{}_b + (\mathcal{R}\cdot t)^a{}_b, \quad (P_{GR}\cdot t)_{ab} \equiv -\frac{1}{2}\xi^2 t_{ab} + \xi^c\xi_{(a}t_{b)c} - \frac{1}{2}\xi_a\xi_b t^c{}_c, \tag{6}$$

where t_{ab} is a symmetric tensor. Then, by means of the gauge transformation $t_{ab} = \hat{t}_{ab} + \xi_{(a}X_{b)}$ with an arbitrary vector X_a, the characteristic equation is reduced into an eigenvalue equation for $\mathcal{R}$ given by (for non-null ξ)

$$\mathcal{R}(x,\xi)\cdot\hat{t} = \frac{1}{2}\xi^2\hat{t}. \tag{7}$$

Hence the problem to find characteristic surfaces is reduced to an eigenvalue problem for $\mathcal{R}$, which can be regarded as a matrix in the $\frac{1}{2}d(d-3)$-dimensional space of physical modes.

In Ref. 2, we applied this method to the following background spacetimes:

- *Spacetime with Killing horizons*: On a Killing horizon $\mathcal{H}$, the Riemann tensor obeys $R_{0i0j} = R_{0ijk} = 0$, where i, j are indices for the angular directions. Employing this property we have found that $\mathcal{H}$ is characteristic and hence it is a boundary of causal domain even in Lovelock theories.

- *Ricci flat type N spacetimes*: Type N spacetimes are algebraically special spacetimes in which the Riemann tensor components vanish except for $R_{1i1j} = \Omega'_{ij}$ ($\Omega'_{ii} = 0$), where we have introduced a null basis by $e_0^a \equiv \ell^a$, $e_1^a \equiv n^a$, $e_i^a \equiv m_i^a$ with null vectors ℓ and n such that $\ell \cdot n = 1$ and m_i being a set of orthonormal spacelike vectors orthogonal to ℓ and n. On this background spacetime, we have found that the principal symbol $P(x, \xi)$ factorizes into the normal vectors contracted with effective metrics:

$$P(x,\xi) = \prod_I G_I^{ab} \xi_a \xi_b, \qquad G_I^{ab} \equiv g^{ab} + \omega_I\, \ell^a \ell^b, \tag{8}$$

  where ω_I depends only on the curvature tensor Ω'_{ij} and may differ from mode to mode. A characteristic cone becomes superluminal if $\omega_I > 0$, and it turned out that there will be always at least one superluminal mode. Despite the existence of such a superluminal mode, causality on this background spacetime is defined in a usual manner using the largest cone but not the light cone.
- *Static, maximally symmetric black holes*: Another examples of the exact solution in Lovelock theories are the static, maximally symmetric black hole solutions,[4,5] and their stability had been investigated intensively in many works.[6–9] The effective metrics on this background spacetime can be read out from the coefficients of the master equations for perturbations clarified in those works. It turned out that the characteristic cones for the tensor, vector and scalar modes coincide with each other in the radial direction but differ in the angular directions, and also that the propagation speed depends on the distance from the black hole and there can be superluminal modes for some parameter choices. Another significant feature was that the hyperbolicity of the effective metric may be violated near horizons of small black holes, where the effect of spacetime curvature becomes large. This phenomena was interpreted as classical instability of black holes in the previous works because exponentially growing mode appears when it happens. We argued that it is more appropriate to interpret it as a loss of well-posedness of the initial value problem, with which a solution describing time evolution does not exist in general and then it does not make sense to consider the initial value problem based on the equation of motion.

3. Shock Formation in Lovelock Theories

By the analysis on characteristic surfaces in Lovelock theories, it was clarified that the propagation speed of gravitational wave varies depending on background spacetime. Such a variable sound speed typically results in nonlinear phenomena, including a shock formation caused by accumulation of waves. Indeed, for a relativistic perfect fluid in $3+1$ dimensions it has been proved that a smooth initial data with small amplitude can cause shock formation within finite time duration.[10]

We examined such a shock formation could occur for gravitational wave in Lovelock theories in Ref. 3.

A shock formation is caused by waveform distortion due to nonlinear effects in wave transport. One of the simplest example of such a shock formation can be found from Burgers' equation

$$u_t + u\,u_x = 0, \tag{9}$$

for which the wave profile $u(x)$ becomes singular within finite time due to the nonlinear term $u\,u_x$. To check if a similar phenomenon could occur in Lovelock theories in a simple manner, we focused on the transport equation of a discontinuity in the second derivative of the metric instead of waves with smooth profile. Such discontinuity must propagate on a characteristic surface in the background spacetime toward which the discontinuity propagates. The discontinuous part of the equation of motion (4) on the characteristic surface Σ is given by

$$[E_{ab}] = \frac{\partial E_{ab}}{\partial(\partial_0^2 g_{cd})}\left[\partial_0^2 g_{cd}\right], \tag{10}$$

where $[f]$ is the discontinuity of f across Σ. This equation implies that the discontinuity $[\partial_0^2 g_{cd}]$ must be proportional to an eigenvector of the principal symbol $P(x,\xi)$, hence it is given by

$$\left[\partial_0^2 g_{cd}\right] = \Pi(x^i)\, r_{cd}, \tag{11}$$

where r_{cd} is the eigenvector of $P(x,\xi)$ and $\Pi(x^i)$ is the amplitude of the discontinuity, which depends on the coordinates x^i on Σ. The transport equation of the discontinuity amplitude Π can be derived by differentiating the equation of motion with respect to x^0 first and then taking the discontinuous part, which results in

$$\dot{\Pi} + N\,\Pi^2 + M\,\Pi = 0, \tag{12}$$

where $\dot{\Pi}$ is the derivative of Π with respect to a parameter along the propagation direction on Σ, and the coefficients N, M are functions constructed from derivatives of the equation of motion. We call the transport equation (12) genuinely nonlinear when the coefficient of the nonlinear term N is generically non-zero. Significance of the nonlinear term N can be seen in the formal solution of Eq. (12):

$$\Pi(s) = \frac{\Pi(0)\, e^{\Phi(s)}}{1 + \Pi(0)\int_0^s N(s')e^{-\Phi(s')}ds'}, \tag{13}$$

where s is the parameter along the propagation direction and $\Phi(s) \equiv \int_0^s M(s')ds'$. When N is not zero, we can choose the initial amplitude of the discontinuity $\Pi(0)$ appropriately to make the denominator of Eq. (13) vanishing at some s. The discontinuity amplitude $\Pi(s)$ will then diverge at such a point, and it corresponds to the formation of a shock. In Lovelock theory, the coefficient N is given by

$$N = -4\sum_{p\geq 2} p(p-1)k_p \delta^{0ikmpr_5\ldots r_{2p}|0jlnps_5\ldots s_{2p}}\Gamma^0_{ij} r_{kl} r_{mn} r_{pq} R_{r_5 r_6 s_5 s_6}\cdots \times R_{r_{2p-1}r_{2p}s_{2p-1}s_{2p}}. \tag{14}$$

It does not vanish for a generic background, hence Lovelock theories are genuinely nonlinear, and then they may suffer from the shock formation.

Having established that Lovelock theories are genuinely nonlinear and suffer from shock formation, it would be now better to find explicit examples of shock formations based on the transport equation (12). It turns out that, although N shown in Eq. (14) does not vanish generically, it does for simple examples of spacetimes and characteristic surfaces including i) flat spacetime, ii) a Ricci flat type N spacetime and a surface along the null direction ℓ, iii) a static spherically symmetric and spherically symmetric surface, and iv) a Killing horizon. Hence we need to look at more complicated system to observe a shock formation. The Ricci-flat type N solution provides such a nontrivial yet simple background solution on which a shock formation could occur. On this background it was shown that, for a discontinuity propagating along the direction opposite from the null direction ℓ, a shock forms within finite time duration for arbitrarily small initial amplitude $\Pi(0)$.

4. Summary and Outlook

In a series of works on Lovelock theories of gravity,[2,3] we investigated causality, hyperbolicity and shock formation of gravitational wave in these theories.

We have shown that, on some background spacetimes, gravitational wave propagates faster than light but still the causality is maintained if we define it with respect to the fastest propagating mode. As shown in this example, the existence of superluminality itself does not immediately imply a problem about causality. Related to this point, Ref. 11 studied Shapiro time advance in Einstein-Gauss-Bonnet theory and argued that such a time advance does not necessarily imply existence of the closed timelike curve, which was used to constrain model parameters of this theory in Ref. 12. We have also shown that the hyperbolicity can be violated in a region where spacetime curvature becomes larger than characteristic scale set by the coupling parameters of the theories. An interesting problem is to check if such a hyperbolicity violation could occur in a time evolution starting from generic smooth initial data, because if a hyperbolicity violation occurs in a generic time evolution it would indicate that the theory is pathological and should be discarded or improved.

It was shown also that Lovelock theories are genuinely nonlinear and then they suffer from shock formation. A shock formation corresponds to a formation of singularity which could be naked, hence it can be interpreted as a violation of the weak cosmic censorship. We however confirmed that a shock does not form in flat spacetime, which may be suggesting that Minkowski spacetime in this theory is stable. In our analysis we looked at the transport of discontinuity in the second derivative of metric to simplify our analysis. It is desirable to check if a shock could arise from a smooth initial data. Since the theory is shown to be genuinely nonlinear, it is expected that a shock formation will occur even when we start from such a smooth data, although we will need new techniques to examine such an expectation.

Lovelock theories can be nontrivial only in higher-dimensional spacetime because the Lovelock part of the Lagrangian does not contribute to dynamics in four dimensions. One way to construct a nontrivial theory in four dimensions is to apply dimensional reduction to Lovelock theories in higher dimensions down to four dimensions, by which the theories can be converted into a four-dimensional theory with scalar fields.[13,14] Phenomena observed in Lovelock theories, such as modifications to causality and shock formation, are expected to be present also in the four-dimensional theories obtained by the dimensional reduction, and possibly also in more general scalar-tensor theories. It would be interesting to pursue this issue for various four-dimensional scalar-tensor theories.[15]

Acknowledgments

This work was supported by the European Research Council grant no. ERC-2011-StG 279363-HiDGR.

References

1. D. Lovelock, "The Einstein tensor and its generalizations," *J. Math. Phys.* **12**, 498 (1971).
2. H. Reall, N. Tanahashi and B. Way, "Causality and Hyperbolicity of Lovelock Theories," *Class. Quant. Grav.* **31**, 205005 (2014)
3. H. S. Reall, N. Tanahashi and B. Way, "Shock Formation in Lovelock Theories," *Phys. Rev. D* **91**, no. 4, 044013 (2015)
4. D. G. Boulware and S. Deser, String Generated Gravity Models, *Phys. Rev. Lett.* **55** (1985) 2656.
5. J. T. Wheeler, Symmetric Solutions to the Maximally Gauss Bonnet Extended Einstein Equations, *Nucl. Phys.* **B273** (1986) 732.
6. T. Takahashi and J. Soda, Stability of Lovelock Black Holes under Tensor Perturbations, *Phys. Rev.* **D79** (2009) 104025
7. T. Takahashi and J. Soda, Instability of Small Lovelock Black Holes in Even-dimensions, *Phys. Rev.* **D80** (2009) 104021
8. T. Takahashi and J. Soda, Master Equations for Gravitational Perturbations of Static Lovelock Black Holes in Higher Dimensions, *Prog. Theor. Phys.* **124** (2010) 911–924
9. T. Takahashi and J. Soda, Catastrophic Instability of Small Lovelock Black Holes, *Prog. Theor. Phys.* **124** (2010) 711–729
10. D. Christodoulou, "The formation of shocks in 3-dimensional fluids", *European Mathematical Society* (2007).
11. G. Papallo and H. S. Reall, "Graviton time delay and a speed limit for small black holes in Einstein-Gauss-Bonnet theory," *JHEP* **1511**, 109 (2015)
12. X. O. Camanho, J. D. Edelstein, J. Maldacena and A. Zhiboedov, "Causality Constraints on Corrections to the Graviton Three-Point Coupling," arXiv:1407.5597 [hep-th].
13. K. Van Acoleyen and J. Van Doorsselaere, "Galileons from Lovelock actions," *Phys. Rev. D* **83**, 084025 (2011) [arXiv:1102.0487 [gr-qc]].
14. C. Charmousis, "From Lovelock to Horndeski's Generalized Scalar Tensor Theory," *Lect. Notes Phys.* **892**, 25 (2015)
15. S. Ohashi and N. Tanahashi, in progress.

Boundary Effect of Anomaly-Induced Action in Curved Spacetime*

Che-Min Shen[a], Keisuke Izumi[b] and Pisin Chen[c]

Leung Center for Cosmology and Particle Astrophysics,
National Taiwan University, Taipei 10617, Taiwan, R.O.C.
[a]*f97222024@ntu.edu.tw*
[b]*izumi@phys.ntu.edu.tw*
[c]*chen@slac.stanford.edu*

The vacuum expectation value of energy momentum tensor is one of the important issues to investigate the quantum effect in curved spacetime. Anomaly-induced action, which gives the stress tensor with the same trace as the trace anomaly, is expected to be a good formalism to be used for computing the VEV of energy momentum tensor. However, in the conventional approach the correspondence between solutions from anomaly-induced action and the related vacuum state is unclear. We showed that after including the boundary contribution, the correspondence becomes clearer. With the modified anomaly-induced action that includes the boundary effect, we are able to compute the energy momentum tensor corresponding to specific vacua naturally.

1. Introduction

The semiclassical approach, quantum field theory in curved spacetimes where the quantum divergences of fields are covariantly renormalized, has been applied widely to study quantum corrections to general relativity.[1] The expectation value of the stress tensor of quantum matter fields can be derived with this procedure, and the result suggests that, even in conformal field theories, a nonzero trace of stress tensor arises through the renormalization. This nonzero trace of stress tensor is called the trace (or conformal) anomaly.[1–3]

The expectation value of the stress tensor of quantum matter fields is in general difficult to compute. One tricky way to tackle with this problem is rebuilding the corresponding *anomaly-induced action*.[4–6] It can be expected and has been checked in some specific cases that in two-dimensional spacetime the anomaly-induced action can exactly describe the stress tensor of quantum field in vacuum states.[7,8] The anomaly-induced action is naturally built in non-local form, and can be localized by introducing an additional auxiliary scalar field.[9,10] Different solutions of the auxiliary scalar fields could describe different quantum states of the original conformal matter field. However, the correspondence between solutions from anomaly-induced action and the associated vacuum state is unclear.

In this article we summarize our recent work.[11] We take the boundary effect into consideration in the discussion of anomaly-induced action, which has been neglected

*Proceedings for *The 2nd LeCosPA Symposium: Everything about Gravity, Celebrating the Centenary of Einsteins General Relativity.* Talk given by Che-Min Shen, on December 17, 2015, Taipei, Taiwan.

in the previous works.[4–8, 12] It can be showed that in this modified anomaly-induced action with the boundary effect, the correspondence becomes clearer and it is able to compute the energy momentum tensor corresponding to specific vacua naturally. Finally, We apply our result to two-dimensional Schwarzschild metric and naturally obtain the stress tensors for the Boulware, Hartle-Hawking and Unruh vacuum states.

2. The analysis of the quantum divergence behavior through path-integral quantization and one-loop effective action

Start with the Lagrangian density of a scalar field in a n-dimensional curved space-time

$$\mathscr{L} = \sqrt{-g}\,\frac{1}{2}\{-g^{\mu\nu}\nabla_\mu\phi\nabla_\nu\phi - (m^2+\xi R)\phi^2\}, \tag{1}$$

where m is mass, ξ is a coupling constant and R is Ricci scalar.

From path-integral formalism, the one-loop effective action can be obtained from

$$e^{-S_{eff}} := \int [D\phi] e^{-S[\phi]}. \tag{2}$$

By using the DeWitt-Schwinger expansion, this effective Lagrangian can be expanded as

$$L_{eff}(x) \approx \frac{1}{2(4\pi)^{n/2}} \sum_{j=0}^{\infty} a_j(x) m^{n-2j}\Gamma(j-\frac{n}{2}) \tag{3}$$

where a_j's are geometric quantities which is written in:

$$a_0(x) = 1, \tag{4}$$

$$a_1(x) = (\frac{1}{6} - \xi)R, \tag{5}$$

$$\ldots .$$

The gamma function $\Gamma(j-n/2)$ within the effective Lagrangian diverges when its argument is non-positive integer, and thus in even dimension, some counter terms should be introduced to renormalize these divergent parts.

In the remaining of this paper, we will focus on conformal scalar field theory, i.e. $\xi(n) := \frac{(n-2)}{4(n-1)}$, and $m = 0$. Due to the conformal symmetry, the classical stress tensor is traceless. However, in even-dimension, due to the counter terms the renormalized stress tensor for the conformal scalar field has a nonzero trace,

$$g^{\mu\nu}\langle T_{\mu\nu}\rangle_{ren} = -\frac{a_k(x)}{(4\pi)^k} \quad \text{with} \quad k = \frac{n}{2}, \quad k \in \mathbb{N}. \tag{6}$$

this violates the conformal symmetry, and thus it is called the trace (or conformal) anomaly.

3. Two-dimensional anomaly-induced action without boundary

According to last section, the divergent parts of the effective action for a conformally coupled scalar field in 2-dimensional spacetime can be derived as:

$$S_{ct} = \frac{-1}{24\pi} \lim_{n\to 2} \int d^2x\sqrt{-g}\frac{R}{(n-2)}. \tag{7}$$

Together with the definition of the Wess-Zumino (WZ) action and anomaly-induced action (S_{anom}):

$$\Gamma_{WZ}[\bar{g},\sigma] := S_{ren}[\bar{g}] - S_{ren}[g] = -\left[S_{ct}[\bar{g}] - S_{ct}[g]\right] := S_{anom}[\bar{g}] - S_{anom}[g], \tag{8}$$

we can obtain a nonlocal anomaly-induced action [a]

$$S_{anom}[g] = \frac{1}{96\pi}\int d^2x \int d^2x' \sqrt{-g}\sqrt{-g'}R(x)D_2(x,x')R(x'), \tag{10}$$

where $\bar{g}_{\mu\nu} := \exp(-2\sigma)g_{\mu\nu}$, and D_2 is the inverse operator of D'Alembert operator which is defined as

$$\Box D_2(x,x') = -\frac{\delta(x-x')}{\sqrt{-g}}. \tag{11}$$

By introducing a real scalar field φ

$$\varphi(x) := \int d^2x' D_2(x,x')R(x') \tag{12}$$

whose field equation is [b]

$$\Box\varphi = -R, \tag{13}$$

the anomaly action (10) can be localized as

$$S_{anom}[g,\varphi] = \frac{1}{96\pi}\int d^2x\sqrt{-g}[g^{\mu\nu}\nabla_\mu\varphi\nabla_\nu\varphi - 2\varphi R]. \tag{14}$$

Then, the corresponding stress tensor can be obtained as

$$\mathrm{T}^{anom}_{\mu\nu} = \frac{1}{24\pi}[-\nabla_\mu\nabla_\nu\varphi + g_{\mu\nu}\Box\varphi - \frac{1}{2}(\nabla_\mu\varphi)(\nabla_\nu\varphi) + \frac{1}{4}g_{\mu\nu}(\nabla_\alpha\varphi)(\nabla^\alpha\varphi)], \tag{15}$$

whose trace consists with the well-known trace anomaly,

$$g^{\mu\nu}\mathrm{T}^{anom}_{\mu\nu} = \frac{1}{24\pi}\Box\varphi = -\frac{1}{24\pi}R. \tag{16}$$

Therefore, it is concluded that scalar field action (14) describes the anomalous contribution.

---

[a]During the derivation, we have imposed the symmetric condition of D_2, i.e. $D_2(x,x') = D_2(x',x)$, and use the relation:

$$2\sqrt{-g}\Box\sigma = 2\sqrt{-\bar{g}}\bar{\Box}\sigma = \sqrt{-\bar{g}}\bar{R} - \sqrt{-g}R. \tag{9}$$

[b]In the previous works for four-dimensional case,[9,10] in order to derive the localized anomaly action, two auxiliary scalar fields which share the same green function (inverse operator) D_4 analogous to D_2 should be introduced. However, according to our recent work,[13] we find that introduce only one auxiliary scalar field is enough to construct the localized S_{anom} in four-dimension.

4. Two-dimensional anomaly-induced action with boundary

In the following, we will introduce timelike boundaries and including the surface terms (i.e. the boundary effect) associated to it which was ignored in the previous works.[4–8,12] It turns out that the boundary effect provides the auxiliary scalar an additional boundary constraint.

The divergent part of effective action including the boundary contribution corresponding to Neumann or Dirichlet boundary condition is derived in Refs. 14, 15, and in two-dimension it has the following form:

$$S_{ct}[g] := \frac{-1}{24\pi}\frac{\int_{\mathcal{M}} d^2x\sqrt{-g}R + 2\int_{\Sigma} d^1x\sqrt{-\gamma}K}{n-2}, \tag{17}$$

where $\mathcal{M}$ is two-dimensional spacetime and Σ is the timelike boundary.

By introducing an arbitrary scalar function $f(x)$ is unity on the boundary and arbitrary elsewhere, we can rewrite the action of the counter terms

$$S_{ct}[g] = \frac{-1}{24\pi}\frac{\int d^2x\sqrt{-g}R + 2\int d^1x\sqrt{-\gamma}fK}{n-2}. \tag{18}$$

Then, the corresponding WZ action can be shown as the form:[16]

$$\begin{aligned}\Gamma_{WZ}[\bar{g},\sigma] &= \frac{1}{24\pi}\lim_{n\to 2}\\ &\quad\times\left[\frac{(\int d^2x\sqrt{-\bar{g}}\bar{R} + 2\int d^1x\sqrt{-\bar{\gamma}}\bar{K}) - (\int d^2x\sqrt{-g}R + 2\int d^1x\sqrt{-\gamma}K)}{n-2}\right]\\ &= -\frac{1}{24\pi}\left\{\int d^2x\sqrt{-\bar{g}}[\sigma\bar{R} - \sigma\bar{\Box}\sigma] + \int d^1x(\sqrt{-\gamma}\sigma K + \sqrt{-\bar{\gamma}}\sigma\bar{K})\right\}\\ &=: S_{anom}[\bar{g}] - S_{anom}[g].\end{aligned} \tag{19}$$

Define a self-adjoint operator L_2^f as

$$L_2^f := (-\Box + \nabla^\mu f n_\mu n_\nu \nabla^\nu), \tag{20}$$

which satisfies a relation analogous to eq.(9)

$$2\sqrt{-g}L_2^f\sigma = 2\sqrt{-\bar{g}}\bar{L}_2^f\sigma = \sqrt{-g}[R + 2\nabla^\mu(n_\mu fK)] - \sqrt{-\bar{g}}[\bar{R} + 2\bar{\nabla}_\mu(\bar{n}^\mu f\bar{K})], \tag{21}$$

we thus can get the non-local anomaly-induced action from Eq.(19)

$$\begin{aligned}S_{anom}[g] = \frac{1}{96\pi}\Bigg[&\int d^2x\int d^2x'\sqrt{-g}\sqrt{-g'}(R(x) + 2\nabla^\mu(n_\mu fK))D_f(x,x')(R(x')\\ &+ 2\nabla'^\mu(n'_\mu f'K')) - 4\int d^2x\sqrt{-g}fK^2\Bigg].\end{aligned} \tag{22}$$

Here, D_f is the symmetric inverse operator of L_2^f, which is defined by

$$L_2^f D_2^f(x,x') = -\frac{\delta(x-x')}{\sqrt{-g}}, \qquad D_2^f(x,x') = D_2^f(x',x). \tag{23}$$

By introducing a real auxiliary scalar field φ

$$\varphi := \int d^2x' \sqrt{-g} D_f(x,x')[R' + 2\nabla'_\mu (n'^\mu f' K')], \tag{24}$$

again the localized version of the anomaly-induced action (22) can be expressed as

$$S_{anom}[g,\varphi] = \frac{1}{96\pi}\left\{ \int d^2x\sqrt{-g}[\varphi L_2^f \varphi - 2\varphi(R(x) + 2\nabla^\mu(n_\mu f K))] \right.$$
$$\left. -4\int d^2x\sqrt{-g} f K^2 \right\}.$$

By taking the limit f approach to the step function, it allow us to grasp the boundary effect. Then, the anomaly-induced action (25) becomes

$$S_{anom}[g,\varphi] \to \frac{1}{96\pi}\{\int d^2x\sqrt{-g}(-\varphi\Box\varphi - 2\varphi R) + \int d^1x\sqrt{\gamma}(\varphi n_\mu \nabla^\mu \varphi - 4\varphi K)\}. \tag{25}$$

It turns out that we have exactly the same action as the previous one (14) expect for the additional boundary terms. The boundary terms have no additional contribution on the stress tensor except on the boundary, and thus the obtained stress tensor in the bulk, $\mathcal{M}$, is still the same as Eq.(15). However, vary this action with φ will provide us not only the original equation of motion

$$\Box\varphi = -R, \tag{26}$$

but also an additional boundary constraint

$$n^\nu \nabla_\nu \varphi = 2K, \qquad x \in \Sigma. \tag{27}$$

This means that there is an additional boundary constraint on φ which was not taken into consideration in the previous works.

As an example, we apply our result to the 2-dim Schwarzschild spacetime. By solving different boundary conditions, finally we can obtain stress tensor for different vacuum states as follows

Hartle-Hawking Vacuum

$$T_{tt} = \frac{1}{24\pi}[\left(\frac{7M^2}{r^4} - \frac{4M}{r^3} + \frac{1}{16M^2}\right) + \frac{{A_H}^2}{64M^2}({R_H}^2 + {T_H}^2)],$$

$$T_{rr} = \frac{1}{24\pi}[\left(1-\frac{2M}{r}\right)^{-2}\left(\frac{1}{16M^2} - \frac{M^2}{r^4}\right) + \frac{{A_H}^2}{64M^2}\left(1-\frac{2M}{r}\right)^{-2}({R_H}^2 + {T_H}^2)],$$

$$T_{tr} = T_{rt} = \frac{1}{24\pi}[\frac{{A_H}^2}{32M^2}\left(1-\frac{2M}{r}\right)^{-1}(T_H R_H)].$$

Boulware Vacuum

$$T_{\mu\nu} = \frac{1}{24\pi}\begin{pmatrix} \frac{-4Mr+7M^2}{r^4} + \frac{A_B^2}{4} & 0 \\ 0 & -\frac{M^2}{r^2(r-2M)^2} + \frac{A_B^2 r^2}{4(r-2M)^2} \end{pmatrix}$$

where A_H, A_B are proportional to temperature of different frames.

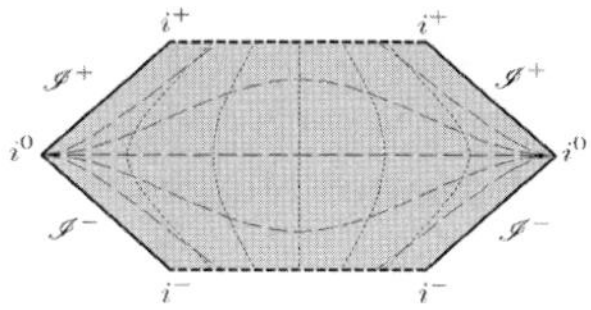

Fig. 1. Region corresponding to Hartle-Hawking vacuum.

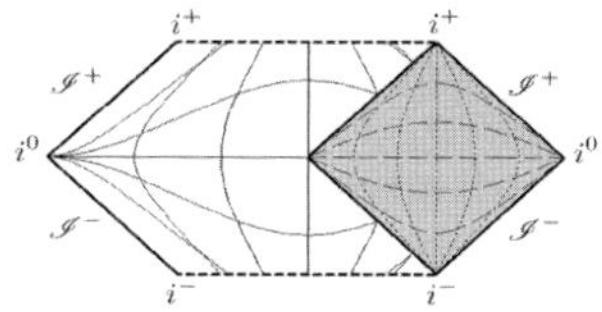

Fig. 2. Region corresponding to the Boulware vacuum.

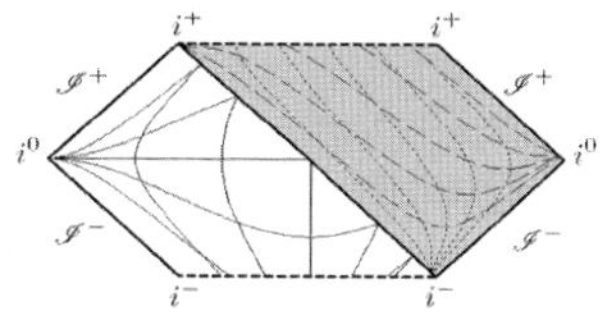

Fig. 3. Region corresponding to the Unruh vacuum.

5. Conclusion

We have shown that after taking boundary effect into consideration, the auxiliary scalar field in anomaly-induced action should satisfy an additional boundary condition. With this modified anomaly-induced action, the correspondence between solutions from anomaly-induced action and the related vacuum state is clear. We thus are able to compute the energy momentum tensor corresponding to specific vacua naturally.

Acknowledgment

The authors would like to thank Je-An Gu for useful comments. K. I. is supported by Taiwan National Science Council under Project No. NSC101-2811-M-002-103 and this work is also supported by Leung Center for Cosmology and Particle Astrophysics, Taiwan, R.O.C.

References

1. N. D. Birrell and P. C. W. Davies, Quantum Fields in Curved Space (Cambridge University Press, Cambridge, England, 1982) Ch6.
2. S. Deser, M. J. Duff and C. J. Isham, Nucl. Phys. B **111**, 45 (1976).
3. M. J. Duff, Nucl. Phys. B **125**, 334 (1977).
4. A. M. Polyakov, Phys. Lett. B **103**, 207 (1981).
5. R. J. Riegert, Phys. Lett. B **134**, 56 (1984).
6. E. S. Fradkin and A. A. Tseytlin, Phys. Lett. B **134**, 187 (1984).
7. R. Balbinot, A. Fabbri and I. L. Shapiro, Nucl. Phys. B **559**, 301 (1999).
8. E. Mottola and R. Vaulin, Phys. Rev. D **74**, 064004 (2006).
9. P. O. Mazur and E. Mottola, Phys. Rev. D **64**, 104022 (2001).
10. I. L. Shapiro and A. G. Zheksenaev, Phys. Lett. B **324**, 286 (1994).
11. C. M. Shen, K. Izumi and P. Chen, Phys. Rev. D **92**, no. 2, 024035 (2015) Addendum: [Phys. Rev. D **92**, no. 4, 049902 (2015)] doi:10.1103/PhysRevD.92.049902, 10.1103/PhysRevD.92.024035 [arXiv:1505.00959 [gr-qc]].
12. C. Barcelo, R. Carballo and L. J. Garay, Phys. Rev. D **85**, 084001 (2012).
13. C. M. Shen, K. Izumi and P. Chen, working in progress.
14. T. P. Branson and P. B. Gilkey, Commun. Part. Diff. Eq. **15**, 245 (1990).
15. I. G. Avramidi, Phys. Atom. Nucl. **56**, 138 (1993) [Yad. Fiz. **56N1**, 243 (1993)].
16. D.V.Fursaev and D.V.Vassilevich, Operators, Geometry and Quanta, Springer, 2011, Sec 8.6

Quantum Gravity Corrections to the Standard Model Higgs in Einstein and R^2 Gravity

Yugo Abe, Masaatsu Horikoshi

Department of Physics, Shinshu University, Asahi3-1-1, Matsumoto, Nagano 390-8621, Japan

Takeo Inami

Department of Physics, National Taiwan University, Taipei 10617, Taiwan, R.O.C.
Theoretical Research Division, Nishina Center, RIKEN, Wako 351-0198, Japan

We evaluate quantum gravity corrections to the standard model Higgs potential $V(\phi)$ a la Coleman-Weinberg and examine the stability question of $V(\phi)$ at scales of Planck mass $M_{\rm Pl}$. We compute the gravity one-loop corrections by using the momentum cut-off in Einstein gravity. The gravity corrections affect the potential in a significant manner for the value of $\Lambda = (1-3)M_{\rm Pl}$. In view of reducing the UV cut-off dependence we also make a similar study in the R^2 gravity.

Keywords: Higgs potential; Coleman-Weinberg; Planck scales; cut off method; R^2 gravity.

1. Introduction

It is curious that the mass of the Higgs boson $m_{\rm H}$ ($= 125.09 \pm 0.24\,{\rm GeV}$), which has recently been discovered at LHC, lies far outside of the mass bound derived from the one-loop radiative corrections.[1,2] This bound arises from the stability condition on the Higgs quartic coupling λ, i.e. $\lambda(\mu) > 0$. The large two-loop corrections come into play and the renormalization group (RG) flows of λ change drastically. The flows are also tangled with the top quark mass $m_{\rm t}$. Some fine-tuning of the parameters, especially that of $m_{\rm t}$, yields the λ barely in accord with the boundary values of the stability bound extended to the scales of Planck mass $M_{\rm Pl}$.[3–5] It implies an interesting possibility that the standard model may be the correct unification theory all the way up to the Planck scales $M_{\rm Pl}$.[6–8] This scenario is consistent with the so far vain results of the susy particle search at the LHC experiment.

It is a common belief that quantum gravity effects should manifest itself at Planck scales, irrespectively of the correct quantum theory of gravity. Hence it is a very urgent problem to study whether incorporating quantum gravity corrections may change drastically the RG analyses with the matter loop corrections alone.[9–11]

In this paper, we calculate the gravity loop corrections to the Higgs potential $V(\phi)$ and examine the property of the Higgs potential at Planck scales $M_{\rm Pl}$. In the Einstein gravity, we have found that there is a significant difference between the Higgs potential with and without gravity corrections. In the previous works,[9,10] the gravity loop corrections to the ϕ^6 and ϕ^8 terms are studied. We consider all such corrections on the RG analysis of $V(\phi)$. This talk contains a brief summary of our recent paper.[12]

2. Gravitational Coleman-Weinberg corrections

In the standard model (SM), one- and two-loop contributions to the bare Higgs mass and the bare couplings have been studied in details.[3–5] We take into account the gravity one-loop effects in addition to the matter loop effects and will see how two contributions compete in the energy region of Planck mass scales $M_{\rm Pl}$.

We derive the Higgs effective potential $V(\phi)$ in the Einstein's gravity theory, by extending the Coleman-Weinberg mechanism[16] to gravity loops. We begin by writing the action for the Higgs field H and the metric $g_{\mu\nu}$

$$S = \int d^4x\sqrt{-g}\left[\frac{2}{\kappa^2}R + g^{\mu\nu}(\partial_\mu H)^\dagger(\partial_\nu H) + m^2 H^\dagger H - \lambda(H^\dagger H)^2 + \cdots\right] \qquad (1)$$

where $\kappa := \sqrt{32\pi G} = \sqrt{32\pi}M_{\rm Pl}^{-1}$ ($M_{\rm Pl} = 1.22\times 10^{19}$GeV). The ellipses show the terms of gauge and fermion fields. Expanding the fields around background fields as $H^\dagger = 1/\sqrt{2}\,(\sigma_1 - i\,\pi_1, \phi + \sigma_2 - i\,\pi_2)$ and $g_{\mu\nu} = \eta_{\mu\nu} + \kappa h_{\mu\nu}$, we evaluate the gravity corrections for the tree level Higgs potential

$$V_{\rm tree} = -\frac{1}{2}m^2\phi^2 + \frac{1}{4}\lambda\phi^4\,. \qquad (2)$$

We take the de Donder gauge fixing term $\mathcal{L}_{\rm gf}$. It is given on the Minkowski background by

$$\mathcal{L}_{\rm gf} = -\eta_{\alpha\beta}\left(\eta^{\mu e}\eta^{\nu\alpha} - \frac{1}{2}\eta^{\mu\nu}\eta^{e\alpha}\right)\left(\eta^{\rho f}\eta^{\sigma\beta} - \frac{1}{2}\eta^{\rho\sigma}\eta^{f\beta}\right)h_{\mu\nu,e}h_{\rho\sigma,f}. \qquad (3)$$

The Higgs and graviton 1-loop corrections to $V_{\rm tree}(\phi)$ is obtained in the cut-off method.[9,10]

$$\begin{aligned}\delta V_{\rm eff} &= \frac{3}{64\pi^2}\left(-m^2+\lambda\phi^2\right)^2\left(\ln\frac{-m^2+\lambda\phi^2}{\Lambda^2} - \frac{3}{2}\right)\\ &\quad + \frac{9}{256\pi^2}\kappa^4\left(-\frac{1}{2}m^2\phi^2 + \frac{1}{4}\lambda\phi^4\right)^2\left(\ln\left(\frac{\kappa^2}{2}\frac{\frac{1}{2}m^2\phi^2 + \frac{1}{4}\lambda\phi^4}{\Lambda^2}\right) - \frac{3}{2}\right)\\ &\quad + \sum_{i=\pm}\frac{C_i^2}{64\pi^2}\left(\ln\frac{C_i}{\Lambda^2} - \frac{3}{2}\right),\end{aligned} \qquad (4)$$

where $C_\pm$ is

$$C_\pm = \frac{1}{2}\left[m_C^2 - m_A^2 \pm \sqrt{(m_C^2 + m_A^2)^2 - 16m_B^4}\right]. \qquad (5)$$

Here $m_A^2 = \frac{\kappa^2}{2}(-\frac{m^2}{2}\phi^2 + \frac{\lambda}{4}\phi^4)$, $m_B^2 = \frac{\kappa}{2}\left(-m^2\phi + \lambda\phi^3\right)$ and $m_C^2 = -m^2 + 3\lambda\phi^2$.

The second and third terms in (4) are due to the graviton loops, which are suppressed by Planck mass at electro-weak scales. Gravity corrections give rise to logarithmic divergent terms of ϕ^6 and ϕ^8. Such higher power terms are not significant at usual energies, but they may become significant at high energy scales, i.e. $M_{\rm Pl}$.

The ϕ^2 and ϕ^4 terms may be renormalized in the usual way. Expanding the third term $\sum_{i=\pm} C_i^2 = m_C^4 + m_A^4 - 8m_B^4$, the counter terms (CT) are found

$$\begin{aligned}\delta V_{\text{CT}} &= \frac{3}{64\pi^2}\left(-m^2+\lambda\phi^2\right)^2 \ln\left(\frac{\Lambda^2}{\mu^2}\right) + \frac{1}{64\pi^2}\left(-m^2+3\lambda\phi^2\right)^2 \ln\left(\frac{\Lambda^2}{\mu^2}\right) \\ &- \frac{\kappa^2}{32\pi^2}\left(m^4\phi^2 - 2\lambda m^2\phi^4\right) \ln\left(\frac{\Lambda^2}{\mu^2}\right) + \frac{5\kappa^4}{512\pi^2} m^4\phi^4 \ln\left(\frac{\Lambda^2}{\mu^2}\right) .\end{aligned} \tag{6}$$

We add these gravitational 1-loop corrections to the tree level potential

$$V_{\text{eff}} = V_{\text{tree}} + \delta V_{\text{eff}} + \delta V_{\text{CT}} . \tag{7}$$

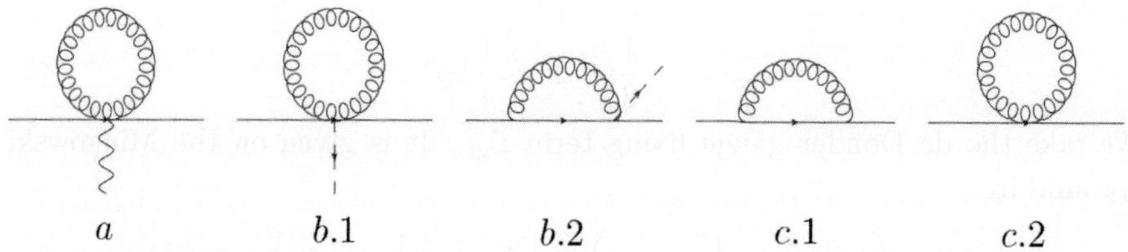

Fig. 1. Gravitational 1-loop diagrams for gauge couplings, Yukawa coupling and anomalous dimension of fermion.

The Higgs potential including the loop corrections can easily be obtained by use of RG. We already know the β functions in SM to the 2-loop corrections.[3,13]

Some graviton loop corrections have recently been computed.[9–11] We have further calculated gravity corrections to other coupling corrections, i.e. gauge and Yukawa couplings, as shown in Fig. 1. β functions and anomalous dimensions due to gravity corrections are

$$\begin{aligned}\beta_{m^2 g} &= \frac{5\kappa^2 m^2}{16\pi^2}\mu^2 - \frac{\kappa^2 m^4}{8\pi^2}, \ \beta_{\lambda g} = \frac{5\kappa^2\lambda}{16\pi^2}\mu^2 - \frac{\kappa^2\lambda m^2}{2\pi^2} - \frac{5\kappa^4 m^4}{64\pi^2}, \\ \beta_{y_t g} &= \frac{\kappa^2}{2\pi^2} y_t \mu^2, \ \gamma_{\phi g} = -\frac{\kappa^2 m^2}{32\pi^2}, \ \gamma_{tg} = \frac{27\kappa^2}{512\pi^2}\mu^2.\end{aligned} \tag{8}$$

3. The potential due to gravity corrections

We solve the RG equation for the coupling constants and V_{eff}, using the β functions and anomalous dimensions due to SM[3] and gravity corrections. We employ the threshold values of these quantities given in the literature[5]

$$
\begin{aligned}
g_y(m_t) &= 0.45187,\ g_2(m_t) \ = \ 0.65354, \\
g_3(m_t) &= 1.1645 - 0.00046\left(\frac{m_t - 173.15}{GeV}\right), \\
y_t(m_t) &= 0.93587 + 0.00557\left(\frac{m_t - 173.15}{GeV}\right) - 0.00003\left(\frac{m_H - 125}{GeV}\right), \\
\lambda(m_t) &= 0.12577 + 0.00205\left(\frac{m_H - 125}{GeV}\right) - 0.00004\left(\frac{m_t - 173.15}{GeV}\right).
\end{aligned}
\tag{9}
$$

We adjust the value of $m^2(m_t)$ so that $V_{\rm eff}(\phi)$gives correct the vacuum, $v = 246$GeV.

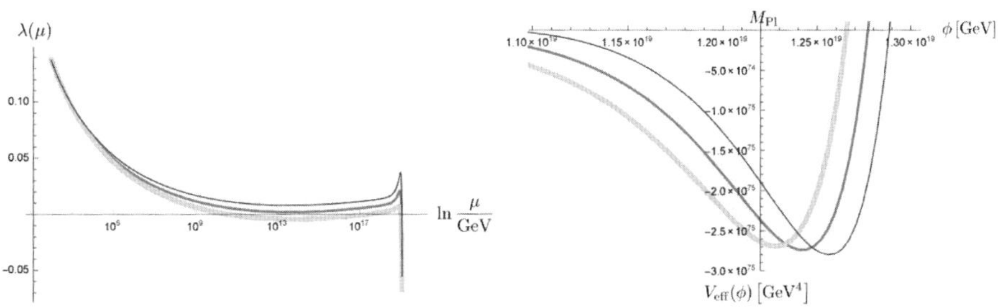

Fig. 2. RG flow of $\lambda(\mu)$ and $V_{\rm eff}(\phi)$ at $\phi \sim M_{\rm Pl}$ These curves correspond to top mass values $m_t = 174, 173, 172$GeV in thick and color order, Green, Red, Blue.

The influence of the gravitational corrections to $\lambda(\mu)$ and $V(\phi)$ is very small, the $\lambda(\mu)$ and $V(\phi)$ are consistent with that in SM, at $\mu, \phi < (O)(10^{17})$GeV.

i) μ-dependence of λ

Gravity corrections are noticeable around $\mu = \mathcal{O}(10^{18}\text{GeV})$, having the effect of increasing the value of λ rapidly, as seen in the Fig. 2. This behavior stops at $\mu = 0.9 \sim 1.0 \times M_{\rm Pl}$, and they start to decrease λ sharply. λ becomes negative at $\mu \sim M_{\rm Pl}$.

ii) $V(\phi)$ near $\phi \sim M_{\rm Pl}$

Regarding $V(\phi)$, gravitational effects begin to be noticeable at $\phi = \mathcal{O}(10^{18}\text{GeV})$, ϕ^6 and ϕ^8 terms become dominant. $\phi < M_{\rm Pl}$, $V(\phi)$ is increasing and is positive. $\phi \sim 0.8 \sim 0.9 \times M_{\rm Pl}$, $V(\phi)$ begins to fall. It stops falling at $\phi = 1.1 \times M_{\rm Pl}$, and takes the minimum there. At $\phi \sim 10^{20}$GeV and larger, $V(\phi)$ is rapidly increasing. However, at such large value of ϕ, more higher loop effects will be dominant, one cannot say anything reliable about graviton loop corrections.

iii) Λ dependence of $V(\phi)$

At $\phi \sim M_{\rm Pl}$ or larger, $\phi^6$and ϕ^8 terms are significant but they depend on the cut-off value Λ, as shown in Fig. 3. $V(\phi)$ take the minimum in each cut-off $\Lambda = 1 \times M_{\rm Pl}, 2 \times M_{\rm Pl}, 3 \times M_{\rm Pl}$. One may safely say that $V_{\rm eff}(\phi)$ take a minimum; the depth of the minimum depends strongly on the cut-off value Λ, however. The value

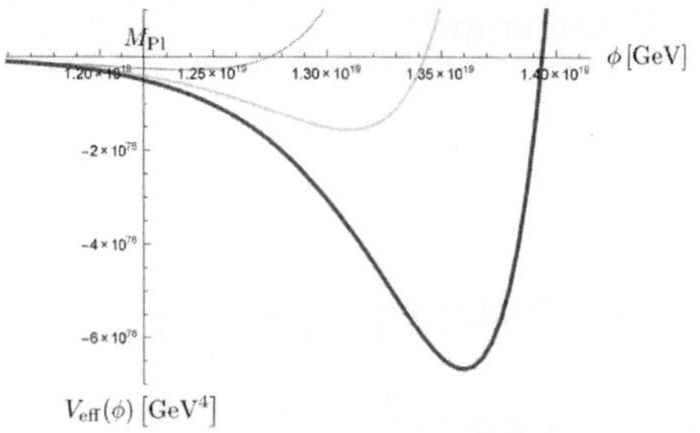

Fig. 3. Plot of $V_{\rm eff}(\phi)$ at $\phi \sim M_{\rm Pl}$, $m_H = 125.09$GeV, $m_t = 173.21$GeV and $\Lambda = 3M_{\rm Pl}, 2M_{\rm Pl}, M_{\rm Pl}$ in thick order. The vertical axis is on the position of $\phi = M_{\rm Pl}$.

of ϕ at the minimum (ϕ_m)increases in direct proportion to cut-off Λ. However Λ dependence of ϕ_m is very mild, ϕ_m stay around $M_{\rm Pl}$. Hence, $V(\phi)$ takes the minimum at $\phi < \Lambda$, but barely do not in $\Lambda = 1 \times M_{\rm Pl}$.

4. Gravity corrections in R^2 gravity

We have evaluated the quantum gravity corrections to Higgs potential $V(\phi)$ in the Einstein gravity in the cut-off method. The resulting $V(\phi)$ apparently depends on the cut-off Λ. It is desirable to study the gravity corrections in some UV renormalizable modified gravity, so that they have no Λ dependence or at most mild Λ dependence. One modest approach is to take R^2 gravity and evaluate R^2 gravity C-W corrections to $V(\phi)$. A certain R^2 gravity, i.e. the action of R^2 term $+(R_{\mu\nu})^2$ term is known to be UV renormalizable.[17,18] R^2 term alone may not be renormalizable but the UV divergence behavior is mild, and this theory tentatively serves our purpose.

$$S = \int d^4x\sqrt{-g}\left[\frac{2}{\kappa^2}\left(R+\alpha R^2\right)+\cdots\right]. \tag{10}$$

In this case, graviton's propagator is $iP_{\mu\nu;\rho\sigma}$, where

$$P_{\mu\nu;\rho\sigma} = \frac{\eta^{\mu\nu}\eta^{\rho\sigma} - \eta^{\mu\rho}\eta^{\nu\sigma} - \eta^{\mu\sigma}\eta^{\nu\rho}}{2k^2} - \frac{1}{4}\frac{\eta^{\mu\nu}\eta^{\rho\sigma}}{k^2+\frac{1}{4\alpha}}. \tag{11}$$

The difference between the propagator of the R^2 gravity and the Einstein gravity is the presence or absence of the second term at (11). Due to the second term of the propagator, we can get the UV finite 1-loop effects about ϕ^6 term

$$-\frac{\kappa^2\lambda^2}{20\pi^4}\ln\left(4\alpha\mu^2\right)\phi^6 . \tag{12}$$

As expected, the cut-off scales did not appear. Hence we can evaluate the ϕ^6 term without setting the cut-off scales value.

5. Conclusions

Evaluating the quantum gravity corrections to the Higgs potential $V(\phi)$ in the Einstein gravity in the cut-off method, we have found a salient difference between

$V(\phi)$ with and without gravity corrections in the region $\phi \gtrsim M_{\rm Pl}$. $V(\phi)$ with gravity corrections takes a minimum at $\phi \sim M_{\rm Pl}$, while $V(\phi)$ without gravity corrections monotonically increase. $V(\phi)$ depends on the cut-off scales, because the Einstein gravity is not renormalizable. As seen in Fig. 3, Λ dependence of the depth of the minimum of $V(\phi)$ is strong. Whereas, the value of ϕ at the minimum depends weakly on the value of Λ, the minimum is exist regardless of the cut-off scales. We may safely guess that qualitatively due to gravity corrections, the Higgs potential takes the minimum around the $\phi \sim M_{\rm Pl}$.

There is a possibility of that Higgs potential does not depend on the cut-off scales if we consider other modified theories of gravity. In a simple case of R^2 gravity (without $R_{\mu\nu})^2$ term, the ϕ^6 term correction does not depend on Λ. Whether more higher terms like ϕ^8 corrections does not either has to be studied further.

Acknowledgment

This talk(M. H.) is supported in part by funding from Shinshu University.

References

1. N. Cabibbo, L. Maiani, G. Parisi and R. Petronzio, Nucl. Phys. B **158**, 295 (1979).
2. M. Spira and P. M. Zerwas, Lect. Notes Phys. **512**, 161 (1998).
3. Y. Hamada, H. Kawai and K. Oda, Phys. Rev. D **87**, 053009 (2013).
4. F. Bezrukov, M. Y. Kalmykov, B. A. Kniehl and M. Shaposhnikov, JHEP **1210**, 140 (2012).
5. G. Degrassi, S. D. Vita, J. Elias-Miro, J. R. Espinosa, G. F. giudice, G. Isidori and A. Strumia, JHEP **1208**, 098 (2012).
6. D. Buttazzo, G. Degrassi, P. P. Giardino, G. F. Giudice, F. Sala, A. Salvio and A. Strumia, JHEP **1312**, 089 (2013).
7. A. Andreassen, W. Frost and M. D. Schwartz, Phys. Rev. Lett. **113**, 24, 241801 (2014).
8. A. V. Bednyakov, B. A. Kniehl, A. F. Pikelner and O. L. Veretin, Phys. Rev. Lett. **115**, 20, 201802 (2015).
9. F. Loebbert and J. Plefka, Mod. Phys. Lett. A **30**, 34, 1550189 (2015).
10. S. Bhattacharjee and P. Majumdar, Nucl. Phys. B **885**, 481 (2014).
11. N. Haba, K. Kaneta, R. Takahashi and Y. Yamaguchi, Phys. Rev. D **91**, 016004 (2015).
12. Y. Abe, M. Horikoshi, T. Inami, arXiv 1602.03792
13. C. Ford, I. Jack and D. R. T. Jones, Nucl. Phys. B **387**, 373 (1992).
14. K. G. Chetyrkin and M. F. Zoller, JHEP **1304**, 091 (2013).
15. A. V. Bednyakov, A. F. Pikelner and V. N. Velizhanin, Nucl. Phys. B **875**, 552 (2013).
16. S. R. Coleman and E. J. Weinberg Phys. Rev. D **7**, 1888 (1973).
17. K. S. Stelle, Phys. Rev. D **16**, 953 (1977).
18. E. S. Fradkin and A. A. Tseytlin, Nucl. Phys. B **201**, 469 (1982).

Modified Gravity inside Astrophysical Bodies

R. Saito* and D. Langlois**
Laboratoire APC, Université Paris Diderot Paris 7,
Paris, 75013, France
**rsaito@apc.in2p3.fr*
***dlangloi@apc.in2p3.fr*

D. Yamauchi
Research Center for the Early Universe, The University of Tokyo,
Tokyo, 113-0033, Japan
yamauchi@resceu.s.u-tokyo.ac.jp

S. Mizuno
Waseda Institute for Advanced Study, Waseda University,
Tokyo, 169-8050, Japan
shuntaro.mizuno@aoni.waseda.jp

J. Gleyzes
Institut de physique théorique, Université Paris Saclay,
CEA, CNRS, Gif-sur-Yvette, 91191, France
Université Paris Sud,
Orsay, 91405, France
jerome.gleyzes@cea.fr

Any infrared modification of gravity, which explains the current cosmic acceleration, should not spoil the successes of general relativity in solar-system observations. In many theories of modified gravity, it is ensured by the Vainshtein mechanism that works near dense sources. Recently, it has been found that the Vainshtein mechanism can be broken inside a dense source, although not outside, in a general class of scalar-tensor theories. In this paper, I will discuss its impact on the density profile of a star, modeling simply it as a polytropic sphere. I will also show the existence of a universal upper bound on the amplitude of this type of modification, independently of the details of the equation of state.

Keywords: Modification of gravity; scalar-tensor theories; stellar structure.

1. Introduction

It is still an unsettled question whether the current cosmic acceleration of the Universe can be explained within the framework of general relativity or gravity is not described by general relativity on cosmological scales. When general relativity is modified on cosmological scales, models of modified gravity often involve a new scalar degree of freedom. While such a scalar degree of freedom is responsible for the cosmic acceleration, it would incidentally mediate a new long-range force, called a fifth force, on small scales. Since such a force is stringently constrained by solar system and terrestrial experiments, it should be screened in the solar system to satisfy these constraints.

Several screening mechanisms are known to work in models of modified gravity. One such an example is the Vainshtein mechanism, which works in, e.g., massive gravity and Horndeski theories.[1–6] In contrast to the tensor modes in general relativity, it can be nontrivial how the scalar degree of freedom mediates the fifth force even in the weak gravity regime. While nonlinear derivative interactions of the tensor modes become important inside the Schwarzschild radius, those of the scalar degree of freedom can be important inside a much larger radius called the Vainshtein radius.

Recently, by analyzing the Vainshtein mechanism in beyond Horndeski theories, it has been shown that the Vainshtein mechanism can be broken inside a region filled with matter, while a deviation from general relativity is not exhibited outside the region.[7] The gravitational force sourced by a spherical object is modified as,

$$\frac{\mathrm{d}\Phi}{\mathrm{d}r} = G_N \left(\frac{\mathcal{M}}{r^2} - \epsilon \frac{\mathrm{d}^2 \mathcal{M}}{\mathrm{d}r^2} \right), \tag{1}$$

in the nonrelativistic regime. Here, $\mathcal{M}(r) \equiv 4\pi \int_0^r r'^2 \rho(r') \mathrm{d}r'$ is the mass enclosed in the sphere of radius r. As mentioned above, the law of gravity is deviated from Newton's law only in a region where the radial derivative of the density, $\mathrm{d}\rho/\mathrm{d}r$, has a non-vanishing value. The amplitude of the deviation is characterized by the parameter ϵ. It is determined by parameters in a model. In the next section, we give how it is related to model parameters in two representative models: beyond Horndeski theories and scalar-tensor theories with a general derivative coupling to matter. However, the main discussion in this paper can be applied to any theory leading to a modification of this type.

Because of its property, the modification of this type is not constrained by experiments on the earth or in space. Therefore, there can be a large modification in a region filled with matter, such as the interior of a star, without conflicting with these experiments. Motivated with this fact, in this paper, we study the impact of the modification (1) on the stellar structure, simply modeling an astrophysical object as a polytropic sphere. After seeing how the parameter ϵ is related to model parameters in the section 2, we derive a *modified* Lane-Emden equation in the section 3. We also show that, independently of the equation of state for matter, there is a universal upper bound on the value of ϵ.

2. Breaking of the Vainshtein mechanism

Before seeing the impact of the modification (1), first, we show how the parameter ϵ is related to model parameters in two representative models: beyond Horndeski theories and scalar-tensor theories with a general derivative coupling to matter.

2.1. *Beyond Horndeski theories*

Horndeski theories provide the most general class of scalar-tensor theories under the assumption that they lead to second-order equations of motion for both the metric and the scalar field. They were believed to be the most general scalar-tensor theories without the occurrence of Ostogradski instabilities. However, it was proposed that they could be further generalized without introducing dangerous degrees of freedom.[9–11]

In Ref. 7, Kobayashi, Watanabe, and Yamauchi have analyzed the Vainshtein mechanism in a subclass of beyond Horndeski theories whose action is given by,

$$S_{\rm bH} = \int \mathrm{d}^4x\, \sqrt{-g}(\mathcal{L}_2 + \mathcal{L}_3 + \mathcal{L}_4 + \mathcal{L}_{4b})\,, \tag{2}$$

with

$$\mathcal{L}_2 = G_2\,, \tag{3}$$
$$\mathcal{L}_3 = G_3 \Box\phi\,, \tag{4}$$
$$\mathcal{L}_4 = G_4 R + G_{4,X}\left[(\Box\phi)^2 - (\nabla_\mu\nabla_\nu\phi)^2\right], \tag{5}$$
$$\mathcal{L}_{4b} = F_4\left\{\nabla^\mu\phi\nabla^\nu\phi\left[\nabla_\mu\nabla_\nu\phi\Box\phi - \nabla_\mu\nabla_\lambda\phi\nabla_\nu\nabla^\lambda\phi\right] + X\left[(\Box\phi)^2 - (\nabla_\mu\nabla_\nu\phi)^2\right]\right\}. \tag{6}$$

Here, each G_a is an arbitrary function of ϕ and X, and $X \equiv -(\nabla\phi)^2/2$. The parameter ϵ is written in terms of the functions G_a $(a = 2, 3, 4)$ as,

$$\epsilon = \frac{\alpha_*^2}{4\pi G_N \Xi}\,, \tag{7}$$

where the various parameters in the expression are defined as,

$$\alpha_* \equiv XF_4\,, \tag{8}$$
$$(8\pi G_N)^{-1} \equiv 2G_4 - 8X(G_{4,X} + XG_{4,XX}) - 4X^2(5F_4 + 2XF_{4,X})\,, \tag{9}$$
$$\Xi \equiv \mathcal{G}(4\alpha_1\alpha_2 - 2\alpha_1\alpha_* + \mathcal{G}\nu) - 2\mathcal{F}\alpha_1\,, \tag{10}$$

and

$$\mathcal{G} \equiv 2(G_4 - 2XG_{4,X})\,, \tag{11}$$
$$\mathcal{F} \equiv 2G_4\,, \tag{12}$$
$$\alpha_1 \equiv G_{4,X} + 2XG_{4,XX} + X(5F_4 + 2XF_{4,X})\,, \tag{13}$$
$$\alpha_2 \equiv G_{4,X} + XF_4\,, \tag{14}$$
$$\nu \equiv G_{4,XX} + 2F_4 + XF_{4,X}\,. \tag{15}$$

Because $\alpha_* = 0$ in Horndeski theories, i.e., models with $F_4 = 0$, the breaking of the Vainshtein mechanism is a specific feature of beyond Horndeski theories.

2.2. *Scalar-tensor theories with a general derivative coupling to matter*

The Lagrangian for beyond Horndeski can be shown to be equivalent to Horndeski Lagrangian with a disformal coupling to matter:[10]

$$\mathcal{L}_{4b}[g_{\mu\nu}] + \mathcal{L}_4[g_{\mu\nu}] + \mathcal{L}_{\text{matter}}[g_{\mu\nu}] = \mathcal{L}_4[\bar{g}_{\mu\nu}] + \mathcal{L}_{\text{matter}}[g_{\mu\nu}]\,, \tag{16}$$

where

$$\bar{g}_{\mu\nu} = g_{\mu\nu} + \Gamma_4(\phi, X)\nabla_\mu\phi\nabla_\nu\phi\,, \tag{17}$$

with the function Γ_4 satisfying

$$\Gamma_{4X} = \frac{F_4}{G_4 - 2XG_{4,X} + X^2F_4}\,. \tag{18}$$

Therefore, scalar-tensor theories with a disformal coupling like (17) should lead to the modification (1). More generally, it can be shown that the modification of this type is exhibited in Horndeski theories with a general derivative coupling as,[12]

$$\mathcal{L} \supset \mathcal{L}_{\text{matter}}[\bar{g}_{\mu\nu}]\,; \quad \bar{g}_{\mu\nu} = \Omega(\phi, X)g_{\mu\nu} + \Gamma(\phi, X)\nabla_\mu\phi\nabla_\nu\phi\,. \tag{19}$$

In this model, the parameter ϵ is related to the amplitude of the derivative coupling:

$$\epsilon \propto (\Omega_{,X} - 2X\Gamma_{,X})^2\,. \tag{20}$$

3. Impact on the stellar structure

Modeling an astrophysical object as a static, spherically symmetric distribution of matter, the hydrostatic equation with the modification of gravity (1) reads,

$$\frac{\mathrm{d}P}{\mathrm{d}r} = -G_N\rho\left(\frac{\mathcal{M}}{r^2} - \epsilon\frac{\mathrm{d}^2\mathcal{M}}{\mathrm{d}r^2}\right)\,. \tag{21}$$

Then, assuming the polytropic relation between the pressure and density,

$$P = K\rho^{1+\frac{1}{n}}\,, \tag{22}$$

we obtain the closed equation for the density, modified Lane-Emden equation,

$$\frac{1}{\xi^2}\frac{\mathrm{d}}{\mathrm{d}\xi}\left[\xi^2\frac{\mathrm{d}}{\mathrm{d}\xi}(\chi - \epsilon\xi^2\chi^n)\right] = -\chi^n\,. \tag{23}$$

Here, we have introduced the dimensionless variables,

$$\xi = r\left\{\frac{(n+1)K\rho_c^{-1+\frac{1}{n}}}{4\pi G_N}\right\}^{-\frac{1}{2}}\,, \quad \rho = \rho_c\chi^n\,, \tag{24}$$

for the density at the center ρ_c.

We have presented the density profiles obtained by numerically solving the modified Lane-Emden equation (23) in Fig. 1(a). Here, the density profiles are plotted for $n = 3$, which gives a models for main sequence stars. For increasing values of ϵ,

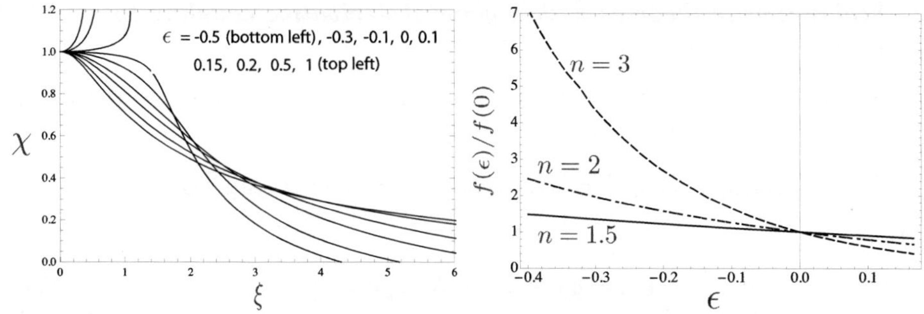

(a) (b)

Fig. 1. (a) density profiles for different values of ϵ ($n = 3$) (b) modification in the mass-radius relation.

the profiles become steeper in the inner region, while it becomes flatter in the outer region. Correspondingly, the mass-radius relation is modified as,

$$M = f(\epsilon)R^{\frac{n-3}{n-1}} \,. \tag{25}$$

The coefficient depends on ϵ, which has been plotted in Fig. 1(b). It is a monotonically decreasing function of ϵ. Therefore, irrespective of the value of n, the mass of an astrophysical object for a fixed radius decreases for increasing values of ϵ.

Finally, it is remarkable that the density profiles blow up and then viable profiles do not exist for large positive values of ϵ. Therefore, these values of ϵ should be excluded. It can be shown that this is true for any physically reasonable equation of state, i.e. the equation of state where the density increases with the pressure. Near the center $r = 0$, the enclosed mass can be expanded as,

$$\mathcal{M} = \frac{4\pi\rho_c r^3}{3} + \mathcal{O}(r^5) \,. \tag{26}$$

Inserting the expansion to the modified gravitational law (1), one obtains,

$$\frac{\mathrm{d}\Phi}{\mathrm{d}r} = (1 - 6\epsilon)\frac{G_N \mathcal{M}}{r^2} \,. \tag{27}$$

Therefore, gravity becomes repulsive for $\epsilon > 1/6$. This means that the pressure has to increase with the radius and then the density does too under the assumption on the equation of state. One can also show that the density gradient cannot vanish at any higher radius.[13] Consequently, the density profile continues to grow for higher radii. We thus conclude that one cannot construct a physically sensible profile when

$$\epsilon > \frac{1}{6} \,. \tag{28}$$

This bound is universal, i.e., it exists independently of the details of the equation of state.

Recently, the lower limit on ϵ has been obtained by studying how the modification (1) changes a minimum mass for hydrogen burning in stars:[14]

$$\epsilon > -0.0068 \,. \tag{29}$$

Combining the universal upper bound (28), ϵ should be in the range:

$$-0.0068 < \epsilon < \frac{1}{6}\,. \tag{30}$$

In conclusion, the modification of the type (1) can be stringently constrained by considering its impact on the stellar structure, while it can evade constraints from experiments on the earth or in space. As a consequence, the generalization to beyond Horndeski and the general derivative coupling to matter (19) are tightly constrained.

Acknowledgement

We thank T. Shigeyama and T. Suda for valuable comments and useful suggestions. R.S. is supported by JSPS Postdoctoral Fellowships for Research Abroad. D.Y. is supported by Grant-in-Aid for JSPS Fellows (No.259800). S.M. is supported by JSPS Grant-in-Aid for Research Activity Start-up No. 26887042 and Waseda University Grant for Special Research Projects (Project number 2014S-191).

References

1. C. de Rham, Massive Gravity, *Living Rev. Rel.* **17**, 7 (2014).
2. T. Narikawa, T. Kobayashi, D. Yamauchi and R. Saito, Testing general scalar-tensor gravity and massive gravity with cluster lensing, *Phys. Rev. D* **87**, 124006 (2013).
3. K. Koyama, G. Niz and G. Tasinato, Effective theory for the Vainshtein mechanism from the Horndeski action, *Phys. Rev. D* **88**, 021502 (2013).
4. R. Kimura, T. Kobayashi and K. Yamamoto, Vainshtein screening in a cosmological background in the most general second-order scalar-tensor theory, *Phys. Rev. D* **85**, 024023 (2012).
5. A. De Felice, R. Kase and S. Tsujikawa, Vainshtein mechanism in second-order scalar-tensor theories, *Phys. Rev. D* **85**, 044059 (2012).
6. R. Kase and S. Tsujikawa, Screening the fifth force in the Horndeski's most general scalar-tensor theories, *JCAP* **1308**, 054 (2013).
7. T. Kobayashi, Y. Watanabe and D. Yamauchi, Breaking of Vainshtein screening in scalar-tensor theories beyond Horndeski, *Phys. Rev. D* **91**, 064013 (2015).
8. G. W. Horndeski, Second-order scalar-tensor field equations in a four-dimensional space, *Int. J. Theor. Phys.* **10**, 363 (1974).
9. J. Gleyzes, D. Langlois, F. Piazza and F. Vernizzi, Healthy theories beyond Horndeski, *Phys. Rev. Lett.* **114**, 211101 (2015).
10. J. Gleyzes, D. Langlois, F. Piazza and F. Vernizzi, Exploring gravitational theories beyond Horndeski, *JCAP* **1502**, 018 (2015).
11. C. Lin, S. Mukohyama, R. Namba and R. Saitou, Hamiltonian structure of scalar-tensor theories beyond Horndeski, *JCAP* **1410**, 071 (2014)
12. R. Saito, D. Yamauchi, and D. Langlois, *in preparation.*
13. R. Saito, D. Yamauchi, S. Mizuno, J. Gleyzes and D. Langlois, Modified gravity inside astrophysical bodies, *JCAP* **1506**, 008 (2015).
14. J. Sakstein, Hydrogen Burning in Low Mass Stars Constrains Scalar-Tensor Theories of Gravity, *Phys. Rev. Lett.* **115**, 201101 (2015).

Tree-Unitarity and Renormalizability in Lifshitz-Scaling Theory as a Toy Model of Hořava's Gravity Theory

Toshiaki Fujimori

Department of Physics, and Research and Education Center for Natural Sciences, Keio University, Hiyoshi 4-4-1, Kanagawa 223-8521, Japan

Takeo Inami

Department of Physics, National Taiwan University, Taipei 10617, Taiwan, R.O.C.
Theoretical Research Division, Nishina Center, RIKEN, Wako 351-0198, Japan

Keisuke Izumi

Departament de Física Fonamental, Institut de Ciències del Cosmos (ICCUB), Universitat de Barcelona, Martí i Franquès 1, E-08028 Barcelona, Spain

Tomotaka Kitamura

Department of physics, Waseda University, Shinjyuku, Tokyo 169-8555, Japan

We study tree-unitarity and renormalizability in Lifshitz-scaling theory, which is characterized by an anisotropic scaling between the spacial and time directions. Due to the lack of the Lorentz symmetry, the conditions for both unitarity and renormalizability are modified from those in relativistic theories. For renormalizability, the conventional discussion of the power counting conditions has to be extended. Because of the dependence of S-matrix elements on the reference frame, unitarity requires stronger conditions than those in relativistic cases. We show that the conditions for unitarity and renormalizabilty are identical as in relativistic theories. We discuss the importance of symmetries for a theory to be renormalizable.

Keywords: Unitarity; renormalizability; Lifshitz scaling; Hořava-Lifshitz gravity.

1. Introduction

A quantum field theory requires both unitarity and renormalizability to be a well-defined theory. They may be implied from each other, which motivated the study of the equivalence between unitarity and renormalizability in gauge theories[1] and gravitational theories.[2] This relation could be helpful to construct a consistent quantum gravity; to check the renormalizability we investigate unitarity and vice versa. It is worth checking whether the relation is a generic property of quantum theories. To avoid the complication due to symmetries, it is better to investigate theories with few symmetries. One of ways to loosen a symmetry is considering non-relativistic theory.

Lifshitz-type theories[3] are characterized by the Lifshitz scaling:

$$t \to b^z t, \qquad x^i \to b x^i \quad (i = 1, \dots d), \tag{1}$$

and manifestly do not have Lorentz symmetry. These theories have a lot of attention in cosmology[4] and some of quantum properties are also investigated.[5–9] Due to the anisotropic scaling (1), the ultraviolet(UV) behaviors are different from those in relativistic theories. This idea has been introduced to gravitational theory, which is called Hořava-Lifshitz (HL) gravity.[10] The application of the HL gravity is widely studied in cosmology (see Ref. 11 for a review), such as the emergence of dark matter as an integration constant[12] and so on. Its quantum properties have not been studied very much.[13] By the power-counting discussion HL gravity is expected to be renormalizable, but the renormalizability still remains uncertain.

We have studied tree-unitarity and renormalizability in theories of a Lifshitz scalar field to investigate the relation between them.[5,6] These thoeries are also useful as toy models of HL gravity theory. The lack of the Lorentz symmetry gives rise to significant modifications to both renormalizability and unitarity. We have shown that, for theories to be renormalizable, the conditions for renormalization have to be extended from the conventional ones. As for tree-level unitarity, since scattering amplitudes have the reference frame dependence, unlike in relativistic cases, stronger conditions are required. We have derived the extended power-counting renomalizability (PRC) conditions and the tree-level unitarity constraints on the forms of self-interaction term in the theory of a single Lifshitz scalar field. We have seen the equivalence between these conditions in cubic and quartic interactions.

We give a brief review of our papers Refs. 5, 6. For details, see these papers. We also discuss a consequence of our argument to the renormalizability of HL gravity.

2. Lifshitz scalar field

The quadratic action for Lifshitz scalar field ϕ can be written in the form

$$S_2 = \int dtd^dx \left[\frac{1}{2}\phi \left\{ \partial_t^2 - f(-\Delta) \right\} \phi \right], \qquad \Delta := \partial^i \partial_i, \tag{2}$$

where $f(-\Delta) = (-\Delta)^z + \alpha(-\Delta)^{z-1} + \dots$ is a polynomial of degree z. The dispersion relation takes the form

$$E = \sqrt{f(p^2)}, \tag{3}$$

where E and p are the energy and the magnitude of the momentum, respectively. This dispersion relation becomes $E \approx p^z$ in the UV regime and is consistent with the Lifshitz scaling (1). Since the asymptotic behavior of propagators is given by

$$\frac{1}{E^2 - f(p^2)} \approx \frac{1}{E^2 - p^{2z}}, \tag{4}$$

UV divergences of loop diagrams are milder than those in relativistic theories for $z > 1$.

Let us introduce the following generic n-point interaction term

$$S_{int} = \lambda \int dtd^dx \, (\partial_x^{a_1}\phi)\,(\partial_x^{a_2}\phi) \cdots (\partial_x^{a_n}\phi), \tag{5}$$

where $a_1, \cdots, a_n$ are non-negative integers. Each ∂_x denotes any of the spatial derivatives ∂_i $(i = 1, \cdots, d)$ and we suppose that this interaction term is designed to keep the d-dimensional spatially rotational symmetry $O(d)$.

3. Renormalizability

The scaling dimension of ϕ can be read from the quadratic term in (2):

$$[\phi] = \frac{d - z}{2}, \tag{6}$$

where we have used the convention that $[E] = z$ and $[p] = 1$. The dimension of the coupling constant λ is obtained as

$$[\lambda] = z + d - \sum_{l=1}^{n} (a_l + [\phi]). \tag{7}$$

In relativistic theories, an interaction term is renormalizable if it has a coupling constant with non-negative dimension: $[\lambda] \geq 0$. On the other hand, in the Lifshitz scalar theory, non-negativity of $[\lambda]$ is not enough for renormalizability.

As an example, let us consider the case with $d = 3$ and $z = 5$, and the following quartic interaction term with $[\lambda] = 0$,

$$S_4 = \lambda \int dt d^3x \, \phi^2 (\Delta^3 \phi)^2. \tag{8}$$

Let us consider an 1-loop diagram with n vertices each of which has two external lines. The term for which two $(\Delta^3\phi)$'s at each vertex correspond to the internal lines gives the leading order contribution. This can be estimated as

$$\int d\omega d^3p \left(\frac{1}{\omega^2 - p^{10}} \right)^n (p^{12})^n \sim \Lambda^{8+2n}, \tag{9}$$

where Λ is a UV cutoff with $[\Lambda] = 1$. For any n, this loop integral diverges as $\Lambda \to \infty$, and thus the interaction term (8) is NOT renormalizable, even though the condition $[\lambda] \geq 0$ is satisfied.

The reason why the interaction term (8) is not renormalizable is as follows. In the calculation of loop integrals, only operators corresponding to the internal lines are involved. Thus, only their dimension is related to the renormalizability. In the above example, the dimension of field is negative ($[\phi] = -1$) and the part of the term $(\Delta^3\phi)^2$ has smaller dimension than the whole of the term $\phi^2(\Delta^3\phi)^2$. Now, the whole of $\phi^2(\Delta^3\phi)^2$ is marginal, and thus the part of the term $(\Delta^3\phi)^2$ is non-renormalizable. Therefor, the criteria of renormalizability is; the dimension of any part of the term has to be larger than the inverse of $[dtd^dx]$.[a] For the proof in generic cases, see Ref. 6.

---

[a]For whole of the term, the case of equality can also be renormalizable. This is the conventional PCR.

4. Tree-unitarity

The unitarity bound is one of the necessary conditions for the untarity of S-matrix elements. The unitarity condition reduces to the optical theorem

$$2\mathrm{Im}\mathcal{M}\,(i \to i) = \sum_X \delta(E - E_X)\delta^d\,(\mathbf{p} - \mathbf{p}_X)\,|\mathcal{M}\,(i \to X)\,|^2, \tag{10}$$

where $\mathcal{M}\,(i \to j)$ is the scattering amplitude, $\{|X\rangle\}$ be a complete orthonormal basis of the Hilbert space:

$$\sum_X |X\rangle\langle X| = 1, \tag{11}$$

i is an element in $\{|X\rangle\}$, and $\sum_X$ denotes the sum with respect to all possible intermediate states. In the following, we ignore unimportant numerical factors. Taking absolute value of eq.(10), the left hand side has to be smaller than or equal to $|\mathcal{M}\,(i \to i)|$. Since the right hand side is the summation of non-negative values, each of them has to be smaller than or equal to the sum. Then, we have the unitarity bound:

$$|\mathcal{M}\,(i \to i)| \geq \delta(E - E_X)\delta^d\,(\mathbf{p} - \mathbf{p}_X)\,|\mathcal{M}\,(i \to j)\,|^2, \tag{12}$$

where j is an element in $\{|X\rangle\}$ and can be the same as i.

The forms of cubic and quartic interaction terms can be constrained with the uniarity bound of two-particle scattering. We take the basis labeled with the total energy and momentum;

$$|E, \mathbf{P}, l\rangle = \int \frac{d^d p_2}{2E_2}\frac{d^d p_2}{2E_2}\delta\,(E_1 + E_2 - E)\,\delta^d\,(\mathbf{p}_1 + \mathbf{p}_2 - \mathbf{P})\;h_l\,(\mathbf{p}_j)\,|\mathbf{p}_1, \mathbf{p}_2\rangle, \tag{13}$$

where $|\mathbf{p}_1, \mathbf{p}_2\rangle$ is the standard asymptotic momentum eigenstates for two-particle states normalized as

$$\int \frac{d^d p_1}{2E_1}\frac{d^d p_2}{2E_2}|\mathbf{p}_1, \cdots, \mathbf{p}_n\rangle\langle \mathbf{p}_1, \cdots, \mathbf{p}_n| = 1, \tag{14}$$

in two-particle basis of subspace, E_1 and E_2 are the energy of particles, and $\{h_l\,(\mathbf{p}_j)\}$ is an orthonormal functions in two-particle subspace with the total energy E and momentum $\mathbf{P}$. In the two-particle subspace, the unitarity bound (12) can be rewritten as

$$|\mathcal{M}\,(E, \mathbf{P}; l \to l')\,| \leq \text{const.} \tag{15}$$

This is a necessary condition for the unitarity. We check the high-energy behavior of this.

In relativistic theories, we can always take the center-of-mass frame, i.e. $\mathbf{P} = \mathbf{0}$. Then, both particles have the same magnitude of momentum. In the high energy limit, the energies of both particles go to infinity. In the theories without Lorentz symmetry, in contrast, states with nonzero total momentum has no relation to those with $\mathbf{P} = \mathbf{0}$. We should also consider the situation where only the energy of one particle goes to infinity and the other is finite. Moreover, this state can transfer to the state where energies of both particle are infinity. Because of these varieties, we have more constraints. From the unitarity bound (15) with the tree-level approximation, we have the same constraints with the extended version of PCR. We omit the derivation of the constraints here. For details, see Ref. 6.

5. Discussion: Importance of Symmetry

We have seen that, if the dimension of fields is non-positive, the extension of PCR is required. This is also true for unitarity, since, as we have demonstrated, the conditions for unitarity are identical to those for renormalizability. In Ref. 5, we have addressed the importance of symmetries for a theory to naturally have renormalizability. We have studied theories with a dimensionless scalar field and shown that in some theories with symmetry, only the conventional conditions of PCR could be enough. The conventional PCR shows that, in theories with dimensionless fields ϕ, marginal interaction terms require the counter terms having all relevant or marginal forms. Because the product of a marginal term and ϕ^n is marginal for any n, we have an infinite number of marginal terms. However, introducing the shift symmetry which is the invariance under $\phi \to \phi$+const., the scalar field ϕ must always appear in combination with space-time derivatives and the dimension of $\partial_\mu \phi$ is positive. Therefore, all possible operators have positive dimension and PCR does NOT need to be extended.

Based on the above discussion, HL gravity with $d = z = 3$ is expected to be renormalizable. If we separately see the interaction terms, HL gravity has terms violating the extended version of PCR. However, because of the symmetry, the possible operators are the three-dimensional curvature R_{ij} and the extrinsic curvature K_{ij} with or without differential operators. These operators have positive dimension, and the number of relevant and marginal terms generated by loop becomes finite. Therefore, the conventional PCR could work well.

The extended version of PCR is important when we introduce matter fields in HL gravity or consider higher z. To keep the scaling of gravitational field, matter fields should satisfy the same scaling, and then they must be dimensionless for $d = z = 3$. Without symmetry of matter sector, the matter fields have to satisfy the extended PCR. Imposing symmetries could naturally lead the theory renormalizable as in the above discussion of HL gravity. For higher z, for instance in the case with $d = 3$ and $z \geq 7$, the dimension of curvature is non-positive, that is, the symmetry does not restrict the possible operator to be positive. In this case, the extended PCR is required.

Acknowledgments

The work of T. F. is supported in part by the Japan Society for the Promotion of Science (JSPS) Grantin-Aid for Scientific Research (KAKENHI Grant No. 25400268) and by the MEXTSupported Program for the Strategic Research Foundation at Private Universities "Topological Science"(Grant No. S1511006). K. I. is supported by FPA2013-46570-C2-2-P, AGAUR 2009-SGR-168, and MDM-2014-0369 of ICCUB (Unidad de Excelencia 'Maria de Maeztu').

References

1. C. H. Llewellyn Smith, Phys. Lett. B **46**, 233 (1973).
   J. M. Cornwall, D. N. Levin and G. Tiktopoulos, Phys. Rev. Lett. **30**, 1268 (1973) [Phys. Rev. Lett. **31**, 572 (1973)].
   J. M. Cornwall, D. N. Levin and G. Tiktopoulos, Phys. Rev. D **10**, 1145 (1974) [Phys. Rev. D **11**, 972 (1975)].
2. F. A. Berends and R. Gastmans, Nucl. Phys. B **88**, 99 (1975).
3. E. Lifshitz, On the theory of Second-Order Phase transitions I II Zh. Eksp.Teor. Fiz. 11 (1941) 255, 269.
4. K. Izumi, T. Kobayashi and S. Mukohyama, JCAP **1010**, 031 (2010) [arXiv:1008.1406 [hep-th]].
5. T. Fujimori, T. Inami, K. Izumi and T. Kitamura, Phys. Rev. D **91**, no. 12, 125007 (2015) [arXiv:1502.01820 [hep-th]].
6. T. Fujimori, T. Inami, K. Izumi and T. Kitamura, arXiv:1510.07237 [hep-th].
7. D. Anselmi and M. Halat, Phys. Rev. D **76**, 125011 (2007) [arXiv:0707.2480 [hep-th]].
   D. Anselmi, JHEP **0802**, 051 (2008) [arXiv:0801.1216 [hep-th]].
   D. Anselmi, Annals Phys. **324**, 874 (2009) [arXiv:0808.3470 [hep-th]].
   D. Anselmi, Annals Phys. **324**, 1058 (2009) [arXiv:0808.3474 [hep-th]].
8. M. Visser, Phys. Rev. D **80**, 025011 (2009) [arXiv:0902.0590 [hep-th]].
   M. Visser, arXiv:0912.4757 [hep-th].
9. M. Colombo, A. E. Gumrukcuoglu and T. P. Sotiriou, arXiv:1410.6360 [hep-th].
   M. Colombo, A. E. Gumrukcuoglu and T. P. Sotiriou, arXiv:1503.07544 [hep-th].
10. P. Horava, Phys. Rev. D **79**, 084008 (2009) [arXiv:0901.3775 [hep-th]].
    P. Horava, JHEP **0903**, 020 (2009) [arXiv:0812.4287 [hep-th]].
11. S. Mukohyama, Class. Quant. Grav. **27**, 223101 (2010) [arXiv:1007.5199 [hep-th]].
12. S. Mukohyama, Phys. Rev. D **80**, 064005 (2009) [arXiv:0905.3563 [hep-th]].
    K. Izumi and S. Mukohyama, Phys. Rev. D **81**, 044008 (2010) [arXiv:0911.1814 [hep-th]].
    K. Izumi and S. Mukohyama, Phys. Rev. D **84**, 064025 (2011) [arXiv:1105.0246 [hep-th]].
13. A. O. Barvinsky, D. Blas, M. Herrero-Valea, S. M. Sibiryakov and C. F. Steinwachs, arXiv:1512.02250 [hep-th].
14. S. R. Das and G. Murthy, "CP**(N-1) Models at a Lifshitz Point," Phys. Rev. D **80**, 065006 (2009) [arXiv:0906.3261 [hep-th]].
15. K. Anagnostopoulos, K. Farakos, P. Pasipoularides and A. Tsapalis, "Non-Linear Sigma Model and asymptotic freedom at the Lifshitz point," arXiv:1007.0355 [hep-th].
16. P. R. S. Gomes, P. F. Bienzobaz and M. Gomes, "Competing interactions and the Lifshitz-type Nonlinear Sigma Model," Phys. Rev. D **88**, 025050 (2013) [arXiv:1305.3792 [hep-th]].
17. T. Fujimori and M. Nitta, JHEP **1510**, 021 (2015) [arXiv:1507.06456 [hep-th]].

Bigravity from Gradient Expansion in DGP 2-Brane Model

Yasuho Yamashita[a] and Takahiro Tanaka[a,b]

[a] *Yukawa Institute for Theoretical Physics, Kyoto University, Kyoto, 606-8502, Japan*
[b] *Department of Physics, Kyoto University, Kyoto, 606-8502, Japan*
yasuho@yukawa.kyoto-u.ac.jp
t.tanaka@tap.scphys.kyoto-u.ac.jp

We discuss how the ghost free bigravity coupled with a single scalar field can be derived from a braneworld setup. We consider DGP two-brane model without radion stabilization and obtain the effective four-dimensional action by solving the bulk configuration for given boundary metrics and substituting back the solution into the action. In the obtained effective theory, two gravitons interact through the Fierz-Pauli mass term and the radion remains as a scalar field, but its coupling to the metrics is non-trivial.

Keywords: Modified gravity; extra dimensions; gravity.

1. Introduction

In general, bigravity, *i.e.*, the gravitational theory that contains two gravitons, was known to be suffered from Boulware-Deser (BD) ghost,[1] which makes the theory unstable in the ultraviolet regime. Recently, de Rham, Gabadadze and Tolley (dRGT) proposed a specific non-derivative interaction in Refs. 5,6 that removes the BD ghost mode.[2–4] The interaction between two metrics different from the dRGT one must be suppressed to obtain a stable model, but we do not know how such fine tuning can be realized. In order to address this fine-tuning problem and also to give a UV completion of the ghost-free bigravity, we try to derive the ghost-free bigravity as a low-energy effective theory from a healthy braneworld setup with two branes.[7,8] This model is free from BD ghost naturally We expect that the effective theory is free from BD ghost and coincides with the ghost-free bigravity, because there is no pathology such as BD ghost in the braneworld setup.

2. Model

We consider braneworld model with two branes, in which the bulk spacetime is compactified by the Z_2 symmetry across each brane. The 4-dimensional effective theory of this model contains one massless graviton and infinite number of massive gravitons whose masses correspond to the momenta tangent to the brane. In order to derive bigravity from 2-brane model, it is necessary to realize the mass hierarchy between the lowest massive mode and the other massive modes. We consider

Dvali-Gabadadze-Porrati (DGP) 2-brane model,[9]

$$S = S_b + \sum_{\sigma=\pm} S_\sigma \,, \tag{1}$$

where

$$S_b := \frac{1}{2}\frac{M_{pl}^2}{2r_c^{(+)}} \int dx^5 \sqrt{-{}^5g} \left({}^5R - \frac{12}{\ell_\Lambda^2} \right) , \tag{2}$$

$$S_+ := \frac{M_{pl}^2}{2} \int d^4x \sqrt{-g_+} \left(R_{(+)} - 2\sigma_+ \right) , \tag{3}$$

$$S_- := \frac{\chi M_{pl}^2}{2} \int d^4x \sqrt{-g_-} \left(R_{(-)} - 2\sigma_- \right) , \tag{4}$$

where ${}^5g_{\mu\nu}$, ${}^5R$, $g^{\pm}_{\mu\nu}$, $R_{(\pm)}$ are 5-dimensional metric, 5-dimensional Ricci scalar, 4-dimensional metrics and Ricci scalars induced on $(\pm)$-branes, respectively. M_{pl} is 4-dimensional Planck mass on $(+)$-brane, while $M_{pl}^2/2r_c^{(+)}$ and χM_{pl}^2 are 5-dimensional bulk Planck mass cubed and 4-dimensional Planck mass squared on the $(-)$-brane, respectively. $\sigma_\pm$ are 4-dimensional cosmological constants on the respective branes, while $6/\ell_\Lambda^2$ is 5-dimensional cosmological constant. Because of the gravitational potential that comes from the Einstein-Hilbert terms localized on the branes, the gravitons are trapped on the respective branes. Then we can keep only the massless and the lowest massive gravitons at low energies while sending the other massive modes away to high energies by tuning the model parameters so that the contributions of the brane-localized Einstein-Hilbert terms dominate the bulk one.[7,10] This tuning is realized by imposing

$$\Delta y / r_c^{(\pm)} \ll 1 \,, \tag{5}$$

where Δy is the brane separation and $r_c^{(-)} := r_c^{(+)}\chi$.

In addition to the spin-2 sector, there exists a spin-0 degree of freedom corresponding to the brane separation, which is called radion. Concerning the radion, it is known that DGP 2-brane model has two branches, the self-accelerating branch and the normal branch. The self-accelerating branch inevitably suffers from either the Higuchi ghost or the radion ghost, while the normal branch is free from ghost.[11,12] In order to choose the ghost-free branch, we should tune the extrinsic curvatures on the $(\pm)$-branes $K_\pm$ so as to satisfy the condition

$$|K_\pm| \lesssim \mathcal{O}\left(1/r_c^{(\pm)} \right) \tag{6}$$

on the respective branes.[7] Assuming that the extrinsic curvature keeps approximately constant in the bulk, this ghost-free condition (6) and the condition (5) restrict the region of the model parameters of our interest to

$$\epsilon := |K \Delta y| \ll 1 \,, \tag{7}$$

Therefore, we expect that ghost-free bigravity and scalar system should be derived as a 4-dimensional effective theory by considering the lowest order truncation of the expansion in $\epsilon := |K\Delta y|$, which is called gradient expansion.[13]

3. Strategy to derive the effective action

Here, we show the strategy to derive the effective action from the action of DGP 2-brane model (1). We solve bulk metric for given boundary metrics on the branes and integrate out the bulk degrees of freedom, which will be simplified by the approximation $|K\Delta y| \ll 1$.

At the lowest order of the gradient expansion, we can set the following metric ansatz:

$$ds^2 = N^2(x)dy^2 + g_{\mu\nu}dx^\mu dx^\nu \,, \tag{8}$$

while the positions of the $(\pm)$-branes are fixed at $y = y^\pm$. Here we imposed $\partial_y N = 0$ and $N^\mu = 0$ using the gauge degrees of freedom. In order to obtain an appropriately interpolating bulk solution for given boundary metrics $g^{(+)}_{\mu\nu}(x)$ and $g^{(-)}_{\mu\nu}(x)$ on the respective branes, we solve the bulk equations

$$\frac{1}{N}\partial_y K_{\mu\nu} = -2K^\rho_\mu K_{\rho\nu} + KK_{\mu\nu} + \frac{4}{\ell^2_\Lambda}g_{\mu\nu} - R_{\mu\nu} + \frac{1}{N}\nabla_\mu\nabla_\nu N \,, \tag{9}$$

$$K_{\mu\nu} = -\frac{1}{2N}\partial_y g_{\mu\nu} \,, \tag{10}$$

$$K^2 - K^\mu_\nu K^\nu_\mu = -\frac{12}{\ell^2_\Lambda} + R \,, \tag{11}$$

$$\nabla_\mu K - \nabla_\nu K^\nu_\mu = 0 \,, \tag{12}$$

where $K_{\mu\nu}$ and $R_{\mu\nu}$ are the extrinsic curvature and the 4-dimensional Ricci tensor evaluated on each y-constant surface, respectively. The covariant differentiation ∇_μ is the one associated with the 4-dimensional metric $g_{\mu\nu}$, and the indices are raised and lowered also using $g_{\mu\nu}$.

Using the gradient expansion, we can expand $g_{\mu\nu}$ around the middle point $\bar{y}$ in the bulk as

$$g^{(\pm)}_{\mu\nu} = \bar{g}_{\mu\nu} + \overline{\partial_y g_{\mu\nu}}\, Ny^\pm + \frac{1}{2}\overline{\partial_y{}^2 g_{\mu\nu}}(Ny^\pm)^2 + \cdots \,, \tag{13}$$

where $\bar{g}_{\mu\nu} := g_{\mu\nu}(\bar{y})$ and we set $\bar{y} = 0$ so that the values of y on the respective branes are related with each other as $y^- = -y^+$. Using the expression (13) and the evolution equation (9), we can obtain $\bar{g}_{\mu\nu}$ and $\bar{K}_{\mu\nu}$ in terms of $g^{(\pm)}_{\mu\nu}$ and Ny^+, and then the Hamiltonian constraint (11) determines Ny^+ in terms of $g^{(\pm)}_{\mu\nu}$. The momentum constraints are automatically imposed by the divergence of the junction conditions on the $(\pm)$-branes[12]

$$K^{(\pm)}_{\mu\nu} - K^{(\pm)}g^{(\pm)}_{\mu\nu} = \pm r^{(\pm)}_c \left(G^{(\pm)}_{\mu\nu} - \sigma_\pm g^{(\pm)}_{\mu\nu}\right) . \tag{14}$$

The effective action written in terms of $g^{(\pm)}_{\mu\nu}$ is obtained from the bulk action

$$S_b = \frac{M^2_{pl}}{2r^{(+)}_c}\oint d^5x\sqrt{-g}\left(R + K^2 - K^\mu_\nu K^\nu_\mu - \frac{12}{\ell^2_\Lambda}\right) , \tag{15}$$

by substituting the bulk solution into the bulk action and integrating the bulk action along y-direction, which is also simplified by the gradient expansion.

4. Result

Here we explicitly derive the 4-dimensional effective action from DGP 2-brane model (1), by taking the leading terms of the expansion in $\epsilon := |KNy^+|$ and assuming $g^{(\pm)}_{\mu\nu} \sim \mathcal{O}(1)$, $\Delta g_{\mu\nu} := g^{(+)}_{\mu\nu} - g^{(-)}_{\mu\nu} \sim \mathcal{O}(\epsilon)$. We want to investigate the parameter region with $|K_\pm| \sim \mathcal{O}\left(1/r_c^{(\pm)}\right)$, in which both radion and the massive gravitons are healthy.[7] We are also interested in the energy regime in which the 4-dimensional momentum squared is comparable to the mass squared of the lowest mode of massive graviton. At the leading order of the gradient expansion, the mass squared of the lowest massive graviton m^2 is estimated as

$$m^2 \sim \frac{1}{r_c^{(+)}\Delta y} + \frac{1}{r_c^{(-)}\Delta y} \sim \mathcal{O}\left(|K|^2/\epsilon\right) . \tag{16}$$

in Ref. 7. Therefore, the Hamiltonian constraint requires that ℓ_Λ^{-2} must be tuned so that $R - 12\ell_\Lambda^{-2} \sim \mathcal{O}\left(K^2\right)$ is much smaller than $R \sim \mathcal{O}\left(|K|^2/\epsilon\right)$. Furthermore, the junction conditions (14) imply that the traceless part of $R_{\mu\nu}$ must be the same order as K^2, and hence $R_{\mu\nu}$ can be written as

$$R_{\mu\nu} = 3\ell_\Lambda^{-2} g_{\mu\nu} + \mathcal{O}\left(|K|^2\right) . \tag{17}$$

From Eqs. (9), (10) and (13), we obtain $\bar{K}_{\mu\nu}$ and $\bar{g}_{\mu\nu}$ as

$$\bar{K}_{\mu\nu} = \frac{\Delta g_{\mu\nu}}{4\Phi} , \tag{18}$$

$$\bar{g}_{\mu\nu} = \tilde{g}_{\mu\nu} + \Phi\nabla_\mu\nabla_\nu\Phi + \frac{\Phi^2}{\ell_\Lambda^2} g_{\mu\nu} + \mathcal{O}\left(\epsilon^2\right) , \tag{19}$$

where

$$\Phi := -Ny^+ , \qquad \tilde{g}_{\mu\nu} := \frac{g^{(+)}_{\mu\nu} + g^{(-)}_{\mu\nu}}{2} . \tag{20}$$

The Hamiltonian constraint (11) at $y = \bar{y}$

$$\bar{R} - \frac{12}{\ell_\Lambda^2} = \frac{\Delta g^2 - \Delta g_{\mu\nu}\Delta g^{\mu\nu}}{16\Phi^2} , \tag{21}$$

where the tensor indices are raised by $\tilde{g}_{\mu\nu}$, determines Φ in terms of $g^{(\pm)}_{\mu\nu}$. Then, the integration of the bulk action (15) along y-direction gives

$$S_b = \frac{M_{pl}^2}{2}\frac{2}{r_c^{(+)}} \int d^4x\sqrt{-\tilde{g}} \left[\frac{\Delta g^2 - \Delta g_{\mu\nu}\Delta g^{\mu\nu}}{16\Phi} + \frac{\Phi}{3}\tilde{\mathcal{L}}^{\mu\nu}\left((\tilde{\nabla}_\mu\Phi)(\tilde{\nabla}_\nu\Phi) - \frac{\Phi^2}{\ell_\Lambda^2}\tilde{g}_{\mu\nu}\right)\right.$$
$$\left. + \mathcal{O}\left(|K|\epsilon^2\right)\right] , \tag{22}$$

where we define $\tilde{\mathcal{L}}^{\mu\nu}$ as

$$\tilde{\mathcal{L}}^{\mu\nu} := \tilde{\nabla}^\mu\tilde{\nabla}^\nu - \left(\tilde{\Box} + \frac{3}{\ell_\Lambda^2}\right)\tilde{g}^{\mu\nu} , \tag{23}$$

where $\tilde{\nabla}$ is defined as the covariant differentiation associated with $\tilde{g}_{\mu\nu}$.[8] In order to reveal the structure of the interaction between two gravitons, it is convenient to treat the radion Φ as an independent variable. Therefore we impose the Hamiltonian constraint with a Lagrange multiplier λ in the effective action:

$$S_b = \frac{M_{pl}^2}{2}\frac{2}{r_c^{(+)}}\int d^4x\sqrt{-\tilde{g}}\left[\frac{\Delta g^2 - \Delta g_{\mu\nu}\Delta g^{\mu\nu}}{16\Phi} + \frac{\Phi}{3}\tilde{\mathcal{L}}^{\mu\nu}\left((\tilde{\nabla}_\mu\Phi)(\tilde{\nabla}_\nu\Phi) - \frac{\Phi^2}{\ell_\Lambda^2}\tilde{g}_{\mu\nu}\right)\right.$$
$$\left. + \lambda\left(\bar{R} - \frac{12}{\ell_\Lambda^2} - \frac{\Delta g^2 - \Delta g_{\mu\nu}\Delta g^{\mu\nu}}{16\Phi^2}\right)\right], \tag{24}$$

where λ is to be determined by the equation of motion for Φ. The equation of motion for Φ is given in an extremely simple form[8] as

$$2\mathcal{L}_\lambda\lambda = \mathcal{L}_\lambda\Phi\,, \tag{25}$$

where

$$\mathcal{L}_\lambda := \left(\tilde{\nabla}_\mu\tilde{\nabla}_\nu\Phi + \frac{\Phi}{\ell_\Lambda^2}\tilde{g}_{\mu\nu}\right)\tilde{\mathcal{L}}^{\mu\nu} + \frac{\Delta g^2 - \Delta g_{\mu\nu}\Delta g^{\mu\nu}}{16\Phi^3}\,. \tag{26}$$

We ignore the homogeneous solution $\mathcal{L}_\lambda\lambda = 0$ by assuming that the energy scale is low enough to treat $\mathcal{L}_\lambda^{-1}$ as a local operator. Substituting $\lambda = \frac{\Phi}{2}$, we obtain the 4-dimensional effective action from the bulk action as

$$S_b = \frac{M_{pl}^2}{2}\frac{2}{r_c^{(+)}}\int d^4x\sqrt{-g}\left[\frac{\Delta g^2 - \Delta g_{\mu\nu}\Delta g^{\mu\nu}}{32\Phi} - \frac{1}{2\ell_\Lambda^2}\Phi^2\left(\Box + \frac{4}{\ell_\Lambda^2}\right)\Phi\right.$$
$$\left. + \frac{\Phi}{2}\left(\tilde{R} - \frac{12}{\ell_\Lambda^2}\right) - \frac{1}{6}\left(\nabla_\mu\Phi\right)\left(\nabla_\nu\Phi\right)\left(\nabla^\mu\nabla^\nu - g^{\mu\nu}\Box - R^{\mu\nu}\right)\Phi\right]$$
$$+\mathcal{O}(\epsilon^2)\,, \tag{27}$$

where ∇ is the covariant differentiation associated with $g_{\mu\nu}$, which is indistinguishable from $g_{\mu\nu}^{(\pm)}$ and $\tilde{g}_{\mu\nu}$ at the leading order of gradient expansion. The interaction between two graviton takes the Fierz-Pauli form. We cannot determine the form of mass interactions at higher order of $\Delta g_{\mu\nu}$. In the effective action (27), $\Phi\tilde{R}$ term can be absorbed into the induced gravity terms by conformal transformations for the boundary metrics $g_{\mu\nu}^{(\pm)}$ and the action for Φ in the second line is a cubic Galileon.[14,15] Therefore, the whole action $S_b + S_+ + S_-$ describes a well-known ghost-free system with two interacting metrics and one scalar.

5. Summary

In this paper, we attempted to derive the ghost-free bigravity action with a single scalar field from DGP 2-brane model with an unstabilized radion. We calculated a 4-dimensional effective action under the approximation that $g_{\mu\nu}^{(+)} - g_{\mu\nu}^{(-)} \sim |KNy^+| \ll 1$, by solving the bulk spacetime for given boundary metrics $g_{\mu\nu}^{(\pm)}$ on $(\pm)$-branes and integrating out the bulk degrees of freedom. As a result, the obtained 4-dimensional effective action contains healthy two gravitons and a scalar field. The scalar field couples to only one of the metrics $g_{\mu\nu}^{(\pm)}$ and its equation of motion does not contain higher-order time derivative. The interaction between two gravitons becomes the

Fierz-Pauli mass term at the leading order of gradient expansion. In order to obtain the mass terms nonlinear in $g^{(+)}_{\mu\nu} - g^{(-)}_{\mu\nu}$, we need to investigate the higher order of the gradient expansion. However, it is difficult to extend our analysis to the higher order of the gradient expansion, because it will produce complex and higher-derivative interaction terms which may correspond to the appearance of other bulk degrees of freedom.

Acknowledgments

YY is supported by the Grant-in-Aid for Japan Society for the Promotion of Science(JSPS) Fellows No. 15J02795 and TT is supported by the Grant-in-Aid for Scientific Research (Nos. 24103006, 24103001, 26287044 and 15H02087).

References

1. D. G. Boulware and S. Deser, Phys. Rev. D **6**, 3368 (1972).
2. S. F. Hassan and R. A. Rosen, JHEP **1202**, 126 (2012).
3. S. F. Hassan and R. A. Rosen, Phys. Rev. Lett. **108**, 041101 (2012).
4. S. F. Hassan and R. A. Rosen, JHEP **1204**, 123 (2012)
5. C. de Rham and G. Gabadadze, Phys. Rev. D **82**, 044020 (2010).
6. C. de Rham, G. Gabadadze and A. J. Tolley, Phys. Rev. Lett. **106**, 231101 (2011).
7. Y. Yamashita and T. Tanaka, JCAP **1406**, 004 (2014).
8. Y. Yamashita and T. Tanaka, arXiv:1510.07551 [hep-th].
9. G. R. Dvali, G. Gabadadze and M. Porrati, Phys. Lett. B **485**, 208 (2000).
10. A. Padilla, Class. Quant. Grav. **21**, 2899 (2004).
11. C. Charmousis, R. Gregory, N. Kaloper and A. Padilla, JHEP **0610**, 066 (2006).
12. K. Izumi, K. Koyama and T. Tanaka, JHEP **0704**, 053 (2007).
13. S. Kanno and J. Soda, Phys. Rev. D **66** (2002) 043526.
14. A. Nicolis, R. Rattazzi and E. Trincherini, Phys. Rev. D **79** (2009) 064036 [arXiv:0811.2197 [hep-th]].
15. C. Deffayet, G. Esposito-Farese and A. Vikman, Phys. Rev. D **79** (2009) 084003 [arXiv:0901.1314 [hep-th]].

The Two Faces of Mimetic Horndeski Gravity: Disformal Transformations and Lagrange Multiplier

Frederico Arroja

Leung Center for Cosmology and Particle Astrophysics, National Taiwan University, No.1, Sec.4, Roosevelt Road, Taipei, 10617 Taipei, Taiwan (R.O.C)
arroja@phys.ntu.edu.tw

After a brief review of the mimetic dark matter paradigm, we generalize its original theory to a very general scalar-tensor theory of gravity and show that generically this theory is invariant under disformal transformations. In the special case of a non-invertible transformation we show that one obtains a generalized mimetic gravity which however contains the same number of derivatives (or less) as the starting theory. We show that these mimetic theories possess a dual formulation in terms of a Lagrange multiplier field. Finally we present one simple cosmological model which can mimic almost any desired expansion history of the Universe by adequately choosing a potential function.

Keywords: Mimetic Horndeski gravity; disformal transformation; Lagrange multiplier.

1. Introduction and brief review of the mimetic dark matter model

This manuscript is based on a talk given on the 17th of December of 2015 by the author in the parallel session "Alternative and modified theories of gravity" of the Second LeCosPA International Symposium "Everything About Gravity – Celebrating the Centenary of Einstein's General Relativity". The slides of the talk can be found online at https://lecospa.ntu.edu.tw/symposium/2015/parallel_session.php?id=9. The talk was entirely based on the work present in Ref. 1.

One of the main motivations for the mimetic dark matter proposal[2] was to find an alternative explanation for the Cold Dark Matter phenomena by considering a different theory of gravity other than General Relativity (GR). In the follow up work, Ref. 3, a generalized model was proposed to also explain the Dark Energy phenomena. In the original mimetic dark matter proposal the field that plays the role of the dark matter is of gravitational origin. This field was introduced via a conformal transformation of the type $g_{\mu\nu} = -w\ell_{\mu\nu}$, where $g_{\mu\nu}$ is the original metric, $\ell_{\mu\nu}$ is an auxiliary metric and w is defined as $w = \ell^{\mu\nu}\partial_\mu\Psi\partial_\nu\Psi$, where Ψ is the new mimetic scalar field. Here we will exactly follow the notation of Ref. 1, in particular we use the mostly plus metric signature. It immediately follows for the previous transformation that $g^{\mu\nu}\partial_\mu\Psi\partial_\nu\Psi = -1$. This is a kinematical constraint that is valid independently of the dynamics. Furthermore, it is easy to see that the theory will be invariant under a Weyl rescaling of the auxiliary metric as $\ell_{\mu\nu} \to \Omega^2(x)\ell_{\mu\nu}$, where $\Omega(x)$ is an arbitrary function, because the theory will be written in terms of the metric $g_{\mu\nu}(\ell_{\alpha\beta}, \Psi)$ which is invariant.

The action of the original mimetic dark matter proposal[2] reads

$$S = \frac{1}{2} \int d^4x \sqrt{-g(\ell_{\mu\nu}, \Psi)} \left[R(g_{\mu\nu}(\ell_{\mu\nu}, \Psi)) + \mathcal{L}_m \right], \quad (1)$$

where $\mathcal{L}_m$ denotes the Lagrangian density of some matter fields. The equations of motion are $G_{\mu\nu} - T_{\mu\nu} = (G-T)\partial_\mu\Psi\,\partial_\nu\Psi, \quad \nabla_\rho\left[(G-T)\partial^\rho\Psi\right] = 0$, where $T_{\mu\nu}$ denotes the energy-momentum tensor of matter, T is its trace and G denotes the trace of the Einstein tensor $G_{\mu\nu}$. ∇_ρ denotes the covariant derivative with respect to $g_{\mu\nu}$ Note that the auxiliary metric does not appear explicitly in the equations but the field Ψ does. One can rewrite the metric equations of motion as $G_{\mu\nu} = T_{\mu\nu} + \tilde{T}_{\mu\nu}$, where the effective energy-momentum tensor $\tilde{T}_{\mu\nu}$ is defined as $\tilde{T}_{\mu\nu} = (G-T)\partial_\mu\Psi\partial_\nu\Psi$. By comparing the previous expression with the energy-momentum tensor of a perfect fluid, i.e. $\tilde{T}_{\mu\nu} = (\rho+P)u_\mu u_\nu + Pg_{\mu\nu}$, where ρ denotes the energy density, P the pressure and u^μ the 4-velocity one can see that by identifying the 4-velocity with $\partial_\mu\Psi$ one obtains that the mimetic fluid behaves as a pressureless fluid with the energy density given by $\rho = G - T$. In the absence of any matter fields, i.e. $T_{\mu\nu} = 0$, one still has a non-zero energy density with zero pressure. This means the mimetic scalar field imitates, or mimics, the behaviour of dust. By generalizing the previous model to include a potential for the scalar field one can have almost any expansion history.[3] For example, one can produce inflation, quintessence and a non-singular bouncing universe. The speed of sound of scalar perturbations in this model is exactly zero on a flat FLRW background.[3] So in order to have a non-zero sound speed, to, for example, successfully generate the primordial inflationary perturbations one needs to introduce higher-derivatives terms in the action.[3]

Since then there have been many works studying different aspects of the original mimetic theory and its generalizations. It is not the purpose of this brief review to cover all of them. Instead we shall briefly summarize the results of some works which are related with the contents of the next sections.

In Ref. 4, the author showed that the reason why the original mimetic theory contains more solutions than GR, which was the starting theory, is because we have performed a non-invertible conformal transformation. He showed that the theory is a conformal extension of GR which is free from ghost instabilities if the energy density of the effective fluid is positive. There may exist caustic instabilities because the fluid elements always follow geodesics. This is a consequence of the mimetic constraint. He also proposed a modification including a vector field which is capable of producing rotational motion. In Ref. 5, it was shown that the mimetic dark matter model has a dual formulation in terms of a Lagrange multiplier field. Deruelle and Rua[6] identified a more general disformal transformation (DT) that leads to the mimetic dark matter model. Generically, Einstein's gravity is invariant under these DT. Finally let us mention that in Ref. 7, a model more general that the original mimetic model of Refs. 2, 3 was proposed and called $\lambda\varphi$-fluid.

2. Mimetic gravity from a disformal transformation

In this work we will be concerned with a DT of the form

$$g_{\mu\nu} = A(\Psi, w)\ell_{\mu\nu} + B(\Psi, w)\partial_\mu\Psi\partial_\nu\Psi, \tag{2}$$

where A and B are arbitrary function of their arguments but should obey some conditions in order to preserve the Lorentzian signature, causality and the inverse metric should exist.[8,9]

By using the inverse function theorem, one can show[1] that the above DT cannot be inverted (for a fixed Ψ) if the B function is related with A as

$$B(\Psi, w) = -\frac{A(\Psi, w)}{w} + b(\Psi), \tag{3}$$

where $b(\Psi)$ is an arbitrary function (with constant sign) of Ψ and it appears as an integration constant. One can also find that the mimetic constraint now reads $b(\Psi)g^{\mu\nu}\partial_\mu\Psi\partial_\nu\Psi = 1$. Let us note that this is a very general result which does not depend on the theory and is a consequence of the transformation only. Also one easily understands why it is not surprising that the mimetic theory will contain additional solutions with respect to the original starting theory. The previous transformations are called mimetic disformal transformations and they were found by Ref. 6 after imposing that the system of equations of motion of the mimetic dark matter model is indeterminate. Let us now generalize their result by performing a DT in a very general scalar-tensor theory of the following type

$$S = \int d^4x\sqrt{-g}\mathcal{L}[g_{\mu\nu}, \partial_{\lambda_1}g_{\mu\nu}, \ldots, \partial_{\lambda_1}\ldots\partial_{\lambda_p}g_{\mu\nu}, \Psi, \partial_{\lambda_1}\Psi, \ldots, \partial_{\lambda_1}\ldots\partial_{\lambda_q}\Psi] \\ + S_m[g_{\mu\nu}, \phi_m], \tag{4}$$

where we assume that the matter field ϕ_m, with action S_m, is uncoupled with Ψ. Here we will omit the details of the calculation, which can be found in Ref. 1, and just state that one can contract the resulting metric ($\ell_{\mu\nu}$) equations of motion with $\ell_{\mu\nu}$ and $\partial_\mu\Psi\partial_\nu\Psi$ to find a system of equations for α_1 and α_2 as

$$M\begin{pmatrix}\alpha_1\\ \alpha_2\end{pmatrix} = 0, \qquad \text{where} \qquad M = \begin{pmatrix} A - w\frac{\partial A}{\partial w} & -w\frac{\partial B}{\partial w} \\ w^2\frac{\partial A}{\partial w} & -A + w^2\frac{\partial B}{\partial w}\end{pmatrix}, \tag{5}$$

with the definitions $\alpha_1 \equiv (E^{\rho\sigma} + T^{\rho\sigma})\ell_{\rho\sigma}$, $\alpha_2 \equiv (E^{\rho\sigma} + T^{\rho\sigma})\partial_\rho\Psi\,\partial_\sigma\Psi$, $E^{\mu\nu} = \frac{2}{\sqrt{-g}}\frac{\delta(\sqrt{-g}\mathcal{L})}{\delta g_{\mu\nu}}$ and $T^{\mu\nu} = \frac{2}{\sqrt{-g}}\frac{\delta(\sqrt{-g}\mathcal{L}_m)}{\delta g_{\mu\nu}}$. The solutions of the previous system depend on whether the determinant of the matrix M is zero or non-zero.

If the determinant of M is not equal to zero then the only solution of the system is $\alpha_1 = \alpha_2 = 0$ and then the equation of motion for $\ell_{\mu\nu}$, Ψ and the matter field ϕ_m reduce to

$$E^{\mu\nu} + T^{\mu\nu} = 0, \qquad \Omega_\Psi = 0, \qquad \Omega_m = 0, \tag{6}$$

where we have defined $\Omega_\Psi = \frac{\delta(\sqrt{-g}\mathcal{L})}{\delta\Psi}$ and $\Omega_m = \frac{\delta(\sqrt{-g}\mathcal{L}_m)}{\delta\phi_m}$. These equations of motion are the same as the equations of motion that one would get by taking the variation with respect to $g_{\mu\nu}$, Ψ and ϕ_m. In other words, a theory of the type (4) is generically, i.e. if $\det(M) \neq 0$, invariant under a disformal transformation of the type (2). On the other hand, if $\det(M) = 0$, then one gets the following differential equation involving the arbitrary functions A and B as $\det(M) = w^2 A \frac{\partial}{\partial w}\left(B + \frac{A}{w}\right) = 0$ which one can easily solve to find that Eq. (3) is a solution. We can see that the condition that the system of equations of motion (5) is indeterminate is the same as the condition for the DT to be non-invertible. In this case, the solution is $\alpha_2 = w\alpha_1$ and the equations of motion read

$$E_{\mu\nu} + T_{\mu\nu} = (E+T)\, b\, \partial_\mu \Psi\, \partial_\nu \Psi, \quad \nabla_\rho[(E+T) b\, \partial^\rho \Psi] - \frac{\Omega_\Psi}{\sqrt{-g}} = \frac{1}{2}(E+T)\frac{1}{b}\frac{db}{d\Psi},$$
$$b(\Psi) g^{\mu\nu} \partial_\mu \Psi \partial_\nu \Psi = 1, \quad \Omega_m = 0. \tag{7}$$

These new equation of motion are different from the ones obtained from taking the variation with respect to the metric $g_{\mu\nu}$, cf. (6). They are the equations of motion of the new mimetic scalar-tensor theory that generalizes the mimetic dark matter model.

3. Mimetic gravity from a Lagrange multiplier

As previously mentioned, Ref. 5 showed that the mimetic dark matter model (recall the action of the original mimetic dark matter model, Eq. (1)) can be formulated in terms of a Lagrange multiplier field without the need to invoke a non-invertible disformal transformation. In this section we will generalize this result to a general scalar-tensor theory (4).

One can show that Eqs. (7) can be derived from the following action

$$S_\lambda = \int d^4x \sqrt{-g} \mathcal{L}[g_{\mu\nu}, \partial_{\lambda_1} g_{\mu\nu}, \ldots, \partial_{\lambda_1} \ldots \partial_{\lambda_p} g_{\mu\nu}, \Psi, \partial_{\lambda_1}\Psi, \ldots, \partial_{\lambda_1} \ldots \partial_{\lambda_q}\Psi]$$
$$+ S_m[g_{\mu\nu}, \phi_m] + \int d^4x \sqrt{-g} \lambda \left(b(\Psi) g^{\mu\nu} \partial_\mu \Psi \partial_\nu \Psi - 1\right), \tag{8}$$

where λ is the Lagrange multiplier field and the function $b(\Psi)$ is the same potential function that appears in the mimetic DT. The Lagrange multiplier field enforces the kinematical constraint. By using Horndeski's identity[10] one can show[1] that the second equation in the first line of (7) can be obtained for the other equations in (7). This has one immediate important consequence, namely, the mimetic theory (8) does not contain more derivatives in the equations of motion than the original theory defined by $\mathcal{L}$. This would be a trivial statement by inspecting the action (8) because the additional term proportional to the Lagrange multiplier does not contain more derivatives than $\mathcal{L}$. However this term does contain one new field, λ. The previous statement about the equal (or less) number of derivatives in the mimetic and original theories is meant to be relevant for Eqs. (7) where the solution for λ has been used to write the equations in terms of $g_{\mu\nu}$, Ψ and ϕ_m only.

4. Simple examples of cosmological models in mimetic Horndeski

In this section we will present one simple example of a mimetic model that possesses interesting cosmological behaviour. See Ref. 1 for another example involving a model with a cubic Galileon term.

For the original theory, namely the Lagrangian $\mathcal{L}$, we choose to take the Horndeski Lagrangian, $\mathcal{L}_H$. The Horndeski theory[10] is known to be the most general 4D local scalar-tensor theory that contain second-order equations of motion and that can be derived from an action. It reads

$$S_H = \int d^4x\sqrt{-g}\mathcal{L}_H = \int d^4x\sqrt{-g}\sum_{n=0}^{3}\mathcal{L}_n, \tag{9}$$

where
$\mathcal{L}_0 = K(X,\Psi)$, $\mathcal{L}_1 = -G_3(X,\Psi)\Box\Psi$, $\mathcal{L}_2 = G_{4,X}(X,\Psi)\left[(\Box\Psi)^2 - (\nabla_\mu\nabla_\nu\Psi)^2\right] + RG_4(X,\Psi)$, $\mathcal{L}_3 = -\frac{1}{6}G_{5,X}(X,\Psi)\left[(\Box\Psi)^3 - 3\Box\Psi(\nabla_\mu\nabla_\nu\Psi)^2 + 2(\nabla_\mu\nabla_\nu\Psi)^3\right] + G_{\mu\nu}\nabla^\mu\nabla^\nu\Psi\, G_5(X,\Psi)$ and $X = -1/2\nabla_\mu\Psi\nabla^\mu\Psi$, $(\nabla_\mu\nabla_\nu\Psi)^2 = \nabla_\mu\nabla_\nu\Psi\nabla^\mu\nabla^\nu\Psi$ and $(\nabla_\mu\nabla_\nu\Psi)^3 = \nabla_\mu\nabla_\nu\Psi\nabla^\mu\nabla^\rho\Psi\nabla^\nu\nabla_\rho\Psi$. The functions $K(X,\Psi)$, $G_3(X,\Psi)$, $G_4(X,\Psi)$ and $G_5(X,\Psi)$ are arbitrary function. The subscript $_{,X}$ denotes the derivative with respect to X. We will now discuss one simple model which is contained in the mimetic Horndeski theory defined by (8) with $\mathcal{L}$ given by $\mathcal{L}_H$ and with $S_m = 0$.

4.1. *A very simple example*

Let us take one of the simplest mimetic Horndeski actions one can imagine as $S_H = \int d^4x\sqrt{-g}\left(\frac{R}{2} + c_2X\right)$, where c_2 is a constant. In a flat FLRW background the independent equations of motion are

$$b(\Psi)\dot{\Psi}^2 + 1 = 0, \qquad 6H^2 + 4\dot{H} + c_2\dot{\Psi}^2 = 0, \tag{10}$$

where H is the Hubble rate defined as $H = \dot{a}/a$, where dot denotes the derivative with respect to cosmic time and a is the scale factor. One can easily check that

$$a(t) = t^{\frac{2}{3(1+\omega)}}, \quad \Psi(t) = \pm\sqrt{-\frac{\alpha}{c_2}}\log\frac{t}{t_0}, \quad b(\Psi) = \frac{c_2}{\alpha}t_0^2 e^{\pm 2\sqrt{-\frac{c_2}{\alpha}}\Psi},$$

is a solution, where t_0 is an integration constant and α is defined as $\alpha = -\frac{8\omega}{3(1+\omega)^2}$, where ω denotes the equation of state. This means that the mimetic field in this model can mimic the expansion history of a perfect fluid with the equations of state ω. In fact, by adjusting the function $b(\Psi)$ one can mimic the expansion history of almost any model. More concretely, by comparing the second equation in (10) with the equation of motion for a perfect fluid, i.e. $6H^2 + 4\dot{H} = -2p$, where p denotes the pressure, one finds that the mimetic field can mimic a perfect fluid with a fixed sign for the pressure. This is a similar fixture to the mimetic models of Refs. 3, 7.

5. Conclusion

In this work, we have generalized previous results obtained for General Relativity only and showed that a very general scalar-tensor theory of the type (4) is generically invariant under a disformal transformation as (2). However, for a particular subset of the DT (satisfying (3)), when the transformation is not invertible, we have shown that the theory that results is different from the original theory and it is a generalization of the mimetic dark matter scenario. These results hold irrespectively of whether the field involved in the transformation is the same as the field in the action or not.[1] We have showed that the mimetic theory possesses a dual formulation in terms of a Lagrange multiplier field. We proposed two simple toy models in the context of the mimetic Horndeski theory that can essentially mimic any desired background expansion history. See also Ref. 1 for an interesting example of promoting the auxiliary metric $\ell_{\mu\nu}$ to the physical metric. We also showed that the mimetic theory contains the same (or less) number of derivatives as the original theory and hence it is free from Ostrogradski instabilities if the original theory was healthy. We finally point out that in Ref. 11 we studied linear perturbations and found that the scalar sound speed is exactly zero in the mimetic Horndeski theory and even in a theory beyond mimetic Horndeski. Furthermore, the comoving curvature perturbation is constant on all scales. This implies that these models need to be extended if they are to be used as models of inflation and the generation of the primordial perturbations.

6. Acknowledgments

We would like to thank Nicola Bartolo, Purnendu Karmakar and Sabino Matarrese for collaboration that lead to the publication 1 on which this manuscript is entirely based. We also would like to thank Keisuke Izumi for organizing the parallel session where this talk was delivered and Pisin Chen and the other members of the LOC for organizing such an enjoyable Symposium. FA is supported by the National Taiwan University (NTU) under Project No. 103R4000 and by the NTU Leung Center for Cosmology and Particle Astrophysics (LeCosPA) under Project No. FI121.

References

1. F. Arroja, N. Bartolo, P. Karmakar, and S. Matarrese, JCAP **1509**, 051 (2015).
2. A. H. Chamseddine and V. Mukhanov, JHEP **1311**, 135 (2013).
3. A. H. Chamseddine, V. Mukhanov, and A. Vikman, JCAP **1406**, 017 (2014).
4. A. Barvinsky, JCAP **1401**, 014 (2014).
5. A. Golovnev, Phys.Lett. **B728**, 39 (2014).
6. N. Deruelle and J. Rua, JCAP **1409**, 002 (2014).
7. E. A. Lim, I. Sawicki, and A. Vikman, JCAP **1005**, 012 (2010).
8. J. D. Bekenstein, Phys.Rev. **D48**, 3641 (1993).
9. D. Bettoni and S. Liberati, Phys.Rev. **D88**, 084020 (2013).
10. G. W. Horndeski, Int.J.Theor.Phys. **10**, 363 (1974).
11. F. Arroja, N. Bartolo, P. Karmakar, and S. Matarrese, (2015), 1512.09374.

Telaparallel Gravity in Five Dimensions

Ling-Wei Luo

Institute of Physics, Chiao Tung University, Hsinchu 300, Taiwan
lwluo@nctu.edu.tw

We study the teleparallel gravity in five-dimensional spacetime with particular discussions on braneworld and Kaluza-Klein (KK) theories. In the braneworld theory, the induced matter formalism of teleparallel gravity in the Friedmann-Lematre-Robertson-Walker (FLRW) cosmology is equivalent to general relativity (GR) due to the same Friedmann equation. In the KK scenario, the dimensional reduction of the 5-dimensional torsion scalar generates a trace-torsion-scalar coupling in the 4D effective Lagrangian, which can lead to an Einstein frame for the non-minimal coupled teleparallel gravity. However, it is still equivalent to GR in the KK scenario up to the total derivative.

Keywords: Teleparallel gravity; braneworld theory; Kaluza-Klein theory.

1. Introduction

The famous five dimensional theory is the Kaluza-Klein (KK) theory, which was first proposed by Kaluza[1] and Klein.[2] In the KK theory, the fifth dimension is always compactified into a circle S^1 with some small scale. Such scale is usually used to explain the problem of the effective Planck constant in four dimensions. If the extra dimension scale is large, such large extra dimension theories are referred to the braneworld theory, given by Arkani-Hamed, Dimopoulos and Dvali.[3] In braneworld theories, the matter fields are localized on the 3-brane embedded into the five dimensional bulk. The well-known *RS-I* and *II* braneworld models were first introduced by Randall and Sundrum (RS) with the special non-factorizable metrics,[4] which leads to the warp extra dimension.

Teleparallel gravity is an alternative gravitational theory considered by Einstein[5] with the absolutely parallelism. The Lagrangian of teleparallel equivalent to general relativity (TEGR) is the quadratic forms of the torsion tensors which construct the torsion scalar T. Recently, the teleparallelism has been used to explain the acceleration of the universe, such as the teleparallel dark energy[6] and $f(T)$[7,8] models.

Teleparallel gravity with an extra dimension has been discussed in the literature.[9–18] In this paper, we review our results obtained in Refs. 14 and 15. We concentrate on the KK and the braneworld scenarios in extra dimension theories. We first give the general geometrical scenario with the projected effect of TEGR on the brane, and then discuss the Friedmann-Lemaître-Robertson-Walker (FLRW) cosmology in the braneworld theory. In the KK theory, we derive the low energy effective action of TEGR in the absence of the electromagnetic field in four-dimension through the the KK reduction. The trace torsion vector and the scalar non-minimal coupled term are generated naturally in the effective action. The Einstein frame for the non-minimal teleparallel gravity can be achieved by including the trace-torsion-scalar coupling. Finally, we study the flat FLRW cosmology and discuss the solution in the KK scenario.

2. Five-Dimensional Geometry

In the four dimensional Weitzenböck geometry T_4, the metric is given by $g = g_{\mu\nu} dx^\mu \otimes dx^\nu$. In the orthonormal frame, the coframe $\vartheta^i = e^i_{\ \mu}\, dx^\mu$, therefore $g_{\mu\nu} = \eta_{ij}\, e^i_{\ \mu}\, e^j_{\ \nu}$ with $\eta_{ij} = \mathrm{diag}(+1,-1,-1,-1)$. The absolute parallelism condition $\nabla_\nu\, e^i_{\ \mu} = \partial_\nu e^i_{\ \mu} - e^i_{\ \rho}\, \Gamma^\rho_{\mu\nu} = 0$ for the vielbein fields $e^i_{\ \mu}$ leads to the Weitzenböck connection $\Gamma^\rho_{\mu\nu} = e_i^{\ \rho} \partial_\nu e^i_{\ \mu}$, which corresponds to the curvature-free condition $R^\sigma{}_{\rho\mu\nu}(\Gamma) = e_i^{\ \sigma} e^j_{\ \rho} R^i{}_{j\mu\nu}(\omega) = 0$ with the special spin connection solution $\omega^i{}_{j\mu} = 0$. As a result, the torsion tensor is $T^i{}_{\mu\nu} \equiv \Gamma^i_{\nu\mu} - \Gamma^i_{\mu\nu} = \partial_\mu e^i_{\ \nu} - \partial_\nu e^i_{\ \mu}$ and the contorsion tensor $K^\rho{}_{\mu\nu} = (-1/2)(T^\rho{}_{\mu\nu} - T_\mu{}^\rho{}_\nu - T_\nu{}^\rho{}_\mu)$. If one does the decomposition of the connection as $\Gamma^\rho_{\mu\nu} = \{^{\ \rho}_{\mu\nu}\} + K^\rho{}_{\mu\nu}$, within the curvature-free condition, the curvature torsion relation can be obtained to be $-\tilde{R}(\{\}) = T - 2\,\tilde{\nabla}_\mu T^\mu$ in the Weitzenböck geometry, where $\tilde{R}(\{\})$ and $\tilde{\nabla}_\mu$ are the scalar curvature and covariant derivative with respect to the Levi-Civita connection $\{^{\ \rho}_{\mu\nu}\}$, respectively, and T is the torsion scalar

$$T \equiv \frac{1}{4}\, T^\rho{}_{\mu\nu}\, T_\rho{}^{\mu\nu} + \frac{1}{2}\, T^\rho{}_{\mu\nu}\, T^{\nu\mu}{}_\rho - T^\nu{}_{\mu\nu}\, T^{\sigma\mu}{}_\sigma = \frac{1}{2}\, T^i{}_{\mu\nu}\, S_i{}^{\mu\nu}\,, \tag{1}$$

with $S_\rho{}^{\mu\nu} \equiv K^{\mu\nu}{}_\rho + \delta^\mu_\rho\, T^{\sigma\nu}{}_\sigma - \delta^\nu_\rho\, T^{\sigma\mu}{}_\sigma$ and $T_\mu := T^\nu{}_{\nu\mu}$ the torsion vector. We can construct the TEGR via the curvature torsion relation

$$S_{\mathrm{TEGR}} = \frac{1}{2\kappa} \int d^4x\, e\, T\,, \tag{2}$$

where $\kappa = 8\pi\, G$ and $e = \det(e^i_{\ \mu})$.

The five-dimensional metric is given by $\bar{g}_{MN} = \bar{\eta}_{IJ}\, e^I_{\ M}\, e^J_{\ N}$, where $\bar{\eta}_{IJ} = \mathrm{diag}(+1,-1,-1,-1,\varepsilon)$ with $\varepsilon = -1$. We take the coordinate frame indices to be $M, N = 0,1,2,3,5$ and $\mu,\nu = 0,1,2,3$ as well as $\alpha,\beta = 1,2,3$, while the orthonormal frame indices $A,B,I,J,K = \hat{0},\hat{1},\hat{2},\hat{3},\hat{5}$ and $i,j,k = \hat{0},\hat{1},\hat{2},\hat{3}$ as well as $a,b = \hat{1},\hat{2},\hat{3}$.

We consider an embedding $T_4 \longrightarrow T_5$. In the Gauss normal coordinate, the metric is given by

$$\bar{g}_{MN} = \begin{pmatrix} g_{\mu\nu}(x^\mu, y) & 0 \\ 0 & \varepsilon\phi^2(x^\mu, y) \end{pmatrix}, \tag{3}$$

where $y = x^5$. It would be easy to start the discussion in the orthonormal frame. The five-dimensional torsion scalar in the orthonormal frame can be decomposed as

$${}^{(5)}T = \bar{T} + \frac{1}{2}\left(\bar{T}_{i\hat{5}j}\, \bar{T}^{i\hat{5}j} + \bar{T}_{i\hat{5}j}\, \bar{T}^{j\hat{5}i}\right) + 2\,\bar{T}^j{}_j{}^i\, \bar{T}^{\hat{5}}{}_{i\hat{5}} - \bar{T}^j{}_{\hat{5}j}\, \bar{T}^{k\hat{5}}{}_k\,, \tag{4}$$

where $\bar{T}$ is the induced four-dimensional torsion scalar. However, the non-vanishing components of vielbein are e^i_μ and $e^{\hat{5}}_5$ only. We then can write down the torsion scalar in the coordinate frame

$$^{(5)}T = \bar{T} + \frac{1}{2}\left(\bar{T}_{\rho 5\nu}\,\bar{T}^{\rho 5\nu} + \bar{T}_{\rho 5\nu}\,\bar{T}^{\nu 5\rho}\right) + 2\,\bar{T}^{\sigma}{}_{\sigma}{}^{\mu}\,\bar{T}^{5}{}_{\mu 5} - \bar{T}^{\nu}{}_{5\nu}\,\bar{T}^{\sigma 5}{}_{\sigma}\,. \tag{5}$$

We note that $\bar{T}^{\rho}{}_{\mu\nu} = T^{\rho}{}_{\mu\nu} = e^{\rho}_{i}(\partial_\mu e^i_\nu - \partial_\nu e^i_\mu)$ a purely four dimensional object.

3. Braneworld Scenario

In the braneworld Scenario, we assume that the bulk metric $\bar{g}$ is spatially flat and maximally symmetric 3-space

$$\bar{g}_{MN} = \mathrm{diag}\left(+1, -a^2(t,y), -a^2(t,y), -a^2(t,y), \varepsilon\,\phi^2(t,y)\right) \tag{6}$$

with $\varepsilon = -1$. The torsion tensors can be calculated by the first Cartan's structure equation $\bar{T}^0 = d\bar{\vartheta}^0 = 0, \bar{T}^i = d\bar{\vartheta}^i = \frac{\dot{a}}{a}\bar{\vartheta}^0 \wedge \bar{\vartheta}^i + \frac{a'}{a\phi}\bar{\vartheta}^5 \wedge \bar{\vartheta}^i, \bar{T}^5 = \frac{\dot{\phi}}{\phi}\bar{\vartheta}^0 \wedge \bar{\vartheta}^5$. Subsequently, the bulk action in the orthonormal frame is

$$S_{\text{bulk}} = \frac{1}{2\kappa_5}\int dvol^5\left(T + \frac{1}{2}(\bar{T}_{i5j}\,\bar{T}^{i5j} + \bar{T}_{i5j}\,\bar{T}^{j5i}) + \frac{2}{\phi}\,e_i(\phi)\,\bar{T}^a - \bar{T}_5\,\bar{T}^5\right), \tag{7}$$

where $\bar{T}_A := \bar{T}^b{}_{bA}$, and the torsion five-form can be read as

$$\bar{\mathcal{T}} = \left[T + \left(\frac{3 + 9\,\varepsilon}{\phi^2}\frac{a'^2}{a^2} + 6\,\frac{\dot{a}}{a}\frac{\dot{\phi}}{\phi}\right)\right] dvol^5\,, \tag{8}$$

where $T = -6H^2$ with $H = \dot{a}/a$. For the perfect fluid on the brane, the energy-momentum four-form can be decomposed into the bulk and brane parts[20] $\bar{\Sigma}_A = \left((\bar{T}^B_A)_{\text{bulk}} + (\bar{T}^B_A)_{\text{brane}}\right)\bar{\star}\bar{\vartheta}_B$, where $\left(\bar{T}^B_A\right)_{\text{brane}} = (\delta(y)/\phi)\,\mathrm{diag}(\rho(t), -P(t), -P(t), -P(t), 0)$ and $\left(\bar{T}^B_A\right)_{\text{bulk}} = (\Lambda_5/\kappa_5)\,\eta^B_A$. By variation with respect to the coframe, we have the four-form equations of motions. We concentrate on the $\hat{0}\hat{0}$-component equation

$$\left(\frac{\dot{a}^2}{a^2} + \frac{\dot{a}}{a}\frac{\dot{\phi}}{\phi}\right) - \frac{1}{\phi^2}\left(\frac{a''}{a} - \frac{a'}{a}\frac{\phi'}{\phi}\right) - \frac{1}{\phi^2}\frac{a'^2}{a^2} = \frac{\kappa_5}{3}\bar{T}_{\hat{0}\hat{0}}\,. \tag{9}$$

The jump of the scale factor is $[a'](t,0) = (-\kappa_5/3\varepsilon)\,\rho\,a_0(t)\,\phi_0(t)$. By imposing the $\mathbb{Z}_2$ symmetry,[19] we have $a'(t,0) = (-\kappa_5/6\varepsilon)\,\rho\,a_0(t)\,\phi_0(t)$. Consequently, the Modified Friedmann equation on the brane is

$$\frac{\dot{a}_0^2(t)}{a_0^2(t)} + \frac{\ddot{a}_0(t)}{a_0(t)} = -\frac{\kappa_5^2}{36}\,\rho(t)(\rho(t) + 3P(t)) - \frac{k_5}{3\phi_0^2(t)}\left(\bar{T}_{55}\right)_{\text{bulk}}\,, \tag{10}$$

which coincides with the result of GR.[20]

4. Kaluza-Klein Theory

We assume that there is no y-dependence (cylindrical condition) and compactify the fifth dimension to S^1 with a small scale r in the five-dimension geometry. The manifold forms a product space $M^4 \times S^1$ and the five-dimensional metric is reduced to

$$\bar{g}_{MN} = \begin{pmatrix} g_{\mu\nu}(x^\mu) & 0 \\ 0 & -\phi^2(x^\mu) \end{pmatrix} . \tag{11}$$

The residual components are $T^\rho{}_{\mu\nu}$ and $\bar{T}^5{}_{\mu 5} = \partial_\mu \phi / \phi$. The five-dimensional torsion scalar becomes

$$^{(5)}T = T + 2\, T^\sigma{}_\sigma{}^\mu\, \bar{T}^5{}_{\mu 5} . \tag{12}$$

After integrating out the fifth dimension, we obtain the four-dimensional effective action

$$S_{\text{eff}} = \frac{1}{2\kappa_4} \int d^4x\, e\, (\phi T + 2\, T^\mu\, \partial_\mu \phi) , \tag{13}$$

where $\kappa_4 = \kappa_5/(2\pi r)$ and $T^\nu{}_{\nu\mu} := T_\mu$ Hence, the trace-torsion-scalar coupling is generated from the KK reduction in the Teleparallelism. Substituting the curvature torsion relation into Eq. (13), we get the effective action as

$$\frac{-1}{2\kappa_4} \int d^4x\, e \Big(\phi \tilde{R}(\{\}) - 2\, \tilde{\nabla}_\mu (\phi T^\mu) \Big) , \tag{14}$$

which is equivalent to the KK effective action in GR up to the total divergence term.

If we redefine the scalar field to a canonical scalar field $\Phi := (\phi/2\kappa_4)^{1/2}$, while the effective action reads as

$$S_{\text{eff}} = \int d^4x\, e \Big(\Phi^2 T + 2\, T^\mu\, \Phi \partial_\mu \Phi \Big) , \tag{15}$$

the trace-torsion-scalar coupling becomes a vector-scalar current coupling.

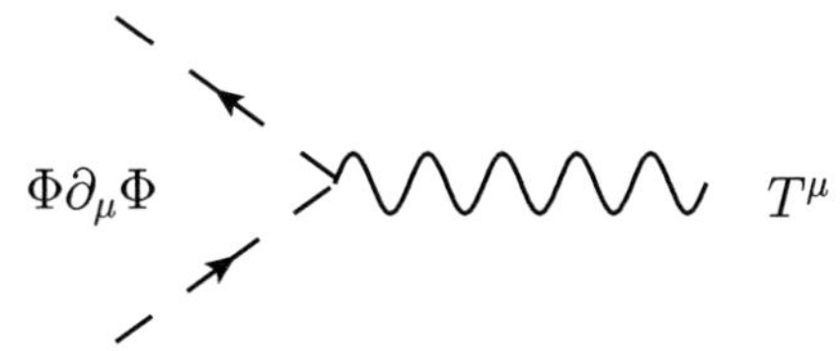

Fig. 1. Vector-scalar current coupling..

We note that by applying the conformal transformation $\tilde{g}_{\mu\nu} = \Omega^2(x)\, g_{\mu\nu}$, the trace torsion and torsion scalar are in terms of the conformal objects as

$$T = \Omega^2\, \tilde{T} - 4\, \Omega\, \tilde{g}^{\mu\nu}\, \tilde{T}_\mu \partial_\nu \Omega + 6\, \tilde{g}^{\mu\nu}\, \partial_\mu \Omega\, \partial_\nu \Omega , \tag{16}$$

$$T_\mu = \tilde{T}_\mu + 3\, \Omega^{-1}\, \partial_\mu \Omega , \tag{17}$$

respectively. By choosing $\phi = \Omega^2$, the effective action can be transformed as

$$S_{\text{eff}} = \int d^4x\, \tilde{e} \left(\frac{1}{2\,\kappa_4} \tilde{T} + \frac{1}{2}\, \tilde{g}^{\mu\nu} \partial_\mu \psi\, \partial_\nu \psi \right), \tag{18}$$

where $\psi = (6/\kappa_4)^{1/2} \ln \Omega$ is the dilaton field. The trace-torsion-scalar coupling leads to an Einstein frame for the non-minimal coupled effective Lagrangian in teleparallel gravity.

The equation of motion of Eq. (13) is

$$\begin{aligned} &\frac{1}{2}\, e_i^\mu \Big(\phi T + 2\, T^\sigma\, \partial_\sigma \phi \Big) - e_i^\rho \Big(\phi\, T^j{}_{\rho\nu}\, S_j{}^{\mu\nu} \Big) \\ &\quad - e_i^\nu \Big(\partial_\sigma \phi\, T^\mu{}_\nu{}^\sigma + \partial_\nu \phi\, T^\mu + \partial^\mu \phi\, T_\nu \Big) \\ &\quad + \frac{1}{e}\, \partial_\nu \Big(e\, (\phi\, S_i{}^{\mu\nu} + e_i^\mu\, \partial^\nu \phi - e_i^\nu\, \partial^\mu \phi) \Big) = \kappa_4\, \Theta_i^\mu \end{aligned} \tag{19}$$

with $\Theta^\mu_\nu = \text{diag}(\rho, -P, -P, -P)$. Hence, the modified Friedmann equations in the flat FLRW universe are obtained

$$3\,\phi\, H^2 + 3\, H\, \dot{\phi} = \kappa_4\, \rho\,, \tag{20}$$

$$3\,\phi\, H^2 + 2\,\dot{\phi}\, H + 2\,\phi\, \dot{H} + \ddot{\phi} = -\,\kappa_4\, P\,. \tag{21}$$

However, the equation of motion of the scalar field ϕ is

$$T - 2\,\partial_\mu T^\mu - 2T^\mu \Gamma^\nu_{\nu\mu} + e\, L_m = 0\,, \tag{22}$$

where L_m is the matter Lagrangian density. We only consider the case in the absence of matter, the equation of motion of ϕ is reduced to

$$a\, \ddot{a} + \dot{a}^2 = 0\,. \tag{23}$$

There are two solutions for the scale factor: (i) the static universe $a(t) = a_s$ and (ii) the accelerated expanding universe $a(t) = b\,\sqrt{t}$, where a_s and b are both constant. In conclusion, the teleparallel gravity in the KK scenario has the same behavior as GR.

5. Summary

We have explored teleparallel gravity with five-dimensional geometry. We have shown that the effect on the lower dimensional manifold is totally determined by a higher dimensional torsion tensor for the TEGR in the braneworld scenario. Braneworld theory of teleparallel gravity in the FLRW cosmology still provides an equivalent viewpoint as Einstein's general relativity. In the FLRW universe, we have found that the accelerated expansion of the universe can be achieved by the effective teleparallel gravity without matter, which is the same as the effective KK scenario in general relativity. The KK reduction of telaparallel gravity generates a trace-torsion-scalar coupling in the effective Lagrangian, which leads to an Einstein frame by the conformal transformation for the non-minimal coupled teleparallel gravity.

References

1. T. Kaluza, Sitzungsber. Preuss. Akad. Wiss. Berlin (Math. Phys.) **1921**, 966 (1921).
2. O. Klein, Z. Phys. **37**, 895 (1926) [Surveys High Energ. Phys. **5**, 241 (1986)].
3. N. Arkani-Hamed, S. Dimopoulos and G. R. Dvali, Phys. Lett. B **429**, 263 (1998); N. Arkani-Hamed, S. Dimopoulos and G. R. Dvali, Phys. Rev. D **59**, 086004 (1999)
4. L. Randall and R. Sundrum, Phys. Rev. Lett. **83**, 3370 (1999); L. Randall and R. Sundrum, Phys. Rev. Lett. **83**, 4690 (1999).
5. A. Einstein, Sitzungsber. Preuss. Akad. Wiss., **1928**(XVII), 217221, (1928); **1929**(X), 156159, (1929).
6. C. Q. Geng, C. C. Lee, E. N. Saridakis and Y. P. Wu, Phys. Lett. B **704**, 384 (2011); C. Q. Geng, C. C. Lee and E. N. Saridakis, JCAP **1201**, 002 (2012); J. A. Gu, C. C. Lee and C. Q. Geng, Phys. Lett. B **718**, 722 (2013); C. Q. Geng, J. A. Gu and C. C. Lee, Phys. Rev. D **88**, 024030 (2013); J. T. Li, Y. P. Wu and C. Q. Geng, Phys. Rev. D **89**, 044040 (2014).
7. R. Ferraro and F. Fiorini, Phys. Rev. D **75**, 084031 (2007).
8. E. V. Linder, Phys. Rev. D **81**, 127301 (2010) [Erratum-ibid. D **82**, 109902 (2010)]; K. Bamba, C. Q. Geng and C. C. Lee, arXiv:1008.4036 [astro-ph.CO]; P. Wu and H. W. Yu, Eur. Phys. J. **C71**, 1552 (2011); K. Bamba, C. Q. Geng, C. C. Lee and L. W. Luo, JCAP **1101**, 021 (2011).
9. V. C. de Andrade, L. C. T. Guillen and J. G. Pereira, Phys. Rev. D **61**, 084031 (2000).
10. A. L. Barbosa, L. C. T. Guillen and J. G. Pereira, Phys. Rev. D **66**, 064028 (2002).
11. F. Fiorini, P. A. Gonzalez and Y. Vasquez, arXiv:1304.1912 [gr-qc].
12. K. Bamba, S. 'i. Nojiri and S. D. Odintsov, Phys. Lett. B **725**, 368 (2013).
13. K. Nozari, A. Behboodi and S. Akhshabi, Phys. Lett. B **723**, 201 (2013).
14. C. Q. Geng, L. W. Luo and H. H. Tseng, Class. Quant. Grav. **31**, 185004 (2014).
15. C. Q. Geng, C. Lai, L. W. Luo and H. H. Tseng, Phys. Lett. B **737**, 248 (2014).
16. A. Behboodi, S. Akhshabi and K. Nozari, Phys. Lett. B **740**, 291 (2015); A. Behboodi and K. Nozari, Phys. Lett. B **750**, 601 (2015).
17. K. Atazadeh and A. Eghbali, Phys. Scripta **90**, no. 4, 045001 (2015).
18. G. G. L. Nashed, Adv. High Energy Phys. **2014**, 830109 (2014).
19. P. Horava and E. Witten, Nucl. Phys. B **460**, 506 (1996).
20. P. Binetruy, C. Deffayet and D. Langlois, Nucl. Phys. B **565**, 269 (2000); P. Binetruy, C. Deffayet, U. Ellwanger and D. Langlois, Phys. Lett. B **477**, 285 (2000).

PARALLEL SESSION

VIII. Cosmology — Early Universe

Summary of Parallel Session: "Cosmology — Early Universe"

Yi-Fu Cai

CAS Key Laboratory for Researches in Galaxies and Cosmology, Department of Astronomy, University of Science and Technology of China, Chinese Academy of Science, Hefei, Anhui 230026, China
yifucai@ustc.edu.cn

Teruaki Suyama

Research Center for the Early Universe (RESCEU), Graduate School of Science, The University of Tokyo, Tokyo 113-0033, Japan
suyama@resceu.s.u-tokyo.ac.jp

This proceeding article summarizes the contents of sixteen talks presented in the the parallel session "Cosmology-Early Universe" in the 2nd LeCosPA International Symposium -Everything about Gravity-, which was held to celebrate the centenary of Einstein's General Relativity in Taipei. Our parallel session focused on the development of early universe cosmology from theoretical viewpoints and then made the outlook of future directions from experimental aspect.

1. Introduction

Rapid progress of observational cosmology in the last two decades has revealed that all the properties of the observed primordial perturbations are consistent with the predictions of inflation, strongly supporting the idea of the rapid accelerated expansion of the Universe in its very early stage. Yet, we do not have concrete picture of the early Universe due to the limited amount of observational data and our ignorance of high energy physics. Thus, further theoretical and observational studies are needed to establish the nature of the early Universe. In light of this situation, there are many directions we can proceed for this purpose. We had sixteen talks in this session (the biggest session in terms of the number of speakers!!) and a wide range of the cosmology is covered by these talks. In order to help any reader figure out what was presented in this session, we decided to classify them into five categories; i) more details of inflation(building of inflation models), ii) period before inflation (bounce scenario, genesis scenario), iii) alternative to inflation (bounce scenario, genesis scenario), iv) inflationary byproducts, v) smoking guns of inflationary cosmology?. In what follows, short summary of each talk is reported based on this classification. Those who want to know more details about each topic are recommended to go to each proceedings paper and references therein.

2. Talks

2.1. *More details of inflation*

Arguably, one of the important goal of inflationary cosmology is to pin down the model of inflation and to embed it in the framework of high energy physics. With the current uncertainty of our knowledge of high energy physics, promising approach is to construct phenomenological models that explain observational data and then examine it if it can be deduced from fundamental theory. In the present subsection we name several significant progresses achieved in our parallel session.

Kazunori Kohri talked about the Starobinsky inflation model which is currently favored by the latest CMB observations. In this model, the gravitational action is modified from the general relativity by adding the R^2 term (R is the Ricci scalar) in the Lagrangian and inflation is caused by this new term. However, it is known that in order for this model to fit the observational data, the dimensionless coefficient in front of the R^2 term must be taken to be a huge value $\sim 10^8$. The main motivation of his work is to provide a natural explanation of such a huge coefficient. For this purpose, ten dimensional gravitational action, which is supposed to be the effective action of the fundamental theory such as string theory, was introduced. It was found that after compactification of the six dimensional space, the Starobinsky model can be nicely recovered just by assuming that the compactification length scale is merely about 30 times bigger than the length scale above which the effective theory describe physics well.

Yuta Hamada talked about the saddle point inflation in string-inspired theory. The underlying assumption made is that spacetime does not respect supersymmetry, which is compatible with the existence of non-supersymmetric vacua in string theory. With this motivation, he considers a ten dimensional effective action consisting of metric and dilaton. It was found that saddle point inflation can be realized after compactifying the theory into four dimension. It was also pointed out that the inflation could be identified with the Higgs field.

Kazunori Nakayama talked about gravitational particle production after inflation with extended gravity models. After inflation, inflaton field oscillates around the bottom of the potential and eventually decays into radiation. In general relativity, during this oscillation period, it is known that the energy density of inflation as well as Hubble parameter vary on Hubble time scale. On the other hand, what he found is that such a picture is drastically altered when the inflaton couples to the Einstein tensor. In this case, both energy density of inflation and the Hubble parameter rapidly oscillate and the gravitational particle production becomes very efficient. Thus, the standard picture of reheating is not justified generically when gravitational sector is modified.

Taku Hayakawa talked about cosmology with heavy Polonyi field. Polonyi model was originally proposed by Polonyi in 1977 as a model for breaking supersymmetry. However, it is known that this model has a serious problem if the model is embedded in cosmology, that is, the success of the Big-Bang Nucleosynthesis is spoiled. To circumvent this problem, he proposes a heavy Polonyi field. In this case, the Polonyi field decays before the formation of light elements. He also shows that dark matter as the decay products of the Polonyi field can be made compatible with the observed abundance of dark matter.

2.2. *Period before inflation*

Since inflation is not the ultimate first period of the history of the Universe, it is very natural to ask what happened before inflation. Two interesting talks were given along this direction.

Taotao Qiu talked about bounce inflation. Since the Universe is expanding, if we trace the Universe backward in time, the size of the Universe eventually shrinks to zero at some time where the Universe becomes singular and description by general relativity breaks down. One interesting possibility to avoid the singularity is to assume that the Universe underwent the bounce before its size reaches zero. However, it is known that two issues must be overcome to achieve the bounce; a violation of the null energy condition and a suppression of the cosmic anisotropy. With the use of the canonical scalar field, violation of the null energy condition involves ghost instability, thus precluding the realization of the bounce. T. Qiu suggested a viable bounce inflation model in which the scalar field has the so-called Galileon term. He showed that those two issues are naturally resolved in his new model.

Tsutomu Kobayashi talked about Galilean creation of the inflationary Universe. He proposed a very interesting idea that the Universe started not from the high energy state but from the empty space, that is, the Minkowski space. As in the case of the bounce model, the null energy condition must be violated in order for the Universe to go from the hot state to the cold state backward in time. This cannot be done without ghost instability in models with canonical scalar field. With the use of the recently proposed beyond Horndeski framework, he succeeded in constructing a viable Galilean genesis model in which no instability arises not only in the genesis phase but also in the transition phase from genesis to inflation.

2.3. *Alternative to inflation*

In order for us to tell with confidence that inflation actually happened in the early Universe, we need to exclude all the other possibilities. This simple argument shows that considering alternatives to inflation is necessary for cosmologists. Two talks were presented in this context.

Sakine Nishi investigated how the matter in the present Universe was created in the Galileon genesis scenario. She showed that the matter is gravitationally produced during the transition phase connecting the genesis phase and the kination phase and the reheating occurs when the matter density becomes greater than the background energy density in the kination phase. She also showed the spectrum of the primordial gravitational waves generated during this scenario. Remarkably, contrary to the standard inflationary scenario, the spectrum is blue and can be interesting target of the terrestrial gravitational wave experiments.

Antonino Marciano talked about superconductive inflation. He considers a possibility of realizing the bounce using the superconducting fermions (Fermionic bounce). The point is that BCS condensation or torsion of the fermionic matter may contribute four fermion term in the energy density and can result in the violation of the null energy condition. He also reported that fermion fluctuation can produce the almost scale-invariant adiabatic scalar perturbation.

2.4. *Inflationary byproducts*

The successes of inflationary cosmology does not only reflect on the enormous model buildings in past decades, but also plentiful phenomena arisen from this very early universe paradigm, such as the great developments of perturbation theory. In the present subsection we name several significant progresses achieved in our parallel session.

Tomo Takahashi brought us into the development of the perturbation theory in the frame of inflationary cosmology in terms of multiple fields. In particular, a class of curvaton models were considered. The speaker began the talk by pointing out that the traditional n_s-r constraint is relatively loose. However, detailed analyses show that there could exist some imprints on the μ distortion at small scales, and therefore, observational constraints on curvaton models can be improved. The study of primordial non-Gaussianities can give rise to stronger limit on the model parameters within curvaton scenarios. Moreover, one can take into account the observational constraint of iso-curvature perturbations that further compress the parameter space of models of multi-field inflation.

Chunshan Lin introduced a new mechanism of generating primordial magnetic fields from the local U(1) symmetry breaking during inflation. The production of cosmic magnetic fields is of observational consideration. A traditional mechanism is to break the conformal symmetry during inflation, but this approach often suffers from issues of strong coupling, back-reaction, iso-curvature perturbation, low-scale inflation, and so on. Therefore, a new mechanism was put forward in which the local U(1) may be softly broken. In the mechanism of the so-called inflationary magnetogenesis, there remains the global U(1) and the scenario can be phenomenologically connected with the information of primordial perturbations, such as primordial non-gaussianities, tensor modes, and so on. Moreover, this interesting mechanism can also provide a theoretical interpretation of the observed baryon asymmetry in our universe.

Yuichiro Tada presented a comprehensive analysis of primordial black holes that might be produced from models of waterfall hybrid inflation. The detailed method is to apply the stochastic system analysis and the δN formalism to derive the non-perturbative algorithm. For the detailed models, the cases of instant waterfall and long water fall have been considered. It can be found that the scale of primordial black holes is in general quite small. As a consequence, depending on the value choices of the abundance parameter ζ_c, either there is no parameter space to allow a detectable primordial black hole or they are too small to survive after a period of evaporation.

Shun Arai talked about the study of inflationary fluctuations within the context of Horava-Lifshitz (HL) gravity. It is well known that the HL gravity is one important theory of modified gravity, which is power-counting renormalisable. Since the anisotropic scaling law in this theory breaks the local Lorentz symmetry, a new degree of freedom appears automatically, which is dubbed as a "scalar graviton". The study of cosmological perturbations seeded by this new scalar degree of freedom was performed within a healthy and non-projectable extension of the HL gravity, which is called BPSH theory. To analyze the growth behavior of the inflaton and scalar graviton fluctuations and then to compare with the CMB data, it becomes observationally possible to constrain the Lorentz-violating parameters.

Chun-Yen Chen showed that the neutrino oscillations and fugacity could make possible connections with early universe. This topic is related to the cosmic phase after inflation, where the BBN was in progress. In this case the neutrino-proton ratio is very sensitive to the BBN process. Consequently, it is suggested that the primordial neutrino fugacity should be considered as one extra cosmological parameter beyond the standard LCDM cosmology due to the effects on the BBN and CMB. The lepton asymmetry for each flavor could also be affected by this newly introduced parameter.

2.5. *Smoking guns of inflationary cosmology?*

Yi Wang raised an important question: How can one falsify thousands of early universe models? And he carried out two possible paths to distinguishing these models, which respectively are the searches for gravitational waves and primordial non-gaussianities. In particular, in order to find the smoking guns that could discriminate various early universe models, it is interesting to introduce a massive field to help to generate features on non-gaussianities of different models including alternatives to inflation. This mechanism was applied into several distinct paradigms of very early universe, which are the regular inflationary cosmology, matter bounce, string gas cosmology, conformal (emergent) universe, and ekpyrotic scenario. It was found that the nonlinear fluctuations induced by a heavy field into primordial power spectrum behave periodically but completely different from others, and thus these cosmological paradigms are in principle distinguishable by confronting those specific patterns imprinted in non-gaussianities with future observations with super high precision. These approximately periodically patterns can be viewed as the ticks of primordial *standard clocks*. An interesting comparison of examining primordial non-gaussianities and gravitational waves was addressed.

Once if cosmologists could theoretically predict particular features for primordial perturbations generated within inflationary cosmology, it becomes urgent to search for these observational signals in forthcoming astronomical/cosmological experiments with high precision. Then, in our session we had Hong Li to give a detailed description of the future CMB experiments in the north part of the Earth. In particular, Hong Li introduced the plan of Ali project at Tibet in the mainland of China. This project will choose Ali area, where we have an observational site at 5100 meters above sea level. The observational goal is to focus on the polarization maps of the CMB by constructing a telescope of which the techniques will be based on other experiments, such as Bicep/Keck Array, SPT, POLARBEAR, Planck, and so on. Considering that the sky coverage will be over 65% at Ali site, we eventually wish to improve observational cosmology into the highest precision.

3. Discussion

Except for a talk by Hong Li, all the talks were theoretical. Currently, we are not yet at the stage of testing those theoretical ideas due to limited amount of observational data. However, not so far in the future, it is almost certain that physical quantities such as 21cm, gravitational waves, CMB spectral distortion, large scale structure and so on will become available as useful new observables. These observables, when measured, will significantly tighten the current theoretical uncertainty of the model of the early Universe. Cosmology has just reached a level of precision science and there is vast domain of research waiting to be explored. Finally, we would like to close this manuscript by quoting a speech made by Prof. Alexei Starobinsky at the closing of this session for guiding and encouraging young scientists. Alexei commented that (after our understanding and re-expressing), so far we cosmologists have made fruitful achievements in the theoretical side of inflationary cosmology, and now, it is the most important to look at early universe physics from observations, and to combine the data with various theories to explore those interesting topics appeared in our parallel session.

Acknowledgments

We would like to thank all the speakers in this session for their wonderful talks. We had a lots of questions and discussions during the session, which made the session more fruitful. We would also like to thank Pisin Chen for giving us the opportunity to attend this nice Symposium and to chair the parallel session of Cosmology-Early Universe. We also acknowledge to Frederica Wang and Yen-Ling Lee as well as all coordinators for various supports which made this session successful. YFC is supported in part by the Chinese National Youth Thousand Talents Program, by the USTC start-up funding (Grant No. KY2030000049) and by the NSFC (Grant No. 11421303).

Gravitational Particle Production after Inflation with Extended Gravity Models

Kazunori Nakayama

Department of Physics, University of Tokyo,
Bunkyo-ku, Tokyo 133-0033, Japan
kazunori@hep-th.phys.s.u-tokyo.ac.jp

We study gravitational particle production in the inflaton oscillation regime after inflation both in Einstein gravity and extended gravity models. In some extended models, the particle production rate is large enough to complete the reheating. We also find that there exists a gradient instability during the inflaton oscillation in some extended models.

Keywords: Inflation; reheating; extended gravity.

1. Introduction

Recently, extended gravity models are drawing much attention partly because it may beautifully explain inflation with a prediction consistent with present observations. A framework to deal with general extended models is the so-called generalized Galileon.[1,2] Although inflation in this general framework has been studied intensively, the reheating era has not been investigated well.

Through a series of works,[3–6] we derived a way to systematically analyze the equation of state of the universe in an inflaton-oscillation dominated era, gravitational particle production and reheating and also a existence of gradient instability.

2. Inflaton oscillation

In this section we derive a useful way to derive background dynamics of coherent inflaton oscillation.

2.1. *Einstein gravity*

In Einstein gravity, the EOM is given by

$$\ddot{\phi} + 3H\dot{\phi} + V' = 0. \tag{1}$$

We consider the limit $m_\phi \gg H$, where $m_\phi^2 \equiv |V'/\phi|_{\phi=\Phi}$ is the effective inflaton mass with Φ being the oscillation amplitude. Although it is difficult to exactly solve this equation, the existence of conserved quantity makes the situation simpler. It is easy to derive the energy conservation equation:

$$\dot{\rho}_\phi + 3H(\rho_\phi + p_\phi) = 0, \tag{2}$$

where

$$\rho_\phi = \frac{1}{2}\dot{\phi}^2 + V, \qquad p_\phi = \frac{1}{2}\dot{\phi}^2 - V. \tag{3}$$

By noting that both ρ_ϕ and p_ϕ are both of $\mathcal{O}(m_\phi^2\phi^2)$, we can immediately see that $\dot{\rho}_\phi \sim H\rho_\phi$, which means that ρ_ϕ is *not* an oscillating quantity with time scale of inflaton oscillation. In this sense, we call it *adiabatic*.

2.2. *Extended gravity*

In general extended gravity models, it is far more non-trivial to find such a conserved or *adiabatic* quantity because the EOM is complicated and the notion of inflaton energy becomes ambiguous. However, we find a very convenient way to derive such an adiabatic invariant for any extended gravity models in the framework of generalized Galileon theories.[1,2] Assuming that the field is homogeneous, the generalized Galileon theories consist of the following Lagrangian:

$$\mathcal{L}_2 = G_2, \tag{4}$$

$$\mathcal{L}_3 = -G_{3\phi}\dot{\phi}^2 - G_{3X}\dot{\phi}^2\ddot{\phi}, \tag{5}$$

$$\mathcal{L}_4 = -6H^2G_4 - 6HG_{4\phi}\dot{\phi} + 6H^2G_{4X}\dot{\phi}^2, \tag{6}$$

$$\mathcal{L}_5 = -3H^2G_{5\phi}\dot{\phi}^2 + H^3G_{5X}\dot{\phi}^3, \tag{7}$$

The EOM is then given by

$$(\mathcal{L}_{\dot{\phi}})^\cdot + 3H\mathcal{L}_{\dot{\phi}} - \mathcal{L}_\phi = 0, \tag{8}$$

$$\mathcal{L} - \dot{\phi}\mathcal{L}_{\dot{\phi}} - H\mathcal{L}_H = 0, \tag{9}$$

$$(\mathcal{L}_H)^\cdot + 3H\mathcal{L}_H - 3\mathcal{L} = 0. \tag{10}$$

From these, it is easily checked that $(\mathcal{L}_H)^\cdot \sim H\mathcal{L}_H$ holds generally. Thus $\mathcal{L}_H$ is not oscillating, but conserved quantity in each inflaton oscillation. Now let us define[6]

$$J \equiv -\frac{\mathcal{L}_H}{6M_P^2}, \tag{11}$$

which we call adiabatic invariant in generalized Galileon theories. In Einstein gravity, it coincides with the energy density of inflaton. This quantity is very useful to analyze the reheating stage, as seen below.

3. Gravitational particle production

Now we estimate the gravitational production rate of massless scalar χ, which couples to ϕ only gravitationally:

$$S = \int d^4x\sqrt{-g}\left(\mathcal{L}(\phi) - \frac{1}{2}(\partial\chi)^2\right). \tag{12}$$

From the general argument of particle production, it is shown that the production rate is estimated to be[5,6]

$$\Gamma_{\phi\to\chi} \sim \frac{q^2 m_\phi^3}{\Phi_c^2} \simeq \frac{\left(\delta\dot{H}\right)^2}{m_\phi \Phi_c^2}, \tag{13}$$

where $q \equiv |\Delta m_\chi^2 / m_\phi^2|$ with Δm_χ^2 being the ϕ-dependent mass term of canonically normalized χ, Φ_c is the inflaton oscillation amplitude in the canonically normalized basis and δH is the oscillating part of the Hubble parameter.

3.1. *Model 1*

We first consider

$$G_2 = X - V(\phi), \tag{14}$$
$$G_4 = f(\phi), \tag{15}$$

where $X = \dot{\phi}^2/2$ and $f(\phi)$ is assumed to be of the form

$$f(\phi) = \frac{1}{2} M_P^2 \left(1 + f_1 \frac{\phi}{M_P}\right). \tag{16}$$

We also assume $|f_1 \phi|/M_P \ll 1$. The adiabatic invariant in this model is calculated as

$$J = H\left(1 + f_1 \frac{\phi}{M_P}\right) + \frac{f_1}{2} \frac{\dot{\phi}}{M_P}. \tag{17}$$

Since this is *not* an oscillating quantity, we can infer the oscillating part of the Hubble parameter as

$$\delta H \simeq -f_1 \frac{\dot{\phi}}{2M_P}. \tag{18}$$

Then, the production rate is estimated as

$$\Gamma_{\phi\to\chi} \sim f_1^2 \frac{m_{\text{eff}}^3}{M_P^2}, \tag{19}$$

which reproduces the result in Refs. 4 and 7.

3.2. *Model 2*

As a second example, we consider a non-minimal derivative coupling model,[8]

$$S = \int d^4x \sqrt{-g} \left[\frac{M_P^2}{2} R - \frac{1}{2}\left(g^{\mu\nu} - \frac{G^{\mu\nu}}{M^2}\right) \partial_\mu \phi \partial_\nu \phi - V(\phi)\right]. \tag{20}$$

This is equivalent to the following assignments:

$$G_2 = X - V(\phi), \tag{21}$$

$$G_4 = \frac{1}{2}M_P^2, \tag{22}$$

$$G_5 = -\frac{\phi}{2M^2}. \tag{23}$$

Let us assume $H \gg M$ so that the non-minimal coupling takes a dominant role in the kinetic term. The adiabatic invariant J is given by

$$J = H\left(1 - \frac{\dot{\phi}^2}{2M_P^2 M^2}\right). \tag{24}$$

From this, the oscillating part of the Hubble parameter is deduced as

$$\delta H \sim \frac{J\dot{\phi}^2}{M_P^2 M^2}. \tag{25}$$

Thus, the decay rate of ϕ is estimated as

$$\Gamma_{\phi\to\chi} \sim \frac{m_{\rm eff}^5 \Phi^2}{M_P^4 M^2}. \tag{26}$$

Note that this can be larger than the Hubble parameter. These examples clearly show that the oscillation of Hubble parameter becomes stronger in extended gravity models and the gravitational particle production becomes very efficient. It is even possible the reheating is completed by the gravitational inflaton decay.

4. Gradient instability

The rapid oscillation of Hubble parameter not only leads to the gravitational inflaton decay into light scalar sector, but also potentially induce *gradient instability.* To make discussion concrete, let us focus on the non-minimal derivative coupling model. By looking at Eq. (20) and noting that $G_{00} \sim H^2$ and $G_{ij} \sim \dot{H}$, we may immediately recognize that the sound speed squared of the inflaton is roughly given by

$$c_s^2 \sim \frac{\dot{H}}{H^2}, \tag{27}$$

whose absolute value is $\sim m_\phi/H$, which can be much larger than unity. Importantly, c_s^2 can take *negative* values in each inflaton oscillation. It means that the shortest wavelength modes grow exponentially. This is known as the gradient instability.

This naive order estimation is actually confirmed by more careful analysis:[5a]

$$c_s^2 = \frac{1}{K}\left[\left(1+\frac{3}{2}\epsilon\right)+\frac{3H^2}{M^2}\left[\left(1+\frac{3}{2}\epsilon\right)+\frac{2}{3F}\epsilon\right]+\frac{6\dot{H}}{M^2}\left(1-\frac{1}{2}\epsilon\right)\right], \tag{28}$$

$$K \equiv \left(1-\frac{1}{2}\epsilon\right)\left(1+\frac{3H^2}{M^2}\frac{1+\frac{3}{2}\epsilon}{1-\frac{1}{2}\epsilon}\right), \quad \epsilon \equiv \frac{\dot{\phi}^2}{M_P^2 M^2}, \quad F = \frac{1-\frac{1}{2}\epsilon}{1-\frac{3}{2}\epsilon}. \tag{29}$$

In the limit $H \gg M$, we have $c_s^2 \sim \dot{H}/H^2$. We do not know what is the exact consequences of gradient instability. It may not invalidate the model, but it surely makes the whole analysis of the reheating process much more complicated. We also stress that this kind of gradient instability might be a general feature of extended gravity models, in which we cannot go to the Einstein frame through conformal transformation.

5. Summary

We have shown that the inflaton dynamics during the reheating is much more violent in some extended gravity models compared with the ordinary Einstein gravity case. The brief summary is as follows:

- Hubble parameter rapidly oscillates in extended gravity models.
- As a result, gravitational production of (non-conformally coupled) particles are very efficient.
- It may also lead to a gradient instability of inflaton itself. This makes the complete analysis of reheating difficult.

References

1. C. Deffayet, X. Gao, D. A. Steer and G. Zahariade, Phys. Rev. D **84**, 064039 (2011) [arXiv:1103.3260 [hep-th]].
2. T. Kobayashi, M. Yamaguchi and J. Yokoyama, Prog. Theor. Phys. **126**, 511 (2011) [arXiv:1105.5723 [hep-th]].
3. R. Jinno, K. Mukaida and K. Nakayama, JCAP **1401**, 031 (2014) [arXiv:1309.6756 [astro-ph.CO]].
4. Y. Ema, R. Jinno, K. Mukaida and K. Nakayama, JCAP **1505**, no. 05, 038 (2015) [arXiv:1502.02475 [hep-ph]].
5. Y. Ema, R. Jinno, K. Mukaida and K. Nakayama, JCAP **1510**, no. 10, 020 (2015) [arXiv:1504.07119 [gr-qc]].

[a]This sound speed is derived both in the gauge $\delta\phi = 0$ and $\zeta = 0$.[5] The former is ill-defined around the point $\dot{\phi} = 0$, but we are concerned with the *short wavelength mode* $|c_s|k \gg m_\phi$, whose timescale of the exponential growth is much shorter than the inflaton oscillation. It is obvious that within a timescale much shorter than the inflaton oscillation, the gauge $\delta\phi = 0$ is well-defined and hence the subtlety related to $\dot{\phi} = 0$ does not matter. The latter gauge is well-defined since the scale factor a is adiabatic and monotonically increasing function of time.

6. Y. Ema, R. Jinno, K. Mukaida and K. Nakayama, JCAP **1510**, no. 10, 049 (2015) [arXiv:1505.04670 [gr-qc]].
7. Y. Watanabe and E. Komatsu, Phys. Rev. D **75**, 061301 (2007) [gr-qc/0612120].
8. C. Germani and A. Kehagias, Phys. Rev. Lett. **105**, 011302 (2010) [arXiv:1003.2635 [hep-ph]].

Bounce Inflation: Towards a Nonsingular Inflationary Universe

Taotao Qiu
Institute of Astrophysics, Central China Normal University, Wuhan 430079, China
qiutt@mail.ccnu.edu.cn

Bounce inflation denotes the scenario that a bouncing process occurs preceding the traditional inflation in the early universe. In this talk I present a new bounce inflation model which can not only realize the background evolution of the bounce and inflationary phases, but also get rid of the problems of "ghost" and "anisotropy" that bothers the usual bounce models. Observable primordial cosmological perturbations can be generated during the inflationary phase after the bounce, therefore can naturally generate the scale-invariant scalar power spectrum and gravitational waves that fits the data. Moreover, the blue-tilted spectrum at large scales generated during the contracting phase can provide characteristic predictions for the bounce inflation scenario.

Keywords: Bounce; inflation; primordial perturbations.

1. Introduction

The standard model of the early universe that consists of "Big Bang" and "Inflation" suffers from the initial singularity. In Hawking et al.'s proof of the Singularity Theorem in 1970's, there should be a singularity point at the very beginning of the early universe.[1] At the singularity, everything blows up and one can not get control of the universe under classical description, making our theory untrustable. Moreover, since the singularity appears before the onset of inflation, it is difficult to remove it within inflation scenario itself. Therefore, it is interesting to ask whether one can find ways to get rid of such a singularity.

In this talk I discuss about the possibility that the universe undergo a contracting phase, where the scale factor $a(t)$ shrinks initially and then, by some mechanism, "bounces" into an expanding one.[2] The whole process can be done non-singularly if $a(t) \neq 0$ at the bouncing point.[3] However, for usual bounce, it may cause new problems. In this talk, I raise two of them, namely the "ghost" and "anisotropy" problems. Then I will propose a new model that can get rid of these problems, and give rise to proper observational signatures. For details please see Ref. 4. For other discussions about early universe problems, see the summary article of Cosmology-Early Universe session of the 2nd LeCosPA Symposium.

The ghost problem will cause quantum instability.[5] In bounce scenario, the Null Energy Condition (NEC) is violated during the bounce, which will invoke ghost modes.[3] However recently, a class of scalar field theory called "Galileons"[6] appears, which contain higher derivatives of the scalar but second order equation of motion, due to the delicate design of the action form. This means the new degrees of freedom is non-dynamical, so there will be no "ghost". For this reason, people become interested in realizing bounce using this theory.[7–9]

The anisotropy problem depends on how the background of the universe evolves in its contracting phase.[10] If our Universe is not exact isotropic at the beginning,[a] as can be described by the Bianchi-IX metric:[11] $ds^2 = -dt^2 + a^2(t)\sum_{i=1}^{3} e^{2\beta_i(t)}dx^{i2}$ with $\sum_{i=1}^{3}\beta_i(t) = 0$, the energy density of anisotropy is defined as $\Delta\rho_{ani} = (\sum_{i=1}^{3}\dot{\beta}_i^2)/2$. According to the Friedmann Equation, $\ddot{\beta}_i + 3H\dot{\beta}_i = 0$, which can be solved to get $\dot{\beta}_i \propto a^{-3}(t)$ and thus $\Delta\rho_{ani} \propto a^{-6}(t)$. This requires that, in order not to have $\Delta\rho_{ani}$ grow faster than the background and destroy the bounce, the matter which drives the bounce must have its EoS no less than unity.

However, such a contraction cannot give rise to scale-invariant power spectrum favored by the data. When the universe has a pre-inflationary contracting, primordial perturbations can be generated either in contracting phase, or in expanding phase. According to the perturbation theory, the equation of motion for the curvature perturbation is:[12]

$$u'' + (c_s^2 k^2 - \frac{z''}{z})u = 0\ ,\ v'' + (k^2 - \frac{a''}{a})v = 0 \tag{1}$$

where $u \equiv z\zeta$, $z \equiv a\sqrt{2\epsilon}$ and ζ is the curvature perturbation. $v \equiv ah$, where h denotes the tensor perturbation. c_s is the sound speed of perturbation, $\epsilon = 3(1+w)/2$ is the slow-roll parameter, and prime denotes derivative with respect to conformal time, $\eta \equiv \int a^{-1}dt$. Eq. (1) can be used to give solutions of ζ and h.

The scalar power spectrum P_ζ and the spectral index n_s are defined as: $P_\zeta \equiv k^3|\zeta|^2/(2\pi^2)$, $n_s = 1 + d\ln P_\zeta/d\ln k$. Similarly, the tensor spectrum and its index are defined as: $P_T \equiv k^3|h|^2/(2\pi^2)$ and $n_T = d\ln P_T/d\ln k$, respectively. According to the solutions of Eq. (1), scale-invariant power spectrum ($n_s \simeq 1$, $n_T \simeq 0$) can be obtained in the case of either inflation ($w \simeq -1$), or matter contraction ($w \simeq 0$),[13] the latter of which leads to anisotropy problem. So we suggest that the bounce is followed by an inflation, which generates the primordial perturbations.

2. The Galileon bouncing inflation model

We consider such a general form of lagrangian as:

$$\mathcal{L} = K(X,\phi) - G(X,\phi)\Box\phi\ , \tag{2}$$

where the last term is introduced by the Galileon theory,[6] with $\Box\phi \equiv g^{\mu\nu}\nabla_\mu\nabla_\nu\phi$. The equation of motion of the Galileon field ϕ are[4]

$$[K_X + 2K_{XX}X - 2(G_\phi + G_{X\phi}X) + 6G_X H\dot{\phi} + 6HG_{XX}\dot{\phi}]\ddot{\phi} - K_\phi$$
$$+3H[K_X - 2(G_\phi - G_{X\phi}X)]\dot{\phi} + [2K_{X\phi} + 6G_X(\dot{H} + 3H^2) - 2G_{\phi\phi}]X = 0\ , \tag{3}$$

and the energy density and pressure are

$$\rho = 2K_X X - K + 3G_X H\dot{\phi}^3 - 2G_\phi X\ ,\ P = K - 2(G_\phi + G_X\ddot{\phi})X \tag{4}$$

respectively.

---

[a] In the statistic point of view, the exact isotropic universe will have zero measure.

In this context, we will consider a rather simple picture, in which the Galileon term only affects on the bouncing point, and can be neglected in the other regions. Our analysis can be greatly simplified if we furtherly assume that the model reduces to a canonical single field model away from the bounce. Therefore, we can take the function forms of $K(X,\phi)$ and $G(X,\phi)$ as:

$$K(X,\phi) = k(\phi)X + t(\phi)X^2 - V^{con} \ , \quad G(X,\phi) = g(\phi)X \ , \tag{5}$$

where $k(\phi) = 1 - 2M_p^4 k_0 (M_p^2 + 2\kappa_1\phi^2)^{-2}$, $t(\phi) = t_0(M_p^2 + 2\kappa_2\phi^2)^{-2}$, $g(\phi) = M_p\gamma(M_p^2 + 2\kappa_2\phi^2)^{-2}$ with k_0, κ_1, t_0, κ_2, γ undetermined parameters.[4] Here we assume that the field ϕ runs monotonically with t, which will be justified by the numerical calculation. Therefore at the regions away from the bounce, we have $k(\phi) \to 1$, $t(\phi) \to 0$, $g(\phi) \to 0$, which induce $K(X,\phi) \simeq X - V$, $G(X,\phi) \simeq 0$, and the model reduces to the canonical scalar field. While around the bounce region, one has $k(\phi) \to 1 - 2k_0$, $t(\phi) \to M_p^{-4} t_0$, $g(\phi) \to M_p^{-3}\gamma$, and the Galileon term becomes important, which helps the bounce to happen.

Moreover, we take the potential term to be:

$$V(\phi) = [1 - \tanh(\lambda_1 \frac{\phi}{\phi_B})]V^{con}(\phi) + [1 + \tanh(\lambda_2 \frac{\phi}{\phi_B})]V^{inf}(\phi) \ , \tag{6}$$

where $V^{con} = -V_0 e^{c\phi}$ and $V^{inf}(\phi) = \Lambda^4[1 - (\phi/\phi_0)^2]^2$, with V_0, and c, Λ, ϕ_0 positive constants, $\lambda_1, \lambda_2 \gg 1$ and ϕ_B is the value of ϕ at bouncing point. The function $1 \pm \tanh(\lambda\phi/\phi_B)$ behaves like a selecting function for large λ, making the potential approach to V^{con}/V^{inf} when $\phi > 0/ < 0$, which drive the universe into an Ekpyrotic-like contraction ($w > 1$)[14] and an inflationary expansion ($w \simeq -1$)[15] respectively. Note that this selecting functions have been applied to construct dark energy models in Ref. 16. The whole potential function $V(\phi)$, as well as the evolutions of ϕ, H and w are numerically calculated and plotted in Fig. 1.

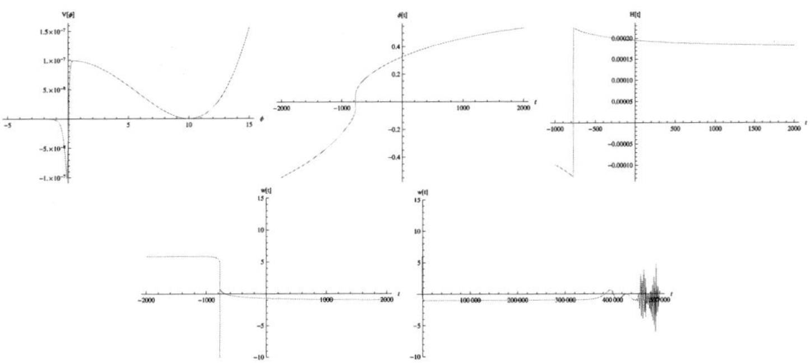

Fig. 1. From left to right: The potential function $V(\phi)$, the evolution of ϕ around bouncing point, the evolution of H around the bouncing point, the evolution of w around the bouncing point, the evolution of w during and after inflation. The horizontal axis is the cosmic time t.

Let's now focus on the perturbations generated in this model. the second order perturbed action of (2) can be written as:

$$\mathcal{S}^{(2)} = \int d\eta d^3 x a^2 \frac{Q}{c_s^2}\left[\zeta'^2 - c_s^2(\partial\zeta)^2\right] , \tag{7}$$

where the explicit form of Q and c_s^2 are

$$Q = 2M_p^4 X[K_X - 2(G_\phi - G_{X\phi}X) + 2(G_X + G_{XX}X)\ddot{\phi} \\ +4HG_X\dot{\phi} - 2G_X^2 X^2/M_p^2]/(M_p^2 H - G_X X\dot{\phi})^2 , \tag{8}$$

$$c_s^2 = \frac{(M_p^2 H - G_X X\dot{\phi})^2}{2M_p^4 X}\{K_X + 2K_{XX}X - 2(G_\phi + G_{X\phi}X) \\ +6H(G_X + G_{XX}X)\dot{\phi} + 6G_X^2 X^2/M_p^2\}^{-1} Q , \tag{9}$$

respectively. The equation of motion for ζ has the same form of (1), with $z = a\sqrt{Q}/c_s$ now. Moreover, since there is no nonminimal coupling between ϕ field and gravity, the equation of motion for gravitational waves remains unaffected.

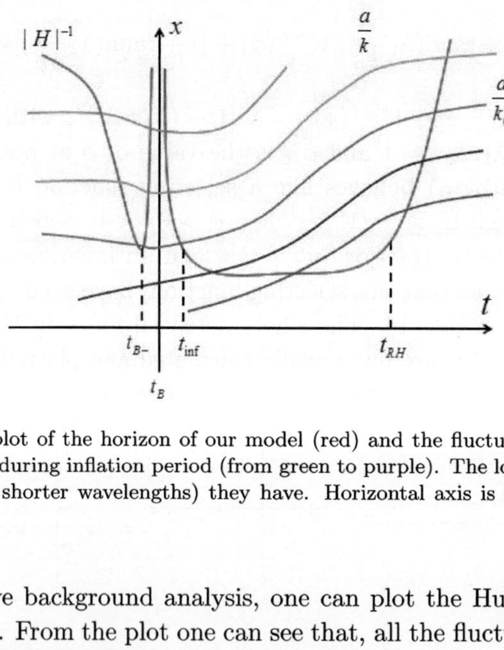

Fig. 2. The sketch plot of the horizon of our model (red) and the fluctuation modes that cross the horizon before or during inflation period (from green to purple). The lower the modes are, the larger momenta (the shorter wavelengths) they have. Horizontal axis is for time, while vertical axis is for space.

From the above background analysis, one can plot the Hubble horizon of our model as in Fig. 2. From the plot one can see that, all the fluctuation modes which are generated inside the Hubble horizon has been divided into two parts. The small k modes exit the horizon before bouncing, while the large k modes exit the horizon after bouncing. Due to such difference, the super-horizon behavior of those two types of modes behaves totally different. Note that in pure bounce or inflation models, all the fluctuation modes are generated and exit horizon within the same background, and will have the same super-horizon behavior.

In all of the regions except near the bounce in our model where ϕ has the canonical approximation, one have $Q \simeq 1$ and $c_s^2 \simeq 1$, so the solution will be the

same as that of a canonical scalar. Due to the page limit, we directly write down the solutions of each era:
Initial Condition:

$$u_k^i \sim Ae^{ic_s k\eta} \ , \ A = (2k)^{-1/2} \ . \tag{10}$$

Contracting Era:

$$u_k \sim (k|\eta - \tilde{\eta}_{B-}|)^{\frac{1}{\epsilon_c - 1}} \ (O) \ , \ \sim \cos(k|\eta - \tilde{\eta}_{B-}|) \ (I) \ , \tag{11}$$

where $\tilde{\eta}_{B-} \equiv \eta_{B-} - [(\epsilon_c - 1)a_{B-}H_{B-}]^{-1}$ and ϵ_c is the slow-varying parameter in contracting phase. I and O denotes "inside horizon solution" and "outside horizon solution", respectively.
Bouncing Era: Since in this era, the universe bounces and the scale factor does not go as power-law any more, we parametrize the scale factor $a(t)$ and the kinetic term X as $a(t) \simeq a_B e^{\alpha t^2/2}$, $X \simeq X_B e^{-\chi t^2}$, where $a_B = a(t = t_B)$ and from the numerics we know $\chi > \alpha$ in our model. Eq. (1) can be expressed as:

$$u_k'' + \bar{c}_s^2 k^2 u_k + (\chi - \alpha)a_B^2 u_k = 0 \ . \tag{12}$$

where $\bar{c}_s^2$ is the parameterized sound speed squared in bouncing era. Moreover, from Fig. 2 one can see that since in this region the Hubble horizon blows up, all the modes will enter into horizon, therefore there will only be inside horizon solution, which is:

$$u_k \sim \cos[\bar{l}(\eta - \eta_B)] \ , \ \bar{l}^2 = \bar{c}_s^2 k^2 + (\chi - \alpha)a_B^2 \ . \tag{13}$$

Expanding Era:

$$u_k \sim (k|\eta - \tilde{\eta}_{B+}|)^{\frac{1}{\epsilon_e - 1}} \ (O) \ , \ \sim \cos(k|\eta - \tilde{\eta}_{B+}|) \ (I) \ . \tag{14}$$

where $\tilde{\eta}_{B+} \equiv \eta_{B+} - [(\epsilon_e - 1)a_{B+}H_{B+}]^{-1}$ and ϵ_e is the slow-varying parameter in expanding phase.

Finally, the solutions in each era is matched by letting ζ and ζ' be continuous between each era, and that of the last (expanding) era gives final results, which generates power spectrum and spectral index observable from the data.

Similar analysis can be done for the tensor fluctuations. The form of solutions in each era are similar to those for scalar perturbations, except that in bouncing phase, due to the absence of X which appears in the scalar solution brought by the nontrivial Q, the solution splits into two branches, namely

$$v_k \sim e^{\bar{l}(\eta - \eta_B)} \ (O) \ , \ \sim \cos[\bar{l}(\eta - \eta_B)] \ (I) \ , \tag{15}$$

and the final tensor spectrum and spectral index can be obtained through the expanding era solution of v (or h). For more details of the calculation, please see Ref. 4.

3. Results

By studying this model, we get several results:

(1) One can get a nearly scale-invariant power spectrum for the perturbations. The newest PLANCK constraint on the scalar spectral index is $n_s = 0.968 \pm 0.006$,[17] while there's also large possibility that the tensor spectrum is also scale-invariant. In our model, the scalar power spectra and the spectral index turns out to be $n_\zeta \simeq 1 + \frac{2\epsilon_e}{\epsilon_e - 1} \simeq 1$, $n_T \simeq \frac{2\epsilon_e}{\epsilon_e - 1} \simeq 0$ for the small-scale ($k > k_0$) perturbations, and $n_\zeta \simeq 1 + \frac{2\epsilon_c}{\epsilon_c - 1} > 1$, $n_T \simeq \frac{2\epsilon_c}{\epsilon_c - 1} > 0$ for the large-scale ($k < k_0$) perturbations. Therefore the small-scale perturbations is nearly scale-invariant, while the large-scale perturbations will have a blue tilt.

(2) It is possible to explain the suppression of the CMB TT spectrum at small-l (large scale) region. For details of small-l suppression, see Ref. 18. Since at large scales the spectrum will be blue-tilted, the power spectrum will be suppressed at large scales and enhanced at small scales.[19] We plot the fitting of the data in Fig. 3.

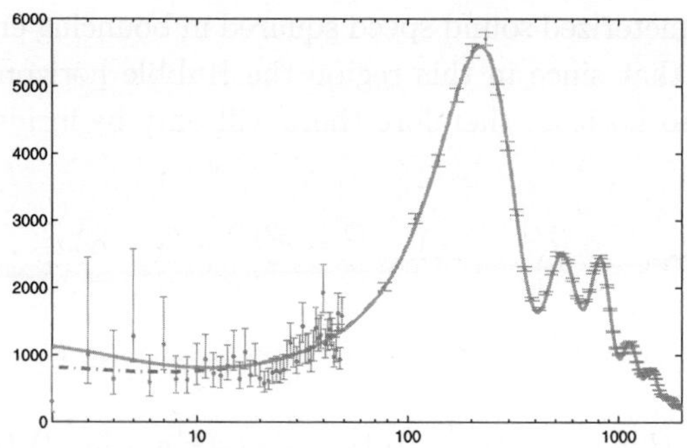

Fig. 3. Best-fit TT power spectra for pure power law (green solid) and our model (dashed) using Planck+WP data. The red points show the Planck data with 1σ errors.

(3) Since when passing through the bouncing region, all the fluctuation modes enter into the horizon, and thus can have oscillating behaviors. This may lead to oscillating signatures in TT spectrum. If such signatures were observed in the future, it will be a support to our model.

(4) The tensor(T)/scalar(S) ratio r can also be calculated from the results of scalar and tensor power spectrum (which are omitted due to the page limit and can be calculated using their definition). We obtain $r \equiv \frac{P_h}{P_\zeta} \simeq \frac{64\epsilon_c \bar{c}_s^2}{(1+\bar{c}_s^2)^2}$ for large k, while $\simeq 16\epsilon_c$ for small k. So for large k modes the T/S ratio also carries information from bounce. Using the observational constraint on r, one could constrain parameters in our model such as $\bar{c}_s^2$. From the observational data from PLANCK, which is roughly about $r \leq 0.11$,[17] one can obtain $\bar{c}_s^2 \leq 5.2 \times 10^{-4}$

for e.g. $\epsilon_c = 3$. However for small k modes the T/S ratio could be large, which will be a smoking gun of our model.

This proceeding is for talk given in the 2nd LeCosPA Symposium and I appreciate LeCosPA, National Taiwan University for hospitality.

References

1. S. W. Hawking and R. Penrose, Proc. Roy. Soc. Lond. A **314**, 529 (1970).
2. M. Novello and S. E. P. Bergliaffa, Phys. Rept. **463**, 127 (2008) [arXiv:0802.1634 [astro-ph]].
3. Y. -F. Cai, T. Qiu, Y. -S. Piao, M. Li and X. Zhang, JHEP **0710**, 071 (2007) [arXiv:0704.1090 [gr-qc]].
4. T. Qiu and Y. T. Wang, JHEP **1504**, 130 (2015) [arXiv:1501.03568 [astro-ph.CO]].
5. S. M. Carroll, M. Hoffman and M. Trodden, Phys. Rev. D **68**, 023509 (2003) [astro-ph/0301273].
6. C. Deffayet, X. Gao, D. A. Steer and G. Zahariade, Phys. Rev. D **84**, 064039 (2011) [arXiv:1103.3260 [hep-th]].
7. T. Qiu, J. Evslin, Y. -F. Cai, M. Li and X. Zhang, JCAP **1110**, 036 (2011) [arXiv:1108.0593 [hep-th]].
8. Y. -F. Cai, D. A. Easson and R. Brandenberger, JCAP **1208**, 020 (2012) [arXiv:1206.2382 [hep-th]].
9. T. Qiu, X. Gao and E. N. Saridakis, Phys. Rev. D **88**, no. 4, 043525 (2013) [arXiv:1303.2372 [astro-ph.CO]].
10. J. K. Erickson, D. H. Wesley, P. J. Steinhardt and N. Turok, Phys. Rev. D **69**, 063514 (2004) [arXiv:hep-th/0312009].
11. C. W. Misner, K. S. Thorne and J. A. Wheeler, *Gravitation*, San Francisco (1973).
12. V. F. Mukhanov, H. A. Feldman and R. H. Brandenberger, Phys. Rept. **215**, 203 (1992).
13. F. Finelli and R. Brandenberger, Phys. Rev. D **65**, 103522 (2002) [hep-th/0112249].
14. J. Khoury, B. A. Ovrut, P. J. Steinhardt and N. Turok, Phys. Rev. D **64**, 123522 (2001) [arXiv:hep-th/0103239];
15. K. A. Olive, Phys. Rept. **190**, 307 (1990).
16. X. F. Zhang and T. Qiu, Phys. Lett. B **642**, 187 (2006) [astro-ph/0603824].
17. P. A. R. Ade *et al.* [Planck Collaboration], arXiv:1502.02114 [astro-ph.CO].
18. P. A. R. Ade *et al.* [Planck Collaboration], arXiv:1506.07135 [astro-ph.CO].
19. Y. -S. Piao, B. Feng and X. -m. Zhang, Phys. Rev. D **69**, 103520 (2004) [hep-th/0310206].

Matter Creation in Generalized Galilean Genesis

Sakine Nishi* and Tsutomu Kobayashi

Department of Physics, Rikkyo University,
Toshima, Tokyo 171-8501, Japan
**sakine@rikkyo.ac.jp*
tsutomu@rikkyo.ac.jp
www.rikkyo.ac.jp

The galilean genesis is known as an alternative to inflation in which the universe starts from Minkowski space-time. We consider the scenario that the phase of universe changes to kination dominant phase from Genesis phase, and we study the matter creation during this phase transition. The temperature at the transition corresponds to the reheating temperature in the inflation scenario. As a result, we obtain the energy density of the created matter and the reheating temperature. Furthermore, this result gives us the shape of the spectrum of the gravitational waves.

Keywords: Cosmology; early universe; modified gravity.

1. Introduction

As Inflation can explain many observational result, it is known as the standard model of the early universe. In this study, we focus on the alternative model called as Galilean Genesis. This model has the scenario the universe started from $a(t) \simeq 1$ The original model of Galilean Genesis is proposed in Ref. 1, and we studied the generalization of Galilean Genesis in Ref. 2. The generalized model is constructed in Horndeski theory.[3,4] It is worth discussing the possibility of the alternative to inflation model can explain the early universe. In this study we consider the phase transition from Genesis to the radiation dominant phase, and discuss the gravitational waves in this scenario. For the matter creation, we use the mechanism of the gravitational particle creation.[5] This study is described in our paper.[6]

2. Horndeski Theory

Horndeski theory is the most general scalar tensor theory.[3,4] This theory gives the field equations which have no third and higher derivative terms. The Lagrangian of this theory is given as

$$\begin{aligned}\mathcal{L} = &G_2(\phi, X) - G_3(\phi, X)\Box\phi + G_4(\phi, X)R \\ &+ G_{4,X}\left[(\Box\phi)^2 - (\nabla_\mu\nabla_\nu\phi)^2\right] + G_5(\phi, X)G^{\mu\nu}\nabla_\mu\nabla_\nu\phi \\ &- \frac{1}{6}G_{5,X}\left[(\Box\phi)^3 - 3\Box\phi(\nabla_\mu\nabla_\nu\phi)^2 + 2(\nabla_\mu\nabla_\nu\phi)^3\right], \qquad (1)\end{aligned}$$

where ϕ is the scalar field, X is the kinetic term of the scalar field, $X := -(1/2)g^{\mu\nu}\partial_\mu\phi\partial_\nu\phi$ and G_i $(i = 2, 3, 4, 5)$ is the arbitrary function of ϕ and X.

3. Generalized Galilean Genesis

The original Genesis model is constructed by Galileon theory, but the generalized model of Galilean Genesis is constructed by defining the arbitrary functions in the action of the Horndeski theory as

$$G_2 = e^{2(\alpha+1)\lambda\phi} g_2(Y), \quad G_3 = e^{2\alpha\lambda\phi} g_3(Y),$$
$$G_4 = \frac{M_{\rm Pl}^2}{2} + e^{2\alpha\lambda\phi} g_4(Y), \quad G_5 = e^{-2\lambda\phi} g_5(Y), \tag{2}$$

where $Y := e^{-2\lambda\phi}X$, g_i $(i = 2,3,4,5)$ is the arbitrary function of Y and α is the arbitrary parameter. This action gives the field equations in Genesis phase $Y = Y_0$ as

$$\mathcal{E} \simeq e^{2(\alpha+1)\lambda\phi} \hat{\rho}(Y_0) \simeq 0, \tag{3}$$
$$\mathcal{P} \simeq 2\mathcal{G}(Y_0)\dot{H} + e^{2(\alpha+1)\lambda\phi} \hat{p}(Y_0) \simeq 0, \tag{4}$$

where

$$\hat{\rho}(Y) := 2Yg_2' - g_2 - 4\lambda Y \left(\alpha g_3 - Yg_3'\right), \tag{5}$$
$$\hat{p}(Y) := g_2 - 4\alpha\lambda Y g_3 + 8(2\alpha+1)\lambda^2 Y(\alpha g_4 - Yg_4'), \tag{6}$$
$$\mathcal{G}(Y) := M_{\rm Pl}^2 - 4\lambda Y \left(g_5 + Yg_5'\right). \tag{7}$$

We can obtain the solutions in $-\infty < t < 0$ from Eq.(3) and Eq.(4) as

$$e^{\lambda\phi} \simeq \frac{1}{\lambda\sqrt{2Y_0}} \frac{1}{(-t)}, \tag{8}$$
$$H \simeq \frac{h_0}{(-t)^{2\alpha+1}} \tag{9}$$
$$a \simeq 1 + \frac{1}{2\alpha} \frac{h_0}{(-t)^{2\alpha}}, \tag{10}$$

where

$$h_0 = -\frac{1}{2(2\alpha+1)(2\lambda^2 Y_0)^{\alpha+1}} \frac{\hat{p}(Y_0)}{\mathcal{G}(Y_0)}. \tag{11}$$

In $t \simeq -\infty$ era the solution of the scale factor is $a(t) \simeq 1$ and we can find our universe started from Minkowski universe in this model. On the other hand, at the end of Genesis $t \simeq 0$ we can find the scale factor diverges. Thus we assume that Genesis ends during the solutions of $t \simeq -\infty$ are applicable.

4. Matter Creation

Let us consider the matter creation in Genesis scenario. At the end of genesis phase $\eta = \eta_*$, we define δ_* as

$$\delta_* := \frac{a(\eta_*)}{a_G} - 1, \tag{12}$$

and we assume $\delta_* \ll 1$. For the created matter, we assume the kination phase as the epoch of after Genesis as

$$\mathcal{L} \simeq \frac{M_{\rm Pl}^2}{2} R + X. \tag{13}$$

In the kination phase, the energy density of χ is $\rho_\chi \propto a^{-4}$. As the energy density of ϕ which generates Genesis phase is $\rho_\phi \propto a^{-6}$, the kination phase terminates and the radiation dominant phase starts at $\rho_\chi = \rho_\phi$. We assume the scalar field χ as the created matter, and its Lagrangian is

$$\mathcal{L}_\chi = -\frac{1}{2} g^{\mu\nu} \partial_\mu \chi \partial_\nu \chi. \tag{14}$$

The matter creation is caused by the interactions between the scalar field χ and gravitation and between the scalar field ϕ and gravitation. The energy density of χ is given as[5]

$$\rho_\chi = -\frac{1}{128\pi^2 a^4} \int_{-\infty}^{\infty} \mathrm{d}\eta_1 \int_{-\infty}^{\infty} \mathrm{d}\eta_2 \ln(m|\eta_1 - \eta_2|) V'(\eta_1) V'(\eta_2), \tag{15}$$

where

$$V(\eta) = \frac{f'' f - (f')^2/2}{f^2}, \quad f(\eta) := a^2(\eta), \tag{16}$$

and the mass parameter m is inserted by dimensional reason. We can find the expansion law generates the matter χ. In the Genesis phase the function $f(\eta)$ is given as

$$f(\eta) \simeq a_G^2 \left[1 + \frac{h_0}{\alpha} \frac{1}{(-a_G \eta)^{2\alpha}} \right] \quad (\eta < \eta_*), \tag{17}$$

with $\eta_* := t_*/a_G$. For the kination phase, the scale factor is $a \propto (\eta + \mathrm{const})^{1/2}$ and the function $f(\eta)$ is given as

$$f(\eta) = c_0 \frac{\eta}{-\eta_*} + c_1. \tag{18}$$

After the calculation we can obtain $V(\eta)$ for two phases, and we can confirm that matter creation is mainly caused in the transition period $\Delta\eta$ because $V'(\eta)$ becomes large value during $\Delta\eta$ so that $V'(\eta)$ of the other epoch can be negligible. From Eq.(15) we obtain the energy density of the created matter as

$$\rho_\chi = \frac{A}{128\pi^2 \alpha^2} \frac{H_*^4}{\delta_*^2} \left(\frac{a_G}{a} \right)^4, \tag{19}$$

where

$$A := (2\alpha + 1)^2 \left[\frac{3}{2} + \ln \left(\frac{1}{m\Delta\eta} \right) \right]. \tag{20}$$

Furthermore, at the time of $\rho_\chi(\eta_R) = \rho_\phi(\eta_R)$, we obtain

$$\frac{a_R}{a_G} = \sqrt{\frac{96\pi^2}{A}} \left(\frac{H_*}{M_{\rm Pl}} \right)^{-\frac{1}{2\alpha+1}} \tilde{h}_0^{\frac{1}{2\alpha+1}}. \tag{21}$$

As the scale factor grows $a_R > a_G$, we can find the condition as

$$\frac{H_*}{M_{\rm Pl}} < \left(\frac{96\pi^2}{A}\right)^{(2\alpha+1)/2} \widetilde{h}_0, \tag{22}$$

where

$$\widetilde{h}_0 := M_{\rm Pl}^{2\alpha} h_0. \tag{23}$$

We can obtain the reheating temperature as

$$T_R = \left(\frac{30}{\pi^2 g_*}\right)^{1/4} \frac{A^{3/4}}{\sqrt{3}(32\pi^2)^{3/4}(2\alpha)^{3/2}} \frac{H_*^2}{M_{\rm Pl}\delta_*^{3/2}}. \tag{24}$$

This result tells us that if we consider the same value of the reheating temperature, the Habble parameter of Genesis is smaller than that of Inflation.

5. Gravitational Waves

From the result of the matter creation, we can discuss the gravitational waves in this scenario. The quadratic action of gravitational waves for Genesis is[2]

$$S_h^{(2)} = \frac{1}{8}\int {\rm d}t{\rm d}^3x\, a^3\mathcal{G}(Y_0)\left[\dot{h}_{ij}^2 - \frac{c_t^2}{a^2}(\partial_k h_{ij})^2\right], \tag{25}$$

and the solution in subhorizon is

$$h_k = \frac{1}{a}\sqrt{\frac{2}{\mathcal{G}c_t k}}e^{-ic_t k\eta}. \tag{26}$$

As the background evolution of the scale factor is $a(\eta) \simeq 1$, this perturbation exit the horizon with almost no evolution, and we can use Eq.(26) in superhorizon. Thus we obtain the power spectrum of the gravitational waves as

$$\mathcal{P}_h^{(p)}(k) = \frac{2k^2}{\pi^2\mathcal{G}c_t a_G^2}, \tag{27}$$

and thus the density parameter of the power spectrum of the gravitational waves is

$$\Omega_{\rm gw} = \Omega_{\rm gw}^{(p)}(k) \times \begin{cases} \dfrac{k_R}{k}\dfrac{k_{\rm eq}^2}{k_R^2}\dfrac{k_0^4}{k_{\rm eq}^4} & (k_R < k < k_*) \\ \dfrac{k_{\rm eq}^2}{k^2}\dfrac{k_0^4}{k_{\rm eq}^4} & (k_{\rm eq} < k < k_R) \\ \dfrac{k_0^4}{k^4} & (k_0 < k < k_{\rm eq}), \end{cases} \tag{28}$$

where

$$\Omega_{\rm gw}^{(p)}(k) = \frac{k^2}{12H_0^2}\mathcal{P}_h^{(p)} = \frac{k^4}{6\pi^2 H_0^2\mathcal{G}c_t a_G^2}. \tag{29}$$

Thus we can describe the behave of the power spectrum of the gravitational waves

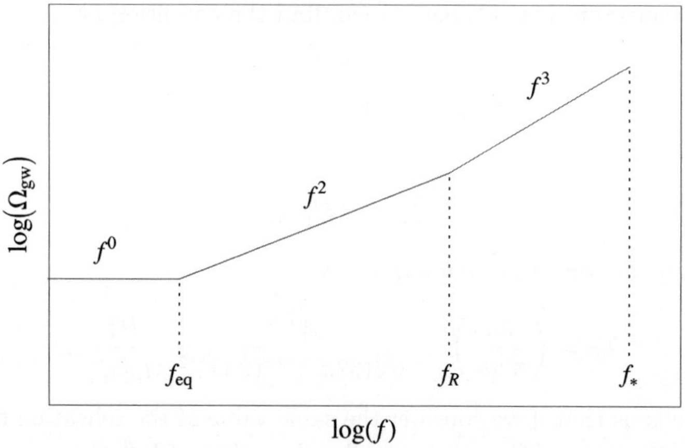

Fig. 1. The schematic picture of the spectrum of GWs in Genesis scenario.

as Fig. 1. Comparing to inflation, we find the behavior of Ω_{gw} for Genesis is different from that for inflation. In inflation Ω_{gw} for $f < f_R$ is given as

$$\Omega_{\rm gw}^{\rm inf} \simeq 10^{-5} \times \left(\frac{H_{\rm inf}}{M_{\rm Pl}}\right)^2 . \tag{30}$$

Recalling the discussion of the reheating temperature, we can find this becomes larger than that in genesis because. In the high frequency region, there is the possibility Ω_{gw} for Genesis becomes larger than that for inflation because the growth in Genesis is rapidly. After the calculations of Eq.(28), we can obtain the value at $f = f_R$ and $f = f_*$

$$\Omega_{\rm gw}(k_R) = 10^{-5} \cdot 3^{-\frac{1}{2+\alpha}} \left(\frac{32\pi^2}{A}\right)^{\frac{1+2\alpha}{2(2+\alpha)}} \left(\frac{\pi^2 g_*}{30}\right)^{\frac{3+2\alpha}{2(2+\alpha)}} \tilde{h}_0^{\frac{1}{2+\alpha}} \left(\frac{T_R}{M_{\rm Pl}}\right)^{\frac{2(3+2\alpha)}{2+\alpha}} \tag{31}$$

$$\Omega_{\rm gw}(k_*) = 10^{-5} \cdot 3^{\frac{2+3\alpha}{2+\alpha}} \left(\frac{32\pi^2}{A}\right)^{\frac{2(1+2\alpha)}{2+\alpha}} \left(\frac{\pi^2 g_*}{30}\right)^{\frac{\alpha}{2+\alpha}} \tilde{h}_0^{\frac{4}{2+\alpha}} \left(\frac{T_R}{M_{\rm Pl}}\right)^{\frac{4\alpha}{2+\alpha}} . \tag{32}$$

6. Conclusions

We have studied the matter creation in Galilean Genesis and the gravitational waves. We have focused on the phase transition from Genesis to the kination phase, and we have discussed the gravitational particle creation of massless scalar field. As a result, we have obtained the energy density of the created matter and the reheating temperature, and we have compared these of Genesis and inflation. This suggests that if we consider the same value of the reheating temperature, the Hubble parameter of Genesis is smaller than that of Inflation. Furthermore, we have discussed the power spectrum of the gravitational waves in this scenario. In this scenario we have obtained the power spectrum different from inflationary scenario.

References

1. P. Creminelli, A. Nicolis and E. Trincherini, "Galilean Genesis: An Alternative to inflation," JCAP **1011**, 021 (2010) [arXiv:1007.0027 [hep-th]].
2. S. Nishi and T. Kobayashi, "Generalized Galilean Genesis," JCAP **1503**, no. 03, 057 (2015) [arXiv:1501.02553 [hep-th]].
3. G. W. Horndeski, "Second-order scalar tensor field equations in a four-dimensional space," Int. J. Theor. Phys. **10**, 363 (1974).
4. T. Kobayashi, M. Yamaguchi and J. Yokoyama, "Generalized G-inflation: Inflation with the most general second-order field equations," Prog. Theor. Phys. **126**, 511 (2011) [arXiv:1105.5723 [hep-th]].
5. L. H. Ford, "Gravitational Particle Creation and Inflation," Phys. Rev. D **35**, 2955 (1987).
6. S. Nishi and T. Kobayashi, "Reheating and Primordial Gravitational Waves in Generalized Galilean Genesis,' arXiv:1601.06561 [hep-th].

Saddle Point Inflation from Non-Supersymmetric String-Inspired Theory

Yuta Hamada
Department of Physics, Kyoto University,
Kyoto, 606-8502, Japan
hamada@gauge.scphys.kyoto-u.ac.jp

In this proceeding, we consider the saddle point inflation scenario inspired by non-supersymmetric string theory. If the scale of the string is close to 10^{16}GeV, the required height of the inflation potential is consistent with one loop suppressed effective potential, which is crucial observation of our work. As an example of the flat potential, we consider the saddle point potential, namely potential has the point where first n derivative vanishes.

Keywords: String theory and inflation.

1. Introduction

The discovered Higgs boson mass is important input in order to speculate high energy physics. Once the Higgs mass is determined, we have fixed all the parameters in the standard model and can lift it up to its ultraviolet cutoff scale. In particular, the quadratically divergent bare Higgs mass is found to be suppressed when the cutoff is at around the Planck scale,[1–3] see also Refs. 4 and 5. Furthermore, the quartic Higgs coupling becomes tiny at the same time, see e.g Refs. 1–3 and 6–8. This opens up new aspect of a possibility of identifying the Higgs field as an inflaton,[9–13] and of non-supersymmetric string theory as Planck scale physics.[14–17] The content of this proceeding is based on our work.[18]

This proceeding is organized as follows. In Sec. 2, we present the general argument in the scenario where the inflaton potential has the point where the first n derivatives vanish. Then we calculate the cosmological predictions of the model. In Sec. 3, we consider the non-supersymmetric heterotic string theory, and argue that the favored string scale is around 10^{16}GeV. In Sec. 4, we summarize our result.

2. Saddle point inflation: general argument

We start with a general potential V as a function of an inflaton field φ. We expand the potential around the inflection point φ_c as $\varphi = \varphi_c + \delta\varphi$:

$$V = \sum_{n=0}^{\infty} \frac{V_c^{(n)}}{n!}\delta\varphi^n = V_c + V_c'\,\delta\varphi + \frac{V_c''}{2}\,\delta\varphi^2 + \frac{V_c'''}{3!}\,\delta\varphi^3 + \cdots . \tag{1}$$

We assume that the first n (≥ 2) derivatives are vanishing at φ_c:

$$V_c' = V_c'' = \cdots = V_c^{(n)} = 0. \tag{2}$$

n	$\varphi_c[M_P]$	$V_c[M_P^4]$	ϵ
2	2.1×10^4	2.0×10^{11}	3.8×10^{17}
4	2.5×10^{-11}	4.1×10^{-49}	7.9×10^{-43}
5	1.2×10^{-7}	2.0×10^{-34}	3.9×10^{-28}
6	1.9×10^{-6}	1.3×10^{-29}	2.5×10^{-23}
7	1.7×10^{-6}	2.8×10^{-27}	5.4×10^{-21}
8	1.6×10^{-5}	6.5×10^{-26}	1.2×10^{-19}
9	2.6×10^{-5}	4.9×10^{-25}	9.4×10^{-19}

The slow roll parameters around the saddle point are obtained as

$$\epsilon := \frac{M_P^2}{2}\left(\frac{V'}{V}\right)^2 = \frac{M_P^2 (V_c^{(n+1)})^2}{2(n!)^2 V_c{}^2}\,\delta\varphi^{2n} + O(\delta\varphi^{2n+1}), \tag{3}$$

$$\eta := M_P^2 \frac{V''}{V} = \frac{M_P^2\, V_c^{(n+1)}}{(n-1)!\, V_c}\,\delta\varphi^{n-1} + O(\delta\varphi^n), \tag{4}$$

$$\zeta^2 := M_P^4 \frac{V'''V'}{V^2} = \frac{M_P^4 (V_c^{(n+1)})^2}{(n-2)!\, n!\, V_c^2}\,\delta\varphi^{n-1} + O(\delta\varphi^n). \tag{5}$$

We see that $\epsilon \ll |\eta|, \zeta^2$. The inflation ends when ϵ becomes of order unity and we define its end point by $\epsilon(\delta\varphi_{\text{end}}) = 1$ to get

$$(\delta\varphi_{\text{end}})^n = \frac{\sqrt{2}\, n!\, V_c}{M_P V_c^{(n+1)}}. \tag{6}$$

The e-folding number N from a given stage of the inflation $\varphi = \varphi_c + \delta\varphi$ until its end $\varphi_{\text{end}} = \varphi_c + \delta\varphi_{\text{end}}$ is

$$\begin{aligned} N = \int_{\varphi_{\text{end}}}^{\varphi} \frac{d\varphi}{M_P^2}\frac{V}{V'} &= \frac{n!}{(n-1)}\frac{V_c}{M_P^2 V_c^{(n+1)}}\left(\frac{1}{(\delta\varphi_{\text{end}})^{n-1}} - \frac{1}{(\delta\varphi)^{n-1}}\right) \\ &\simeq \frac{n!}{(n-1)}\frac{V_c}{M_P^2 |V_c^{(n+1)}| |\delta\varphi|^{n-1}}, \end{aligned} \tag{7}$$

where we have approximated $|\delta\varphi_{\text{end}}| \gg |\delta\varphi|$ in the last step. Then the COBE normalization at a point that would give the e-folding N becomes:

$$\frac{V}{\epsilon\, M_P^4} = 2\left[\frac{(n-1)^{2n}}{(n!)^2} N^{2n}\left(\frac{V_c}{M_P^2}\right)^{n-3} (V_c^{(n+1)})^2\right]^{\frac{1}{n-1}} = 5.3 \times 10^{-7}. \tag{8}$$

We see that $V \simeq V_c$ is constrained from $\epsilon \ll 1$ by

$$V_c \ll 5.3 \times 10^{-7} M_P^4. \tag{9}$$

The slow roll parameters at the same point are

$$\epsilon = \frac{1}{2M_P^2}\left[\frac{n!}{(n-1)^n}\frac{1}{N^n}\frac{V_c}{M_P^2 |V_c^{(n+1)}|}\right]^{\frac{2}{n-1}}, \quad \eta = -\frac{n}{(n-1)N}, \quad \zeta^2 = \frac{n}{(n-1)N^2}. \tag{10}$$

It appears that the spectral index n_s and its running $dn_s/d\ln k$ in Eq. (10) change their values discretely with n. This is due to the small $\delta\varphi$ expansion made in obtaining the e-folding (7). Taking into account the next order term, we can explicitly check that the limit of $V_c^{(n+1)} \to 0$ continuously connects the case n to $n+1$.

The cosmological observables, namely the spectral index n_s, the tensor-to-scalar ratio r, and the running index $dn_s/d\ln k$:

$$n_s = 1 - 6\epsilon + 2\eta \simeq 1 - \frac{2n}{(n-1)N}, \tag{11}$$

$$r = 16\epsilon, \tag{12}$$

$$\frac{dn_s}{d\ln k} = -16\epsilon\eta + 24\epsilon^2 + 2\zeta^2 \simeq \frac{n}{(n-1)N^2}, \tag{13}$$

are constrained by the Planck 2015 data[19]

$$0.954 < n_s < 0.980, \qquad r < 0.168, \qquad -0.03 < \frac{dn_s}{d\ln k} < 0.007, \tag{14}$$

at the 95% CL.[a] The e-folding number

$$N_* = 62 - \ln\left(\frac{10^{16}\text{GeV}}{V_{\text{end}}^{1/4}}\right) \simeq 64 + \frac{1}{4}\ln\epsilon \tag{15}$$

corresponds to the stage of inflation observed by the Planck experiment.

Then, let us examine the possibilities of the saddle point Higgs inflation. We take

$$V_c \simeq \lambda_c \varphi_c^4, \qquad V_c^{(n+1)} \simeq (n+1)!\,\lambda_c \varphi_c^{3-n}, \tag{16}$$

and we can solve Eq. (8) for φ_c,

$$\varphi_c = \left(\frac{12\pi^2\delta}{\lambda_c}\right)^{1/2+1/(n-3)} \frac{1}{N(n-1)} \left(\frac{1}{N^3(n-1)^3(n+1)}\right)^{1/(n-3)} M_p. \tag{17}$$

Putting $N = 60, \delta = 2.2\times10^{-9}, \lambda_c = 10^{-6}, M_p = 2.4\times10^{18}$GeV, we obtain the following values in Table 2. Here we assume that the first n derivatives take zero due to, for example, the balance between φ^4 and higher dimensional terms.

3. Saddle point inflation from string theory

In this section, we consider the inflection point inflation starting from non-supersymmetric heterotic string. Here we assume that the tree level potential of the inflaton is absent, which means that, for example, the inflaton comes from the extra component of the gauge field/metric. In such a case, the dominant contribution to

[a] To give the most conservative bound, here we employ the constraint from the Planck TT+lowP data.

the potential is the one loop one. This is suppressed compared to the string scale as follows:

$$\int \frac{d^dk}{(2\pi)^d} = \frac{S_{d-1}}{2(2\pi)^d} \int dk^2 (k^2)^{\frac{d}{2}-1}. \tag{18}$$

For $d = 10$, we obtain the following loop factor

$$C_{\text{loop}} \equiv \frac{S_{d-1}}{2(2\pi)^d} = \frac{2\pi^5}{\Gamma(5)} \frac{1}{2(2\pi)^{10}} \simeq 1.3 \times 10^{-7}. \tag{19}$$

In fact, the 10 dimensional cosmological constant of $SO(16) \times SO(16)$ heterotic string theory is calculated as[20, 21]

$$\Lambda_{SO(16)\times SO(16)} \simeq 3.9 \times 10^{-6} M_s^{10}. \tag{20}$$

We believe that this estimation would be valid if the ten dimensional effective field theory picture is trusted such as large extra dimension compactification. By assuming that the tree potential of the Higgs vanishes, the effective action below the string scale becomes

$$\begin{aligned} S = & \frac{M_s^8}{g_s^2} \int d^{10}x \sqrt{g} A(\varphi) \mathcal{R} + \frac{M_s^8}{g_s^2} \int d^{10}x \sqrt{g} B(\varphi) (\partial\varphi)^2 \\ & + C_{\text{loop}} M_s^{10} \int d^{10}x \sqrt{g} V(\varphi) + \cdots \\ = & \frac{M_s^8}{g_s^2} V_6 \int d^4x \sqrt{g} A(\varphi) \mathcal{R} + \frac{M_s^8}{g_s^2} V_6 \int d^4x \sqrt{g} B(\varphi) (\partial\varphi)^2 \\ & + C_{\text{loop}} M_s^{10} V_6 \int d^4 \sqrt{g} V(\varphi) + \cdots , \end{aligned} \tag{21}$$

Here φ is the dimensionless inflaton field, g_s is the string coupling, $V(\varphi)$ is the one loop potential, and V_6 is the internal volume. $A(\varphi)$ and $B(\varphi)$ is the order one functions:

$$A(\varphi) = 1 + a_2 g_s^2 \varphi^2 + \cdots , \qquad B(\varphi) = 1 + b_2 g_s^2 \varphi^2 + \cdots , \tag{22}$$

with a_2 and b_2 being order one constants. Next let us move to the Einstein frame. Namely, we refine the metric in such a way that $A(\varphi)$ becomes 1. In the Einstein frame, we have

$$S = M_P^2 \int d^4x \sqrt{g} \mathcal{R} + M_P^2 \int d^4x \sqrt{g} C(\varphi) (\partial\varphi)^2 + C_{\text{loop}} g_s^2 M_P^2 M_s^2 \int d^4x \sqrt{g} U(\varphi), \tag{23}$$

where

$$M_P^2 = \frac{M_s^2}{g_s^2} (M_s^6 V_6), \qquad C(\varphi) = 1 + c_2 g_s^2 \varphi^2 + \cdots , \tag{24}$$

where c_2 is the order one coefficient. In terms of the canonical field χ, this becomes

$$S = M_P^2 \int d^4x \sqrt{g} \mathcal{R} + M_P^2 \int d^4x \sqrt{g} (\partial\chi)^2 + C_{\text{loop}} g_s^2 M_P^2 M_s^2 \int d^4x \sqrt{g} W(\chi). \tag{25}$$

Here $W(\chi)$ is the order one function. The argument so far is the quite general. In the following, we assume that the potential has the critical point where the first n derivatives vanish. We consider that this is happened by some mechanism beyond the ordinary local field theory such as multiple point criticality principle[22] and maximum entropy principle.[23–27] By performing a fine tuning we have a saddle point for $W(\chi)$. Therefore, we take

$$W = -(1-\chi)^{n+1} + 1. \tag{26}$$

as a simple possibility. Then, from Eq. (10), we get

$$\epsilon = \frac{1}{18N^4}, \frac{1}{64N^3}, \frac{1}{18\times 15^{3/2}N^{8/3}}, \frac{1}{64\sqrt{6}N^{5/2}}, \frac{1}{50\times 35^{2/5}N^{12/5}}, \tag{27}$$

for $n = 2,3,4,5,6$. Furthermore, Eq. (27) and the COBE normalization fixes the $C_{\rm loop} g_s^2 M_P^2 M_s^2$. In the Table 3, we present the predictions of the cosmological parameters taking $C_{\rm loop} = 10^{-7}, N = 60$.

n	$V[M_P^4]$	n_s	ϵ	$g_s M_s[M_P]$
2	2.2×10^{-15}	$0.933\cdots$	4.3×10^{-9}	1.5×10^{-4}
3	3.8×10^{-14}	0.95	7.2×10^{-8}	6.1×10^{-4}
4	8.6×10^{-14}	$0.955\cdots$	1.7×10^{-7}	9.3×10^{-4}
5	1.2×10^{-13}	$0.95833\cdots$	2.3×10^{-7}	1.1×10^{-3}
6	1.4×10^{-13}	0.96	2.6×10^{-7}	1.2×10^{-3}

Eq. (14) implies that $n \geq 4$ is favored. The tensor to scalar ratio is very small compared to the current limit. Depending on the value of n, $g_s M_P$ takes 4×10^{14} GeV to 3×10^{15} GeV. If g_s is $\mathcal{O}(0.1)$ parameter, the result means that the M_s is around 10^{16} GeV. This is the reasonable scale.

4. Summary

We have considered the inflation scenario inspired by non-supersymmetric string theory, which is motivated by the recent LHC result. We have shown that if the string scale is close to the GUT scale, 10^{16}GeV, the successful inflation can be realized assuming that the inflaton potential has the point where first n derivatives vanish.

Acknowledgments

We thank Hikaru Kawai and Kiyoharu Kawana for fruitful collaboration.

References

1. Y. Hamada, H. Kawai and K.-y. Oda, Bare Higgs mass at Planck scale, *Phys. Rev.* **D87**, p. 053009 (2013), [Erratum: Phys. Rev.D89,no.5,059901(2014)].

2. F. Jegerlehner, The Standard model as a low-energy effective theory: what is triggering the Higgs mechanism?, *Acta Phys.Polon.* **B45**, 1167 (2014).
3. F. Jegerlehner, The hierarchy problem of the electroweak Standard Model revisited (2013).
4. M. Einhorn and D. Jones, The Effective potential and quadratic divergences, *Phys.Rev.* **D46**, 5206 (1992).
5. D. Jones, The quadratic divergence in the Higgs mass revisited, *Phys.Rev.* **D88**, p. 098301 (2013).
6. M. Holthausen, K. S. Lim and M. Lindner, Planck scale Boundary Conditions and the Higgs Mass, *JHEP* **1202**, p. 037 (2012).
7. F. Bezrukov, M. Y. Kalmykov, B. A. Kniehl and M. Shaposhnikov, Higgs Boson Mass and New Physics, *JHEP* **1210**, p. 140 (2012).
8. G. Degrassi, S. Di Vita, J. Elias-Miro, J. R. Espinosa, G. F. Giudice *et al.*, Higgs mass and vacuum stability in the Standard Model at NNLO, *JHEP* **1208**, p. 098 (2012).
9. F. L. Bezrukov and M. Shaposhnikov, The Standard Model Higgs boson as the inflaton, *Phys.Lett.* **B659**, 703 (2008).
10. Y. Hamada, H. Kawai and K.-y. Oda, Minimal Higgs inflation, *PTEP* **2014**, p. 023B02 (2014).
11. Y. Hamada, H. Kawai, K.-y. Oda and S. C. Park, Higgs inflation still alive, *Phys.Rev.Lett.* **112**, p. 241301 (2014).
12. F. Bezrukov and M. Shaposhnikov, Higgs inflation at the critical point, *Phys. Lett.* **B734**, 249 (2014).
13. Y. Hamada, H. Kawai, K.-y. Oda and S. C. Park, Higgs inflation from Standard Model criticality, *Phys. Rev.* **D91**, p. 053008 (2015).
14. M. Blaszczyk, S. G. Nibbelink, O. Loukas and S. Ramos-Sanchez, Non-supersymmetric heterotic model building (2014).
15. Y. Hamada, T. Kobayashi and S. Uemura, Standard Model-like D-brane models and gauge couplings, *Nucl. Phys.* **B897**, 563 (2015).
16. Y. Hamada, H. Kawai and K.-y. Oda, Eternal Higgs inflation and the cosmological constant problem, *Phys. Rev.* **D92**, p. 045009 (2015).
17. S. Abel, K. R. Dienes and E. Mavroudi, Towards a nonsupersymmetric string phenomenology, *Phys. Rev.* **D91**, p. 126014 (2015).
18. Y. Hamada, H. Kawai and K. Kawana, Saddle point inflation in string-inspired theory, *PTEP* **2015**, p. 091B01 (2015).
19. P. A. R. Ade *et al.*, Planck 2015 results. XX. Constraints on inflation (2015).
20. L. J. Dixon and J. A. Harvey, String Theories in Ten-Dimensions Without Space-Time Supersymmetry, *Nucl. Phys.* **B274**, 93 (1986).
21. L. Alvarez-Gaume, P. H. Ginsparg, G. W. Moore and C. Vafa, An O(16) x O(16) Heterotic String, *Phys. Lett.* **B171**, p. 155 (1986).
22. C. Froggatt and H. B. Nielsen, Standard model criticality prediction: Top mass 173 $\pm$ 5-GeV and Higgs mass 135 $\pm$ 9-GeV, *Phys.Lett.* **B368**, 96 (1996).
23. H. Kawai and T. Okada, Solving the Naturalness Problem by Baby Universes in the Lorentzian Multiverse, *Prog.Theor.Phys.* **127**, 689 (2012).
24. H. Kawai, Low energy effective action of quantum gravity and the naturalness problem, *Int.J.Mod.Phys.* **A28**, p. 1340001 (2013).
25. Y. Hamada, H. Kawai and K. Kawana, Evidence of the Big Fix, *Int.J.Mod.Phys.* **A29**, p. 1450099 (2014).
26. K. Kawana, Reconsideration of the Coleman's Baby Universe (2014).
27. Y. Hamada, H. Kawai and K. Kawana, Weak Scale From the Maximum Entropy Principle, *PTEP* **2015**, p. 033B06 (2015).

Cosmology with a Heavy Polonyi Field

Taku Hayakawa

Institute for Cosmic Ray Research, The University of Tokyo, 5-1-5 Kashiwanoha, Kashiwa, Chiba 277-8582, Japan
taku1215@icrr.u-tokyo.ac.jp

I consider a cosmologically consistent scenario with a heavy Polonyi field. The Polonyi field with a mass of $\mathcal{O}(100)$ TeV decays before the Big-Bang Nucleosynthesis (BBN) and can avoid the severe constraint from the BBN. However, the abundance of the Lightest Supersymmetric Particle (LSP) produced from the decay often exceeds the observed dark matter density. In my scenario, the dark matter density is obtained by the LSP abundance with an aid of entropy production, and baryon asymmetry is generated by the Affleck-Dine mechanism. It is found that the observed baryon-to-dark matter ratio of $\mathcal{O}(0.1\text{-}1)$ is naturally explained in sequestering models with a (pseudo-)Nambu-Goldstone Boson (NGB).

Keywords: Polonyi problem; baryon asymmetry; dark matter.

1. Introduction

Cosmological observations have revealed the existence of the dark matter (DM) and the baryon asymmetry, while their origins remain unknown for a long time. We need a cosmologically consistent scenario explaining both the baryon asymmetry and the DM abundance ($\Omega_B/\Omega_{\rm DM} = 0.18$).[1]

Since those origins cannot be explained in the framework of the Standard Model (SM), there should be underlying physics beyond the SM. Supersymmetry (SUSY)[2] is one of the most interesting models since it achieves the unification of the SM gauge couplings and can relax the hierarchy problem. In addition, supersymmetric models with a conserved R parity predict the stability of the Lightest Supersymmetric Particle (LSP), which becomes a good candidate for DM. Moreover, SUSY extensions of the SM contain a lot of flat directions with $B-L$ charges,[3] which can produce $B-L$ asymmetry.[4] In the early universe, one of the flat directions, which I call the Affleck-Dine field, may receive an angular kick from SUSY breaking and R symmetry breaking effects and rotates in its complex plane, which corresponds to the generation of the $B-L$ asymmetry. The $B-L$ asymmetry is converted into the baryon asymmetry through the sphaleron process.[5,6] This mechanism, known as "the Affleck-Dine (AD) mechanism",[4,7] can produce baryon number more effectively than most baryogenesis scenarios. For these advantages, I focus on supersymmetric theories in this paper.

The SUSY must be spontaneously broken since superparticles have not been discovered yet. The simplest SUSY breaking model is the Polonyi model[8] in which an F-term of an elementary singlet field Z breaks the SUSY in the hidden sector.[a] This model is attractive because of its simplicity. However, such a singlet field Z, called the Polonyi field, may cause a cosmological difficulty.[9] After inflation, the Polonyi field begins to oscillate around the true minimum with an amplitude of the order of the Planck scale. The energy density of its coherent oscillation immediately dominates the universe after the inflaton decays. Since the Polonyi field very weakly interacts with particles in the observable sector, its late-time decay upsets the standard scenario of the Big-Bang Nucleosynthesis (BBN), which is referred to as "the Polonyi problem".

In this paper, I consider the case where the Polonyi field is heavy enough to decay before the BBN, i.e., $m_Z \sim \mathcal{O}(100)$TeV, where m_Z is a mass of the Polonyi field. The mass of the Polonyi field is assumed to be of the order of the gravitino mass $m_{3/2}$ for generic Kähler potentials. I construct a cosmologically consistent scenario in the presence of the heavy Polonyi field. In my scenario, the baryon asymmetry is created by dynamics of the AD field which takes the vacuum expectation value (VEV) of the order of the Planck scale in the early universe. Both the Polonyi density and the baryon asymmetry are diluted by late-time entropy production, for example, by the thermal inflation.[10] After the entropy production, the Polonyi field decays into LSPs which explain the observed DM abundance. I show that the baryon-to-DM ratio is simply determined by the LSP mass and the branching fraction of the decay of the Polonyi field into superparticles. The correct baryon-to-DM ratio is obtained when the LSP mass is of $\mathcal{O}(1)$ TeV and the branching fraction of the decay of the Polonyi into superparticles is of $\mathcal{O}(10^{-3})$.

In order to realize the branching fraction of $\mathcal{O}(10^{-3})$, I consider sequestering models with a (pseudo-)Nambu-Goldstone boson (NGB). In this model, the Polonyi field mainly decays into NGBs, but their contribution to the DM abundance is negligible since the NGB is much lighter than the LSP (as is the case with the QCD axion) or decays into SM particles. On the other hand, the decay of the Polonyi into superparticles is suppressed for the sequestered potential, and the branching ratio is of $\mathcal{O}(10^{-3})$. The DM abundance is determined by the LSP abundance produced through the Polonyi decay.[b]

2. Affleck-Dine Mechanism without Superpotential

The Affleck-Dine mechanism[4] is a promising candidate for the baryogenesis in cosmological scenarios with dilution. To obtain the large field value, I assume that the AD field does not appear in the superpotential.[c] In this case, SUSY breaking effects including the A-terms are provided by the Kähler potential. The potential

---

[a]In the Polonyi model, the SUSY breaking scale can be easily obtained by dynamical transmutation, by assigning a vanishing R charge to the Polonyi field and breaking the R symmetry by gaugino condensation.

[b]The detailed analysis is found in Ref. 11.

[c]For example, a $U(1)_R$ symmetry can prohibit appearance of the AD field in the superpotential.

for the AD field Φ is given by

$$V(\Phi) = m_\phi^2|\Phi|^2 - \frac{m_{3/2}^2}{nM_{\rm pl}^{n-2}}(a_n\Phi^n + {\rm h.c.}) + c_4 m_{3/2}^2\frac{|\Phi|^4}{M_{\rm pl}^2} + \cdots, \tag{1}$$

where m_ϕ and $M_{\rm pl}$ denote the soft scalar mass of the AD field and the reduced Planck mass($\simeq 2.4\times10^{18}$ GeV), respectively. a_n and $c_4(>0)$ are $\mathcal{O}(1)$ dimensionless parameters.

When $m_\phi \ll m_{3/2}$ and $c_4 = 0$, there exists charge/color breaking minima smaller than the Planck scale because of the relatively large A-terms.[12] The AD field needs to avoid dropping the minima during its evolution. The quartic term, however, lifts the potential near the Planck scale, and the global minima disappear.

In addition, I assume that the AD field has a negative Hubble induced mass term so that it takes a large field value at the onset of its oscillation. Due to this negative mass term, the AD field takes its field value of the order of the Planck scale until $H(t) \simeq m_{3/2}$. When $H(t) \lesssim m_{3/2}$, the position of the local minimum of the AD field, which is determined by a balance between the negative Hubble induced mass term and the positive quartic term, becomes smaller than the Planck scale. Since the position is quickly driven towards the origin, the AD field cannot track the local minimum and starts to roll down to the origin when $H(t) \simeq m_{3/2}$ (for details, see Ref. 13).

I consider the case where the density of the Polonyi field is diluted by the entropy production because LSPs produced from the Polonyi field tends to be overproduced. In that case, the baryon asymmetry is also diluted. Assuming that the inflaton decays after the onset of the oscillation of the AD field, the yield of the baryon number is estimated as

$$Y_B \equiv \frac{n_B}{s} \simeq \frac{2}{23}\frac{1}{\Delta}\frac{T_{\rm inf}}{m_{3/2}}\left(\frac{|\Phi_{\rm osc}|}{M_{\rm pl}}\right)^2, \tag{2}$$

where $T_{\rm inf}$ denotes the temperature when the inflaton decays, and Δ denotes a dilution factor.

3. Baryon-to-DM Ratio

First, I estimate the DM abundance. When the Polonyi field starts its oscillation before the inflaton decays, the Polonyi energy density-to-entropy ratio after the inflaton decay is given by

$$\frac{\rho_Z}{s} \simeq \frac{T_{\rm inf}}{8}\left(\frac{z_0}{M_{\rm pl}}\right)^2, \tag{3}$$

where s denotes the entropy density, and z_0 represents the oscillation amplitude of Z, which is expected to be of the order of the Planck scale. Then, the LSP-to-entropy ratio is estimated as

$$\frac{\rho_{\rm LSP}}{s} = m_{\rm LSP}\frac{2{\rm Br}_{\rm SUSY}}{\Delta}\frac{n_Z}{s_i} \simeq \frac{{\rm Br}_{\rm SUSY}}{\Delta}\frac{T_{\rm inf}m_{\rm LSP}}{4m_Z}\left(\frac{z_0}{M_{\rm pl}}\right)^2, \tag{4}$$

where $\mathrm{Br}_{\mathrm{SUSY}}$ denotes a branching fraction for the Polonyi decay into two superparticles,[d] and m_{LSP} denotes the LSP mass. The number of the produced superparticles is almost equal to that of the LSPs due to the R-parity conservation. Note that the pair annihilation between LSPs is not efficient since the Polonyi field decays after the dilution. I assume that decay products other than the LSPs do not contribute to the DM abundance. Hence, the DM abundance is obtained by the nonthermally produced LSPs.

Next, let me compare the DM abundance with the baryon asymmetry created by the AD mechanism. The ratio of the density of the Polonyi field to the $B-L$ number remains the same after they begin their oscillations since the densities of both components decrease as a^{-3}. From Eqs. (2) and (4), I obtain the following relation:

$$\begin{aligned}\frac{\Omega_B}{\Omega_{\mathrm{LSP}}} &= \frac{8}{23}\frac{1}{\mathrm{Br}_{\mathrm{SUSY}}}\frac{m_p m_Z}{m_{\mathrm{LSP}} m_{3/2}}\left(\frac{|\Phi_{\mathrm{osc}}|}{z_0}\right)^2 \\ &\simeq 0.33\left(\frac{10^{-3}}{\mathrm{Br}_{\mathrm{SUSY}}}\right)\left(\frac{1\,\mathrm{TeV}}{m_{\mathrm{LSP}}}\right)\left(\frac{|\Phi_{\mathrm{osc}}|}{z_0}\right)^2, \end{aligned} \tag{5}$$

where m_p represents the proton mass ($m_p \simeq 0.938\,\mathrm{GeV}$). Note that z_0 and $|\Phi_{\mathrm{osc}}|$ are of the order of the Planck scale. One can find that the baryon-to-DM ratio is determined by the LSP mass and the branching fraction of the decay of the Polonyi into superparticles. Assuming that the LSP mass is of $\mathcal{O}(1)$ TeV, $\mathrm{Br}_{\mathrm{SUSY}}$ is required to be of $\mathcal{O}(10^{-3})$ in order to realize the observed value, $\Omega_B/\Omega_{\mathrm{DM}} \simeq 0.18$.

4. Sequestering Model and Decay Process of the Polonyi Field

In order to suppress the branching fraction into superparticles ($\mathrm{Br}_{\mathrm{SUSY}} \simeq \mathcal{O}(10^{-3})$), I consider the so-called sequestering model,[14,15,e] in which the SUSY breaking sector is sequestered from the visible sector. I also assume that the SM gauge sector does not directly couple to the hidden sector. In this setup, gaugino masses vanish at the tree level, and the quantum corrections to the gaugino masses arise at loop-suppressed level, which mainly come from the anomaly mediation.[15,16] The lightest gaugino is the neutral wino with a mass of $\mathcal{O}(1)$ TeV when $m_{3/2} \simeq \mathcal{O}(100)$ TeV. This is compatible with my scenario with the neutral wino LSP.

In the sequestering models, the Polonyi field mainly decays into gluinos and gluons through the anomaly-mediated effects when the decay into 2 gravitinos is kinetically forbidden ($m_Z < 2m_{3/2}$). If it is the leading process, however, the Polonyi field becomes long-lived, and the constraint from the BBN is again severe even with $m_Z \simeq \mathcal{O}(100)$ TeV. I need some other efficient decay processes. Note

[d] Gravitinos are not produced from the Polonyi decay assuming that the decay is kinematically forbidden ($m_Z < 2m_{3/2}$). The abundance of gravitinos produced during the reheating becomes negligible after the dilution.

[e] The sequestering model has been introduced in the context of extra dimension.[15] It is also realized in a four dimensional strongly coupled CFT.[17,18]

that those decay processes should not yield large DM abundance. As a suitable decay process, I consider the decay of the Polonyi field into a (pseudo-)NGB.

The rate of the loop-suppressed decay is much smaller than that of the tree-level decay, and I obtain the branching fraction of the Polonyi decay into superparticles as

$$\mathrm{Br}_{\mathrm{SUSY}} = \frac{\Gamma(Z \to 2 \text{ superparticles})}{\Gamma(Z \to 2 \text{ axions})} \sim 1 \times 10^{-3}. \tag{6}$$

Since the (pseudo-)NGB mass is typically much smaller than the LSP mass, the abundance of (pseudo-)NGBs produced from the Polonyi decay is negligible compared with the LSP abundance. Therefore, the DM abundance is determined by the abundance of the decay products of the suppressed decay into superparticles. From Eqs. (5) and (6), it is found that the observed baryon-to-DM ratio of $\Omega_B/\Omega_{\mathrm{DM}} \simeq 0.18$ is explained in the sequestering model with the (pseudo-)NGB.

5. Summary

In this paper, I have considered the cosmological consistent scenario in the presence of the heavy Polonyi field. When the Polonyi field is heavier than $\mathcal{O}(100)$ TeV, it decays before the BBN, and the constraint from the BBN becomes much milder. LSPs produced by the decay of the Polonyi are diluted by entropy production. With the entropy production, the most promising candidate for the baryogenesis is the AD mechanism.

In my scenario, the DM is explained by the LSPs produced from the decay of the Polonyi field, and the baryon asymmetry is created by the AD mechanism without superpotential. I have shown that the baryon-to-DM ratio is simply determined by the LSP mass (m_{LSP}) and the branching fraction of the decay of the Polonyi into DM ($\mathrm{Br}_{\mathrm{SUSY}}$). The observed ratio $\Omega_B/\Omega_{\mathrm{DM}} \simeq 0.18$ is realized for the LSP mass of $\mathcal{O}(1)$ TeV and the branching fraction of $\mathcal{O}(10^{-3})$. The Polonyi density is diluted before its decay by the entropy production, for example, by thermal inflation.

In general, the branching fraction, $\mathrm{Br}_{\mathrm{SUSY}}$, is of $\mathcal{O}(1)$ when the Polonyi field is connected with the observable sector through non-renormalizable interactions. In order to suppress $\mathrm{Br}_{\mathrm{SUSY}}$ ($\simeq \mathcal{O}(10^{-3})$), I have considered the sequestering model with a (pseudo-) NGB, which can be identified with the QCD axion. In this model, the Polonyi field decays into NGBs at the tree level, which do not contribute to the DM abundance. On the other hand, it decays into superparticles mainly through anomaly-induced interactions and hence is suppressed compared with the decay into NGBs.

Acknowledgments

This work is supported by the Program for the Leading Graduate Schools, MEXT, Japan.

References

1. K. A. Olive *et al.* [Particle Data Group Collaboration], Chin. Phys. C **38**, 090001 (2014).
2. For reviews, see H. P. Nilles, Phys. Rept. **110**, 1 (1984); S. P. Martin, Adv. Ser. Direct. High Energy Phys. **21**, 1 (2010) [Adv. Ser. Direct. High Energy Phys. **18**, 1 (1998)] [hep-ph/9709356].
3. T. Gherghetta, C. F. Kolda and S. P. Martin, Nucl. Phys. B **468**, 37 (1996) [hep-ph/9510370].
4. I. Affleck and M. Dine, Nucl. Phys. B **249**, 361 (1985).
5. V. A. Kuzmin, V. A. Rubakov and M. E. Shaposhnikov, Phys. Lett. B **155**, 36 (1985).
6. S. Y. Khlebnikov and M. E. Shaposhnikov, Nucl. Phys. B **308**, 885 (1988); J. A. Harvey and M. S. Turner, Phys. Rev. D **42**, 3344 (1990).
7. M. Dine, L. Randall and S. D. Thomas, Nucl. Phys. B **458**, 291 (1996) [hep-ph/9507453].
8. J. Polonyi, Budapest preprint KFKI-1977-93 (1977).
9. G. D. Coughlan, W. Fischler, E. W. Kolb, S. Raby and G. G. Ross, Phys. Lett. B **131**, 59 (1983).
10. K. Yamamoto, Phys. Lett. B **168**, 341 (1986); D. H. Lyth and E. D. Stewart, Phys. Rev. D **53**, 1784 (1996) [hep-ph/9510204].
11. K. Harigaya, T. Hayakawa, M. Kawasaki and M. Yamada, arXiv:1601.02140 [hep-ph].
12. M. Kawasaki, T. Watari and T. Yanagida, Phys. Rev. D **63**, 083510 (2001) [hep-ph/0010124].
13. K. Harigaya, M. Ibe, M. Kawasaki and T. T. Yanagida, JCAP **1511**, no. 11, 003 (2015) [arXiv:1507.00119 [hep-ph]].
14. K. Inoue, M. Kawasaki, M. Yamaguchi and T. Yanagida, Phys. Rev. D **45**, 328 (1992).
15. L. Randall and R. Sundrum, Nucl. Phys. B **557**, 79 (1999) [hep-th/9810155].
16. G. F. Giudice, M. A. Luty, H. Murayama and R. Rattazzi, JHEP **9812**, 027 (1998) [hep-ph/9810442].
17. M. A. Luty and R. Sundrum, Phys. Rev. D **65**, 066004 (2002) [hep-th/0105137]; Phys. Rev. D **67**, 045007 (2003) [hep-th/0111231].
18. M. Ibe, K.-I. Izawa, Y. Nakayama, Y. Shinbara and T. Yanagida, Phys. Rev. D **73**, 015004 (2006) [hep-ph/0506023]; Phys. Rev. D **73**, 035012 (2006) [hep-ph/0509229].

Aspects of Multi-Field Models of Inflation

Tomo Takahashi
Department of Physics, Saga University, Saga 840-8502, Japan
tomot@cc.saga-u.ac.jp

We discuss multi-field models of inflation, such as the curvaton, modulated reheating scenario, as the origin of density fluctuations. After summarizing the current status of such models from the viewpoint of observations of the scalar spectral index, the tensor-to-scalar ratio and so on, we investigate the CMB spectral μ distortion in such a model, which captures the property of density fluctuations on small scales. We find that, given the current constraint on the spectral index and the tensor-to-scalar ratio, the predicted distribution of the value of μ in multi-field models could be broader than that in the standard single-field inflation model and larger values of μ could be realized. Some issues of isocurvature fluctuations in multi-field models are also discussed.

Keywords: Inflation; primordial fluctuations; cosmic microwave background.

1. Introduction

It is now widely believed that the Universe experienced a very rapid expansion in the very early Universe, called inflation. Although current cosmological observations well probe the primordial fluctuations to give a severe constraint on inflationary models,[1,2] we still do not know yet its precise mechanism or the model which is realized in the nature. One of the key quantities of inflationary models is the tensor-to-scalar ratio r, which is now severely constrained as $r \leq 0.07$ (95 % C.L.).[2] Therefore, inflation models with small tensor-to-scalar ratio may be favored. Although one can construct a model with small r in the framework of the standard single-field inflation models such as the new inflation or of the extension of gravity such as the Starobinsky model,[3] small r can be easily realized in multi-field models of inflation, in particular, the so-called spectator field models such as the curvaton model,[4–6] modulated reheating scenario[7,8] and so on. In this talk, we focus on such multi-field inflation models. First we discuss its current status of such models, then we study its prediction for the CMB spectral μ distortion, which could be useful to test the inflationary models. Furthermore, in multi-field models of inflation, there exist multiple sources of density fluctuations, hence isocurvature fluctuations can be generated. Since current cosmological observations put a stringent constraint on the fraction of isocurvature fluctuations,[1] if a sizable amount of isocurvature fluctuations is generated, such a model would be excluded by the constraint. We also discuss this issue briefly, focusing on the curvaton model.

2. Mixed inflaton and spectator field models

Here we summarize the prediction of multi-field models of inflation for the spectral index n_s and the tensor-to-scalar ratio r. In the following, we particularly have the spectator field models in mind such as the curvaton and modulated reheating scenarios. Usually, when one discusses the spectator field models, it is usually assumed that only the spectator field is responsible for the generation of density fluctuations. However, even in such models, the inflaton should also exist to drive the inflationary expansion, hence in general, fluctuations from the inflaton can also contribute to primordial fluctuations. Therefore we work in the framework of mixed inflaton and spectator field models where fluctuations from the inflaton and the spectator field can both contribute to primordial density fluctuations.

In this kind of model, the power spectrum of primordial density fluctuations is given by the sum of two contributions from the inflaton ϕ and the spectator field σ:

$$\begin{aligned}\mathcal{P}_\zeta^{(\rm total)} &= \mathcal{P}_\zeta^{(\phi)} + \mathcal{P}_\zeta^{(\sigma)} \\ &= A_s^{(\phi)}(k_{\rm ref})\left(\frac{k}{k_{\rm ref}}\right)^{n_s^{(\phi)}-1} + A_s^{(\sigma)}(k_{\rm ref})\left(\frac{k}{k_{\rm ref}}\right)^{n_s^{(\sigma)}-1},\end{aligned} \tag{1}$$

where $\mathcal{P}_\zeta^{(\phi)}$ and $\mathcal{P}_\zeta^{(\sigma)}$ are power spectra for the inflaton and the spectator field, respectively. $A_s^{(\phi)}(k_{\rm ref})$ and $A_s^{(\sigma)}(k_{\rm ref})$ are their amplitudes at the reference scale $k_{\rm ref}$. For each power spectrum, we can define the spectral index $n_s^{(\phi)}$ and $n_s^{(\sigma)}$ to describe its scale dependence as in the same manner as the single-source (single-field) case, which are respectively given by

$$n_s^{(\phi)} - 1 = -6\epsilon + 2\eta_\phi, \qquad n_s^{(\sigma)} - 1 = -2\epsilon + 2\eta_\sigma, \tag{2}$$

where ϵ, η_ϕ and η_σ are the slow-roll parameters defined as

$$\epsilon = -\frac{\dot{H}}{H^2}, \qquad \eta_\phi = \frac{1}{3H^2}\frac{d^2V}{d\phi^2}, \qquad \eta_\sigma = \frac{1}{3H^2}\frac{d^2V}{d\sigma^2}. \tag{3}$$

We note that $n_s^{(\phi)}$ is given by the same formula as that in the standard case, on the other hand, the expression for $n_s^{(\sigma)}$ is a bit different from $n_s^{(\phi)}$ since σ behaves as an isocurvature field during inflation.[9,10]

The tensor mode is not modified since the spectator field does not affect the inflationary dynamics. Therefore, the tensor power spectrum is given by the usual formula:

$$\mathcal{P}_T = \frac{8}{M_{\rm pl}^2}\left(\frac{H_{\rm inf}}{2\pi}\right)^2. \tag{4}$$

Then the tensor-to-scalar ratio in the mixed model can be written as

$$r = \frac{\mathcal{P}_T}{\mathcal{P}_\zeta^{(\rm total)}} = \frac{\mathcal{P}_T}{\mathcal{P}_\zeta^{(\phi)} + \mathcal{P}_\zeta^{(\sigma)}} = \frac{16\epsilon}{1+R}, \tag{5}$$

where we have defined the ratio of the power spectra between those from the spectator field and the inflaton by R:

$$R \equiv \frac{\mathcal{P}_\zeta^{(\sigma)}}{\mathcal{P}_\zeta^{(\phi)}}. \tag{6}$$

The limits of $R \to 0$ and $R \to \infty$ correspond to the case where fluctuations from the inflaton and the spectator field completely dominate in the (total) power spectrum, respectively.

As one can see from Eq. (5), when R is large, in which the contribution from the spectator field dominates in the power spectrum, the tensor-to-scalar ratio is suppressed. Therefore, small tensor-to-scalar ratio can be easily accommodated in multi-field models of inflation. In Fig. 1, we show the constraint on ϵ and R from current CMB observations of Planck,[11] high ℓ data from Atacama Cosmology Telescope (ACT) and South Pole Telescope (SPT)[12–14] and joint analysis of BICEP2/KECK and Planck (BKP).[15] Since current observations just put the upper bound on the tensor-to-scalar ratio, one cannot obtain a significant constraint on R.

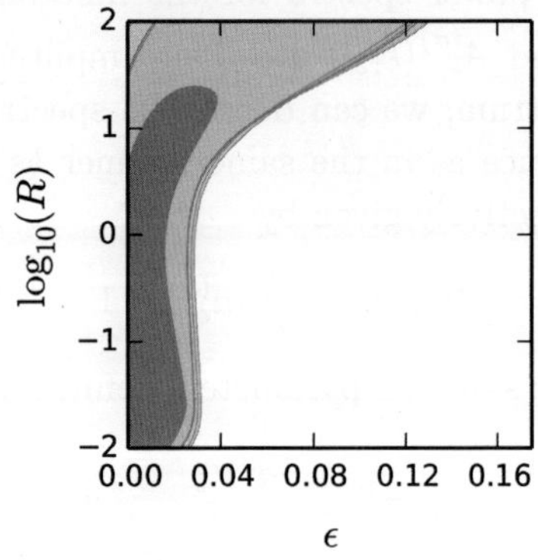

Fig. 1. Constraints on the ϵ–R plane from Planck alone (green), Planck+high ℓ data (gray), Planck+BKP (red) and Planck+BKP+high ℓ data (blue). The figure is adopted from Enqvist et al.[16] See Enqvist et al.[16] for the detail.

3. CMB spectral μ distortion in the mixed models

Now we discuss the CMB spectral μ distortion in the mixed inflaton and spectator field models (for the details of the CMB μ distortion, see Refs. 17–21). In general, the scale dependences of those power spectra differ. Hence it is possible that the contribution of one source (e.g., $\mathcal{P}_\zeta^{(\phi)}$) dominates on large scales, on the other hand, that from the other source (e.g., $\mathcal{P}_\zeta^{(\sigma)}$) does on small scales. Although observations of CMB can give a stringent constraint on the scale dependence or the spectral index of primordial power spectrum, they probe those only on large scales. Therefore relatively flat or blue-tilted spectrum is possible on small scales, which could generate a large value of μ.

In Fig. 2, scatter plots of μ are shown, which are superimposed on the CMB constraints on n_s and r for the cases with a standard single-field inflation and a general mixed inflaton and spectator field. Although the μ distortion for the standard case has already been studied in the literature,[17,18] we show it for comparison. In the right panel which shows the case with a general mixed inflaton and spectator field, n_s in the horizontal axis corresponds to the "effective" spectral index which is defined as

$$n_s^{(\mathrm{eff})} - 1 = \frac{1}{1+R}(n_s^{(\phi)} - 1) + \frac{R}{1+R}(n_s^{(\sigma)} - 1). \tag{7}$$

As one can see from the figure, large values of μ can be generated in multi-field models (mixed model). This tendency can also be observed by looking at 1D posterior predictive distribution of μ which is shown in Fig. 3. For the detail, see Enqvist et al.[16]

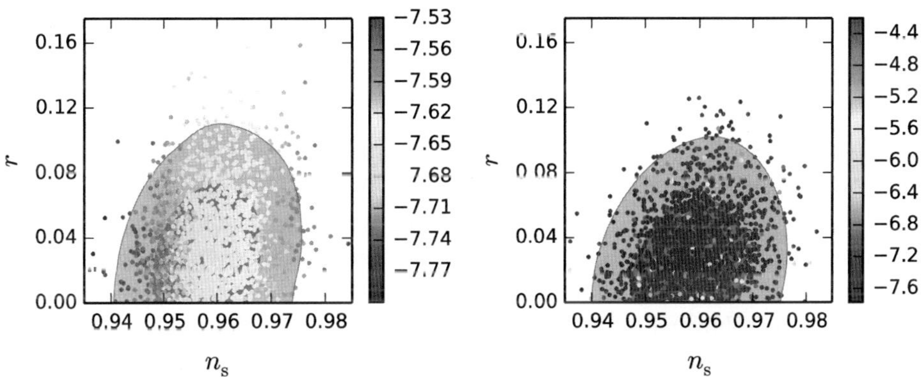

Fig. 2. Shaded regions are 1σ and 2σ allowed regions from current observations for the standard single-field case (left) and a general mixed inflaton and spectator field case (right). Scatter plots of μ are superimposed on the constraints. Color legend indicates the value of $\log_{10}\mu$. The figure is adopted from Enqvist et al.[16] See Enqvist et al.[16] for the detail.

4. Isocurvature fluctuations in the curvaton model

In multi-field models of inflation, there exist multiple sources of fluctuations and, in such a case, residual isocurvature fluctuations can be generated. Since current observations of CMB give a very severe constraint on isocurvature fluctuations, when sizable isocurvature fluctuations are generated, such a model would be excluded.

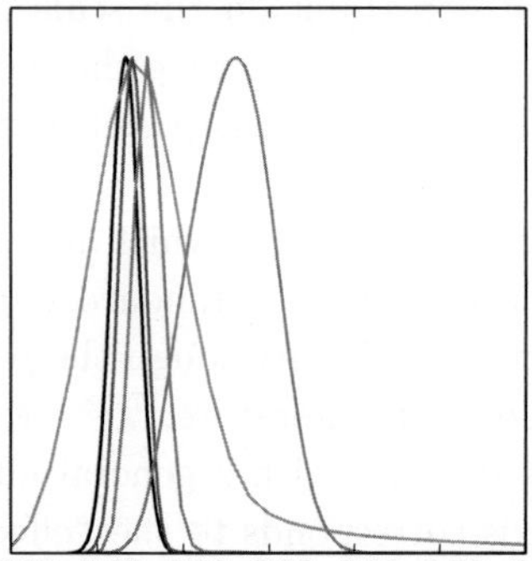

Fig. 3. 1D posterior predictive distribution of μ given the current constraint on n_s and r for the standard single-field inflaton case (black), the general mixed case (red), the mixed case with $(\eta_\phi, \eta_\sigma) = (0.2, 0)$ (purple), the mixed case with $\eta_\sigma = 0$ (blue), and the mixed case with $(r, \eta_\sigma) = (0.005, 0)$ (green). The figure is adopted from Enqvist et al.[16] See Enqvist et al.[16] for the detail.

In the curvaton model, it has been discussed that large isocurvature fluctuations are generated when cold dark matter (CDM) is created before the curvaton decay.[22–26] In fact, this tends to be the case given the constraint on non-Gaussianity. Planck data provides the limit on the local-type non-linearity parameter as $f_{\rm NL}^{\rm (local)} = 0.8 \pm 5.0$ (68 % C.L.),[27] which can be translated into the constraint on the energy fraction of the curvaton at the time of its decay, denoted as $r_{\rm dec}$, as $r_{\rm dec} > 0.19$ (95 % C.L.).[27] This bound means that the curvaton should (almost) dominate the Universe at its decay and be long-lived. Hence the epoch of DM creation tends to be earlier than the decay time of the curvaton, which would be problematic for the model since too large CDM isocurvature fluctuations could be generated in such a case.

However, when the curvaton is very long-lived, dilute plasma from the curvaton decay can be dominant in radiation component (radiation consists of that from the inflaton and the curvaton decays). If we consider WIMP DM, when dilute plasma is the dominant component in radiation, DM freezes out from the thermal bath of the dilute plasma whose fluctuations originate from the curvaton. If this is the case, there is no isocurvature fluctuations between CDM and radiation since fluctuations of both of these come from the curvaton ones. Therefore the model is not excluded by observational constraints even if DM is created (freezes out) before the curvaton decay. This consideration indicates that isocurvature constraint is not so severe than previously considered. For the details of this issue, we refer the readers to Kitajima et al.[28]

5. Summary

We discussed multi-field models of inflation, in particular, having the so-called spectator field models in mind. From the viewpoint of current cosmological observations where the upper bound on the tensor-to-scalar ratio has become severer, multi-field models of inflation would be interesting to study since the tensor-to-scalar ratio in the model tends to be suppressed. We investigated the CMB spectral μ distortion in such multi-field models, which can serve as a critical test for the model. We also briefly discussed the issues of isocurvature fluctuations, focusing the curvaton model. By carefully studying the model, isocurvature fluctuations may not be so problematic than previously thought. In light of these considerations, multi-field models of inflation may should more be detailedly explored.

Acknowledgments

This work is partially supported by JSPS KAKENHI Nos. 15H05888 and 15K05084.

References

1. P. A. R. Ade *et al.* [Planck Collaboration], arXiv:1502.02114 [astro-ph.CO].
2. P. A. R. Ade *et al.* [BICEP2 and Keck Array Collaborations], Phys. Rev. Lett. **116**, no. 3, 031302 (2016) [arXiv:1510.09217 [astro-ph.CO]].
3. A. A. Starobinsky, Phys. Lett. B **91**, 99 (1980).
4. K. Enqvist and M. S. Sloth, Nucl. Phys. B **626**, 395 (2002) [hep-ph/0109214].
5. D. H. Lyth and D. Wands, Phys. Lett. B **524**, 5 (2002) [hep-ph/0110002].
6. T. Moroi and T. Takahashi, Phys. Lett. B **522**, 215 (2001) [Phys. Lett. B **539**, 303 (2002)] [hep-ph/0110096].
7. G. Dvali, A. Gruzinov and M. Zaldarriaga, Phys. Rev. D **69**, 023505 (2004) [astro-ph/0303591].
8. L. Kofman, astro-ph/0303614.
9. C. T. Byrnes and D. Wands, Phys. Rev. D **74**, 043529 (2006) [astro-ph/0605679].
10. T. Kobayashi, F. Takahashi, T. Takahashi and M. Yamaguchi, JCAP **1310**, 042 (2013) [arXiv:1303.6255 [astro-ph.CO]].
11. P. A. R. Ade *et al.* [Planck Collaboration], Astron. Astrophys. **571**, A15 (2014) [arXiv:1303.5075 [astro-ph.CO]].
12. J. Dunkley *et al.*, JCAP **1307**, 025 (2013) [arXiv:1301.0776 [astro-ph.CO]].
13. R. Keisler *et al.*, Astrophys. J. **743**, 28 (2011) [arXiv:1105.3182 [astro-ph.CO]].
14. C. L. Reichardt *et al.*, Astrophys. J. **755**, 70 (2012) [arXiv:1111.0932 [astro-ph.CO]].
15. P. A. R. Ade *et al.* [BICEP2 and Planck Collaborations], Phys. Rev. Lett. **114**, 101301 (2015) [arXiv:1502.00612 [astro-ph.CO]].
16. K. Enqvist, T. Sekiguchi and T. Takahashi, arXiv:1511.09304 [astro-ph.CO].
17. J. Chluba, A. L. Erickcek and I. Ben-Dayan, Astrophys. J. **758**, 76 (2012) [arXiv:1203.2681 [astro-ph.CO]].
18. J. B. Dent, D. A. Easson and H. Tashiro, Phys. Rev. D **86**, 023514 (2012) [arXiv:1202.6066 [astro-ph.CO]].
19. W. Hu, D. Scott and J. Silk, Astrophys. J. **430**, L5 (1994) [astro-ph/9402045].
20. J. Chluba and R. A. Sunyaev, Mon. Not. Roy. Astron. Soc. **419**, 1294 (2012) [arXiv:1109.6552 [astro-ph.CO]].

21. J. Chluba, R. Khatri and R. A. Sunyaev, Mon. Not. Roy. Astron. Soc. **425**, 1129 (2012) [arXiv:1202.0057 [astro-ph.CO]].
22. T. Moroi and T. Takahashi, Phys. Rev. D **66**, 063501 (2002) [hep-ph/0206026].
23. D. H. Lyth, C. Ungarelli and D. Wands, Phys. Rev. D **67**, 023503 (2003) [astro-ph/0208055].
24. D. H. Lyth and D. Wands, Phys. Rev. D **68**, 103516 (2003) [astro-ph/0306500].
25. T. Moroi and T. Takahashi, Phys. Lett. B **671**, 339 (2009) [arXiv:0810.0189 [hep-ph]].
26. T. Takahashi, M. Yamaguchi and S. Yokoyama, Phys. Rev. D **80**, 063524 (2009) [arXiv:0907.3052 [astro-ph.CO]].
27. P. A. R. Ade *et al.* [Planck Collaboration], arXiv:1502.01592 [astro-ph.CO].
28. N. Kitajima, D. Langlois, T. Takahashi and S. Yokoyama in preparation.

Probing the Primordial Universe using Massive Fields

Xingang Chen[1,2], Mohammad Hossein Namjoo[1,2] and Yi Wang[3]

[1] *Institute for Theory and Computation, Harvard-Smithsonian Center for Astrophysics, 60 Garden Street, Cambridge, MA 02138, USA*
[2] *Department of Physics, The University of Texas at Dallas, Richardson, TX 75083, USA*
[3] *Department of Physics, The Hong Kong University of Science and Technology, Clear Water Bay, Kowloon, Hong Kong, P.R.China*

Inflation models are numerous. It is extremely difficult, if possible at all, to identify the actual underlying inflation model of our primordial universe. Thus, for the purpose of proving/falsifying inflation and using inflation to probe new physics, model independent approaches are crucial. Massive fields play a uniquely important role in those missions. This short review is based on a talk by one of the authors (YW) in the 2nd LeCosPA Symposium.

Keywords: Quasi-single-field inflation; non-Gaussianity; inflation; alternative to inflation.

1. Introduction

Inflation is the leading paradigm of the very early universe cosmology. During inflation the universe expands almost exponentially as $a \sim e^{Ht}$, where H is Hubble parameter and is slowly varying. Such a simple scenario explains horizon and flatness puzzles in the hot big bang cosmology and predicts primordial density fluctuations which turn out to be the seeds of the cosmic microwave background (CMB) anisotropies and the large scale structure (LSS).

What would be a logical route to systematically investigate the inflationary physics? A naïve approach would be copying what collider particle physicists have done.

In particle physics, building on the general principles of quantum field theory, the particle physics standard model is established in the 1960s. In the 50 years that follows, the standard model is tested to agree with experiments extremely well. The standard model now provides a reference model for the search of new physics, which is referred to as beyond standard model physics.

If the similar thing would about to happen for inflation, one would first build inflation models, and afterwards test those models and pick the right one from experiments. Once the standard model of inflation is established, one can then use the model as a standard reference to probe new physics during inflation.

Unfortunately, such an approach for inflation fails badly, simply because there are too many inflation models, way too many compared to the number of observables that one typically expects to observe.

For example, in Encyclopædia Inflationaris,[1] 74 models (potentials) of single field slow roll inflation are summarized. Even before addressing the completeness of this list, one can easily come up with inflation models with a summation of some those potential terms resulting in a combinatorial number of new models. On top of that, one can add scale dependent sharp features,[2–5] non-Bunch-Davies initial conditions,[6–9] generalized kinetic terms such as Galileons,[10] and so on and so forth.

This is more or less the story for single scalar field inflation. One can in addition consider quasi-single field,[11–13] multi-field, and generalize to higher spin fields.[14] As a result, the possibilities of inflation models is practically infinite. (Beyond those practically infinite possibilities of inflation, there are also all kinds of alternative to inflation scenarios, which we shall address later.) Although lots of experiments are on going or being planned, we are still very unlikely to pin down a standard model for inflation. So the attempt to copy the theoretical development of particle physics very likely leads to an unfortunate end here.

Why such a "logical" route fails? This is related to the simple $a \sim e^{Ht}$ nature of inflation. Ugly, bad and good consequences arise from here. As the ugly consequence, this is why there are so many inflationary models. Any potential satisfying slow roll conditions can inflate. Although WMAP and Planck marks the era of ruling out simple models such as $\lambda\phi^4$, the possibility is still widely open for many models mentioned above. As the bad consequence, those models predict very similar observational signatures, which only differ by 1% (the order of slow roll parameters) or smaller. This adds to the extreme difficulty in distinguishing inflationary models observationally.

Fortunately, there also exists a good consequence: As long as one postulates $a \sim e^{Ht}$, model independent results can be obtained and one can use such model independent results to test inflation in a model independent way, and probe new high energy physics model independently based on the scenario of inflation, instead of a particular inflation model.

A nice example – and the only example for a long time in the literature – is gravitational waves.[15] Gravitational waves is the universal prediction of inflation. Once observed, it is a non-trivial test of inflation, and indicates new physics – the quantum fluctuation of the metric degree of freedom.

Recently, another example of such is realized – the massive fields during inflation.[11–13,16–20] The massive fields predict unique signature in observations. Once observed, the expansion history of the inflationary universe can be reconstructed and thus it is a direct probe of inflation. Moreover, the detection of new massive particles during inflation is an invaluable hint for the development of high energy physics.

2. Massive fields during inflation

The inflaton field is not the only field existing during inflation. There exists plenty of massive fields during inflation. Those massive fields can come from:

- IR uplifting. There are many fields in the particle physics standard model. The flat space mass spectrum of those fields are negligibly small in the context of a typical inflation model. However, on an inflationary background, those fields get masses due to the Gibbons-Hawking temperature $M \sim gT \sim gH/(2\pi)$, where g is a coupling constant (or its square root) depending on the details of the particle.

- UV completion. A quantum field theory with UV divergence needs UV completion. Inflation especially requires UV completion because even a dimension 6 operator $V(\phi)\phi^2/M_p^2$ contributes part of the inflaton mass $\delta m \sim H$ and thus spoils the slow roll potential. In the viewpoint of effective field theory, new massive modes arise at the UV completion scale. For example, in string theory, there are various massive modes coming from moduli spaces, compactifications, stringy excitations, and so on. Those massive modes are typically below or at the string scale.
- Supersymmetry breaking. It is often believed that the inflationary energy scale is high and there is supersymmetry at such high energy scales. However, supersymmetry must be broken during inflation because of the quasi-de Sitter geometry. Thus the light fields whose mass is protected by supersymmetry obtain a correction of order Hubble.[13,21]

In the long wavelength limit, a massive field oscillates as $(-\tau)^{\pm i\mu}$, where τ is the conformal time, and $\mu \equiv \sqrt{M^2/H^2 - \alpha^2}$, where α depends on spin. For spin 0, 1/2 and 1, α takes value 3/2, 0 and 1/2 respectively. When $M/H < \alpha$ (which applies only for scalars and vectors), the massive field behaves as an over-damped oscillator and shall not oscillate in the IR. However, when $M/H > \alpha$, the Hubble friction cannot stop the field from oscillating. And fermions oscillate in the IR regardless of their mass. We focus on the $M/H > \alpha$ case in the following, and the results of the $M/H < \alpha$ case can be obtained by an analytical continuation.

The contributions from the massive fields, imprinting onto the primordial density fluctuations, can be classified into two types, namely local and non-local ones.[17] The non-local contribution originates from thermal particle production and cannot be be integrated out or mimicked by a single inflaton effective theory. The comparison between the local and non-local contribution is shown in Table 1.

name	origin	analyticity	integrate out	size
local	vacuum	analytic	can	$1/\mu^2$
non-local	thermal	non-analytic	cannot	$e^{-\pi\mu}$

Fig. 1. Comparison of different imprints on density fluctuation from a massive field.

3. Detecting new massive particles in the cosmological collider

The key mission for collider particle physics is searching for new particles. The major technique of new particle detection is through its resonance peaks. Things are similar here for the primordial universe.[11–13,17] An under-damped massive field oscillates as $\sigma \sim \exp(iMt)$. Thus the three point function between this field and two curvature scalar fields contains

$$\langle \zeta_{\mathbf{k}}^2 \sigma \rangle \supset \int f(\tau) e^{i(Mt-2k\tau)} d\tau \propto e^{i(Mt_*-2k\tau_*)} , \tag{1}$$

where $f(\tau)$ is a slowly varying function of time. As τ_* (or t_*) satisfying $k/a(\tau_*) = M/2$, a resonance is triggered and the density fluctuation is amplified;[5] and we have used the saddle point approximation in the second step. The effect of the σ field converts to density fluctuations either through gravitational or direct coupling. At the time of resonance, the phase of the massive field quantum oscillation is recorded in the above integral. This oscillatory behavior in k is the characteristic signal from massive fields. In single field effective field theory, we expect the correlation functions to be analytic in k thus such terms cannot appear. Note that the cases of $M/H < \alpha$ and $M/H > \alpha$ are related by an analytic continuation.

4. Towards proving/falsifying the inflation scenario

It is interesting to view the $\langle \zeta_{\mathbf{k}}^2 \sigma \rangle$ correlation in a broader sense. While inflation is the leading paradigm of the primordial universe, there are possible alternatives. The massive fields can be considered as a clock, whose ticks and tocks are uniform in physical time and labelling the different modes in density perturbations with its phases.[18–20] Depending on whether this oscillation of massive fields is classical or quantum-mechanical, we have the classical or quantum standard clocks, respectively. Solving the resonance condition $a(t_*) = 2k/M$ for t_* or τ_*, one can see that the phase of the correlation function (1) as a function of k is proportional to the inverse function of $a(t_*)$ or $a(\tau_*)$, so the evolution of the scale factor of the primordial universe is directly recorded in the astrophysical observables.

5. Summary and discussions

We would like to conclude this proceeding paper by comparing the physical implication of inflationary massive fields with inflationary gravitational waves, because unlike the other observables both of them can be used to distinguish the inflation from alternative scenarios model-independently. We refer them as "standard clock signals" versus "gravitational waves":

- Both exist in Nature model-independently. For standard clock signals, we refer to the quantum standard clock signal[20] in this statement.
- Both predict characteristic observational features.
  - The gravitational waves are imprinted on the CMB as the B-mode polarization, which is hard to mimic by other primordial physics (though not impossible, for example, primordial non-decaying vector fluctuations).
  - The standard clock signals are imprinted in all kinds of density fluctuations as special types of oscillatory features. Classical standard clocks predict unique

scale-dependent oscillatory signals in power spectrum and correlatedly in non-Gaussianites. Quantum standard clocks predict shape-dependent oscillatory signals in non-Gaussianities, which are protected by the analyticity of quantum field theory.

- Both test the inflation scenario in a model independent way, with some caveats.
  - The conventional wisdom is that, the amplitude H^2/M_p^2 (and its tilt) of the gravitational wave directly probes the Hubble scale (and its time dependence), and by observing a scale-invariant gravitational waves one proves inflation. However, there are a few caveats. First, conservation of the amplitude of the gravitational wave outside the horizon is assumed. However, this is not true in contracting universes, where the growing mode dominates over the constant mode. As a result, matter contraction,[22,23] cannot be distinguished from inflation by the amplitude and tilt of the gravitational waves. Moreover, non-trivial time evolution (even parametric resonance) of the super-horizon gravitational waves can also arise in massive gravity.[24] Second, vacuum initial condition is assumed, and by relaxing it the string gas cosmology[25] may produce the same amplitude (and see Refs. 26 and 27 for a discussion of tilt). Moreover, non-Bunch Davies initial conditions or sources from the matter sector (for example, particle production[28] or enhanced non-linear perturbations[29]) can also change the amplitude and tilt of the gravitational waves. So among the variety of alternative-to-inflation scenarios in the literature, the gravitational wave mainly distinguishes inflation from the ekpyrosis scenario.[30] Even between inflation and ekpyrosis, it is worth noting that the gravitational wave is a one sided test – seeing gravitational waves supports inflation and rules out ekpyrosis but not seeing gravitational waves shall not support ekpyrosis or rule out inflation.
  - The standard clock signal is model independent. The oscillation frequency does not depend on initial condition or super-horizon evolution of the field (though the amplitude depends on those details and is model dependent). It is shown that the clock signal distinguishes the inflation, ekpyrosis, string gas and matter bounce scenarios by predicting a unique signal pattern for each scenario. The frequency of the massive field oscillation can be affected by time dependent parameters in the Lagrangian. A fine tuned $m(t)$ may let the clock signal record a deformed evolution history of the primordial universe. These are interesting topics for future investigations.
- Both indicate significant new physics if detected.
  - Gravitational waves indicate the quantized fluctuations of gravity. Although in theory its existence is expected, an actual discovery would pin down the energy scale of inflation, which is new physics.
  - Standard clock signals is a probe of massive fields. Similarly, although in theory their existence is expected, an actual discovery would tell us the detailed information about the particle spectrum and coupling, along with, perhaps

even more importantly, the recorded evolutionary history of the primordial universe.

- Unfortunately, both are hard to detect.
  - The amplitude of gravitational waves is of order H^2/M_p^2. The Planck suppression make it hard to detect, especially for low scale inflation models.[31,32] Even worse, the gravitational waves behaves as radiation and decays after returning to the horizon. Thus we have to rely on observations on large scales, and the number of modes is limited. Also, there can be contamination from astrophysical sources such as polarized emissions from dust.
  - The quantum standard clock signals can be suppressed by two factors, the mass and the coupling. When the mass of the field is large, the suppression factor is $\exp(-\pi M/H)$.[17] Thus probing very massive fields is exponentially hard. Although it's natural to expect fields with $M \sim H$, we do not know it for sure. Coupling-wise, direct couplings generically predict much larger signals than gravitational coupling, giving rise to $f_{NL} \sim \epsilon$ or much larger.[12,13] These signals may be observable in future experiments. The classical standard clock signals are easier to observe but its existence is model-dependent. Observation-wise, the squeezed limit non-Gaussianity gives the sharpest clock signals, but can only make use of a small subset of available modes. On the other hand, oscillatory signals may be more difficult to analyze but less susceptible to contaminations from astrophysical sources.

As well-known, the primordial gravitational waves is a very important probe of the scenario type of the primordial universe. On the other hand, recent researches have uncovered another set of phenomena induced by massive fields, which could be used to achieve the same and complimentary science goals. The study is in its early stage and there are many works need to be done.

Acknowledgements

YW thanks organizers of the 2nd LeCosPA Symposium, especially Pisin Chen and Yifu Cai for invitation. XC and MHN are supported in part by the NSF grant PHY-1417421. YW is supported by the CRF Grants of the Government of the Hong Kong SAR under HKUST4/CRF/13G.

References

1. J. Martin, C. Ringeval and V. Vennin, Encyclopædia Inflationaris, *Phys. Dark Univ.* **5-6**, 75 (2014).
2. A. A. Starobinsky, Spectrum of adiabatic perturbations in the universe when there are singularities in the inflation potential, *JETP Lett.* **55**, 489 (1992), [Pisma Zh. Eksp. Teor. Fiz.55,477(1992)].
3. J. A. Adams, B. Cresswell and R. Easther, Inflationary perturbations from a potential with a step, *Phys. Rev.* **D64**, p. 123514 (2001).

4. X. Chen, R. Easther and E. A. Lim, Large Non-Gaussianities in Single Field Inflation, *JCAP* **0706**, p. 023 (2007).
5. X. Chen, R. Easther and E. A. Lim, Generation and Characterization of Large Non-Gaussianities in Single Field Inflation, *JCAP* **0804**, p. 010 (2008).
6. R. Easther, B. R. Greene, W. H. Kinney and G. Shiu, Inflation as a probe of short distance physics, *Phys. Rev.* **D64**, p. 103502 (2001).
7. X. Chen, M.-x. Huang, S. Kachru and G. Shiu, Observational signatures and non-Gaussianities of general single field inflation, *JCAP* **0701**, p. 002 (2007).
8. H. Jiang and Y. Wang, Towards the physical vacuum of cosmic inflation (2015).
9. H. Jiang, Y. Wang and S. Zhou, On the initial condition of inflationary fluctuations (2016).
10. T. Kobayashi, M. Yamaguchi and J. Yokoyama, G-inflation: Inflation driven by the Galileon field, *Phys. Rev. Lett.* **105**, p. 231302 (2010).
11. X. Chen and Y. Wang, Large non-Gaussianities with Intermediate Shapes from Quasi-Single Field Inflation, *Phys. Rev.* **D81**, p. 063511 (2010).
12. X. Chen and Y. Wang, Quasi-Single Field Inflation and Non-Gaussianities, *JCAP* **1004**, p. 027 (2010).
13. D. Baumann and D. Green, Signatures of Supersymmetry from the Early Universe, *Phys. Rev.* **D85**, p. 103520 (2012).
14. A. Golovnev, V. Mukhanov and V. Vanchurin, Vector Inflation, *JCAP* **0806**, p. 009 (2008).
15. A. A. Starobinsky, Spectrum of relict gravitational radiation and the early state of the universe, *JETP Lett.* **30**, 682 (1979), [Pisma Zh. Eksp. Teor. Fiz.30,719(1979)].
16. X. Chen and Y. Wang, Quasi-Single Field Inflation with Large Mass, *JCAP* **1209**, p. 021 (2012).
17. N. Arkani-Hamed and J. Maldacena, Cosmological Collider Physics (2015).
18. X. Chen, Primordial Features as Evidence for Inflation, *JCAP* **1201**, p. 038 (2012).
19. X. Chen, M. H. Namjoo and Y. Wang, Models of the Primordial Standard Clock, *JCAP* **1502**, p. 027 (2015).
20. X. Chen, M. H. Namjoo and Y. Wang, Quantum Primordial Standard Clocks (2015).
21. E. J. Copeland, A. R. Liddle, D. H. Lyth, E. D. Stewart and D. Wands, False vacuum inflation with Einstein gravity, *Phys. Rev.* **D49**, 6410 (1994).
22. D. Wands, Duality invariance of cosmological perturbation spectra, *Phys. Rev.* **D60**, p. 023507 (1999).
23. F. Finelli and R. Brandenberger, On the generation of a scale invariant spectrum of adiabatic fluctuations in cosmological models with a contracting phase, *Phys. Rev.* **D65**, p. 103522 (2002).
24. C. Lin and M. Sasaki, Resonant Primordial Gravitational Waves Amplification, *Phys. Lett.* **B752**, 84 (2016).
25. R. H. Brandenberger and C. Vafa, Superstrings in the Early Universe, *Nucl. Phys.* **B316**, p. 391 (1989).
26. Y. Wang and W. Xue, Inflation and Alternatives with Blue Tensor Spectra, *JCAP* **1410**, p. 075 (2014).
27. M. He, J. Liu, S. Lu, S. Zhou, Y. Cai, Y. Wang and R. Brandenberger, To appear.
28. S. Mukohyama, R. Namba, M. Peloso and G. Shiu, Blue Tensor Spectrum from Particle Production during Inflation, *JCAP* **1408**, p. 036 (2014).
29. M. Biagetti, M. Fasiello and A. Riotto, Enhancing Inflationary Tensor Modes through Spectator Fields, *Phys. Rev.* **D88**, p. 103518 (2013).

30. J. Khoury, B. A. Ovrut, P. J. Steinhardt and N. Turok, The Ekpyrotic universe: Colliding branes and the origin of the hot big bang, *Phys. Rev.* **D64**, p. 123522 (2001).
31. P. W. Graham, D. E. Kaplan and S. Rajendran, Cosmological Relaxation of the Electroweak Scale, *Phys. Rev. Lett.* **115**, p. 221801 (2015).
32. H. Jiang, T. Liu, S. Sun and Y. Wang, Echoes of Inflationary Particle Phase Transitions in the CMB (2015).

Cosmological Implications of Fubini Type Instanton

Bum-Hoon Lee[a], Wonwoo Lee*, and Daeho Ro[b]

*[a] *Center for Quantum Spacetime, Sogang University, Seoul 04107, Korea*
[ab] *Department of Physics, Sogang University, Seoul 04107, Korea*
[a] *Asia Pacific Center for Theoretical Physics, Pohang 37673, Korea*
[a] *bhl@sogang.ac.kr*
* *warrior@sogang.ac.kr*
[b] *dhro@sogang.ac.kr*

A Fubini instanton is a bounce solution which describes the decay of a vacuum state located at the top of the tachyonic potential via the tunneling without a barrier. We investigate various types of Fubini instantons both without gravity and with gravity. We present numerically Fubini bounce solutions in curved space. We discuss the cosmological implications of Fubini type instanton in the early universe. They could affect the cosmological large scale structure formation and CMB.

Keywords: Fubini instanton; tunneling without a barrier.

1. Introduction

A bounce solution was originally introduced to describe the decay of metastable vacuum state without gravity.[1,2] A decay rate per unit time per unit volume is given by $\Gamma/V = Ae^{-B}$, in which the exponent B is the Euclidean action and the pre-exponential factor A is obtained in terms of the eigenvalues of second variational derivation of Euclidean action.[3] The mechanism of false vacuum decay was developed with gravity,[4] which only consider the flat space of true or false vacuum. It was extended into the cases of arbitrary vacuum energy.[5] Later, the possible types of true vacuum bubbles in de Sitter (dS) space and general background spaces were studied in more detail.[6]

In the equation of motion for scalar field without gravity, there exists a viscous damping term that makes an ordinary bounce solution not oscillate. When the gravity taken into account, the situation is abruptly changes. Depending on the background space, the role of damping term can be changed into an anti-damping term if the dS background is included during the vacuum decay. Then, there can exists an oscillating bounce solution. The exponent of decay rate can be obtained by a path integral and the dominant contribution comes from the ordinary bounce solution, but the oscillating bounce solution with arbitrary number of oscillation still contribute to that.[7] The oscillating bounce solution was found indeed in dS background numerically.[8] Recently, the oscillating bounce solution on the symmetric double-well potential was studied in general background space and the condition for the existence of oscillating bounce was also studied.[9] There was a study about the oscillating solution as homogeneous tunneling channels.[10]

The Fubini instanton[12] describes the decay of a metastable vacuum state through the tunneling on a negative quartic potential. The conformal scalar field theory allows the existence of one-parameter Euclidean solution. If the conformal invariance is broken by a mass term, there cannot exists a bounce solution.[13] The Fubini instanton represents the tunneling without a potential barrier between the top of potential and an arbitrary state. There were intensive studies about the tunneling without a barrier in flat space[14] and curved space.[15] Recently, the vacuum decay from the flat to Anti-de Sitter (AdS) space was studied as a tunneling without a barrier.[16] Furthermore, the analysis of Fubini instanton is extended into the many scalar field solution.[17]

When the gravity is taken into the model of Fubini instanton, the conformal invariance of scalar field theory is broken and the one-parameter solution might be disappear or changed. We have shown that the existence of Fubini instanton in the presence Einstein gravity for the first time with an initially flat and AdS background and Z_2-symmetric solutions with an initially dS background.[18] The oscillating solutions under the potential with a mass term was also studied.[19] We were also investigated the oscillating Fubini instantons in the presence of Einstein gravity with AdS and dS background and constructed the parametric phase diagram for each type of solutions.[20] For the stability issue on the tachyonic top, we were discussed the issue with quadratic potential.[20]

In Sec. 2, we set up the basic framework for this paper. In Sec. 3, we represent the various types of solutions and parametric phase diagram. The results are discussed in Sec. 4. This article is prepared for the proceedings of 2nd LeCosPA Symposium in NTU, Taiwan, Dec 2015. The main work of this article is based on our previous works.[18,20]

2. Tunneling without a barrier without gravity

We consider the action in the absence of gravity

$$S = \int_{\mathcal{M}} \sqrt{-g} d^4x \left[-\frac{1}{2} \nabla_\alpha \Phi \nabla^\alpha \Phi - U(\Phi) \right] , \tag{1}$$

where $g = \det \eta_{\mu\nu}$, $\eta_{\mu\nu} = diag(-1, 1, 1, 1)$ is the Minkowski metric, and the potential is a polynomial function and has sufficiently flat top.

The important reason why the tunneling without a barrier is considered is that the field theory has an infinite number of degrees of freedom, so the potential energy for the rolling particle is not simple $U(\Phi)$, but rather the functional

$$V[\Phi] = \int d^3x \left[\frac{1}{2} (\nabla\Phi)^2 + U(\Phi) \right] , \tag{2}$$

as pointed out in Ref. 14. The gradient term can gives rise to the effective barrier in the potential energy. Thus, the tunneling can occur even without a barrier.

We consider the tunneling process in Euclidean space, in which the solution has $O(4)$ symmetry. The equation of motion is given by

$$\frac{d^2\Phi}{d\eta^2}+\frac{3}{\eta}\frac{d\Phi}{d\eta}=-\frac{d(-U)}{d\Phi}\,, \tag{3}$$

where $\eta(=\sqrt{\tau^2+x^2})$ plays the role of the evolution parameter in Euclidean space and the second term in the left-hand side plays the role of a damping term. We impose the boundary conditions as follows:

$$\left.\frac{d\Phi}{d\eta}\right|_{\eta=0}=0 \;\text{ and }\; \Phi|_{\eta=\infty}=0\,. \tag{4}$$

If the potential has only a tachyonic quartic self-interaction term, there exists an instanton.[12] The so-called Fubini instanton corresponds to the bounce solution representing the decay of the background vacuum state under tunneling instead of rolling in the presence of a tachyonic potential. The tunneling occurs from the maximum of the potential, which is a vacuum state, to any arbitrary state, belonging to the tunneling without a barrier. The analytic solution in this case has the form $\Phi=\sqrt{\frac{8}{\lambda}\frac{b}{\eta^2+b^2}}$, where η is the radial length in Euclidean space, b is any arbitrary length scale which characterizes the size of the instanton and is related to the initial value Φ_o.

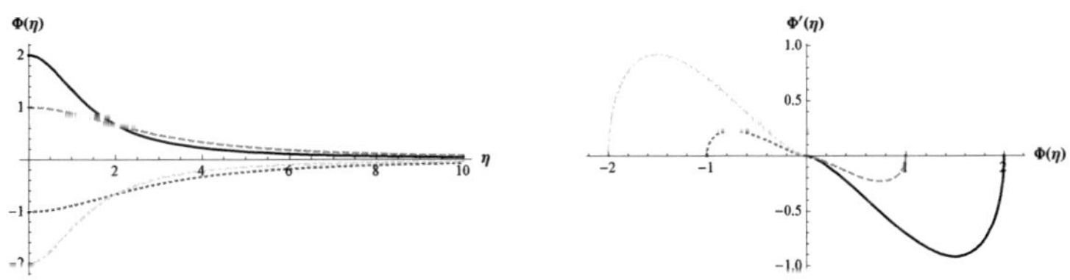

(a) Solutions for Φ (b) Solutions for Φ vs. Φ'

Fig. 1. (Color online) Fubini solutions with $\lambda=1$ and corresponding values of b with respect to $\Phi_o=2,\,1,\,-1$, and -2.

Figure 1 presents the characteristic behavior of the analytic solutions in terms of the value of the parameter b.

If the potential has both a tachyonic quartic and a quadratic term, the conformal invariance is broken and any solution with a finite action is forbidden by scaling argument. In other worlds, the particle can not have enough energy to reach the hill overcome the barrier near $\Phi=0$ because the damping term has got a large value due to a large value of Φ' near the initial point.[13]

If the potential is given by[14]

$$U(\Phi)=\begin{cases}-\frac{\lambda}{4}\Phi^4, & |\Phi|<\hat{\Phi}\\ -K(|\Phi|-\hat{\Phi})-\frac{\lambda}{4}\hat{\Phi}^4, & |\Phi|>\hat{\Phi},\end{cases} \tag{5}$$

where K and $\hat{\Phi}$ are constants. Fortunately, there exists an analytic solution which is given by

$$\Phi(\eta) = \begin{cases} -\frac{K}{8}\eta^2 + \hat{\Phi}(2-\gamma), & \eta < \sqrt{\frac{8\hat{\Phi}}{K}(1-\gamma)} \\ \frac{8\hat{\Phi}^2}{K}\frac{1}{(\eta^2+8\hat{\Phi}\gamma/K)}, & \eta > \sqrt{\frac{8\hat{\Phi}}{K}(1-\gamma)}, \end{cases} \tag{6}$$

where $\gamma = \lambda\hat{\Phi}^3/K$. Figure 2 presents the characteristic behavior of the analytic solutions with $\hat{\Phi} = 1$ and $K = 2$.

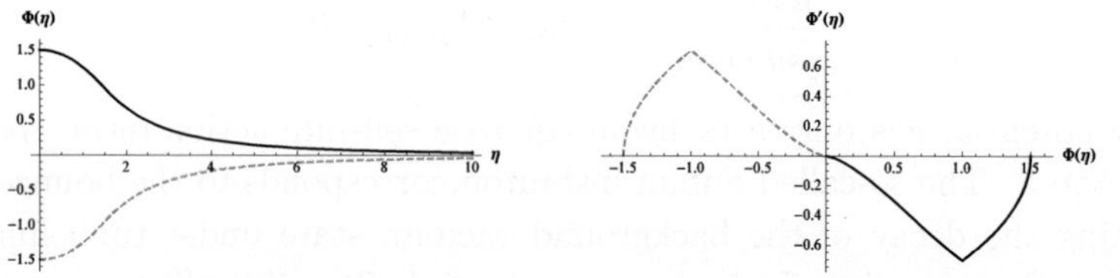

(a) Solutions for Φ (b) Solutions for Φ vs. Φ'

Fig. 2. (Color online) Solutions for Eq. (5) with parameters $\lambda = 1$, $\hat{\Phi} = 1$, and $K = 2$.

3. Tunneling without a barrier with gravity

We consider the action in the presence of gravity

$$S = \int_{\mathcal{M}} \sqrt{-g}d^4x \left[\frac{R}{2\kappa} - \frac{1}{2}\nabla_\alpha\Phi\nabla^\alpha\Phi - U(\Phi)\right] + \oint_{\partial\mathcal{M}} \sqrt{h}d^3x\frac{K - K_o}{\kappa}, \tag{7}$$

where $g = \det g_{\mu\nu}$, $\kappa \equiv 8\pi G$, R denotes the scalar curvature of the spacetime $\mathcal{M}$, and h is the determinant of the first fundamental form. K and K_o are the traces of the second fundamental form of the boundary $\partial\mathcal{M}$ for the metric $g_{\mu\nu}$ and $\eta_{\mu\nu}$, respectively. The second term on the right-hand side is the so-called York-Gibbons-Hawking boundary term.[21,22]

The geometry is written as

$$ds^2 = d\eta^2 + \rho(\eta)^2\left[d\chi^2 + \sin^2\chi\left(d\theta^2 + \sin^2\theta d\phi^2\right)\right]. \tag{8}$$

The field equations are given by

$$\Phi'' + \frac{3\rho'}{\rho}\Phi' = -\frac{d(-U)}{d\Phi}, \quad \rho'' = -\frac{\kappa}{3}\rho\left(\Phi'^2 + U\right), \tag{9}$$

and the Hamiltonian constraint is given by

$$\rho'^2 - 1 - \frac{\kappa\rho^2}{3}\left(\frac{1}{2}\Phi'^2 - U\right) = 0, \tag{10}$$

where the prime denotes differentiation with respect to η. Figure 3 presents the characteristic behavior of the analytic solutions with $\hat{\Phi} = 1$ and $K = 2$.

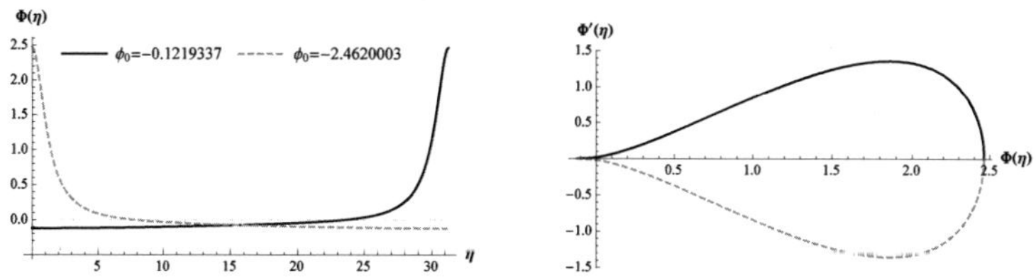

(a) Solutions for Φ (b) Solutions for Φ vs. Φ'

Fig. 3. (Color online) Solutions for Eq. (5) with parameters $\lambda = 1$, $\hat{\Phi} = 1$, and $K = 2$.

4. Discussion

The issue on the instability of a tachyonic vacuum is very subtle. There are three possible modes, i.e. quantum fluctuation, the typical decay mode representing the homogeneous rolling on the potential, and the tunneling without a barrier[14,15] as inhomogeneous transition as shown in Fig. 4. We did not consider quantum fluctuation in the present paper. The possibility may depend on both the curvature of the tachyonic top and the cosmological constant. If the shape of the potential near its top is very sharp we could not expect the existence of the tunneling without a barrier. If the shape near its top is flat enough, there exist two decay modes for the rolling and the tunneling. If so there is a competition between them. As pointed out in Ref. 15, if the size of a solution is much smaller than the horizon size, the decay mode by the tunneling is distinguishable from others, while if the sizes are comparable, the decay mode by the tunneling becomes indistinguishable from others. To analyze which mode is dominant, one may need to estimate the decay time for each mode. If the decay time for the rolling is less than that for the tunneling the rolling becomes the dominant decay mode, and vice versa. In general all modes mix together. All modes may affect the cosmological large scale structure formation and CMB. If the tunneling would have occurred in some place as well as the rolling representing the inflation would have occurred homogeneously except for that some place in the early universe, the reheating processes for each event would have been different. The some place would have small number of particles than others, therefore the temperature of the some place is lower than others.

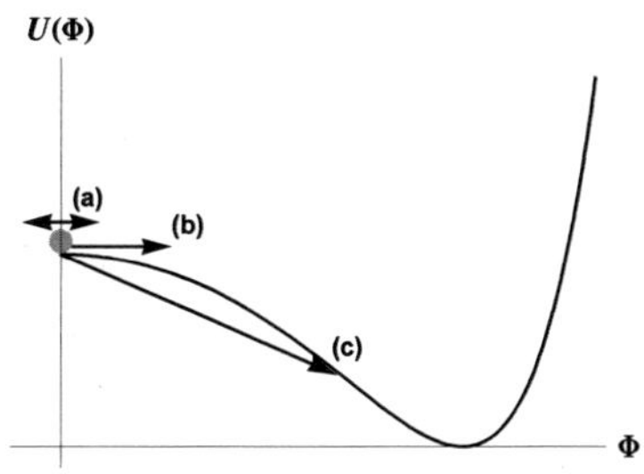

Fig. 4. (Color online) Three decay modes (a) quantum fluctuation, (b) rolling, and (c) tunneling.

Acknowledgements

BHL was supported by the National Research Foundation of Korea(NRF) grant funded by the Korea government(MSIP) (No. 2014R1A2A1A01002306). WL was supported by Basic Science Research Program through the National Research Foundation of Korea(NRF) funded by the Ministry of Education, Science and Technology(2012R1A1A2043908).

References

1. S. Coleman, The Fate of the False Vacuum. 1. Semiclassical Theory, *Phys. Rev. D* **15**, 2929 (1977); **16**, 1248(E) (1977).
2. I. Yu. Kobzarev, L. B. Okun, and M. B. Voloshin, Bubbles in Metastable Vacuum, *Yad. Fiz.* **20**, 1229 (1974)[*Sov. J. Nucl. Phys.* **20**, 644 (1975)].
3. C. G. Callan, Jr., and S. Coleman, The Fate of the False Vacuum. 2. First Quantum Corrections, *Phys. Rev. D* **16**, 1762 (1977).
4. S. Coleman and F. De Luccia, Gravitational Effects on and of Vacuum Decay, *Phys. Rev. D* **21**, 3305 (1980).
5. S. Parke, Gravity and the decay of the false vacuum, *Phys. Lett. B* **121**, 313 (1983).
6. B.-H. Lee and W. Lee, Vacuum bubbles in a de Sitter background and black hole pair creation, *Classical Quantum Gravity* **26**, 225002 (2009).
7. J. Zinn-Justin, *Path Integrals in Quantum Mechanics* (Oxford University Press, Oxford, 2010).
8. J. C. Hackworth and E. J. Weinberg, Oscillating bounce solutions and vacuum tunneling in de Sitter spacetime, *Phys. Rev. D* **71**, 044014 (2005).
9. B.-H. Lee, C. H. Lee, W. Lee, and C. Oh, Oscillating instanton solutions in curved space, *Phys. Rev. D* **85**, 024022 (2012).
10. B.-H. Lee, W. Lee, and D.-h. Yeom, Oscillating instantons as homogeneous tunneling channels, *Int. J. Mod. Phys. A* **28** 1350082 (2013).
11. L. Battarra, G. Lavrelashvili, and J.-L. Lehners, Zoology of instanton solutions in flat potential barriers, *Phys. Rev. D* **88**, 104012 (2013).
12. S. Fubini, A New Approach to Conformal Invariant Field Theories, *Nuovo Cim. A* **34**, 521 (1976).
13. I. Affleck, On Constrained Instantons, *Nucl. Phys. B* **191**, 429 (1981).
14. K. Lee and E. J. Weinberg, Tunneling without barriers, *Nucl. Phys. B* **267**, 181 (1986).
15. K. Lee, Tunneling Without Barriers in Curved Space-time, *Nucl. Phys. B* **282**, 509 (1987).
16. S. Kanno, M. Sasaki, and J. Soda, Tunneling without barriers with gravity, *Classical Quantum Gravity* **29**, 075010 (2012).
17. A. Aravind, B. S. DiNunno, D. Lorshbough and S. Paban, Analyzing multifield tunneling with exact bounce solutions, *Phys. Rev. D* **91**, 025026 (2015).
18. B.-H. Lee, W. Lee, C. Oh, D. Ro, and D.-H. Yeom, Fubini instantons in curved space, *JHEP*, **1306**, 003 (2013).
19. B.-H. Lee, W. Lee, D. Ro, and D.-H. Yeom, Generalized Fubini instantons of a self-gravitating scalar field, *J. Korean Phys. Soc.* **65**, 884 (2014).

20. B.-H. Lee, W. Lee, D. Ro, and D.-H. Yeom, Oscillating Fubini instantons in curved space, *Phys. Rev. D* **91**, 124044 (2015).
21. J. W. York, Jr., Role of conformal three geometry in the dynamics of gravitation, *Phys. Rev. Lett.* **28**, 1082 (1972).
22. G. W. Gibbons and S. W. Hawking, Action Integrals and Partition Functions in Quantum Gravity, *Phys. Rev. D* **15**, 2752 (1977).

Growing Isocurvature in the Lifshitz Regime of Horava Gravity

Shun Arai

Department of Physics and Astrophysics, Nagoya University,
Nagoya 464-8602, Japan,
arai.shun@a.mbox.nagoya-u.ac.jp

We study the evolution of scalar perturbations in inflationary epoch with a single Lifshitz scalar in the context of the BPSH theory, which generalizes the original non-projectable Horava Lifshitz gravity. We consistently solve the evolution of the coupled scalar graviton and the inflaton fluctuation and derive the isocurvature perturbation. We find that the isocurvature mode grows in high energy.

Keywords: Horava gravity; inflation.

1. Introduction

The General Relativity theory of gravity (GR) has still remained the most successful theory of gravity which has passed the all kind of observational tests ever considered. We have well known, however, it is quite a big deal that the GR conflicts theoretically with quantum theory; Non-renormalizability arises in the GR. Recently P.Horava has proposed a promising model so called Horava-Lifshitz gravity,[1] which describes a new scaling law is diffrent between space and time. Specifically (D + 1)-dimensional spacetime scales anisotropically $t \to b^z t$, $\mathbf{x} \to b\mathbf{x}$, what is called "Lifshitz scaling". Here z is the dynamical exponent describing the anisotropy between space and time. When $z \geq D$ the Lagrangian becomes power-counting renormalizable. While such a spacetime loses the local Lorentz invariance. This feature charactorizes the HL gravity. The current observational constraints have been obtained that the Lorentz invariance is preserved up to TeV scale by the gravitational anomaly in the solar system,[5] or gamma ray observations.[6,7] The Lorentz invariance violation causes another interesting feature that the perturbed HL gravity has a spin zero mass-less mode, called "scalar graviton" or "Khronon". The previous studies have revealed that the behaviour of the scalar in the fixed background with the Lifshitz scaling, and an isocurvature mode in the Einstein Aether theory which coincides with a Iow energy limit of the BPSH theory,[4] one of the healthy extension of Horava Lifshitz gravity free from theoretical instabilities.[2,3]

In our study, We consider the evolution of scalar perturbations in inflationary epoch with a single Lifshitz scalar in the context of the BPSH theory in the (3+1)-dimensional spacetime. We consistently solve the evolution of the coupled scalar graviton and the inflaton fluctuation and derive the isocurvature perturbation. We find that the isocurvature mode grows in high energy. This paper is organized as follows. In section 2 we set up our model. In section 3 we mention shortly our results of the isocurvature mode in the case $z = 3$. We sum up our paper in discussion about on-going work.

2. Set up

We set up the action

$$\mathcal{L}_{HG} = N\sqrt{\gamma}\Bigg\{\frac{M_*^2}{2}\left[\frac{1}{\alpha_1}K_{ij}K^{ij} - \frac{1}{\alpha_2}K^2\right] - \frac{1}{2M_*^2}\left[\frac{(\nabla_i R_{jk})^2}{\gamma_1} + \frac{(\nabla_i R)^2}{\gamma_2} + \frac{\Delta R \nabla_i a^i}{\gamma_3} - \frac{a_i \Delta^2 a^i}{\gamma_4}\right]\Bigg\}, \tag{1}$$

$$\mathcal{L}_{inf} = N\sqrt{\gamma}\left\{\frac{(\dot{\Phi} - N^i\partial_i\Phi)^2}{2N^2} + \varkappa_3\frac{\Phi\Delta^3\Phi}{2M_*^4} - V(\Phi)\right\}, \tag{2}$$

where the notations are:

$$ds^2 = (N^2 - N_iN^i)dt^2 - 2N_i dt dx^i - \gamma_{ij}dx^i dx^j\ , \tag{3}$$

$$K_{ij} = \frac{\dot{\gamma}_{ij} - \nabla_i N_j - \nabla_j N_i}{2N}\ , \tag{4}$$

$$a_i = \frac{\partial_i N}{N}\ , \tag{5}$$

$$R_{ij} \quad \text{— 3-dimensional Ricci tensor,} \tag{6}$$

$$\Delta \quad \text{— 3-dimensional Laplacian .} \tag{7}$$

and Φ is an Lifshitz type inflaton. Then we define model parameters. Firstly we introduce an mass scale at which the Lorentz invariance is broken.

$$M_*^2 = M_P^2\alpha_1\ , \tag{8}$$

where M_P is the reduced Planck mass. The experimental data require $\alpha_{1,2} \ll 1$ (In detail, see the paper[3]). We will also assume the following hierarchy for other couplings: $\gamma_{1,2,3} \sim \alpha_1$ and $\gamma_4 \lesssim 1$. This will ensure that the transition towards the UV form of the dispersion relations happens at $k \sim \omega \sim M_*$ both for the spin-2 and scalar gravitons. Indeed, the dispersion relation of the spin-2 modes is affected only by the first terms in each square bracket and to be modified at M_* it must contain $\gamma_1 \sim \alpha_1$. While the necessary requirements for the scalar mode can be read off from Eqs. (5.1), (5.3), (5.4) of the paper.[3] Finally, we assume that $\varkappa_3 \sim 1$.

3. Isocurvature

We define an curvature perturbation ζ

$$\zeta \equiv \mathcal{R} - \frac{H}{\dot{\Phi}}\varphi. \tag{9}$$

Here $\mathcal{R}$ is a spatial curvature perturbation which corresponds to a scalar graviton and φ is an inflaton fluctuation. After ζ is frozen, inflaton perturbation φ is divided into a constant mode and an isocurvature mode. In this moment the Lagrangian of φ is given by

$$a^{-3}\mathcal{L}_\varphi \simeq \frac{1}{2}\left[\frac{\alpha_1}{\varepsilon_1}\Omega_3\dot{\varphi}^2 - \frac{\omega_{\mathcal{R}}^2}{\varepsilon_1}a^{-2}\varphi^2\right] - \frac{2M_*^2}{\alpha_1}\frac{H}{\dot{\Phi}}\omega_{\mathcal{R}}^2 a^{-2}\zeta\varphi\,, \tag{10}$$

where p is an momentum in Fourier space, a is a scale factor, H is a Hubble parameter, and $\mathcal{H}$ is a comoving Hubble parameter. If we consider the case that $z = 3$ is dominant, Ω_1, Ω_3 and $\omega_{\mathcal{R}}^2/\mathcal{H}^2$ is

$$\Omega_1 \sim 1\,, \tag{11}$$

$$\Omega_3 \sim \frac{1}{\gamma_4}\left(\frac{p}{\mathcal{H}}\right)^2\left(\frac{p}{aM_*}\right)^4, \tag{12}$$

$$\frac{\omega_{\mathcal{R}}^2}{\mathcal{H}^2} \sim \alpha_1\left(\frac{3}{\gamma_1}+\frac{8}{\gamma_2}+\frac{3}{\gamma_3}-\varepsilon_1\right)\left(\frac{p}{\mathcal{H}}\right)^2\left(\frac{p}{aM_*}\right)^4, \tag{13}$$

and then we obtain the equation of motion

$$\ddot{\varphi} + (-3+2\varepsilon_1-\varepsilon_2)H\dot{\varphi} + m^2\varphi + \sqrt{\frac{2\varepsilon_1}{\alpha_1}}M_* m^2\zeta = 0\,. \tag{14}$$

where

$$m^2 \equiv \gamma_4\left(\frac{3}{\gamma_1}+\frac{8}{\gamma_2}+\frac{3}{\gamma_3}-\varepsilon_1\right)H^2\,. \tag{15}$$

With $\mathcal{O}(\varepsilon^2)$ accuracy we obtain the isocurvature mode $\varphi_{\rm iso}$

$$\varphi_{\rm iso} = a^{(3-\varepsilon_1+\varepsilon_2)/2}\left(C_1 a^{i\frac{m}{H}} + C_2 a^{-i\frac{m}{H}} + \mathcal{O}(\varepsilon^2)\right)\,. \tag{16}$$

C_1 and C_2 are determined by the initial condition when ζ is frozen. We find that $\varphi_{\rm iso}$ grows when the higher derivative at $z = 3$ is dominant.

4. Conclusions

We have calculated perturbations analytically in the context of the non projectable version of Horava gravity. Then we have found the isocurvature perturbation grows with the scale factor during the regime when $z = 3$. We are now studying on the curvature power spectrum and the coeficients of the isocurvature evaluated at the time ζ become constant. And we have to test them with the CMB observations such as PLANCK and we expect to obtain much precise constraint of the parameters in the Horava-Lifshitz gravitational theory.

Acknowledgements

I thank Y.Urakawa for useful discussions. This work is in part supported by MEXT Grant-in- Aid for Scientific Research on Innovative Areas, No. 15H05890.

References

1. P. Horava, Phys. Rev. D **79**, 084008 (2009) [arXiv:0901.3775 [hep-th]].
2. D. Blas, O. Pujolas and S. Sibiryakov, Phys. Rev. Lett. **104**, 181302 (2010) [arXiv:0909.3525 [hep-th]].
3. D. Blas, O. Pujolas and S. Sibiryakov, JHEP **1104**, 018 (2011) [arXiv:1007.3503 [hep-th]].

4. C. Armendariz-Picon, N. F. Sierra and J. Garriga, JCAP **1007**, 010 (2010) [arXiv:1003.1283 [astro-ph.CO]].
5. M.C.Will 2005 Living Rev.Rel.9 3
6. M.Ackermann et al. [Fermi GBM/LAT Collaborations] 2009 Nature 462 331
7. J.Albert et al. [MAGIC Collaboration and Other Contributors Collaboration] 2008 Phys. Lett. B 668 253

PARALLEL SESSION

IX. Holographic Principle, AdS/CFT, and Applications

Summary of Parallel Session: "Holographic Principle, AdS/CFT, and Applications"

Nakwoo Kim

Physics Department, Kyung Hee University,
Seoul, 02447, Republic of Korea
nkim@khu.ac.kr

This article summarizes the presentations of the parallel session "Holographic Principle, AdS/CFT, and Applications", and discuss the recent developments and the current status of this topic.

1. Introduction

The parallel session on *"Holographic Principle, AdS/CFT, and Applications"* was organized by Keun-Young Kim and Janet Hung. Unfortunately Keun-Young Kim could not attend the conference and I gave the summary presentation instead, on Dec. 18 2015.

The parallel session was held on Dec. 15 2015, and there were 10 short presentations (20 minutes each, given in the following chronological order).

- **Nakwoo Kim**, Bulk local operators and D-instantons in pp-wave holography[1]
- **Wei Li**, Generalized Conformal Anomalies in Higher Spin Theory[2]
- **Bogeun Gwak**, Testing the validity of cosmic censorship in the rotating anti-de Sitter black hole[3]
- **Shin Nakamura**, Application of AdS/CFT Correspondence to Studies on Non-equilibrium Steady States[4]
- **Shu Lin**, Holographic thermalization of strongly coupled field theory in two dimension[5]
- **Yun-Long Zhang**, Rindler Fluid and Its Properties[6]
- **Chanyong Park**, Thermodynamic law from the entanglement entropy bound[7]
- **Xian-Hui Ge**, Lower bound of conductivity from Holography[8]
- **Feng-Li Lin**, Entanglement renormalization and integral geometry[9]
- **Dong-Hoon Kim**, Analogies between Gravity and Quantum Entanglement: Viewed from Information Geometry

2. Brief Summary of Past Development and Current Status of AdS/CFT

Before we talk about each talk and what was reported, let me briefly provide an overview of recent progress made in AdS/CFT.[10] After it was first proposed in 1997, there have been a huge number of works and subsequently the field diversified. So it will be useful to locate ourselves in this long and winding road.

The original motivation of AdS/CFT correspondence of course goes back to the black hole entropy formula saying that it is proportional to the horizon area, not the volume. It got huge boost by a concrete proposal of AdS/CFT by J. Maldacena in 1997. There, we have a very concrete two theories related through holography. One is the $\mathcal{N} = 4$ supersymmetric Yang-Mills theory in four dimensions, the other IIB string theory in the background of $AdS_5 \times S^5$. They are both maximally supersymmetric theories and thanks to the huge symmetry we can do many quantitative checks of holographic principle. Namely, we do explicit computations on both sides and match the results and find agreement. That means it is usually useful to use string theory construction of gauge theory observables, and restrict to quantities which are preserved by supersymmetry, and also crucially, we take the large N limit of gauge field theory. The quantities compared include central charge, partition function, Wilson loop expectation values etc. Initially the checks were done mostly in $\mathcal{N} = 4$ super Yang-Mills, but after 2008 it was discovered how to describe multiple M2-branes, so the program was extended to M-theory and dual three dimensional field theory. Crucially the peculiar behavior of $N^{3/2}$ scaling of degrees of freedom was checked using localization formula,[11] and agreement established including the correct coefficient. For a similar check with M5-branes and their N^3 scaling, see e.g. Gang et al.[12,13] Perhaps this quantitative and precision holography is nearing the completion.

On the other hand some of the string theoriests got more ambitious and speculated wildly. They assumed, without rigorous proof or quantitative evidence, that any asymptotically AdS classical gravity should be dual to some strongly-coupled system. They don't need to be necessarily gauge field theory in the large N limit.

It was initially applied to QCD, and the main physical phenomenon in mind was the quark gluon plasma and the thermalization of relativistic high energy collider etc. Then more recently this approach was applied to superfluid or superconductor. Important landmark works include the Sakai Sugimoto model in 2004,[14] and Hartnoll, Herzog and Horowitz's holographic superconductor in 2008.[15] Some people are skeptical about this kind of bottom-up approach, since sometimes the string theory construction is not clear, there is stability issue due to lack of supersymmetry, no perturbative control and so on. But at least it is true that this program gave us better understanding black holes in AdS, especially when and how the no-hair theorem can be avoided.

In the rest of this report I will briefly summarize each talk. For logical coherence I classified the presentations into four categories: bottom-up holography, extending the applicability of holography, fundamental questions concerning holography principle, and quantum information theoretical implications of AdS/CFT.

3. Summary of Presentations

3.1. *Bottom-up holography*

There were two speakers whose research took the bottom-up approach towards holographic principle.

Shin Nakamura talked about non-equilibrium physics using holography. Most quantitative and precise research of AdS/CFT so far involves conformal field theory spectrum etc. But of course at the end of the day we would like to understand real physics, where there is no supersymmetry or integrability. But more importantly, we would like to understand time-dependent physics. But it is quite hard in holography since the gravity equations of motion are nonlinear partial differential equations. Probably an easier but still important problem is steady state among non-equilibrium cases. More concretely, Nakamura considered a probe D-brane withing black hole geometry. The D-brane dynamics is expected to give quark sector physics. The black hole background provides heat bath at a certain temperature T. Then one discovers that the fluctuation modes of D-brane leads to a notion of horizon on D-brane worldvolume. One turns on electric field on the D-brane, and there is energy flow towards *horizon* on the boundary. This phenomenon is an example of analogue black hole, since the D-brane dynamics itself is not a gravity theory.

Then Xian-Hui Ge talked about his work with S-J Sin and S-F Wu. They have more concrete physical phenomenon in mind, to explain using holography. Their goal was to construct a holographic model which exhibits linear T-resistivity, like Luttinger liquid. They performed a general analysis using AdS-Einstein-Maxwell-scalar system and calculated holographic resistivity.

$$S = \int d^{d+1}x\sqrt{-g}\left[\frac{1}{16\pi G}\left(R - \frac{1}{2}(\partial\phi)^2 + V(\phi) - \frac{1}{2}\sum_{i=1}^{p-1} Y_i(\phi)(\partial\psi_i)^2\right)\right.$$
$$\left. - \frac{1}{4g_{d+1}^2} Z(\phi)F^2\right]. \tag{1}$$

3.2. *Extending the applicability of holography principle in a fundamental way*

There are three presentations which belong to this category.

Wei Li talked about conformal anomalies in higher spin theory, a work done in collaboration with S. Theisen. It is a basic property of string theory that it contains particles with arbitrarily high spin, as higher oscillation modes. Maybe this feature is essential in quantum gravity to provide UV completion. Now, the ordinary AdS/CFT dictionary says that the Yang-Mills coupling is some positive power of curvature radius divided by the string length scale. So, including the higher spin fields mean we probe weak/weak duality relation instead of the usual strong/weak duality.

Wei Li considered a particular realization of higher spin theory, which is Chern-Simons theory with $sl(N,R) \times sl(N,R)$ gauge group. Of course this is a natural extension of the three dimensional AdS gravity, in the first order formalism. Then she computed the conformal anomaly of the theory of the dual field theory. The quantity computed is strictly speaking an extension of Weyl anomaly, since the Virasoro algebra is extended to W-algebra in this version of higher spin theory. Aiming to provide the extended holography dictionary, she computed anomaly in several different gauge choices in higher spin field theory. The gauge choices include Fefferman-Graham, PBH transformation etc.

Shu Lin's presentation was about time dependent process. The thermalization mechanism is an important subject in relativistic heavy ion collision. In AdS/CFT, thermalization is dual to black hole formation. But solving the full-fledged time-dependent process in gravity is very hard, so Shu Lin used a special metric ansatz called Vaidya metric.

$$ds^2 = \frac{-f dv^2 - 2dvdz + dx^2}{z^2} \tag{2}$$

This is a special choice and solving the time-evolution in the light-cone time is manageable. Shu Lin computed the correlation functions holographically, using analytically obtained solutions for the metric.

Shu Lin argued that previous works had ignored existence initial long-range correlation. But in the work reported, variou correlation functions are explicitly given and the enhanced correlation is quantified. It was also checked that there is no log correction of the thermalization time.

Yun-Long Zhang talked about Rindler fluid, work done with R-G Cai and Q Yang. It is well known that the black hole horizon behaves like viscous fluid. And this is the motivation for fluid/gravity duality. He imposed the so-called Petrov I condition and derived the Navier-Stokes equation from the Einstein's equation for the Rindler horizon. Then he performed derivative expansion and extracted the transport coefficients and checked that the results are the same as the fluid/gravity results.

3.3. *Foundation of holography*

Then there are two talks which I would categorize as pertinent to the foundational problems of holographic principle.

Bogeun Gwak talked about cosmic censorship in anti-de Sitter black hole background. AdS black holes are dual to conformal field theory at finite temperature, with possibly a few additional charges such as angular momentum, particle density, magnetic field etc. Gwak studied cosmic censorship in particular for rotation black holes in AdS. The tool employed is particle's geodesic motion and it was verified that the results are consistent with black hole thermodynamics.

Chanyong Park's talk was about entanglement thermodynamics. Entanglement entropy has been the object of intense study in AdS/CFT circle recently. It is von

Neumann entropy of *reduced density matrix* of a subsystem. It is a non-local object and can be used as an order parameter. It is quite difficult to compute in general quantum field theory, mainly due to non-locality. But Ryu and Takayanagi provided a simple geometrical prescription in AdS/CFT, as the minimal area of codimension-2 surface which is homologous to the boundary region in the field theory spacetime we are interested in. Since it is entropy, this selection of a subsystem naturally gives rise to a notion of emergent thermodynamics, with the first law $S_{ent} = T_{ent}E$. Park studied the entanglement thermodynamics of charged AdS black hole with scalar hair, and verified T_{ent} is inversely proportional to the size of entangling region.

3.4. *AdS/CFT geometry and quantum information*

Nakwoo Kim talked about bulk local operators in pp-wave holography. It is motivated by the surface/state (SS) correspondence by Miyaji and Takayanagi, which is in turn motivated by AdS/MERA correspondence which was first pointed out by B. Swingle. MERA, or multiscale entanglement renormalization ansatz, was proposed by G. Vidal in 2005 and it is an efficient numerical technique for critical systems. The unique property of MERA is that it uses a tensor called disentangler, which removes short-distance entanglement, in addition to the usual coarse-graining process. Then Swingle pointed out that MERA network is naturally a discretized version of AdS space. More recently, Miyaji and Takayanagi came up with surface/state correspondence that a density matrix can be assigned to arbitrary surface, open or closed, located within bulk not nessarily anchored on the boundary.

Kim talked about his work done with Hyeonjoon Shin. They applied SS duality to pp-wave holography where the holographic screen is obscure and SS duality might be useful in particular. It was argued that D-instantons are natural candidates for bulk local operators. It was shown that D-instanton overlap functions are consistent with pp-wave metric, as they should be.

Feng-Li Lin's talk was about entanglement renormalization and integral geometry. The work was done with X Huang. There is notion of integral geometry, which is the geometry of a space where a point corresponds to a geodesic. They studied the renormalization process in the AdS/MERA, which corresponds to a small change of end points of the geodesic curve. Then the change was identified with MERA tensors, isometry (coarse-grainer) and disentangler.

Dong-Hoon Kim talked about the similarity of gravity physics and quantum entanglement. MERA, which is a version of tensor network, is understood also as quantum circuit. Given a probability distribution, Fisher-Rao metric gives a Riemannian metric.

$$g_{\mu\nu}(\Theta) = \int dX P(X|\Theta)\partial_\mu \ln P(X|\Theta)\partial_\nu P(X|\Theta) \tag{3}$$

He calculated Fisher metric from the 2-particle scattering wavefunction in one dimension, and made an observation that it is exactly three-dimensional hyperbolic space. He also discussed how to derive area law for entropy etc.

4. Comments and Future Prospects

At the 2nd LeCosPA meeting 2015, we heard about a number of interesting recent developments on AdS/CFT correspondence. Of course what is the most exciting part for gravity should be the implication of AdS/CFT on the true nature of strongly coupled general relativity. Theorists with adventurous minds will keep pushing the frontier in following years.

References

1. N. Kim and H. Shin, Surface/state correspondence and bulk local operators in pp-wave holography, *Phys. Lett.* **B751**, 508 (2015).
2. W. Li and S. Theisen, Some aspects of holographic W-gravity, *JHEP* **08**, p. 035 (2015).
3. B. Gwak and B.-H. Lee, Cosmic Censorship of Rotating Anti-de Sitter Black Hole with a Probe (2015).
4. H. Hoshino and S. Nakamura, Effective temperature of nonequilibrium dense matter in holography, *Phys. Rev.* **D91**, p. 026009 (2015).
5. S. Lin, Holographic thermalization with initial long range correlation, *Phys. Rev.* **D93**, p. 026007 (2016).
6. R.-G. Cai, Q. Yang and Y.-L. Zhang, Petrov type I Condition and Rindler Fluid in Vacuum Einstein-Gauss-Bonnet Gravity, *JHEP* **12**, p. 147 (2014).
7. C. Park, Thermodynamic law from the entanglement entropy bound (2015).
8. X.-H. Ge, S.-J. Sin and S.-F. Wu, Lower Bound of Electrical Conductivity from Holography (2015).
9. X. Huang and F.-L. Lin, Entanglement renormalization and integral geometry, *JHEP* **12**, p. 081 (2015).
10. J. M. Maldacena, The Large N limit of superconformal field theories and supergravity, *Int.J.Theor.Phys.* **38**, 1113 (1999).
11. N. Drukker, M. Marino and P. Putrov, From weak to strong coupling in ABJM theory, *Commun. Math. Phys.* **306**, 511 (2011).
12. D. Gang, N. Kim and S. Lee, Holography of Wrapped M5-branes and Chern-Simons theory, *Phys. Lett.* **B733**, 316 (2014).
13. D. Gang, N. Kim and S. Lee, Holography of 3d-3d correspondence at Large N, *JHEP* **04**, p. 091 (2015).
14. T. Sakai and S. Sugimoto, Low energy hadron physics in holographic QCD, *Prog. Theor. Phys.* **113**, 843 (2005).
15. S. A. Hartnoll, C. P. Herzog and G. T. Horowitz, Building a Holographic Superconductor, *Phys. Rev. Lett.* **101**, p. 031601 (2008).

Bulk Local Operators and D-Instantons in pp-Wave Holography

Nakwoo Kim

Physics department, Kyung Hee University
Seoul, 02447, Republic of Korea
nkim@khu.ac.kr

We apply the surface/state correspondence proposal of Miyaji *et al.* to IIB string theory on pp-waves. We propose that the bulk local operators can be realized by D-instantons. In line with ordinary AdS/CFT correspondence, the bulk local operators in pp-waves also create a hole, or a boundary, in the dual gauge theory as pointed out by H. Verlinde, and by Y. Nakayama and H. Ooguri.

Keywords: Holography; surface/state correspondence.

1. Introduction

Recently there appeared an interesting proposal called surface/state correspondence[1–3] which aims to generalize the holography principle.[4] It is partly motivated by the observation that a particular tensor network prescription, called MERA (Multiscale Entanglement Renormalization Ansatz),[5] apparently realizes the emergence of holographic spacetime through a real space renormalization.[6] The idea of surface/state correspondence is based on the holographic entanglement entropy conjecture,[7] and extends it by assigning a density matrix not only to a spacelike minimal surface ending on the boundary but also to any surface in the bulk. Moving and deforming the surface in the bulk should correspond to performing the coarse-graining procedure: dubbed as "disentangler" and "isometry" in MERA literature. If possible, this generalized holography relation can be also applied to bulk geometrics without a boundary, including the Minkowski and de Sitter spacetime. In particular, when the surface shrinks to a point the dual state should be bulk local operators which have been studied by *e.g.* Hamilton et al.[8]

In this report we consider the maximally supersymmetric IIB pp-wave,[9] and its gauge/gravity correspondence by Berenstein, Maldacena and Nastase (BMN).[10] The advantage is obviously twofold. The string spectrum is exactly solvable in the lightcone gauge, and the proposed mapping to large R-charge operators has been extensively studied and confirmed. On the other hand, just like the flat background the pp-wave does not have a clear boundary and the notion of holography has remained rather ambiguous.[11–13] Thus we reckon the pp-wave/BMN relation provides an excellent setup to test the surface/state correspondence in more detail and also in the full string theory, potentially including even the string interaction through string field theory.[14]

Our main point is that instantonic D-branes are the "surfaces" which are mapped to gauge theory "states" through the BMN relation to gauge operators. Especially for the bulk local operators we use D-instantons. D-instantons probing the bulk geometry is a very well known idea in AdS/CFT: they are dual to Yang-Mills instantons, of which the moduli space contains AdS_5 precisely.[15] As usual, it is expected that the supersymmetry and modular invariance of 1/2-BPS D-instantons will protect them against perturbative corrections and string interactions.

We take a modest first step in this work. Following mainly the discussions in Refs. 3,16,17, we first discuss how the pp-wave symmetry algebra should constrain the candidate bulk local operators, and also illustrate how to extract the information of the pp-wave metric from D-instantons. Along the way we also argue that the D-instantons as bulk local operators are "boundary creating" operators on the boundary, just as proposed in Refs. 16,17. We conclude with discussions on possible future avenues.

Disclaimer: This proceeding article is based on the work of this author with H. Shin[18] and should not be considered as an original work.

2. IIB pp-wave and BMN correspondence

In this section we give a brief review of IIB pp-wave geometry[9] and its gauge/gravity duality relation.[10] The metric of IIB maximally supersymmetric pp-wave is given as $(I = 1, 2, \cdots, 8)$

$$ds^2 = 2dx^+dx^- - \mu^2(x^I)^2(dx^+)^2 + dx^I dx^I \,. \tag{1}$$

Note that the apparent $SO(8)$ symmetry of the metric is broken into $SO(4)\times SO(4)$ due to the nonvanishing RR 5-form $F_{+1234} = F_{+5678} = 2\mu$. This background is related to $AdS_5 \times S^5$ through the Penrose limit, where one takes a null geodesic and blows up the spacetime around it. Morally speaking the first $\mathbb{R}^4$ is understood as Euclidean spacetime, and the second $\mathbb{R}^4$ is the (infinitely magnified) four-dimensional space around a great circle in S^5.

The total bosonic symmetry group is 30 dimensional: there are kinematical generators $P^+, P^I, J^{+I}, M^{ij}, M^{i'j'}$ and a dynamical generator P^- (here and below $i, j = 1, 2, 3, 4$ and $i', j' = 5, 6, 7, 8$). $M^{ij}, M^{i'j'}$ are the usual rotation generators of $SO(4) \times SO(4)$. The relevant commutation relations are

$$\begin{aligned} &[P^-, P^I] = -i\mu^2 J^{+I}, \quad [P^-, J^{+I}] = iP^I, \\ &[P^I, J^{+J}] = -i\delta^{IJ} P^+ . \end{aligned} \tag{2}$$

P^+ is a central element and commutes with all other generators. It is easy to see that P^-, P^I, J^{+I} comprise Heisenberg algebras when we define $a^I = (\mu J^{+I} + iP^I)/\sqrt{2\mu P^+}$ and $a^{I\dagger} = (\mu J^{+I} - iP^I)/\sqrt{2\mu P^+}$. In that case $-P^-/\mu$ becomes the number operator, since $[P^-, a^I] = \mu a^I$. The isometry generators can be of course more explicitly written in terms of string modes in lightcone gauge which will be explained below.

The proposal of Ref. 10 by Berenstein, Maldacena and Nastase (BMN) relates a certain class of long operators in $\mathcal{N} = 4$ supersymmetric Yang-Mills (SYM) theory to the string spectrum in pp-wave background. The relevant operators, which we will call BMN operators, carry a large $U(1)_R$ charge J, so the conformal dimension Δ is also large. The BMN correspondence equates $-P^-/\mu = \Delta - J$ as a duality relation.

The Green-Schwarz superstring theory of pp-waves in lightcone gauge turns out to be a free theory of 8 massive scalars and 8 massive fermions.[19] The quantum spectrum starts with a unique vacuum state $|0;p^+\rangle$, where p^+ is lightcone momentum and a free parameter. There is a tower of string modes: first the zero modes α_0^I, θ_0 and their conjugates. For $n \geq 1$, $\alpha_n^I, \tilde{\alpha}_n^I$ are the right and the left moving bosonic modes in $SO(8)$ vector representation, while $\theta_n, \tilde{\theta}_n$ are the right and the left moving fermionic oscillators in spinor representation of $SO(8)$. The string vacuum $|0;p^+\rangle$ is the Fock vacuum with lightcone energy $P^- = 0$. Raising the level of a bosonic or a fermionic zero mode by one will increase $\Delta - J$ by one precisely. On the other hand, adding the occupation number of the n-th mode by one increases $\Delta - J$ by $\sqrt{1 + n^2/(\alpha' p^+ \mu)^2}$. The usual level matching condition also applies, and the total occupation number of left- and right-moving modes should be the same for physical states.

The string spectrum beautifully matches that of the large R-charge operators in $\mathcal{N} = 4$ SYM. The vacuum state of the string theory $|0;p^+\rangle$ corresponds to $\mathrm{Tr}(Z^J)$, where Z is one of the complex scalar fields in $\mathcal{N} = 4$ SYM with $\Delta = J = 1$. These are the only kind of $\mathcal{N} = 4$ operators which satisfy $\Delta = J$ classically and also in the interacting theory. The eight bosonic fields on the worldsheet correspond to: 4 remaining scalar fields ϕ^i $(i = 1, \cdots, 4)$ of SYM, and covariant derivatives D_μ along 4 Euclidean spacetime directions. And the eight fermionic directions come from the half of the set of fermion fields in $\mathcal{N} = 4$ SYM with $\Delta = 3/2, J = 1/2$. The BMN excited states are long operators where the aforementioned "impurity" are inserted in the "string" of Z fields. Excitation by zero modes are constructed as totally symmetric combinations: for instance (up to a normalization constant),

$$a_0^{i\dagger} a_0^{j'\dagger} |0;p^+\rangle \longleftrightarrow \sum_l \mathrm{Tr}(Z^l (D_i Z) Z^{J-l} \phi^{j'}) \tag{3}$$

and in an analogous way with fermions. These totally symmetrized combinations are also 1/2-BPS just as $\mathrm{Tr}Z^J$, so the conformal dimensions are protected. It agrees with the string side computation that $\Delta - J$ is exactly integral. For non-zero modes, the BMN proposal is that one should now consider impurities with "momentum". It goes like for instance

$$\begin{aligned} &a_n^{i'\dagger} a_n^{j'\dagger} |0;p^+\rangle \\ &\longleftrightarrow \sum_l e^{2\pi i n l/J} \mathrm{Tr}(Z^l \phi^{i'} Z^k \phi^{j'} Z^{J-k}) . \end{aligned} \tag{4}$$

Then the string spectrum predicts that the SYM operator should receive quantum corrections for conformal dimension so that $\Delta - J = 2\sqrt{1 + n^2/(\alpha' p^+ \mu)^2}$. This is justified from the explicit perturbative computations in SYM, upon the duality identification[10]

$$\frac{1}{(\alpha' p^+ \mu)^2} = \frac{g_{YM}^2 N}{J^2} . \tag{5}$$

This proposal was extensively analyzed and confirmed to higher orders in perturbation theory, for a review see e.g. Ref. 20.

3. Bulk local operators as D-instantons

Let us now consider how the isometry algebra should restrict the putative BMN operator assignment in the bulk of IIB pp-wave. The maximally supersymmetric pp-wave is homogeneous, so we may start with any convenient point: let us choose $x^+ = x^- = x^I = 0$. Obviously the little group is then generated by $J^{+I}, M^{ij}, M^{i'j'}$. The candidate string states are easily constructed. For the Fock vacuum $|0\rangle$ (in string theory realization of course this is the unique string ground state $|0; p^+\rangle$) satisfying $a^I|0\rangle = 0$, the "position" eigenstate $J^{+I}|B\rangle = 0$ is given by

$$|B\rangle = e^{\frac{1}{2}\sum_I (a^{I\dagger})^2} |0\rangle . \tag{6}$$

In the string theory realization of course the symmetry requirement alone does not lead to a unique string state. There are fermionic modes as well, but more importantly the kinematical generators only involve the zero-modes and we may consider excitation by higher string modes. This is akin to the conformal field theory consideration in Refs. 3, 16, 17 , that in general the symmetry requirement for boundary states is satisfied by a linear combination of Ishibashi states. It is a natural guess that in string theory the bulk local state $|B\rangle$ should be D-instantons constructed as coherent states in closed string theory. Indeed, IIB pp-waves do allow 1/2-BPS D-instantons[21] and we propose them to be the bulk local operators. The idea is that we have bulk local operator expressions in closed string theory, which then can be mapped SYM operators through BMN proposal.

For the standard AdS/CFT correspondence, it was argued that the local bulk operators are mapped to *boundary creation* operators through a unitary transformation.[3,16,17] It was in particular emphasized in Ref. 17 that the mapping is mediated through dilatation (i.e. Hamiltonian in the radial quantization) by the imaginary unit. The spherical boundaries of the holes created on the CFT side are put at a constant time slice.

For the holography of pp-waves, we recall that the light-cone coordinate x^+ becomes time both on the string worldsheet and presumably also for the dual BMN sector. One way to see it is to rewrite the metric as follows.[13]

$$\begin{aligned} ds^2 = {} & 2dx^+dx^- + dr_1^2 + r_1^2(-\mu^2(dx^+)^2 + d\Omega_{3(1)}^2) \\ & + dr_2^2 + r_2^2(-\mu^2(dx^+)^2 + d\Omega_{3(2)}^2) . \end{aligned} \tag{7}$$

It is implied here that the radial coordinates of $\mathbb{R}^4 \times \mathbb{R}^4$ are the holographic coordinates, and the boundary includes two copies of $\mathbb{R} \times S^3$. We note that in the surface/state correspondence[1,3] the holes are codimension-two spacelike hypersurfaces in the bulk. From the rearrangement in (7) we infer that it is a natural guess that the hole is at $x^+ = 0$, extended along x^-, and at $r = \infty$. This subspace is invariant under the same symmetry algebra of the point of our interest: J^{+I} and $SO(4) \times SO(4)$ rotations. It is amusing to see that, in general,

$$|\text{A hole at } x^+ = x_0^+\rangle = e^{ix_0^+ P^-} |B\rangle \,. \tag{8}$$

This is the central point of this note: namely, the bulk local operator and the spherical boundary are again mapped through the time evolution in the boundary theory in pp-wave holography.

4. Discussion

We have studied the IIB pp-wave D-instantons in the light of the surface/state correspondence.[2] Admittedly we only addressed the kinematics and zero-mode sector of the string theory so far, and it is desirable to probe genuinely stringy aspects. It will be also very nice if one can find MERA-type unitary transformations between the various instantonic D-brane constructions in Ref. 21.

Probably the biggest difference between the holographic entangling surface and D-branes is that the latter is dynamical. In other words they not only represent a density matrix, but can also induce a change in vacuum density matrix. It will be very interesting if we can relate the gravitational backreaction of D-instantons using the recent proposals on holographic Fisher information metric.[22,23]

The study of M-theory pp-waves and their relation to a massive matrix theory[10] is also an interesting subject. The kinematical consideration should straightforwardly carry through, in terms of the abelian part of the BMN matrix model. On the other hand, there is no systematic way of constructing M-branes in the BMN matrix level at quantum level. And according to the supergravity probe analysis in Ref. 24 there are no supersymmetric instantons in $D = 11$ pp-waves, so we have a more challenging problem.

Acknowledgments

We thank Hyeonjoon Shin for enjoyable collaboration. We are partly supported by APCTP, and NRF grants funded by the Korea government with grant No. 2015R1D1A1A09059301.

References

1. M. Miyaji, S. Ryu, T. Takayanagi and X. Wen, Boundary States as Holographic Duals of Trivial Spacetimes, *JHEP* **1505**, p. 152 (2015).
2. M. Miyaji and T. Takayanagi, Surface/State Correspondence as a Generalized Holography, *PTEP* **2015**, p. 073B03 (2015).
3. M. Miyaji, T. Numasawa, N. Shiba, T. Takayanagi and K. Watanabe, cMERA as Surface/State Correspondence in AdS/CFT (2015).

4. J. M. Maldacena, The Large N limit of superconformal field theories and supergravity, *Int.J.Theor.Phys.* **38**, 1113 (1999).
5. G. Vidal, Entanglement Renormalization, *Phys. Rev. Lett.* **99**, p. 220405 (2007).
6. B. Swingle, Entanglement Renormalization and Holography, *Phys. Rev.* **D86**, p. 065007 (2012).
7. S. Ryu and T. Takayanagi, Holographic derivation of entanglement entropy from AdS/CFT, *Phys. Rev. Lett.* **96**, p. 181602 (2006).
8. A. Hamilton, D. N. Kabat, G. Lifschytz and D. A. Lowe, Local bulk operators in AdS/CFT: A Holographic description of the black hole interior, *Phys.Rev.* **D75**, p. 106001 (2007).
9. M. Blau, J. M. Figueroa-O'Farrill, C. Hull and G. Papadopoulos, A New maximally supersymmetric background of IIB superstring theory, *JHEP* **01**, p. 047 (2002).
10. D. E. Berenstein, J. M. Maldacena and H. S. Nastase, Strings in flat space and pp waves from N=4 superYang-Mills, *JHEP* **0204**, p. 013 (2002).
11. S. R. Das, C. Gomez and S.-J. Rey, Penrose limit, spontaneous symmetry breaking and holography in PP wave background, *Phys.Rev.* **D66**, p. 046002 (2002).
12. E. Kiritsis and B. Pioline, Strings in homogeneous gravitational waves and null holography, *JHEP* **0208**, p. 048 (2002).
13. R. G. Leigh, K. Okuyama and M. Rozali, P P waves and holography, *Phys.Rev.* **D66**, p. 046004 (2002).
14. D. Berenstein and H. Nastase, On light cone string field theory from superYang-Mills and holography (2002).
15. M. Bianchi, M. B. Green, S. Kovacs and G. Rossi, Instantons in supersymmetric Yang-Mills and D instantons in IIB superstring theory, *JHEP* **08**, p. 013 (1998).
16. H. Verlinde, Poking Holes in AdS/CFT: Bulk Fields from Boundary States (2015).
17. Y. Nakayama and H. Ooguri, Bulk Locality and Boundary Creating Operators (2015).
18. N. Kim and H. Shin, Surface/state correspondence and bulk local operators in pp-wave holography, *Phys. Lett.* **B751**, 508 (2015).
19. R. R. Metsaev and A. A. Tseytlin, Exactly solvable model of superstring in Ramond-Ramond plane wave background, *Phys. Rev.* **D65**, p. 126004 (2002).
20. D. Sadri and M. M. Sheikh-Jabbari, The Plane wave / superYang-Mills duality, *Rev.Mod.Phys.* **76**, p. 853 (2004).
21. M. R. Gaberdiel and M. B. Green, The D instanton and other supersymmetric D branes in 2B plane wave string theory, *Annals Phys.* **307**, 147 (2003).
22. M. Miyaji, T. Numasawa, N. Shiba, T. Takayanagi and K. Watanabe, Gravity Dual of Quantum Information Metric (2015).
23. N. Lashkari and M. Van Raamsdonk, Canonical Energy is Quantum Fisher Information (2015).
24. N. Kim and J.-T. Yee, Supersymmetry and branes in M theory plane waves, *Phys. Rev.* **D67**, p. 046004 (2003).

Testing the Validity of Cosmic Censorship in the Rotating Anti-de Sitter Black Hole

Bogeun Gwak
Center for Quantum Spacetime, Sogang University, Seoul 04107, Korea
rasenis@sogang.ac.kr

Bum-Hoon Lee
Center for Quantum Spacetime, Sogang University, Seoul 04107, Korea
Department of Physics, Sogang University, Seoul 04107, Korea
Asia Pacific Center for Theoretical Physics, Pohang 37673, Korea
Department of Physics, Postech, Pohang 37673, Korea
bhl@sogang.ac.kr

We review our investigation of the cosmic censorship of a rotating black hole by particle absorption. The energy and angular momentum of particles are assumed to change the conserved quantities of a black hole. When particles are added, the mass of the extremal black hole increases more than its angular momentum, so that the angular momentum of the black hole remains within its extremal bound. Therefore, the cosmic censorship of the black hole is valid.

Keywords: Black hole; cosmic censorship; thermodynamics.

1. Introduction

In general relativity, black holes have a singularity inside their event horizon. This singularity cannot be observed through cosmic censorship. Rotating or charged black holes have a bound for this conserved quantity in which black holes have horizons. Beyond these bounds, the singularity appears without the horizon. In this case, the cosmic censorship is invalid. The possibility of a particle overspinning or overcharging a black hole has been studied to test the cosmic censorship in asymptotically flat[1–14] and anti-de Sitter (AdS) spacetime.[15–21]

The temperature of a black hole can be defined from the quantum radiation called Hawking radiation.[22,23] With the removal of rotational energy, natural processes, such as particle absorption, the Kerr black hole mass increase.[24–26] This mass has a relationship with Bekenstein-Hawking entropy.[27,28] The thermodynamics of a black hole can be defined using its temperature and entropy.

The gravity theory defined in bulk AdS spacetime has duality with the conformal field theory (CFT) defined on the spacetime boundary, which is called AdS/CFT correspondence.[29–32] Thermodynamic properties, such as the temperature of the black hole in the bulk, appear in the dual CFT on the boundary.[33] The phases of the CFT can be interpreted by the stabilities of the black holes,[34–36] and the configurations of the strings have a physical interpretation in the dual CFT as a giant magnon and spike.[37,38] Under AdS/CFT duality, the microscopic origin of the black hole entropy is explained as the asymptotic growth of states described by the CFT[18, 39–43] in the (2+1)-dimensional Bañados-Teitelboim-Zanelli (BTZ) black hole.[44, 45] The BTZ black hole entropy is equivalent to the Cardy formula using diffeomorphism operators.[46–48]

In this proceeding, we will briefly review cosmic censorship in particle absorption. Changes in the black hole after it absorbs the particle will be described.

2. Review of Testing Cosmic Censorship in Particle Absorption

In this section, we will briefly review about the cosmic censorship in the particle absorption in Refs. 13, 18, 20, 21. Having the mass parameter M and spin parameter a, the rotating black hole in the asymptotically AdS spacetime is

$$ds^2 = -\frac{\Delta_r}{\rho^2}\left(dt - \frac{a\sin^2\theta}{\Xi}d\phi\right)^2 + \frac{\rho^2}{\Delta_r}dr^2 + \frac{\Delta_\theta \sin^2\theta}{\rho^2}\left(a\,dt - \frac{r^2+a^2}{\Xi}d\phi\right)^2 ,$$
$$+\frac{\rho^2}{\Delta_\theta}d\theta^2 , \quad \Delta_r = (r^2+a^2)(1+\frac{r^2}{\ell^2}) - 2Mr , \tag{1}$$
$$\rho^2 = r^2 + a^2\cos^2\theta , \quad \Delta_\theta = 1 - \frac{a^2}{\ell^2}\cos^2\theta , \quad \Xi = 1 - \frac{a^2}{\ell^2} ,$$

where the AdS radius ℓ. The function Ξ should be a positive value. The black hole mass M_B and angular momentum J_B are defined in non-rotating frame,

$$M_B = \frac{M}{\Xi^2} , \quad J_B = \frac{Ma}{\Xi^2} . \tag{2}$$

When a black hole absorbs a particle, the conserved quantities of the particle move to those of the black hole. To describe the relationship between particle and black hole momenta, the equations for particle motion are obtained from the first-order equations of motion. The Hamilton-Jacobi method with separability[49, 50] is appropriate for finding the equations of motion. The Hamiltonian and Hamilton-Jacobi action are

$$\mathcal{H} = \frac{1}{2}g^{\mu\nu}p_\mu p_\nu , \quad S = \frac{1}{2}m^2\lambda - Et + L\phi + S_r(r) + S_\theta(\theta) , \tag{3}$$

where the metric and particle momenta are $g_{\mu\nu}$ and p_μ. This describes the particle of mass m, energy E, and angular momentum L. The function $S_r(r)$ and $S_\theta(\theta)$ can be obtained using separate variable $\mathcal{K}$,

$$S_r(r) = \sqrt{\frac{\mathcal{K}}{\Delta_r} + \frac{1}{\Delta_r^2}\left(a\Xi L - (r^2+a^2)E\right)^2 - \frac{m^2r^2}{\Delta_r}} , \tag{4}$$
$$S_\theta(\theta) = \sqrt{-\frac{\mathcal{K}}{\Delta_\theta} - \frac{1}{\Delta_\theta^2}\left(\Xi L\csc\theta - aE\sin\theta\right)^2 - \frac{m^2a^2\cos^2\theta}{\Delta_\theta}} .$$

By eliminating the separate variables in the equations of motion, we can rewrite the particle energy equation in terms of the momenta for a given location. The particle will be assumed to be absorbed into the black hole when it passes through the horizon of the black hole, so we will use the particle energy equation at the horizon r_h. After the black hole absorbs the particle, the conserved quantities of the black hole will be infinitesimally changed by those of the particle. The particle energy and angular momentum change the black hole mass and angular momentum. To match the conserved quantities of the particle and the black hole, we redefine the particle energy and angular momentum in a non-rotating frame.[20,51,52] In a non-rotating frame, the conserved quantities of the black hole and particle are related

$$E = dM_B \,, \quad L = dJ_B \,. \tag{5}$$

The change in the black hole mass is obtained in terms of the change in the black hole angular momentum and particle radial momentum p^r ,

$$E = \frac{a\left(1+\frac{r_h^2}{\ell^2}\right)}{r_h^2+a^2} L + \frac{\rho_h^2}{r_h^2+a^2}|p^r| \,. \tag{6}$$

Now, using Eq. (6), we will test the validity of cosmic censorship in the extremal black hole. The location of the extremal black hole is the minimum point of the function $\Delta_r(r)$, so

$$\Delta_r\Big|_{r=r_h} = 0 \,, \quad \frac{\partial \Delta_r}{\partial r}\Big|_{r=r_h} = 0 \,. \tag{7}$$

After the particle absorption, the mass and angular momentum of the black hole infinitesimally change as much as those of the particle. The shape of the function $\Delta_r(r)$ is also changed under the effects. The value of the changed minimum point $r_h + \delta r_e$ always becomes a negative value, so

$$\Delta_r(r_h+\delta r_e) = \frac{\partial \Delta_r}{\partial M_B}\Big|_{r=r_h} \delta M_B + \frac{\partial \Delta_r}{\partial J_B}\Big|_{r=r_h} \delta J_B + \frac{\partial \Delta_r}{\partial r}\Big|_{r=r_h} \delta r_e < 0 \,. \tag{8}$$

Therefore, the outer horizon still exists, and the cosmic censorship is valid for the extremal black hole. This also implies that the black hole has two horizons, an outer one and an inner one. The existence of two horizons can be described from the analysis of the function $\Delta_r(r)$. The function $\Delta_r(r)$ has only one extremum, which is the minimum. The second derivative of function $\Delta_r(r)$ with respect to r is always a positive function, such as

$$\frac{\partial^2 \Delta_r}{\partial r^2} = 2 + \frac{2a^2}{\ell^2} + \frac{12r^2}{\ell^2} > 0 \,, \tag{9}$$

which means that the first derivative of function $\Delta_r(r)$ with respect to r monotonically increases. The first derivative of function $\Delta_r(r)$ is

$$\frac{\partial \Delta_r}{\partial r} = -2M + 2r + \frac{2a^2 r}{\ell^2} + \frac{4r^3}{\ell^2} \,. \tag{10}$$

The first derivative of function $\Delta_r(r)$ has a negative value at $r = 0$ and becomes positive infinity at $r \to +\infty$, so there is only one solution in $0 < r < \infty$. The solution must be the minimum, because the second derivative of function $\Delta_r(r)$ is always positive as shown in Eq. (9). The value of the function $\Delta_r(r)$ at $r = 0$ is always positive such as,

$$\Delta_r(r = 0) = a^2 > 0. \tag{11}$$

After particle absorption, the minimum value of the function $\Delta_r(r)$ becomes negative, as shown in Eq. (8), so we can expect that one more solution exists between $r = 0$ and the minimum point. This is the inner horizon. Moreover, the function $\Delta_r(r)$ still has another solution corresponding to the outer horizon between the minimum point and $r \to \infty$. Therefore, two solutions, corresponding to the outer and inner horizons, satisfy $\Delta_r(r) = 0$. After particle absorption, the extremal black hole becomes non-extremal. These can be also shown as Fig. 1.

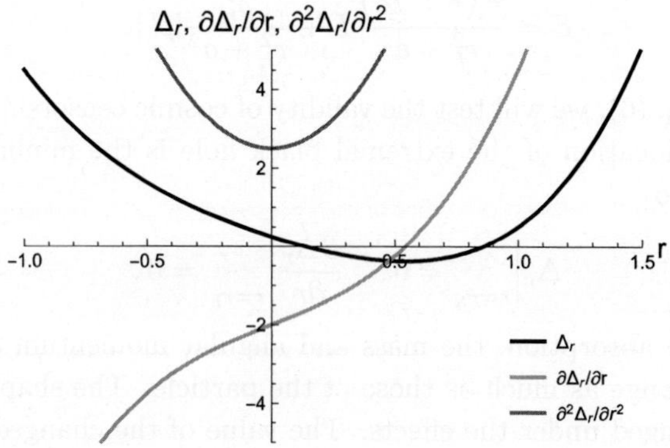

Fig. 1. The graphs for $M = 1$, $a = 0.5$, and $\ell = 1$. The function $\Delta_r(r)$ is black solid line, and the first derivative of $\Delta_r(r)$ is red solid line, and the second derivative of $\Delta_r(r)$ is blue solid line. The function $\Delta_r(r)$ has only one extremum corresponding to the minimum and gives two horizons for this negative minimum.

Therefore, the extremal black hole mass increases more than its angular momentum under the particle absorption process. Due to this effect, the extremal black hole becomes non-extremal, so the cosmic censorship of the extremal black hole is valid.

3. Summary

We reviewed the cosmic censorship in AdS black holes in Refs. 13, 20, 21, 18. The Hamilton-Jacobi equations are solved using separate variable method, and then the particle energy equation at the horizon of the black hole relates the particle energy to its conserved quantities. We assumed that the particle is absorbed when it touches the horizon of the black hole. The particle infinitesimally changes the black hole mass and angular momentum under the satisfaction of the particle energy equation. Using this effect, we test the validity of cosmic censorship when the conserved quantities of the black hole satisfy the extremal condition. After particle absorption, the minimum value of the function $\Delta_r(r)$ becomes a negative value, so the mass of the black hole increases more than its angular momentum by a particle conserved quantity. The extremality of the black hole is broken, and then the black hole has two horizons. Therefore, the horizon of the black hole still exists, and the cosmic censorship of the extremal black hole is valid under particle absorption.

Acknowledgments

This research was supported by Basic Science Research Program through the National Research Foundation of Korea(NRF) funded by the Ministry of Science, ICT & Future Planning(NRF-2015R1C1A1A02037523) and the National Research Foundation of Korea(NRF) grant funded by the Korea government(MSIP)(No. 2014R1A2A1A010).

References

1. R. M. Wald, *Annals Phys.* **82**, 548 (1974).
2. T. Jacobson and T. P. Sotiriou, *Phys. Rev. Lett.* **103**, 141101 (2009).
3. E. Barausse, V. Cardoso and G. Khanna, *Phys. Rev. Lett.* **105**, 261102 (2010).
4. E. Barausse, V. Cardoso and G. Khanna, *Phys. Rev. D* **84**, 104006 (2011).
5. M. Colleoni and L. Barack, *Phys. Rev. D* **91**, 104024 (2015).
6. M. Colleoni, L. Barack, A. G. Shah and M. van de Meent, *Phys. Rev. D* ***92**, no. 8,* 084044 (2015).
7. V. E. Hubeny, *Phys. Rev. D* **59**, 064013 (1999).
8. S. Isoyama, N. Sago and T. Tanaka, *Phys. Rev. D* **84**, 124024 (2011).
9. R. C. Myers and M. J. Perry, *Annals Phys.* **172**, 304 (1986).
10. M. Bouhmadi-Lopez, V. Cardoso, A. Nerozzi and J. V. Rocha, *Phys. Rev. D* **81**, 084051 (2010).
11. J. Doukas, *Phys. Rev. D* **84**, 064046 (2011).
12. A. Saa and R. Santarelli, *Phys. Rev. D* **84**, 027501 (2011);
13. B. Gwak and B. H. Lee, *Phys. Rev. D* **84**, 084049 (2011).
14. S. Gao and Y. Zhang, *Phys. Rev. D* **87**, no. 4, 044028 (2013).
15. J. V. Rocha and V. Cardoso, *Phys. Rev. D* **83**, 104037 (2011).
16. J. V. Rocha, R. Santarelli and T. Delsate, *Phys. Rev. D* **89**, no. 10, 104006 (2014).
17. J. V. Rocha and R. Santarelli, *Phys. Rev. D* **89**, no. 6, 064065 (2014).
18. B. Gwak and B. H. Lee, *Class. Quant. Grav.* **29**, 175011 (2012).
19. B. McInnes and Y. C. Ong, JCAP **1511**, no. 11, 004 (2015).
20. B. Gwak and B. H. Lee, accepted in *JCAP*, arXiv:1509.06691 [gr-qc].
21. B. Gwak and B. H. Lee, arXiv:1510.08215 [gr-qc].
22. S. W. Hawking, *Commun. Math. Phys.* **43**, 199 (1975).
23. S. W. Hawking, *Phys. Rev. D* **13**, 191 (1976).
24. D. Christodoulou, *Phys. Rev. Lett.* **25**, 1596 (1970).
25. D. Christodoulou and R. Ruffini, *Phys. Rev. D* **4**, 3552 (1971).
26. L. Smarr, *Phys. Rev. Lett.* **30**, 71 (1973).
27. J. D. Bekenstein, *Phys. Rev. D* **7**, 2333 (1973).
28. J. D. Bekenstein, *Phys. Rev. D* **9**, 3292 (1974).
29. J. M. Maldacena, *Int. J. Theor. Phys.* **38**, 1113 (1999) [*Adv. Theor. Math. Phys.* **2**, 231 (1998)].

30. S. S. Gubser, I. R. Klebanov and A. M. Polyakov, *Phys. Lett. B* **428**, 105 (1998).
31. E. Witten, *Adv. Theor. Math. Phys.* **2**, 253 (1998).
32. O. Aharony, S. S. Gubser, J. M. Maldacena, H. Ooguri and Y. Oz, *Phys. Rept.* **323**, 183 (2000).
33. E. Witten, *Adv. Theor. Math. Phys.* **2**, 505 (1998).
34. B. Gwak and B. H. Lee, *Phys. Rev. D* **91**, no. 6, 064020 (2015).
35. W. K. Ahn, B. Gwak, B. H. Lee and W. Lee, *Eur. Phys. J. C* **75**, no. 8, 372 (2015).
36. B. Gwak, B. H. Lee and D. Ro, arXiv:1509.03493 [gr-qc].
37. B. Gwak, B. H. Lee, K. L. Panigrahi and C. Park, *JHEP* ***0904***, 071 (2009).
38. B. Gwak, B. H. Lee and C. Park, *Phys. Rev. D* **87**, no. 8, 086002 (2013).
39. J. W. York, Jr., *Phys. Rev. D* **28**, 2929 (1983).
40. W. H. Zurek and K. S. Thorne, *Phys. Rev. Lett.* **54**, 2171 (1985).
41. V. P. Frolov and I. Novikov, *Phys. Rev. D* **48**, 4545 (1993).
42. F. Larsen and F. Wilczek, *Phys. Lett. B* **375**, 37 (1996).
43. A. Strominger, *JHEP* **9802**, 009 (1998).
44. M. Banados, C. Teitelboim and J. Zanelli, *Phys. Rev. Lett.* **69**, 1849 (1992).
45. M. Banados, M. Henneaux, C. Teitelboim and J. Zanelli, *Phys. Rev. D* **48**, 1506 (1993) [*Erratum-ibid. D* **88**, no. 6, 069902 (2013)].
46. J. L. Cardy, *Nucl. Phys. B* **270**, 186 (1986).
47. J. D. Brown and M. Henneaux, *Commun. Math. Phys.* **104**, 207 (1986).
48. O. Coussaert and M. Henneaux, *Phys. Rev. Lett.* **72**, 183 (1994).
49. B. Carter, *Commun. Math. Phys.* **10**, 280 (1968).
50. M. Vasudevan and K. A. Stevens, *Phys. Rev. D* **72**, 124008 (2005).
51. M. M. Caldarelli, G. Cognola and D. Klemm, *Class. Quant. Grav.* **17**, 399 (2000).
52. G. W. Gibbons, M. J. Perry and C. N. Pope, *Class. Quant. Grav.* **22**, 1503 (2005).

Application of AdS/CFT Correspondence to Studies on Non-Equilibrium Steady States

S. Nakamura

Department of Physics, Chuo University,
Tokyo, 112-8551, Japan
nakamura@phys.chuo-u.ac.jp

The general relativity is not only a theory of gravity but also a useful tool to analyze non-equilibrium physics by virtue of the AdS/CFT correspondence. In the gravity dual of a non-equilibrium steady state, an analogue black hole on a D-brane provides basic characteristics of the non-equilibrium steady state.

Keywords: AdS/CFT; non-equilibrium steady state.

1. Introduction

The general relativity was born 100 years ago as a theory of gravity. However, the general relativity, in modern point of view, is not only a theory for gravity but also for non perturbative description of gauge theories. This idea came from superstring theory as the AdS/CFT correspondence.[1,2]

The AdS/CFT correspondence is a map between a strongly-interacting gauge theory and a supergravity theory on a curved spacetime. When we take the large-N_c limit of the gauge theory, where N_c is the rank of the gauge group, the supergravity can be treated as a classical theory. The string-theory correction to the supergravity is also negligible when we take the strong-coupling limit of the gauge theory. The AdS/CFT correspondence has been paid much attention as a powerful tool to analyze a quantum gauge theory.

In this talk, I would like to emphasize that the AdS/CFT correspondence is also useful to statistical physics, especially to non-equilibrium physics.[3] In this talk, I will overview my recent work on the applications of the AdS/CFT correspondence to non-equilibrium physics. The readers can consult with Refs. 4–7 for more details.

2. Difficulties in Non-Equilibrium Physics

We are surrounded by many phenomena which are out of equilibrium. Studies on non-equilibrium physics is thus very important, but construction of non-equilibrium statistical physics is still a challenge in modern physics.

Here, "non-equilibrium" means that the principle of detailed balance is broken in the system we study. When the principle of detailed balance is broken, the distribution function is not guaranteed to be a standard Maxwell-Boltzmann distribution (or Bose/Fermi distribution, depending on the statistics at the quantum level). When the distribution function is unknown, expectation values of physical quantities are hardly computable with a conventional method. Is it possible to obtain the expectation values without knowing the distribution function explicitly?

We find one such example in gravitational systems. As we know, black holes have thermodynamics.[8] A black hole has macroscopic variables such as Hawking temperature, entropy and so on. We can compute the free energy of the black hole from the Euclidean on-shell Einstein-Hilbert action. These quantities "merely" come from the classical solution to the Einstein's equation. Somehow, gravity yields the macroscopic quantities "automatically."

The AdS/CFT correspondence relates the gravity and quantum gauge theories. We can prepare a non-equilibrium system by using the gauge particles. Then, we can map the system into a gravitational setup where we can analyze the macroscopic quantities without knowing the distribution function explicitly.

3. Non-Equilibrium Steady States in Gravity Dual

Non-equilibrium states can be classified into two categories: time-dependent states and steady states. The latter is referred to non-equilibrium steady state (NESS). Here, steady state means that the macroscopic variables do not depend on time. We consider NESS in this talk.

A representative of NESS is a heater with a constant flow of current along a constant electric field, located in a large volume of air. In the heater, the electric power is converted into the heat, which means that the system is out of equilibrium. However, the heat accumulated in the heater can be constant since the surplus heat is flowing out to the air, a heat bath. When the energy coming into the heater from the power supply and the dissipation going into the heat bath are in balance, the heater realizes NESS.

Let us construct NESS in terms of an $SU(N_c)$ gauge theory. Suppose that we have quarks and gluons. We apply an external electric field acting only on the global quark charge. The gluons are neutral with respect to the quark charge. Then quarks start moving, but their velocity reaches a constant owing to the friction between the quark and the gas of the gluons. Here, we assume that the heat capacity of the gluon sector is large enough comparing to that of the quark sector. This can be realized by taking the large-N_c limit, since the degrees of freedom of quarks and gluons are proportional to N_c and $N_c^2 - 1$, respectively. The degrees of freedom of gluons is N_c times that of the quarks and is sufficiently large at the large-N_c limit. The gluon sector is a good heat bath for the quarks at this limit.

In this talk, we employ a $(3+1)$-dimensional supersymmetric QCD as a system of quarks and gluons. The gluon sector of the supersymmetric QCD is $SU(N_c)$ $\mathcal{N} = 4$ supersymmetric Yang-Mills (SYM) theory at the large-N_c limit. We also take the large-λ limit where λ is the 't Hooft coupling of the SYM theory. The quark sector is the hypermultiplet of $\mathcal{N} = 2$ supersymmetric gauge theory.

The gravity dual of this system is known as the D3-D7 system.[9] The gluon sector is mapped to a 5-dimensional Schwarzschild black hole in an asymptotically

AdS spacetime. The metric of the black hole is given by

$$ds^2 = -\frac{1}{z^2}\frac{(1 - z^4/z_H^4)^2}{1 + z^4/z_H^4}dt^2 + \frac{1 + z^4/z_H^4}{z^2}\left(dx^2 + d\vec{y}^2\right) + \frac{dz^2}{z^2}, \tag{1}$$

where z is the radial coordinate of the black hole. The horizon is located at $z = z_H$ and the boundary is at $z = 0$. The Hawking temperature which corresponds to the temperature of the gluon sector is given by $T = \sqrt{2}/(\pi z_H)$. $x = x^1$ and $\vec{y} = (x^2, x^3)$ denote the 3-dimensional spatial directions. The bulk geometry is a 10-dimensional spacetime. The remaining 5-dimensional part is an S^5 whose metric is given by $d\Omega_5^2 = d\theta^2 + \sin^2\theta d\psi^2 + \cos^2\theta d\Omega_3^2$, where $0 \le \theta \le \pi/2$ and $d\Omega_d$ is the volume element of the unit d-dimensional sphere. The radius of the S^5 is taken to 1.

The gravity dual of the quark sector is realized as a probe D7-brane embedded in the black hole geometry. The action of the D7-brane is

$$S = -T_{\rm D7}\int d^8\xi\sqrt{-\det[\gamma_{\alpha\beta} + F_{\alpha\beta}]}, \tag{2}$$

where $\gamma_{\alpha\beta}$ is the induced metric on the D7-brane whose worldvolume coordinates are ξ^α. $F_{\alpha\beta}$ is the field strength of the $U(1)$ gauge field on the D7-brane. $T_{\rm D7}$ is the tension of the D7-brane. We take the string tension to be 1 for simplicity. The D7-brane is wrapped on an S^3 part of the S^5. We choose our coordinates in such a way that the S^3 part is specified by $\psi = 0$. Then $\cos\theta(z)$ represents the radius of the S^3 part of the D7-brane at a given position z.

We consider a system at finite quark-charge density D. The quark-charge density is introduced by turning on an "electric field" along the radial direction z on the D7-brane.[10–13] More explicitly, the quark-charge density is taken into account by the vector potential on the D7-brane as

$$A_t = \mu - \frac{1}{2}\frac{D}{\mathcal{N}}z^2 + O(z^4), \tag{3}$$

where μ is the quark-charge chemical potential. $\mathcal{N} = 2\pi^2 T_{\rm D7}$, and $\mathcal{N} = (2\pi)^{-4}\lambda N_c$ in our convention. At this stage, the system is still in thermal equilibrium.

Let us apply an external electric field E in the x direction on the D7-brane.[14] We employ a gauge where the vector potential is given by

$$A_x = -Et + h(z). \tag{4}$$

As a response to E, the system may have a current density J. According to the GKP-Witten prescription of the AdS/CFT correspondence, J is related to $h(z)$ as

$$h(z) = \frac{1}{2}\frac{J}{\mathcal{N}}z^2 + O(z^4). \tag{5}$$

Under the presence of the electric field, the D7-brane can be unstable. This is understood by looking at the on-shell action

$$S = -\mathcal{N}\int dz dt \cos^6\theta g_{xx}^{5/2}\sqrt{\frac{-g_{tt}g_{zz}(g_{tt}g_{xx} + E^2)}{g_{tt}g_{xx}^3\cos^6\theta + \frac{g_{tt}D^2 + g_{xx}J^2}{\mathcal{N}^2}}}. \tag{6}$$

Notice that $g_{tt} \leq 0$ when $0 \leq z \leq z_H$. More precisely, $g_{tt} = 0$ at $z = z_H$ whereas it negatively diverges at $z = 0$. Then $g_{tt}g_{xx} + E^2$ goes across zero at some point, say $z = z_*$, between the horizon and the boundary. This makes the action to be complex and the system is unstable. In the presence of the instability, the system can never be a steady state. Only the way to make the action real is that we adjust J in such a way that $g_{tt}g_{xx}^3 \cos^6\theta + \frac{g_{tt}D^2 + g_{xx}J^2}{\mathcal{N}^2}$ also flips the sign at the same location $z = z_*$, simultaneously. Now, E and J cannot be independent of each other, and they have to follow

$$J = \sqrt{\frac{N_c^2 T^2}{16\pi^2}\sqrt{e^2+1}\cos^6\theta(z_*) + \frac{d^2}{e^2+1}}, \tag{7}$$

where $e = E/(\frac{\pi}{2}\sqrt{\lambda}T^2)$ and $d = /(\frac{\pi}{2}\sqrt{\lambda}T^2)$.[14]

This is how we compute the non-linear conductivity of the system. Note that we do not need to employ any perturbative method, and we can analyze beyond the linear-response regime.

4. Effective Temperature of NESS

It is interesting to note that an analogue black hole emerges on the worldvolume of the D7-brane. The small fluctuations of the fields on a D-brane obey the open-string metric[15] $s_{\alpha\beta}$ that is given by

$$s_{\alpha\beta} = \gamma_{\alpha\beta} - (F\gamma^{-1}F)_{\alpha\beta}. \tag{8}$$

This governs the fluctuations in the NESS. The presence of E, J, and D turn on the off-diagonal components of $s_{\alpha\beta}$. After an appropriate coordinate transformation from $(t, x) \to (\tilde{t}, \tilde{x})$,[7] the open-string metric at the vicinity of $z = z_*$ can be diagonalized in such a way that

$$ds^2 = -a(z-z_*)^2 d\tilde{t}^2 + \frac{b}{(z-z_*)^2}dz^2 + \tilde{g}_{xx}d\tilde{x}^2 + \tilde{g}_{yy}d\tilde{y}^2, \tag{9}$$

where a and b are constants given by

$$a = \left.\frac{\mathcal{N}^2 g_{xx}^2 \cos^6\theta(-g_{tt}g_{xx})'}{\mathcal{N}^2 g_{xx}^3\cos^6\theta + D^2}\right|_{z=z_*}, \tag{10}$$

$$b = \left.\frac{\mathcal{N}^2 g_{xx}^3\cos^6\theta(-g_{tt}g_{xx})}{\left(\frac{-g_{tt}}{g_{xx}}(\mathcal{N}^2 g_{xx}^3\cos^6\theta + D^2)\right)' g_{xx}}\right|_{z=z_*}. \tag{11}$$

Here, prime denotes the differentiation with respect to z. Note that $z = z_*$ plays the role of the horizon of the analogue black hole. We can compute the Hawking temperature (T_{eff}) of the analogue black hole on the D-brane[16,17] as

$$T_{\text{eff}} = \frac{1}{4\pi}\sqrt{\frac{a}{b}}. \tag{12}$$

What is the physical meaning of this temperature? For example, a scalar fluctuation on the D-brane obeys a Klein-Gordon equation on the geometry given by (8). This is parallel to a scalar field on the genuine black hole geometry except that the temperature is given by $T_{\rm eff}$ instead of the heat-bath temperature T. This suggests that the relationship between the correlation functions is given by[6, 17–20]

$$G^{\rm sym}(\omega) = -\coth(\omega/2T_{\rm eff}){\rm Im}G^{\rm R}(\omega), \tag{13}$$

where $G^{\rm sym}(\omega)$ and $G^{\rm R}(\omega)$ are the Fourier transform of the symmetrized Wightman function and the retarded Green's function, respectively, of the fluctuating modes. Now, the fluctuations in the NESS are characterized not by the heat-bath temperature, but by $T_{\rm eff}$ which we call the effective temperature.[21] In general, $T_{\rm eff}$ is different from T at the order of E^2. This means that they can be distinguished only when we are beyond the linear-response regime. $T_{\rm eff}$ can be higher than T, but interestingly, it can be lower than T as well.[6,7] $T_{\rm eff}$ measured by a single test particle dragged in a heat bath can also be lower than the heat-bath temperature in gravity-dual models,[6, 18–20] and a consistency of the behaviours of $T_{\rm eff}$ between the conductors and the dragged particles is confirmed in Ref. 7.

5. Another Information from the Analogue Black Hole

When we consider the system at finite quark charge, we find another interesting feature of the NESS. The spectrum of the Hawking radiation of the analogue black hole on the D7-brane obeys the distribution function proportional to

$$\exp\left\{-\frac{\omega - \mu_x k_x}{T_{\rm eff}}\right\}, \tag{14}$$

where ω and k_x are the energy and the momentum in the x direction of the mode in consideration, and $\mu_x = -s_{tx}/s_{xx}|_{z=z_*}$.[22] In our setup, $s_{tx} = -F_{tz}\gamma^{zz}F_{zx}$ which is non-vanishing only when both D and J are non-zero. Appearance of μ_x is owing to the off-diagonal component s_{tx} of the open-string metric, which resembles us of the Kerr black hole.[16] μ_x can be understood as the velocity of the analogue black hole in the x direction. What is the meaning of μ_x in the NESS?

Now the fluctuation around the NESS is characterized by the effective temperature $T_{\rm eff}$ and by the "chemical potential" μ_x for the mode with k_x. The presence of μ_x breaks the rotational symmetry of the distribution. However, if we boost the system at the velocity μ_x along the x direction, one finds that (14) is transformed into the form without μ_x. This means that, if the observer has the velocity μ_x with respect to the heat bath, the fluctuation still looks to be isotropic. Our NESS is associated with the velocity μ_x that characterizes the system. More detailed analysis and consideration will be reported in Ref. 22.

Acknowledgments

The author thanks the organizers of the Second CosPA International Symposium for invitation. The work reported in this talk was supported in part by the Grant-in-Aid for Scientific Research on Innovative Areas "Fluctuation & Structure" (No. 26103517) from the Ministry of Education, Culture, Sports, Science, and Technology of Japan; Chuo University Personal Research Grant, and the visiting professorship of Institute for Solid State Physics, University of Tokyo.

References

1. J. M. Maldacena, The Large N limit of superconformal field theories and supergravity, *Adv. Theor. Math. Phys.* **2**, 231 (1998).
2. S. S. Gubser, I. R. Klebanov and A. M. Polyakov, Gauge theory correlators from non-critical string theory, *Phys. Lett. B* **428**, 105 (1998); E. Witten, Anti-de Sitter space and holography, *Adv. Theor. Math. Phys.* **2**, 253 (1998).
3. V. E. Hubeny and M. Rangamani, A Holographic view on physics out of equilibrium, *Adv. High Energy Phys.* **2010**, 297916 (2010).
4. S. Nakamura, Negative Differential Resistivity from Holography, *Prog. Theor. Phys.* **124**, 1105 (2010).
5. S. Nakamura, Nonequilibrium Phase Transitions and Nonequilibrium Critical Point from AdS/CFT, *Phys. Rev. Lett.* **109**, 120602 (2012).
6. S. Nakamura and H. Ooguri, Out of Equilibrium Temperature from Holography, *Phys. Rev. D* **88**, 126003 (2013).
7. H. Hoshino and S. Nakamura, Effective temperature of nonequilibrium dense matter in holography, *Phys. Rev. D* **91**, 026009 (2015).
8. See, for example, R. M. Wald, *General Relativity*, (Univ. Chicago Press, Chicago, 1984).
9. A. Karch and E. Katz, Adding flavor to AdS / CFT, *JHEP* **0206**, 043 (2002).
10. N. Horigome and Y. Tanii, Holographic chiral phase transition with chemical potential, *JHEP* **0701**, 072 (2007).
11. K. Y. Kim, S. J. Sin and I. Zahed, Dense hadronic matter in holographic QCD, *J. Korean Phys. Soc.* **63**, 1515 (2013).
12. S. Nakamura, Y. Seo, S. J. Sin and K. P. Yogendran, A New Phase at Finite Quark Density from AdS/CFT, *J. Korean Phys. Soc.* **52**, 1734 (2008).
13. S. Kobayashi, D. Mateos, S. Matsuura, R. C. Myers and R. M. Thomson, Holographic phase transitions at finite baryon density, *JHEP* **0702**, 016 (2007).
14. A. Karch and A. O'Bannon, Metallic AdS/CFT, *JHEP* **0709**, 024 (2007).
15. N. Seiberg and E. Witten, String theory and noncommutative geometry, *JHEP* **9909**, 032 (1999).
16. K. -Y. Kim, J. P. Shock and J. Tarrio, The open string membrane paradigm with external electromagnetic fields, *JHEP* **1106**, 017 (2011).
17. J. Sonner and A. G. Green, Hawking Radiation and Non-equilibrium Quantum Critical Current Noise, *Phys. Rev. Lett.* **109**, 091601 (2012).
18. S. S. Gubser, Momentum fluctuations of heavy quarks in the gauge-string duality, *Nucl. Phys. B* **790**, 175 (2008).
19. J. Casalderrey-Solana and D. Teaney, Transverse Momentum Broadening of a Fast Quark in a N=4 Yang Mills Plasma, *JHEP* **0704**, 039 (2007).
20. U. Gursoy, E. Kiritsis, L. Mazzanti and F. Nitti, Langevin diffusion of heavy quarks in non-conformal holographic backgrounds, *JHEP* **1012**, 088 (2010).
21. For discussions on the effective temperature in condensed matter physics, see for example, L. F. Cugliandolo, The Effective Temperature, *J. Phys. A: Math. Theor.* 44, 483001 (2011).
22. H. Hoshino and S. Nakamura, in progress.

Holographic Thermalization with Initial Long Range Correlation

Shu Lin

School of Physics and Astronomy, Sun Yat-Sen University,
Guangzhou, 510275, China
linshu8@mail.sysu.edu.cn

We studied the evolution of Wightman correlator in a thermalizing state using holography. We gave a prescription for calculating the correlator in coordinate space. For equal-time correlator $\langle O(v,x)O(v,0)\rangle$, we obtained an enhancement factor v^2 due to long range correlation present in the initial state. This was missed by previous studies based on geodesic approximation. We also studied spatially integrated correlator and presented evidence indicating that our intuition from equilibrium physics is not necessarily true for a far-from-equilibrium state. We also calculated radiation spectrum of particle weakly coupled to O and found low frequency mode thermalizes faster than high frequency mode.

Keywords: Thermalization; CFT; holography.

1. Introduction

Thermalization is a phenomenon where a quantum state evolves toward thermal equilibrium. It emerges in different physical systems such as heavy ion collisions, cold atom gas etc. Thermalization phenomenon often occurs in systems which are strongly correlated, making theoretical description difficult. The AdS/CFT provides a useful approach to study off-equilibrium field theory problem at strong coupling. In this talk, I will discuss the thermalization of 2D conformal field theory (CFT). While general results on thermalization of 2D CFT have been obtained by CFT techniques by Cardy and Calabrese (CC), our setting differs from theirs in that our initial state has long range correlation. As we will see, the long range correlation in the initial state does have an imprint on the evolution of correlators: for equal-time correlator, it gives enhancement factor missed by previous holographic studies using geodesic approximation. Furthermore, it also provides evidence indicating that our intuition from equilibrium physics is not necessarily true for a far-from-equilibrium state. In particular, the low momentum physics should not be equated with long distance physics.

2. Correlator in a thermalizing state

We will study the evolution of a thermalizing state modeled by AdS$_3$-Vaidya background

$$ds^2 = \frac{-f dv^2 - 2dvdz + dx^2}{z^2}, \tag{1}$$

with $f = 1 - mz^2\theta(v)$ and $dv = dt - \int_0^z \frac{dz'}{f(z')}$. Here v is a ingoing lightcone coordinate. The dual field state is initially a vacuum state. At $v = 0$, there is an instantaneous injection of energy density, which triggers a thermalization process. The end point of the evolution is a thermal state with temperature $T = \sqrt{m}/2\pi$. We will fix the units by setting $m = 1$. As a probe of the thermalization process, we study Wightman correlator of scalar operator$O$ dual to dilaton in gravity background. The operator has scaling dimension two. The correlator in initial (vacuum) state $iG_0^>(2|1)$ and final (thermal) state $iG_{\text{th}}^>(2|1)$ are given by

$$iG_0^>(2|1) = \frac{2}{\pi} \frac{1}{\left(-(v_2 - v_1 - i\epsilon) + x_{21}\right)^2},$$
$$iG_{\text{th}}^>(2|1) = \frac{1}{2\pi} \frac{1}{\left(-\cosh(v_2 - v_1 - i\epsilon) + \cosh x_{21}\right)^2}. \tag{2}$$

where we use 1 and 2 to denote collectively coordinates of point 1 and 2 respectively. We note the power law dependence in both $v_2 - v_1$ and $x_2 - x_1$ in $iG_0^>(2|1)$, indicating long range correlation in the initial state. $iG_{\text{th}}^>(2|1)$ decays exponentially at large spatial and temporal separation, characterizing Debye screening in final state.

The calculation of Wightman correlator in the thermalizing state is formulated as an initial value problem. We proceed by calculating the bulk Wightman correlator:[1]

$$\Delta G^>(4|3) = \int dz_1 dx_1 dz_2 dx_2 \Delta G^>(2|1) \overleftrightarrow{D}^{v1} \overleftrightarrow{D}^{v2} G_{\text{th}}^R(3|1) G_{\text{th}}^R(4|2), \tag{3}$$

where $\Delta G^>(4|3) = G^>(4|3) - G_{\text{th}}^>(4|3)$ is the difference between bulk correlators in Vaidya background and BTZ background and $\Delta G^>(2|1) = G_0^>(2|1) - G_{\text{th}}^>(2|1)$ is the difference between bulk correlators in empty AdS and BTZ background. $\Delta G^>(2|1)$ is the initial value for $\Delta G^>(4|3)$. It is to be evaluated on the trajectory of the shell $v_2 = v_1 = 0$. The symbol $\overleftrightarrow{D}^v = \sqrt{-g} g^{vz} \overleftrightarrow{\partial}_z$ is a two-way differential operator. $iG_{\text{th}}^R(3|1)$ and $iG_{\text{th}}^R(4|2)$ are retarded bulk-bulk propagators in BTZ. The resulting bulk correlator $\Delta G^>(4|3)$ gives us the boundary Wightman correlator:[2]

$$\Delta\langle G^>(4|3)\rangle \to z_4^2 z_3^2 \Delta\langle O(v_4, x_4) O(v_3, x_3)\rangle, \quad \text{as} \quad z_4, z_3 \to 0. \tag{4}$$

Note that we have used regularized initial value, so the resulting correlator is also regularized. This is necessary as the unregularized initial value is singular on the trajectory of the shell. By construction, $\Delta\langle O(v_4, x_{43}) O(v_3, 0)\rangle$ is the difference of correlation in thermalizing state minus the counterpart in thermal state:

$$\Delta\langle O(v_4, x_4) O(v_3, x_3)\rangle = \langle O(v_4, x_{43}) O(v_3, 0)\rangle - \langle O(v_4, x_{43}) O(v_3, 0)\rangle_{\text{th}}. \tag{5}$$

(2) can be evaluated analytically in certain kinematic regime and numerically in any regime. Below we will present results on both correlators in coordinate space and spatially integrated correlator.

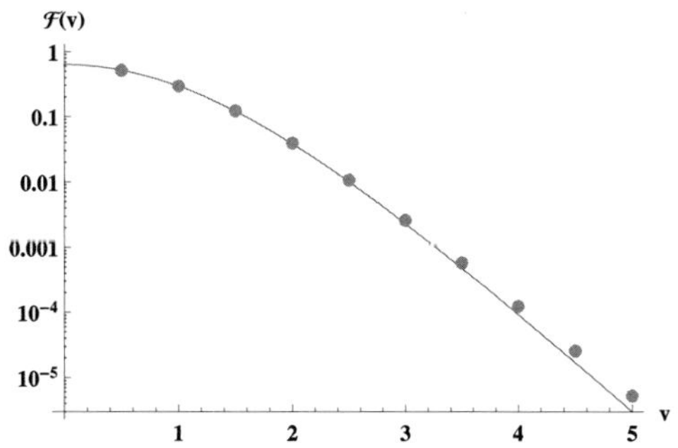

Fig. 1. $\mathcal{F}(v) = x^4\Delta\langle O(v,x)O(v,0)\rangle$ as a function of v for $x = 15$. The analytic fitting function is given by (7). It fits very well in a wide range of v.

2.1. *Correlator in coordinate space*

We will present results on equal-time correlator $\Delta\langle O(v,x)O(v,0)\rangle$, which has a clear physical interpretation as a measure of time evolution of spatial correlation. More general results can be found in Ref. 3.The early time limit of equal-time correlator is given by

$$\Delta\langle O(v,x)O(v,0)\rangle \xrightarrow{v\to 0} \frac{2}{\pi}\frac{1}{x^4} - \frac{1}{2\pi}\frac{1}{(-1+\cosh x)^2}. \tag{6}$$

The first term is a power law decay, indicating long range correlation in vacuum. The second term is an exponential decay, characteristic of Debye screening. Only the first term evolves with time. We consider two limits of equal-time correlator: large separation $x \gg 1$ and small separation $x \ll 1$. At large separation $x \gg 1$, the second term is negligible. $\Delta\langle O(v,x)O(v,0)\rangle$ simply measures decay of vacuum correlation $1/x^4$. This motivates the parameterization $\Delta\langle O(v,x)O(v,0)\rangle = \mathcal{F}(v)/x^4$. The function can be obtained analytically as

$$\mathcal{F}(v) = \frac{18}{\pi}\left(\frac{v\coth v - 1}{\sinh^2 v}\right)^2. \tag{7}$$

We have confirmed the parameterization and the analytic function (7) by numerical results in Figure 1. At small separation $x \ll 1$, there is partial cancellation between the first and second terms in (6). The leading remaining result scales as $1/x^2$. Restricting ourselves to early time regime, we expect the following scaling $\Delta\langle O(v,x)O(v,0)\rangle = \mathcal{G}(v/x)/x^2$. This is indeed confirmed by numerical results shown in Figure 2.

It is informative to compare our results with general results obtained by CC.[4] In their case, the equal-time correlator $\langle O(v,x)O(v,0)\rangle$ in a thermalizing state is given by

$$\langle O(v,x)O(v,0)\rangle \sim e^{-4\pi v/2\tau_0}, \quad x > 2v. \tag{8}$$

We identify $2\tau_0 = \pi$. Note that (8) does not contain power law dependence on x. The reason as pointed out in Ref. 5, is the absence of long range correlation in

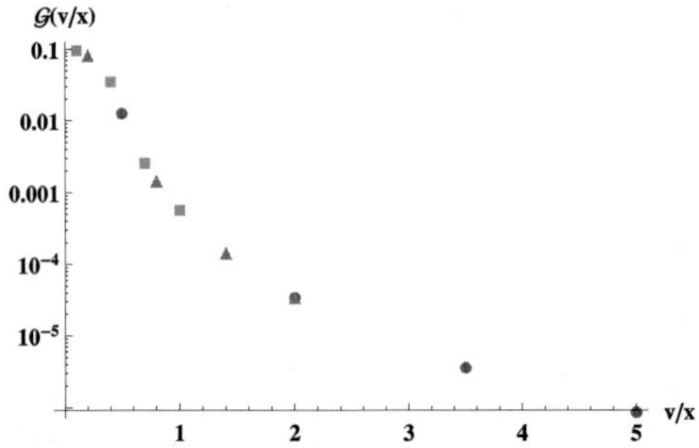

Fig. 2. $\mathcal{G}(v/x) = x^2\Delta\langle O(v,x)O(v,0)\rangle$ as a function of v/x for $x = 1/50$(blue point), $x = 1/20$(purple triangle) and $x = 1/10$(brown square). Points with different x collapse on the same curve. The tail is numerically consistent with a power law $\mathcal{G}\left(\frac{v}{x}\right) \sim \left(\frac{x}{v}\right)^4$.

their initial state. In the presence of initial long range correlation, it follows from (7) that $\langle O(v,x)O(x,0)\rangle \sim v^2 e^{-4v}/x^4$ for $v \gg 1$. The power law dependence $1/x^4$ is reminiscent of long range correlation. In addition, we also find an enhancement factor v^2, which is missed in Ref. 5 using a geodesic approximation. We argue that the enhancement factor v^2 is also due to the long range correlation in the initial state: as $\langle O(v,x)O(v,0)\rangle$ receives contribution from regions in the backward light-cone, two factors of v arise from the distance of the correlated regions propagating the initial state correlation to point x and 0.

2.2. *Spatially integrated correlator*

Next we look at spatially integrated correlator defined by

$$\int dx_{43}\Delta\langle O(v_4,x_4)O(v_3,x_3)\rangle, \tag{9}$$

which is the mode with $k = 0$. We are able to obtain simple analytic results for two limiting cases. For $v_3 \ll 1$, $v_4 \gg 1$, we obtain

$$\int dx_{43}\Delta\langle O(v_4,x_4)O(v_3,x_3)\rangle \simeq \frac{3\sqrt{2}}{32\sqrt{v_3}}v_4 e^{-2v_4}. \tag{10}$$

We readily notice the same enhancement factor and exponential decay factor in the dependence on v_4. For $v_3, v_4 \ll 1$.

$$\int dx_{43}\Delta\langle O(v_4,x_4)O(v_3,x_3)\rangle \simeq \frac{3\pi}{128\sqrt{v_3 v_4}}. \tag{11}$$

The dependence on v_3 and v_4 are of particular interest. In fact we can show that the dominant contribution to (11) comes from the integration of short distance with $x_{43} \ll 1$. This is in contrast to our intuition in equilibrium physics, which would tell us that the low momentum ($k = 0$) physics is equivalent to long distance physics. (11) provides evidence that our intuition in equilibrium physics might be incorrect for a far-from-equilibrium state.

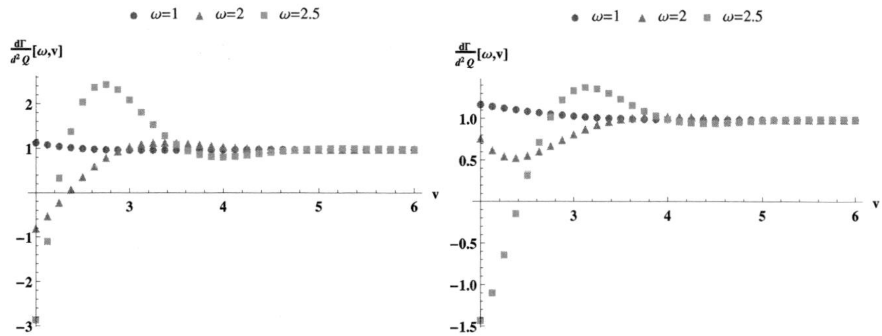

Fig. 3. Emission rates $\frac{d\Gamma}{d^2Q}(\omega, v)$ as a function of v in the thermalizing state (left) and in an instantaneous thermalization scenario (right). Both rates are in unit of thermal emission rate. The symbols represent $\omega = 1$ (blue point), $\omega = 2$ (purple triangle) and $\omega = 2.5$ (brown square). All the rates approach thermal limit at late time. We observe a hierarchy among spectra with different frequencies, with lower frequency mode tends to thermal spectrum faster. The high frequency mode shows more oscillations in relaxing to thermal spectrum.

As an application, we calculate a physical observable: emission spectrum of particles weakly coupled to operator O. We can draw analogy with dilepton emission: we can regard O as current, and radiated particles as dilepton. The coupling constant g_O between radiated particle and O is small like the electromagnetic coupling e. With an abuse of terminology, we will refer to the radiated particle as dilepton and the field created by O as photon. The differential rate of dilepton at time v is defined by

$$\frac{d\Gamma(\omega, v)}{d^2Q} = g_O^2 \Big[\int_{-\infty}^{v} dX^0 e^{i\omega(X^0 - v)}|_{Y^0=v} + \int_{-\infty}^{v} dY^0 e^{i\omega(v - Y^0)}|_{X^0=v} \Big] \times \int d(X^1 - Y^1) e^{-ik(X^1 - Y^1)} \langle i|O(Y)O(X)|i\rangle. \tag{12}$$

Here $|i\rangle$ is the thermalizing state. We have calculated the regularized version of $\int d(X^1 - Y^1) e^{-ik(X^1-Y^1)} \langle i|O(Y)O(X)|i\rangle$ for $k = 0$. To obtain the full result, we need to add back the thermal part. To compare the emission rate with different frequencies, we normalize the rate by dividing out the thermal rate. As a separate reference, we also calculate the emission rate, assuming instantaneous thermalization of the state at $v = 0$. We present our results for $\omega = 1/2$, $\omega = 1$ and $\omega = 2.5$ in Figure 3. we do observe interesting hierarchy in frequency: low frequency mode tends to appear thermal faster than high frequency mode. Previous studies have shown that short distance physics tends to thermalize faster than long distance physics. Our results can be viewed as a complementary picture to this, although in a counter intuitive way.

3. Summary

In summary, we calculated the evolution of Wightman correlator in a thermalizing state modeled by AdS_3-Vaidya background. The thermalizing starts with vacuum state with long range correlation and end with a thermal state with Debye screening. We gave a prescription for calculating Wightman correlator in coordinate space without using any approximation. We obtained results on both correlator in coordinate space and spatially integrated correlator. We found interesting imprint of the long range correlation in the results: For equal-time correlator $\langle O(v,x)O(v,0)\rangle$, we obtained an enhancement factor v^2, which was missed by previous studies based on geodesic approximation. We also studied spatially integrated Wightman correlator and showed evidence against our intuition from equilibrium physics. The dominant contribution to low momentum ($k=0$) physics comes from short distance physics for a far-from-equilibrium state. As an application, we also calculated radiation spectrum of particle weakly coupled to O and found lower frequency mode approaches thermal spectrum faster than high frequency mode.

References

1. S. Caron-Huot, P. M. Chesler and D. Teaney, Fluctuation, dissipation, and thermalization in non-equilibrium AdS_5 black hole geometries, *Phys. Rev.* **D84**, p. 026012 (2011).
2. V. Keranen and P. Kleinert, Non-equilibrium scalar two point functions in AdS/CFT, *JHEP* **04**, p. 119 (2015).
3. S. Lin, Holographic thermalization with initial long range correlation, *Phys. Rev.* **D93**, p. 026007 (2016).
4. P. Calabrese and J. L. Cardy, Time-dependence of correlation functions following a quantum quench, *Phys. Rev. Lett.* **96**, p. 136801 (2006).
5. J. Aparicio and E. Lopez, Evolution of Two-Point Functions from Holography, *JHEP* **12**, p. 082 (2011).

Rindler Fluid and Petrov Type I Condition

Qing Yang

State Key Laboratory of Theoretical Physics, Institute of Theoretical Physics, Chinese Academy of Sciences, Beijing 100190, China

Yun-Long Zhang

Department of Physics, National Taiwan University, Taipei 10617, Taiwan
zhangyunlong001@gmail.com

This proceeding overviews the relation between Rindler Fluid and Petrov Type I condition. Rindler fluid is a special fluid dual to vacuum Einstein gravity, which lives on the time-like hypersurface in an accelerating frame. Through imposing the Petrov type I condition and Hamiltonian constraint, we show that the stress tensor of Rindler fluid can be recovered with correct first order and second order transport coefficients, which lead to an interesting recurrence relation.

Keywords: Rindler fluid; Petrov type I condition.

1. Introduction

The holographic duality between gravity and one lower dimensional fluid has attracted much attention over the past years. Recently, the authors in Ref. 1 consider the gravitational fluctuations confined inside a finite cutoff surface outside a horizon, and in this case the dual fluid lives on this hypersurface. The Dirichlet condition on the cutoff surface and the regularity on the horizon are imposed. This procedure has also been generalized to the asymptotically flat[2,3] and de Sitter[4] spacetimes. In particular, a systematical method to reconstruct the solution of vacuum Einstein gravity to an arbitrary order has been presented in both of the non-relativistic and relativistic hydrodynamic expansions.[5–8] And the Rindler fluid is a special fluid dual to vacuum Einstein gravity, which lives on the time-like hypersurface in an accelerating frame.

It is interesting to note that, instead of imposing the regularity condition on the horizon, imposing the Petrov type I condition on a hypersurface in near-horizon limit is alternatively introduced in Ref. 9. The Petrov type I condition gives $p(p+1)/2$ constraints on the Brown-York stress tensor T_{ab}, which leads to $p+1$ independent variables. These variables are exactly the degrees of freedom of a fluid in $p+1$ dimensions. They have shown that combining the Petrov type I condition with Hamiltonian and momentum constraints can lead to the incompressible Navier-Stokes equation for the dual fluid on the cutoff surface in the near-horizon limit. Some further generalizations and discussions can be seen in Ref. 10–17.

However, it is found in Ref. 18 that the condition holds up to order ϵ^3 and is broken at order ϵ^4 in the non-relativistic hydrodynamic expansion. While those Petrov type I condition violated terms contain only the third order terms of the derivative expansion parameter ∂ of relativistic fluid if an improved frame is taken. This motivates us to check the Petrov type I condition for the solution of vacuum Einstein equations in the relativistic hydrodynamic expansion. It is shown in Ref. 19 that the condition holds up to the second order of the derivative expansion parameter ∂, by using the vacuum solution available to this order in Ref. 6. Turn the logic around, it is shown that imposing the Petrov type I condition and Hamiltonian constraint on a finite cutoff surface, the stress tensor of the dual relativistic fluid can be fixed up to the second order of the derivative expansion. This relation is also generalised to the fluid dual to vacuum Einstein-Gauss-Bonnet gravity in Ref. 20. In this proceeding, we overview the relation between Rindler Fluid and Petrov Type I condition in vacuum Einstein gravity.

2. Rindler Fluid and Its Properties

To study the Rindler fluid dual to the vacuum Einstein gravity, we begin with the Einstein-Hilbert action on a $(p+2)$ dimensional Lorentz manifold $\mathcal{M}$, with appropriate surface term

$$S = \frac{1}{16\pi G_{p+2}} \int d^{p+2}x\sqrt{-g}(R - 2\Lambda) + S_{\partial\mathcal{M}}. \tag{1}$$

Varying this action with respect to the metric $g_{\mu\nu}$ yields the vacuum Einstein field equations,

$$G_{\mu\nu} = 0, \qquad G_{\mu\nu} \equiv R_{\mu\nu} - \frac{1}{2}Rg_{\mu\nu}, \quad \mu,\nu = 0,1,...,p+1. \tag{2}$$

The $p+2$ dimensional Rindler metric

$$\mathrm{d}s^2_{p+2} = -r\mathrm{d}\tau^2 + 2\mathrm{d}\tau\mathrm{d}r + \delta_{ij}\mathrm{d}x^i\mathrm{d}x^j, \quad i,j = 1,...,p, \tag{3}$$

is an exact solution of the field equations (2). On a timelike hypersurface Σ_c with $r = r_c$, the induced metric is intrinsic flat,

$$\mathrm{d}s^2_{p+1} = \gamma_{ab}\mathrm{d}x^a\mathrm{d}x^b = -r_c\mathrm{d}\tau^2 + \mathrm{d}x_i\mathrm{d}x^i\ , \quad a,b = 0,1,...,p\,. \tag{4}$$

And after setting $16\pi G_{p+2} = 1$, the Brown-York stress tensor of Einstein gravity on the cutoff surface Σ_c can be written as[21]

$$\hat{T}_{ab} = -2\left(K_{ab} - K\gamma_{ab}\right), \tag{5}$$

where K_{ab} is the extrinsic curvature of the hypersurface Σ_c , and we omit the ambiguous up to a constant multiple of the induced metric.

In order to study the dual fluid on the hypersurface Σ_c , one introduces the $(p+1)$ independent parameters $u^a = \gamma_v(1, v^i)$ and $\mathbb{p}$, which are slowly varying functions of $x^a = (\tau, x^i)$. Here γ_v is fixed through $\gamma_{ab}u^a u^b = -1$. Keep the induced metric on a timelike hypersurface Σ_c flat and impose the regularity on the future horizon, the solution of vacuum Einstein field equations (2) up to the second order in the derivative expansion is given by,[22]

$$ds^2_{p+2} = g_{\mu\nu}dx^\mu dx^\nu = -2\mathrm{p}u_a dx^a dr + g_{ab}dx^a dx^b, \tag{6}$$

$$g_{ab} = g^{(0)}_{ab} + g^{(1)}_{ab} + g^{(2)}_{ab} + O(\partial^3). \tag{7}$$

The superscript indices $(0), (1), (2), \ldots$ on the metric components g_{ab} denote the order of the derivative of the velocity u^a and pressure p with respect to the transverse space-time coordinates x^a: $\partial \sim \varepsilon$, where we have introduced the small parameter $\varepsilon \ll 1$. The leading order term of g_{ab} in the derivative expansion is given by

$$g^{(0)}_{ab} = [1 - \mathrm{p}^2(r - r_c)]u_a u_b + h_{ab}, \tag{8}$$

where the projection tensor $h_{ab} \equiv \gamma_{ab} + u_a u_b$. We can read off the horizon position through $r_h = r_c - 1/\mathrm{p}^2$ with $g^{(0)}_{ab}$ in the case of equilibrium state. The first order term in g_{ab} in the derivative expansion is

$$g^{(1)}_{ab} = 2\mathrm{p}(r - r_c)\left[(D \ln \mathrm{p})u_a u_b + 2a_{(a}u_{b)}\right], \tag{9}$$

where $D \equiv u^c\partial_c$ and the acceleration $a^a \equiv u^b\partial_b u^a$. The second order term in g_{ab} in the derivative expansion is

$$\begin{aligned}
u^c u^d g^{(2)}_{cd} &= 2(r - r_c)\mathcal{K}_{cd}\mathcal{K}^{cd} + \frac{1}{2}\mathrm{p}^2(r - r_c)^2\left(\mathcal{K}_{cd}\mathcal{K}^{cd} + 2a_c a^c\right) + \frac{1}{2}\mathrm{p}^4(r - r_c)^3\Omega_{cd}\Omega^{cd},\\
h^c_a u^d g^{(2)}_{cd} &= -2(r - r_c)h^c_a\partial_d\mathcal{K}^d_c + \mathrm{p}^2(r - r_c)^2\left[h^b_a\partial_c\mathcal{K}^c_b - (\mathcal{K}_{ad} + \Omega_{ad})a^d\right],\\
h^c_a h^d_b g^{(2)}_{cd} &= 2(r - r_c)\left(-\mathcal{K}_a{}^c\mathcal{K}_{cb} + 2\mathcal{K}_{c(a}\Omega^c{}_{b)} - 2h^c_a h^d_b D\mathcal{K}_{cd}\right) - \mathrm{p}^2(r - r_c)^2\Omega_{ac}\Omega^c{}_b.
\end{aligned} \tag{10}$$

Here the fluid shear and vorticity are defined as

$$\mathcal{K}_{ab} \equiv h^c_a h^d_b\partial_{(c}u_{d)}, \quad \Omega_{ab} \equiv h^c_a h^d_b\partial_{[c}u_{d]}. \tag{11}$$

And one also needs to consider the following constraints

$$\begin{aligned}
\partial_a u^a &= 2\mathrm{p}^{-1}\mathcal{K}_{ab}\mathcal{K}^{ab} + O(\partial^3),\\
a_a + D^\perp_a \ln \mathrm{p} &= 2\mathrm{p}^{-1}h^c_a\partial_b\mathcal{K}^b_c + O(\partial^3),
\end{aligned} \tag{12}$$

with $D^\perp_a \equiv h^c_a\partial_c$, so that the metric (6) solves the vacuum Einstein field equations (2) up to the second order in the derivative expansion.

We can project the Weyl tensor on the hypersurface Σ_c and define $\mathrm{P}_{ij} \equiv 2C_{(\ell)i(\ell)j}|_{\Sigma_c}$. In Ref. 9, $\mathrm{P}_{ij} = 0$ is named as Petrov type I condition and P_{ij} can be rewritten in terms of the extrinsic curvature K_{ab} of Σ_c by employing the Gauss-Codazzi equations. Notice that K_{ab} can be expressed in terms of the Brown-York stress tensor through $\hat{T}_{ab} = 2(K\gamma_{ab} - K_{ab})$. We have $\mathrm{P}_{ij} = m_i{}^a m_j{}^b\mathbb{P}_{ab}$ where

$$\begin{aligned}
4\mathbb{P}_{ab} &= h^m_a h^n_b\left[\left(\hat{T}_{mc}\hat{T}_{nd} - \hat{T}_{mn}\hat{T}_{cd}\right)u^c u^d - \hat{T}_{mc}\hat{T}^c{}_n - 4u^c\partial_c\hat{T}_{mn} + 4u^c\partial_{(m}\hat{T}_{n)c}\right]\\
&\quad + p^{-2}\left[\hat{T}(\hat{T} + p\hat{T}_{cd}u^c u^d) + 4p\,u^c\partial_c\hat{T}\right]h_{ab}.
\end{aligned} \tag{13}$$

With the bulk metric in (6), the dual stress tensor can be expanded in the following form

$$\hat{T}_{ab} = \hat{T}^{(0)}_{ab} + \hat{T}^{(1)}_{ab} + \hat{T}^{(2)}_{ab} + O(\partial^3), \tag{14}$$

and these terms are obtained in Ref. 6 as

$$\begin{aligned}
\hat{T}^{(0)}_{ab} &= \mathbb{p} h_{ab}, \\
\hat{T}^{(1)}_{ab} &= \zeta'(u^c\partial_c \ln \mathbb{p}) u_a u_b - 2\eta \mathcal{K}_{ab}, \\
\hat{T}^{(2)}_{ab} &= \mathbb{p}^{-1}\Big\{ \big[d_1 \mathcal{K}_{ab}\mathcal{K}^{ab} + d_2 \Omega_{ab}\Omega^{ab} + d_3 (u^c\partial_c \ln \mathbb{p})^2 \\
&\quad + d_4 u^c \partial_c (u^d \partial_d \ln \mathbb{p}) + d_5 h^{cd}(\partial_c \ln \mathbb{p})(\partial_d \ln \mathbb{p}) \big] u_a u_b \\
&\quad + \big[c_1 \mathcal{K}_{ac}\mathcal{K}^c{}_b + c_2 \mathcal{K}_{c(a}\Omega^c{}_{b)} + c_3 \Omega_{ac}\Omega^c{}_b + c_4 h^c_a h^d_b \partial_c \partial_d \ln \mathbb{p} \\
&\quad + c_5 \mathcal{K}_{ab}(u^c \partial_c \ln \mathbb{p}) + c_6 (h^c_a \partial_c \ln \mathbb{p})(h^d_b \partial_d \ln \mathbb{p}) \big] \Big\}.
\end{aligned} \tag{15}$$

Here the first and second order transport coefficients are

$$\begin{aligned}
\zeta' &= 0, \qquad \eta = 1, \\
d_1 &= -2, \quad d_2 = d_3 = d_4 = d_5 = 0, \\
c_1 &= -2, \quad c_2 = c_3 = c_4 = c_5 = -c_6 = -4\,.
\end{aligned} \tag{16}$$

Considering the equilibrium entropy density $s = 4\pi$, one obtains the universal ratio of shear viscosity over entropy density as $\eta/s = 1/4\pi$.[23]

The momentum constraint $2G_{\mu b} n^\mu|_{\Sigma_c} = 0$, which leads to the conservation of the stress tensor $\partial^a \hat{T}_{ab} = 0$, gives the constraint equations (12), while the Hamiltonian constraint $2G_{\mu\nu} n^\mu n^\nu|_{\Sigma_c} = 0$ leads to $4\mathbb{H} \equiv p\hat{T}_{ab}\hat{T}^{ab} - \hat{T}^2 = 0$, which can be viewed as the equation of state for the dual fluid. In addition, one can show that the trace of the stress tensor satisfies $\hat{T} = p\mathbb{p} + O(\partial^3)$.

Putting the stress tensor (14) into the expression (13), we then obtain $\mathbb{P}_{ab} = O(\partial^3)$, which implies $\mathrm{P}_{ij} = O(\partial^3)$. Thus we show that the Petrov type I condition $\mathrm{P}_{ij} = 0$ is satisfied up to ∂^2 by using the stress tensor of the dual relativistic fluid.

3. From Petrov Type I Condition to Rindler Fluid

In this section we turn the logic around, and show that through imposing the Hamiltonian constraint and Petrov type I condition on a finite cutoff surface, the stress tensor of Rindler fluid can be fixed up to the second order of the derivative expansion, without using the details of the bulk metric[19] .

Firstly, one can introduce an undetermined symmetric stress tensor $\hat{T}_{ab}$, and it satisfies $h^b_a \hat{T}_{bc} u^c = 0$, where u^a is regarded as the relativistic fluid velocity. Then the stress tensor can be decomposed as $\hat{T}_{ab} = \mathbb{e} u_a u_b + \Pi_{ab}$, where

$$\mathbb{e} \equiv \hat{T}_{ab} u^a u^b, \quad \Pi_{ab} \equiv h^c_a h^d_b \hat{T}_{cd}. \tag{17}$$

The Hamiltonian constraint becomes $\mathbb{H} = 0$, where

$$4\mathbb{H} \equiv p\,(\mathbb{e}^2 + \Pi_{ab}\Pi^{ab}) - \hat{T}^2, \tag{18}$$

and $\hat{T} = -\mathbb{e} + \Pi_{ab}h^{ab}$. The Petrov type I condition can be generalized as $\mathbb{P}_{ab} = 0$, where

$$4\mathbb{P}_{ab} \equiv -\mathbb{e}\Pi_{ab} - \Pi_{ac}\Pi^c{}_b - 4h_a^c h_b^d(u^e\partial_e\Pi_{cd}) - 4\Pi_{(a}{}^c h_{b)}^d\partial_d u_c$$
$$- 4\mathbb{e}\mathcal{K}_{ab} + p^{-2}\big[\hat{T}(\hat{T} + p\,\mathbb{e}) + 4p\,u^c\partial_c\hat{T}\big]h_{ab}. \tag{19}$$

Expanding the stress tensor in terms of the derivative expansion parameter ∂ as

$$\mathbb{e} = \mathbb{e}^{(0)} + \mathbb{e}^{(1)} + \mathbb{e}^{(2)} + O(\partial^3),$$
$$\Pi_{ab} = \Pi_{ab}^{(0)} + \Pi_{ab}^{(1)} + \Pi_{ab}^{(2)} + O(\partial^3), \tag{20}$$

and we identify $\mathbb{e}^{(0)} = 0$, $\Pi_{ab}^{(0)} = \mathbb{p}h_{ab}$ from the zeroth order Brown-York stress tensor in (15). Then through

$$\mathbb{H}^{(1)} = 0 \Rightarrow \mathbb{e}^{(1)} = 0, \tag{21}$$
$$\mathbb{P}_{ab}^{(1)} = 0 \Rightarrow \Pi_{ab}^{(1)} = -2\mathcal{K}_{ab}, \tag{22}$$

we can fix the stress tensor at the first order. With these,

$$\mathbb{H}^{(2)} = 0 \Rightarrow \mathbb{e}^{(2)} = -2\mathbb{p}^{-1}\mathcal{K}_{ab}\mathcal{K}^{ab}, \tag{23}$$
$$\mathbb{P}_{ab}^{(2)} = 0 \Rightarrow \Pi_{ab}^{(2)} = \mathbb{p}^{-1}\big[-2\mathcal{K}_{ac}\mathcal{K}^c{}_b - 4\mathcal{K}_{c(a}\Omega^c{}_{b)} - 4\Omega_{ac}\Omega^c{}_b$$
$$- 4h_a^c h_b^d\partial_c\partial_d\ln\mathbb{p} - 4\mathcal{K}_{ab}(u^c\partial_c\ln\mathbb{p})$$
$$+ 4(h_a^c\partial_c\ln\mathbb{p})(h_b^d\partial_d\ln\mathbb{p})\big], \tag{24}$$

we can then fix the second order terms in the stress tensor. In the above procedure we have chosen the isotropy gauge that there is no higher order corrections to the pressure $\mathbb{p}$. Thus, up to the second order, we obtain the total stress tensor of the dual relativistic fluid as

$$\hat{T}_{ab} = \mathbb{e}^{(2)}u_a u_b + \mathbb{p}h_{ab} + \Pi_{ab}^{(1)} + \Pi_{ab}^{(2)}. \tag{25}$$

It is identical to the Brown-York stress tensor in (14) which is calculated from the whole metric (6).

4. Conclusion

As pointed out in Ref. 9, the Petrov type I condition is expected to be equivalent to the regularity condition on the future horizon of the spacetime, and it gives the constraint on the dual theory from gravity. We have indeed shown that imposing the Petrov type I condition is mathematically much simpler than imposing the regularity requirement, because one no longer needs to solve the perturbation equations of bulk gravity. To summarise, we overview the properties of Rindler Fluid dual to vacuum Einstein gravity, and regard the Petrov type I condition as an interesting recurrence relation. Through imposing the Petrov type I condition and Hamiltonian constraint on a finite cutoff surface, the stress tensor of the Rindler fluid can be fixed up to the second order of the derivative expansion. The resulting stress tensor identically matches the one calculated from the solution of vacuum Einstein equations.

Acknowledgments

This work is supported in part by National Natural Science Foundation of China (No.10821504, No.11035008, and No.11375247). We thank Prof. Rong-Gen Cai and Li Li for collaborations in this series of studies.

References

1. I. Bredberg, C. Keeler, V. Lysov and A. Strominger, JHEP **1103**, 141 (2011) [arXiv:1006.1902 [hep-th]].
2. I. Bredberg, C. Keeler, V. Lysov and A. Strominger, JHEP **1207**, 146 (2012) [arXiv:1101.2451 [hep-th]].
3. I. Bredberg and A. Strominger, JHEP **1205**, 043 (2012) [arXiv:1106.3084 [hep-th]].
4. D. Anninos, T. Anous, I. Bredberg and G. S. Ng, JHEP **1205**, 107 (2012) [arXiv:1110.3792 [hep-th]].
5. G. Compere, P. McFadden, K. Skenderis and M. Taylor, JHEP **1107**, 050 (2011) [arXiv:1103.3022 [hep-th]].
6. G. Compere, P. McFadden, K. Skenderis and M. Taylor, JHEP **1203**, 076 (2012) [arXiv:1201.2678 [hep-th]].
7. C. Eling, A. Meyer and Y. Oz, JHEP **1205**, 116 (2012) [arXiv:1201.2705 [hep-th]].
8. A. Meyer and Y. Oz, JHEP **1307**, 090 (2013) [arXiv:1304.6305].
9. V. Lysov and A. Strominger, arXiv:1104.5502 [hep-th].
10. T. -Z. Huang, Y. Ling, W. -J. Pan, Y. Tian and X. -N. Wu, JHEP **1110**, 079 (2011) [arXiv:1107.1464 [gr-qc]].
11. T. -Z. Huang, Y. Ling, W. -J. Pan, Y. Tian and X. -N. Wu, Phys. Rev. D **85**, 123531 (2012) [arXiv:1111.1576 [hep-th]].
12. C. -Y. Zhang, Y. Ling, C. Niu, Y. Tian and X. -N. Wu, Phys. Rev. D **86**, 084043 (2012) [arXiv:1204.0959 [hep-th]].
13. X. Wu, Y. Ling, Y. Tian and C. Zhang, Class. Quant. Grav. **30**, 145012 (2013) [arXiv:1303.3736 [hep-th]].
14. B. Wu and L. Zhao, Nucl. Phys. B **874**, 177 (2013) [arXiv:1303.4475 [hep-th]].
15. Y. Ling, C. Niu, Y. Tian, X. N. Wu and W. Zhang, Phys. Rev. D **90**, 043525 (2014) [arXiv:1306.5633 [gr-qc]].
16. V. Lysov, arXiv:1310.4181 [hep-th].
17. B. Wu and L. Zhao, Class. Quant. Grav. **31**, 105018 (2014) [arXiv:1401.6487 [hep-th]].
18. R. -G. Cai, L. Li, Q. Yang and Y. -L. Zhang, JHEP **1304**, 118 (2013) [arXiv:1302.2016 [hep-th]].
19. R. G. Cai, Q. Yang and Y. L. Zhang, Phys. Rev. D **90**, no. 4, 041901 (2014) [arXiv:1401.7792 [hep-th]].
20. R. G. Cai, Q. Yang and Y. L. Zhang, JHEP **1412**, 147 (2014) [arXiv:1408.6488 [hep-th]].
21. J. D. Brown and J. W. York, Jr., Phys. Rev. D **47**, 1407 (1993) [gr-qc/9209012].
22. C. Eling, A. Meyer and Y. Oz, JHEP **1208**, 088 (2012) [arXiv:1205.4249 [hep-th]].
23. G. Chirco, C. Eling and S. Liberati, JHEP **1108**, 009 (2011) [arXiv:1105.4482 [hep-th]].

Brief Review on the Entanglement Entropy Bound

Chanyong Park

Asia Pacific Center for Theoretical Physics, Pohang, 790-784, Korea
Department of Physics, Postech, Pohang, 790-784, Korea
chanyong.park@apctp.org

In this note, we briefly summarize our recent work on the entanglement entropy bound. In a quantumly entangled system, the non-negativity of the relative entropy leads to the entanglement entropy bound. When the entanglement entropy bound is saturated, a quantum system satisfies the thermodynamics-like law with an appropriately defined entanglement temperature. We show that the saturation of the entanglement entropy bound accounts for a universal feature of the entanglement temperature proportional to the inverse of the system size.

Keywords: Entanglement entropy; relative entropy; entanglement entropy bound.

1. Introduction

In black hole physics, the Bekenstein bound has been proposed from a thought experiment.[1] When an object is absorbed into a black hole, it describes that the increased thermal entropy of a black hole should be bounded from the absorbed energy. This Bekenstein bound can be further generalized to an entanglement entropy bound which is also valid in quantum systems. When a ground state is excited, the entanglement entropy bound implies that the increased entanglement entropy is bounded by the excitation energy similar to the Bekenstein bound. This entanglement entropy bound has been derived from the non-negativity of the relative entropy.[2,3]

In this note, we briefly summarize the entanglement entropy of an excited state and the universality of its thermodynamics-like law.[4] To do so, we take into account a charged AdS black brane geometry with peculiar properties.[5–7] This geometry allows for the dual field theory to have a Fermi sea and massless fluctuations on the Fermi surface. Using the holographic renormalization technique, its thermodynamic properties have been studied from the boundary stress tensor.[8] Interestingly, it has been shown that the trace of the boundary stress tensor does not vanish even though it has an AdS asymptote. This fact implies that matter of the dual field theory is non-conformal. On the dual geometry of a strongly interacting fermioninc system, we investigate the entanglement entropy bound which accounts for the universality of the entanglement temperature.

2. Entanglement entropy bound

In order to describe the entanglement entropy bound, let us first define a relative entropy which is independent of the renormalization scheme. When two states are in the same Hilbert space, the relative entropy gives rise to a fundamental statistical measure of their distance. When two reduced density matrices are given by ρ_1 and ρ_0, the relative entropy $S(\rho_1|\rho_0)$ is defined as[3]

$$S(\rho_1|\rho_0) \equiv \mathrm{Tr}\,(\rho_1 \log \rho_1) - \mathrm{Tr}\,(\rho_1 \log \rho_0). \tag{1}$$

Here we can identify ρ_0 with the reduced density matrix of a ground or thermal state, while ρ_1 is one for a quantumly or thermally excited state. Using the variations of the modular Hamiltonian and entanglement entropy

$$\Delta \langle K \rangle = \mathrm{Tr}\,(\rho_1 K) - \mathrm{Tr}\,(\rho_0 K) \quad \text{and} \quad \Delta S = S(\rho_1) - S(\rho_0), \tag{2}$$

the relative entropy can be reexpressed as

$$S(\rho_1|\rho_0) = \Delta \langle K \rangle - \Delta S. \tag{3}$$

The relative entropy is usually non-negative, which leads to the following inequality

$$\Delta \langle K \rangle \geq \Delta S. \tag{4}$$

This is the entanglement entropy bound holding for any region in QFT and implies that the increase of the entanglement entropy is bounded by the increased energy.

3. Dual of a fermionic system

Recently, numerous charged dilatonic black brane solutions have been extensively studied for understanding the Fermi surface.[9] It has been shown that a Fermi surface and massless fluctuations on it can be described by the following dual gravity[5]

$$S = \frac{1}{2\kappa^2} \int d^5x \sqrt{-g} \left[\mathcal{R} - \frac{1}{4} e^{4\phi} F_{\mu\nu} F^{\mu\nu} - 12 \partial_\mu \phi \partial^\mu \phi + \frac{1}{R^2} \left(8 e^{2\phi} + 4 e^{-4\phi} \right) \right], \tag{5}$$

where $R = 1$ corresponds to an AdS radius. This action shows that there is a local minimum at $\phi = 0$ where the scalar potential reproduces the five-dimensional AdS cosmological constant. If there exists a nontrivial ϕ approaching to zero at the asymptotic boundary, there exists an asymptotic AdS solution. Solving equations of motion gives rise to the following solution

$$\begin{aligned} ds^2 &= r^2 e^{2A(r)} \left(-f(r) dt^2 + d\vec{x}^2 \right) + \frac{e^{2B(r)}}{r^2 f(r)} dr^2, \\ A &= A_t dt, \end{aligned} \tag{6}$$

with

$$\begin{aligned} A(r) &= \frac{1}{3} \log \left(1 + \frac{Q^2}{8mr^2} \right), \quad B(r) = -\frac{2}{3} \log \left(1 + \frac{Q^2}{8mr^2} \right), \\ \phi(r) &= \frac{1}{6} \log \left(1 + \frac{Q^2}{8mr^2} \right), \quad f(r) = 1 - \frac{m}{r^4 \left(1 + \frac{Q^2}{8mr^2} \right)^2}, \\ A_t &= 2\kappa^2 \mu - \frac{Q}{2r^2 \left(1 + \frac{Q^2}{8mr^2} \right)}, \end{aligned} \tag{7}$$

where m, μ and Q indicate the charged black brane's mass, chemical potential and charge density respectively. The extremal limit supports normal modes of massless charged fermions so that the geometry we consider is regarded as the dual of Fermi liquid.[5] In the grand canonical ensemble, the internal energy satisfying the first law of thermodynamics is given by[8]

$$E = \frac{3\pi^4 V_3}{2\kappa^2} T_H^4 + 6\pi^2\kappa^2 V_3 T_H^2 \mu^2 + \frac{14}{3}\kappa^6 V_3 \mu^4. \tag{8}$$

4. Thermodynamics-like law from the entanglement entropy bound

In recent studies,[10–12] it has been found that the entanglement entropy of an excited state in a strip region follows the thermodynamics-like law after defining an appropriate entanglement temperature. Intriguingly, the entanglement temperature shows a universal feature inversely proportional to the system size. In this note, we try to explain this universality by using the entanglement entropy bound. According to a simple dimension counting in a relativistic QFT, we can guess that the increased modular Hamiltonian is proportional to the increased energy

$$\Delta \langle K \rangle = \lambda \, l \Delta E \tag{9}$$

with a non-universal numerical factor λ. If it is correct, we can see that the entanglement temperature is universally proportional to the inverse of the system size

$$T_E = \frac{1}{\lambda l}. \tag{10}$$

Since the entanglement entropy bound is universal, the universality of the entanglement temperature naturally appears in the thermodynamics-like law regardless of the shape of the entangling surface.

In general, the modular Hamiltonian is not known except several simple cases. One of them is the case with a spherical entangling surface. Now, let us consider the entanglement entropy contained in a ball-shaped region. Parameterizing a disk with a radius l as

$$0 \le \rho \le l. \tag{11}$$

and considering z as a function of ρ, the action for the minimal surface is reduced to

$$A = \Omega_2 \int_0^{l-l_*} d\rho \, \frac{e^{2A}\rho^2}{z^3} \sqrt{e^{2A} + \frac{e^{2B}}{f} z'^2}, \tag{12}$$

where l_* in the upper limit is introduced to denote a UV cutoff and the prime means a derivative with respect to ρ. The equation of motion for z reads

$$\begin{aligned} 0 = {} & \rho z f z'' + 2 z z'^3 e^{2B-2A} - 4\rho z f z'^2 A' + \rho z f z'^2 B' - \frac{1}{2}\rho z z'^2 f' \\ & - 3\rho z f^2 e^{2A-2B} A' + 3\rho f^2 e^{2A-2B} + 3\rho f z'^2 + 2 z f z'. \end{aligned} \tag{13}$$

Since μl and $T_H l$ have small values in a UV region, one can expand z as follows

$$z(\rho) = z_0(\rho) + \kappa^4\mu^2 l^2 z_1(\rho) + \kappa^8\mu^4 l^4 z_2(\rho) + T_H^4 l^4 z_3(\rho) + \kappa^4\mu^2 T_H^2 l^4 z_4(\rho) + \cdots\,, \tag{14}$$

where the ellipsis indicates higher order corrections.

At leading order, the action is exactly reduced to that obtained from a pure AdS space and its solution has already been known as[13,14]

$$z_0(\rho) = \sqrt{l^2 - \rho^2}. \tag{15}$$

Around this known solution, the first correction caused by the deformation are governed by z_1. At l^2 order, z_1 is given by

$$z_1 = -\frac{2\rho^2(2l^2-\rho^2)}{3l^2\sqrt{l^2-\rho^2}} + \frac{(l-\rho)^2 c_1}{\rho\sqrt{l^2-\rho^2}} + \frac{c_2}{\sqrt{l^2-\rho^2}}. \tag{16}$$

Since the subsystem we consider is located at $z = 0$, all higher order functions should vanish at $\rho = l$. This constraint fixes c_2 to be $c_2 = \frac{2l^2}{3}$. In addition, the smoothness of the minimal surface at the turning point, $z_1' = 0$ at $\rho = 0$, determines the remaining integral constant to be $c_1 = 0$.

At l^4 order, z_2, z_3 and z_4 satisfying equations of motion are given by

$$\begin{aligned}
z_2 &= \frac{2(5l^4\rho^2 + 2l^2\rho^4 - 3\rho^6)}{45l^4\sqrt{l^2-\rho^2}} + \frac{(l-\rho)^2 c_3}{\rho\sqrt{l^2-\rho^2}} + \frac{c_4}{\sqrt{l^2-\rho^2}}, \\
z_3 &= \frac{\pi^4(5l^4\rho^2 - 4l^2\rho^4 + \rho^6)}{10l^4\sqrt{l^2-\rho^2}} + \frac{(l-\rho)^2 c_5}{\rho\sqrt{l^2-\rho^2}} + \frac{c_6}{\sqrt{l^2-\rho^2}}, \\
z_4 &= \frac{2\pi^2(5l^4\rho^2 - 4l^2\rho^4 + \rho^6)}{5l^4\sqrt{l^2-\rho^2}} + \frac{(l-\rho)^2 c_7}{\rho\sqrt{l^2-\rho^2}} + \frac{c_8}{\sqrt{l^2-\rho^2}}.
\end{aligned} \tag{17}$$

Imposing again that all higher order functions should be zero at $\rho = l$, a half of unknown integral constants are determined to be

$$c_2 = \frac{2l^2}{3}\ ,\quad c_4 = -\frac{8l^2}{45}\ ,\quad c_6 = -\frac{\pi^4 l^2}{5}\quad \text{and}\quad c_8 = -\frac{4\pi^2 l^2}{45}. \tag{18}$$

In addition, the smoothness of the minimal surface at the turning point, $z_1' = z_2' = z_3' = z_4' = 0$ at $\rho = 0$, yields $c_1 = c_3 = c_5 = c_7 = 0$. Substituting above solutions into the action, the minimal area up to l^4 order gives rise to

$$A = \Omega_2 \int_0^{l-l_*} d\rho \left[\frac{l\rho^2}{(l^2-\rho^2)^2} - \frac{8}{15}\kappa^8 l\mu^4\rho^2 + \frac{3}{5} l\rho^2\left(\pi^2 T_H^2 + 2\kappa^4\mu^2\right)^2\right]. \tag{19}$$

Here l_* is associated with the UV cutoff of the z coordinate denoted by ϵ. The perturbative solution we found determines their relation up to higher order corrections

$$l_* = \frac{\epsilon^2}{2l}\left[1 + \left\{\frac{1}{4l^2} - \frac{4\kappa^4}{3}\mu^2 - \frac{8l^2}{45}\kappa^8\mu^4 + \frac{l^2}{5}\left(\pi^2 T_H^2 + 2\kappa^4\mu^2\right)^2\right\}\epsilon^2 + \cdots\right]. \tag{20}$$

Using this relation, the entanglement entropy finally becomes

$$S = \frac{\pi l^2 \Omega_2}{\kappa^2 \epsilon^2} + \frac{\pi \Omega_2}{\kappa^2} \log\left(\frac{\epsilon}{l}\right) - \frac{\pi \Omega_2}{2\kappa^2}(1 + 2\log 2)$$
$$+ \frac{4\pi l^2 \Omega_2}{3\kappa^2} \kappa^4 \mu^2 - \frac{8\pi l^4 \Omega_2}{45\kappa^2} \kappa^8 \mu^4 + \frac{\pi l^4 \Omega_2}{5\kappa^2} \left(\pi^2 T_H^2 + 2\kappa^4 \mu^2\right)^2 . \tag{21}$$

This is the entanglement entropy of the excited state with the chemical potential. At a given chemical potential, the increased entanglement entropy up to l^4 order is given by

$$\Delta S = \frac{\pi^5 l^4 \Omega_2}{5\kappa^2} T_H^4 + \frac{4\pi^3 \kappa^4 l^4 \Omega_2}{5\kappa^2} \mu^2 T_H^2, \tag{22}$$

which is the exact same as the increased modular Hamiltonian. the saturation of the entanglement entropy bound leads to the thermodynamics-like law

$$\Delta \langle K \rangle = \Delta S = \frac{\Delta E}{T_E}, \tag{23}$$

with

$$T_E = \frac{5}{2\pi l}. \tag{24}$$

This result shows that the entanglement temperature is universally proportional to the inverse of the system size as expected.

5. Conclusion

When an object is absorbed into a black hole, the Bekenstein bound has been proposed to explain the increase of the thermal entropy. In this note, we have investigated the generalized entanglement entropy bound for a holographic fermion system with a Fermi surface. Intriguingly, the entanglement entropy bound is originated from the non-negativity of the relative entropy and can be applied to a quantum system unlike the Bekenstein bound. Rewriting the entanglement entropy bound in terms of the system energy instead of the modular Hamiltonian, it is equivalent to the Bekenstein bound except that it is working even in a quantum system. Recently, it has been shown that the entanglement temperature satisfying the thermodynamics-like law has a universal feature inversely proportional to the system size.[10] We showed that the almost saturated entanglement entropy bound can account for the universality of the entanglement temperature.

Acknowledgments

C. Park was supported by Basic Science Research Program through the National Research Foundation of Korea funded by the Ministry of Education (NRF-2013R1A1A2A10057490) and also by the Korea Ministry of Education, Science and Technology, Gyeongsangbuk-Do and Pohang City.

References

1. J. D. Bekenstein, Phys. Rev. D **7**, 2333 (1973); J. D. Bekenstein, Phys. Rev. D **9**, 3292 (1974); J. D. Bekenstein, Phys. Rev. D **23**, 287 (1981).
2. H. Casini, Class. Quant. Grav. **25**, 205021 (2008) [arXiv:0804.2182 [hep-th]].
3. D. D. Blanco, H. Casini, L. Y. Hung and R. C. Myers, JHEP **1308**, 060 (2013) [arXiv:1305.3182 [hep-th]].
4. C. Park, arXiv:1511.02288 [hep-th].
5. S. S. Gubser and F. D. Rocha, Phys. Rev. D **81**, 046001 (2010) [arXiv:0911.2898 [hep-th]].
6. S. S. Gubser and I. Mitra, JHEP **0108**, 018 (2001) [hep-th/0011127].
7. R. A. Davison, M. Goykhman and A. Parnachev, JHEP **1407**, 109 (2014) [arXiv:1312.0463 [hep-th]].
8. C. Park, Adv. High Energy Phys. **2014**, 565219 (2014) [arXiv:1405.1490 [hep-th]].
9. B. Gouteraux and E. Kiritsis, JHEP **1112**, 036 (2011) [arXiv:1107.2116 [hep-th]]; X. Dong, S. Harrison, S. Kachru, G. Torroba and H. Wang, JHEP **1206**, 041 (2012) [arXiv:1201.1905 [hep-th]]; S. Kulkarni, B. H. Lee, C. Park and R. Roychowdhury, JHEP **1209**, 004 (2012) [arXiv:1205.3883 [hep-th]]; S. Kulkarni, B. H. Lee, J. H. Oh, C. Park and R. Roychowdhury, JHEP **1303**, 149 (2013) [arXiv:1211.5972 [hep-th]]; C. Park, Adv. High Energy Phys. **2013**, 389541 (2013) [arXiv:1209.0842 [hep-th]]; C. Park, Phys. Rev. D **91**, no. 12, 126003 (2015) [arXiv:1501.02908 [hep-th]].
10. J. Bhattacharya, M. Nozaki, T. Takayanagi and T. Ugajin, Phys. Rev. Lett. **110**, no. 9, 091602 (2013) [arXiv:1212.1164].
11. E. Bianchi and R. C. Myers, Class. Quant. Grav. **31**, 214002 (2014) [arXiv:1212.5183 [hep-th]].
12. M. Nozaki, T. Numasawa, A. Prudenziati and T. Takayanagi, Phys. Rev. D **88**, no. 2, 026012 (2013) [arXiv:1304.7100 [hep-th]].
13. S. Ryu and T. Takayanagi, Phys. Rev. Lett. **96**, 181602 (2006) [hep-th/0603001]; S. Ryu and T. Takayanagi, JHEP **0608**, 045 (2006) [hep-th/0605073].
14. C. Park, Phys. Rev. D **92**, no. 12, 126013 (2015) [arXiv:1505.03951 [hep-th]]; K. K. Kim, O. K. Kwon, C. Park and H. Shin, Phys. Rev. D **90**, no. 4, 046006 (2014) [arXiv:1404.1044 [hep-th]]; K. K. Kim, O. K. Kwon, C. Park and H. Shin, Phys. Rev. D **90**, no. 12, 126003 (2014) [arXiv:1407.6511 [hep-th]].

Entanglement Renormalization and Integral Geometry

Feng-Li Lin

Department of Physics, National Taiwan Normal University,
Taipei, 116, Taiwan
linfengli@phy.ntnu.edu.tw

We adopt the integral geometry to study the holographic entanglement entropy in AdS_3 and realize the dual picture of entanglement renormalization of multi-scale entanglement renormalization ansatz (MERA).

Keywords: Holographic entanglement entropy; MERA; integral geometry.

1. Introduction

Ever since the seminal work by the Ryu and Takayanagi,[1,2] the relation between geometry in the bulk and the entanglement of the boundary state has been studied extensively, see e.g. Refs. 3,4. However, a working prescription on building the bulk spacetime from boundary entanglement remains much desired.

A framework to achieve the above goal is an efficient algorithm proposed for evaluating the ground states of many-body systems by evolving the RG flow through the network of local unitary operations (LUs). This algorithm is called the multi-scale entanglement renormalization ansatz, denoted as MERA[5,6] and is a revised version of real space renormalization of quantum state by also removing local entanglement (called disentanglers in MERA) besides coarse graining (called isometries in MERA). A keen observation in Ref. 7 about the resemblance between the geometric structures of quantum circuit for MERA and the AdS space initiates further elaborated works[8,9] along this direction, the so-called AdS/MERA correspondence.

To build the connection between the bulk geometry and boundary entanglement renormalization in more microscopic sense, one needs some bulk geometric picture of the individual entangled pair so that the action of LUs of MERA on it generates the bulk line element along the RG trajectory. Besides, the entanglement entropy for some region can be understood as the additive sum of the contribution from the entangled pairs with one end inside and one outside. This provides a microscopic view point of entanglement entropy, a concept called entanglement contour as introduced in Ref. 10.

The recent developments in differential entropy[11–13] offers powerful tools ready for this purpose. Especially, a special way of discussing the distance measured in 3D hyperbolic space by the the integral geometry, e.g. Ref. 14, leads Ref. 15 to express the differential entropy as the volume of the kinematic space of the geodesics intersecting the target bulk curve. Each point in the kinematic space by itself corresponds to a geodesic with two boundary endpoints, which can be intuitively thought as a pair of entangled parties, and their contribution to the differential entropy is weighted by the volume measure of the kinematic space, i.e., the entanglement contour.

In this work, we adopt the framework introduced in Ref. 15 to explore the local and microscopic understanding of the relation between bulk line element and the entanglement renormalization under the quantum state RG flow of the boundary CFT. The additive feature of the entanglement contour enables us to study the entanglement renormalization of an individual entangled pair in a RG step.

2. Review of integral geometry and entanglement contour

Motivated by AdS/MERA, a more refined picture in the context AdS/CFT proposed recently[16,17] is the so-called surface/state correspondence (SS-duality), which we will adopt for our discussions in this work. The proposal states that any co-dimensional two convex surface Σ is dual to a holographic quantum state described by density matrix $\rho(\Sigma)$. Especially, if Σ is closed, it is dual to a pure state $\rho(\Sigma) = |\psi(\Sigma)\rangle\langle\psi(\Sigma)|$. In the context of AdS/CFT, the bulk space can be foliated into congruence of these convex surfaces, and different surfaces/states are connected by a network of LUs which can be interpreted as renormalization flow if one follows the UV to IR directions. The LUs of MERA are utilized to disentangle the IR-irrelevant degrees of freedom (d.o.f.) from the the IR-relevant ones. By keeping only the IR-relevant d.o.f., the RG flow can be treated as a unitary transformation. The feature of disentangling the IR-relevant and irrelevant d.o.f. makes the RG flow as a process of entanglement renormalization.

In the framework of integral geometry, the volume of a q-dimensional object M^q in a d-dimensional symmetric space $\mathcal{M}$ can be measured by r-dimensional planes. More precisely, the volume of M^q can be expressed in terms of an integral over an auxiliary space $\mathcal{K}$ called kinematic space, which consists of all the r-planes. In the context of AdS/CFT, we apply integral geometry to a d-dimensional spatial slice of the stationary bulk spacetime. In AdS_3 ($d = 2$), the planes we need to consider are geodesics ($r = 1$) and we are mainly interested in the bulk curves ($q = 1$), whose lengths can be written in terms of integrals via the Crofton's formula (G is the Newton constant of bulk gravity)

$$\frac{\sigma(\gamma)}{4G} = \frac{1}{4}\int_{\gamma\cap\Gamma\neq\emptyset} N(\gamma\cap\Gamma)\,\epsilon_{\mathcal{K}}\,, \tag{1}$$

where $N(\gamma \cap \Gamma)$ denotes intersection number between the bulk curve γ and oriented geodesics Γ ending on the boundary, and $\epsilon_{\mathcal{K}}$ denotes the volume form of the kinematic space $\mathcal{K}$.

In the case of pure AdS_3 the measure follows from that of the spacetime isometry group and has a natural entropic interpretation

$$\epsilon_{\mathcal{K}}(u,v) = \frac{\partial^2 S(u,v)}{\partial u \partial v} du \wedge dv\,. \tag{2}$$

The function $S(u,v)$ is the length (divided by $4G$) of the geodesic with u,v being the boundary end points, which are now also the coordinates of the kinematic space. According to Ryu-Takayanagi (RT) formulation,[1,2] $S(u,v)$ is the holographic entanglement for the the boundary interval $[u,v]$.

Based on (1), the holographic entanglement entropy for a surface state at $z=\tilde{z}$ horizontal surface of AdS$_3$ ($ds^2=\frac{dz^2-dt^2+dx^2}{z^2}$) of the interval $I_\ell := [\,\tilde{x},\tilde{x}+\ell\,]$ can be rewritten as the sum of contribution associated with spatially separated pairs of points

$$S_{EE}(\ell;\tilde{z})=\frac{1}{2}\int_{-\infty}^{\tilde{x}}\mathrm{d}u\int_{\tilde{x}}^{\tilde{x}+l}\mathrm{d}v\frac{\partial^2 S(u,v;\tilde{z})}{\partial u\partial v}+\frac{1}{2}\int_{\tilde{x}+\ell}^{\infty}\mathrm{d}v\int_{\tilde{x}}^{\tilde{x}+l}\mathrm{d}u\frac{\partial^2 S(u,v;\tilde{z})}{\partial u\partial v}\,. \quad (3)$$

In the above either u or v is inside I_ℓ but not both. The relation to the differential entropy is given by

$$\int_{\gamma\cap\Gamma\neq\emptyset}\frac{\partial^2 S(u,v)}{\partial u\partial v}\mathrm{d}u\wedge\mathrm{d}v=-\oint\frac{\partial S(u,v(u))}{\partial u}\mathrm{d}u\,. \quad (4)$$

Based on (3) and (4) we can define the long-distance entanglement contour (of a point) for latter usage, which only includes the contour of pairs separated by distance larger than some length scale $\ell_{\geq}$, i.e.,

$$\Theta(\ell_{\geq},\tilde{x};\tilde{z})\mathrm{d}x:=\frac{1}{2}\mathrm{d}x\Big(\frac{\partial S(\tilde{x}-\ell_{\geq},v;\tilde{z})}{\partial v}|_{v=\tilde{x}}-\frac{\partial S(u,\tilde{x}+\ell_{\geq};\tilde{z})}{\partial u}|_{u=\tilde{x}}\Big)\,. \quad (5)$$

We note that this quantity by definition is another entanglement contour since $\bar{s}(i\cup j)=\bar{s}(i)+\bar{s}(j)$.

3. Integral geometrical view of entanglement renormalization

Now we like to consider how the entanglement contour renormalizes along the RG flow from the integral geometry point of view. We notice that the change of the cutoff line from $z=z_1<\tilde{z}$ to $z=z_2<z$ will not affect the measure of the kinematic space (number density of geodesics) at $z=\tilde{z}$ as long as $z_1,z_2<\tilde{z}$

$$\frac{\partial^2 S(u(z_1),v(z_1);z_1)}{\partial u\partial v}\mathrm{d}u(z_1)\wedge\mathrm{d}v(z_1)=\frac{\partial^2 S(u(z_2),v(z_2);z_2)}{\partial u\partial v}\mathrm{d}u(z_2)\wedge\mathrm{d}v(z_2)\,, \quad (6)$$

where $u(z),v(z)$ are end-point coordinates defined on different slices. Obviously, the integrated form of (6) also holds. By Crofton's formula, It implies that the differential entropy of some bulk curve (within the region of $z<\tilde{z}$) is invariant as we move the running RG scale from $z=z_1<\tilde{z}$ to $z=z_2<\tilde{z}$.

Moreover, as we can see from Fig. 1 the distance $|u(z)-v(z)|$ between the end points of the same geodesic shrinks along the RG flow. In other words, the same correlation is seen as for shorter scale from the viewpoint of an IR observer.

To derive some concrete result from the fact of (6), we consider renormalization of the long-distance entanglement contour (of a point) (5) in homogeneous space, i.e. holographically dual to pure AdS$_3$. For a surface/state $\psi(\tilde{z})$, the entanglement entropy between a line element around the point at $(\tilde{x},\tilde{z})$ and its complement

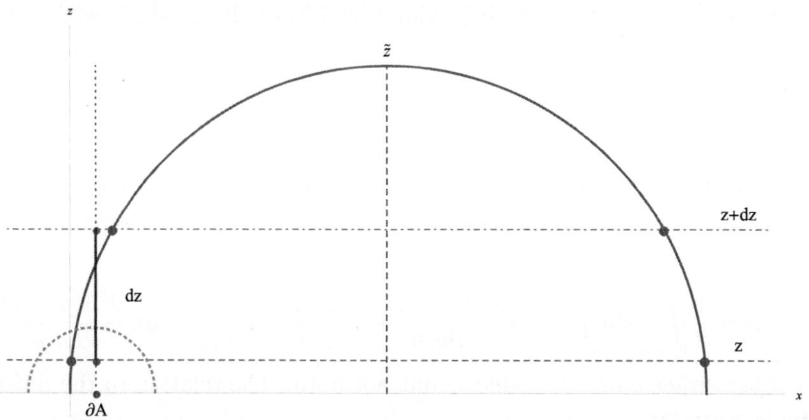

Fig. 1. The long-distance entanglement is invariant under RG flow but it is reshuffled to shorter scale. Two situations are shown here. The first is the blue geodesic which intersects both surface states at z and $z + dz$. This implies that the entanglement contour for the intersecting pair of points is just reshuffled but not removed. However, the entanglement entropy is reduced (and turned into line element dz) as the left end point is dragged across the entangling surface (the intersection points of the vertical bar with the horizontal lines). The second situation is the red-dotted geodesic for which the intersecting points shrink to none as the surface state is pushed up from z to $z + dz$. The first situation is similar to isometry operation in MERA, while the second is to disentangler.

is $\Theta(0, \tilde{x}; \tilde{z})$ as already given by (5). Due to the assumed homogeneity along x-direction, $\Theta(\ell, \tilde{x}; z)$ is independent of $\tilde{x}$ (thus we omit the $\tilde{x}$ in $\Theta(\ell, \tilde{x}; z)$ hereafter) so that we can add up the this contour on a horizontal closed line (by imposing periodic boundary condition $\tilde{x} \sim \tilde{x} + L_x$) to arrive its length (or the entanglement entropy) $\Theta(0; \tilde{z})L_x$. On the other hand, the length can be seen as the differential entropy for the surface/state $\psi(z)$, i.e., using (4) and the fact that (5) is given by $\int \partial_{\Delta x} S(\Delta x) d\tilde{x} = L_x \Theta(2\sqrt{\tilde{z}^2 - z^2}; z)$. The equivalence of the above two different views is guaranteed by the same integral geometric construction and then it yields

$$\Theta(2\sqrt{\tilde{z}^2 - z^2}; z) = \Theta(0; \tilde{z}) \; . \tag{7}$$

Eq. (7) implies that the long-distance entanglement (more precisely, the part captured by the contour) is reshuffled into shorter scale as one flows the surface state from the UV state $\psi(0)$ to $\psi(z)$.

This result of (7) can be understood holographically as follows. As shown in Fig. 1, the entanglement contour for the differential entropy associated with the line element around the point $(\tilde{x}, \tilde{z})$ is manifested as the geodesic with $(\tilde{x}, \tilde{z})$ as its highest point. The locations of the pair of points are then specified by its intersection with the cutoff surface for the UV reference state. As long as $z \leq \tilde{z}$, the entanglement contour remains after the RG step from $\psi(0)$ to $\psi(z)$ but with the separation of the end points change from $2\tilde{z}$ to $2\sqrt{\tilde{z}^2 - z^2}$. As the contour of all the other pairs contributed to $\Theta(2\sqrt{\tilde{z}^2 - z^2}; z)$ are manifested in a larger geodesic than the above one so that they also remain after the above RG step. Thus the long-distance entanglement contour remains the same after the RG running as long as the running scale $z < \tilde{z}$.

In some sense, this reshuffling is similar to the isometry operations of MERA, for which the entanglement of a pair are not removed but rescaled under coarse-graining into that of a new pair.[a] This is exactly what we have shown in Fig. 1. On the other hand, there are situations in which the entanglement contour of a pair of points is removed under RG flow, as shown in (the red-dotted geodesic of) Fig. 1. The entanglement entropy reduced in this case is also turned into (a proportion of) the length of the line element as predicted by the Crofton's formula (1). This is similar to the disentangler operations in MERA.

In fact, the above entanglement renormalization can be encoded in the following RG equation for the entanglement contour, i.e.,

$$\frac{d}{dz}\Theta(2\sqrt{\tilde{z}^2 - z^2}; z) = 0\,. \tag{8}$$

Finally we would like to remark on how the RT formula can be realized from the renormalization of entanglement contour. Both types of LUs (disentangler and isometry) can be visualized as segments of geodesics (red and blue respectively in Fig. 1) chopped off by the surfaces of RG, which are the analogues of the links (disentangler/isometry) connecting neighboring sites on the MERA network. Taking both types of geodesics (long and short) into account, the Crofton's formula (1) turns into the familiar form in MERA

$$\text{length of curve } \gamma = \#(\text{LUs cut})\,. \tag{9}$$

Moreover, the entanglement entropy of an interval A is bounded by the number of cuts (LUs that reduce the density matrix ρ_A to the one with no entanglement)

$$S_A \lesssim \#(\text{LUs cut}) = \text{length of curve } \gamma \text{ ending on } \partial A, \tag{10}$$

which is precisely the MERA version of RT formula.

4. Conclusion

The central theme of this study is to understand bulk geometry from the entanglement of the boundary field theory, or vice versa. To achieve our goal, we propose a new physical interpretation for the measure in the kinematic space of integral geometry as the entanglement (contour) associated with the two end points of geodesics on the boundary. We then show that along the RG flow the entanglement contour evolves as if under the influence of local unitary operations in MERA (or rather cMERA). Moreover, a RG equation for the long-distance entanglement contour is

[a] On a MERA network, a similar process is realized as transferring of entanglement to the sites on the next layer, which is part of the isometry operation. We note however that the correspondence is not precise because a site in general carries entanglement to both sides and hence the entanglement entropy is not always reduced when the site is dragged across ∂A.

derived, which could be used as a guide for deriving the bulk dynamics. Our result is the refinement of the AdS/MERA duality by looking into the renormalization of the entanglement contour for a pair of points, which is a new type of measure of spatial entanglement.

Acknowledgments

This work is supported by MoST grant: 103-2112-M-003 -001 -MY3 and 103-2811-M-003 -024. Also, thank Xing Huang for collaboration.

References

1. S. Ryu and T. Takayanagi, "Holographic derivation of entanglement entropy from AdS/CFT," Phys. Rev. Lett. **96**, 181602 (2006) [hep-th/0603001].
2. S. Ryu and T. Takayanagi, "Aspects of Holographic Entanglement Entropy," JHEP **0608**, 045 (2006) [hep-th/0605073].
3. M. Van Raamsdonk, "Comments on quantum gravity and entanglement," arXiv:0907.2939 [hep-th].
4. M. Van Raamsdonk, "Building up spacetime with quantum entanglement," Gen. Rel. Grav. **42**, 2323 (2010) [Int. J. Mod. Phys. D **19**, 2429 (2010)] [arXiv:1005.3035 [hep-th]].
5. G. Vidal, "A class of quantum many-body states that can be efficiently simulated", Phys. Rev. Lett. **101**, 110501 (2008). [arXiv:quant-ph/0610099]
6. G. Evenbly, G. Vidal, "Algorithms for entanglement renormalization", Phys. Rev. B **79**, 144108 (2009). [arXiv:0707.1454 [cond-mat.str-el]].
7. B. Swingle, "Entanglement Renormalization and Holography," Phys. Rev. D **86**, 065007 (2012) [arXiv:0905.1317 [cond-mat.str-el]].
8. G. Evenbly, G. Vidal, "Tensor network states and geometry", J. Stat. Phys. (2011) 145:891-918, arXiv:1106.1082 [quant-ph].
9. M. Nozaki, S. Ryu and T. Takayanagi, "Holographic Geometry of Entanglement Renormalization in Quantum Field Theories," JHEP **1210**, 193 (2012) [arXiv:1208.3469 [hep-th]].
10. Y. Chen and G. Vidal, "Entanglement contour," J. Stat. Mech. (2014) P10011. [arXiv:1406.1471 [cond-mat.str-el]].
11. V. Balasubramanian, B. D. Chowdhury, B. Czech, J. de Boer and M. P. Heller, "Bulk curves from boundary data in holography," Phys. Rev. D **89**, no. 8, 086004 (2014) [arXiv:1310.4204 [hep-th]].
12. B. Czech, X. Dong and J. Sully, "Holographic Reconstruction of General Bulk Surfaces," JHEP **1411**, 015 (2014) [arXiv:1406.4889 [hep-th]].
13. B. Czech and L. Lamprou, "Holographic definition of points and distances," Phys. Rev. D **90**, no. 10, 106005 (2014) [arXiv:1409.4473 [hep-th]].
14. L. Santaló, "Integral geometry and geometric probability," Cambridge University Press, 1976.
15. B. Czech, L. Lamprou, S. McCandlish and J. Sully, "Integral Geometry and Holography," arXiv:1505.05515 [hep-th].
16. M. Miyaji, S. Ryu, T. Takayanagi and X. Wen, "Boundary States as Holographic Duals of Trivial Spacetimes," JHEP **1505**, 152 (2015) [arXiv:1412.6226 [hep-th]].
17. M. Miyaji and T. Takayanagi, "Surface/State Correspondence as a Generalized Holography," arXiv:1503.03542 [hep-th].

PARALLEL SESSION

X. Black Hole Evaporation and Information Loss Paradox

Summary of Parallel Session: "Black Hole Evaporation and Information Loss Paradox"

Yen Chin Ong
Nordita, KTH Royal Institute of Technology & Stockholm University, Roslagstullsbacken 23, SE-106 91 Stockholm, Sweden
yenchin.ong@nordita.org

Dong-Han Yeom
Leung Center for Cosmology and Particle Astrophysics, National Taiwan University, Taipei 10617, Taiwan
innocent.yeom@gmail.com

We summarize the presentations of the parallel session "Black hole evaporation and information loss paradox", and provide some thoughts on the open problems and future prospects of the field.

Keywords: Black hole, information loss paradox.

1. Introduction

The parallel session on *"Black hole evaporation and information loss paradox,"* jointly organized by us, was held on 17th December 2015. The presentations were:

- **Pei-Ming Ho**, *The absence of horizon in black hole formation*
- **Wen-Yu Wen**, *Black hole remnant in asymptotic anti-de Sitter space*
- **Michael R. R. Good**, *Black mirrors: a one-to-one correspondence between moving mirrors and black holes*
- **Dong-han Yeom**, *Information loss problem and the roles of instantons*
- **Guillem Doménech**, *Stationary bubbles and their tunneling channels toward trivial geometry*
- **Yen Chin Ong**, *Black hole: the interior spacetime*
- **Naritaka Oshita**, *Unruh radiation produced by a uniformly accelerating charged particle in thermal random motions*
- **Yao-Chieh Hu**, *Two interpretations on thin-shell instantons*

In addition, there were a few relevant plenary session presentations and review talks:

- **William G. Unruh**, *Firewalls: what and why the issue*
- **Don N. Page**, *Naked firewalls*
- **Pisin Chen**, *Decelerating plasma mirror to investigate black hole information loss paradox*

– **Abhay Ashtekar**, *Quantum gravity: An overview*

In this review, we shall first summarize these presentations. Each of them is either a possible candidate resolution of the information loss paradox, or could provide some clues towards said resolution. We then comment on the possible future research directions and open issues in this field.

2. Summary of Presentations

2.1. *Candidates of resolutions*

In this session, there were many suggestions on the resolution of the information loss problem.[1] In order to classify these candidates of resolutions, as a reference we shall list the usual assumptions of black hole complementarity:[2]

(1) *Unitarity* holds throughout the entire process of black hole evaporation;

(2) *No drama* for an in-falling observer in accordance to general relativity;

(3) *Local quantum field theory* is correct, at least outside the stretched horizon of a sufficiently massive black hole;

(4) *The statistical entropy of a black hole is proportional to the black hole area*: $\log N = \mathcal{A}/4$, where N is the number of states of a black hole and $\mathcal{A}$ is the area of the black hole horizon;

To this we add an often implicit assumption that:

(5) *There is an observer* who can extract information from the Hawking radiation.[a]

If we accept the various arguments – such as the firewall claim – that black hole complementarity is inconsistent,[4,5] then we need to drop at least one of these assumptions. Therefore, we have the following options (We denote the negation of the aforementioned assumptions with a "¬" sign):

¬(1) "Non-unitarity" is not a serious problem: it need not violate energy-momentum conservation **[Unruh]**.[6] For example, it could simply mean that there are "extra places" the information could go to, such as the singularity or a baby universe.

¬(2) Either general relativity breaks down not only at the singularity but at a larger scale and gives rise to new effects at the horizons, or that when the backreaction of Hawking radiation is properly taken into account (such as in the approach of **[Ho]**[7]) the horizon never formed (However, Unruh

[a] This observer does not have to be an asymptotic observer. In view of the firewall proposal, We have in mind an observer whose role is to collect and attempt to decode the highly scrambled information encoded in the Hawking radiation.[3]

emphasized that black holes should form in his plenary talk), or that no-drama is violated by the presence of a firewall. The firewall view is criticized in, e.g., **[Page]**.[8]

¬(3) Local quantum field theory is violated, e.g., there may be non-local quantum effects **[Page]**.[9]

¬(4) $\log N \neq \mathcal{A}/4$. Regarding this possibility, there are various ways to realize the idea,[10] for example,

(i) Information can be compressed by the quantum gravitational regime, i.e. the entropy bound may be violated at late time **[Ashtekar]**. See also Ref.11. Hence, the resolution of the singularity may be sufficient, or perhaps *essential* to explain the information loss problem.

(ii) A small Planck scale remnant can store all the information (related to **[Wen]**[12,b]).

(iii) The interior of a black hole may store sufficiently large amount of information, e.g., via a "naturally large volume" or a bubble universe **[Ong]**.[13]

¬(5) There is no observer who can read information from the Hawking radiation; information does not disappear but semi-classical observers are not able to read the information **[Yeom, Domènech, Hu]**.[14–16]

2.2. *How can we obtain some insights?*

Some presentations in this parallel session were related to methods that one might apply to obtain some insights on black hole evaporation and the information loss paradox. The common theme is that we can investigate analog models of black hole evaporation, both at the theoretical level as well as their possible experimental realizations. We learned from the various presentations that:

(1) There are many correspondences between the physics of moving mirrors and that of radiating black holes **[Good]**.[17] The moving mirror approach is a well-known tool in the literature but has yet to be fully exploited to study black hole evaporation.

(2) In order to detect Unruh radiation, we need to distinguish pure Unruh effects from thermal random motions **[Oshita]**.[18]

(3) One can build a plasma mirror from laser experiments. This can provide a model that mimics black hole evaporation in the laboratory **[Chen]**.[19]

---

[b]Note that, in [Wen]'s talk, he discussed whether remnants in anti-de Sitter space can be more stable than thermal radiations. We do not intend to imply that he supports the remnant picture as a resolution of the information loss paradox.

3. Open Problems and Future Prospects

It is best to have in mind some crucial questions while one attempts to resolve the information loss paradox. The first important aspect is the roles played by the horizon, the singularity, and the interior spacetime of a black hole. Let us elaborate:

(1) *Horizon:* In general relativity, a black hole is defined by its event horizon, beyond which nothing, not even light, can escape from. Other notions of horizons, such as apparent/trapping horizons, have also been discussed in the recent literature. An important and crucial question is whether a black hole, defined in the strict sense by the event horizon, really exist? Apparent horizons are not without problems since they are hypersurface dependent. There exist attempts to prevent the information loss issue from ever arising, by arguing that event horizons do not form (and in some of these approaches, not even apparent horizons exist, see, e.g., **[Ho]**[7]).

(2) *Singularity:* One often assumes that singularities in general relativity can be cured by quantum gravity. There are two issues here worth emphasizing: firstly, we do not have any evidence, other than wishful thinking, that the singularity can indeed be cured. If quantum gravity fails to cure the singularity,[10] then perhaps information just falls through the edge of spacetime into the singularity. This possibility was emphasized in the plenary talk by Unruh. Even if quantum gravity successfully cures the singularity, it might be that the information simply falls into whatever it is that replaces the singularity. This is related to the second point we wish to raise here: can the information loss paradox be understood without the full knowledge of quantum gravity (including the full understanding of singularities)?

(3) *Interior Spacetime:* The information loss paradox is often discussed from the point of view of the exterior observers. Perhaps some insights can be gained by investigating the issue from the point of view of an infalling observer instead. Does the dynamical interior of the black hole play any role at all in somehow "storing the information"? Such a question is interesting considering the fact that in some sense the black hole interior spacetime has a very large spatial volume **[Ong]**,[13,20–22] although a recent investigation suggests that the associated entropy of the large volume is only proportional to the surface area.[23] A related issue, as raised by Unruh in his plenary talk, is: how do we understand the "bag-of-gold" type of geometry, especially in the context of AdS/CFT correspondence? That is, can the boundary probe the interior of the large volume of such a nontrivial geometry? This question[24] was previously also discussed in the literature in the context of the so-called "monster" spacetime,[25–27] but there is still some room for further study.

More fundamentally, there are some rather philosophical questions that are worth emphasizing. The most important issue is: *what do we mean by resolv-*

ing the information loss paradox? The "preferred" approach in the literature seems to be that the *external* observers should be able to, at least in principle, recover all the information after the black hole has evaporated away. The alternative solutions, such as having the information somehow locked inside a remnant[10] or goes into a baby universe (or simply falls into the singularity if one takes general relativity at face value), remain somewhat unpopular. Note that one is not as bothered by, say, information from the very early cosmos that in principle could leak into the various bubble universes in the eternal inflationary scenario.

Indeed, this issue is closely related to another equally important question: *what do we really mean by unitarity?* As pointed out in quite a few talks in this parallel session **[Yeom, Domènech, Hu]**,[14–16] perhaps we are still thinking "too classically." That is, information loss only occurs if one looks at semi-classical picture in which matter is quantized but there is still one definite – albeit evolving – spacetime geometry. It is possible that all that is required for unitarity is that the information is conserved at the level of wavefunctions, and that we should instead consider superpositions of different geometries (and topologies). In such a picture, information is loss as far as semi-classical observers (who only see one definite geometry) are concerned,[28] but unitarity remains intact at the fundamental level.

Lastly, although one wishes to find *the* resolution of the information loss paradox, one should keep in mind the possibility that perhaps there is *no unique resolution.* Afterall, there exist many types of black holes with various topologies and hairs. One often finds that a proposed resolution works for one type of black hole but not for another, more complicated, type. Perhaps Nature does respect unitarity and information is always conserved, but there are many ways to ensure this, instead of a simple elegant mean. Furthermore, even a specific type of black hole can be formed from many different initial conditions. As Stephen Fulling pointed out in his talk during the Hawking Radiation Conference held in Stockholm in August 2015, perhaps different initial conditions lead to black holes with different end states. For example, some might have a "bag-of-gold" while some have relatively "trivial" interiors. It is possible that some black holes might completely evaporate but some remain as a remnant.

The information loss paradox is indeed a long-standing problem in black hole physics. It is quite unfortunate that we have not made much progress in this subject after 40 years. The firewall controversy demonstrates that our understanding of black hole physics is even less than previously thought. In view of this, hidden assumptions are starting to be examined, and further understanding would certainly be made. Hopefully by the 200th anniversary of general relativity, we would have the answer ready.

References

1. S. W. Hawking, Breakdown of predictability in gravitational collapse, *Phys. Rev. D* **14**, 2460 (1976).

2. L. Susskind, L. Thorlacius and J. Uglum, The stretched horizon and black hole complementarity, *Phys. Rev. D* **48**, 3743 (1993) [arXiv:hep-th/9306069].
3. D. Harlow and P. Hayden, Quantum Computation vs. Firewalls, JHEP **06**, 085 (2013) [arXiv:1301.4504 [hep-th]].
4. D. Yeom and H. Zoe, Semi-classical black holes with large N re-scaling and information loss problem, *Int. J. Mod. Phys. A* **26**, 3287 (2011) [arXiv:0907.0677 [hep-th]].
5. A. Almheiri, D. Marolf, J. Polchinski and J. Sully, Black holes: complementarity or firewalls?, *JHEP* **1302**, 062 (2013) [arXiv:1207.3123 [hep-th]].
6. W. G. Unruh and R. M. Wald, On evolution laws taking pure states to mixed states in quantum field theory, *Phys. Rev. D* **52**, 2176 (1995) [hep-th/9503024].
7. P. M. Ho, The absence of horizon in black-hole formation, [arXiv:1510.07157 [hep-th]].
8. P. Chen, Y. C. Ong, D. N. Page, M. Sasaki and D. Yeom, Naked firewalls, [arXiv:1511.05695 [hep-th]].
9. D. N. Page, Excluding black hole firewalls with extreme cosmic censorship, *JCAP* **1406**, 051 (2014) [arXiv:1306.0562 [hep-th]].
10. P. Chen, Y. C. Ong and D. Yeom, Black hole remnants and the information loss paradox, *Phys. Rept.* **603**, 1 (2015) [arXiv:1412.8366 [gr-qc]].
11. A. Ashtekar and M. Bojowald, Black hole evaporation: a paradigm, *Class. Quant. Grav.* **22**, 3349 (2005) [arXiv: gr-qc/0504029].
12. W. Y. Wen and S. Y. Wu, Black hole remnant in asymptotic anti-de Sitter space, *Eur. Phys. J. C* **75**, no. 12, 608 (2015) [arXiv:1509.01305 [hep-th]].
13. Y. C. Ong, Never judge a black hole by its area, *JCAP* **1504**, no. 04, 003 (2015) [arXiv:1503.01092 [gr-qc]].
14. M. Sasaki and D. Yeom, Thin-shell bubbles and information loss problem in anti de Sitter background, *JHEP* **1412**, 155 (2014) [arXiv:1404.1565 [hep-th]].
15. P. Chen, G. Doménech, M. Sasaki and D. Yeom, Stationary bubbles and their tunneling channels toward trivial geometry, [arXiv:1512.00565 [hep-th]].
16. P. Chen, Y. C. Hu and D. Yeom, Two interpretations on thin-shell instantons, [arXiv:1512.03914 [hep-th]].
17. M. R. R. Good and Y. C. Ong, Signatures of energy flux in particle production: a black hole birth cry and death gasp, *JHEP* **1507**, 145 (2015) [arXiv:1506.08072 [gr-qc]].
18. N. Oshita, K. Yamamoto and S. Zhang, Unruh radiation produced by a uniformly accelerating charged particle in thermal random motions, [arXiv:1509.03038 [hep-th]].
19. P. Chen and G. Mourou, Accelerating plasma mirrors to investigate black hole information loss paradox, [arXiv:1512.04064 [gr-qc]].
20. M. Christodoulou and C. Rovelli, How big is a black hole?, Phys. Rev. D **91**, 064046 (2015) [arXiv:1411.2854 [gr-qc]].
21. I. Bengtsson and E. Jakobsson, Black holes: their large interiors, Mod. Phys. Lett. A, **30**, 1550103 (2015) [arXiv:1502.01907 [gr-qc]].
22. Y. C. Ong, The persistence of the large volumes in black holes, Gen. Relativ. Gravit. **47**, 88 (2015) [arXiv:1503.08245 [gr-qc]].
23. B-C. Zhang Entropy in the interior of a black hole and thermodynamics, Phys. Rev. D **92**, 081501(R) (2015) [arXiv:1510.02182 [gr-qc]].
24. D. Marolf, Black holes, AdS, and CFTs Gen. Rel. Grav. **41**, 903 (2009) [arXiv:0810.4886 [gr-qc]].
25. S.D.H. Hsu and D. Reeb, Black hole entropy, curved space and monsters Phys. Lett. B **658**, 244 (2008) [arXiv:0706.3239 [hep-th]].

26. S.D.H. Hsu and D. Reeb, Monsters, black holes and the statistical mechanics of gravity, Mod. Phys. Lett. A **24**, 1875 (2009) [arXiv:0908.1265 [gr-qc]].
27. Y. C. Ong and P. Chen, The fate of monsters in anti-de Sitter spacetime, JHEP **07**, 147 (2013) [arXiv:1304.3803 [hep-th]].
28. S. D.H. Hsu, Macroscopic superpositions and black hole unitarity, [arXiv:1302.0451 [hep-th]].

BTZ Remnant in Quantum Gravity

Wen-Yu Wen

Department of Physics and Center for High Energy Physics
Chung Yuan Christian University, Chung Li City, Taiwan

Leung Center for Cosmology and Particle Astrophysics
National Taiwan University, Taipei 106, Taiwan
wenw@cycu.edu.tw

In this letter, we investigate remnant solutions as candidates for the BTZ black hole ground state after it receives correction from loop quantum effect or noncommutative geometry. Though the remnant solutions are energetically disfavored in comparison to thermal AdS below the Hawking-Page phase transition temperature, we argue that they are still possible to be found in the overcooled phase if strings were present and its implication is discussed.

Keywords: Black hole remnant; quantum gravity; noncommutative geometry.

1. Introduction

It was shown that for a Schwarzschild black hole evaporating massless particles, the Hawking-Bekenstein area law[1] receives the logarithmic correction, $S = 4\pi M^2 - 4\pi\sigma \log(M + \sigma)$ for one-loop corrected temperature $T = (8\pi M)^{-1}(1 + \sigma(M^2)^{-1})$, where σ is the conformal anomaly. In particular, a Planck size remnant is implied for $\sigma < 0$.[2] There were complaints about remnants,[3] mainly due to extraordinary large amount of entropy confined within a tiny volume. Nevertheless, if the Hawking radiation were not exactly thermal, the retained information would have been released through the evaporation process. One could have expected the existence of a remnant with zero or small amount of residual entropy. In this case, instead to help resolve the information loss paradox, the remnant solution might still answer the call for the hypothetically fundamental physics at Planck scale. In general, remnant solutions can be easily found in those theories such as noncommutative geometry, doubly special relativity, or generalized uncertainty principle (GUP). Readers are directed to a recent review for instance.[4]

On the other hand, it is well known that the three dimensional Banados-Teitelboim-Zanelli (BTZ) black hole system has two distinct phases:[5] the black hole with mass $M > 0$ above the Hawking-Page temperature and thermal AdS with $M = -1$ below that temperature.[6] It is unclear for us whether this thermal AdS phase is still energetically favored at low temperature after the BTZ receives correction from loop quantum effect or noncommutative geometry. In other words, there might be a possibility that the black hole would have already stopped evaporating and stayed as a warm remnant before it enters the thermal phase.

To proceed our discussion, one needs to assume that both first law of thermodynamics and logarithmic correction to the black hole entropy are valid within the energy range in our discussion. The extrapolation of both relations to the limit of Planck size maybe too naive, especially for our ignorance of complete theory of quantum gravity. However, our strategy is to compare free energy of remnant and that of thermal AdS around the Hawking-Page temperature $T_{HP} = 1/2\pi l$. This corresponds to the energy scale $\mathcal{O}(l^{-1})$, which is still far from the Planck or string scale, it should be reasonable to assume the laws of thermodynamics are valid and the black hole size around $\mathcal{O}(l)$ can be treated as classical and static background.

This letter is organized as follows. We review the two loop-corrected entropy and free energy for BTZ black hole and discuss its phase transition to thermal AdS In Sec. 2. In Sec. 3, we compute the entropy and free energy for BTZ black hole in the noncommutative space. In Sec. 4, we discuss possible scenarios around phase transition if stringy excitation is considered.

2. Quantum correction to surface gravity and the emergence of the remnant

To justify which phase is energetically favored at specific temperature, we would like to compare free energy of BTZ and that of vacuum. Since we expect a vacuum without nonzero angular momentum, we will only consider nonrotating BTZ black hole. With that being said, we will not inspect solutions such as extremal BTZ solution or exotic BTZ.[7] Although the phase diagram would become rich and interesting by including those solutions, they cannot tunnel into AdS vacuum as long as conservation of angular momentum is respected. The nonrotating BTZ solution has the metric

$$ds^2 = -(-M + \frac{r^2}{l^2})dt^2 + (-M + \frac{r^2}{l^2})^{-1}dr^2 + r^2 d\phi^2 \tag{1}$$

with horizon $r_+ = l\sqrt{M}$.

The quantum correction to the Hawking temperature has following form

$$T_H^{[1]} = \frac{r_+}{2\pi l^2}(1 + \sum_i \alpha_i \frac{\hbar^i}{r_+^i})^{-1}. \tag{2}$$

In generic, the coefficients α_i will depend on species of particles included in the i-th loop perturbative correction.[2] In this letter we will study a general form of correction:

$$T_H^{[2]} = \frac{r_+}{2\pi l^2}\frac{1 + \beta\frac{\hbar}{r_+}}{1 + \alpha\frac{\hbar}{r_+}}. \tag{3}$$

We only keep one-loop correction in the denominator for simplicity. The numerator can be regarded as quantum correction to the ADM mass. Following the first

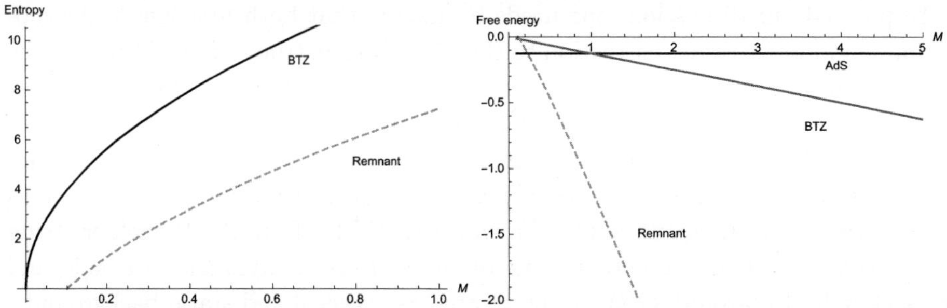

Fig. 1. (Left) A numerical simulation of entropy for BTZ blakc hole (black solid) and loop-corrected remnant with $\alpha < 0$ and $\beta > 0$ (red dashed). (Right) A numerical simulation of free energy for BTZ blakc hole (blue solid), quantum corrected remnant (red dashed), and thermal AdS (black solid). At low temperature or small mass limit, the thermal AdS is always energetically favored than BTZ and remnant. The remnant mass $m_c = .1$ in the unit of $G = \hbar = l = 1$ for the choice of $\alpha = -0.01$ and $\beta = 0.1$.

law of thermodynamics, the entropy receives corresponding correction as follows:

$$S_{BH}^{[2]} = \frac{1}{8G}\int \frac{dM}{T_H^{[2]}} = \frac{\pi r_+}{2G} + (\alpha - \beta)\hbar\frac{\pi}{2G}\ln(r_+ + \beta\hbar) + C, \tag{4}$$

where C refers to the constant of integration. The conformal field theory (CFT) calculation also implies a similar logarithmic correction with coefficient $\alpha = -\frac{3G}{\pi\hbar}$ and $\beta = 0$ in the above expression.[8] However, we remark that the temperature in the CFT computation was not modified according to (2). Therefore, it is more appropriate to think the logarithmic correction we will discuss here comes from a different origin and therefore α and β could vary for different models of quantum gravity. Nevertheless, the expression (4) could be UV incomplete in some choices of parameters since it suffers from the divergence as $M \to 0$. Without knowing much about gravity at the Planck scale, we would like to assume the entropy takes the same expression as (4) and should be regularized, if necessary, to vanish at some critical mass m_c, which can be regarded as the remnant mass. The vanishing entropy garantees the uniqueness of ground state. In this letter, we focus on a class of solution where $\alpha < 0$, $\beta > 0$ and C is chosen to be

$$C = -\frac{\pi l\sqrt{m_c}}{2G} - (\alpha - \beta)\hbar\frac{\pi}{2G}\ln(l\sqrt{m_c} + \beta\hbar) \tag{5}$$

At last, the quantum corrected free energy can be computed as

$$F_{BH}^{[2]} = \frac{M}{8G} - T_H^{[2]}S_{BH}^{[2]} \tag{6}$$

We simulate the entropy and free energy of thermal AdS, BTZ and quantum-corrected remnant and plot them in the Fig. 1. It has shown that for quantum corrected BTZ, entropy vanishes at the critical mass m_c. We also found that with quantum correction, the Hawking-Page transition happens at slightly smaller M

and the black hole turns into the AdS thermal vacuum before a remnant could possibly form. This result can be partially understood by absorbing those quantum corrections in (2) into a definition of *effective* horizon size $r_+^{eff} \equiv r_+ + (\beta - \alpha)\hbar - \cdots$ for $r_+ \gg l_p$. That is, the quantum fluctuation makes the horizon look bigger than its classical value, therefore more stable (for large black hole). However, unless the quantum corrected black hole stop radiation before it hits the Hawking-Page temperature, it eventually becomes energetically disfavored and decayed into the thermal vacuum. We conclude that qauntum-corrected BTZ remnant is unstable just like the small BTZ black hole. Free energy for 2-loop corrected Hawking temperature defined in the (2) has been computed and there the thermal AdS is also energetically favored.[9]

3. BTZ remnant in the noncommutative space

One resolution to the singularity in the general relativity is to introduce the noncommutative space. In particular, the noncommutative BTZ metric were contructed in Refs. 10–13 and one version reads:

$$d\hat{s}^2 = -[\frac{(r^2 - r_+^2)}{l^2} - \frac{\theta B}{2l^2}]dt^2 + \frac{dr^2}{l^2 r^2}[(r^2 - r_+^2 - \theta B)r^2 + \frac{\theta B}{2} r_+^2] + (r^2 + \frac{\theta B}{2})d\phi^2, \quad (7)$$

where θ is the noncommutative parameter and B is the magnetic flux in the corresponding Chern-Simons theory. Using the Hamilton-Jacob method of tunneling,[14] the corrected Hawking temperature was obtained

$$\hat{T}_H = \frac{\hat{r}_+}{2\pi l^2}\sqrt{1 - \frac{\theta B M l^2}{2\hat{r}_+{}^4}}, \quad (8)$$

for the corrected horizon $\hat{r}_+{}^2 = Ml^2 + \frac{\theta B}{2}$. The remnant mass is that which makes (8) vanish. In the limit $\frac{\theta B}{Ml^2} \ll 1$, one obtains $m_c \simeq \frac{\theta B}{2l^2}$ and the corrected entropy after regularization reads:

$$\hat{S}_{BH} = \frac{\pi l}{2G}\sqrt{M} - \frac{\pi \theta^2 B^2}{96 G l^3 \sqrt{M^3}} - \frac{\pi l}{2G}\sqrt{m_c} + \frac{\pi \theta^2 B^2}{96 G l^3 \sqrt{m_c^3}} \quad (9)$$

The free energy is also obtained through $\hat{F} = \frac{M}{8G} - \hat{T}_H \hat{S}_{BH}$. We plot its entropy and free energy in the Fig. 2.

4. Discussion

We have investigated remnant solutions as candidates for the BTZ black hole ground state after it receives correction from loop quantum effect and noncommutative geometry. Though the remnant solutions are energetically disfavored in comparison to thermal AdS below the Hawking-Page phase transition temperature, in the previous

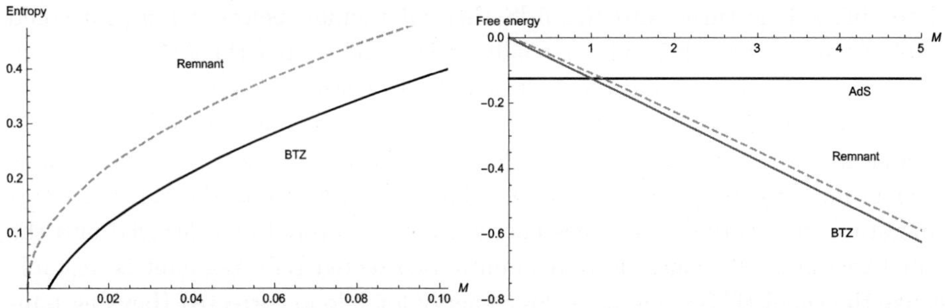

Fig. 2. (Left) A numerical simulation of entropy for BTZ blakc hole (black solid) and noncommutative remnant with $\theta B = 0.01$ (red dashed). (Right) A numerical simulation of free energy for BTZ blakc hole (blue solid), noncommutative remnant (red dashed), and thermal AdS (black solid). At low temperature or small mass limit, the thermal AdS is always energetically favored than BTZ and remnant.

work,[9] we have argued that remnant is still possible to be found in the overcooled phase[15] if strings were present where the momentum mode could stablize the Atick-Witten tachyon winding mode.[16]

Acknowledgments

The author is grateful to the hospitality of YITP and LeCosPA for fruitful discussion with Hirotaka Irie and Pisin Chen during the early stage of this project. WYW is also grateful to the hospitality of Osaka University for useful feedbacks from Koji Hashimoto, Norihiro Iizuka, Satoshi Yamaguchi and Ohta Nobuyoshi. This work is supported in parts by the Taiwan's Ministry of Science and Technology (grant No. 102-2112-M-033-003-MY4) and National Center for Theoretical Science in Taiwan.

References

1. J. D. Bekenstein, "Black holes and entropy," Phys. Rev. D **7**, 2333 (1973).
2. D. V. Fursaev, "Temperature and entropy of a quantum black hole and conformal anomaly," Phys. Rev. D **51**, 5352 (1995) [hep-th/9412161].
3. L. Susskind, "Trouble for remnants," hep-th/9501106.
4. P. Chen, Y. C. Ong and D. h. Yeom, "Black Hole Remnants and the Information Loss Paradox," arXiv:1412.8366 [gr-qc].
5. M. Banados, C. Teitelboim and J. Zanelli, "The Black hole in three-dimensional spacetime," Phys. Rev. Lett. **69**, 1849 (1992) [hep-th/9204099].
6. M. Eune, W. Kim and S. -H. Yi, "Hawking-Page phase transition in BTZ black hole revisited," JHEP **1303**, 020 (2013) [arXiv:1301.0395 [gr-qc]].
7. P. K. Townsend and B. Zhang, "Thermodynamics of Exotic Baados-Teitelboim-Zanelli Black Holes," Phys. Rev. Lett. **110**, no. 24, 241302 (2013) [arXiv:1302.3874 [hep-th]].
8. S. Carlip, "Logarithmic corrections to black hole entropy from the Cardy formula," Class. Quant. Grav. **17**, 4175 (2000) [gr-qc/0005017].
9. W. Y. Wen and S. Y. Wu, "Black hole remnant in asymptotic Anti-de Sitter space," Eur. Phys. J. C **75**, no. 12, 608 (2015) doi:10.1140/epjc/s10052-015-3841-7 [arXiv:1509.01305 [hep-th]].

10. M. Banados, O. Chandia, N. E. Grandi, F. A. Schaposnik and G. A. Silva, "Three-dimensional noncommutative gravity," Phys. Rev. D **64**, 084012 (2001) doi:10.1103/PhysRevD.64.084012 [hep-th/0104264].
11. H. C. Kim, M. I. Park, C. Rim and J. H. Yee, "Smeared BTZ Black Hole from Space Noncommutativity," JHEP **0810**, 060 (2008) doi:10.1088/1126-6708/2008/10/060 [arXiv:0710.1362 [hep-th]].
12. E. Chang-Young, D. Lee and Y. Lee, "Noncommutative BTZ Black Hole in Polar Coordinates," Class. Quant. Grav. **26**, 185001 (2009) doi:10.1088/0264-9381/26/18/185001 [arXiv:0808.2330 [hep-th]].
13. R. Jahangir and K. Saifullah, "Thermodynamics of noncommutative BTZ black hole," arXiv:1101.6073 [gr-qc].
14. M. A. Anacleto, F. A. Brito, A. G. Cavalcanti, E. Passos and J. Spinelly, "Quantum correction to the entropy of noncommutative BTZ black hole," arXiv:1510.08444 [hep-th].
15. M. Berkooz, Z. Komargodski and D. Reichmann, "Thermal AdS(3), BTZ and competing winding modes condensation," JHEP **0712**, 020 (2007) [arXiv:0706.0610 [hep-th]].
16. J. J. Atick and E. Witten, "The Hagedorn Transition and the Number of Degrees of Freedom of String Theory," Nucl. Phys. B **310**, 291 (1988).

Reflections on a Black Mirror

Michael R. R. Good

Physics Department, Nazarbayev University,
Astana, Republic of Kazakhstan
michael.good@nu.edu.kz

A black mirror is an accelerated boundary that produces particles in an exact correspondence to an evaporating black hole. We investigate the spectral dynamics of the particle creation during the formation process.

Keywords: Black hole evaporation, moving mirror, dynamical Casimir effect.

To understand particle creation from black holes, it is essential to consider the time-dependent formation phase. Recently, it was shown[1,2] that there is an exact correspondence between a black hole and a moving mirror. That is, for a massless minimally coupled scalar field in a (1+1)D spacetime in which a black hole forms from the collapse of a null shell, and for the quantum amplification of modes for this field which results from an accelerating moving mirror with a particular trajectory in (1+1)D, the particle creation is the same. The moving mirror in this case is called a *black mirror*.

First, before we investigate the black mirror's early-time particle creation, allow us to take a moment to inquire about a system which emits thermal radiation forever.[3] In this case there is no time dependence for the particle creation. This system, unlike the black mirror, is in equilibrium. The particles residing in each frequency mode are Planckian:

$$N_\omega = \frac{1}{e^{2\pi\omega/\kappa} - 1}, \tag{1}$$

with temperature $T = \kappa/2\pi$. By construction, the immutable constant emission of energy ($F = \kappa^2/48\pi$) of this Carlitz-Willey[3] moving mirror is echoed by the eternal constant emission of Planckian distributed particles. The relevant beta Bogoliubov coefficient for this eternally thermal moving mirror (henceforth called the *equilibrium mirror*) is

$$\beta_{\omega\omega'} = -\frac{1}{2\pi\kappa}\sqrt{\frac{\omega}{\omega'}}e^{-\pi\omega/2\kappa}\left(\frac{\omega'}{\kappa}\right)^{-i\omega/\kappa}\Gamma\left[\frac{i\omega}{\kappa}\right]. \tag{2}$$

One can see the form of the Planckian distribution of Eq. 1 by complex conjugating:

$$|\beta_{\omega\omega'}|^2 = \frac{1}{2\pi\kappa\omega'}\frac{1}{e^{2\pi\omega/\kappa} - 1}. \tag{3}$$

Albeit, there is a pre-factor of $1/2\pi\kappa\omega'$ which is responsible for the total infinite production of particles, residing in mode ω, when one integrates over ω':

$$\langle N_\omega\rangle = \int_0^\infty d\omega' |\beta_{\omega\omega'}|^2 . \tag{4}$$

To avoid infinite particle production, and to remove this pre-factor in Eq. 3, one most localize the production of particles in time and frequency. Following Hawking,[4] one uses orthonormal and complete wave packets to packetize and localize the beta Bogoliubov coefficient in Eq. 2. Then, one has packetized beta Bogoliubov coefficients,[5–7]

$$\beta_{jn\omega'} = \frac{1}{\sqrt{\epsilon}} \int_{j\epsilon}^{(j+1)\epsilon} d\omega \, e^{\frac{2\pi i \omega n}{\epsilon}} \beta_{\omega\omega'} \,. \tag{5}$$

The values of n and j designate the time and frequency bins respectfully. The total number of particles in any particular packet is then,

$$\begin{aligned} \langle N_{jn} \rangle &= \int_0^\infty d\omega' |\beta_{jn,\omega'}|^2 \,, \\ &= \int_0^\infty d\omega' \int_{j\epsilon}^{(j+1)\epsilon} \frac{d\omega_1}{\sqrt{\epsilon}} \int_{j\epsilon}^{(j+1)\epsilon} \frac{d\omega_2}{\sqrt{\epsilon}} e^{2\pi i(\omega_1-\omega_2)n/\epsilon} \beta_{\omega_1\omega'} \beta^*_{\omega_2\,\omega'} \,, \\ &= \frac{1}{\epsilon} \int_{j\epsilon}^{(j+1)\epsilon} d\omega_1 \int_{j\epsilon}^{(j+1)\epsilon} d\omega_2 e^{2\pi i(\omega_1-\omega_2)n/\epsilon} \langle N_{\omega_1\omega_2} \rangle \,, \end{aligned} \tag{6}$$

where

$$\langle N_{\omega_1\omega_2} \rangle \equiv \int_0^\infty d\omega' \beta_{\omega_1\omega'} \beta^*_{\omega_2\omega'} \,. \tag{7}$$

The particles are localized in the range of frequencies $j\epsilon \leqslant \omega \leqslant (j+1)\epsilon$. They peak in the range of times $(2\pi n - \pi)/\epsilon \leqslant u \leqslant (2\pi n + \pi)/\epsilon$. Further details about these wave packets are discussed in Fabbri[7] and Good-Anderson-Evans.[5] In the case of the equilibrium mirror, the packetized particle count is found[a] exactly:

$$\langle N_{\omega_1\omega_2} \rangle = \frac{1}{e^{\omega_2/T} - 1} \delta(\omega_2 - \omega_1). \tag{8}$$

To see how there is no time dependence, one integrates with the Dirac delta function by plugging Eq. 8 in the last line of Eq. 6. This causes the time dependence to cancel out because the n time bin values, in the term $e^{2\pi i(\omega_1-\omega_2)n/\epsilon}$, are negated by the Dirac delta function. Therefore, the particle count, for the equilibrium mirror, becomes eternally time independent, $\langle N_{jn} \rangle \to \langle N_j \rangle$:

$$\langle N_j \rangle = \frac{1}{\epsilon} \int_{j\epsilon}^{(j+1)\epsilon} d\omega_1 \frac{1}{e^{\omega_1/T} - 1}. \tag{9}$$

This integral is easy to do and the final result is:

$$\langle N_j \rangle = \frac{1}{2\bar{\epsilon}} \log\left[\sinh\left(j+1\right)\bar{\epsilon}) \operatorname{csch}\left(j\bar{\epsilon}\right)\right] - \frac{1}{2} \tag{10}$$

[a]This delta function comes about with help from using the integral: $\int_0^\infty d\omega' \omega'^{i(\omega_2-\omega_1)-1} = 2\pi\delta(\omega_2 - \omega_1)$.

where $\bar{\epsilon} \equiv \epsilon/2T$, or alternatively expressed,

$$\langle N_j \rangle = \frac{T}{\epsilon} \log \left[\frac{e^{(j+1)\epsilon/T} - 1}{e^{j\epsilon/T} - 1} \right] - 1 \,. \tag{11}$$

This is the exact Planckian distribution. It is this expression, Eq. 11, which will allow us to numerically confirm, the late-time thermal emission of the black mirror. This is the reason for investigating the equilibrium mirror first. The equilibrium emission will help us explore non-equilibrium emission.

It is easy to check, that in an expansion of ϵ, one can obtain the more familiar Planckian form. Keeping the center frequency, $\omega_j \equiv (j + 1/2)\epsilon$, constant, and doing the series expansion, then $\langle N_j \rangle \to \langle N_{\omega_j} \rangle$, and Eq. 11 becomes the usual Planckian distribution:

$$\langle N_j \rangle = \langle N_{\omega_j} \rangle + O(\epsilon^2), \tag{12}$$

where

$$\langle N_{\omega_j} \rangle \equiv \frac{1}{e^{\omega_j/T} - 1} \tag{13}$$

and $T = \kappa/2\pi$. Eq. 13 is the discrete counterpart of Eq. 1 and for it to be accurate, one requires a small $\epsilon < 1$ wave packet parametrization. The exact Planckian distribution, Eq. 11, will be more useful than Eq. 13, as an anchor for investigating the time evolution of non-thermal systems, which eventually evolve into thermal equilibrium.

Do we have an intuition about the exact Planckian distribution, Eq. 11? What choice to make for ϵ as an intuitive parametrization for the wave packets? Our preference is that it should be small, where Eq. 13 holds, yet not so small, that time resolution is too coarse. Fine-grained frequency *and* time resolution is required to analyze the spectral dynamics of the formation to a black hole. To this end, we ask what is the typical angular frequency, ϵ_T for a Hawking quanta in thermal equilibrium? Yan[8] asserts that the general thermal de Broglie wavelength is:

$$\lambda_T = \frac{2\sqrt{\pi}}{T} \left(\frac{\Gamma(d/2+1)}{\Gamma(d+1)} \right)^{1/d} , \tag{14}$$

using a massless linear dispersion relation with space dimensions d, and $\hbar = c = k_B = 1$. Knowing this value, we could pick a packetization ϵ that is smaller, $\epsilon < \epsilon_T$, in order to resolve well in frequency, yet not be so small that we can't also have reasonably fine-grained time resolution for a non-eternal process. In our $1+1$ dimensional setting, Eq. 14 gives,

$$\lambda_T = \frac{\pi}{T}, \quad \text{and} \quad f_T \lambda_T = 1 \to \frac{\epsilon_T}{2\pi} \frac{\pi}{T} = 1 \quad \to \quad \epsilon_T = 2T. \tag{15}$$

We are then justified in picking $\epsilon < \epsilon_T = 2T$, as a wave packet parametrization for ϵ for good frequency resolution. Note that $\bar{\epsilon} \equiv \epsilon/2T = \epsilon/\epsilon_T$. Let us pick

$$\epsilon_{gr} = T \ln \frac{1+\sqrt{5}}{2} = T \ln \phi, \tag{16}$$

where $\phi = 0.0765$ is the Golden Ratio, such that $T \ln \phi \approx T/2$. This means that our choice of ϵ_{gr} is about 4 times smaller than the general thermal de Broglie angular frequency, Eq. 15, $\epsilon_T = 2T$. It also means a 5% relative error between Eq. 13 and Eq. 11. That is, 5% error between the approximate familiar Planckian form and the exact Planckian form, when $j = 1$.

However, the actual intuitive usefulness of our choice of $\epsilon = \epsilon_{gr}$, is that it gives on average, *one particle per packet*, in the lowest non-diverging $j = 1$ frequency bin, for $\langle N_j \rangle$:

$$\langle N_{j=1} \rangle = 1. \tag{17}$$

This is checked by solving Eq. 11, using $j = 1$ and $\epsilon = \epsilon_{gr}$. This also happens to be the same assumption as supposing that the total amount of particles detected on average is exactly 2 particles, that is:

$$\sum_{j=2}^{\infty} \langle N_j \rangle = 1, \tag{18}$$

or, of course,

$$\sum_{j=1}^{\infty} \langle N_j \rangle = 2. \tag{19}$$

In this note, up until now, we have been looking at the equilibrium mirror. We developed an intuitive tool set to understand the thermal equilibrium where particle emission is discrete and one particle per packet is observed. Now we are in a better position to time resolve the early-time, non-thermal history of the black mirror. The black mirror trajectory is given by[1,3]

$$z(t) = -t - \frac{W(2e^{-2\kappa t})}{2\kappa}, \tag{20}$$

where W is the Lambert W function, and the horizon is taken to be on the $t = -x$ null ray. It is plotted in Fig. (1) and it begins at rest, at $z = +\infty$ and at late times asymptotically approaches the speed of light. At late times this trajectory is very similar to the equilibrium mirror[b]. The black mirror has particles in a non-thermal distribution at early times and they only become Planckian at late times.

The black mirror has beta Bogoliubov coefficients,

$$\beta_{\omega\omega'} = \frac{\sqrt{\omega\omega'}}{2\pi\kappa(\omega + \omega')} e^{-\pi\omega/2\kappa} \left(\frac{\omega + \omega'}{\kappa}\right)^{-i\omega/\kappa} \Gamma\left[\frac{i\omega}{\kappa}\right], \tag{21}$$

with different form than equilibrium mirror beta coefficients of Eq. 2. The total number of particles produced with frequency ω is Eq. 4 and therefore complex conjugating, one finds

$$|\beta_{\omega\omega'}|^2 = \frac{\omega'}{2\pi\kappa(\omega + \omega')^2} \frac{1}{e^{2\pi\omega/\kappa} - 1}. \tag{22}$$

[b]See Good-Anderson-Evans[5] or Good[9] for the explicit equilibrium $z(t)$ trajectory.

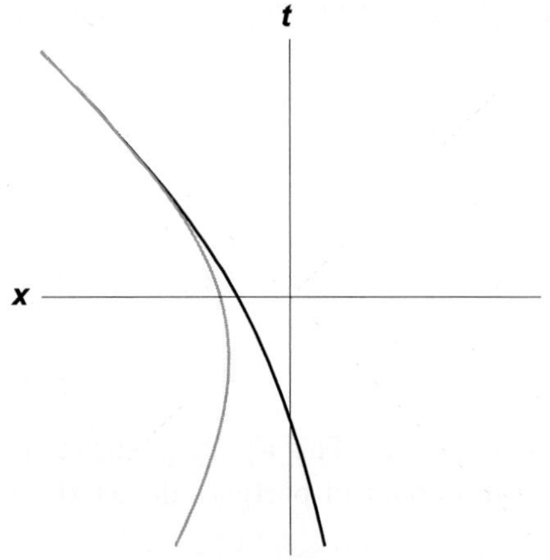

Fig. 1. The black mirror in a spacetime diagram (the black trajectory) is overlaid with the equilibrium mirror (the gray trajectory). The accelerations are time dependent. The horizon is at $t = -x$. We have set $\kappa = 1$.

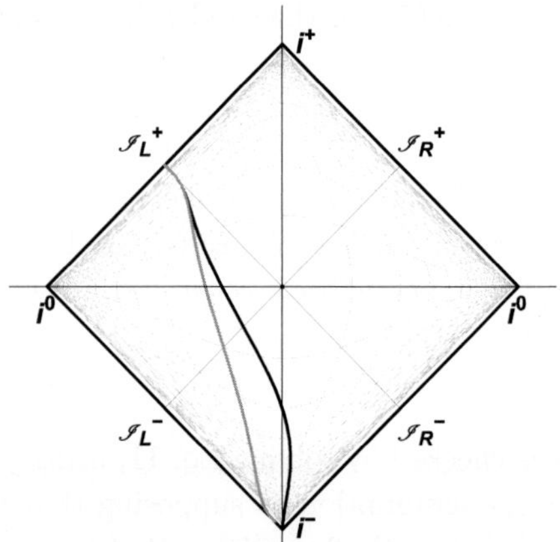

Fig. 2. The two mirrors in a Penrose diagram. Notice that both start asymptotically inertial. The black mirror starts at rest, while the equilibrium mirror starts at the speed of light. $\kappa = 1$.

Like the equilibrium mirror, we also see the explicit Planckian signature. It's easy to see that in the late time limit where $\omega' \gg \omega$, then Eq. 22 goes to Eq. 3,

$$|\beta_{\omega\omega'}|^2 \sim \frac{1}{2\pi\kappa\omega'}\frac{1}{e^{2\pi\omega/\kappa} - 1}, \tag{23}$$

and there is a thermal distribution of particles with temperature $T = \kappa/2\pi$. Late time thermality was found 40 years ago by Davies-Fulling[10,11] for particular late-time behaved mirrors.

The black mirror beta can only be exactly analytically packetized at late times. This is because the exact Planckian result, Eq. 11, holds. Early-time resolution is only tractable numerically by using wave packets of Eq. 5 on Eq. 21. As a check, one numerically confirms the late-time exact analytic Planckian distribution, Eq. 11. The reason for first studying the equilibrium mirror before tackling the black mirror head-on should now be clear: since we know a priori that the black mirror produces late-time Planckian particles, we can anchor the late-time numerical precision and accuracy goals against the exact Planckian distribution to gain confidence for the early-time, non-thermal radiation. A plot of the time dependence of the particle production for the frequency band $\epsilon = \epsilon_{gr}$ and $\kappa = j = 1$ is given in Fig. (3). The radiation emission is monotonic, i.e. here there is no black hole birth cry.[6] Since black mirror radiation is black hole evaporation, the black mirror provides a window into the early-time formation phase.

Acknowledgments

MRRG thanks Paul Anderson, Yen Chin Ong and Charles Evans for fruitful conversations. MRRG is supported by the Nazarbayev University Social Policy grant.

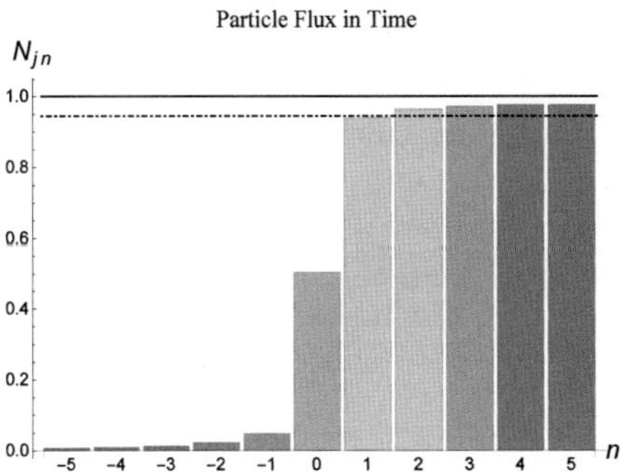

Fig. 3. Black mirror particle flux in time. The solid black line at the top is the number of particles produced in an exact Planckian distribution, Eq. 11, $\langle N_{j=1} \rangle = 1$. The dotted-dashed line is the Planckian distribution, Eq. 13. Here $j = 1$, $\epsilon = T \ln \phi$, $\kappa = 1$.

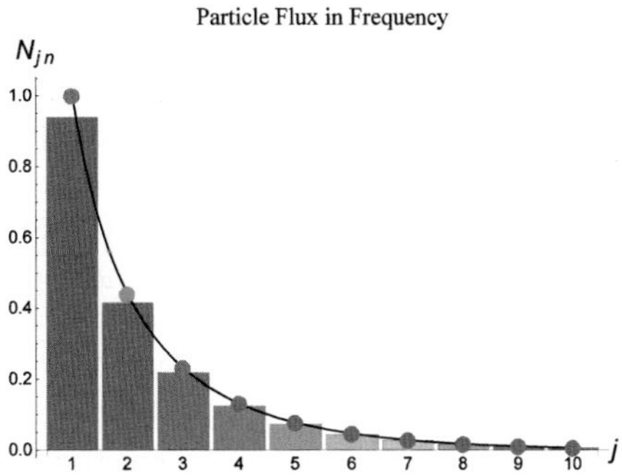

Fig. 4. Black mirror particle flux in frequency. The black curve at the top of the plot is the Planckian distribution as a function of the frequency, Eq. 11. Note that the packet ($j = n = 1$) is in both Figs. Here $n = 1$, $\epsilon = T \ln \phi$, $\kappa = 1$.

References

1. P. R. Anderson, M. R. R. Good, and C. R. Evans, "Black Hole - Moving Mirror I: An Exact Correspondence," to appear in the Proceedings of the 14th Marcel Grossmann Meeting, (2015) arXiv:1507.03489 [gr-qc]
2. M. R. R. Good, P. R. Anderson and C. R. Evans, "Black Hole - Moving Mirror II: Particle Creation," to appear in the Proceedings of the 14th Marcel Grossmann Meeting, (2015) arXiv:1507.05048 [gr-qc].
3. R. D. Carlitz and R. S. Willey,"Reflections On Moving Mirrors," Phys. Rev. D **36**, 2327 (1987).
4. S. Hawking, "Particle Creation by Black Holes", Commun. Math. Phys. **43** (1975) 199.
5. M. R. R. Good, P. R. Anderson and C. R. Evans,"Time Dependence of Particle Creation from Accelerating Mirrors," Phys. Rev. D **88**, 025023 (2013) [arXiv:1303.0756 [gr-qc]].
6. M. R. R. Good and Y. C. Ong, JHEP **1507**, 145 (2015) [arXiv:1506.08072 [gr-qc]].
7. A. Fabbri and J. Navarro-Salas,"Modeling black hole evaporation," London, UK: Imp. Coll. Pr. (2005)
8. Z. Yan, "General thermal wavelength and its applications,' European Journal of Physics, 21, 6, (2000).
9. M. R. R. Good, "On Spin-Statistics and Bogoliubov Transformations in Flat Spacetime With Acceleration Conditions," Int. J. Mod. Phys. A **28**, 1350008 (2013) [arXiv:1205.0881 [gr-qc]].
10. P. C. W. Davies, S. A. Fulling, "Radiation from a Moving Mirror in Two-Dimensional Space-Time Conformal Anomaly," Proc. Roy. Soc. Lond. A **348** (1976) 393.
11. P. C. W. Davies, S. A. Fulling, "Radiation from Moving Mirrors and from Black Holes," Proc. Roy. Soc. Lond. A **356** (1977) 237.

Information Loss Problem and Roles of Instantons*

Dong-Han Yeom

Leung Center for Cosmology and Particle Astrophysics,
National Taiwan University, Taipei 10617, Taiwan
innocent.yeom@gmail.com

In order to understand the information loss problem, non-perturbative effects will do significant roles. Instantons are in general helpful for this purpose. There are various and rich thin-shell instantons and tunneling channels that eventually result a trivial geometry without a singularity nor an event horizon. We further discuss that there are some subtle examples in asymptotic de Sitter cases that need further investigations.

Keywords: Information loss paradox; Euclidean path-integral approach; instantons.

1. Introduction: effective loss of information

In order to resolve the information loss problem,[1] various ideas have been suggested in the literature (Fig. 1).[2] However, it is also fair to say that various ideas have their own problems. If the black hole evolution is not unitary, then in itself it can cause a serious problem.[3] If Hawking radiation contains information,[4] then we finally reach the inconsistency.[5,6] If information should be retained by another objects, the problem is that we still do not have a generic argument to realize this.[2] Then what does this mean? If information should be conserved, but it is not attached by Hawking radiation nor by a remnant, then where is the information? One interesting way to explain this dilemma is to think that the wave function of the universe contains information; however, semi-classical observers, like us, cannot access the information since we should see a specific semi-classical geometry[7] ('question mark' in Fig. 1). We name this idea as the *effective loss of information.*

Then, how can we realize this idea? In order to do this, we need to understand the wave function of the universe,[8] while this is a genuinely quantum gravitational problem. Even though there is no consensus on quantum theory of gravity, the Euclidean path integral approach[9] can be a highway to understand, at least, the skeleton of the wave function of the universe, by using instantons.

2. Euclidean path integral approach

According to the Euclidean path integral approach, the ground state wave function[9] that satisfies the Wheeler-DeWitt equation[8] is presented by the following form:

$$\langle f|i\rangle = \int_{i\to f_j} \mathcal{D}g_{\mu\nu}\mathcal{D}\phi \; e^{-S_{\mathrm{E}}[g_{\mu\nu},\phi]}, \tag{1}$$

*Proceedings for *The 2nd LeCosPA Symposium: Everything about Gravity, Celebrating the Centenary of Einstein's General Relativity.* Talk on December 17, 2015, Taipei, Taiwan.

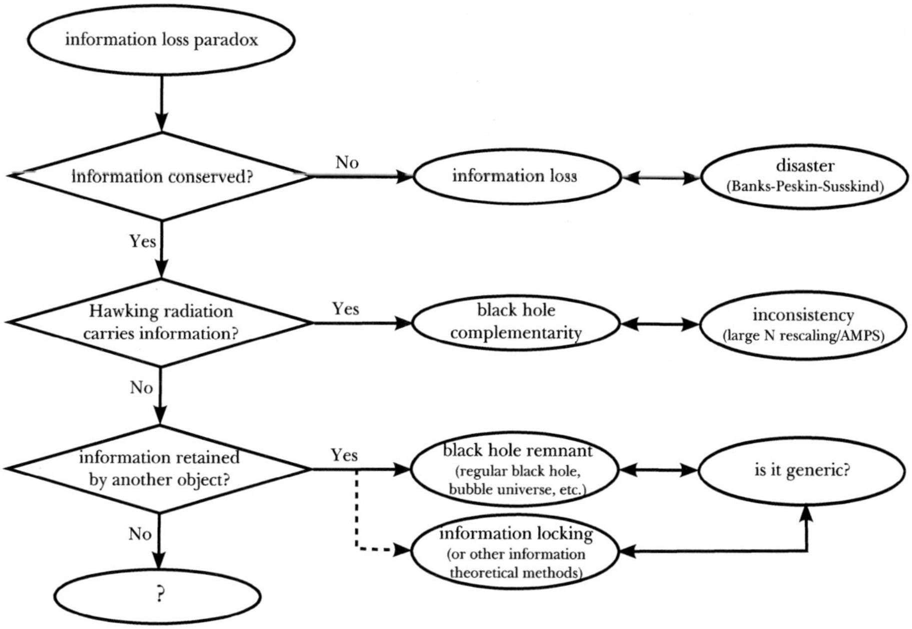

Fig. 1. Flowchart on resolutions of the information loss problem.

where $|i\rangle$ is the initial state, $|f\rangle = \sum_j a_j|f_j\rangle$ is the final state as a superposition of classical states $\{|f_j\rangle\}$ with probability weights $\{a_i\}$, S_{E} is the Euclidean action, and we sum over all geometries and field combinations that connect from $|i\rangle$ to a certain classical state $|f_j\rangle$.

Since this path integral is not very easy to do, we approximate this by summing over only on-shell solutions, or instantons, relying on the steepest-descent approximation. Then we can simplify

$$\langle f|i\rangle \simeq \sum_{i\to f_j} e^{-S_{\mathrm{E}}^{\text{on-shell}}}. \tag{2}$$

If there exists a history with non-zero probability that connects from $|i\rangle$ to $|f_k\rangle$ and if this history has no event horizon nor singularity, i.e., if this has a *trivial geometry*, then information will be conserved through the geometry.[10,11] Hence, in this sense, the wave function of the universe has information. However, since its probability can be highly suppressed due to the entropy cost of the instanton, perhaps the most probable history is the non-trivial geometry that is estimated by the perturbative quantum field theory in a curved spacetime; and in this sense, the semi-classical observer will lose information.[7]

Now the task is to find such an instanton that mediates a trivial geometry. Still we do not have a generic solution, but we can find various interesting examples using the thin-shell approximation.

3. Examples of thin-shell instantons

We consider a spacetime with the metric, $ds_{\pm}^2 = -f_{\pm}(R)dT^2 + f_{\pm}(R)^{-1}dR^2 + R^2d\Omega^2$, where a thin-shell locates at r: outside the shell is $r < R$ (denoted by $+$) and inside the shell is $R < r$ (denoted by $-$). The thin-shell satisfies the metric $ds^2 = -dt^2 + r^2(t)d\Omega^2$. We assume that the outside and inside the shell look like $f_{\pm}(R) = 1 - 2M_{\pm}/R - R^2/\ell_{\pm}^2$, where $M_+ > 0$ and $M_- = 0$ are the mass parameters of each region and $\ell_{\pm}^2 = 3/8\pi U_{\pm}$ denotes the amount of the vacuum energy ($\ell_{\pm}^2$ can be chosen negative in order to present negative vacuum energy).

The equation of motion of the thin-shell is determined by the junction equation:[12] $\epsilon_- \sqrt{\dot{r}^2 + f_-(r)} - \epsilon_+ \sqrt{\dot{r}^2 + f_+(r)} = 4\pi r\sigma$, where $\epsilon_{\pm} = \pm 1$ denotes the direction of the outward normal direction and σ is a tension parameter which is a constant for a scalar field case, while in general this can be a function of r and should satisfy physical constraints.

3.1. *Asymptotic anti-de Sitter and Minkowski*

For asymptotic anti-de Sitter or Minkowski cases, many models have been observed. For example,

- *Sasaki and Yeom*[7] considered true vacuum bubbles in anti-de Sitter background.
- *Lee, Lee and Yeom*[13] considered a magnetic shell in 3-dimensional anti-de Sitter background. Due to the magnetic field, there appeared much diverge tunneling channels.
- *Chen, Domènech, Sasaki and Yeom*[14] considered vacuum bubbles with various tensions that can be in principle justified by physical matters. Some of these examples can show an initial condition that is free from the white hole initial singularity.
- *Chen, Hu and Yeom*[15] considered instanton creation from a no-shell combination to a two-shell combination. The probability can be well-defined, not only relying on the Euclidean method,[16] but also from the Hamiltonian method.[17]

These kind of examples should be generalized in the sense that *(i) it should not rely on vacuum structures* and *(ii) it should not rely on the thin-shell approximation.* We remain these topics as future research directions.

3.2. *Asymptotic de Sitter and subtle issues*

For asymptotic de Sitter cases, there appears further subtle issues. In this section, we see more details.

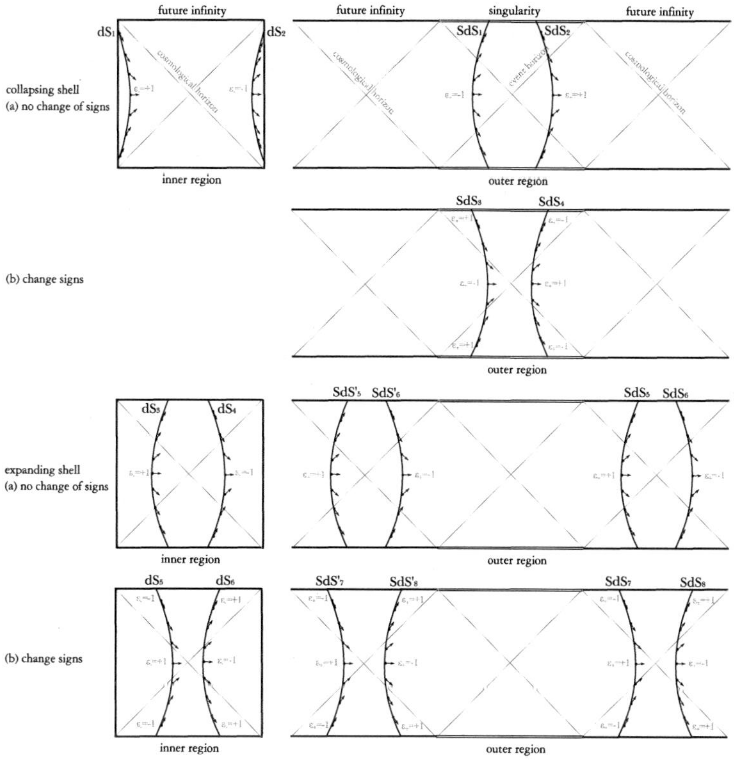

Fig. 2. Summary of the shell dynamics for symmetric solutions.

If the tension is a constant, then after simple calculations, we can reduce the junction equation by $\dot{r}^2 + V(r) = 0$, where

$$V(r) = f_+(r) - \frac{\left(f_-(r) - f_+(r) - 16\pi^2\sigma^2 r^2\right)^2}{64\pi^2\sigma^2 r^2} \tag{3}$$

$$= 1 - \left(\frac{1}{\ell_+^2} + \frac{\mathcal{B}^2}{64\pi^2\sigma^2}\right) r^2 - 2M_+\left(1 + \frac{\mathcal{B}}{32\pi^2\sigma^2}\right)\frac{1}{r} - \frac{M_+^2}{16\pi^2\sigma^2}\frac{1}{r^4} \tag{4}$$

and $\mathcal{B} \equiv -\ell_-^{-2} + \ell_+^{-2} - 16\pi^2\sigma^2$.

By analyzing the equation of motion, we can classify the shell dynamics as well as possible tunneling channels[18] (Fig. 2). In terms of tunneling processes toward a trivial geometry, we illustrate two important processes.

(i) *Disappearance of a black hole.* A collapsing shell can tunnel to a bouncing shell with out-going energy.[16] This was investigated by Gregory-Moss-Withers[16] (Fig. 3).

(ii) *Sweep of a bubble.* For some parameter spaces, a tunneling process such as $dS_3 - SdS_6$ or $dS_3 - SdS'_6$ is possible (Fig. 4). The latter was investigated by Farhi-Guth-Guven (FGG)[19] and Fischler-Morgan-Polchinksi,[17] while the former was investigated by Aguirre-Johnson (AJ).[18]

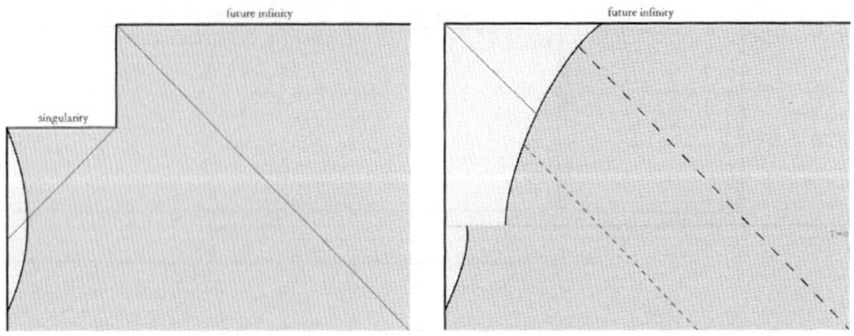

Fig. 3. Left: Usual gravitational collapse and evaporation due to Hawking radiation. Right: Tunneling from $dS_1 - SdS_2$ to $dS_3 - SdS_5$. Black dashed curve is for the false vacuum bubble, while blue dashed curve is for the true vacuum bubble.

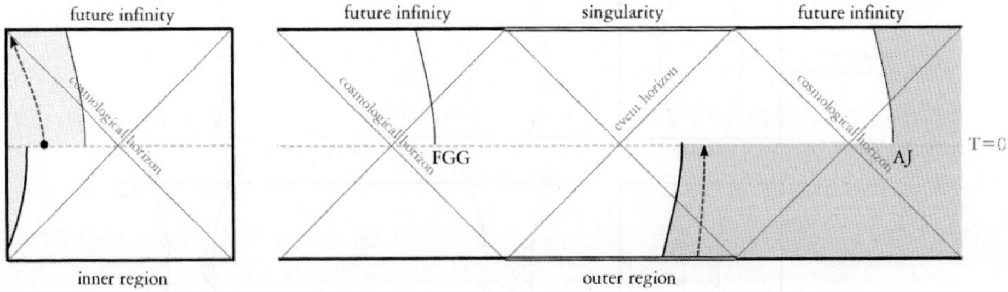

Fig. 4. Tunneling from $dS_1 - SdS_2$ to $dS_3 - SdS_6$ (AJ) and $dS_3 - SdS_6'$ (FGG). The blue dashed arrow is a typical world-line of an observer.

The Aguirre-Johnson tunneling can change to a trivial geometry at least for a local observer (blue dashed arrow in Fig. 4). However, in this case, a more correct expression is that the shell sweeps the observer and hence the observer suddenly locates inside the bubble. Hence, *the singularity should be located outside the causal patch of the observer.* Only if we restrict the path integral inside the observer-dependent causal patch, we can say that this tunneling induces a trivial geomtery.

4. Future perspectives

Non-perturbative effects will shed some lights on the information loss problem and instantons will help to understand this. We have found various thin-shell instantons, but we need to generalize further: we need to find instantons that result trivial geometries (i) without relying on vacuum structures and (ii) without relying on the thin-shell approximation.

For de Sitter cases, there is an interesting process, so-called the Aguirre-Johnson tunneling. This contributes the cosmological horizon scale. Can this mean that a unitary observer should see a superposition of states or uncertainty around the cosmological horizon? In any case, unitarity of the de Sitter space is less clear in the literature. This may cause subtle issues and we need further investigations.

Acknowledgment

The author would like to thank Pisin Chen, Guillem Doménech, Yao-Chieh Hu, Bum-Hoon Lee, Wonwoo Lee, Yen Chin Ong, Don N. Page, and Misao Sasaki for comments and stimulated discussions. This work is supported by Leung Center for Cosmology and Particle Astrophysics (LeCosPA) of National Taiwan University (103R4000).

References

1. S. W. Hawking, Breakdown of predictability in gravitational collapse, *Phys. Rev. D* **14**, 2460 (1976).
2. P. Chen, Y. C. Ong and D. Yeom, Black Hole Remnants and the Information Loss Paradox, *Phys. Rept.* **603**, 1 (2015).
3. T. Banks, L. Susskind and M. E. Peskin, Difficulties for the Evolution of Pure States Into Mixed States, *Nucl. Phys. B* **244**, 125 (1984).
4. L. Susskind, L. Thorlacius and J. Uglum, The stretched horizon and black hole complementarity, *Phys. Rev. D* **48**, 3743 (1993).
5. D. Yeom and H. Zoe, Semi-classical black holes with large N re-scaling and information loss problem, *Int. J. Mod. Phys. A* **26**, 3287 (2011);
   S. E. Hong, D. Hwang, E. D. Stewart and D. Yeom, The Causal structure of dynamical charged black holes, *Class. Quant. Grav.* **27**, 045014 (2010);
   D. Yeom and H. Zoe, Constructing a Counterexample to the Black Hole Complementarity, *Phys. Rev. D* **78**, 104008 (2008).
6. A. Almheiri, D. Marolf, J. Polchinski and J. Sully, Black Holes: Complementarity or Firewalls?, *JHEP* **1302**, 062 (2013).
7. M. Sasaki and D. Yeom, Thin-shell bubbles and information loss problem in anti de Sitter background, *JHEP* **1412**, 155 (2014).
8. B. S. DeWitt, Quantum Theory of Gravity. 1. The Canonical Theory, *Phys. Rev.* **160**, 1113 (1967).
9. J. B. Hartle and S. W. Hawking, Wave Function of the Universe, *Phys. Rev. D* **28**, 2960 (1983).
10. J. M. Maldacena, Eternal black holes in anti-de Sitter, *JHEP* **0304**, 021 (2003).
11. S. W. Hawking, Information loss in black holes, *Phys. Rev. D* **72**, 084013 (2005).
12. W. Israel, Singular hypersurfaces and thin shells in general relativity, *Nuovo Cim. B* **44S10**, 1 (1966) [*Nuovo Cim. B* **48**, 463 (1967)].
13. B. H. Lee, W. Lee and D. Yeom, Dynamics of magnetic shells and information loss problem, *Phys. Rev. D* **92**, no. 2, 024027 (2015).
14. P. Chen, G. Doménech, M. Sasaki and D. Yeom, Stationary bubbles and their tunneling channels toward trivial geometry, arXiv:1512.00565 [hep-th].
15. P. Chen, Y. C. Hu and D. Yeom, Two interpretations on thin-shell instantons, arXiv:1512.03914 [hep-th].
16. R. Gregory, I. G. Moss and B. Withers, Black holes as bubble nucleation sites, *JHEP* **1403**, 081 (2014).
17. W. Fischler, D. Morgan and J. Polchinski, Quantization of False Vacuum Bubbles: A Hamiltonian Treatment of Gravitational Tunneling, *Phys. Rev. D* **42**, 4042 (1990).
18. A. Aguirre and M. C. Johnson, Two tunnels to inflation, *Phys. Rev. D* **73**, 123529 (2006).
19. E. Farhi, A. H. Guth and J. Guven, Is It Possible to Create a Universe in the Laboratory by Quantum Tunneling?, *Nucl. Phys. B* **339**, 417 (1990).

Stationary Bubbles: Information Loss Paradox?

Guillem Domènech* and Misao Sasaki†

Yukawa Institute for Theoretical Physics,
Kyoto University, Kyoto 606-8502, Japan
**guillem.domenech@yukawa.kyoto-u.ac.jp*
†misao@yukawa.kyoto-u.ac.jp

The main purpose of this work is to build classically stationary bubbles, within the thin-shell formalism, which are unstable under quantum effects; they either collapse into a black hole or expand. Thus, the final state can be thought of a superposition of geometries. We point out that, from a quantum mechanical point of view, there is no issue with a loss of information in such configuration. A classical observer sees a definite geometry and, hence, finds an effective loss of information. Although it does not cover all possible cases, we emphasise the role of semi-classical gravitational effects, mediated by instatons, in alleviating/solving the information loss paradox.

Keywords: Black holes; information loss paradox; Euclidean gravity.

1. Introduction

Black holes are truly interesting objects in General Relativity but not free from problems. In particular, the information loss paradox[1] is a long standing question without a definitive answer. It has been shown that one cannot hold on to general relativity, semi-classical quantum field theory, area-entropy relation, and unitarity all at the same time.[4,5]

Although there are several proposals with their own share of issues,[2–12] here we tackle the information loss paradox from the euclidean path integral approach and the wave function of the universe.[13] There, one sums over all possible geometries and configurations, formally given by

$$\psi(\phi_i, g_i|\phi_f, g_f) = \int \mathcal{D}\phi\mathcal{D}g \; e^{iS[\phi,g]} \,, \tag{1}$$

but in practise one cannot deal with the whole path integral. Nevertheless, there is an interesting proposal by Maldacena[14] and Hawking[2] who argued that if one of those histories has a trivial topology, i.e. no horizon nor singularity, then information is conserved in the whole wave function. However, a classical observer who sees a definite geometry experiences an *effective loss of information.* In this work, we construct concrete examples supporting this point of view. Although one cannot build them in general, we show that there is a wide range of parameters that allow such configurations.

We have in mind the picture of a star collapse, where a stable star suffers for any reason a phase transition and undergoes gravitational collapse, perhaps forming a black hole. We model this behaviour within the thin shell formalism and look for a

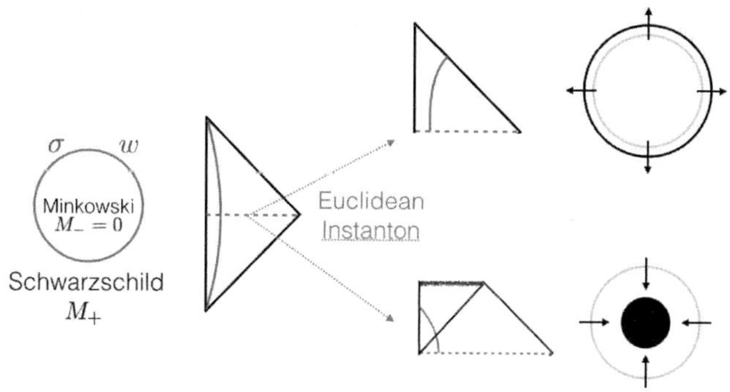

Fig. 1. Rough sketch of the set up. A stationary thin shell, constant r blue curve in the left hand side, tunnels into a black hole, right-down, or expands to infinity, right-up. Those two outcomes are possible and hence the final state is a combination of two geometries.

stationary bubble solutions. To do that a simple tension does not suffice and thus we modify it at high energies regime to obtain the desired solution.

The idea of this work is as follows. As schematically shown in Figure 1, we want to build a stationary bubble, as much as a toy model for a star collapse with a regular interior, which the final geometry, black hole or expanding shell, is due to semi-classical effects. Thus, in Section 2 we review the thin-shell formalism, in Section 3 we build a stationary thin shell solution and discuss the implications to the information loss paradox. Finally, in Section 4 we summarize and conclude our work.

2. Thin-shell junction conditions

By virtue of Israel junction conditions[15] one can glue two solutions of Einstein equations, allowing for a thin shell which accounts for any jump of the extrinsic curvature at the junction. In particular, assume a spherically symmetric system where the metric is given by

$$ds^2_\pm = -f_\pm(R)dT^2 + \frac{1}{f_\pm(R)}dR^2 + R^2 d\Omega^2 \qquad \left(f_\pm(R) = 1 - \frac{2M_\pm}{R} + \frac{R^2}{\ell^2_\pm}\right), \quad (2)$$

where $\pm$ refer to outside ($R > r$) and inside ($R < r$) the shell, $M_\pm$ is the mass parameter and $\ell_\pm$ is the AdS radius. Then the junction condition yields

$$\epsilon_- \sqrt{\dot{r}^2 + f_-(r)} - \epsilon_+ \sqrt{\dot{r}^2 + f_+(r)} = 4\pi r \sigma \,, \quad (3)$$

where σ is the tension of a perfect fluid thin shell and $\epsilon_\pm$ is the sign of the extrinsic curvature out/inside respectively. On top of that we have the conservation of energy, relating surface energy density σ and pressure λ, which given a constant equation of state $w = \lambda/\sigma$ leads us to

$$\sigma(r) = \frac{\sigma_0}{r^{2(1+w)}} \,. \quad (4)$$

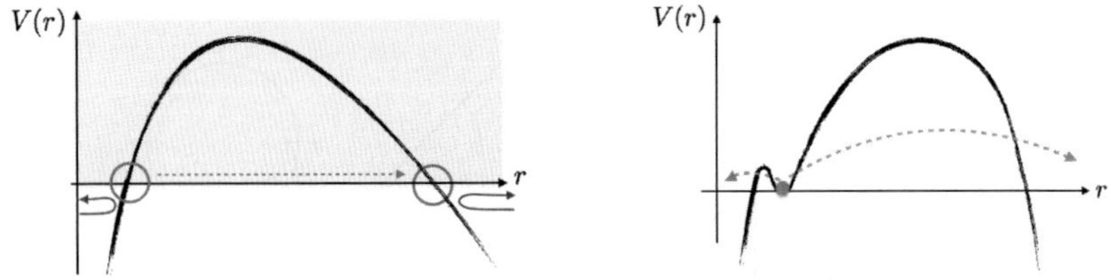

Fig. 2. Schematic form of the effective potential. Left hand side shows the general behaviour for true vacuum bubbles in AdS background with a constant thin shell tension. Right hand side shows the desired behaviour we want to achieve: a stationary bubble.

With this simple set of equations one can find very interesting examples.[16,17] In most cases though, a simple choice of the tension yields only two possible outcomes; collapse or expansion. In the next section we show some sufficient requirements for the tension so as to obtain a stationary thin shell or bubble. Before that, let us remind the reader that the situation gets more interesting when semi-classical effects come into play.[18,19]

2.1. *Effective potential and thin shell tunnelling*

One can grasp a more physically intuitive picture by rewriting the junction condition Eq. (3) as a particle with zero energy moving in a 1-D potential, that is:[20]

$$\dot{r}^2 + V(r) = 0, \quad V(r) = f_+(r) - \frac{\left(f_-(r) - f_+(r) - 16\pi^2\sigma^2 r^2\right)^2}{64\pi^2\sigma^2 r^2}. \tag{5}$$

This form is particularly useful to understand the dynamics of the thin shell with radius r. For example, for true vacuum bubbles in an AdS background with a constant tension,[11] the potential generally takes the form as in the left side of Figure 2. It is easy to see that the shell either unavoidably collapse or expands. Now, if our model parameters have a region with $V(r) > 0$, we have two classically disconnected regions. On one hand, the shell starts from $r = 0$, expands, bounces and collapses back to $r = 0$. On the other hand, the shell starts from $r = \infty$, collapses, bounces and expands. However, as it is well know from quantum mechanics, a particle can cross a classically forbidden region by a quantum tunnelling. In our case, the thin shell can tunnel,[19] for example, from an expanding solution from $r = 0$ to an expanding solution to $r = \infty$, with a exponentially suppressed yet non-vanishing probability. The main drawback of the latter example though, is that the shell starts from a white hole at $r = 0$ and thus one could question its initial conditions. For this reason, let us modify the latter example and look for an stationary bubble as an initial condition, see right of Figure 2.

Before ending this section, let us recall that the sign of the extrinsic curvature dictates the location of the thin shell in the Penrose diagram. We require the extrinsic curvature to be positive for r larger than the outer horizon so as to be located at the right hand side of the diagram. This is ensured by requiring $\sqrt{f_- - f_+} > 4\pi\sigma r$,

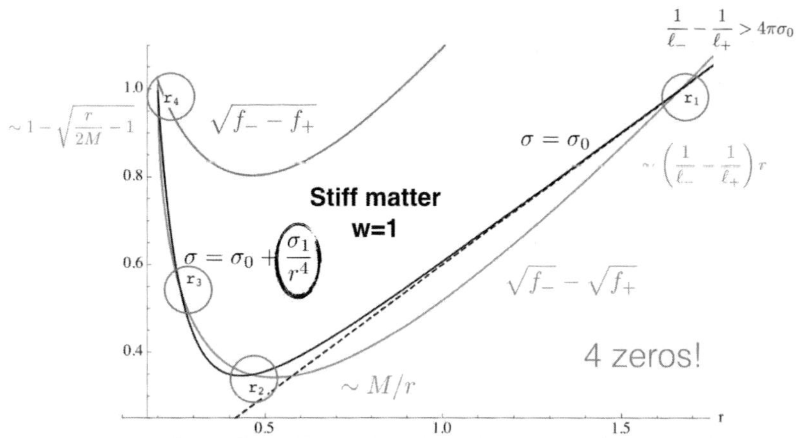

Fig. 3. Counting zeros of the potential. The red line is $\sqrt{f_-} - \sqrt{f_+}$ and the approximate behaviour in each zone is written in red. Black lines are $4\pi\sigma r$, black dotted for a constant tension and black curve for the modified behaviour $\sim 1/r^4$. Any crossing with the blue line indicates a change in the extrinsic curvature sign. The parameters of this plot are as follows ($c = \hbar = G = 1$): $M_+ = 0.1$, $M_- = 0$, $\ell_+ = \infty$, $\ell_- = 2$, $4\pi\sigma_0 = 0.3$ and $4\pi\sigma_1 = 0.0067$.

as one can see from recovering the signs of the extrinsic curvature, that is

$$\beta_\pm(r) \equiv \frac{f_-(r) - f_+(r) \mp 16\pi^2\sigma^2 r^2}{8\pi\sigma r} = \epsilon_\pm \sqrt{\dot{r}^2 + f_\pm(r)}\,. \tag{6}$$

We show this requrement as a blue line in all forthcoming plots.

3. Stationary bubbles and superposition of geometries

Let us intuitively explain how the tension σ should behave at high energies (small r) regime so as to achieve a classically stable configuration. For this purpose, it is useful to count the number of zeros in the effective potential, by ploting the junction condition Eq.(3) setting $\dot{r} = 0$, rather than the effective potential form. For example, for an AdS true vacuum bubble one generally obtains the behaviour shown in Figure 3. Notice that for a constant tension, which corresponds to the black dotted line, we obtain two intersections with the red line and, therefore, the potential has two zeros and takes the form of left Figure 2.

Now, we want to modify this solution so as to obtain four zeros in the effective potential. In other words, we want four intersections with the red line in total. From Figure 3, it is readily seen that the tension must be steep enough in order to overcome the $1/r$ behaviour of the red line curve. We choose to add stiff matter with an equation of state $w = 1$, which leads to a $1/r^4$ behaviour of the tension. See the black line in Figure 3. This is obviously not the only possibility but the simplest modification, as far as we understand.

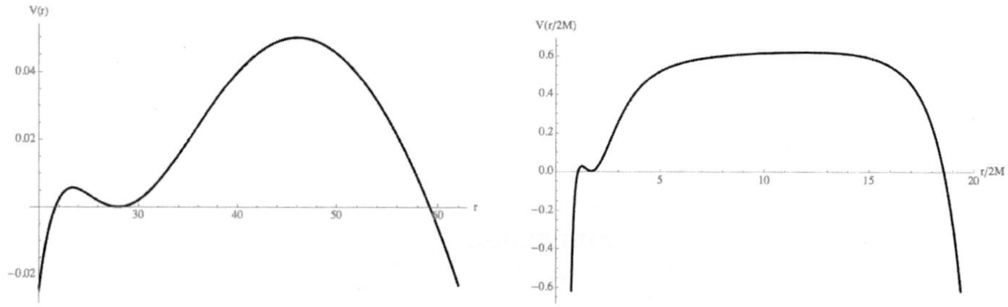

Fig. 4. Concrete examples of the potential. On the left hand side the parameters are: $M_+ = 10$, $M_- = 0$, $\ell_+ = \infty$, $\ell_- = 40$, $4\pi\sigma_0 = 0.016$ and $4\pi\sigma_1 = 5230$. On the right hand side we chose: $\ell_\pm = \infty$, $\bar{\sigma}_0 = 0.87$, $\mu_1 = 1/2$, $\mu_2 = 1/3$, $\bar{r}_1 = 1$ and $\bar{r}_2 = 20$.

The same procedure can be applied in the case of a thin shell in a Minkowski background, although one must tune more the tension.[11] In particular, we use:

$$w(\bar{r}) = \begin{cases} \frac{1}{2}\left(\tanh\left[\mu_1\left(\bar{r}_1 - \bar{r}\right)\right] + 1\right) & \bar{r} < \bar{r}_m \\ \frac{1}{6}\left(\tanh\left[\mu_2\left(\bar{r} - \bar{r}_2\right)\right] + 1\right) & \bar{r} > \bar{r}_m \end{cases}, \tag{7}$$

where $\bar{r} = r/2M$, $\bar{\sigma} \equiv 8\pi M\sigma$, μ_i and r_i respectively control the steepness and position of the transition, and r_m is the matching radius in an intermediate region. One can easily find a parameter region where three or four zeros are obtained and, therefore, find a stationary solution. A concrete example is shown in Figure 4.

So far we showed examples of stationary bubbles, which we think of a toy model of a star collapse. Now, it is time to discuss the implications to the information loss paradox. In fact, one can numerically compute the probabilities[19] and find that the probability of tunnelling into a black hole is exponentially higher than the probability to tunnel to an expanding shell. Still, the latter is non-vanishing and therefore there is a path in the whole wave function of the universe where the information is recovered. That implies that a classical observer who sees a definite geometry, the most probable one, finds an *effective loss of information.*

4. Conclusion

We studied the information loss paradox in the context of euclidean path integrals. Using the thin shell formalism, we built classically stable stationary bubbles. We showed the existence of such examples for true vacuum bubbles in asymptotically flat and AdS backgrounds and also in Minkowski spacetime. They could be thought of as a toy model of a star collapse. Afterwards, we considered that such bubbles are not stable under quantum effects and eventually they tunnel, mediated by instantons, either to a collapsing shell that forms a black hole or to an expanding shell that expands all the way to infinity. In terms of the wave function, the final geometry is a composition of two possible final states. However, a classical observer only sees a definite geometry and, most likely, finds a black hole configuration since it is exponentially favoured with respect to the expanding solution. Thus, a classical observer might find an effective loss of information, while the whole information is encoded in the wave function of the universe. Although our work is based in a collection of examples, we want to emphasise that in such cases the role of semi-classical effects in solving the information loss paradox is essential. It might be helpful, for example, in cases such as eternal black holes or extremal charged black holes.

Acknowledgments

GD and MS would like to thank P.Chen and D.h.Yeom for many fruitful discussions. This work was supported in part by MEXT KAKENHI Grant Number 15H05888.

References

1. S. W. Hawking, Phys. Rev. D **14**, 2460 (1976).
2. S. W. Hawking, Phys. Rev. D **72**, 084013 (2005) [arXiv:hep-th/0507171]; S. W. Hawking, arXiv:1401.5761 [hep-th].
3. L. Susskind, L. Thorlacius and J. Uglum, Phys. Rev. D **48** (1993) 3743 [hep-th/9306069].
4. D. Yeom and H. Zoe, Phys. Rev. D **78**, 104008 (2008) [arXiv:0802.1625 [gr-qc]]; P. Chen, Y. C. Ong and D. Yeom, JHEP **1412**, 021 (2014) [arXiv:1408.3763 [hep-th]].
5. A. Almheiri, D. Marolf, J. Polchinski and J. Sully, JHEP **1302**, 062 (2013) [arXiv:1207.3123 [hep-th]]; A. Almheiri, D. Marolf, J. Polchinski, D. Stanford and J. Sully, JHEP **1309**, 018 (2013) [arXiv:1304.6483 [hep-th]].
6. W. G. Unruh and R. M. Wald, Phys. Rev. D **52** (1995) 2176 [hep-th/9503024].
7. D. Hwang, B.-H. Lee and D. Yeom, JCAP **1301**, 005 (2013) [arXiv:1210.6733 [gr-qc]]; W. Kim, B.-H. Lee and D. Yeom, JHEP **1305**, 060 (2013) [arXiv:1301.5138 [gr-qc]].
8. A. Ashtekar and M. Bojowald, Class. Quant. Grav. **22** (2005) 3349 [gr-qc/0504029]; P. M. Ho, arXiv:1510.07157 [hep-th].
9. D. N. Page, JCAP **1406** (2014) 051 [arXiv:1306.0562 [hep-th]]; P. Chen, Y. C. Ong, D. N. Page, M. Sasaki and D. Yeom, arXiv:1511.05695 [hep-th].
10. P. Chen, Y. C. Ong and D. h. Yeom, Phys. Rept. **603** (2015) 1 [arXiv:1412.8366 [gr-qc]]; W. Y. Wen and S. Y. Wu, Eur. Phys. J. C **75** (2015) 12, 608 [arXiv:1509.01305 [hep-th]]; Y. C. Ong, JCAP **1504** (2015) 04, 003 [arXiv:1503.01092 [gr-qc]].
11. M. Sasaki and D. Yeom, JHEP **1412**, 155 (2014) [arXiv:1404.1565 [hep-th]]; P. Chen, G. Domènech, M. Sasaki and D. h. Yeom, arXiv:1512.00565 [hep-th]; P. Chen, Y. C. Hu and D. h. Yeom, arXiv:1512.03914 [hep-th].
12. S. W. Hawking, M. J. Perry and A. Strominger, arXiv:1601.00921 [hep-th]. A. Averin, G. Dvali, C. Gomez and D. Lust, arXiv:1601.03725 [hep-th].
13. J. B. Hartle and S. W. Hawking, Phys. Rev. D **28**, 2960 (1983).
14. J. M. Maldacena, JHEP **0304**, 021 (2003) [arXiv:hep-th/0106112].
15. W. Israel, Nuovo Cim. B **44S10** (1966) 1
16. R. J. Adler, J. D. Bjorken, P. Chen and J. S. Liu, Am. J. Phys. **73**, 1148 (2005) [gr-qc/0502040].
17. R. B. Mann and J. J. Oh, Phys. Rev. D **74** (2006) 124016 [gr-qc/0609094].
18. E. Farhi, A. H. Guth and J. Guven, Nucl. Phys. B **339**, 417 (1990); W. Fischler, D. Morgan and J. Polchinski, Phys. Rev. D **42**, 4042 (1990).
19. S. Ansoldi, A. Aurilia, R. Balbinot and E. Spallucci, Class. Quant. Grav. **14** (1997) 2727 [gr-qc/9706081].
20. S. K. Blau, E. I. Guendelman and A. H. Guth, Phys. Rev. D **35**, 1747 (1987).

Black Hole: The Interior Spacetime

Yen Chin Ong

Nordita, KTH Royal Institute of Technology & Stockholm University, Roslagstullsbacken 23, SE-106 91 Stockholm, Sweden
yenchin.ong@nordita.org

The information loss paradox is often discussed from the perspective of the observers who stay outside of a black hole. However, the interior spacetime of a black hole can be rather nontrivial. We discuss the open problems regarding the volume of a black hole, and whether it plays any role in information storage. We also emphasize the importance of resolving the black hole singularity, if one were to resolve the information loss paradox.

Keywords: Black hole volume, information loss paradox.

1. What's Inside a Black Hole?

The information loss paradox remains unresolved 40 years after Hawking first pointed out the problem.[1] Consider a cloud of matter in a pure state that collapses to form a black hole. The common viewpoint is that, if physics is indeed unitary, exterior observers should be able to recover a pure state after the black hole has completely evaporated. Indeed, according to black hole complementarity,[2] an exterior observer and an in-falling observer can observe different events, but each see a completely self-consistent, unitary-preserving, physics. It may therefore be sufficient to discuss the information loss paradox solely from the point of view of the exterior observers. Nevertheless, the fate of the information, as seen from a comoving observer that falls into the black hole, remains an interesting problem. To answer this question, one has to first ask the obvious question: "*what is inside a black hole?*" We shall discuss some possibilities.

For simplicity, let us focus our attention on the Schwarzschild black hole, described by the following metric, in the units $G = c = 1$,

$$g[\text{Sch}] = -\left(1 - \frac{2M}{r}\right) \mathrm{d}t^2 + \left(1 - \frac{2M}{r}\right)^{-1} \mathrm{d}r^2 + r^2(\mathrm{d}\theta^2 + \sin^2\theta\, \mathrm{d}\phi). \quad (1)$$

A textbook on general relativity typically mentions that one can analytically continue the Schwarzschild manifold to the Kruskal-Szekeres manifold, which contains another asymptotically flat region inside the black hole, on the other side of the Einstein-Rosen bridge. There are at least two issues with this picture. The first is well known: in a realistic gravitational collapse, one does not seriously expect that the resulting black hole would contain another universe inside it. The second issue is that, in general, analytic continuations are not unique. If we drop some conditions such as vacuum, then other analytic continuations exist.[3] (Analyticity is a rather strong condition, if one drops this and considers smooth continuations only, then even more extensions are possible.) This means that the interior geometries of two black holes can be vastly different even if both of them are exactly Schwarzschild as seen from the outside.

Let us first consider the simplest scenario in which the black hole is formed from gravitational collapse and is thus "one-sided", i.e., does not harbor another asymptotically flat region. Can we say something definitive about its interior geometry? In particular, what is its volume?

This is a rather tricky question to answer. Although the area of the black hole event horizon, which by virtue of being a null surface, is a well-defined geometric invariant independent of the choice of spacelike hypersurfaces, the same is not true for the spatial 3-volume inside the black hole.[4] In fact, various definitions of black hole volumes have been proposed (see e.g. Ref.5), including definition that is more thermodynamical than geometrical.[6,7] It is also noted that an invariant way of defining volume does exist, albeit this is not a proper volume.[8] What we are interested in is the question: if we look at, say, the black hole at the center of the Milky Way, how much volume does it contain?

Since the proper volume is hypersurface-dependent, there is no unique answer to this. What one can ask, however, is the *largest* volume possible. Assuming the Schwarzschild geometry, Christodoulou and Rovelli showed that[9] a cross section of the event horizon taken at late times bounds a spatial volume which grows with time as

$$\text{Vol.} \sim 3\sqrt{3}\pi M^2 v, \tag{2}$$

where v is the advanced time, and M the mass of the black hole. This volume corresponds to the maximal slice[10,11] at $r = 3M/2$, which we shall refer to as the "CR-volume". More explicitly, this is obtained via the volume integral

$$\text{Vol.} \sim \int^v \int_{S^2} \max\left[r^2\sqrt{\frac{2M}{r} - 1}\right] \sin\theta \, d\theta \, d\phi \, dv. \tag{3}$$

The time dependence is not so surprising since the interior geometry of a Schwarzschild black hole is, unlike the exterior geometry, *not* static.

Christodoulou and Rovelli estimated that Sagittarius A*, the supermassive black hole at the center of our galaxy, contains sufficient space to fit a *million* solar systems, despite its areal radius being only a factor of 10 or so larger than the distance from Earth to the Moon. Taking into account the rotation of Sagittarius A* and repeating the calculation using the Kerr metric, Bengtsson and Jakobsson showed that this estimate does not change by much,[12] and so there is a lot of room inside a black hole! (Such volume for asymptotically locally anti-de Sitter black hole has also been investigated.[13]) It is therefore tempting to ask whether this large volume can shed some light on the information loss paradox.

2. Black Hole Volume and Information Storage

To answer this question, one must first ask: How much do we trust the naive spherically symmetric Schwarzschild geometry to continue to hold in the interior spacetime?

Near the singularity it is conjectured that spacetime should become highly chaotic (BKL singularity[14]), and one would not be able to trust the aforementioned naive volume integral to continue to hold. However, this complication should not arise until the very late part of the evolution. Therefore, let us for the moment assume that the Schwarzschild geometry is a sufficiently good approximation and there is indeed a large volume inside a black hole, at least for some time during its evolution under Hawking evaporation. The question is then: how does Hawking evaporation affect this volume? Does it shrink together with the horizon area?

For the most part of the evolution, the mass loss is well-approximated by the thermal loss equation, which is given by

$$\frac{\mathrm{d}M}{\mathrm{d}v} \approx -\alpha a \sigma T^4 = -\alpha a \left(27\pi M^2\right)\left(\frac{\hbar}{8\pi M}\right)^4 = -\frac{\mathcal{C}}{M^2}, \quad \mathcal{C} > 0, \tag{4}$$

where $a = \pi^2/(15\hbar^3)$ is the radiation constant (which is $4/c$ times the Stefan-Boltzmann constant), and α is the greybody factor, which depends on the number of particle species emitted by the Hawking radiation. It is found in Ref.15 that somewhat counter-intuitively, the CR-volume continues to increase even though the black hole is losing mass, and consequently, its area is decreasing. This is easy to see, for the volume integral is

$$V(v) \sim 3\sqrt{3}\pi \int^v M^2 \,\mathrm{d}v. \tag{5}$$

(We remark that the volume is no longer asymptotically linear in v, as it would be when the mass of the black hole is a constant.) So by the Fundamental Theorem of Calculus, one immediately sees that

$$\frac{\mathrm{d}V}{\mathrm{d}v} \sim 3\sqrt{3}\pi M^2(v) \geqslant 0. \tag{6}$$

As discussed in Ref.15, this could imply that at sufficiently late time, one either have a "bag-of-gold" type of geometry, which is still connected to the exterior asymptotically flat spacetime via a throat, or that the throat could pinch off entirely and the interior spacetime becomes an isolated universe on its own. This is consistent with the results of Ref.16.

In view of this, it is perhaps tempting to conjecture that information can be stored in the CR-volume despite the shrinking of the black hole area. However, this appears not to be the case since the actual entropy content associated with the CR-volume, S_{CR}, turns out to be proportional to the *area* (though the coefficient is not the same as the Bekenstein-Hawking area formula), instead of the volume.

Explicitly, Zhang[17] showed that

$$S_{\mathrm{CR}} \sim \frac{3\sqrt{3}}{\mathcal{C}(90\pi \times 8^4)} A. \tag{7}$$

Zhang also argued that the thermodynamics associated with the entropy in the CR volume is caused by the vacuum polarization near the horizon.

3. The Importance of Understanding the Singularities

One crucial issue that we have not discussed thus far is the singularity inside a black hole. For a Schwarzschild black hole this is particularly important, since it is spacelike and lies in the future of anything that falls into the hole. It is well known that the maximum proper time before an in-falling object eventually terminates at the singularity is πM. This means that if the singularity is not resolved, information would eventually "fall off the edge of spacetime" at the singularity. More precisely, there is an entanglement between particles behind the horizon, and particles that remain in the exterior of it. However, the one behind the horizon ultimately gets destroyed at the singularity – hence the information loss paradox.

This means that having a large spacelike volume *by itself* does not resolve the information loss paradox. Instead, one has to understand whether the singularity is indeed resolved in quantum gravity, and if so how. Although the common viewpoint is that the singularity is a sign that general relativity breaks down and will be cured by a full working theory of quantum gravity, *it may not be.*

Only if there is no singularity, then a black hole remnant with a huge interior volume or a baby universe may be a viable candidate to resolve the information loss paradox. This proposal is of course not without problems. The readers should refer to Ref.18 for detailed discussion, but let us mention one of the obvious problem: If information is indeed contained in the large volume in some way or another (i.e. not necessarily in the CR volume), this seems to violate the common wisdom that the Bekenstein-Hawking entropy – which is proportional to the area – should count the number of states of the black hole degrees of freedom. Nevertheless, there remains a possibility that the common wisdom is incorrect. For more discussion, see Refs.18, 19, 20, 21, 22.

4. Conclusion: Pay Attention to Geometry

To conclude, general relativity is a *geometric* theory of gravity. We should therefore pay more attention to the spacetime geometry, whether it is the exterior geometry or the interior one. Let us not be biased by the fact that we are exterior observers. Perhaps the interior spacetime has nontrivial geometry that allows information to be stored despite the shrinkage of the black hole horizon during Hawking evaporation.

In addition, the volume of (two-sided) black holes has recently gained some attention in the context of holography. To be more specific, there might exist a volume/complexity relation, such that the computational complexity[23] $C(t)$ of a certain quantum state, as a function of some proper time t, goes like

$$C(t) \propto \mathrm{Vol}(\Sigma_t), \tag{8}$$

where Σ_t is a codimension-one space-like section of the anti-de Sitter bulk with extremal volume. See Ref.24 and the references therein for detail.

Despite the potentially important roles of black hole volumes, we should not ignore the singularities, since if they remain unresolved by quantum gravity information can be lost by simply getting destroyed there.

In the context of the firewall controversy,[25] it has been proposed by Susskind that perhaps the firewall, if it exists, is just the singularity of the black hole that has "migrated" to the horizon, due to the volume inside a black hole gradually disappearing as the entanglement between the interior and exterior spacetime gets broken[26] at sufficiently late times. This further exemplifies the importance of understanding both the volume, as well as the singularity, of a black hole.

It is of course possible that the information loss paradox is a question that can be resolved without knowing the details whether – or how – quantum gravity cures the black hole singularity, but again, *it may not be.*

References

1. S. W. Hawking, Breakdown of predictability in gravitational collapse, *Phys. Rev. D* **14**, 2460 (1976).
2. L. Susskind, L. Thorlacius and J. Uglum, The stretched horizon and black hole complementarity, *Phys. Rev. D* **48**, 3743 (1993) [arXiv:hep-th/9306069].
3. J. M. M. Senovilla, Singularity theorems and their consequences (Review), *Gen. Rel. Grav.* **30**, 701 (1998).
4. B. S. DiNunno and R. A. Matzner, The volume inside a black hole, *Gen. Rel. Grav.* **42** 63, (2010) [arXiv:0801.1734 [gr-qc]].
5. W. Ballik and K. Lake, The vector volume and black holes, *Phys. Rev. D* **88**, 104038 (2013) [arXiv:1310.1935 [gr-qc]].
6. M. Cvetic, G. W. Gibbons, D. Kubiznak and C. N. Pope, Black hole enthalpy and an entropy inequality for the thermodynamic volume, *Phys. Rev. D* **84**, 024037 (2011) [arXiv:1012.2888 [hep-th]].
7. B. P. Dolan, D. Kastor, D. Kubiznak, R. B. Mann and J. Traschen, Thermodynamic volumes and isoperimetric inequalities for de Sitter black holes, *Phys. Rev. D* **87**, 104017 (2013) [arXiv:1012.2888 [hep-th]].
8. M. K. Parikh, The volume of black holes, *Phys. Rev. D* **73**, 124021 (2006) [arXiv:hep-th/0508108].
9. M. Christodoulou and C. Rovelli, How big is a black hole?, *Phys. Rev. D* **91**, 064046 (2015) [arXiv:1411.2854 [gr-qc]].
10. B. L. Reinhart, Maximal foliations of extended Schwarzschild space, *J. Math. Phys.* **14**, 719 (1973).
11. F. Estabrook, H. Wahlquist, S. Christensen, B. DeWitt, L. Smarr and E. Tsiang, Maximally slicing a black hole, *Phys. Rev. D* **7**, 2814, (1973).
12. I. Bengtsson and E. Jakobsson, Black holes: their large interiors, *Mod. Phys. Lett. A*, **30**, 1550103 (2015) [arXiv:1502.01907 [gr-qc]].

13. Y. C. Ong, Never judge a black hole by its area, *JCAP* **1504**, no. 04, 003 (2015) [arXiv:1503.01092 [gr-qc]].
14. V. A. Belinskii, I. M. Khalatnikov and E. M. Lifshitz, Oscillatory approach to a singular point in the relativistic cosmology, *Adv. Phys.* **19**, 525 (1970).
15. Y. C. Ong, The persistence of the large volumes in black holes, *Gen. Relativ. Gravit.* **47**, 88 (2015) [arXiv:1503.08245 [gr-qc]].
16. R. Parentani and T. Piran, The internal geometry of an evaporating black hole, *Phys. Rev. Lett.* **73**, 2805 (1994).
17. B-C. Zhang, Entropy in the interior of a black hole and thermodynamics, *Phys. Rev. D* **92**, 081501(R) (2015) [arXiv:1510.02182 [gr-qc]].
18. P. Chen, Y. C. Ong and D. Yeom, Black hole remnants and the information loss paradox, *Phys. Rept.* **603**, 1 (2015) [arXiv:1412.8366 [gr-qc]].
19. D. Marolf, Black holes, AdS, and CFTs, *Gen. Rel. Grav.* **41**, 903 (2009) [arXiv:0810.4886 [gr-qc]].
20. S. D. H. Hsu and D. Reeb, Black hole entropy, curved space and monsters, *Phys. Lett. B* **658**, 244 (2008) [arXiv:0706.3239 [hep-th]].
21. S. D. H. Hsu and D. Reeb, Monsters, black holes and the statistical mechanics of gravity, *Mod. Phys. Lett. A* **24**, 1875 (2009) [arXiv:0908.1265 [gr-qc]].
22. Y. C. Ong and P. Chen, The fate of monsters in anti-de Sitter spacetime, *JHEP* **07**, 147 (2013) [arXiv:1304.3803 [hep-th]].
23. L. Susskind, Computational complexity and black hole horizons, [arXiv:1402.5674 [hep-th]].
24. J. L. F. Barbon and J. Martin-Garcia, Holographic complexity of cold hyperbolic black holes, *JHEP* **11**, 181 (2015) [arXiv:1510.00349 [hep-th]].
25. A. Almheiri, D. Marolf, J. Polchinski and J. Sully, Black holes: complementarity or firewalls?, *JHEP* **1302**, 062 (2013) [arXiv:1207.3123 [hep-th]].
26. L. Susskind, Singularities, firewalls, and complementarity, [arXiv:1208.3445 [hep-th]].

Unruh Radiation Produced by a Uniformly Accelerating Charged Particle in Thermal Random Motions

Naritaka Oshita

Department of Physics, Graduate School of Science, The University of Tokyo, Bunkyo-ku, Tokyo 113-0033, Japan, and Research Center for the Early Universe (RESCEU), Graduate School of Science, The University of Tokyo, Bunkyo-ku, Tokyo 113-0033, Japan
naritaka@resceu.s.u-tokyo.ac.jp

Kazuhiro Yamamoto

Department of Physical Science, Graduate School of Science, Hiroshima University, Higashi-Hiroshima 739-8526, Japan, and Hiroshima Astrophysical Science Center, Hiroshima University, Higashi-Hiroshima 739-8526, Japan

Sen Zhang

Okayama Institute for Quantum Physics, Kyoyama 1-9-1, Kita-ku, Okayama 700-0015, Japan

In this study, we investigate the signature of the Unruh effect in quantum radiation from an accelerated charged particle interacting with vacuum fluctuations. Because a charged particle in uniformly accelerated motion exhibits thermal random motion around the classical trajectory because of the Unruh effect, its quantum radiation might be termed Unruh radiation. We show that the energy flux of the quantum radiation is negative and that its amplitude is smaller than the classical Larmor radiation by a factor of a/m, where a is the acceleration and m is the mass of the particle. The total radiation flux of the classical Larmor radiation and the quantum radiation is positive; therefore, the quantum radiation appears to suppress the total radiation. Interestingly, the results are consistent with the prediction for the quantum correction to classical Larmor radiation, which were obtained using a different approach.

Keywords: Hawking radiation; Unruh effect; laser cosmology.

1. Introduction

Hawking radiation refers to the theoretically predicted black hole emission of thermal radiation with a temperature of $T_{\rm BH} = 1/8\pi GM$, where M is the mass of the black hole and G is the gravitational constant.[1] The detection of Hawking radiation is extremely difficult because the Hawking temperature of a black hole with solar mass $M_\odot \sim 2 \times 10^{30}$ [kg] is of the order of $T_{\rm BH} \sim 10^{-7}$ [K], which is much smaller than the temperature of the cosmic microwave background $T_{\rm CMB} \simeq 2.7$ [K]. The Unruh effect refers to the prediction that a uniformly accelerated observer sees the Minkowski vacuum as a thermal bath with a temperature of $T_{\rm U} = a/2\pi$, where a is the acceleration of the observer.[2]

The Unruh effect is related to the Hawking radiation via the principle of equivalence. Therefore, any experimental verification of the Unruh effect will be useful as an analogy of the Hawking radiation. P. Chen and T. Tajima pointed out the possibility of testing the Unruh effect using an extremely intense laser.[3] An electron accelerated by the laser will see the Minkowski vacuum as a thermal bath; then, radiation might be produced via the Unruh effect, which is referred to as Unruh radiation. Using a laser with an intensity of the order of 10^{25} [W/cm^2],[4] the Unruh temperature reaches $T_{\rm U} \simeq 10^3$ [K], which is much higher than room temperature. Thus, Unruh radiation is interesting as evidence of the Unruh effect.

Possible signatures of Unruh radiation have been theoretically investigated by several authors. The authors in Refs.3, 5, 6 investigated the Unruh radiation produced by scattering between an accelerated electron and the Rindler particle of the thermal bath originating from the Unruh effect. However, the importance of quantum interference in predicting the Unruh radiation has been pointed out;[7–12] we take this interference into account in our theoretical investigation of the signature of the Unruh radiation.[13]

The paper is organized as follows. In Sec. 2, we review our results in Ref.13. Adopting a model consisting of a charged particle and an electromagnetic field, we calculate the energy flux while taking the quantum interference term into account. In Sec. 3, we show that the results in Sec. 2 are consistent with the first-order quantum correction to the Larmor radiation, which was previously obtained using a different approach in Refs.16–20.

2. Quantum radiation produced by an accelerated charged particle

We adopt a model consisting of a point particle (with a mass m and a charge e) and an electromagnetic field $A^\mu(x)$, which are coupled to each other. The action of the model is given by $S_{\rm tot} = S_{\rm P} + S_{\rm EM} + S_{\rm int}$ with $S_{\rm P} = -m\int d\tau\sqrt{\eta_{\mu\nu}\dot{z}^\mu\dot{z}^\nu}$, $S_{\rm EM} = -\int d^4x F^{\mu\nu}F_{\mu\nu}/4$, and $S_{\rm int}(z, A^\mu) = -e\int d\tau \int d^4x \delta^4_{\rm D}(x - z(\tau))\dot{z}^\mu(\tau)A_\mu(x)$, where $F_{\mu\nu} = \partial_\mu A_\nu - \partial_\nu A_\mu$ is the field strength and $\eta_{\mu\nu}$ is the metric of the Minkowski spacetime, for which we follow the convention $(+ - - -)$. The equations of motion are derived from the action as follows:

$$m\ddot{z}_\mu = e(\partial_\mu A_\nu - \partial_\nu A_\mu)\dot{z}^\nu + f_\mu(A_{\rm BG}), \tag{1}$$

$$\partial_\mu\partial^\mu A^\nu = e\int d\tau \dot{z}^\nu(\tau)\delta^4_{\rm D}(x - z(\tau)), \tag{2}$$

$$\text{with } \partial_\mu A^\mu = 0 \text{ (gauge condition)}, \tag{3}$$

where we introduce the external force f_μ that accelerates the particle with acceleration a using a uniform electromagnetic field $A_{\rm BG}$. The solution of Eq. (2) is given by

$$A_\mu(x) = A_{{\rm BG}\mu}(x) + A_{{\rm h}\mu}(x) + e\int d\tau G_{\rm R}(x, z(\tau))\dot{z}_\mu(\tau), \tag{4}$$

where $A^\mu_{\rm h}$ satisfies $\partial_\nu\partial^\nu A^\mu_{\rm h} = 0$, and $G_{\rm R}(x, y)$ represents the retarded Green's function that satisfies $\partial_\mu\partial^\mu G_{\rm R}(x, y) = \delta^4_{\rm D}(x - y)$. Note that the background part of the electromagnetic field $A_{\rm BG}(x)$ does not contribute to the radiation energy flux, but instead, simply uniformly accelerates the particle. We introduce the inhomogeneous solution of the electromagnetic field $A^\mu_{\rm inh}(x)$ as follows

$$A^\mu_{\rm inh}(x) \equiv e\int d\tau G_{\rm R}(x, z(\tau))\dot{z}^\mu(\tau). \tag{5}$$

Decomposing the trajectory of the particle $z_\mu = z_\mu(\tau)$ into two parts

$$z^\mu = \bar{z}^\mu + \delta z^\mu \tag{6}$$

$$\bar{z}^\mu = (a^{-1}\sinh(a\tau), a^{-1}\cosh(a\tau), 0, 0), \tag{7}$$

and substituting (4) into (1), we obtain the linearized equation of (1),

$$m\delta\ddot{z}^i(\tau) = \frac{e^2}{6\pi}(\delta\dddot{z}^i - a^2\delta\dot{z}^i) + e(\eta^{i\nu}\dot{\bar{z}}^\alpha - \eta^{i\alpha}\dot{\bar{z}}^\nu)\partial_\nu A_{{\rm h}\alpha}(x)|_{x=z(\tau)}. \tag{8}$$

In the following discussion, we omit the third-order time derivative term of the radiation reaction force $(e^2/6\pi)\delta\dddot{z}^i$ in (8), because the contribution of this term to the perturbation δz^i is of the order of $\mathcal{O}((a/m)^2)$,[a] as mentioned in the paper by S. Zhang.[15] Moreover, this term corresponds to the effect caused by the short-distance dynamics, whose scale is of the order of e^2/m, which is much smaller than the Compton length $1/m$.

To obtain the energy flux of the Unruh radiation, we first calculate the two-point correlation function,

$$\begin{aligned}&\langle A^\mu(x)A^\nu(y)\rangle - \langle A^\mu_{\rm h}(x)A^\nu_{\rm h}(y)\rangle - A^\mu_{\rm BG}(x)A^\nu_{\rm BG}(y)\\ &= \langle A^\mu_{\rm h}(x)A^\nu_{\rm inh}(y)\rangle + \langle A^\mu_{\rm inh}(x)A^\nu_{\rm h}(y)\rangle + \langle A^\mu_{\rm inh}(x)A^\nu_{\rm inh}(y)\rangle,\end{aligned} \tag{9}$$

where $\langle A^\mu_{\rm h}(x)A^\nu_{\rm h}(y)\rangle$ represents the Wightman function for vacuum fluctuations of the field and $A^\mu_{\rm BG}$ does not contribute to the energy flux; therefore, we subtracted it from the two-point function. The inhomogeneous solution is written as

$$A^\mu_{\rm inh}(x) = \frac{e\dot{z}^\mu(\tau^x_-)}{4\pi\rho(\tau^x_-)} \tag{10}$$

with $\rho(x^x_-) \equiv \dot{z}_\mu(\tau^x_-)(x^\mu - z^\mu(\tau^x_-))$, where τ^x_- satisfies the relation $(x - z(\tau^x_-))^2 = 0$ and the trajectory $z(\tau^x_-)$ is in the R-region (see the left panel of Fig. 1). Using the expansion in (6), we obtain $\rho(x) = \rho_0(x) + \delta\rho(x)$, where ρ_0 and $\delta\rho$ are defined as $\rho_0(x) \equiv \dot{\bar{z}}_\mu(\tau^x_-)x^\mu$ and $\delta\rho(\tau^x_-) \equiv \delta\dot{z}_\mu(\tau^x_-)(x - \bar{z}^\mu(\tau^x_-))$, respectively. Then, we obtain the formula for the inhomogeneous part $A^\mu_{\rm inh}(x)$

$$A^\mu_{\rm inh}(x) = \frac{e}{4\pi\rho_0(x)}(\dot{\bar{z}}^\mu(\tau^x_-) - E^\mu_{(-)i}(x)\delta\dot{z}^i(\tau^x_-)), \tag{11}$$

where we define $E^{\mu i}_\mp(x)$ as $E^{\mu i}_\mp(x) \equiv \eta^{\mu i} - \frac{\dot{\bar{z}}^\mu(\tau^x_\mp)x^i}{\rho_0(x)}$ and $\tau^x_\pm$ satisfies the relation $(x - \bar{z}(\tau^x_\pm))^2 = 0$ (see the left pane of Fig. 1). Using (11), we obtain the formula

[a]We need to impose the condition $a \ll m$ so as to avoid the boiling vacuum that is due to the Schwinger effect. Therefore, a/m is much smaller than unity.

(see Ref.13 for details)

$$[\langle A^{\mu}_{\rm h}(x)A^{\nu}_{\rm inh}(y)\rangle + \langle A^{\mu}_{\rm inh}(x)A^{\nu}_{\rm h}(y)\rangle + \langle A^{\mu}_{\rm inh}(x)A^{\nu}_{\rm inh}(y)\rangle]_S$$
$$= \left(\frac{e}{4\pi}\right)^2 \frac{\dot{z}^{\alpha}(\tau^x_-)}{\rho_0(x)}\frac{\dot{z}^{\beta}(\tau^x_-)}{\rho_0(x)} + \left[\frac{e}{4\pi\rho_0(x)}\frac{e}{4\pi\rho_0(y)}\frac{E^{\beta}_{(-)i}(y)E^{\alpha i}_{(+)}(x)}{2m}I_2(x,y)\right.$$
$$\left.+\frac{e}{4\pi\rho_0^3(x)}\frac{e}{4\pi\rho_0(y)}E^{\beta}_{(-)i}(y)x^i\eta^{\alpha A}a\epsilon_A^{\ A'}x_{A'}\frac{i}{2m}(I_1(x,y)-I_3(x,y))\right]$$
$$+[(x,\alpha)\leftrightarrow(y,\beta)], \tag{12}$$

where the functions $I_1(x,y)$, $I_2(x,y)$ and $I_3(x,y)$ are defined as

$$I_1(x,y) = -\frac{i}{2\pi\sigma} + \frac{i}{\pi}\log\left(1+e^{-a|\tau^y_- - \tau^x_+|}\right) + \frac{i}{\pi}a(\tau^y_- - \tau^x_+)\theta(\tau^y_- - \tau^x_+)$$
$$+\mathcal{O}(\sigma), \tag{13}$$
$$I_2(x,y) = -\frac{a}{\pi}\frac{1}{e^{a(\tau^x_+)}+1} + \mathcal{O}(\sigma), \tag{14}$$
$$I_3(x,y) = -\frac{i}{2\pi\sigma} + \frac{i}{\pi}\log\left(1-e^{-a|\tau^y_- - \tau^x_-|}\right) + \frac{i}{\pi}a(\tau^y_- - \tau^x_-)\theta(\tau^y_- - \tau^x_-)$$
$$+\mathcal{O}(\sigma), \tag{15}$$

for the F-region $x^0 > |x^1|$, where the dimensionless parameter σ is defined as $\sigma \equiv e^2a/6\pi m$. Using the formulas of the energy-momentum tensor $T_{0\mu} = -(A_{\alpha,0} - A_{0,\alpha})(A^{\alpha}_{,\mu} - A_{\mu}^{;\alpha})$ and the energy flux $f = -\Sigma^3_{i=1}T_{0i}n^i$ ($n^i \equiv x^i/\sqrt{\Sigma x^i x^i} \equiv x^i/r$), we obtain the energy flux expressed as $f(x) = f^C(x) + f^Q(x)$ at a large distance (i.e., $r \to \infty$). $f^C(x)$ and $f^Q(x)$ are the classical part (the Larmor radiation) and the quantum part (the Unruh radiation), respectively, defined by

$$f^C = \left(\frac{e}{4\pi}\right)^2 a^2\frac{1}{r^2}\frac{1}{\sin^4\theta}G(q), \qquad f^Q = \left(\frac{e}{4\pi}\right)^2\frac{a^3}{2\pi m}\frac{1}{r^2}\frac{1}{\sin^4\theta}F(q), \tag{16}$$

with

$$G(q) = \frac{1-P^2}{(1+q^2)^2}, \tag{17}$$
$$F(q) = \frac{1}{(1+q^2)^3}\left[6P(2P^2-1)\left\{\log a\epsilon - \log\left(1+e^{-a|\tau_- - \tau_+|}\right)\right.\right.$$
$$\left.-a(\tau_- - \tau_+)\theta(\tau_- - \tau_+)\right\} + \frac{2P}{(a\epsilon)^2}$$
$$\left.+2\frac{(3-e^{a(\tau_+-\tau_-)})(2-e^{a(\tau_+-\tau_-)}(9-e^{a(\tau_+-\tau_-)}))}{(1+e^{a(\tau_+-\tau_-)})^3}\right], \tag{18}$$

where we defined $\epsilon = |\tau^x_- - \tau^y_-|$ and $P = -\tanh a(\tau_+ - \tau_-)/2 = q/\sqrt{1+q^2}$ with $q = a(t - r - 1/2a^2r)/\sin\theta$. Note that $F(q)$ contains two terms that diverge in the coincidence limit $y \to x$. This divergence derives from the point particle approximation.[12,14] The approximation can be removed by taking into account the finite-size effect of the particle. Here, for simplicity, we omit the divergent terms. This procedure does not change our conclusions, as long as the cutoff value is of the order of $a\epsilon = \mathcal{O}(1)$. The right panel of Fig. 1 plots $G(q)$ and $F(q)$ as functions of q. We see that $F(q)$ is typically negative; thus, the quantum part of the radiation flux is typically negative. However, the total radiation flux of the classical part and the quantum part is positive; therefore, the quantum radiation appears to suppress the total radiation. This property is consistent with the predictions of the model based on the massless scalar field.[12]

3. Quantum correction to the Larmor radiation

The quantum correction to the Larmor radiation has been investigated using quantum field theory for a time dependent background.[16–20] The field theoretical approach to the quantum radiation from an accelerated charge is different from that described in the previous section. However, the results are consistent with those presented in the previous section. In the previous section, we adopted the unit $\hbar = 1$; here, we explicitly include $\hbar$.

According the results in Refs.16, 18, in the non-relativistic limit of the velocity of a charged particle, $|\mathbf{v}| \ll 1$, and the energy of the quantum Larmor radiation is expressed as the combination of the classical Larmor radiation $E^{(0)}$ of the zeroth order of $\hbar$, and the quantum part $E^{(1)}$ of the first order of $\hbar$, $E = E^{(0)} + E^{(1)}$ where

$$E^{(0)} = \frac{e^2}{6\pi}\int dt(\dot{v}^1(t))^2, \tag{19}$$

$$E^{(1)} = \frac{e^2\hbar}{6\pi^2 m}\int dt\int dt'\frac{\ddot{v}^1(t)\dot{v}^1(t') - \dot{v}^1(t)\ddot{v}^1(t')}{t-t'}, \tag{20}$$

where dots indicate differentiation with respect to t or t'. We may use the relations $t = \bar{z}^0(\tau) = a^{-1}\sinh a\tau$, $\bar{z}^1(\tau) = a^{-1}\cosh a\tau$, and $v^1(t) = \tanh a\tau$, for evaluating the radiation energy. We roughly estimate the order of the radiation rate as

$$\frac{dE^{(0)}}{dt} \simeq \frac{e^2a^2}{6\pi}, \tag{21}$$

$$\frac{dE^{(1)}}{dt} \sim -\frac{\hbar e^2 a^3}{6\pi^2 m}. \tag{22}$$

Note that the first-order correction to the Larmor radiation (22) is negative and that its amplitude is of the order of $\sim \hbar e^2 a^3/m$; these two statements are consistent with the results of (16). Therefore, the quantum radiation obtained in the previous section is related to the quantum correction to the Larmor radiation. The quantum correction to the Larmor radiation (20) is non-local in time, which might be useful for understanding the quantum radiation that is due to the Unruh effect.

Acknowledgements

We would like to thank J. Yokoyama, T. Suyama, and K. Fukushima for helpful discussions. We also thank Prof. P. Chen for critical discussions that took place at the beginning of this study.

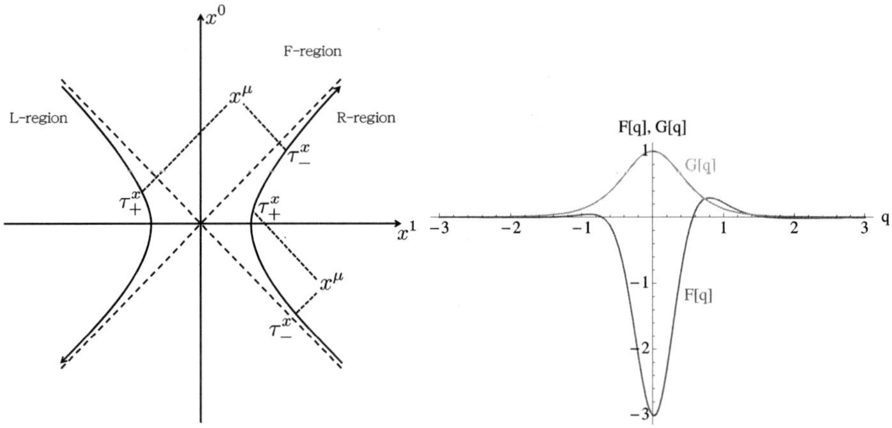

Fig. 1. (Left) The hyperbolic curve in the R-region is the trajectory of a uniformly accelerating particle. The hyperbolic curve in the L-region is the hypothetical trajectory obtained from an analytic continuation of the true trajectory. For an observer at point x^μ in the R-region, τ_-^x is defined by the proper time of the particle's trajectory intersecting with the past light cone; τ_+^x is similarly defined by the future light cone. For an observer in the F-region, τ_+^x is the proper time of the hypothetical trajectory in the L-region intersecting with the past light cone. (Right) Functions $G(q)$ (red line) and $F(q)$ (blue line) with divergent terms omitted. The sign of $G(q)$ is positive; the sign of $F(q)$ is typically negative.

References

1. S. W. Hawking, Commun. Math. Phys. **43**, 199 (1975). [Erratum-ibid. **46**, 206 (1976).]
2. W. G. Unruh, Phys. Rev. D **14**, 870 (1976).
3. P. Chen and T. Tajima, Phys. Rev. Lett. **83**, 256 (1999).
4. P.G. Thirolf, et al., Eur. Phys. J. D **55**, 379 (2009).
5. R. Schutzhold, et al., Phys. Rev. Lett. **97**, 121302 (2006).
6. R. Schutzhold, G. Schaller, and D. Habs, Phys. Rev. Lett. **100**, 091301 (2008).
7. D. J. Raine, D. W. Sciama, and P. G. Grove, Proc. R. Soc. Lond. A **435**, 205 (1991).
8. A. Raval, B. L. Hu, and J. Anglin, Phys. Rev. D **53**, 7003 (1996).
9. S-Y. Lin, B. L. Hu, Phys. Rev. D **73**, 124018 (2006).
10. S. Iso, K. Yamamoto, and S. Zhang, Prog. Theor. Exp. Phys. 063B01 (2013).
11. S. Iso, Y. Yamamoto, and S. Zhang, Phys. Rev. D **84**, 025005 (2011).
12. N. Oshita, K. Yamamoto, and S. Zhang, Phys. Rev. D **92**, 045027 (2015).
13. N. Oshita, K. Yamamoto and S. Zhang, to be published in Phys. Rev. D, arXiv:1509.03038 [hep-th].
14. N. Oshita, K. Yamamoto, and S. Zhang, Phys. Rev. D **89**, 124028 (2014).
15. S. Zhang, Prog. Theor. Exp. Phys. **12**, 123A01 (2013).
16. A. Higuchi and P. J. Walker, Phys. Rev. D **80**, 105019 (2009).
17. H. Nomura, M. Sasaki, and K. Yamamoto, J. Cosmol. Astropart. **11**, 013 (2006).
18. K. Yamamoto and G. Nakamura, Phys. Rev. D **83**, 045030 (2011).
19. R. Kimura, G. Nakamura, and K. Yamamoto, Phys. Rev. D **83**, 045015 (2011).
20. G. Nakamura and K. Yamamoto, Int. J. Mod. Phys. A **27**, 1250142 (2012).

Thin-Shell Instanton Tunneling: Something-to-Something or Nothing-to-Something?*

Pisin Chen[a], Yao-Chieh Hu[b] and Dong-Han Yeom[c]

Leung Center for Cosmology and Particle Astrophysics,
National Taiwan University, Taipei 10617, Taiwan, R.O.C.
[a]*pisinchen@phys.ntu.edu.tw*
[b]*r04244003@ntu.edu.tw*
[c]*innocent.yeom@gmail.com*

There exist two interpretations of instantons in the literature. The first interpretation regards instanton as divider between the initial and final hypersurfaces. The Coleman-De Luccia instanton is one such an example. The second interpretation, proposed by Brown and Weinberg, considers instanton as connector between the initial and final hypersurfaces. In this proceedings, we try to suitably and intuitively argue that these two interpretations are complementary to each other under certain conditions. Furthermore, we demonstrate that the decay rate obtained from the Euclidean treatment and the Hamiltonian treatment both are consistent with each other, which may help to dissolve some concerns about the validity of regularization technique employed in the treatment of the cusp singularity of instantons. Based on these, we argue that instantons can be a sensible tool to address the information loss problem.

Keywords: Information loss paradox; thin-shell instanton; bubble nucleation; black hole tunneling.

1. Introduction

Instanton is an interesting and extremely important object in the Euclidean path-integral approach to quantum field theory, which describes nonperturbative phenomenon of a system. In this article we summarize our recent work[1] on interpretations of instantons, and in particular, its connection with black hole physics.

By interpretation, we actually refer to the technical issue of how an instanton, which begins with the Euclidean spacetime, can be analytically continued to the eventual Lorentzian one. In the literature, it is known that there are two interpretations of instanton that can be summarized as follows. For simplicity, we shall refer to these two interpretations as *Nothing-to-Something* and *Something-to-Something*, where

- *Nothing-to-Something interpretation*: the initial and final hypersurfaces are separated by an instanton.[2]
- *Something-to-Something interpretation*: the initial and final hypersurfaces are connected by an instanton.[3]

*Proceedings for *The 2nd LeCosPA Symposium: Everything about Gravity, Celebrating the Centenary of Einsteins General Relativity.* Talk given by Yao-Chieh Hu, on December 17, 2015, Taipei, Taiwan.

So far the investigation of instanton solutions in the models of Coleman and De Luccia[2] and Brown and Weinberg[3] have been based on O(4) symmetry. In our work, we go further to study with the spherical symmetry that can include objects like black holes.

Besides the issues of interpretations, both Euclidean and Hamiltonian approaches have been invoked in the analysis of the something-to-something interpretation. On the contrary, for the nothing-to-something interpretation, there was no article that used the Hamiltonian treatment to calculate the decay rate. To make sure the consistency between two interpretations, it is worthwhile to check that the Hamiltonian approach also works well for the nothing-to-something interpretation. We therefore test the consistency of the Euclidean treatment by computing the decay rate from the Hamiltonian treatment.

2. Two interpretations—de Sitter space

Here we list the pros and cons of the two interpretations with $O(4)$ symmetry.

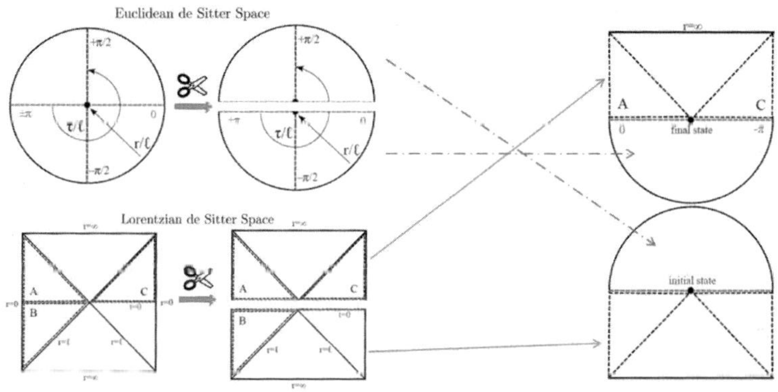

Fig. 1. Pictorial illustration for the nothing-to-something interpretation. One can easily see that the glued manifold is smoothly connected and maximally extended but contains some unphysical region beyond the Hubble radius.

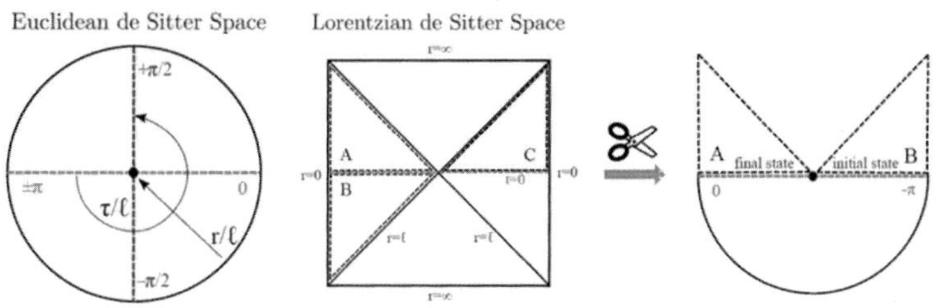

Fig. 2. Pictorial illustration for the something-to-something interpretation. It was shown by Brown and Weinberg[3] that once we identify the initial hypersurface with the final hypersurface between Euclidean and Lorentzian manifolds, a consistent interpretation is possible.

2.1. *Nothing-to-something*

As we mentioned before, the nothing-to-something interpretation is the case where the initial and final hypersurfaces are separated by the instanton solution (Fig. 1). This is the traditional interpretation that smoothly glues and maximally extends manifold, which is of course mathematically natural. In this interpretation, however, the manifold contains a region that lies beyond the cosmological horizon and hence inaccessible.

2.2. *Something-to-something*

For the something-to-something interpretation, in which the initial and final hypersurfaces are connected by the instanton solution, the tunneling process takes place in one causal patch. This interpretation is therefore more physically intuitive. However, there are still some mathematically technical issues; the manifold is not maximally extended and it is difficult to apply it to some important cases such as including the anti-de Sitter space.

3. Two interpretations—spherical symmetry

For our own interests in black hole physics, we move to the spherical symmetry and testify whether it is possible to interpret in both ways. In this section, we will briefly review the dynamics of a thin-shell bubble. Then, we will introduce both Euclidean and Hamiltonian methods in order to describe the tunneling process.

3.1. *Dynamics of thin-shell bubbles*

Let us consider a true vacuum bubble of the scalar field ϕ with the following action:[1]

$$S = \int_M \sqrt{-g}d^4x \left[\frac{\mathcal{R}}{16\pi} - \frac{1}{2}\nabla^\mu\phi\nabla_\mu\phi - U(\phi)\right] + \int_{\partial M} \sqrt{-h}d^3x \left[\frac{\mathcal{K}-\mathcal{K}_0}{8\pi}\right], \quad (1)$$

where $\mathcal{R}$ is the Ricci scalar, ϕ is a scalar field, $U(\phi)$ is the potential of the scalar field, and $\mathcal{K}$ and $\mathcal{K}_0$ are Gibbons-Hawking boundary terms for a given metric and the Minkowski metric, respectively. With the spherical symmetry, one can easily obtain the equation of motion of the bubble wall by using the thin-shell approximation: after straightforward calculations, we obtain the following form of the equation:[1]

$$\dot{r}^2 + V(r) = 0, \quad (2)$$

where the effective potential $V(r)$ is a function of several physical quantities, such as the mass and the surface energy density of the black hole.

3.2. *Two approaches toward tunneling process*

There are two approaches to treat the tunneling process. The first one is the Euclidean method, for example Farhi-Guth-Guven method[4] that relies on the Euclidean path integral. The second one is the Hamiltonian method proposed by

Fischler-Morgan-Polchinski,[5] where the integration is performed over the constant-time hypersurfaces. In the literature, both methods are adopt under the something-to-something interpretation. However, both methods can be interpreted as nothing-to-something as long as the Euclidean time period is properly chosen, that is $\tau_{r_1 \to r_2} = 8\pi M$ (Fig. 3). We can use the *true vacuum decay* as a demonstration as follows.

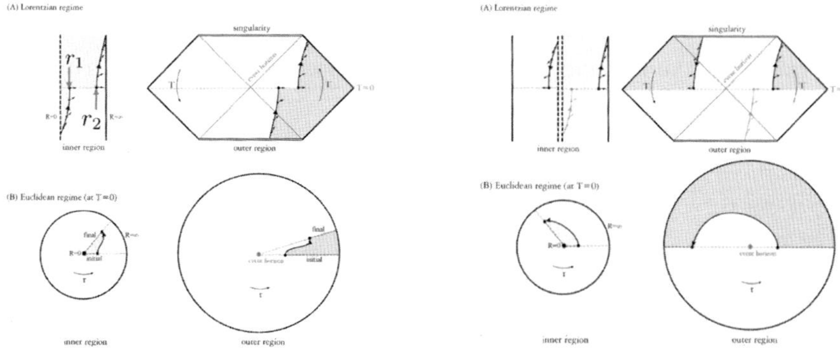

Fig. 3. In the case of the true vacuum decay problem in the Minkowski background, the inside region has the anti-de Sitter geometry and the outside region has the Schwarzschild geometry, where where the something-to-something interpretation is on the left and the nothing-to-something interpretation is on the right.

For the nothing-to-something interpretation, though the Euclidean method is verified by Gregory-Moss-Withers,[6] the Hamiltonian method has not been clarified. Therefore, we perform a consistency check as follows.

4. Consistency check

Both Farhi-Guth-Guven (FGG)[4] and Fischler-Morgan-Polchinski (FMP)[5] tunnelings give the same decay rates. These methods are therefore equally valid. A question then naturally arises— Can one also find a coherent decay rate between Euclidean and Hamiltonian approaches for nothing-to-something interpretation?

Table 1. Comparison table of our consistency checks.

	Something-to-Something	Nothing-to-Something
Euclidean Approach (FGG)	FGG[4]	GMW[6]
Hamiltonian Approach (FMP)	FMP[5]	CHY[1]

4.1. *Decay rates in the Euclidean approach*

In the Euclidean approach, the decay rate is given as

$$\Gamma \propto e^{-2B}, \tag{3}$$

where $B = S_{\mathrm{E}}(\text{instanton}) - S_{\mathrm{E}}(\text{background})$ is the tunneling exponent and the Euclidean action takes the form

$$S_{\mathrm{E}} = -\int_{M} \sqrt{+g}d^4x \left[\frac{\mathcal{R}}{16\pi} - \frac{1}{2}\nabla^{\mu}\phi\nabla_{\mu}\phi - U(\phi)\right] - \int_{\partial M} \sqrt{+h}d^3x \left[\frac{\mathcal{K}-\mathcal{K}_0}{8\pi}\right]. \tag{4}$$

The decay rate with the nothing-to-something interpretation has been derived by Gregory-Moss-Withers,[6] with

$$2B = \frac{\mathcal{A}_i - \mathcal{A}_f}{4} + 2\int_{r_1}^{r_2} drr \left| \cos^{-1}\left(\frac{f_+ + f_- - 16\pi^2\sigma^2 r^2}{2\sqrt{f_+ f_-}}\right)\right|, \tag{5}$$

where $f_+ \equiv 1 - 2M/r$, $f_- \equiv 1 + r^2/l^2$, and $\mathcal{A}_{i,f}$ are areas of horizons. Note that the first part of the right hand side corresponds to the entropy term which came from the regularization technique around the cusp singularity.

4.2. *Decay rates in the Hamiltonian approach*

In the Hamiltonian approach, the decay rate is given by[5]

$$\Gamma \propto e^{2i(\Sigma_f - \Sigma_i)}, \tag{6}$$

where $\Sigma_f - \Sigma_i$ is the difference between the integrals on initial and final space-like hypersurfaces (Fig. 4). Following the nothing-to-something interpretation, we[1] computed the action integration and we have shown that the integration takes the form

$$\left[\int_0^{\eta_1-\epsilon} d\eta(\ldots) + \int_{\eta_1+\epsilon}^{\eta_2-\epsilon} d\eta(\ldots) + \int_{\eta_2-\epsilon}^{\eta_2+\epsilon} d\eta(\ldots)\right] + \left[\int_{\eta_1-\epsilon}^{\eta_1+\epsilon} + \int_{\eta_2-\epsilon}^{\eta_2+\epsilon}\right], \tag{7}$$

where the first parenthesis was referred to as the *volume-integration* and the second the *shell-integration*.

We have showed that[1]

$$\left[\int_0^{\eta_1-\epsilon} d\eta(\ldots) + \int_{\eta_1+\epsilon}^{\eta_2-\epsilon} d\eta(\ldots) + \int_{\eta_2-\epsilon}^{\eta_2+\epsilon} d\eta(\ldots)+\right] = \frac{\mathcal{A}_i - \mathcal{A}_f}{4} \tag{8}$$

and

$$\left[\int_{\eta_1-\epsilon}^{\eta_1+\epsilon} + \int_{\eta_2-\epsilon}^{\eta_2+\epsilon}\right] = 2\int_{r_1}^{r_2} drr \left| \cos^{-1}\left(\frac{f_+ + f_- - 16\pi^2\sigma^2 r^2}{2\sqrt{f_+ f_-}}\right)\right|. \tag{9}$$

Thus, we have demonstrated that not only the Euclidean but also the Hamiltonian method work well for the nothing-to-something interpretation. Therefore, we can safely say that these two interpretations are complementary with each other.

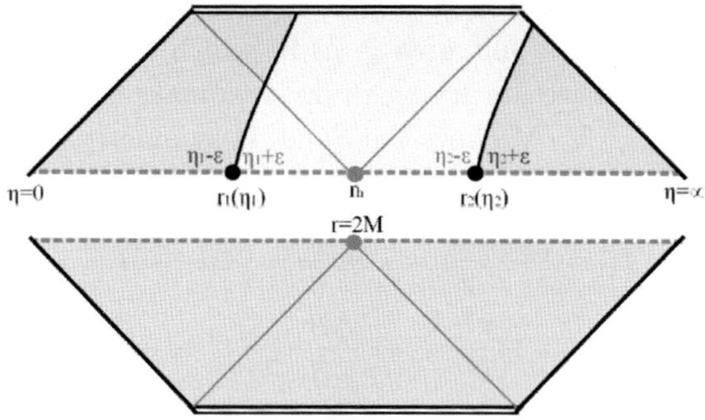

Fig. 4. Integration taken over spacelike hypersurfaces.

5. Conclusion

We have shown that the two interpretations on instantons are complementary each other under some constraint conditions. In addition, by checking consistency between the Euclidean method and the Hamiltonian method, we have verified that the regularization technique of Gregory-Moss-Withers is valid.

Acknowledgment

YH would like to thank Hsu-Wen Chiang and Chun-Yen Chen for beneficial discussions. PC and YH are supported by Taiwan's Ministry of Science and Technology (MOST 104-2112-M-002-024-MY3). DY is supported by Leung Center for Cosmology and Particle Astrophysics (LeCosPA) of National Taiwan University (103R4000).

Appendix A. Details on the consistency check

Let us choose the natural units $G = \hbar = c = 1$. The integration from the Euclidean action gives the following form[1]

$$\int drr \left[\cos^{-1}\left(\frac{\beta_+}{\sqrt{f_+}}\right) - \cos^{-1}\left(\frac{\beta_-}{\sqrt{f_-}}\right)\right]. \tag{A.1}$$

By using the identity

$$\cos^{-1}\alpha - \cos^{-1}\beta = \cos^{-1}\left[\alpha\beta + \sqrt{(1-\alpha^2)(1-\beta^2)}\right],$$

this is consistent with the second term in Eq. (5).

In the Hamiltonian tunneling exponent,[5] the *shell-integration* is given as

$$\hat{F}[r_2 - r_1]$$
$$\propto \int_{r_1}^{r_2} drr \left[\cos^{-1}\left(\frac{2M-\mu^2r^3(\lambda-1)}{2\mu r^2\sqrt{1-\frac{\Lambda_I r^2}{3}}}\right) - \cos^{-1}\left(\frac{2M-\mu^2r^3(\lambda+1)}{2\mu R^2\sqrt{1-\frac{2M}{r}}}\right)\right], \tag{A.2}$$

where $\lambda \equiv \Lambda_I/3\mu^2$ is XXX and Λ_I is XXX. As claimed at the beginning of FMP,[5] the shell energy density per unit area is $\mu/4\pi$ which also stands for the surface tension $\sigma = \mu/4\pi$. What we are trying to do is to make sure (A.1) and (A.2) are equivalent. From FMP,[5] $\mathbf{V}\cdot\mathbf{V}$ obeys the $SO(1,1)$ symmetry which is independent of angles. Thus, one can identify the notations of FMP as follows:

$$f_+ = 1 - \frac{2M}{r} = 1 - \frac{2M}{r}, \quad M \equiv M, \tag{A.3}$$

$$f_- = 1 + \frac{r^2}{\ell^2} = 1 - \frac{\Lambda_I r^2}{3}, \quad \ell^2 \equiv -\frac{\Lambda_I}{3}, \tag{A.4}$$

where the Schwarzschild radius is $r_s = 2M$ and the de Sitter radius is $r_d = \sqrt{3/\Lambda_I}$. Thus from (A.1), together with $\sigma = \mu/4\pi$ and $\lambda = \Lambda_I/3\mu^2$,

$$\frac{\beta_+}{\sqrt{f_+}} = \frac{\frac{f_- - f_+ - 16\pi^2\sigma^2 r^2}{8\pi\sigma r}}{\sqrt{1-\frac{2M}{r}}} = \frac{(1-\frac{\Lambda_I r^2}{3}) - (1-\frac{2M}{r}) - 16\pi^2(\frac{\mu}{4\pi})^2 r^2}{8\pi\sigma r(\frac{\mu}{4\pi})\sqrt{1-\frac{2M}{r}}} \tag{A.5}$$

$$= \frac{2M - \mu^2 r^3\Lambda - \mu^2 r^3}{2\mu r^2\sqrt{1-\frac{2M}{r}}} = \frac{2M - \mu^2 r^3(\lambda+1)}{2\mu r^2\sqrt{1-\frac{2M}{r}}}. \tag{A.6}$$

Similarly,

$$\frac{\beta_-}{\sqrt{f_-}} = \frac{2M - \mu^2 r^3(\lambda-1)}{2\mu r^2\sqrt{1-\frac{\Lambda_I r^2}{3}}}. \tag{A.7}$$

Thus, we have safely connected (A.1) and (A.2).

References

1. Pisin Chen, Yao-Chieh Hu, and Dong-han Yeom. "Two interpretations on thin-shell instantons." arXiv preprint arXiv:1512.03914 (2015).
2. Sidney Coleman, and Frank De Luccia. "Gravitational effects on and of vacuum decay." Physical Review D 21.12 (1980): 3305.
3. Adam R. Brown, and Erick J. Weinberg. "Thermal derivation of the Coleman-De Luccia tunneling prescription." Physical Review D 76.6 (2007): 064003.
4. Edward Farhi, Alan H. Guth, and Jemal Guven. "Is it possible to create a universe in the laboratory by quantum tunneling?." Nuclear Physics B 339.2 (1990): 417-490.
5. Willy Fischler, Daniel Morgan, and Joseph Polchinski. "Quantization of false-vacuum bubbles: A Hamiltonian treatment of gravitational tunneling." Physical Review D 42.12 (1990): 4042.
6. Ruth Gregory, Ian G. Moss, and Benjamin Withers. "Black holes as bubble nucleation sites." Journal of High Energy Physics 2014.3 (2014): 1-27.

PARALLEL SESSION

XI. Mathematical Structure of GR

Summary of Parallel Session: "Mathematical Structure of GR"

Chiang-Mei Chen

Department of Physics, National Central University, Chungli 32001, Taiwan
cmchen@phy.ncu.edu.tw

James M. Nester

Department of Physics, National Central University, Chungli 32001, Taiwan
Leung Center for Cosmology and Particle Astrophysics, National Taiwan University, Taipei 10617, Taiwan
nester@phy.ncu.edu.tw

This is a summary of the presentations delivered to the parallel session "Mathematical Structure of GR". There were five talks covering various topics related to the theory of general relativity.

The parallel section "Mathematical Structure of GR" was held on 17 December 2015. The list of the five presentations is

- **Jong Hyuk Yoon**, *Hamiltonian Reduction of Einstein's Equations in (2+2) Formalism under no Isometry Assumptions*
- **Seung Hun Oh**, *Some Exact Solutions in (2+2) Hamiltonian Reduction*
- **Gyula Fodor**, *Localized Oscillating Configurations Formed by Real Scalar Fields*
- **Jerzy Lewandowski**, *Isolated Horizons, Near Horizon Geometries: How to Find a New Exact Solution to the Einstein's Equations*
- **Chiang-Mei Chen**, *Covariant Hamiltonian Boundary Term: Reference and Quasi-local Quantities*

The five talks cover diverse topics except for the first two successive presentations. The main results are very briefly summarized as follows.

The first topic presented in the parallel section is about (2+2) dimensional reduction.[1,2] The first talk given by **Jong Hyuk Yoon** reviews the (2+2) fibration of four dimensional generic spacetimes without isometries, in which the physical Hamiltonian is expressed in terms of the conformal 2-metric and its conjugate momentum. In this formalism there is a natural time parameter given by the area of the two dimensional spacelike hypersurface. In addition, there are certain privileged coordinates of the Newman-Unti type in which the constraints of GR can be expressed in a more "convenient" form. A member of the audience suggested that this formalism may be useful for investigating the membrane paradigm for black holes.

The follow up talk by **Seung Hun Oh** shows some implications of (2+2) Hamiltonian reduction addressing canonical quantization, in particular for Einstein-Rosen waves. It was argued that the (2+2) reduction is a generalization of the Hamiltonian reduction of cylindrical gravitational waves done by Kuchař in 1971. The (2+2) reduction was also applied to some other well-known solutions including Schwarzschild.

The second topic is about the localized oscillating configurations of scalar fields.[3,4] In the talk of **Gyula Fodor**, a particular configuration, a spherically-symmetric star like object formed from a real scalar field, called oscillaton, was discussed. Oscillatons are oscillating objects which are extremely long living and stable, generating a small amount of radiation at a slow rate. The numerical and analytical results for an one parameter family of oscillatons were presented for both the zero and negative cosmological constant cases.

The third topic considered the relation between isolated horizons and near horizon geometries.[5–7] In the talk of **Jerzy Lewandowski**, two mathematical frameworks associated with black holes, namely isolated horizons and near horizon geometries, were discussed. A derivation of the near horizon geometries from an isolated horizon was presented. The relation between these two frameworks can be helpful for finding a class of exact solutions to the Einstein's equations.

The last topic in the parallel section concerned the quasi-local quantities in general relativity and the associated reference choice.[8–10] In the talk of **Chiang-Mei Chen**, the covariant Hamiltonian formulation was discussed. The Hamiltonian includes a boundary term which determines both the value of the conserved quantities and the boundary conditions. The boundary term depends not only on the dynamical variables but also on their reference values; the latter determine the ground state (having vanishing quasi-local quantities). To determine the reference metric 4D isometric matching and extremizing the energy are proposed.

References

1. J. H. Yoon, "New Hamiltonian formalism and quasilocal conservation equations of general relativity," Phys. Rev. D **70**, 084037 (2004) [gr-qc/0406047].
2. J. H. Yoon, "Hamiltonian structure and constraint algebra in the (2+2) formalism," Class. Quant. Grav. **31**, no. 4, 045005 (2014).
3. G. Fodor, P. Forgacs and M. Mezei, "Mass loss and longevity of gravitationally bound oscillating scalar lumps (oscillatons) in D-dimensions," Phys. Rev. D **81**, 064029 (2010) [arXiv:0912.5351 [gr-qc]].
4. P. Grandclement, G. Fodor and P. Forgacs, "Numerical simulation of oscillatons: extracting the radiating tail," Phys. Rev. D **84**, 065037 (2011) [arXiv:1107.2791 [gr-qc]].
5. J. Lewandowski and T. Pawlowski, "Extremal isolated horizons: A Local uniqueness theorem," Class. Quant. Grav. **20**, 587 (2003) [gr-qc/0208032].

6. T. Pawlowski, J. Lewandowski and J. Jezierski, "Space-times foliated by Killing horizons," Class. Quant. Grav. **21**, 1237 (2004) [gr-qc/0306107].
7. J. Jezierski and B. Kaminski, "Towards uniqueness of degenerate axially symmetric Killing horizon," Gen. Rel. Grav. **45**, 987 (2013) [arXiv:1206.5136 [gr-qc]].
8. J.-L. Liu, C.-M. Chen and J.-M. Nester, "Quasi-local energy and the choice of reference," Class. Quant. Grav. **28**, 195019 (2011) [arXiv:1105.0502 [gr-qc]].
9. G. Sun, C.-M. Chen, J.-L. Liu and J. M. Nester, "An Optimal Choice of Reference for the Quasi-Local Gravitational Energy and Angular Momentum," Chin. J. Phys. **52**, 111 (2014) [arXiv:1307.1039 [gr-qc]].
10. C.-M. Chen, J. M. Nester and R.-S. Tung, "Gravitational energy for GR and Poincare gauge theories: A covariant Hamiltonian approach," Int. J. Mod. Phys. D **24**, 1530026 (2015) [arXiv:1507.07300 [gr-qc]].

Hamiltonian Reduction of Einstein's Equations in the (2+2) Formalism under No Isometry Assumptions

J. H. Yoon

School of Physics, Konkuk University,
120 Neungdong-ro, Gwangjin-gu, Seoul 05029, Korea
yoonjh@konkuk.ac.kr

I apply the Hamiltonian reduction procedure to spacetimes of 4 dimensions without isometries in the (2+2) formalism and find privileged spacetime coordinates in which the physical Hamiltonian is solely expressed in terms of the conformal two metric and its conjugate momentum. Physical time is the area element of the spatial cross-section of null hypersurfaces, and the physical radial coordinate is defined by *equipotential* surfaces on a given spacelike hypersurface of constant physical time. The physical Hamiltonian is local and positive in the privileged coordinates. I present the complete set of Hamilton's equations and find that they coincide with the Einstein's equations written in the privileged coordinates. This shows that the Hamiltonian reduction is self-consistent and respects general covariance.

Keywords: Problem of time; Hamiltonian reduction; 2+2 formalism; constraint equations.

1. Introduction

I apply the standard Hamiltonian reduction procedure to general spacetimes of 4 dimensions under no symmetry assumption in the (2+2) formalism. The area element of the spatial cross-section of null hypersurfaces is identified as the physical time, and the physical radial coordinate is defined by equipotential surfaces on a given spacelike hypersurface of constant physical time. The remaining two spacelike coordinates are chosen as orthogonal to the surface of constant physical radial coordinate on that spacelike hypersurface. I present the fully-reduced physical Hamiltonian in these privileged coordinates, which is a local function of the conformal two metric and its conjugate momentum. The momentum constraints are also solved; they are simply *defining* equations of the physical linear momentum.

2. The Action

Let me recall that the metric in the (2+2) decomposition of 4 dimensional spacetimes can be written as[1–7]

$$ds^2 = 2dudv - 2hdu^2 + \tau\rho_{ab}\left(dy^a + A_+{}^a du + A_-{}^a dv\right) \times \left(dy^b + A_+{}^b du + A_-{}^b dv\right). \quad (1)$$

The vector fields $\hat{\partial}_\pm$ defined as

$$\hat{\partial}_\pm := \partial_\pm - A_\pm{}^a \partial_a, \quad (2)$$

where

$$\partial_+ = \partial/\partial u, \quad \partial_- = \partial/\partial v, \quad \partial_a = \partial/\partial y^a \quad (a = 2,3), \tag{3}$$

are horizontal vector fields orthogonal to the two-dimensional spacelike surface N_2 generated by ∂_a. The inner products of the horizontal vector fields are given by

$$< \hat{\partial}_+, \hat{\partial}_+ >= -2h, \quad < \hat{\partial}_+, \hat{\partial}_- >= 1, \quad < \hat{\partial}_-, \hat{\partial}_- >= 0. \tag{4}$$

These equations tell us that $\hat{\partial}_+$ has a norm $-2h$, which can be either positive, negative, or zero, depending on the sign of h, and that $\hat{\partial}_-$ is a null vector field with v as an affine parameter. In this paper, I choose the sign $-2h > 0$ so that $v =$ constant is a spacelike hypersurface. The metric on N_2 is $\tau\rho_{ab}$, where τ is the area element of N_2 and ρ_{ab} is the conformal two metric with $\det \rho_{ab} = 1$.

As was shown in Ref. 4, the Einstein's equations can be obtained from the variational principle of the following action integral:

$$S = \int dv du d^2y \{\pi_\tau \dot{\tau} + \pi_h \dot{h} + \pi_a \dot{A}_+^{\ a} + \pi^{ab}\dot{\rho}_{ab} - \text{“1”} \cdot C_- - \text{“0”} \cdot C_+ - A_-^{\ a} C_a\}, \tag{5}$$

where the overdot $\dot{} = \partial_-$, and “1”, “0”, and $A_-^{\ a}$ are Lagrange multipliers that enforce the constraints $C_- = 0$, $C_+ = 0$, and $C_a = 0$, which are given by

$$\begin{aligned} C_- :=& \frac{1}{2}\pi_h\pi_\tau - \frac{h}{4\tau}\pi_h^2 - \frac{1}{2\tau}\pi_h D_+\tau + \frac{1}{2\tau^2}\rho^{ab}\pi_a\pi_b - \frac{\tau}{8h}\rho^{ab}\rho^{cd}(D_+\rho_{ac})(D_+\rho_{bd}) \\ & - \frac{1}{2h\tau}\rho_{ab}\rho_{cd}\pi^{ac}\pi^{bd} - \frac{1}{2h}\pi^{ac}D_+\rho_{ac} - \tau \mathrm{R}_{(2)} + D_+\pi_h - \partial_a(\tau^{-1}\rho^{ab}\pi_b) \\ =& 0, \end{aligned} \tag{6}$$

$$\begin{aligned} C_+ :=& \pi_\tau D_+\tau + \pi_h D_+ h + \pi^{ab}D_+\rho_{ab} - 2D_+(h\pi_h + D_+\tau) \\ & + 2\partial_a(h\tau^{-1}\rho^{ab}\pi_b + \rho^{ab}\partial_b h) \\ =& 0, \end{aligned} \tag{7}$$

$$C_a := \pi_\tau\partial_a\tau + \pi_h\partial_a h + \pi^{bc}\partial_a\rho_{bc} - 2\partial_b(\rho_{ac}\pi^{bc}) - D_+\pi_a - \partial_a(\tau\pi_\tau) = 0, \tag{8}$$

respectively. Here, $\mathrm{R}_{(2)}$ is the scalar curvature of N_2, and the diffN_2-covariant derivative of a tensor density q_{ab} with weight s is defined as in Ref. 4

$$\begin{aligned} D_+ q_{ab} :=& \partial_+ q_{ab} - [A_+,\ q]_{\mathrm{L}ab} \\ =& \partial_+ q_{ab} - A_+^{\ c}\partial_c q_{ab} - q_{cb}\partial_a A_+^{\ c} - q_{ac}\partial_b A_+^{\ c} - s(\partial_c A_+^{\ c}) q_{ab}, \end{aligned} \tag{9}$$

where $[A_+,\ q]_{\mathrm{L}ab}$ is the Lie derivative of q_{ab} along $A_+ := A_+^{\ a}\partial_a$. For instance, the diff$N_2$-covariant derivatives of the area element τ and the conformal metric ρ_{ab}, which are a scalar density and a tensor density with weight 1 and -1 with respect to the diffN_2 transformations, respectively, are given by

$$D_\pm \tau = \partial_\pm\tau - A_\pm^{\ a}\partial_a\tau - (\partial_a A_\pm^{\ a})\tau, \tag{10}$$

$$D_\pm \rho_{ab} = \partial_\pm\rho_{ab} - A_\pm^{\ c}\partial_c\rho_{ab} - \rho_{cb}\partial_a A_\pm^{\ c} - \rho_{ac}\partial_b A_\pm^{\ c} + (\partial_c A_\pm^{\ c})\rho_{ab}. \tag{11}$$

On the other hand, h is a scalar under the diffN_2 transformations, whose diffN_2-covariant derivative is given by

$$D_{\pm}h = \partial_{\pm}h - A_{\pm}^{\ a}\partial_a h, \tag{12}$$

and the diffN_2-covariant field strength $F_{+-}^{\ \ a}$ is defined as

$$\begin{aligned} F_{+-}^{\ \ a} &:= \partial_+ A_-^{\ a} - \partial_- A_+^{\ a} - [A_+,\ A_-]^a \\ &= \partial_+ A_-^{\ a} - \partial_- A_+^{\ a} - A_+^{\ b}\partial_b A_-^{\ a} + A_-^{\ b}\partial_b A_+^{\ a}. \end{aligned} \tag{13}$$

The diffN_2-covariant derivatives of the conjugate momenta π_τ, π_h, π_a, and π^{ab}, which are tensor densities of weights $0, 1, 1$ and 2, respectively, are defined similarly. Because ρ_{ab} has a unit determinant, its derivatives automatically satisfy the following conditions

$$\rho^{bc}\partial_{\pm}\rho_{bc} = \rho^{bc}\partial_a\rho_{bc} = \rho^{bc}D_{\pm}\rho_{bc} = 0. \tag{14}$$

3. The Privileged Spacetime Coordinates

Let me define a *potential* function R and its conjugate momentum π_R as

$$\partial_+ R := -h\pi_h, \quad \pi_R = -\partial_+ \ln(-h), \tag{15}$$

respectively.[8] The transformation from $\{h, \pi_h\}$ to $\{R, \pi_R\}$ is a *canonical* transformation, as it changes the action integral by total derivatives only. If the constraints $C_+ = 0$ and $C_a = 0$ are imposed and the Lagrange multiplier is chosen such that $A_-^{\ a} = 0$, then the action in Eq. (5) becomes

$$S = \int dvdud^2y\{\pi_\tau\dot{\tau} + \pi_R\dot{R} + \pi_a\dot{A}_+^{\ a} + \pi^{ab}\dot{\rho}_{ab} - C_-\} + \text{ total derivatives,} \tag{16}$$

where the constraint C_- becomes

$$\begin{aligned} C_- = &-\frac{1}{2h}\pi_\tau\partial_+ R - \frac{1}{4h\tau}(\partial_+ R)^2 + \frac{1}{2h\tau}(D_+\tau)(\partial_+ R) - \frac{1}{h}D_+(\partial_+ R) \\ &+\frac{1}{h}(\partial_+ R)\partial_+ \ln(-h) - \frac{1}{h}(\partial_+ R)A_+^{\ a}\partial_a \ln(-h) - \tau R^{(2)} + \frac{1}{2\tau^2}\rho^{ab}\pi_a\pi_b \\ &-\partial_a(\tau^{-1}\rho^{ab}\pi_b) - \frac{1}{2h\tau}\rho_{ab}\rho_{cd}\pi^{ac}\pi^{bd} - \frac{\tau}{8h}\rho^{ab}\rho^{cd}(D_+\rho_{ac})(D_+\rho_{bd}) \\ &-\frac{1}{2h}\pi^{ac}D_+\rho_{ac} = 0. \end{aligned} \tag{17}$$

The function h still appears in Eq. (17), but because h is related to R and π_R by Eq. (15), the constraint function C_- given by Eq. (17) may be viewed as a function of the new variables $(\tau, Q^I; \pi_\tau, \Pi_I)$, where $Q^I = (R, A_+^{\ a}, \rho_{ab})$ and $\Pi_I = (\pi_R, \pi_a, \pi^{ab})$. Notice that the first term in Eq. (17) is linear in π_τ, and that all the remaining terms are independent of π_τ. Thus, the equation of motion for τ is given by

$$\dot{\tau} = \int dud^2y\frac{\delta C_-}{\delta\pi_\tau} = -\frac{1}{2h}\partial_+ R. \tag{18}$$

Now, recall that $\tau = \tau(v, u, y^a)$. If this equation is solved for v, then v may be viewed as a function of (τ, u, y^a) and consequently, $(R,\ A_+{}^a,\ \rho_{ab})$ may be regarded as functions of (τ, u, y^a). Therefore, one has

$$\dot{R} = \dot{\tau}\partial_\tau R, \quad \dot{A}_+{}^a = \dot{\tau}\partial_\tau A_+{}^a, \quad \dot{\rho}_{ab} = \dot{\tau}\partial_\tau \rho_{ab}, \quad C_- = -(\frac{2h}{\partial_+ R})\dot{\tau}\ C_-, \tag{19}$$

where I used the coordinate conditions $\partial u/\partial v = \partial y^a/\partial v = 0$. Then, the action in Eq. (16) becomes

$$\begin{aligned} S &= \int dv du d^2 y \dot{\tau}\{\pi_\tau + \pi_R \partial_\tau R + \pi_a \partial_\tau A_+{}^a + \pi^{ab}\partial_\tau \rho_{ab} + (\frac{2h}{\partial_+ R})C_-\} \\ &= \int d\tau du d^2 y\{\pi_R \partial_\tau R + \pi_a \partial_\tau A_+{}^a + \pi^{ab}\partial_\tau \rho_{ab} - C_{(1)}\} \\ &= \int d\tau du d^2 y\{\sum_I \Pi_I \partial_\tau Q^I - C_{(1)}\}. \end{aligned} \tag{20}$$

Here, $dv\dot{\tau}$ is replaced by $d\tau$ in the second line, and $C_{(1)}$ is given by

$$C_{(1)} = -(\frac{2h}{\partial_+ R})C_- - \pi_\tau = -\pi_\tau, \tag{21}$$

where the constraint $C_- = 0$ was imposed in the second equality.

The second step in the Hamiltonian reduction consists of *identifying* arbitrarily specifiable coordinates u and y^a as

$$u = R, \quad y^a = Y^a \tag{22}$$

such that the two-dimensional "shift" vector $A_+{}^a$ is zero, i.e.,

$$A_+{}^a = 0. \tag{23}$$

For the class of spacetimes whose spatial topology is either $N_2 \times \mathbb{R}$ or $N_2 \times S^1$, where N_2 is a compact two-dimensional space with the genus g, the condition in Eq. (23) can always be met by relabeling N_2 such that Y^a = constant is normal to u = constant at each point on Σ_τ. If N_2 is continued to labeled this way, then the "shift" $A_+{}^a$ can always be made zero for the class of spacetimes under consideration. Thus, I will work in the coordinates $X^A := (\tau, R, Y^a)$, which satisfy the coordinate conditions

$$\frac{\partial X^A}{\partial X^B} = \delta_B^A. \tag{24}$$

Then, it follows from Eq. (18) that

$$\dot{\tau} = -\frac{1}{2h} > 0, \tag{25}$$

which means that τ increases monotonically along the out-going null vector field.

Hamilton's equations of motion follow from the variational principle of the action integral in Eq. (20):

$$\partial_\tau Q^I = \int_{\Sigma_\tau} du d^2 y \frac{\delta C_{(1)}}{\delta \Pi_I}|_{u=R, y^a=Y^a}, \quad \partial_\tau \Pi_I = -\int_{\Sigma_\tau} du d^2 y \frac{\delta C_{(1)}}{\delta Q^I}|_{u=R, y^a=Y^a}, \tag{26}$$

where $Q^I = (R, A_+^{\ a}, \rho_{ab})$, $\Pi_I = (\pi_R, \pi_a, \pi^{ab})$, and Σ_τ is a spacelike hypersurface defined by $\tau =$ constant.

4. Einstein's Equations in the Privileged Spacetime Coordinates

In the following, I present the Einstein's equations in the privileged spacetime coordinates (τ, R, Y^a) obtained through the Hamiltonian reduction.

Evolution equations I

$$\frac{\partial R}{\partial \tau} = 0 \Rightarrow \tau \mathrm{R}_{(2)} = \frac{1}{2}\tau^{-2}\rho^{ab}\pi_a\pi_b - \partial_a(\tau^{-1}\rho^{ab}\pi_b) \quad \text{(topological censorship)} \tag{27}$$

$$\frac{\partial A_+^{\ a}}{\partial \tau} = 0 \Rightarrow \tau^{-1}\pi_a = -\partial_a \ln(-h) \quad \text{(superpotential)} \tag{28}$$

$$\frac{\partial \tau}{\partial \tau} = 1 \quad \text{(trivial)} \tag{29}$$

$$\frac{\partial \ln(-h)}{\partial \tau} = H - \frac{1}{\tau} \quad \text{(superpotential)} \tag{30}$$

$$\frac{\partial \pi_a}{\partial \tau} = 2\tau^{-1}\pi_a + (\pi^{bc} + \frac{\tau}{2}\rho^{bd}\rho^{ce}\partial_R\rho_{de})\partial_a\rho_{bc} - \partial_b(2\pi^{bc}\rho_{ac} + \tau\rho^{bc}\partial_R\rho_{ac}) \tag{31}$$

$$\frac{\partial \pi_\tau}{\partial \tau} = \frac{1}{2}\tau^{-2} + \tau^{-2}\rho_{ab}\rho_{cd}\pi^{ac}\pi^{bd} - \frac{1}{4}\rho^{ab}\rho^{cd}(\partial_R\rho_{ac})(\partial_R\rho_{bd}) - 2\tau^{-2}\partial_a(h\rho^{ab}\pi_b) \tag{32}$$

Constraint equations

$$C_- = 0 \Rightarrow \pi_\tau = -H + 2\partial_R \ln(-h) \quad \text{(definition of physical Hamiltonian)} \tag{33}$$

$$C_+ = 0 \Rightarrow \pi_R = -\pi^{ab}\partial_R\rho_{ab} \quad \text{(definition of physical momentum)} \tag{34}$$

$$C_a = 0 \Rightarrow \tau^{-1}\pi_a = -\pi^{bc}\partial_a\rho_{bc} + 2\partial_b(\pi^{bc}\rho_{ac}) - \tau\partial_a(H + \pi_R) \tag{35}$$

(definition of physical momentum)

Superpotential $\ln(-h)$

$$-\partial_R \ln(-h) = \pi_R \tag{36}$$

Integrability conditions

$$\partial_R(\tau^{-1}\pi_a) = \partial_a\pi_R, \quad \partial_\tau\pi_R = -\partial_R H, \quad \partial_\tau(\tau^{-1}\pi_a) = -\partial_a H \tag{37}$$

Evolution equations II: Equations of motion of physical variables

$$\frac{\partial \rho_{ab}}{\partial \tau} = \int_{\Sigma_\tau} dR d^2Y \frac{\delta C^*_{(1)}}{\delta \pi^{ab}}, \quad \frac{\partial \pi^{ab}}{\partial \tau} = -\int_{\Sigma_\tau} dR d^2Y \frac{\delta C^*_{(1)}}{\delta \rho_{ab}}, \tag{38}$$

where

$$C^*_{(1)} := C_{(1)}|_{u=R, y^a=Y^a} = -\pi_\tau = H - 2\partial_R \ln(-h) \tag{39}$$

The metric in these privileged coordinates

$$ds^2 = -4hdRd\tau - 2hdR^2 + \tau\rho_{ab}dY^a dY^b \tag{40}$$

In the above equations, H is defined by

$$H = \frac{1}{\tau}\rho_{ab}\rho_{cd}\pi^{ac}\pi^{bd} + \frac{1}{4}\tau\rho^{ab}\rho^{cd}(\partial_R\rho_{ac})(\partial_R\rho_{bd}) + \pi^{ac}\partial_R\rho_{ac} + \frac{1}{2\tau}. \tag{41}$$

5. Discussion

Let me mention that the whole set of equations summarized in Section 4 is *identical to the vacuum Einstein's equations*

$$\mathrm{R}_{AB} = 0 \tag{42}$$

for spacetimes whose metric is given by Eq. (40). Thus, the whole procedure of the Hamiltonian reduction proposed discussed in this article respects general covariance, as it must, even though the final theory is written in the privileged coordinates.

The integral of Eq. (27) over a closed two surface N_2 becomes

$$\int_{N_2} d^2Y \tau^{-2} \rho^{ab} \pi_a \pi_b = 16\pi(1-g) \geq 0, \tag{43}$$

where g is the genus of N_2. This identity states that, as long as the out-going null hypersurface forms a congruence of null geodesics which admits a cross-section, the spatial topology of that null hypersurface is either a two sphere or a torus.[9] This is a remarkably simple proof of *topological censorship*, as it does not rely on assumptions such as global hyperbolicity, asymptotic conditions, energy conditions, and so on, either inside or outside the null hypersurface, which are normally assumed in the literature.[10–12]

Acknowledgments

This work was prepared in part by the support of National Research Foundation of Korea (NRF-2015R1D1A1A01059407).

References

1. R. Sachs, *Proc. Roy. Soc. A* **270**, 103 (1962).
2. R. A. d'Inverno and J. Stachel, *J. Math. Phys.* **19** (12), 2447 (1978).
3. Y. M. Cho, Q. H. Park, K. S. Soh, and J. H. Yoon, *Phys. Lett. B* **286**, 251 (1992).
4. J. H. Yoon, *Phys. Rev. D* **70**, 084037 (2004).
5. J. H. Yoon, *Class. Quant. Grav.* **31**, 045005 (2014).
6. J. H. Yoon, *J. Korean Phys. Soc.* **64**, 192 (2014).
7. J. H. Yoon, *J. Korean Phys. Soc.* **64**, 631 (2014).
8. K. V. Kuchař, *Phys. Rev. D* **4**, 955 (1971).
9. T. Pawlowski, J. Lewandowski, and J. Jezierski, `http://www.arxiv.org/pdf/gr-qc/0306107`.
10. J. L. Friedman, K. Schleich, and D. M. Witt, *Phys. Rev. Lett.* **71**, 1486 (1993).
11. P. T. Chruściel and R. M. Wald, *Class. Quantum Grav.* **11**, L147 (1994).
12. G. J. Galloway, K. Schleich, D. M. Witt, and E. Woolgar, *Phys. Rev. D* **60**, 104039 (1999).

Schwarzschild Solution in the (2+2) Hamiltonian Reduction

S. H. Oh

Research Institute of Basic Sciences, Konkuk University,
120 Neungdong-ro, Gwangjin-gu, Seoul 05029, Korea
thereisnoi@naver.com

Jong Hyuk Yoon

School of Physics, Konkuk University, 120 Neungdong-ro,
Gwangjin-gu, Seoul 05029, Korea
yoonjh@konkuk.ac.kr

In this article, it will be discussed how we get the Einstein-Rosen wave spacetime and the Schwarzschild solution by solving the (2+2) Hamilton's equations. And the local Hamiltonian and the field momentum densities corresponding to those solutions will be presented.

Keywords: Hamiltonian reduction; 2+2 formalism; Schwarzschild solution.

1. Introduction

Recently one of the authors proposed a Hamiltonian reduction of the Einstein's theory without isometry, called the (2+2) Hamiltonian reduction.[1,2] It was shown that the Einstein's theory can be transformed into a constraint-free field theory in a privileged coordinate system with some topological restriction on spacelike hypersurfaces. In this privileged coordinates, one has a non-zero Hamiltonian density so that the ordinary Hamiltonian formalism works. Classically this means that one can get the solutions to the Einstein's equations by solving the (2+2) Hamilton's equations instead of the Einstein's equations.[3] As a simple application, the Einstein-Rosen wave solution will be obtained in this way after a short review of the (2+2) Hamiltonian reduction. And then the Schwarzschild solution will be obtained in a similar way.

2. Review of the (2+2) Hamiltonian Reduction

Geometrically our spacetime can be regarded as a fiber bundle E_4 which consists of a 2-dimensional Lorentzian base manifold and its 2-dimensional fiber N_2. If its spacelike hypersurfaces have the topology of $\mathbb{R} \times N_2$ or $\mathbb{S}^1 \times N_2$, then it can be described in a privileged coordinate system (τ, R, Y^1, Y^2) with the following metric;[1]

$$ds^2 = -2h(2dRd\tau + dR^2) + \tau\rho_{ab}dY^a dY^b, \tag{1}$$

with $a, b = 1, 2$. Here ρ_{ab} is called the conformal two-metric tensor on N_2, and its determinant is 1 so that τ can be interpreted as the area element of these surfaces.

In this sense, we call τ *the area time*. We take a sign convention $h < 0$ in order to keep the constant τ surfaces to be spacelike. All functions h and ρ_{ab} are allowed to depend on all coordinates since we have no isometry.

Based on these settings, one has the non-zero Hamiltonian density $\mathcal{H}$ and physical momentum densities π_R and π_a for $a = 1, 2$ as following;

$$\mathcal{H} = \tau^{-1}\rho_{ab}\rho_{cd}\pi^{ac}\pi^{bd} + \frac{1}{4}\tau\rho^{ab}\rho^{cd}(\partial_R\rho_{ac})(\partial_R\rho_{bd}) + \pi^{ac}\partial_R\rho_{ac} + \frac{1}{2\tau}, \tag{2}$$

$$\pi_R = -\pi^{ab}\partial_R\rho_{ab}, \tag{3}$$

$$\tau^{-1}\pi_a = -\pi^{bc}\frac{\partial}{\partial Y^a}\rho_{bc} + 2\frac{\partial}{\partial Y^b}(\pi^{bc}\rho_{ac}) - \frac{\partial}{\partial Y^a}\{\tau(\mathcal{H} + \pi_R)\}, \tag{4}$$

where π^{ab} is the conjugate momentum of ρ_{ab}, and is traceless, i.e., $\rho_{ab}\pi^{ab} = 0$. Remark that they are solely expressed in terms of ρ_{ab} and π^{ab}. Indeed ρ_{ab} has two degrees of freedom corresponding to the true physical degrees of freedom of gravitational fields. The Hamilton's equations of this system are given by,

$$\frac{\partial}{\partial\tau}\rho_{ab} = 2\tau^{-1}\rho_{ac}\rho_{bd}\pi^{cd} + \partial_R\rho_{ab}, \tag{5}$$

$$\frac{\partial}{\partial\tau}\pi^{ab} = -2\tau^{-1}\rho_{cd}\pi^{ac}\pi^{bd} + \partial_R\pi^{ab} - \frac{\tau}{2}\rho^{ai}\rho^{bj}\rho^{ck}(\partial_R\rho_{ic})(\partial_R\rho_{jk}) + \frac{\tau}{2}\rho^{ac}\rho^{bd}(\partial_R^2\rho_{cd}) \tag{6}$$

Once one imposes an appropriate initial data, ρ_{ab} and π^{ab} will be uniquely determined by the above equations. And then we can figure out h from the superpotential equations,

$$\partial_\tau \ln(\;h) = \mathcal{H} \quad \tau^{-1}, \tag{7}$$

$$\partial_R \ln(-h) = -\pi_R, \tag{8}$$

$$\partial_a \ln(-h) = -\tau^{-1}\pi_a. \tag{9}$$

This completes the solving process of the Einstein's equations in the (2+2) Hamiltonian reduction.

3. Cylindrically Symmetric Spacetime

To illustrate a simple application of (2+2) Hamiltonian reduction, let us consider the following simple ansatz,

$$\rho_{ab} = \begin{pmatrix} f & 0 \\ 0 & 1/f \end{pmatrix}, \qquad \pi^{ab} = \begin{pmatrix} \pi_F/f & 0 \\ 0 & -\pi_F f \end{pmatrix}. \tag{10}$$

And let us assume that h, f and π_F depend only on τ and R. It means we have two spacelike Killing vectors $\frac{\partial}{\partial Y^1}$ and $\frac{\partial}{\partial Y^2}$. Then the Hamilton's equations (5) and (6) become,

$$\partial_\tau F = \frac{4}{\tau}\pi_F + \partial_R F, \tag{11}$$

$$\partial_\tau \pi_F = \frac{\tau}{4}\partial_R^2 F + \partial_R \pi_F, \tag{12}$$

where $F := 2\ln f$. The superpotential equation (9) is automatically satisfied due to the equation (4). The other two superpotential equations (7) and (8) are,

$$\partial_\tau \ln(-h) = -\frac{1}{2\tau} + \frac{2}{\tau}\pi_F^2 + \frac{\tau}{8}(\partial_R F)^2 + \pi_F(\partial_R F), \tag{13}$$

$$\partial_R \ln(-h) = \pi_F(\partial_R F). \tag{14}$$

From the equation (11), π_F can be expressed in terms of F as,

$$\pi_F = \frac{\tau}{4}(\partial_\tau - \partial_R)F. \tag{15}$$

Eliminating π_F, the equations (12), (13), and (14) can be written as

$$\partial_\tau^2 F + \frac{1}{\tau}(\partial_\tau - \partial_R)F - 2\partial_\tau\partial_R F = 0, \tag{16}$$

$$\partial_\tau \ln(-h) = -\frac{1}{2\tau} + \frac{\tau}{8}(\partial_\tau F)^2, \tag{17}$$

$$\partial_R \ln(-h) = \frac{\tau}{4}(\partial_\tau F)(\partial_R F) - \frac{\tau}{4}(\partial_R F)^2. \tag{18}$$

Now (16) is a linear second order PDE, so it can be easily solved if one imposes an initial condition. Once F, equivalently f, is fixed from (16), then (17), and (18) will give the function h so that our spacetime will be uniquely determined.

In order to clarify what this system describes, let us change the coordinates as,

$$t(\tau, R) = \tau + R, \quad \rho(\tau, R) = \tau, \quad Y^1 = \phi, \quad Y^2 = z. \tag{19}$$

And let us define new functions Γ and Ψ by

$$\Psi = -\ln(f/\rho) + i\pi, \qquad \Gamma = \ln(-2\rho h/f) + i\pi. \tag{20}$$

Then the privileged line element will be transformed into

$$ds^2 = e^{\Gamma-\Psi}(dt^2 - d\rho^2) - \rho^2 e^{-\Psi} d\phi^2 - e^{\Psi} dz^2. \tag{21}$$

Here complex constants were introduced to keep our sign convention $h < 0$. These constants don't change the physics, but they just change the signature of our spacetime from $(-,+,+,+)$ to $(+,-,-,-)$. This trick is required since the physical time τ was chosen to be the area element of a 2-dimensional spacelike surface. And now the equations (16), (17), and (18) become,

$$\partial_t^2\Psi - \frac{1}{\rho}\partial_\rho\Psi - \partial_\rho^2\Psi = 0, \tag{22}$$

$$\partial_t\Gamma = \rho(\partial_t\Psi)(\partial_\rho\Psi), \tag{23}$$

$$\partial_\rho\Gamma = \frac{\rho}{2}\{(\partial_t\Psi)^2 + (\partial_\rho\Psi)^2\}. \tag{24}$$

This is nothing but the Einstein-Rosen waves.[4] Therefore we had obtained the Einstein-Rosen waves by solving the Hamilton's equations with a simple ansatz.

4. Schwarzschild Spacetime

As is well-known, the Schwarzschild solution can be written in the form of the canonical Weyl metric. In order to use the canonical Weyl metric as a bridge between our privileged metric and the Schwarzschild metric, we should find a transformation from the (2+2)-privileged coordinates to the canonical Weyl coordinates. To do so, let us take the same ansatz (10) in the previous section and assume that f and h and π_F depend only on τ and R. And let us do a complex coordinate transformation from (τ, R, Y^1, Y^2) to (t, ρ, ϕ, z) such that

$$t = Y^1, \quad \rho = \tau, \quad \phi = iY^2, \quad z = i(\tau + R). \tag{25}$$

and define two functions $\gamma(\rho, z)$ and $\psi(\rho, z)$ as,

$$\psi = \frac{1}{2}\ln(\rho f), \quad \gamma = \frac{1}{2}\ln(-2\rho f h). \tag{26}$$

Then the priviledged line element becomes the canonical Weyl form,

$$ds^2 = e^{2\psi}dt^2 - e^{2(\gamma-\psi)}(d\rho^2 + dz^2) - \rho^2 e^{-2\psi}d\phi^2. \tag{27}$$

And the equations (16), (17), and (18) become,

$$\partial_\rho^2\psi + \frac{1}{\rho}\partial_\rho\psi + \partial_z^2\psi = 0, \tag{28}$$

$$\partial_\rho\gamma = \rho[(\partial_\rho\psi)^2 - (\partial_z\psi)^2] \tag{29}$$

$$\partial_z\gamma = 2\rho(\partial_\rho\psi)(\partial_z\psi). \tag{30}$$

Note that the canonical Weyl metric represents the Schwarzschild solution if the following relations hold.

$$\gamma = \frac{1}{2}\ln\frac{(r_+ + r_-)^2 - 4m^2}{4r_+r_-}, \tag{31}$$

$$\psi = \frac{1}{2}\ln\frac{r_+ + r_- - 2m}{r_+ + r_- + 2m}, \tag{32}$$

where $r_\pm = [\rho^2 + (z \pm m)^2]^{\frac{1}{2}}$ and m denotes the Schwarzschild mass. Using the equations from (25) to (32), one can straightforwardly obtain an expression of Schwarzschild metric in the (2+2) privileged coordinates as following,

$$ds^2 = -2h\,(2dRd\tau + dR^2) + \tau\rho_{ab}\,dY^a dY^b\,,$$

where

$$\rho_{ab} = \begin{pmatrix} f & 0 \\ 0 & 1/f \end{pmatrix}, \quad f = \frac{1}{\tau}\frac{\sqrt{\xi+\sqrt{\xi^2+\eta^2}} - \sqrt{2}m}{\sqrt{\xi+\sqrt{\xi^2+\eta^2}} + \sqrt{2}m},$$

$$h = -\frac{1}{4\sqrt{\xi^2+\eta^2}}\left(\sqrt{\xi + \sqrt{\xi^2+\eta^2}} + \sqrt{2}m\right)^2. \tag{33}$$

Here ξ and η are functions of τ and R,

$$\xi = m^2 - 2R\tau - R^2, \quad \eta = 2m(\tau + R)\,. \tag{34}$$

5. Discussion

Since the Einstein's theory was deparametrized by fixing the coordinates and solving the constraint equations, it is not necessary to consider the concept of quasi-local energy and momenta of gravitational fields in the (2+2) reduction formalism. Instead one has the local Hamiltonian density $\mathcal{H}$ and the local field momentum densities π_R and π_a as well. It is straightforward to calculate them using (2), (3), and (4).

Now let us analyze the cylindrically symmetric system of gravitational field in the viewpoint of the (2+2) reduction formalism. Since the system has a single degree of freedom F, the configuration space would consist of (F, π^F). Then the reduced action becomes,

$$\begin{aligned} S &= \int d\tau dR d^2Y \, \{\pi^{ab}\partial_\tau \rho_{ab} - \mathcal{H}(F, \partial_R F, \pi_F; \tau)\} \\ &= \int d\tau dR d^2Y \, \{\pi_F \partial_\tau F - \mathcal{H}^C(F, F', \pi_F; \tau)\}, \end{aligned} \tag{35}$$

with the Hamiltonian density,

$$\mathcal{H}^C = \frac{1}{2\tau} + \frac{2}{\tau}\pi_F^2 + \frac{\tau}{8}(\partial_R F)^2 + \pi_F(\partial_R F). \tag{36}$$

Now we have the non-zero Hamiltonian density so that the ordinary Hamiltonian formalism can be applied. Actually the evolution equations of canonical variables (11) and (12) can be derived from the above action. The field momentum densities carried by the Einstein-Rosen waves are,

$$\pi_R^C = -\pi_F(\partial_R F), \tag{37}$$

$$\pi_a^C = 0, \tag{38}$$

for $a = 1, 2$. This is a natural consequence since $\frac{\partial}{\partial Y^1}$ and $\frac{\partial}{\partial Y^2}$ are Killing vectors of the system. Note that all components of the metric tensor is determined by F. Classically, therefore, F can be interpreted as a news function of the cylindrically symmetric spacetime proposed by H. Bondi et al. in 1962.[5]

On the other hand, the Hamiltonian density and the momentum densities of the Schwarzschild spacetime are given by,

$$\mathcal{H}^S = \frac{1}{2\tau} + \frac{\tau}{2}\Big(\frac{2\sqrt{2}m((m^2+R^2)(R+2\tau) - R\sqrt{\xi^2+\eta^2}\,)}{\sqrt{\xi^2+\eta^2}\sqrt{\xi+\sqrt{\xi^2+\eta^2}}(\xi+\sqrt{\xi^2+\eta^2}-2m^2)} - \frac{1}{\tau}\Big)^2, \tag{39}$$

$$\pi_R^S = \frac{2(\tau+R)}{\sqrt{\xi^2+\eta^2}}\Big(\frac{\sqrt{\xi+\sqrt{\xi^2+\eta^2}} - \sqrt{2}m}{\sqrt{\xi+\sqrt{\xi^2+\eta^2}}} - \frac{R^2+2R\tau+m^2}{\sqrt{\xi^2+\eta^2}}\Big), \tag{40}$$

$$\pi_a^S = 0, \tag{41}$$

for $a = 1, 2$, and ξ and η are functions of τ and R defined by (34). So we have a time-dependent Hamiltonian density of the Schwarzschild spacetime, since we had chosen a different spacelike slice from the ordinary choices. Indeed, the direction of time was not chosen to be a Killing direction.

6. Conclusion

The (2+2) reduction formalism has remarkable advantages since it enables us to analyze gravitational systems in the standard process of Hamiltonian mechanics. Of course Hamiltonian physics gives a natural way to quantize physical systems. But the reduction formalism has important meanings as a classical theory. At first, the local Hamiltonian density and the local field momentum densities can be straightforwardly calculated if the metric of a gravitational system is expressed in the privileged coordinates. And there is no need to consider the stability of constraint equations since all constraint equations were solved from the beginning. In addition, all physical degrees of freedom are isolated in the conformal two-metric ρ_{ab} and its conjugate momentum π^{ab}, and their evolutions are solely determined by themselves through the Hamilton's equations. So the (2+2) reduction formalism can be applied to classical analysis of gravitational systems in a minimal way.

Acknowledgement

This paper was supported by Konkuk University in 2015.

References

1. J. H. Yoon, Hamiltonian Reduction for Einstein's Equations in the (2+2) Formalism, *J. Korean Phys. Soc.* **64**, 192 (2014).
2. J. H. Yoon, Consistency of Hamiltonian Reduction of Einstein's equations, *J. Korean Phys. Soc.* **65**, 926 (2014).
3. J. H. Yoon, Hamiltonian Reduction and Exact Solutions to Einstein's Equations, *J. Korean Phys. Soc.* **65**, 935 (2014).
4. A. Einstein and N. Rosen, On Gravitational Waves, *J. Franklin Inst.* **223**, 43 (1937).
5. H. Bondi, M. G. J. van der Burg and A. W. K. Metzner, Waves from Axi-Symmetric Isolated Systems, *Proc. R. Soc. Lond. A* **269**, 21 (1962).

When Isolated Horizons met Near Horizon Geometries

Jerzy Lewandowski[a], Adam Szereszewski[b], and Piotr Waluk[c]
Faculty of Physics, University of Warsaw,
ul. Pasteura 5, 02–093 Warsaw, Poland
[a] *Jerzy.Lewandowski@fuw.edu.pl*
[b] *Adam.Szereszewski@fuw.edu.pl*
[c] *Piotr.Waluk@fuw.edu.pl*
www.fuw.edu.pl

There are two mathematical relativity frameworks generalizing the black hole theory: the theory of isolated horizons (IH) and the theory of near horizon geometries (NHG). We outline here and discuss the derivation of the NHG from the theory of IH by composing spacetimes from IH. The simplest but still quite general class of solutions to Einstein's equations of this type defines spacetimes foliated by Killing horizons emanating from extremal horizons. That derivation, clearly being a link between the two frameworks, seems to be unknown to the NHG researchers and is hardly acknowledged in reviews on the IH. This lecture was a contribution to the Mathematical Structures session of the 2nd LeCosPA International Symposium "Everything about Gravity" celebrating the centenary of Einstein's General Relativity on December 14–18, 2015 in Taipei.

Keywords: Isolated horizon; near horizon geometry; Kundt spacetime.

1. Introduction

A family of solutions to 4d-Einstein's equations was constructed in Ref. 1 from the following data:

- S — a 2 dimensional manifold diffeomorphic to a sphere,
- $g = g_{AB}dx^A dx^B$ — a metric tensor on S,
- $\omega = \omega_A dx^A$ — a differential 1-form on S,

such that the following (tensor) equation is satisfied:

$$D_{(A}\omega_{B)} + \omega_A\omega_B - \frac{1}{2}R_{AB} = 0, \tag{1}$$

where D_A and R_{AB} are, respectively, the torsion free covariant derivative and the Ricci tensor, defined on S by the metric g_{AB}.

The corresponding solution to the vacuum Einstein equations is a metric tensor $g_{\mu\nu}$ defined on

$$S \times \mathbb{R} \times \mathbb{R}$$

in terms of coordinates

$$(x^\mu) = (x^A, v, u),$$

in the following way

$$g_{\mu\nu}dx^\mu dx^\nu \;:=\; g_{AB}dx^A dx^B \;-\; 2du\left(dv - 2v\omega - \frac{1}{2}v^2(D_A\omega^A + 2\omega^A\omega_A)du\right) \quad (2)$$

2. Foliation by Killing horizons

What is special about each of the solutions (2) is that the corresponding spacetime is foliated by Killing horizons. They are the surfaces

$$u \;=\; \text{const} \quad (3)$$

of the topology $S \times \mathbb{R}$ and

$$v \;=\; 0. \quad (4)$$

The Killing vector corresponding to each value $u = u_0$ is

$$K^{(u_0)} \;=\; v\partial_v - (u - u_0)\partial_u. \quad (5)$$

This was exactly the aim of performing the construction in Ref. 1: to compose a spacetime out of horizons.

An unexpected property, though, was the emergence of an extremal Killing horizon

$$v \;=\; 0 \quad (6)$$

of the Killing vector

$$K \;=\; \partial_u. \quad (7)$$

All the horizons (3) foliating the spacetime emanate from that extremal horizon.

The 1-form $\omega_A dx^A$, considered on a spacial slice

$$v \;=\; \text{const}$$

of any of the horizons (3), coincides with minus the rotation 1-form potential, namely

$$\nabla_A K^{(u_0)} \;=\; -\omega_A K^{(u_0)}. \quad (8)$$

Considered on each spacial slice

$$u \;=\; \text{const}$$

of the extremal horizon (6), on the other hand, the 1-form $\omega_A dx^A$ coincides exactly with its rotation 1-form potential, that is, the pullback of the spacetime covariant derivative $\nabla_\mu K$ to the slice is

$$\nabla_A K \;=\; \omega_A K. \quad (9)$$

The equation (1) coincides with the vacuum constraint that has to be satisfied by every induced metric tensor g_{AB} and rotation 1-form potential ω_A on an extremal Killing or, more generally, isolated horizon, at which the vacuum Einstein equations hold.[2] The equation was studied in Ref. 3 for compact 2d-surfaces. It was proven therein that the only axisymmetric solutions (g, ω) admitted by S diffeomeomorphic to S^2 are those provided by the horizons of extremal Kerr solutions. This may be interpreted as a quasi-local version of the black hole uniqueness theorem. Those solutions correspond to the Horowitz limit[4] of the extremal Kerr spacetimes, also known as the Kundt class solutions of the Petrov type D.[5] The equation (1) was farther studied in Refs. 6, 7. However, intriguingly enough, still very little is known about a generic, axially non-symmetric case. It is not even known whether there exists any non-axisymmetric solution to eq. (1) on S diffeomorphic to 2-sphere. The other compact 2d-S cases are either excluded — the higher genus case — or trivialized by the equation (1): on S diffeomorphic to 2-torus, the only solutions are flat g and $\omega = 0$. It might be possible, though, that inclusion of a non-zero cosmological constant in the calculations would loosen these restrictions somehow. More specifically, as the argument relies on the definiteness of $\omega_A\omega^A - \Lambda$. A negative cosmological constant yet strengthens that property, and even eliminates the 2-torus case, allowing only the 2-sphere. However, a positive cosmological constant destroys the positive definiteness and we have no argument against the higher genus.

Above, Einstein's vacuum equations can be replaced by the Einstein–Maxwell vacuum equations. The extremal horizon constraint (1) is then suitably generalized by the presence of the Maxwell field. The uniqueness theorem still holds for the resulting extremal horizon constraint equation, with the Kerr solutions replaced by the Kerr–Newman solution.[3] The suitable generalization of the metrics (2) is also available.[1] Those metrics still have the form

$$g_{\mu\nu}(x,u,v)dx^\mu dx^\nu = g_{AB}(x)dx^A dx^B - 2du\left(dv - 2v\omega_A(x)dx^A - \frac{1}{2}v^2 H(x)du\right), \tag{10}$$

where the function $H(x)$ depends now on both $\omega_A(x)$ and the Maxwell field.[1]

3. Near horizon limit

Yet more generally, metric tensors of the form (10), with arbitrary $g_{AB}(x)$, $\omega_A(x)$, and $H(x)$ are called near horizon geometries.[8] The vacuum Einstein equations imposed on such a metric are equivalent to (1,2). In the case with matter, the equation (1) is generalized to

$$D_{(A}\omega_{B)} + \omega_A\omega_B - \frac{1}{2}R_{AB} = \frac{1}{2}\mathcal{R}_{AB}, \tag{11}$$

where the new term $\mathcal{R}_{AB}$ is the pullback of the spacetime Ricci tensor to a spacial slice of an extremal Killing horizon.[10] The corresponding change in (2) amounts to a suitable modification of the function $H(x)$.[8]

The class of metric tensors (10) was derived by a neat argument[9] from the general class of metric tensors admitting an extremal Killing horizon. Let us recall. Suppose a spacetime metric g admits a Killing horizon. Locally, in the neighbourhood of the horizon, g can be written as

$$g = g_{AB}(v,x)dx^A dx^B - 2du\left(dv - 2v\omega_A(v,x)dx^A + (v\kappa H_0(v,x) + \frac{1}{2}v^2 H(v,x))du\right). \quad (12)$$

The Killing vector is

$$K = \partial_u$$

and the Killing horizon is the surface

$$v = 0.$$

To remove the ambiguity of κ, we demand that $H_0\big|_{v=0} \equiv -1$. Apart from that, g_{AB}, ω_A, H_0, and H are *a priori* arbitrary (differentiable) functions of the variables (v, x^A). Under such circumstances, the κ in (12) is a constant which coincides with the surface gravity of the Killing vector K at the horizon (this actually holds also for a wide class of non-vacuum metrics, satisfying an energy inequality $T_{\alpha\beta}K^\alpha K^\beta \geq 0$ at the horizon). The horizon is extremal if and only if

$$\kappa = 0.$$

Then, the metric (12) admits the (pointwise) limit

$$\epsilon \to 0$$

of the transformation

$$f_\epsilon : (v, u, x^A) \mapsto (\epsilon v, \frac{1}{\epsilon}u, x^A). \quad (13)$$

This is a generalization of the Horowitz limit mentioned before. The result is

$$\lim_{\epsilon\to 0} f_\epsilon^* g = g_{AB}(0,x)dx^A dx^B - 2du\left(dv - 2v\omega_A(0,x)dx^A + \frac{1}{2}v^2 H(0,x))du\right). \quad (14)$$

As the limit of the transformation (13), the resulting metric is f_ϵ invariant, that is, in addition to the original Killing vector field

$$K = \partial_u \quad (15)$$

it also has a second Killing vector field

$$K^{(0)} = v\partial_v - u\partial_u. \quad (16)$$

Notice, that the surface $v = 0$ is a Killing horizon also for $K^{(0)}$. It is, however, not an *extremal* horizon with respect to this second Killing field.

Moreover, every surface $u = u_0$ is the Killing horizon of a Killing vector field

$$K^{(0)} + u_0 K.$$

4. Conclusions, outlook

Another generalization of (1, 2) is also possible. Consider a spacetime foliated by non-expanding horizons. Let the foliation be defined by

$$u = \text{const.} \tag{17}$$

Then, a distinguished null vector field ℓ tangent to all the horizons is

$$\ell^{\mu} = g^{\mu\nu} u_{,\nu}. \tag{18}$$

On every spacial section S of a leaf

$$u = u_0$$

of the foliation, the rotation 1-form potential $-\omega_A$ is defined by

$$\nabla_A \ell = -\omega_A \ell. \tag{19}$$

Suppose the spacetime satisfies vacuum Einstein's equations. Then, a constraint equation implied by Einstein's equations is again(!)

$$D_{(A}\omega_{B)} + \omega_A \omega_B - \frac{1}{2} R_{AB} = 0 \tag{20}$$

(we use the same notation as in (1)). It has to be satisfied on every leaf of the foliation independently (given a leaf, the independence of choice of slice S is automatic, and follows from the 0th low of non-expanding horizons). So, it takes the mathematical form of the extremal horizon constraint even though there is no visible extremal horizon in the spacetime (and perhaps none at all). This mystery is yet to be understood.

Constructing the family of solutions (2) to Einstein's equations by composing isolated horizons not only was a link between the theories of isolated horizons and theories of near horizon geometries (surprisingly, the paper[1] is unknown in the living reviews of those theories[8, 11]), but it also still contains intriguing mysteries. Throughout the paper we were assuming that the spacetime dimension is 4. Most of the reasoning generalizes to higher dimensions[8–10] and there is still potential for new results. A non-axially symmetric solution to the equation (1) would produce a new solution to Einstein's equations. Understanding the mysterious emergence of the extremal horizon constraint in the context of the spacetimes composed from non-extremal isolated horizons could also prove interesting.

Acknowledgments

We would like to thank Maciej Dunajski for pointing out the reference.[9] This work was partially supported by the Polish National Science Centre grant No. 2015/17/B/ST2/02871.

References

1. T. Pawłowski, J. Lewandowski, J. Jezierski, Spacetimes Foliated by Killing Horizons, *Class. Quant. Grav.* **21** (2004), 1237–1252, arXiv:gr-qc/0306107.
2. A. Ashtekar, C. Beetle, J. Lewandowski, Geometry of Generic Isolated Horizon, *Class. Quant. Grav.* **19** (2002), 1195–1225, arXiv:gr-qc/0111067.
3. J. Lewandowski, T. Pawłowski, Extremal Isolated Horizons: A Local Uniqueness Theorem, *Class. Quant. Grav.* **20** (2003), 587–606, arXiv:gr-qc/0208032.
4. J. M. Bardeen, G. T. Horowitz, The Extreme Kerr Throat Geometry: A Vacuum Analog of $AdS(2) \times S^2$, *Phys. Rev. D* **60** (1999), 104030, arXiv:hep-th/9905099.
5. H. Stephani, D. Kramer, M. MacCallum, C. Hoenselaers, E. Herlt, Exact Solutions to Einsteins Field Equations, Cambridge University Press, 2003.
6. J. Jezierski, On the Existence of Kundt's Metrics and Degenerate (or Extremal) Killing Horizons, *Class. Quant. Grav.* **26** (2009), 035011, arXiv:0806.0518 [gr-qc].
7. J. Jezierski, B. Kaminski, Towards Uniqueness of Degenerate Axially Symmetric Killing Horizon, *Gen. Relativ. Gravit.* **45** (2016), 987–1004, arXiv:1206.5136 [gr-qc].
8. H. K. Kunduri, J. Lucietti, Classification of Near-Horizon Geometries of Extremal Black Holes, *Living Rev. Rel.* **16** (2013), 8, http://www.livingreviews.org/lrr-2013-8, arXiv:abs/1306.2517.
9. H. S. Reall, Higher Dimensional Black Holes and Supersymmetry, *Phys. Rev. D* **68** (2003), 024024, arXiv:hep-th/0211290.
10. J. Lewandowski, T. Pawłowski, Quasi-local Rotating Black Holes in Higher Dimension: Geometry, *Class. Quant. Grav.* **22** (2005), 1573–1598, arXiv:gr-qc/0410146.
11. A. Ashtekar, B. Krishnan, Isolated and Dynamical Horizons and Their Applications, *Living Rev. Rel.* **7** (2004), 10, http://www.livingreviews.org/lrr-2004-10, arXiv:gr-qc/0407042.

Covariant Hamiltonian Boundary Term: Reference and Quasi-Local Quantities

Gang Sun[1], Chiang-Mei Chen[2], Jian-Liang Liu[3] and James M. Nester[2,4,5]

[1] *Institute of Theoretical Physics, Beijing University of Technology, Beijing, 100136, China*
[2] *Department of Physics, National Central University, Chungli 32001, Taiwan*
[3] *Department of Mathematics, Shantou University, Guangdong 515063, China*
[4] *Graduate Institute of Astronomy, National Central University, Chungli 32001, Taiwan*
[5] *Leung Center for Cosmology and Particle Astrophysics, National Taiwan University, Taipei 10617, Taiwan*
sungang@bjut.edu.cn, cmchen@phy.ncu.edu.tw, liujl_mc@163.com, nester@phy.ncu.edu.tw

The Hamiltonian for dynamic geometry generates the evolution of a spatial region along a vector field. It includes a boundary term which determines both the value of the Hamiltonian and the boundary conditions. The value gives the quasi-local quantities: energy-momentum, angular-momentum and center-of-mass. The boundary term depends not only on the dynamical variables but also on their reference values; the latter determine the ground state (having vanishing quasi-local quantities). For our preferred boundary term for Einstein's GR we propose 4D isometric matching and extremizing the energy to determine the reference metric and connection values.

Keywords: Quasi-local quantity; reference; covariant Hamiltonian.

1. Introduction

Although the global *total* energy-momentum is well defined (for spaces with suitable asymptotic regions), for any gravitating system — and hence for all real *physical* systems — the localization of energy-momentum is still an outstanding fundamental problem.[1,2] Unlike all matter and other interaction fields, the gravitational field itself has *no* proper energy-momentum density. In view of the fact that energy-momentum is conserved, and that sources *exchange* energy-momentum *locally* with the gravitational field, one expects some kind of "local description" of the energy-momentum density of gravity itself. But all attempts at constructing such an expression led only to reference frame dependent quantities, generally referred to as *pseudotensors*.[3] Physically this can be understood as a consequence of *Einstein's equivalence principle*: gravity cannot be detected at a point. The energy-momentum of gravity — and thus for all physical systems is inherently *non-local.* The modern idea is *quasi-local*: energy-momentum is associated with a closed surface bounding a region.[4]

2. Covariant Hamiltonian Formalism

The *first order* Lagrangian[5] for an *f-form field* φ and its *conjugate momentum* p is given by

$$\mathcal{L} = \mathcal{L}(d\varphi; \varphi, p) = d\varphi \wedge p - \Lambda(\varphi, p). \tag{1}$$

The variation (with respect to φ and p independently)

$$\delta\mathcal{L} = d(\delta\varphi \wedge p) + \delta\varphi \wedge \frac{\delta\mathcal{L}}{\delta\varphi} + \frac{\delta\mathcal{L}}{\delta p} \wedge \delta p \tag{2}$$

gives the equations of motion, with $\varsigma := (-1)^f$,

$$\frac{\delta\mathcal{L}}{\delta p} := d\varphi - \partial_p\Lambda = 0, \qquad \frac{\delta\mathcal{L}}{\delta\varphi} := -\varsigma dp - \partial_\varphi\Lambda = 0. \tag{3}$$

Diffeomorphism invariance (in terms of the Lie derivative $\pounds_N = di_N + i_N d$) leads an identity for *any vector* N

$$di_N\mathcal{L} \equiv \pounds_N\mathcal{L} \equiv d(\pounds_N\varphi \wedge p) + \pounds_N\varphi \wedge \frac{\delta\mathcal{L}}{\delta\varphi} + \frac{\delta\mathcal{L}}{\delta p} \wedge \pounds_N p. \tag{4}$$

From this one gets a conserved "translational current" 3-form:

$$\mathcal{H}(N) := \pounds_N\varphi \wedge p - i_N\mathcal{L}, \qquad -d\mathcal{H}(N) \equiv \pounds_N\varphi \wedge \frac{\delta\mathcal{L}}{\delta\varphi} + \frac{\delta\mathcal{L}}{\delta p} \wedge \pounds_N p. \tag{5}$$

Note that $\mathcal{H}(N)$ is not unique:

$$\mathcal{H}' = \mathcal{H} + d\mathcal{B}' \quad \Rightarrow \quad d\mathcal{H} = d\mathcal{H}'. \tag{6}$$

Furthermore it can be written in the form, $\mathcal{H}(N) = N^\mu\mathcal{H}_\mu + d\mathcal{B}(N)$ then

$$d\mathcal{H}(N) = d[N^\mu\mathcal{H}_\mu + d\mathcal{B}(N)] \equiv dN^\mu \wedge \mathcal{H}_\mu + N^\mu d\mathcal{H}_\mu \tag{7}$$

$\Rightarrow$ $\mathcal{H}_\mu$ vanishes "on shell".

Hence for gravitating systems the Noether translational "charge" — *energy-momentum* — is *quasi-local*, it is given by the integral of the boundary term, $\mathcal{B}(N)$. But this boundary term can be completely modified to any value. However, the Hamiltonian approach tames the ambiguity. Quasi-local quantities are determined only by the surface integral

$$E(N) = \int_\Sigma \mathcal{H}(N) = \int_\Sigma [N^\mu\mathcal{H}_\mu + d\mathcal{B}(N)] = \oint_{\partial\Sigma} \mathcal{B}(N). \tag{8}$$

The two parts of the Hamiltonian have distinct roles: The 3-form part $N^\mu\mathcal{H}_\mu$ generates the equations of motion. As mentioned, for diffeomorphic invariant theories it has vanishing value. The Hamiltonian generally also includes a boundary term $\mathcal{B}(N)$: (i) it determines the values of the quasi-local quantities, and (ii) it determines the boundary conditions.[6,7]

2.1. *Quasi-Local Quantities*

The Hamiltonian boundary terms determines the values of the quasi-local quantities:

- Energy is given by a suitable *timelike* displacement;
- Linear momentum is obtained from a *spatial* translation;
- Angular momentum from a suitable *rotational* displacement;

- A spacetime displacement which is asymptotically a *boost* will give the center-of-mass moment.

Our Noether analysis has revealed that $\mathcal{B}(N)$ can be adjusted, changing the conserved value to a new value. However the variational principle contains an additional (largely overlooked) feature which distinguishes all of these choices. The boundary variation principle, i.e. the boundary term in the variation, tells us what to hold fixed on the boundary — it determines the *boundary conditions.*[3]

The different Hamiltonian boundary terms are each associated with distinct boundary conditions. As in thermodynamics or electrostatics there are various "energies" which correspond to how the system interacts with the outside through its boundary. In general (in particular for gravity) it is necessary (in order to guarantee functional differentiability of the Hamiltonian on the phase space with the desired boundary conditions) to adjust the boundary term $\mathcal{B}(N) = i_N\varphi \wedge p$ which is naturally inherited from the Lagrangian (1). The variation of the Hamiltonian implies

$$\delta\mathcal{H}(N) \equiv -\delta\varphi \wedge \pounds_N p + \pounds_N \varphi \wedge \delta p + d i_N(\delta\varphi \wedge p) - i_N \left(\delta\varphi \wedge \frac{\delta\mathcal{L}}{\delta\varphi} + \frac{\delta\mathcal{L}}{\delta p} \wedge \delta p\right). \quad (9)$$

There is a freedom for modifying the boundary term $\mathcal{B}(N) \to \mathcal{B}'(N)$. Moreover, a reference configuration, $\bar{\varphi}$ and $\bar{p}$, (which determines the ground state) is essential especially for gravity, in particular to allow the desired phase space asymptotics.

With $\Delta\varphi := \varphi - \bar{\varphi}$, $\Delta p := p - \bar{p}$, we found two boundary choices (essentially *Dirichlet* and *Neumann*) which have the indicated covariant boundary terms in $\delta\mathcal{H}$:

$$\mathcal{B}_\varphi = i_N\varphi \wedge \Delta p - \varsigma\Delta\varphi \wedge i_N\bar{p} \quad \Longrightarrow \quad i_N(\delta\varphi \wedge \Delta p), \quad (10)$$

$$\mathcal{B}_p = i_N\bar{\varphi} \wedge \Delta p - \varsigma\Delta\varphi \wedge i_N p \quad \Longrightarrow \quad -i_N(\Delta\varphi \wedge \delta p). \quad (11)$$

We also found two other physical interesting choices:

$$\mathcal{B}_{\text{dynamics}} = i_N\bar{\varphi} \wedge \Delta p - \varsigma\Delta\varphi \wedge i_N\bar{p} \quad \Longrightarrow \quad \varsigma\delta\varphi \wedge i_N\Delta p - i_N\Delta\varphi \wedge \delta p, \quad (12)$$

$$\mathcal{B}_{\text{constraint}} = i_N\varphi \wedge \Delta p - \varsigma\Delta\varphi \wedge i_N p \quad \Longrightarrow \quad i_N\delta\varphi \wedge \Delta p - \varsigma\Delta\varphi \wedge i_N\delta p. \quad (13)$$

Let us look at the following two applications.

2.2. *Applications: Electromagnetism and General Relativity*

The first order Lagrangian 4-form for the source free $U(1)$ gauge field one-form A and its conjugate momentum H is

$$\mathcal{L}_{\text{EM}} = dA \wedge H - \frac{1}{2}{}^\star H \wedge H. \quad (14)$$

The pair of first order equations are

$$dH = 0, \qquad dA - {}^\star H = 0. \quad (15)$$

These are just the vacuum Maxwell equations with ${}^{\star}H = F := dA$; hence $H = -{}^{\star}F$ and $d{}^{\star}F = 0$. The natural reference in electromagnetism is $\bar{A} = 0$, $\bar{H} = 0$. The best boundary choice is $\mathcal{B}_{\text{dynamics}}$, which vanishes for this reference choice.

The first order Lagrangian for Einstein's (vacuum) gravity theory is

$$\mathcal{L}_{\text{GR}} = R^{\alpha}{}_{\beta} \wedge \eta_{\alpha}{}^{\beta}, \tag{16}$$

where the curvature 2-form is $R^{\alpha}{}_{\beta} := d\Gamma^{\alpha}{}_{\beta} + \Gamma^{\alpha}{}_{\gamma} \wedge \Gamma^{\gamma}{}_{\beta}$, and the dual basis 2-form is $\eta^{\alpha\beta} := *(\vartheta^{\alpha} \wedge \vartheta^{\beta})$. For gravity itself, two different choices of boundary condition correspond to the quasi-local expressions which asymptotically give (a) the ADM energy, (b) the Bondi energy and, moreover, (c) the Bondi flux: the celebrated outgoing flux plus an incoming flux.[8] Our general formalism with $\varphi \to \Gamma^{\alpha}{}_{\beta}$ and $p \to \eta_{\alpha}{}^{\beta}$ gives 4 quasi-local expressions. There is a distinguished energy expression with a very desirable property: it corresponds to imposing boundary conditions on a *manifestly covariant object*:

$$\mathcal{B}_{\vartheta}(\mathbf{N}) := \frac{1}{16\pi}\left(\Delta\Gamma^{\alpha}{}_{\beta} \wedge i_N\eta_{\alpha}{}^{\beta} + \bar{D}^{\alpha}N^{\beta}\Delta\eta_{\alpha\beta}\right). \tag{17}$$

The associated energy flux expression is

$$\pounds_N\mathcal{H}_{\vartheta} \simeq di_N\left(\Delta\Gamma^{\alpha\beta} \wedge \pounds_N\eta_{\alpha\beta}\right), \tag{18}$$

and the natural reference in gravity for the asymptotic flat spacetime is the Minkowski spacetime:

$$\bar{g}_{\mu\nu} = \text{diag}(-1, +1, +1, +1). \tag{19}$$

3. Reference Choice

Now let us turn to how to select the reference; effectively one should embed the 2-boundary into Minkowski space.[9,10] In a neighborhood of the desired spacelike boundary 2-surface S, 4 smooth functions $y^i = y^i(x^{\mu})$, $i = 0, 1, 2, 3$ with $dy^0 \wedge dy^1 \wedge dy^2 \wedge dy^3 \neq 0$ define a Minkowski reference:

$$\bar{g} = -(dy^0)^2 + (dy^1)^2 + (dy^2)^2 + (dy^3)^2. \tag{20}$$

The reference connection is

$$\bar{\Gamma}^{\alpha}{}_{\beta} = x^{\alpha}{}_{i}(\bar{\Gamma}^{i}{}_{j}y^{j}{}_{\beta} + dy^{i}{}_{\beta}) = x^{\alpha}{}_{i}dy^{i}{}_{\beta}, \tag{21}$$

where $dy^i = y^i{}_{\alpha}dx^{\alpha}$ and $dx^{\alpha} = x^{\alpha}{}_{j}dy^j$ with vanishing Minkowski reference connection coefficients. N^{μ} is a *translational Killing field* of the Minkowski reference, then the second quasi-local term vanishes. Our quasi-local expression then takes the form

$$\mathcal{B}(N) = N^k x^{\mu}{}_{k}(\Gamma^{\alpha}{}_{\beta} - x^{\alpha}{}_{j}\, dy^{j}{}_{\beta}) \wedge \eta_{\mu\alpha}{}^{\beta}. \tag{22}$$

To determine the reference choice $y^i{}_{\mu}$ in terms of quasi-spherical foliation adapted coordinates t, r, θ, ϕ, the isometric matching on the 2-surface implies

$$g_{AB} = \bar{g}_{AB} = \bar{g}_{ij}y^i_A y^j_B = -y^0_A y^0_B + \delta_{ab}y^a_A y^b_B, \quad a, b = 1, 2, 3;\ A, B = 2, 3 = \theta, \phi, \tag{23}$$

where the reference metric on the dynamical space has the components $\bar{g}_{\mu\nu} = \bar{g}_{ij}y^i{}_\mu y^j{}_\nu$. From a classic closed 2-surface into $\mathbb{R}^3$ embedding theorem, we expect that—as long as one restricts S and $y^0(x^A)$ such that on S

$$g'_{AB} := g_{AB} + y^0_A y^0_B \tag{24}$$

is convex—one can prove that there is a unique isometric embedding. (But, unfortunately, there is no explicit formula.)

3.1. *4D Isometric Matching*

Complete 4D isometric matching on S has 10 constraints:[9]

$$g_{\mu\nu}|_S = \bar{g}_{\mu\nu}|_S = \bar{g}_{ij}y^i{}_\mu y^j{}_\nu|_S. \tag{25}$$

There are 12 embedding functions on the constant t, r 2-surface:

$$y^i(\Rightarrow y^i_\theta, y^i_\phi), \quad y^i_t, \quad y^i_r. \tag{26}$$

The 10 constraints split into 3 for the already discussed 2D isometric matching: $g_{\theta\theta}, g_{\theta\varphi}, g_{\varphi\varphi}$ which constrain the 4 y^i; 3 normal bundle algebraic quadratic expressions: g_{tt}, g_{tr}, g_{rr}; and 4 mixed linear algebraic expressions: $g_{t\theta}, g_{t\varphi}, g_{r\theta}, g_{r\varphi}$. The 2D isometric matching can be regarded as a given y^0 uniquely determining y^1, y^2, y^3 on S. The remaining 7 algebraic equations can be regarded as finding all the other embedding variables in terms of y^i and $y^0{}_r$ on S. Thus one can take $y^0, y^0{}_r$ as the embedding control variables. Geometrically $y^0{}_r$ controls a boost in the plane normal to S.

3.2. *An Optimal Choice*

One can regard the value of the boundary term as a measure of the difference between the dynamical boundary values and the reference boundary values. However, how to find the "best matched" reference geometry? Because 12 embedding variables are subject to 10 isometric conditions, one will obtain the best matched reference geometry as long as one can obtain the two unknown variables.[11–13]

For a given S there are 2 different quantities which can be considered: $m^2 = -\bar{g}^{ij}p_i p_j$ and $E(N, S)$. For the latter there are 2 different ways to fix N. The critical points are distinguished: (1) The critical points of m^2. This determines the reference up to Poincaré transformations. (2) The critical points of $E(\partial_T, S)$. (If $m^2 > 0$, this may equivalent to (1).) (3) The critical points of $E(N, S)$ for a given dynamical vector field N. (Afterward one could find the extreme choice of N.)

The first two approaches lead to quasi-local quantities associated with S, the third alternative gives a quasi-local energy associated with an observer.[14,15] Based on some physical and practical computational arguments, it is reasonable to expect a unique solution. For our quasi-local values for axisymmetric solutions including Kerr see Ref. 13.

4. Summary

For any gravitating system — and hence for all *physical* systems — the *localization* of energy-momentum is an outstanding problem. We've displayed the relation between covariant Hamiltonian boundary term and the quasi-local quantities. For gravitating systems, we have obtained four quasi-local energy-momentum expressions; each is associated with a physically distinct, and geometrically clear, boundary condition. With the "*best matched*" reference, we have a satisfactory way of fixing the Hamiltonian boundary term quasi-locally for locally Poincaré gauge invariant gravity including GR. This in particular gives a way of resolving the ambiguities in determining the quasi-local energy-momentum of classical physical systems.

Acknowledgement

C.M.C. was supported by the Ministry of Science and Technology of the R.O.C. under the grant MOST 102-2112-M-008-015-MY3.

References

1. L. B. Szabados, *Living Rev. Relativ.* **12**, 4 (2009).
2. C.-M. Chen, J. M. Nester and R.-S. Tung, *Int. J. Mod. Phys. D* **24**, 1530026 (2015) [arXiv:1507.07300 [gr-qc]].
3. C.-C. Chang, J. M. Nester and C.-M. Chen, *Phys. Rev. Lett.* **83**, 1897–1901 (1999) [gr-qc/9809040].
4. R. Penrose, *Proc. R. Soc. London* A **381**, 53–63 (1982).
5. K. Kuchař, *J. Math. Phys.* **17** 801–820 (1976).
6. C.-M. Chen, J. M. Nester and R.-S. Tung, *Phys. Lett. A* **203**, 5–11 (1995) [gr-qc/9411048].
7. C.-M. Chen and J. M. Nester, *Class. Quantum Grav.* **16**, 1279–1304 (1999) [gr-qc/9809020].
8. C.-M. Chen, J. M. Nester and R.-S. Tung, *Phys. Rev. D* **72**, 104020 (2005) [gr-qc/0508026].
9. J.-L. Liu, C.-M. Chen and J.-M. Nester, *Class. Quant. Grav.* **28**, 195019 (2011) [arXiv:1105.0502 [gr-qc]].
10. M.-F. Wu, C.-M. Chen, J.-L. Liu and J. M. Nester, *Gen. Rel. Grav.* **44**, 2401 (2012) [arXiv:1206.0506 [gr-qc]].
11. C.-M. Chen, J.-L. Liu, J. M. Nester and M.-F. Wu, *Phys. Lett. A* **374**, 3599 (2010) [arXiv:0909.2754 [gr-qc]].
12. M.-F. Wu, C.-M. Chen, J.-L. Liu and J. M. Nester, *Phys. Rev. D* **84**, 084047 (2011) [arXiv:1109.4738 [gr-qc]].
13. G. Sun, C.-M. Chen, J.-L. Liu and J. M. Nester, *Chin. J. Phys.* **52**, 111 (2014) [arXiv:1307.1039 [gr-qc]].
14. J. M. Nester, C.-M. Chen, J.-L. Liu and G. Sun, Springer Proc. Phys. **157**, 177 (2014) [arXiv:1210.6148 [gr-qc]].
15. G. Sun, C.-M. Chen, J.-L. Liu and J. M. Nester, *Chin. J. Phys.* **53**, 110107 (2015).

APPENDIX

CONFERENCE COMMITTEES AND STAFF

International Advisory Committee

DACG EXCO members

M. Sasaki (Chair)(YITP), R.-G. Cai (IPT, CAS), K. Choi (KAIST), W. J. Couch (AAO), B. Dawson (U. of Adelaide), L. Dunne (U. of Canterbury), Y. Gong (HUST), S.-I. Inutsuka (Nagoya U.), G. Kang (KISTI), S.P. Kim (Kusnsan U.), S. Kim (Ewha U.), M. Mori (Ritsumeikan U.), T. Tanaka (Tokyo U.), R. Volkas (U. of Melbourne), X.-F. Wang (Tsinghua U.), S. Wyithe (U. of Melbourne), J. Yokoyama (RESCEU), S. Zhang (IHEP, CAS), Z.-H. Zhu (Beijing Normal U.)

Local Organizing Committee

P. Chen (Chair) (LeCosPA), C.-M. Chen (NCU), J.-W. Chen (NTU), C.-S. Chu (Tsinghua U.), C.-Q. Geng (Tsinghua U.), J.-A. Gu (NTU), X.-G. He (NTU), K. Hosomichi (NTU), H.-L. Yu (ASIOP)

Staff

Y.-L. Lee, S.-T. Yang, H.-H. Wang, Y.-K. Tsai.

LIST OF PARTICIPANTS

Arai, Shun	Nagoya University	falcon.olzq@gmail.com
Arroja, Frederico	LeCosPA, National Taiwan University	arroja@phys.ntu.edu.tw
Ashtekar, Abhay	Institute for Grav & the Cosmos, Penn State	ashtekar@gravitu.psu.edu
Cai, Rong-Geng	Inst. of Theo. Phys., Chinese Acad. of Sciences	cairg@itp.ac.cn
Cai, Yifu	University of Science and Technology of China	yifucai@ustc.edu.cn
Chang, Po-Wen	National Taiwan University	r02222057@ntu.edu.tw
Chen, Che-Yu	LeCosPA, National Taiwan University	b97202056@gmail.com
Chen, Chian-Shu	Institute of Physics, Academia Sinica	chianshu@gmail.com
Chen, Chiang-Mei	National Central University	cmchen@phy.ncu.edu.tw
Chen, Chin-Hao	LeCosPA, National Taiwan University	chchen@phys.ntu.edu.tw
Chen, Chun-Yen	LeCosPA, National Taiwan University	R02222076@ntu.edu.tw
Chen, Pisin	LeCosPA, National Taiwan University	pisinchen@phys.ntu.edu.tw
Chiang, Hsu-Wen	LeCosPA, National Taiwan University	b98202036@ntu.edu.tw
Chiueh, Tzihong	Dept. of Physics, National Taiwan University	ntulecospa@ntu.edu.tw
Cho, Hee-Sek	Korea Inst. of Science and Tech. Information	chohs1439@gmail.com
Chu, Chong-Sun	NCTS and National Tsing Hua University	cschu@phys.nthu.edu.tw
De Felice, Antonio	Yukawa Inst. for Theo. Phys., Kyoto University	antonio.defelice@yukawa.kyoto-u.ac.jp
Di Serego Alighieri, Sperello	INAF - Osservatorio Astrofisico di Arcetri	sperello@arcetri.astro.it
Di Virgilio, Angela	INFN	angela.divirgilio@pi.infn.it
Domènech, Guillem	Yukawa Inst. for Theo. Phys., Kyoto University	guillem.domenech@yukawa.kyoto-u.ac.jp
Donà, Pietro	Fudan University	pietro_dona@fudan.edu.cn
Ellis, Jonathan	King's College London/CERN	John.Ellis@cern.ch
Feng, Wei-Xiang	National Tsing Hua University	mh062467306@gmail.com
Fodor, Gyula	Paris Observatory	gyula.fodor@obspm.fr
Ge, Je-An	LeCosPA, National Taiwan University	f3202019@ms16.hinet.net
Ge, Xian-Hui	Department of Physics, Shanghai University	gexh@shu.edu.cn
Geng, Chao-Qiang	National Tsinghua University	geng@phys.nthu.edu.tw
Gondolo, Paolo	University of Utah	paolo.gondolo@utah.edu
Good, Michael	Nazarbayev University	michael.good@nu.edu.kz
Gwak, Bogeun	Sogang University/CQUeST	rasenis@sogang.ac.kr
Hamada, Yuta	Kyoto University	hamada@gauge.scphys.kyoto-u.ac.jp
Hayakawa, Taku	Institute for Cosmic Ray Research	taku1215@icrr.u-tokyo.ac.jp
Hayashinaka, Takahiro	RESCEU, University of Tokyo	hayashinaka@resceu.s.u-tokyo.ac.jp
Ho, Pei-Ming	Dep. of Phys., National Taiwan University	pmho@phys.ntu.edu.tw
Hong, Jeong-Pyong	Inst. for Cosmic Ray Research, U. of Tokyo	hjp0731@icrr.u-tokyo.ac.jp
Horava, Petr	Berkeley Center for Theoretical Physics	horava@berkeley.edu
Horikoshi, Masaatsu	Shinshu University	14st307b@shinshu-u.ac.jp
Hosomichi, Kazuo	Dep. of Phys., National Taiwan University	hosomiti@phys.ntu.edu.tw
Hsu, Shih-Ying	LeCosPA, National Taiwan University	cktheonly@gmail.com
Hu, Yao-Chieh	LeCosPA, National Taiwan University	r04244003@ntu.edu.tw

Huang, Da	National Tsinghua University	dahuang@phys.nthu.edu.tw
Huang, Jiang-Jung	LeCosPA, National Taiwan University	njinee@hotmail.com
Hung, Ling Yan	Fudan University	lyhung@fudan.edu.cn
Inami, Takeo	National Taiwan University	inami@phys.chuo-u.ac.jp
Izumi, Keisuke	Barcelona University, ICC	keisuke.aset@gmail.com
Kalogeropoulos, Nikos	Weill Cornell Medicine - Qatar / Cornell U.	nik2011@qatar-med.cornell.edu
Kang, Jae Hwan	Stanford University	jaehwan@stanford.edu
Kerr, Matthew	Australia Telescope National Facility	matthew.kerr@gmail.com
Kim, Dong-Hoon	Ewha Womans University	ki1313@yahoo.com
Kim, Keun-Young	Gwangju Inst.of Science and Technology (GIST)	fortoe@gmail.com
Kim, Nakwoo	Kyung Hee University	nkim@khu.ac.kr
Kim, Sang Pyo	Kunsan National University	sangkim@kunsan.ac.kr
Kitamoto, Hiroyuki	Kyoto University	kitamoto@tap.scphys.kyoto-u.ac.jp
Kobayashi, Tsutomu	Rikkyo University	tsutomu@rikkyo.ac.jp
Kohri, Kazunori	KEK	kohri@post.kek.jp
Koyama, Yoji	National Center for Theoretical Sciences	koyama811@gmaillcom
Kuo, Chao-Lin	Standford University	clkuo@stanford.edu
Kuroyanagi, Sachiko	Nagoya University	skuro@nagoya-u.jp
Lee, Bum-Hoon	Sogang University	bhl@sogang.ac.kr
Lee, Chung-Chi	National Center for Theoretical Sciences	g9522545@oz.nthu.edu.tw
Lee, Wonwoo	Center for Quantum Spacetime, Sogang U.	warrior@sogang.ac.kr
Lewandowski, Jerzy	University of Warsaw	jerzy.lewandowski@fuw.edu.pl
Li, Hong	Institute of high energy physics	hongli@ihep.ac.cn
Li, Wei	Institute of Theoretical Physics- CAS	u.duality@gmail.com
Lin, Chunshan	Yukawa Inst. for Theo. Phys., Kyoto University	chunshan.lin@ipmu.jp
Lin, Feng-Li	National Taiwan Normal University	linfengli@phy.ntnu.edu.tw
Lin, Guey-Lin	Inst. of Phys., National Chiao-Tung University	showchiung@gmail.com
Lin, Shih-Yuin	National Changhua University of Education	sylin156@gmail.com
Lin, Shu	Sun Yat-Sen University	linshu8@mail.sysu.edu.cn
Lin, Tzu-Lu	Dept. of Phys., National Taiwan University	liampassion6626@gmail.com
Lin, Yao-Yu	LeCosPA, National Taiwan University	joshualin24@gmail.com
Liu, Tsungche	LeCosPA, National Taiwan University	tcliu@ntu.edu.tw
Luo, Ling-Wei	Inst. of Phy., National Chiao Tung University	lwluo@nctu.edu.tw
Marciano, Antonino	Fudan University	antonino.marciano@gmail.com
Mase, Keiichi	Chiba University	mase@hepburn.s.chiba-u.ac.jp
Misonoh, Yosuke	Waseda University	misonoh@aoni.waseda.jp
Nakamura, Shin	Chuo University	nakamura@phys.chuo-u.ac.jp
Nakayama, Kazunori	University of Tokyo	kazunori@hep-th.phys.s.u-tokyo.ac.jp
Nam, Jiwoo	LeCosPA, National Taiwan University	jwnam@phys.ntu.edu.tw
Namba, Ryo	Kavli IPMU	ryo.namba@ipmu.jp
Nester, James	National Central U. & LeCosPA NTU	jmnester@gmail.com
Ni, Wei-Tou	National Tsing Hua University	weitou@gmail.com
Nishi, Sakine	Rikkyo University	sakine@rikkyo.ac.jp
Noumi, Toshifumi	Hong Kong U. of Science and Technology	iasnoumi@ust.hk
Nugier, Fabien	LeCosPA, National Taiwan University	fnugier@ntu.edu.tw

Ogawa, Hiromu	Rikkyo University	jh.ogawa@rikkyo.ac.jp
Oh, Seung Hun	Konkuk University	thereisnoi@naver.com
Ong, Yen Chin	Nordita	ongyenchin@gmail.com
Oshita, Naritaka	University of Tokyo	naritaka@resceu.s.u-tokyo.ac.jp
Paganis, Stathes	National Taiwan University	paganis@phys.ntu.edu.tw
Page, Don	University of Alberta	profdonpage@gmail.com
Pan, Sheau-Shi	Industrial Technology Research Institute	Sheau.Shi.Pan@itri.org.tw
Parikh, Maulik	Arizona State University	Maulik.Parikh@asu.edu
Park, Chanyong	Sogang University	cyong21@sogang.ac.kr
Park, Jiwon	Kunsan National University	pjwsjh@naver.com
Pu, Hung-Yi	ASIAA	hypu@asiaa.sinica.edu.tw
Qiu, Taotao	Central China Normal University	qiutt@mail.ccnu.edu.cn
Ripa, Jakub	LeCosPA, National Taiwan University	jripa@ntu.edu.tw
Romano, Antonio Enea	UDEA	aer@phys.ntu.edu.tw
Sadakazu, Haino	Academia Sinica	sadakazu.haino@cern.ch
Saito, Ryo	AstroParticle and Cosmology laboratory	rsaito@apc.in2p3.fr
Sakai, Norisuke	Keio University	norisuke.sakai@gmail.com
Sasaki, Misao	Yukawa Inst. for Theo.Phys., Kyoto University	misao@yukawa.kyoto-u.ac.jp
Sekiguchi, Yuichiro	Toho University	y.sekiguchi@sci.toho-u.ac.jp
Shen, Che-Min	LeCosPA, National Taiwan University	f97222024@ntu.edu.tw
Shiao, Jerry	LeCosPA, National Taiwan University	jshiao@phys.ntu.edu.tw
Shiu, Gary	University of Wisconsin-Madison & HKUST	shiu@physics.wisc.edu
Soo, Chopin	National Cheng Kung University	cpsoo@mail.ncku.edu.tw
Starobinsky, Alexei	Landau Institute For Theoretical Physics RAS	alstar@landau.ac.ru
Su, Meng	MIT	mengsu@mit.edu
Sun, Sichun	IAS, HKUST	sichun@uw.edu
Suyama, Teruaki	University of Tokyo	suyama@resceu.s.u-tokyo.ac.jp
Tada, Yuichiro	Kavli IPMU	yuichiro.tada@ipmu.jp
Takahashi, Tomo	Saga University	tomot@cc.saga-u.ac.jp
Tanahashi, Norihiro	University of Cambridge	N.Tanahashi@damtp.cam.ac.uk
Tsai, Ho-Chin	National Taiwan Normal University	tsaihochin@gmail.com
Tsujikawa, Shinji	Tokyo University	shinji@rs.kagu.tus.ac.jp
Tye, Henry	HKUST IAS	iastye@ust.hk
Unruh, William	University of British Columbia	unruh@physics.ubc.ca
Van Putten, Maurice	Sejong University	mvp@sejong.ac.kr
Wang, I-Chin	National Cheng Kung University	icwcmb@gmail.com
Wang, Min-Zu	National Taiwan University	mwang@phys.ntu.edu.tw
Wang, Qing	Tsing Hua University	wangq@mail.tsinghua.edu.cn
Wang, Shih-Hao	LeCosPA, National Taiwan University	wsh4180@gmail.com
Wang, Yi	HKUST	phyw@ust.hk
Wen, Wen-Yu	Chun Yuan Christian University	steve.wen@gmail.com
Wong, Henry	Academia Sinica	htwong@phys.sinica.edu.tw
Wu, Yi-Peng	Academia Sinica	ypwu@phys.sinica.edu.tw

Yamashita, Yasuho	Yukawa Inst.for Theo. Phys., Kyoto University	yasuho@yukawa.kyoto-u.ac.jp
Yang, Hyun Seok	Sogang University	ya19683@gmail.com
Yau, Shing Tung	Harvard University	yassist@math.ntu.edu.tw
Ye, Fang	National Central University	shivaandalice@gmail.com
Yeom, Dong-han	LeCosPA, National Taiwan University	innocent.yeom@gmail.com
Yokoyama, Jun'ichi	RESCEU, University of Tokyo	yokoyama@resceu.s.u-tokyo.ac.jp
Yoon, Jong Hyuk	Konkuk University	yoonjh@konkuk.ac.kr
Yu, Hoi Lai	Institute of Physics, Academia Sinica	hlyu@phys.sinica.edu.tw
Zhang, Fan	Beijing Normal University	fnzhang@bnu.edu.cn
Zhang, Ying	Xi'an Jiaotong University	hepzhy@mail.xjtu.edu.cn
Zhang, Yun-Long	National Taiwan University	zhangyunlong001@gmail.com

CONFERENCE PROGRAM

Monday, December 14

08:00-09:00	Registration	
09:00-10:45	**Plenary Session**	**Chair: Pisin Chen**
09:00-09:35	Shung-Tung Yau	"Quasi-Local Mass in General Relativity"
09:35-10:10	Abhay Ashtekar	"Isolated Systems and Gravitational Waves with a Positive Cosmological Costant"
10:10-10:45	Jim Nester	"Gravity: A Gauge Theory Perspective"
10:45-11:15	Conference photo& Coffee break	
11:15-12:25	**Plenary Session**	**Chair: Ronald Taam**
11:15-11:50	Matthew Kerr	"From Binaries to Backgrounds: Pulsar Timimg and General Relativity"
11:50-12:25	Wei-Tou Ni	"Sensitivities of Gravitational-Wave Detection—A Century Outlook"
12:25-13:30	Lunch Break	
13:30-15:30	**Parallel Session-CF**	**Chair: Chong-Sun Chu& Gary Shiu**
13:30-13:50	Yoji Koyama	"Dilaton, IR Effects in De-Sitter Space and IR-Driven Inflation"
13:50-14:10	Toshifumi Noumi	'"Effective Field Theory of Anisotropic Inflation"
14:10-14:30	Fang Ye	"Enhancing the Axion Field Range through Mixings"
14:30-14:50	Ryo Namba	"Scale-Dependent Gravitational Waves from a Rolling Axion"
14:50-15:10	Hiroyuki Kitamoto	"Stochastic Dynamics of Infrared Fluctuations in Accelerating Universe"
15:10-15:30	Sichun Sun	"Echoes of Inflationary Particle Phase Transition in the CMB"
15:30-16:00	Coffee Break	
16:00-17:30	**Parallel Session-EQ-I**	**Chair: Antonino Marciano**
16:00-16:55	Abhay Ashtekar	"Quantum Gravity: An Overview"
16:55-17:15	Pietro Dona	"Asymptotic Safety in an Interacting System of Gravity and Scalar Matter"
17:15-17:30	Hsu-Wen Chiang	"Quantized Geodesic with Local Coordinate Difference Operator"
13:30-15:30	**Parallel Session – DD**	**Chairs: Chao-Qiang Geng & Kaixuan Ni**
13:30-14:00	Meng Su	"Future of Indirect Search of Dark Matter Particle from the Gamma-Ray Sky"
14:00-14:30	Paolo Gondolo	"Review of Dark Matter Theory"
14:30-14:50	Da Hung	"Cosmic Ray Excesses From Multi-Component Dark Matter"
14:50-15:10	Jeong-Pyong Hong	"Charged Q-Balls in Gauge Mediated SUSY Breaking Models"

15:10-15:30	Yao-Yu Lin	" Impact of Gravitational Slingshot of Dark Matter on Galactic Halo Profiles"
15:30-16:00	Coffee Break	
15:40-16:00	Antonio Enea Romano	"H_0 Tension and the Effects of Local Structure"
16:00-16:30	Antonio De Felice	" The Minimal Theory of Massive Gravity"
16:30-16:50	Chung-Chi Lee	" Unified Models of Inflation and Dark Energy"
16:50-17:10	Yi-Peng Wu	" The Fate of Dark Energy with Nonminimal Coupling to Gravity"
17:10-17:30	Fabien Nugier	" Cosmology with an Adapted System of Coordinates"
13:30-17:30	**Parallel Session – GW**	**Chair: Arroja Frederico**
13:30-14:00	Shinji Tsujikawa	" Inflationary Gravitational Waves in the Effective Field Theory of Modified Gravity"
14:00-14:25	Sachiko Kuroyanagi	" Gravitational Waves from Second-Order Scalar Perturbations"
14:25-14:50	Dong-Hoon Kim	" Gravitational Waves from Bubble Collisions in the Early Universe"
14:50-15:10	Hee-Suk Cho	" Search for Gravitational Waves from Binary Black Hole Mergers"
15:10-15:30	Fan Zhang	"Electromagnetic Counterparts to Gravitational Waves from Binary Black Hole Mergers"
15:30-16:00	Coffee Break	
16:00-16:30	Teruaki Suyama	" Primordial Black Holes as a Novel Probe of Primordial Gravitational Waves"
16:30-17:00	Chunshan Lin	" Gravitational Waves Signal of Massive Gravities"
17:00-17:30	Maurice Van Putten	"Directed Searches for Broadband Extended Gravitational-Wave Emission in Energetic Core-Collapse Supernovae"
13:30-17:30	**Parallel Session – PA**	**Chairs: Jakub Ripa & Min-Zu Wang**
13:30-14:00	Sadakazu Haino	" Particle Physics and Cosmology with Alpha Magnetic Spectrometer"
14:00-14:30	Keiichi Mase	" The Dawn of Neutrino Astrophysics"
14:30-14:50	Tsung-Che Liu	" Profile of Third Flight of the Antarctic Impulsive Transient Antenna (ANITA) in 2015"
14:50-15:10	Chin-Hao Chen	"Searching Ultra-High Energy Cosmic Neutrinos and Flavor Identification with ARA Detector"
15:10-15:30	Wang Shih-Hao	"Feasibility of Determining Diffuse Ultra-High Energy Cosmic Neutrino Flavor Ratio through ARA Neutrino Observatory"
15:30-16:00	Coffee Break	
16:00-16:30	Guey-Lin Lin	"Thermal Transport of Dark Matter in the Sun"
16:30-16:50	Jakub Ripa	"Gamma-Ray Bursts and Their Relation to Astroparticle Physics and Cosmology"

16:50-17:10	Dong-Hoon Kim	"General Relativistic Effects on Pulsar Radiation"
17:10-17:30	Hayashinaka Takahiro	"QED Correction of the Radiation from the Magnetors"
17:40-18:15	**Plenary Session**	**Chair: Tzihong Chiueh**
17:40-18:15	Yuichiro Sekiguchi	"Numerical Relativity: Solving Einstein's Equation with Computer"
18:30-19:30	Welcome Reception	
19:30-20:30	Opening Night Concert	

Tuesday, December 15

08:00~09:00	Registraion	
09:00-10:45	**Plenary Session**	**Chair: Petr Horava**
09:00-09:35	William Unruh	"Firewalls: What and Why the Issue"
09:35-10:10	Don Page	"Naked Firewalls"
10:10-10:45	Pisin Chen	"Decelerating Plasma Mirror to Investigate Black Hole Information Loss Paradox"
10:45-11:15	Coffee Break	
11:15-12:25	Plenary Session	Chair: John Ellis
11:15-11:50	Alexei Starobinsky	"Inflation: Present Status and Future Prospects"
11:50-12:25	Henry Tye	"A Discrepancy/Puzzle in Higgs Physics"
12:25-13:30	Lunch Break	
13:30-15:15	Parallel Session –OT	Chairs: Wei-Tou Ni & Sheau-shi Pan
13:30-14:00	Sperllo di Senego Alighieri	"Gaining Confidence on General Relativity with Cosmic Polarization Rotation"
14:00-14:20	Wei-Tou Ni	"New Constraints on Cosmic Polarization Rotation Including SPTpol B-Mode Polarization Observations"
14:20-14:50	Angela di Virgilio	"GINGER, An array of RingLasers to test General relativity"
14:50-15:15	Hung-Yi Pu	"Observing the Black Hole Shadow of M87 and the Greenland Telescope Project: GR Test in the Strong Gravity Regime"
15:30-16:00	Coffee Break	
16:00-17:30	**Parallel Session –AM-I**	**Chair: Keisuke Izumi**
16:00-16:30	Norihiro Tanahashi	"Causality, Hyperbolicity and Shock Formation in Lovelock Theories"
16:30-17:00	Pei-Ming Ho	"Geometry of Area without Length"
17:00-17:30	Chung-Chi Lee	"Poincare Gauge Theory of Gravity"
13:30-17:30	**Parallel Session –CE-I**	**Chairs: Yi-Fu Cai & Teruaki Suyama**
13:30-14:00	Kazunori Kohri	"Reinterpretation of the Starobinsky-Type Inflation Models"
14:00-14:30	Kazunori Nakayama	"Gravitational Particle Production after Inflation with Extended Gravity Models"
14:30-15:00	Taotao Qiu	"Bounce Inflation: A Nonsingular Inflationary Universe"

15:00-15:30	Antonino Marciano	"Superconductive Cosmology"
15:30-16:00	Coffee Break	
16:00-16:30	Tsutomu Kobayashi	"Galilean Creation of the Inflationary Universe"
16:30-16:50	Sakine Nishi	"Matter Creation in Generalized Galilean Genesis"
16:50-17:10	Yuta Hamada	"Saddle Point inflation in String-Inspired Theory"
17:10-17:30	Taku Hayakawa	"Cosmology with Heavy Polonyi Field"
13:30-17:30	**Parallel Session –HP**	**Chairs: Keun-Young Kim & Janet Hung**
13:30-13:50	Nakwoo Kim	"Bulk Local Operators and D-Instantons in pp-wave Holography"
13:50-14:10	Wei Li	"Generalized Conformal Anomalies in Higher Spin Theory"
14:10-14:30	Bogeun Gwak	"Testing the Validity of Cosmic Censorship in the Rotating Anti-de Sitter Black Hole"
14:30-14:50	Shin Nakanura	"Application of AdS/CFT Correspondence to Studies on Non-Equilibrium Steady States"
14:50-15:10	Shu Lin	"Holographic Thermalization of Strongly Coupled Field Theory in Two Dimension"
15:10-15:30	Yun-Long Zhang	"Rindler Fluid and Its Properties"
15:30-16:00	Coffee Break	
16:00-16:20	Chanyong Park	"Thermodynamic Law from the Entanglement Entropy Bound"
16:20-16:40	Xian Hui Ge	"Lower Bound of Quantum Critical Conductivity from Holography"
16:40-17:00	Lin Feng-Li	"Interplay between Entanglement and Geometry"
17:00-17:20	Dong-Hoon Kim	"Analogies between Gravity and Quantum Entanglement: Viewed from Information Geometry"
13:30-17:30	Parallel Session –EQ-II	Chair: Antonino Marciano
13:30-14:25	Jerzy Lewandowski	"Canonical Quantum Gravity"
14:25-14:55	Antonino Marciano	"Yang-Mills Vacuum and the Planck Scale in Reduced Loop Quantum Gravity"
14:55-15:25	Ho-Lai Yu	"Intrinsic Time Quantum Gravity"
15:25-15:45	Coffee Break	
15:45-16:15	Hyun Seok Yang	"Cosmic Inflation from Emergent Spacetime Picture"
16:15-16:45	Yifu Cai	"Cosmology of f (T) Teleparallel Gravity"
16:45-17:15	Nikolaos Kalogeropoulos	"Non-Additive Entropies in ... Gravity ?"
17:15-17:30	Yosuke Misonoh	"Black Holes and Thunderbolt Singularities with Lifshitz Scaling Terms"
17:40-18:15	**Plenary Session**	**Chair: Misao Sasaki**
17:40-18:15	Sang Pyo Kim	"Schwinger Effect, Hawking Radiation and Unruh Effect"

Wednesday, December 16

08:00-9:00	Registration	
09:00-09:35	**Plenary Session**	**Chair: Bum-Hoon Lee**
09:00~09:35	Misao Sasaki	"Conformal Frames in Cosmology: Frame Dependence vs Observational Equivalence-"
09:35-17:00	Excursion :Taipei-Yilan Tour	
17:40-21:00	Banquet at Formosa Pearl (Yilan)	

Thursday, December 17

09:00-10:45	**Plenary Session**	**Chair: Hoi-Lai Yu**
09:00-09:35	Petr Hořava	"Surprises with Non-relativistic Naturalness"
09:35-10:10	Maulik Parikh	"Two Roads to the Null Energy Condition"
10:10-10:45	Chopin Soo	"Quantum Geometrodynamics with Intrinsic Time Development"
10:45-11:15	Conference Photo & Coffee Break	
13:30-17:00	**Parallel Session-AM-II**	**Chair: Keisuke Izumi**
13:30-13:50	Che-Min Shen	"Boundary Effect of Anomaly-Induced Action in Curved Spacetime"
13:50-14:10	Masaatsu Horikoshi	"R^2 Gravitational Coleman-Weinberg Corrections to the Standard Model Higgs at Planck Scales"
14:10-14:30	Che-Yu Chen	"Cosmological Singularities in the Eddington-Inspired-Born-Infeld Theory and its Possible Extension"
14:30-14:55	Ryo Saito	"Modified Gravity Inside Astrophysical Bodies"
14:55-15:20	Keisuke Izumi	"Tree-Unitarity and Renormalizability in Lifshitz-Scaling Theory as a Toy Model of Horava's Gravity Theory"
15:30-15:40	Break	
15:40-15:55	Yasuho Yamashita	"Bigravity from Gradient Expansion"
15:55-16:20	Frederico Arroja	"The Two Faces of Mimetic Horndeski Gravity: Disformal Transformations and Lagrange Multiplier"
16:20-16:40	Hiroma Ogawa	"Instability of Hairy Black Holes in Shift-Symmetric Horndeski Theories"
16:40-17:00	Ling-Wei Luo	"Teleparallel Gravity in Five Dimension"
13:30-17:00	**Parallel Session-BH**	**Chairs: Yen Chin Ong & Dong-han Yeom**
13:30-14:00	Pei-Ming Ho	"The Absence of Horizon in Black-Hole Formation"
14:00-14:30	Wen-Yu Wen	"Black Hole Remnant in Asymptotic Anti-De Sitter Space"
14:30-14:55	Michael Good	"Black Mirrors: A One-to-One Correspondence between Moving Mirrors and Black Holes."
14:55-15:20	Dong-Han Yeom	"Information Loss Problem and Roles of Instantons"

15:20-15:45	Guillem Domènech Fuertes	"Stationary Bubbles and their Tunneling Channels toward Tivial Geometry"
15:45-16:10	Yen Chin Ong	"Black Hole: The Interior Spacetime"
16:10-16:35	Naritaka Oshita	"Unruh Radiation Produced by a Uniformly Accelerating Charged Particle in Thermal Random Motions"
16:35-17:00	Hu Yao-Chieh	"Two Interpretations on Thin-Shell Instantons"
13:30-16:50	**Parallel Session-CE-II**	**Chairs: Yi-Fu Cai & Teruaki Suyama**
13:30-14:00	Tomo Takahashi	"Aspects of Multi-Field Models of Inflation"
14:00-14:30	Yi Wang	"Quantum Primordial Standard Clocks -- Towards an Observational Proof of Inflation"
14:30-15:00	Chunshan Lin	"Inflationary Magnetogenesis with Broken Local U (1) Symmetry"
15:00-15:30	Hong Li	"The CMB Ali Project at Tibet"
15:30-15:50	Yuichiro Tada	"Can Massive Primordial Black Holes be Produced in Mild Waterfall Hybrid Inflation?"
15:50-16:10	Wonwoo Lee	"Cosmological Implications of Fubini Type Instanton"
16:10-16:30	Shun Arai	"Inflationary Perturbations in Lifshitz Regime of Horava Gravity"
16:30-16:50	Chen Chun-Yen	"Cosmological Singularities in the Eddington-Inspired-Born-Infeld theory and its Possible Extension"
13:30-15:30	**Parallel Session-MS**	**Chair: Chiang-Mei Chen**
13:30-14:00	Jong Hyuk Yoon	"Hamiltonian Reduction of Einstein's Equations in (2+2) Formalism Under no Isometry Assumptions"
14:00-14:30	Seung-Hun Oh	"Some Exact Solutions in (2+2) Hamiltonian Reduction"
14:30-15:00	Gyula Fodor	"Localized Oscillating Configurations Formed by Real Scalar Fields"
15:10-15:40	Jerezy Lewandowski	"Isolated Horizons, Near Horizon Geometries: How to Fins a New Exact Solution to the Einstein's Equations."
15:00-15:30	Chiang-Mei Chen	"Covariant Hamiltonian Boundary Term: Reference and Quasi-local Quantities"
17:00-18:00	Coffee Hour: Christmas Celebration	
18:00-18:35	**Plenary Session**	**Chair: Juin-Huei Proty Wu**
18:00-18:35	Chao-Lin Kuo	"The Search for Inflationary Gravitational Waves: the BICEP/Keck Program"

Friday, December 18

09:00-10:45	**Plenary Session**	**Chair: Wei-Shu Hou**
09:00-09:35	Rong-Gen Cai	"Holographic Magnetism from General Relativity"
09:35-10:10	Henry Wong	"Direct Experimental Searches of Dark Matter"

10:10-10:45	Jiwoo Nam	"Ultra-High Energy Neutrinos and Cosmic Rays via Radio; ANITA, ARA, and TAROGE"
10:45-11:15	Coffee Break	
11:15-12:25	Plenary Session	Chair: William Unruh
11:15-11:50	Tzihong Chiueh	"Extremely Light Bosonic Dark Matter, Wave DM"
11:50-12:25	Jonathan Ellis	"Testing Models of Inflation: How is Einstein's Gravity Modified?"
12:25-13:30	Lunch Break	
13:30-18:00	**Plenary Session**	**Chair: Pisin Chen**
13:30-13:50	MS: Jim Nester	
13:50-14:02	OT: Angela Di Virgilio	
14:02-14:30	AM: Keisuke Izumi	
14:30-14:50	GW: Frederico Arroja	
14:50-15:30	CE: Yifu Cai & Teruaki Suyama	
15:30-16:00	Coffee Break	
16:00-16:20	BH: Yen Chin Ong & Dong-han Yeom	
16:20-16:40	HP: Nakwoo Kim	
16:40-16:52	CF: Chong-Sun Chu	
16:52-17:20	EQ: Antonino Marciano	
17:20-17:40	DD: Chao-Qiang Geng	
17:40-18:00	PA: Jakub Ripa	
18:00-18:15	Closing Remark	Pisin Chen

Second LeCosPA Symposium Conference Photo, December 14, 2015, Taipei Taiwan

Opening Remarks

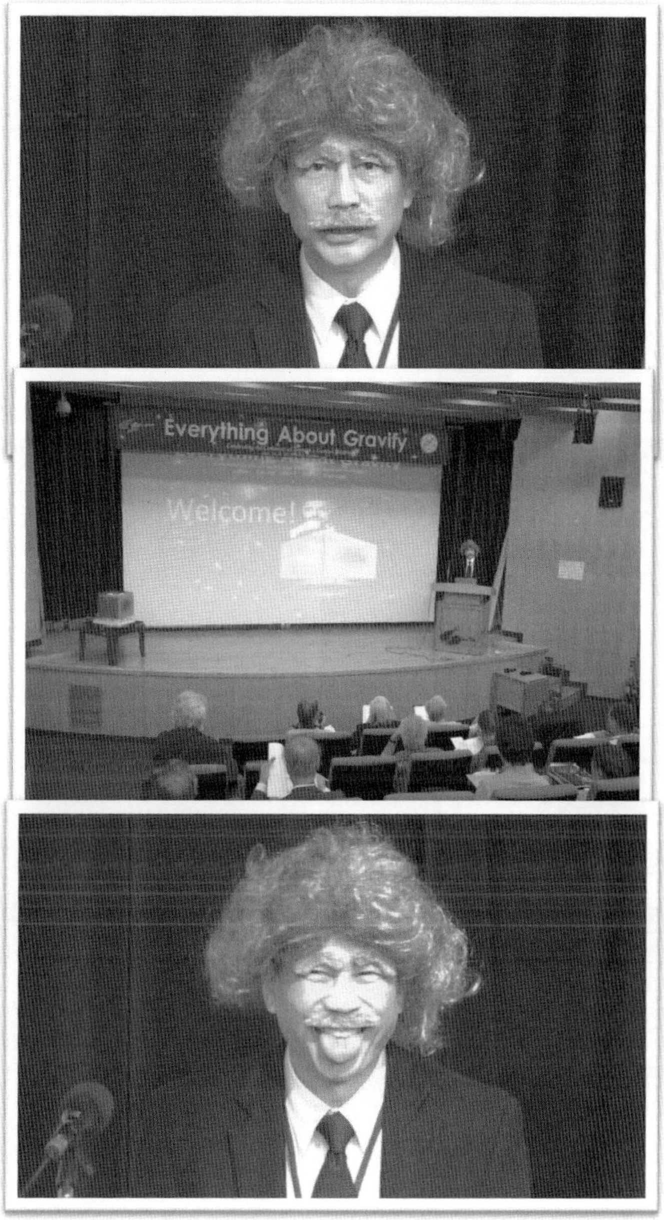

Plenary Sessions

Plenary Sessions

Plenary Sessions

Plenary Sessions

Parallel Sessions

Parallel Sessions

Parallel Sessions

Parallel Sessions

Parallel Sessions

Parallel Sessions

Welcome Reception

Welcome Reception

Opening Night Concert

Taipei-Yilan Excursion

Banquet at Formosa Pearl Restaurant, Yilan

Coffee Hour Christmas Celebration

Coffee Hour Christmas Celebration

Symposium Staff

Author Index